Measurement Systems: Application and Design

Measurement Systems: Application and Design

Ernest O. Doebelin

Department of Mechanical Engineering
The Ohio State University

McGraw-Hill Book Company

New York St. Louis San Francisco
Toronto London Sydney

Measurement Systems: Application and Design

Library of Congress Catalog Card Number 66-18475

17335

234567890 MP 732106987

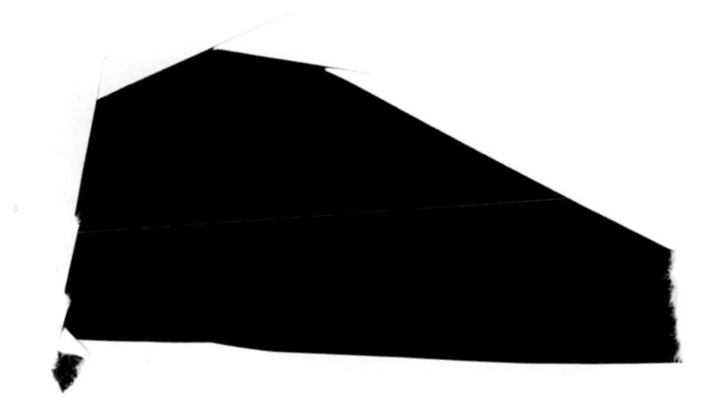

Preface

The title of this book, "Measurement Systems: Application and Design," is sufficiently broad that one can read into it a wide range of possibilities with regard to content and approach. The author would like to use the preface to define them more sharply, in addition to presenting the way in which the material has been used and the purpose he feels it can serve in engineering education and practice.

Since measurement in one form or another is used regularly by all sorts of people in all sorts of jobs, one must first restrict the scope. This material has been used in connection with courses and laboratories in the mechanical engineering curriculum at The Ohio State University and is thus biased toward this audience. Sufficient material at a suitable level is included for two courses: an introductory treatment useful for a required undergraduate course and more advanced considerations suitable for an elective course at the advanced undergraduate or beginning graduate level. The inclusion of both types of material in a single text is in part a recognition of the variation in curricula from school to school. By including a wide scope of material it is hoped that the individual instructor will be able to select topics and place emphasis so as to best utilize the previous preparation of his particular students.

The area of measurement is, of course, closely allied to that of laboratory teaching, and since this facet of engineering education is the subject of some controversy, there is certainly room for a wide variety of approaches. One can consider the general field to be composed of two parts: the hardware of measurement and the techniques of experimentation. It is difficult completely to separate the two, and some, by choice or because of pressure of time, will design courses to treat both types of material concurrently. At the author's school, three required courses are devoted to this general subject. The first is aimed mainly at developing an understanding of the operating principles of measurement hardware and the problems involved in the analysis, design, and application of such equipment. "Application" here (and in the book title) is not construed to encompass the detailed planning of comprehensive experiments but rather is limited mainly to consideration of the disturbing effect of the measuring instrument on the measured system and the influence of extraneous system variables on the instrument output. The method of presentation of this material is by 1-hr lectures each week and one 4-hr laboratory session each week. Lecturing is divided about equally between generalized theory (static characteristics, dynamic characteristics, etc.) and specific measurement hardware (vibration pickups, recorders, etc.) which may not be covered in the laboratory experiments. Laboratory experiments are preceded by about 1 hr of lecturing specific to the particular measurement area treated in that experiment. Experiments will not here be discussed in detail; however, the philosophy is that of having a relatively small number of experiments so as to allow sufficient time for adequate penetration and understanding. It is felt that it is impossible to cover all the significant hardware. Therefore experiments, while dealing with important specific areas, are designed mainly to serve as vehicles for the illustration of general concepts.

Following the course just described are two courses devoted to experimental analysis of engineering problems. These initially devote lecture time to development of systematic methods of planning, executing, and evaluating experiments. Then small groups of students undertake original projects which involve theoretical analysis, experiment planning, equipment design and construction, measurement-system design, experiment execution, evaluation, and report writing.

For the first course mentioned above, selected material from this book serves as text, while for the second and third courses it becomes a valuable reference. In addition to these three required courses, an additional elective in measurement systems, which essentially extends in breadth and depth from the first required course, is offered. This course has 3 hr of lecture and one 2-hr laboratory period per week. The objective is to develop increased competence in both the design and use of measurement equipment. This is implemented by consideration of both more advanced general concepts and also more sophisticated specific hardware. Experiments are designed mainly to provide familiarity with actual instrumentation equipment and problems involved in its use. Material from this book again serves as the text for this course.

Some explanation of the use of the word "system" in the title of this book may be in order. While the term "system engineering" has come to mean, at least to some, the planning and implementation of complex schemes on a grand scale, no such meaning is intended here. In fact, the author subscribes to the view that one man's component may be another man's system and that this varied use of the word is not objectionable. Thus to the designer of a large-scale data system, who essentially selects hardware from that available to achieve a compatible arrangement that meets the specified requirements, a tape recorder is legitimately considered a component. However, a designer of tape recorders would certainly insist that his machine is a most complex electromechanical system. Since this book is addressed to both these classes of application, the title word "system" is intended to be interpreted in either way, as appropriate.

The word "design" is used in the title to emphasize that many mechanical engineers not only use instrumentation but also are engaged in designing it. While the design of an electronic amplifier has perhaps only limited mechanical aspects (packaging, shock mounting, cooling, manufacturing, etc.), much other equipment, particularly transducers, is as much mechanical as electrical. Although highly specialized electrical aspects of electromechanical system design are handled by electrical engineers, the electrical background of mechanical engineers is adequate to allow them to treat many electrical problems that are closely coupled to the mechanical aspects. Design is intended to consider not only the problems of individual components but also the assembly of available components (transducers, amplifiers, recorders, etc.) into a compatible system capable of meeting required specifications.

Some important features of the text include the following:

1. Consideration of measurement as applied to research and development operations and also to monitoring and control of industrial and military systems and processes.
2. A generalized treatment of error-compensating techniques.
3. Treatment of dynamic response for all types of inputs: periodic, transient, and random, on a uniform basis, utilizing frequency response.
4. Detailed consideration of problems involved in interconnecting components.
5. Discussion, including numerical values, of standards for all important quantities. These give the reader a feeling for the ultimate performance currently achievable.
6. Quotation of detailed numerical performance specifications of actual instruments.
7. Inclusion of significant material on important specific areas such as sound measurement, heat-flux sensors, gyroscopic instruments, hot-wire anemometers, digital methods, random signals, mass flowmeters, amplifiers, and the use of feedback principles.

Since material for both introductory and advanced courses is included, it may be helpful to indicate the division. This is somewhat arbitrary since material considered advanced for a course at a given level in one curriculum might be thought elementary in another. For example, students who have had a course in dynamic systems analysis could cover the material on dynamic characteristics very rapidly. Thus the individual instructor must make the necessary selection. For those who would appreciate some assistance, the author offers the following suggestions.

Chapters 1 and 2 can be covered quite quickly and easily and should be included in an introductory course. In Chap. 3 the material on the proofs of the generalized loading equations, which starts at Fig. 3.29, can be omitted in a first course. The material on dynamic response utilizing Fourier transforms for transient response (Fig. 3.74) and on amplitude-modulated and -demodulated signals (Figs. 3.81 to 3.90) can also be omitted, as can the treatment of random signals. In the remainder of this chapter the sections Requirements on the Instrument Transfer Function to Ensure Accurate Measurement and Experimental Determination of Measurement-system Parameters should be retained and the others omitted.

In Chap. 4 we begin to deal with specific devices, and the choice of topics to be omitted becomes less clear and somewhat a matter of personal preference.

Sections 4.10 to 4.12 might reasonably be omitted and, in the remaining material, emphasis put on standards and calibration, potentiometers, strain gages, differential transformers, piezoelectric transducers, and accelerometers. Section 5.7 can be omitted, as can Secs. 6.6, 6.7, and 6.10. In Chap. 7 the material on hot-wire anemometers, electromagnetic flowmeters, and ultrasonic flowmeters can be left out. Radiation methods in Chap. 8 can be considerably cut, and heat-flux sensors omitted. Section 9.1 is the only essential part of Chap. 9. Section 10.1 is essential and would, in fact, usually be assigned much earlier, since it is needed in Chap. 4. Section 10.2 can be eliminated as can the hydraulic filter and statistical averaging of Sec. 10.3, and also Secs. 10.6 to 10.13. Chapter 11 may be omitted but most instructors will wish to cover all of Chap. 12, perhaps early in the course if an associated laboratory requires the use of recording equipment. Since Chap. 13 is short, entirely descriptive, and designed as a unifying conclusion, it should be included in an introductory course.

The author would like to acknowledge his appreciation to other workers in this field whose contributions are evidenced by the voluminous references and bibliography. Production of the manuscript was made as painless as possible by the faultless typing of Mrs. Maxine Fitzgerald. The contributions of students over the past 10 years to my understanding of methods of presenting material should not be minimized. Finally, the forbearance of a long-suffering wife and family is gratefully acknowledged.

Ernest O. Doebelin

Contents

part II Measuring Devices

part III Manipulation, Transmission, and Recording of Data

Measurement Systems: Application and Design

part I

General Concepts

1 Types of applications of measurement instrumentation

1.1 Introduction

As background for our later detailed study of measuring instruments and their characteristics, it will be useful first to discuss in a general way the uses to which such devices are put. We here choose to classify these applications according to the following scheme:

1. Monitoring of processes and operations
2. Control of processes and operations
3. Experimental engineering analysis

Each of these classes of application will now be described in more detail.

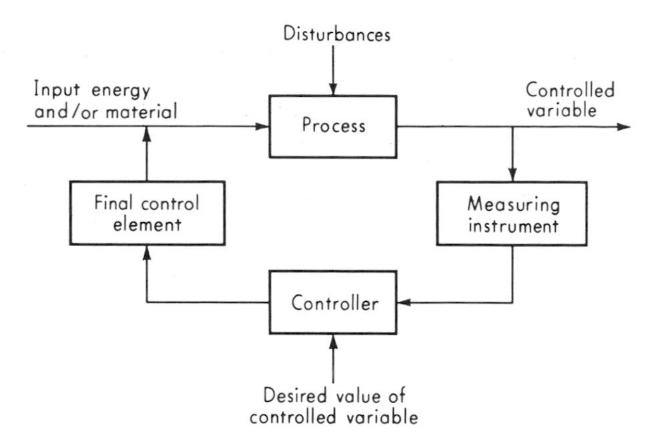

Fig. 1.1. *Feedback-control system.*

1.2

Monitoring of Processes and Operations Certain applications of measuring instruments may be characterized as having essentially a monitoring function. The thermometers, barometers, and anemometers used by the Weather Bureau serve in such a capacity. They simply indicate the condition of the environment, and their readings do not serve any control functions in the ordinary sense. Similarly, water, gas, and electric meters in the home keep track of the quantity of the commodity used so that the cost to the user may be computed. The film badges worn by workers in radioactive environments serve to monitor the cumulative exposure of the wearer to radiations of various types.

1.3

Control of Processes and Operations Another extremely important type of application for measuring instruments is that in which the instrument serves as a component of an automatic control system. A functional block diagram illustrating the operation of such a system is shown in Fig. 1.1. It is clear that in order to control any variable by such a "feedback" scheme it is first necessary to measure it; thus all such control systems must incorporate at least one measuring instrument.

Examples of this type of application are endless. A familiar one is the typical home-heating system employing some type of thermostatic control. A temperature-measuring instrument (often a bimetallic element) senses the room temperature, thus providing the information necessary for proper functioning of the control system. Much more sophisticated examples are found among the aircraft and missile control

1. Often give results that are of general use rather than for restricted application.
2. Invariably require the application of simplifying assumptions. Thus not the actual physical system but rather a simplified "mathematical model" of the system is studied. This means the theoretically predicted behavior is *always* different from the real behavior.
3. In some cases, may lead to complicated mathematical problems. This has blocked theoretical treatment of many problems in the past. Today, increasing availability of high-speed computing machines allows theoretical treatment of many problems that could not be so treated in the past.
4. Require only pencil, paper, computing machines, etc. Extensive laboratory facilities are not required. (Some computers are very complex and expensive, but they can be used for solving all kinds of problems. Much laboratory equipment, on the other hand, is special-purpose and suited only to a limited variety of tasks.)
5. No time delay engendered in building models, assembling and checking instrumentation, and gathering data.

Fig. 1.2. *Features of theoretical methods.*

systems. Here a single control system may require information from many measuring instruments such as pitot-static tubes, angle-of-attack sensors, thermocouples, accelerometers, altimeters, and gyroscopes.

The reader, in attempting to classify applications within his own experience according to the three categories of Sec. 1.1, may find instances where the distinction between monitoring, control, and analysis functions is not clear-cut. Thus the category decided upon may depend somewhat on one's point of view. The data obtained by the Weather Bureau, for instance, serve mainly in a monitoring function for the average person. For fruit growers, however, a report of cold weather may act in a control sense because it signals them to turn on smudge pots and apply other anti-frost measures. Also, present weather data for large areas are correlated and analyzed to form the basis of short- and long-range weather predictions, so that one could say the instruments are supplying data for an engineering analysis. Once one recognizes the possibility of a variety of interpretations, depending on the point of view, the apparent looseness of the classification should not cause any difficulty.

1.4

Experimental Engineering Analysis In solving engineering problems, two general methods are available: theoretical and experimental.

1. Often give results that apply only to the specific system being tested. However, techniques such as dimensional analysis may allow some generalization.
2. No simplifying assumptions necessary if tests are run on an actual system. The true behavior of the system is revealed.
3. *Accurate* measurements necessary to give a true picture. This may require expensive and complicated equipment. *The characteristics of all the measuring and recording equipment must be thoroughly understood.*
4. Actual system or a scale model required. If a scale model is used, similarity of all significant features must be preserved.
5. Considerable time required for design, construction, and debugging of apparatus.

Fig. 1.3. *Features of experimental methods.*

Many problems require the application of both methods. The relative amount of each employed depends on the nature of the problem. Problems on the frontiers of knowledge often require very extensive experimental studies since adequate theories are not yet available. Theory and experiment should thus be thought of as complementing each other, and the engineer who takes this attitude will, in general, be a more effective problem solver than one who neglects one or the other of these two approaches.

It may be helpful to summarize quickly the salient features of the theoretical and the experimental methods of attack. This is done in Figs. 1.2 and 1.3.

In considering the application of measuring instruments to problems of experimental engineering analysis, it may be helpful to have at hand a classification of the types of problems encountered. This classification may be accomplished according to several different plans, but one which the author has found meaningful is given in Fig. 1.4.

1.5

Conclusion Whatever the nature of the application, intelligent selection and use of measurement instrumentation depend on a broad knowledge of what is available and how the performance of the equipment may be best described in terms of the job to be done. New equipment is continuously being developed, but there are certain basic devices that have proved their usefulness in broad areas and will undoubtedly be widely used for many years. A representative cross section of such devices is discussed in this text. These devices are of great interest in

1. Testing the validity of theoretical predictions based on simplifying assumptions.
   Example: frequency-response testing of mechanical linkage for resonant frequencies.
2. Formulation of generalized empirical relationships in situations where no adequate theory exists.
   Example: determination of friction factor for turbulent pipe flow.
3. Determination of material, component, and system parameters; variables; and performance indices.
   Examples: determination of yield point of a certain alloy steel, speed-torque curves for an electric motor, thermal efficiency of a steam turbine.
4. Study of phenomena with hopes of developing a theory.
   Example: electron microscopy of metal fatigue cracks.
5. Solution of mathematical equations by means of analogies.
   Example: solution of shaft torsion problems by measurements on soap bubbles.

Fig. 1.4. *Types of experimental-analysis problems.*

themselves; they also serve as the vehicle for the presentation and development of general techniques and principles needed in handling problems in measurement instrumentation. In addition, these general concepts will be useful in treating any new devices that may be developed in the future.

The treatment is also intended to be on a level that will be of service not only to the user but also to the designer of measurement instrumentation equipment. There are two main reasons for this emphasis. One is that much experimental equipment (including measurement instruments) is often "homemade," especially in smaller companies where the high cost of specialized gear cannot always be justified. The other reason is that the instrument industry is a large and growing one which utilizes many engineers in a design capacity. While the general techniques of mechanical and electrical design as applied to *machines* are also applicable to instruments, in many cases a rather different point of view is necessary in instrument design. This is due, in part, to the fact that the design of machines is mainly concerned with considerations of *power* and *efficiency* whereas instrument design almost completely neglects these areas and concerns itself with the acquisition and manipulation of *information.* Since a considerable number of engineering graduates will work in the instrument industry, their education should include treatment of the most significant aspects of this area.

The planning, execution, and evaluation of experiments are barely touched on in this text. The author feels this extremely important material logically *follows* a treatment of measurement instrumentation. Fortunately, excellent texts treating these matters are available.[1]

Problems

1.1 By consulting various technical journals in the library, find accounts of experimental studies carried out by engineers or scientists. Find three such articles, reference them completely, explain briefly what was accomplished, and attempt to classify them according to one or more categories of Fig. 1.4.

1.2 Give three specific examples of measuring-instrument applications in each of the following areas:

a. Monitoring of processes and operations
b. Control of processes and operations
c. Experimental engineering analysis.

1.3 Compare and contrast the experimental and the theoretical approaches to the following problems:

a. What is the tolerable vibration level to which astronauts may safely be exposed in launch vehicles?
b. Find the relationship between applied force F and resulting friction torque T_f in the simple brake of Fig. P1.1.

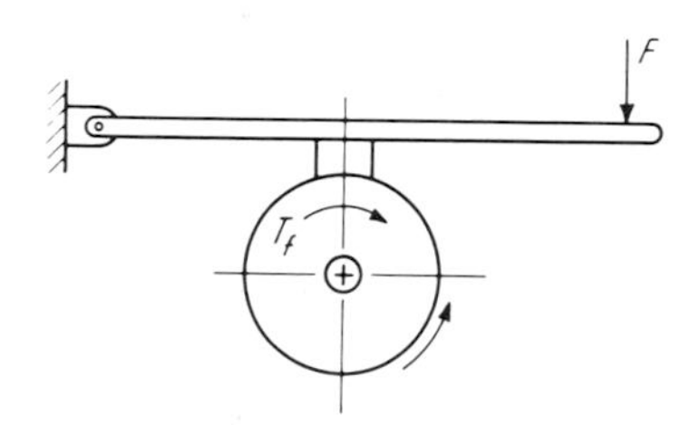

Fig. P1.1

c. Find the location of the center of mass of the rocket shown in Fig. P1.2 if the shapes, sizes, and materials of all the component parts are known.

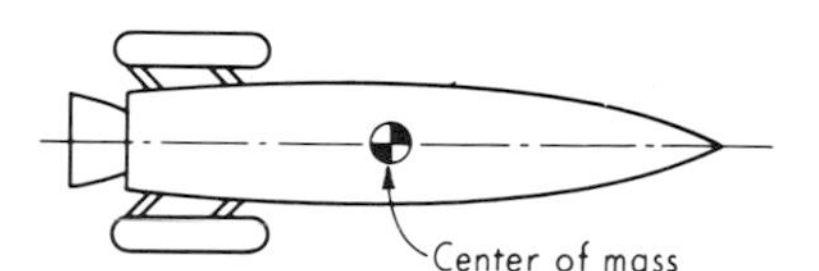

Fig. P1.2

d. At what angle with the horizontal should a projectile be launched to achieve the greatest horizontal range?

Bibliography

books

1. K. S. Lion: "Instrumentation in Scientific Research," McGraw-Hill Book Company, New York, 1959.

[1] Hilbert Schenck, Jr., "Theories of Engineering Experimentation," McGraw-Hill Book Company, New York, 1961.

2. C. F. Hix and R. P. Alley: "Physical Laws and Effects," John Wiley & Sons, Inc., New York, 1958.
3. P. K. Stein: "Measurement Engineering," Stein Engineering Services, Inc., Phoenix, Ariz., 1964.
4. C. S. Draper, Walter McKay, and Sidney Lees: "Instrument Engineering," vols. I, II, III, McGraw-Hill Book Company, New York, 1955.
5. R. H. Cerni and L. E. Foster: "Instrumentation for Engineering Measurement," John Wiley & Sons, Inc., New York, 1962.
6. W. M. Cady: "Physical Measurements in Gas Dynamics and Combustion," Princeton University Press, Princeton, N.J., 1954.
7. E. B. Wilson, Jr.: "An Introduction to Scientific Research," McGraw-Hill Book Company, New York, 1952.
8. R. C. Dove and P. H. Adams: "Experimental Stress Analysis and Motion Measurement," Charles E. Merrill Books, Inc., Columbus, Ohio, 1964.
9. D. Bartholomew: "Electrical Measurements and Instrumentation," Allyn and Bacon, Inc., Boston, 1963.
10. E. Frank: "Electrical Measurement Analysis," McGraw-Hill Book Company, New York, 1959.
11. T. G. Beckwith and W. L. Buck: "Mechanical Measurements," Addison-Wesley Publishing Company, Inc., Reading, Mass., 1961.
12. N. H. Cook and E. Rabinowicz: "Physical Measurement and Analysis," Addison-Wesley Publishing Company, Inc., Reading, Mass., 1963.
13. D. P. Eckman: "Industrial Instrumentation," John Wiley & Sons, Inc., New York, 1950.
14. G. L. Tuve: "Mechanical Engineering Experimentation," McGraw-Hill Book Company, New York, 1961.
15. D. M. Considine (ed.): "Process Instruments and Controls Handbook," McGraw-Hill Book Company, New York, 1957.
16. Transducer Compendium, Instrument Society of America, Pittsburgh, Pa.

periodicals

1. *The Review of Scientific Instruments*
2. *Journal of Scientific Instruments* (Great Britain)
3. *Transactions of Instrument Society of America*
4. *Experimental Mechanics*
5. *Measurement Techniques* (USSR; English translation)
6. *Instruments and Experimental Techniques* (USSR; English translation)
7. *Industrial Laboratory* (USSR; English translation)
8. *Instruments and Control Systems*
9. *Control Engineering*
10. *Journal of Instrument Society of America*
11. *Archiv für Technisches Messen* (Germany)
12. *Journal of Research of the National Bureau of Standards*
13. *Transactions of the Society of Instrument Technology* (Great Britain)
14. *Electromechanical Design*

2 Generalized configurations and functional descriptions of measuring instruments

2.1 *The Functional Elements of an Instrument*

It is possible and desirable to describe both the operation and the performance (degree of approach to perfection) of measuring instruments and associated equipment in a generalized way without recourse to specific physical hardware. The operation can be described in terms of the functional elements of instrument systems, and the performance is defined in terms of the static and dynamic performance characteristics. This section develops the concept of the functional elements of an instrument or instrument system.

If one examines diverse physical instruments with a view toward

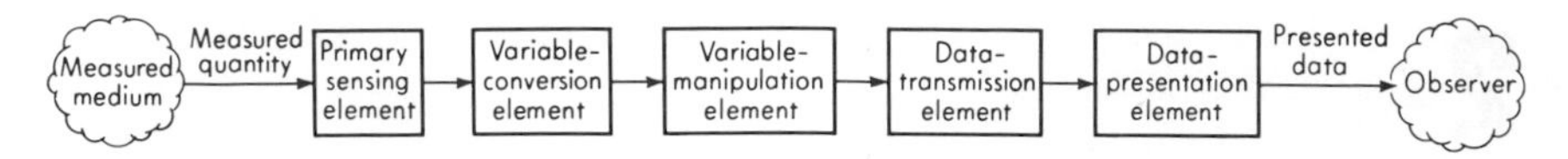

Fig. 2.1. *Functional elements of an instrument or measurement system.*

generalization, he soon recognizes in the elements of the instruments a recurring pattern of similarity with regard to function. This leads to the concept of breaking down instruments into a limited number of types of elements according to the generalized function performed by the element. This breakdown can be made in a number of ways, and no standardized universally accepted scheme is at present in use. We shall present one such scheme which may be of help to the reader in understanding the operation of any new instrument with which he may come in contact and also in planning the design of a new instrument.

Consider the diagram of Fig. 2.1, which represents a possible arrangement of functional elements in an instrument and includes *all* the basic functions considered necessary for a description of any instrument. The *primary sensing element* is that which first receives energy from the measured medium and produces an output depending in some way on the measured quantity. It is important to note that an instrument *always* extracts some energy from the measured medium. Thus the measured quantity is *always* disturbed by the act of measurement, making a perfect measurement theoretically impossible. Good instruments are designed to minimize this effect but it is always present to some degree.

The output signal of the primary sensing element is some physical variable, such as a displacement or a voltage. For the instrument to perform the desired function, it may be necessary to convert this variable to another more suitable variable while preserving the information content of the original signal. An element that performs such a function is called a *variable-conversion element.* It should be noted that every instrument need not include a variable-conversion element while some require several. Also, the "elements" we speak of are *functional* elements, not physical elements. That is, Fig. 2.1 shows an instrument neatly separated into blocks, which may lead the reader to think of the physical apparatus as being precisely separable into subassemblies performing the specific functions shown. This is, in general, not the case; a specific piece of hardware may perform *several* of the basic functions, for instance.

In performing its intended task, an instrument may require that a signal represented by some physical variable be manipulated in some way. By manipulation we here mean specifically a change in numerical value according to some definite rule but a preservation of the physical nature of the variable. Thus an electronic amplifier accepts a small voltage signal as input and produces an output signal that is also a voltage but is some constant times the input. An element that performs such a function will be called a *variable-manipulation element.* Again, the reader should not be misled by Fig. 2.1. A variable-manipulation element does not necessarily *follow* a variable-conversion element; it may precede it, appear elsewhere in the chain, or not appear at all.

When the functional elements of an instrument are actually physically separated, it becomes necessary to transmit the data from one to another. An element performing this function is called a *data-transmission element.* It may be as simple as a shaft and bearing assembly or as complicated as a telemetry system for transmitting signals from missiles to ground equipment by radio.

If the information about the measured quantity is to be communicated to a human being for monitoring, control, or analysis purposes, it must be put into a form recognizable by one of the human senses. An element that performs this "translation" function is called a *data-presentation element.* This function includes the simple *indication* of a pointer moving over a scale and also the *recording* of a pen moving over a chart. Indication and recording may also be performed in discrete increments (rather than smoothly) as exemplified by an optical flat used for measuring flatness of surfaces by light-interference principles and an electric typewriter for recording numerical data. While the majority of instruments communicate with people through the medium of the visual sense, the use of other senses such as hearing and touch is certainly conceivable. Certain methods of recording may present the data in a form not directly detectable by human senses. The magnetic tape recorder is a noteworthy example. For such a case, suitable instruments for extracting the stored information at any desired time and converting it to a form intelligible to man are required.

Before going on to some illustrative examples, let it be emphasized again that Fig. 2.1 is intended as a vehicle for presenting the concept of functional elements and not as a physical schematic of a generalized instrument. A given instrument may involve the basic functions in any number and combination; they need not appear in the order of Fig. 2.1. A given physical component may serve several of the basic functions.

As an example of the above concepts, consider the rudimentary pressure gage of Fig. 2.2. One of several possible valid interpretations is as follows: The primary sensing element is the piston, which also serves the function of variable conversion since it converts the fluid pressure (force per unit area) into a resultant force on the piston face. Force is transmitted by the piston rod to the spring, which converts force into a proportional displacement. This displacement of the piston rod is magnified (manipulated) by the linkage to give a larger pointer displace-

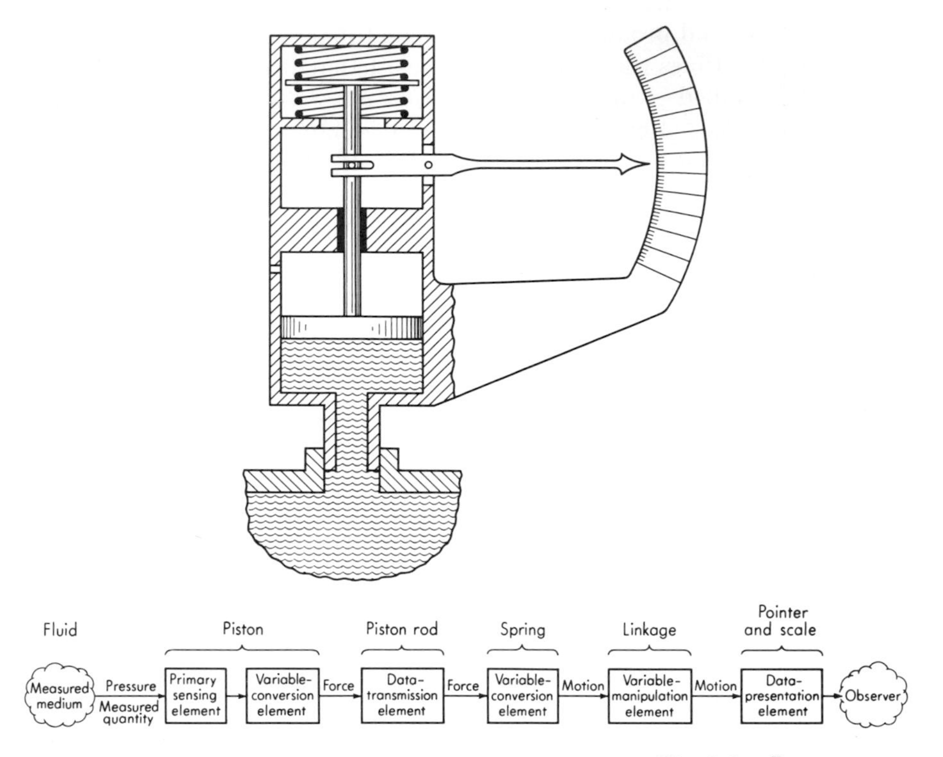

Fig. 2.2. *Pressure gage.*

ment. The pointer and scale indicate the pressure, thus serving as data-presentation elements. If it were necessary to locate the gage at some distance from the source of pressure, a small tube could serve as a data-transmission element.

Figure 2.3 depicts a pressure-type thermometer. The liquid-filled bulb acts as primary sensor and variable-conversion element since a temperature change results in a pressure buildup within the bulb, because of the constrained thermal expansion of the filling fluid. This pressure is transmitted through the tube to a Bourdon-type pressure gage which converts pressure to displacement. This displacement is manipulated by the linkage and gearing to give a larger pointer motion. A scale and pointer again serve for data presentation.

A remote-reading shaft-revolution counter is shown in Fig. 2.4. The microswitch sensing arm and the camlike projection on the rotating shaft serve both a primary sensing function and a variable-conversion function since rotary displacement is converted to linear displacement. The microswitch contacts also serve for variable conversion, converting a

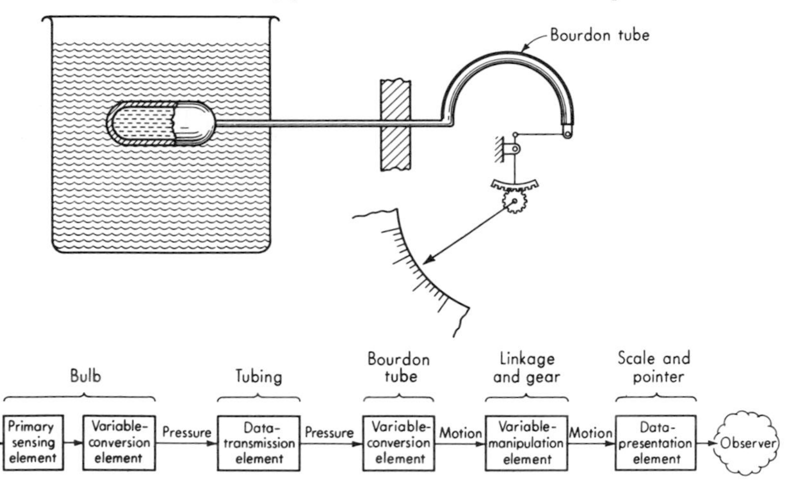

Fig. 2.3. *Pressure thermometer.*

mechanical oscillation into an electrical oscillation (a sequence of voltage pulses). These voltage pulses may be transmitted relatively long distances over wires to a solenoid. The solenoid reconverts the electrical pulses into mechanical reciprocation of the solenoid plunger which serves as input to a mechanical counter. The counter itself involves variable conversion (reciprocating to rotary motion), variable manipulation (rotary motion to decimalized rotary motion), and data presentation.

As a final example, let us examine Fig. 2.5 which illustrates schematically a D'Arsonval galvanometer as used in oscillographs. A time-varying voltage to be recorded is applied to the ends of the two wires

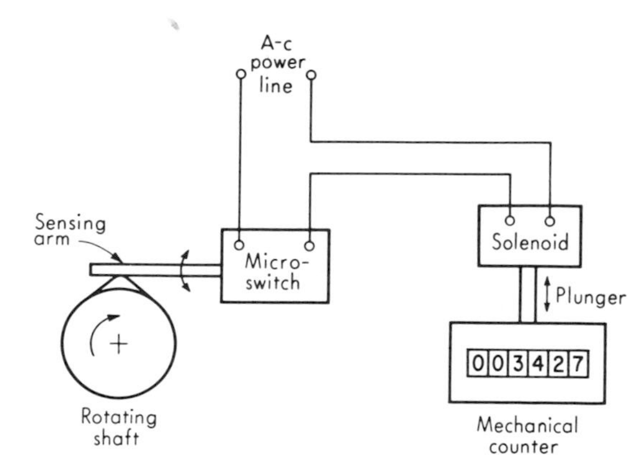

Fig. 2.4. *Digital revolution counter.*

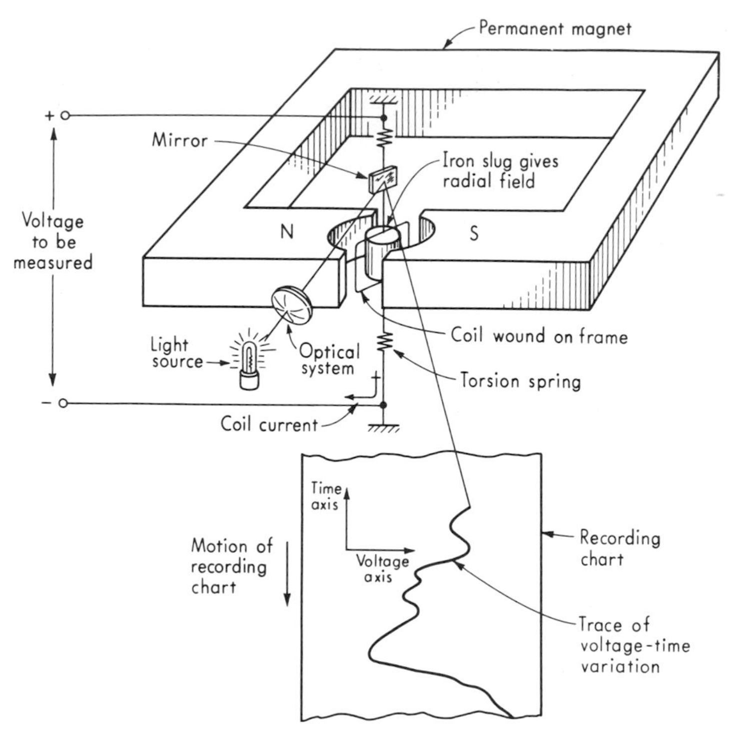

Fig. 2.5. *D'Arsonval galvanometer.*

which transmit the voltage to a coil made up of a number of turns wound on a rigid frame. This coil is suspended in the field of a permanent magnet. The resistance of the coil converts the applied voltage to a proportional current (ideally). The interaction between the current and the magnetic field produces a torque on the coil, giving another variable conversion. This torque is converted to an angular deflection by the torsion springs. A mirror rigidly attached to the coil frame converts the frame rotation into the rotation of a light beam which it reflects. The light-beam rotation is twice the mirror rotation, giving a motion magnification. The reflected beam intercepts a recording chart made of photosensitive material which is moved at a fixed and known rate, giving a time base. The combined horizontal motion of the light spot and the vertical motion of the recording chart generate a graph of voltage versus time. The "optical lever arm" (the distance from the mirror to the recording chart) has a motion-magnifying effect since the spot displacement per unit mirror rotation is directly proportional to it.

In this instrument the coil and magnet assembly would probably be considered as the primary sensing element since the lead wires (which serve a transmission function) are not really part of the instrument and the coil resistance (which acts in a variable-conversion function) is an intrinsic part of the coil. In any case, the assignment of precise names to specific components is not nearly as important as the recognition of the basic functions necessary to the successful operation of the instrument. By concentrating on these functions and the various physical devices available for accomplishing them, we develop our ability to synthesize new combinations of elements leading to new and useful instruments. This ability is fundamental to all instrument design.

2.2

Active and Passive Transducers Once certain basic functions common to all instruments have been identified, it is then in order to see if it is possible to make some generalizations on *how* these functions may be performed. One such generalization is concerned with energy considerations. In performing any of the general functions indicated in Fig. 2.1, a physical component may act as an *active transducer* or a *passive transducer*.

A component whose output energy is supplied entirely or almost entirely by its input signal is commonly called a *passive transducer*. The output and input signals may involve energy of the same form (say both mechanical) or there may be an energy conversion from one form to another (say mechanical to electrical). (In much technical literature the term transducer is restricted to devices involving energy *conversion*, but, conforming to the dictionary definition of the term, we do not make this restriction.)

An *active transducer*, on the other hand, has an auxiliary source of power which supplies a major part of the output power while the input signal supplies only an insignificant portion. Again, there may or may not be a conversion of energy from one form to another.

In all the examples of Sec. 2.1 there is only one active transducer, the microswitch of Fig. 2.4; all other components are passive transducers. The power to drive the solenoid comes not from the rotating shaft but from the a-c power line, an auxiliary source of power. Some further examples of active transducers may be in order. The electronic amplifier shown in Fig. 2.6 furnishes a good one. The element supplying the input-signal voltage e_i need supply only a negligible amount of power since almost no current is drawn, owing to negligible grid current and a high R_g. However, the output element (the load resistance R_L) receives significant current and voltage and thus power. This power must be supplied by the plate battery E_{bb}, the auxiliary power source. Thus the input *controls* the output but does not actually supply the output power.

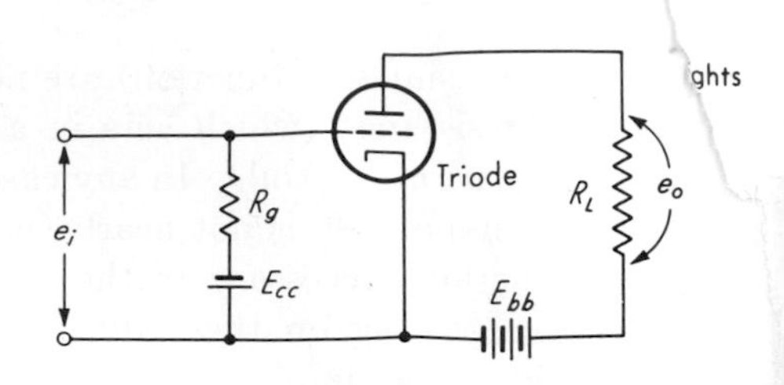

Fig. 2.6. Triode amplifier.

Another active transducer of great practical importance, the *instrument servomechanism*, is shown in simplified form in Fig. 2.7. This is actually an instrument *system* made up of components, some of which are passive transducers and some active transducers. When considered as an entity, however, with input voltage e_i and output displacement x_o, it meets the definition of an active transducer and is profitably thought of as such. The purpose of this device is to cause the motion x_o to follow the variations of the voltage e_i in a proportional manner. Since the motor torque is proportional to the error voltage e_e, it is clear that the system can be at rest only if e_e is zero. This occurs only when $e_i = e_{sl}$; since e_{sl} is proportional to x_o, this means that x_o must be proportional to e_i in the

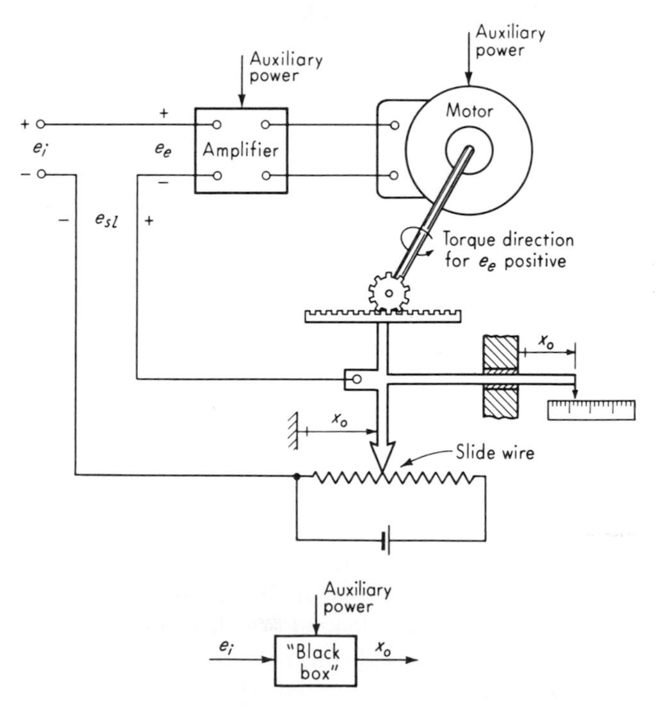

Fig. 2.7. Instrument servomechanism.

tic case. If e_i varies, x_o will tend to follow it, and by proper design ccurate "tracking" of e_i by x_o should be possible.

2.3

Analog and Digital Modes of Operation It is possible further to classify how the basic functions may be performed by turning attention to the continuous or discrete nature of the signals that represent the information. Signals that vary in a continuous fashion and can take on an infinity of values in any given range are called *analog* signals; the devices that produce such signals are called analog devices. (This is strictly in a macroscopic sense since all physical effects become discrete in atomistic considerations.) In contrast, signals that vary in discrete steps and can thus take on only a finite number of different values in any given range are described as *digital* signals; the devices that produce such signals are called digital devices.

The majority of present-day measuring instruments are of the analog type. The only digital device illustrated in this text up to this point is the revolution counter of Fig. 2.4. This is clearly a digital device since it is impossible for this instrument to indicate, say, 0.79 revolution; it measures only in steps of one revolution. The importance of digital instruments is increasing, perhaps mainly because of the increasing use of digital computers in both data-reduction and automatic control systems. Since the digital computer works only with digital signals, any information supplied to it must be in digital form. The computer's output is also in digital form. Thus any communication with the computer at either the input or output end must be in terms of digital signals. Since most present-day measurement and control apparatus is of an analog nature it is necessary to have both *analog-to-digital converters* (at the input to the computer) and *digital-to-analog converters* (at the output of the computer). These devices (which are discussed in more detail in a later chapter) serve as "translators" that enable the computer to communicate with the outside world, which is largely of an analog nature. Effort is being expended to develop both measuring and control devices that are *inherently* digital in nature so that the somewhat complex converters will not be needed.

2.4

Null and Deflection Methods A useful classification with regard to the mode of operation of instruments separates devices by their operation on a null or a deflection principle. In a *deflection-type* device the measured quantity produces some physical effect that engenders a similar

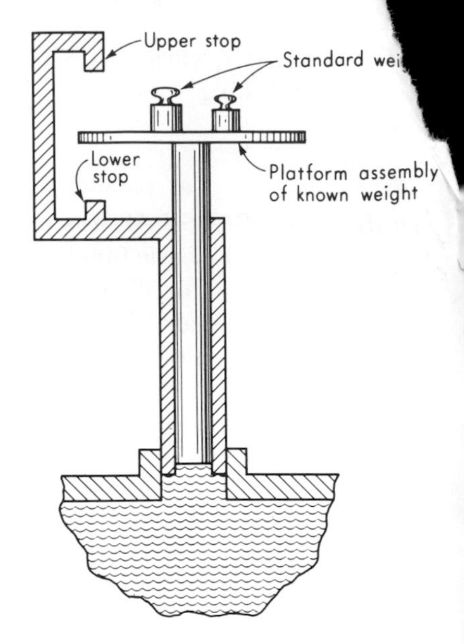

Fig. 2.8. Dead-weight pressure gage.

but opposing effect in some part of the instrument. The opposing effect is closely related to some variable (usually a mechanical displacement or deflection) that can be directly observed by some human sense. The opposing effect increases until a balance is achieved, at which point the "deflection" is measured and the value of the measured quantity inferred from this. The pressure gage of Fig. 2.2 exemplifies this type of device since the pressure force engenders an opposing spring force due to an unbalance of forces on the piston rod (called the force-summing link) which causes a deflection of the spring. As the spring deflects, its force increases; thus a balance will be achieved at some deflection if the pressure is within the design range of the instrument.

In contrast to the deflection-type device, a *null-type* device attempts to maintain deflection at zero by suitable application of an effect opposing that generated by the measured quantity. Necessary to such an operation are a detector of unbalance and a means (manual or automatic) of restoring balance. Since deflection is kept at zero (ideally), determination of numerical values requires accurate knowledge of the magnitude of the opposing effect. A pressure gage operating on a null principle is depicted in simplified form in Fig. 2.8. By adding the proper standard weights to the platform of known weight, the pressure force on the face of the piston may be balanced by gravitational force. The condition of force balance is indicated by the platform remaining at rest between the upper and lower stops. Since the weights and the piston area are all known, the unknown pressure may be computed.

Upon comparing the null and deflection methods of measurement exemplified by the pressure gages described above, we note that, in the deflection instrument, accuracy depends on the calibration of the spring whereas in the null instrument it depends on the accuracy of the standard weights. In this particular case (and also for most measurements in general) the accuracy attainable by the null method is of a higher level than that of the deflection method. One reason for this is that the spring is not in itself a primary standard of force but must be calibrated by standard weights, whereas in the null instrument a *direct* comparison of the unknown force with the standard is achieved. Another advantage of null methods is the fact that, since the measured quantity is balanced out, the detector of unbalance can be made very sensitive because it need cover only a small range around zero. Also the detector need not be calibrated since it must detect only the presence and direction of unbalance and not the amount. On the other hand, a deflection instrument must be larger, more rugged, and thus less sensitive if it is to measure large magnitudes.

The disadvantages of null methods appear mainly in dynamic measurements. Let us consider the pressure gages again. The difficulty in keeping the platform balanced for a fluctuating pressure should be apparent. The spring-type gage suffers not nearly so much in this respect. By use of automatic balancing devices (such as the instrument servomechanism of Fig. 2.7) the speed of null methods may be improved considerably, and instruments of this type are of great importance.

2.5

Input-Output Configuration of Measuring Instruments and Instrument Systems Before going on to discuss instrument performance characteristics, it is desirable to develop a generalized configuration which brings out the significant input-output relationships present in all measuring apparatus. A scheme suggested by Draper, McKay, and Lees[1] is presented in somewhat modified form in Fig. 2.9. Input quantities are classified into three categories: desired inputs, interfering inputs, and modifying inputs. *Desired inputs* represent the quantities that the instrument is specifically intended to measure. *Interfering inputs* represent quantities to which the instrument is unintentionally sensitive. A desired input produces a component of output according to an input-output relation symbolized by F_D, where F_D denotes the mathematical operations necessary to obtain the output from the input.

[1] C. S. Draper, Walter McKay, and Sidney Lees, "Instrument Engineering," vol. III, p. 58, McGraw-Hill Book Company, New York, 1955.

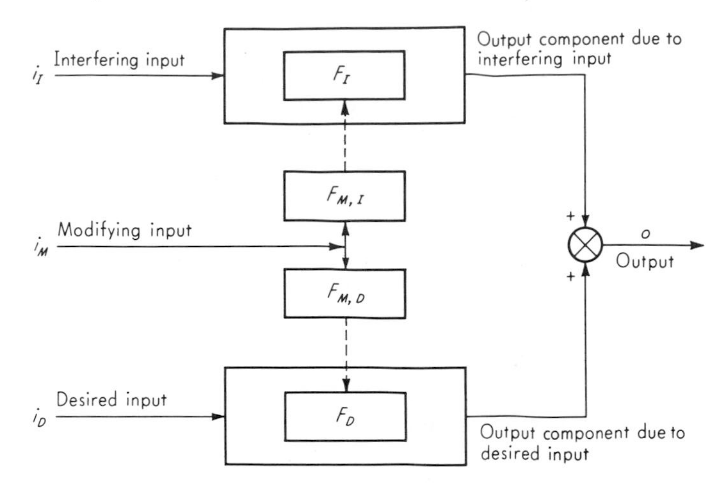

Fig. 2.9. Generalized input-output configuration.

The symbol F_D may represent different concepts, depending on the particular input-output characteristic that is being described. Thus F_D might be a constant number K giving the proportionality constant relating a constant static input to the corresponding static output for a linear instrument. For a nonlinear instrument a simple constant is not adequate to relate static inputs and outputs; an algebraic or transcendental *function* is required. To relate dynamic inputs and outputs, differential equations are necessary. If a description of the output "scatter," or dispersion, for repeated equal static inputs is desired, a statistical distribution function of some kind is needed. The symbol F_D encompasses all such concepts. The symbol F_I serves a similar function for an interfering input.

The third class of inputs might perhaps be thought of as being included among the interfering inputs, but a separate classification is actually more significant. This classification is that of modifying inputs. *Modifying inputs* are the quantities that cause a change in the input-output relations for the desired and interfering inputs; that is, they cause a change in F_D and/or F_I. The symbols $F_{M,I}$ and $F_{M,D}$ represent (in the appropriate form) the specific manner in which i_M affects F_I and F_D. These symbols, $F_{M,I}$ and $F_{M,D}$, are again to be interpreted in the same general way as F_I and F_D were.

The block diagram of Fig. 2.9 illustrates the above concepts. The circle with a cross in it is a conventional symbol for a *summing device.* The two plus signs as shown indicate that the output of the summing device is the instantaneous algebraic sum of its two inputs. Since an instrument system may have several inputs of each of the three types and

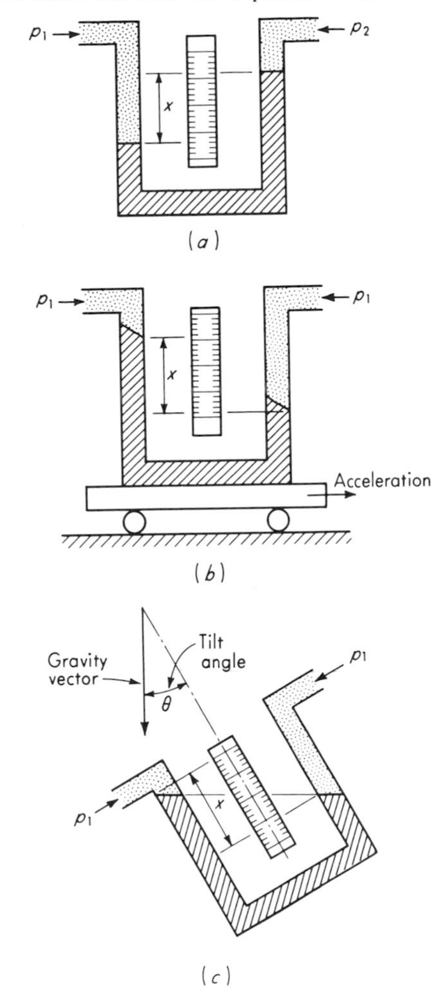

Fig. 2.10. *Spurious inputs for manometer.*

also several outputs, it may be necessary to draw more complex block diagrams than in Fig. 2.9. This extension is, however, straightforward.

The above concepts can be clarified by means of specific examples. Consider the mercury manometer used for differential-pressure measurement as shown in Fig. 2.10*a*. The desired inputs are the pressures p_1 and p_2 whose difference causes the output displacement x which can be read off the calibrated scale. Figure 2.10*b* and *c* shows the action of two possible interfering inputs. In Fig. 2.10*b* the manometer is mounted on some vehicle that is accelerating. A simple analysis will show that there will be an output x even though the differential pressure might be zero. Thus if one is trying to measure pressures under such circumstances an error

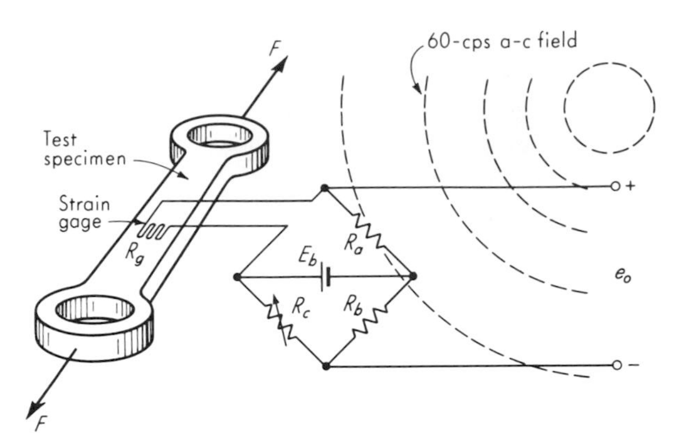

Fig. 2.11. *Interfering input for strain-gage circuit.*

will be engendered because of the interfering acceleration input. Similarly, in Fig. 2.10*c*, if the manometer is not properly aligned with the gravity vector it may give an output signal x even though no pressure difference exists. Thus the tilt angle θ is an interfering input. (It is also a modifying input.)

Modifying inputs for the manometer include ambient temperature and gravitational force. Ambient temperature manifests its influence in a number of ways. First, the calibrated scale changes length with temperature; thus the proportionality factor relating $p_1 - p_2$ to x is modified whenever temperature varies from its basic calibration value. Also, the density of mercury varies with temperature, again leading to a change in the proportionality factor. A change in gravitational force due to changes in location of the manometer, such as moving it to another country or putting it aboard a spaceship, leads to a similar modification in the scale factor. It should be noted that the effects of *both* the desired and interfering inputs may be modified by the modifying inputs.

As another example, consider the electrical-resistance strain-gage setup shown in Fig. 2.11. The gage consists of a fine wire grid of resistance R_g firmly cemented to the specimen whose unit strain ϵ at a certain point is to be measured. When strained, the gage's resistance changes according to the relation

$$\Delta R_g = (GF)R_g\epsilon \tag{2.1}$$

where $\Delta R_g \triangleq$ change in gage resistance, ohms (2.2)

$GF \triangleq$ gage factor, dimensionless (2.3)

$R_g \triangleq$ gage resistance when unstrained, ohms (2.4)

$\epsilon \triangleq$ unit strain, in./in. (2.5)

The resistance change is proportional to the strain; thus if we could measure the resistance, we could compute the strain. The resistance is measured by using the Wheatstone-bridge arrangement shown. When no load F is present, the bridge is balanced (e_o set to zero) by adjusting R_c. Application of load causes a strain, a ΔR_g, and thus unbalances the bridge, causing an output voltage e_o which is proportional to ϵ and can be measured on a meter or oscilloscope. The voltage e_o is given by

$$e_o = -(GF)R_g\epsilon E_b \frac{R_a}{(R_g + R_a)^2} \tag{2.6}$$

The desired input here is clearly the strain ϵ which causes a proportional output voltage e_o. One interfering input which often results in trouble in such apparatus is the 60-cycle field caused by nearby power lines, electric motors, etc. This field induces voltages in the strain-gage circuit, causing output voltages e_o even when the strain is zero. Another interfering input is the gage temperature. If this varies, it causes a change in gage resistance that will cause a voltage output even if there is no strain. Temperature has another interfering effect since it causes a differential expansion of the gage and the specimen which gives rise to a strain ϵ and a voltage e_o even though no force F has been applied. Temperature also acts as a modifying input since the gage factor is sensitive to temperature. The battery voltage E_b is another modifying input. Both these are modifying inputs since they tend to change the proportionality factor between the desired input ϵ and the output e_o or between an interfering input (gage temperature) and output e_o.

Methods of correction for interfering and modifying inputs. In the design and/or use of measuring instruments a number of methods for nullifying or reducing the effects of spurious inputs are available. We shall briefly describe some of the most widely used.

The *method of inherent insensitivity* proposes the obviously sound design philosophy that the elements of the instrument should *inherently* be sensitive only to the desired inputs. While this is not usually entirely possible, the simplicity of this approach encourages one to consider its application wherever feasible. In terms of the general configuration of Fig. 2.9, this approach requires that somehow F_I and/or $F_{M,D}$ be made as nearly equal to zero as possible. Thus, even though i_I and/or i_M may exist, they cannot affect the output. As an example of the application of this concept to the strain gage of Fig. 2.11, we might try to find some gage material that exhibits an extremely low temperature coefficient of resistance while retaining its sensitivity to strain. If such a material can be found, the problem of interfering temperature inputs is at least partially solved. Similarly, in mechanical apparatus that must maintain

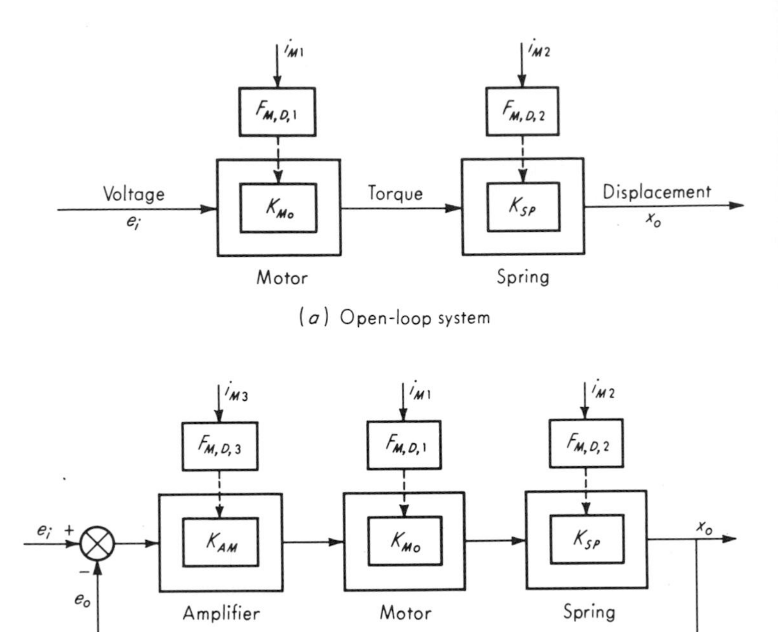

Fig. 2.12. *Use of feedback to reduce effect of spurious inputs.*

accurate dimensions in the face of ambient-temperature changes, the use of a material of very small temperature coefficient of expansion (such as the alloy Invar) may be helpful.

The *method of high-gain feedback* is exemplified by the system shown in Fig. 2.12*b*. Suppose we wish to measure a voltage e_i by applying it to a motor whose torque is applied to a spring, causing a displacement x_o which may be measured on a calibrated scale. By proper design, the displacement x_o might be made proportional to the voltage e_i according to

$$x_o = (K_{Mo}K_{SP})e_i \qquad (2.7)$$

where K_{Mo} and K_{SP} are appropriate constants. This arrangement, shown in Fig. 2.12*a*, would be called an open-loop system. If modifying inputs i_{M1} and i_{M2} exist, they cause changes in K_{Mo} and K_{SP} which lead to errors in the relation between e_i and x_o. These errors are in *direct proportion* to the changes in K_{Mo} and K_{SP}. Suppose, instead, we construct a system as in Fig. 2.12*b*. Here the output x_o is measured by the feedback device which produces a voltage e_o proportional to x_o. This voltage is subtracted from the input voltage e_i and the difference applied to an amplifier which drives the motor and thereby the spring to produce x_o. We may write

$$(e_i - e_o)K_{AM}K_{Mo}K_{SP} = (e_i - K_{FB}x_o)K_{AM}K_{Mo}K_{SP} = x_o \tag{2.8}$$

$$e_iK_{AM}K_{Mo}K_{SP} = (1 + K_{AM}K_{Mo}K_{SP}K_{FB})x_o \tag{2.9}$$

$$x_o = \frac{K_{AM}K_{Mo}K_{SP}}{1 + K_{AM}K_{Mo}K_{SP}K_{FB}}\, e_i \tag{2.10}$$

Suppose now that we design K_{AM} to be very large (a "high-gain" system), so that $K_{AM}K_{Mo}K_{SP}K_{FB} \gg 1$. Then

$$x_o \approx \frac{1}{K_{FB}}\, e_i \tag{2.11}$$

The significance of Eq. (2.11) is that the effect of variations in K_{Mo}, K_{SP}, and K_{AM} (due to modifying inputs i_{M1}, i_{M2}, and i_{M3}) on the relation between input e_i and output x_o has been made negligible. *We now require only that K_{FB} stay constant (unaffected by i_{M4}) in order to maintain constant input-output calibration as shown by Eq.* (2.11).

The reader may question whether much really has been gained by this somewhat elaborate scheme since we have merely transferred the requirements for stability from K_{Mo} and K_{SP} to K_{FB}. In actual practice, however, this method often leads to great improvements in accuracy. One reason for this is that, since the amplifier supplies most of the power needed, the feedback device can be designed with low power-handling capacity. This in general leads to greater accuracy and linearity in the feedback-device characteristics. Also, the input signal e_i need carry only negligible power; thus the feedback system extracts less energy from the measured medium than the corresponding open-loop system. This, of course, results in less distortion of the measured quantity due to the presence of the measuring instrument. Finally, if the open-loop chain consists of several (perhaps many) devices, each susceptible to its own spurious inputs, *all* these bad effects can be negated by the use of high amplification and a stable and accurate feedback device.

Before passing on to other methods, it should be mentioned that application of the feedback principle is not without its own peculiar problems. The main one is that of dynamic instability, wherein excessively high amplification leads to destructive oscillations. The study of the design of feedback systems is a whole field in itself, and many texts treating this subject are available.[1]

The *method of calculated output corrections* requires that one be able to measure or estimate the magnitudes of the interfering and/or modifying inputs and that one know quantitatively how they affect the output. With this information available, it is possible to calculate corrections which may be added to or subtracted from the indicated output so as to leave (ideally) only that component associated with the desired input. Thus, in the manometer of Fig. 2.10, the effects of temperature on the calibrated scale's length and on the density of mercury may both be quite accurately computed if the temperature is known. The local gravitational acceleration is also known for a given elevation and latitude so that this effect may also be corrected by calculation. Although theoretically applicable to any form of input, the method of calculated output corrections is probably most used for inputs that are essentially constant.

The *method of signal filtering* is based on the possibility of introducing certain elements ("filters") into the instrument which in some fashion block the spurious signals so that their effects on the output are removed or reduced. The filter may be applied to any suitable signal in the instrument, be it input, output, or intermediate signal. The concept of signal filtering is shown schematically in Fig. 2.13 for the cases of input filtering and output filtering. The application to intermediate signals should be obvious. In Fig. 2.13*a* the inputs i_I and i_M are caused to pass through filters whose input-output relation is (ideally) zero. Thus i_I' and i_M' are zero even if i_I and i_M are not zero. The concept of output filtering is illustrated in Fig. 2.13*b*. Here the output o, though really one signal, is thought of as a superposition of o_I (output due to interfering input), o_D (output due to desired input), and o_M (output due to modifying input). If it is possible to construct filters that selectively block o_I and o_M but allow o_D to pass through, this may be symbolized as in Fig. 2.13*b* and results in o' consisting entirely of o_D.

The filters necessary in the application of this method may take several forms; they are best illustrated by examples. If a filter is put directly in the path of a spurious input, it can be designed (ideally) to block completely the passage of the signal. If, however, the filter is inserted at a point where the signal contains both desired and spurious components, it must be designed to be selective; that is, it must pass the desired components essentially unaltered while effectively suppressing all others.

It is often necessary to attach delicate instruments to structures

[1] E. O. Doebelin, "Dynamic Analysis and Feedback Control," McGraw-Hill Book Company, New York, 1962.

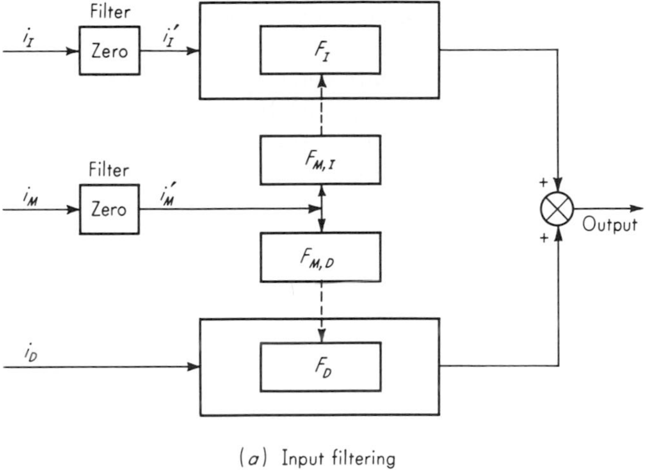

(a) Input filtering

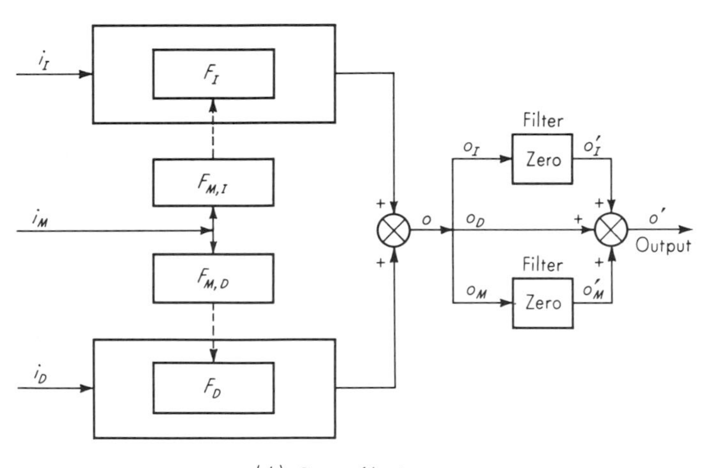

(b) Output filtering

Fig. 2.13. *General principle of filtering.*

that vibrate. Electromechanical devices for navigation and control of aircraft or missiles are outstanding examples. Figure 2.14*a* shows how the interfering vibration input may be filtered out by use of suitable spring mounts. The mass-spring system is actually a mechanical filter which passes on to the instrument only a negligible fraction of the motion of the vibrating structure.

The interfering tilt-angle input to the manometer of Fig. 2.10*c* may be effectively filtered out by means of the gimbal-mounting scheme of Fig. 2.14*b*. If the gimbal bearings are essentially frictionless, the rotations θ_1 and θ_2 cannot be communicated to the manometer; thus it always hangs vertical.

In Fig. 2.14*c* the thermocouple reference junction is shielded from ambient-temperature fluctuations by means of thermal insulation. Such an arrangement acts as a filter for temperature or heat-flow inputs.

The strain-gage circuit of Fig. 2.14*d* is shielded from the interfering 60-cps field by enclosing it in a metal box of some sort. This solution of the problem corresponds to filtering the interfering *input*. Another possible solution, which corresponds to selective filtering of the *output*, is shown in Fig. 2.14*e*. For this approach to be effective, it is essential that the frequencies in the desired signal occupy a frequency range considerably separated from those in the undesired component of the signal.

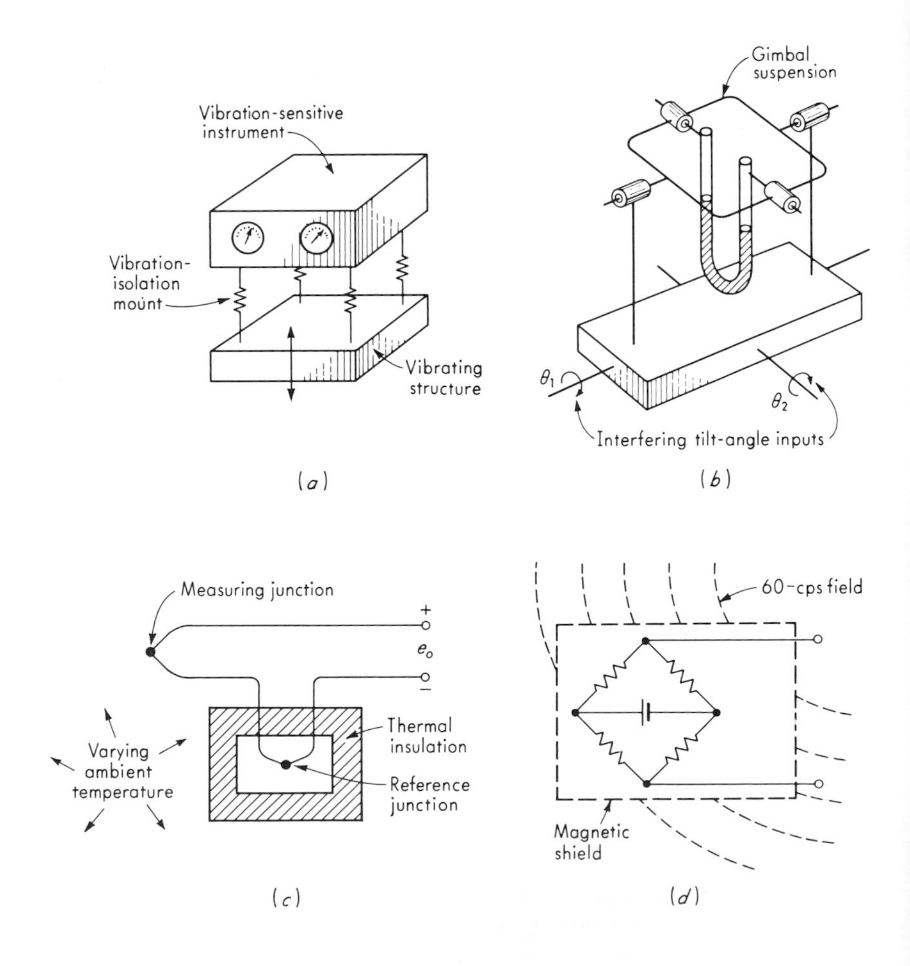

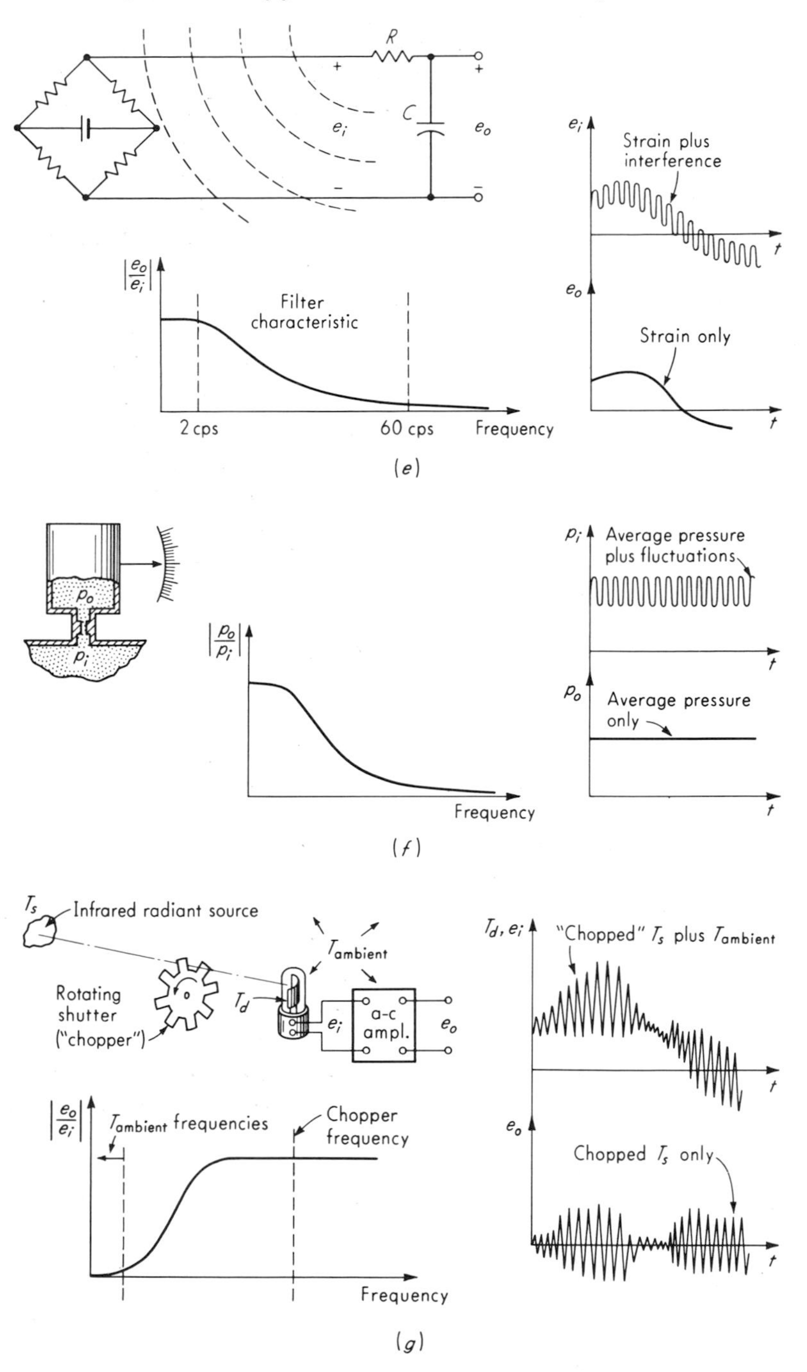

Fig. 2.14. *Examples of filtering.*

In the present example, suppose the strains to be measured are mainly steady and will never vary more rapidly than 2 cps. It is then possible to insert a simple *RC* filter, as shown, that will pass the desired signals but almost completely block the 60-cps interference.

Figure 2.14*f* shows the pressure gage of Fig. 2.2 modified by the insertion of a flow restriction between the source of pressure and the piston chamber. Such an arrangement is useful, for example, if one wishes to measure only the average pressure in a large air tank that is being supplied by a reciprocating compressor. The pulsations in the air pressure may be smoothed by the pneumatic filtering effect of the flow restriction and associated volume. The variation of the output-input amplitude ratio $|p_o/p_i|$ with frequency is similar to that for the electrical *RC* filter of Fig. 2.14*e*. Thus steady or slowly varying input pressures are accurately measured while rapid variations are strongly attenuated. The flow restriction may be in the form of a needle valve, allowing easy adjustment of the filtering effect.

A "chopped" radiometer is shown in simplified form in Fig. 2.14*g*. The purpose of this device is to sense the temperature T_s of some body in terms of the infrared radiant energy that it emits. The emitted energy is focused on a detector of some sort and causes the temperature T_d of the detector and thus its output voltage e_i to vary. The difficulty with such devices is that the ambient temperature, as well as T_s, affects T_d. This effect is serious since the radiant energy to be measured causes very small changes in T_d; thus small ambient drifts can completely mask the desired input. An ingenious solution to this problem interposes a rotating shutter between the radiant source and the detector so that the desired input is "chopped," or modulated, at a known frequency. This frequency is chosen to be much higher than the frequencies at which ambient drifts may occur. The output signal e_i of the detector thus is a superposition of slow ambient fluctuations and a high-frequency wave whose amplitude varies in proportion to variations in T_s. Since the desired and interfering components are thus widely separated in frequency, they may be selectively filtered. In this case one desires a filter that rejects constant and slowly varying signals but faithfully reproduces rapid variations. Such a characteristic is typical of an ordinary a-c amplifier, and since amplification is necessary in such instruments in any case, the use of an a-c amplifier as shown solves two problems at once.

In summing up the method of signal filtering, it may be said that, in general, it is usually possible to design filters of mechanical, electrical, thermal, pneumatic, etc., nature which separate signals according to their frequency content in some specific manner. Figure 2.15 summarizes the most common useful forms of such devices.

The *method of opposing inputs* consists of intentionally introducing

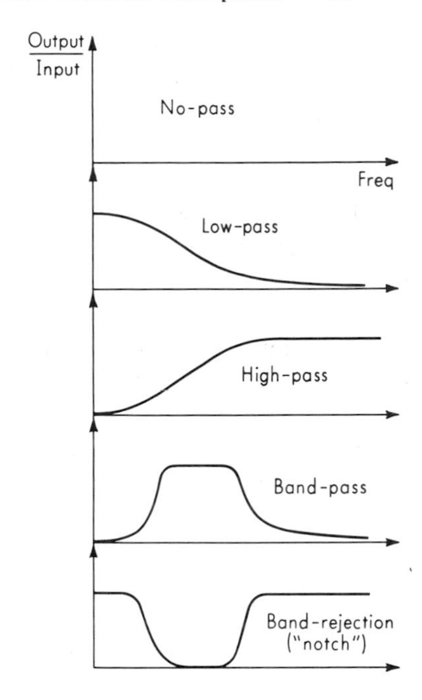

Fig. 2.15. *Basic filter types.*

into the instrument interfering and/or modifying inputs that tend to cancel the bad effects of the unavoidable spurious inputs. Figure 2.16 shows schematically the concept for interfering inputs. The extension to modifying inputs should be obvious. The intentionally introduced input is designed so that the signals o_{I1} and o_{I2} are essentially equal but act in opposite sense; thus the net contribution $(o_{I1} - o_{I2})$ to the output

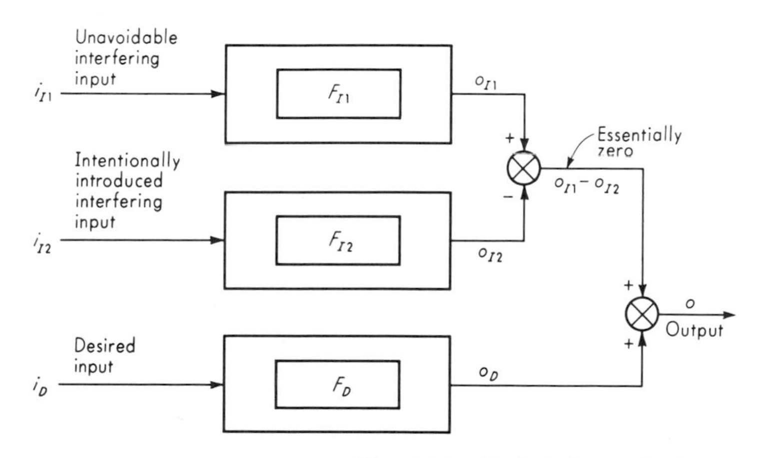

Fig. 2.16. *Method of opposing inputs.*

is essentially zero. This method might actually be considered as a variation on the method of calculated output corrections; however, the "calculation" and application of the correction are achieved automatically owing to the structure of the system, rather than by numerical calculation by a human operator. Thus the two methods are similar; however, the distinction between them is a worthwhile one since it helps to organize one's thinking in inventing new applications of these generalized correction concepts.

Some examples of the method of opposing inputs are shown in Fig. 2.17. A millivoltmeter, as shown in Fig. 2.17*a*, is basically a *current-sensitive* device. However, as long as the total circuit resistance is constant, its scale can be calibrated in voltage since voltage and current are proportional. A modifying input here is the ambient temperature, since it causes the coil resistance R_{coil} to change, thereby changing the propor-

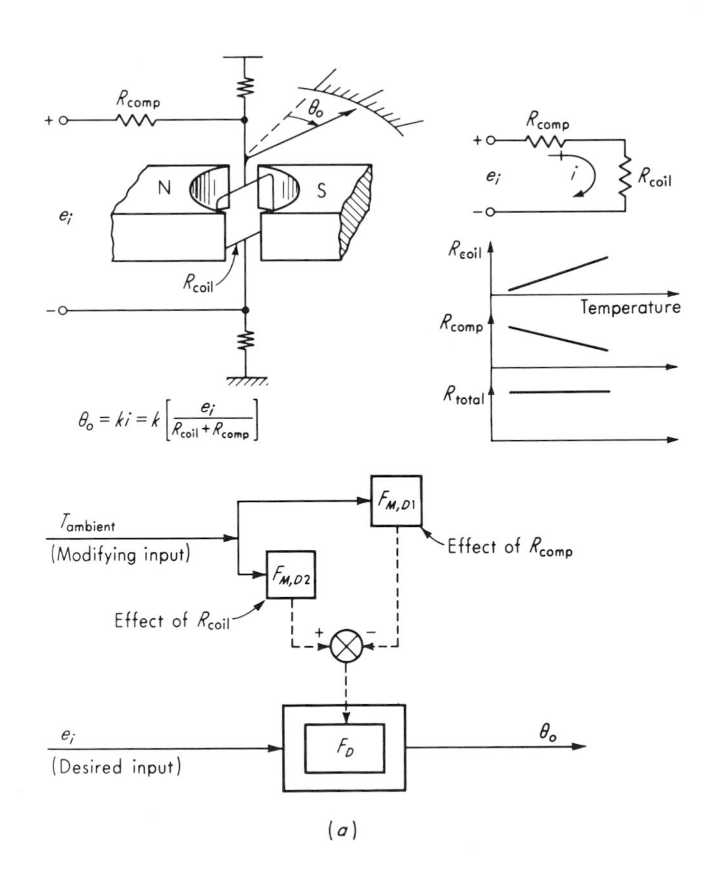

(*a*)

tionality factor between current and voltage. To correct for this error, the compensating resistance R_{comp} is introduced into the circuit and its material carefully chosen to have a temperature coefficient of resistance *opposite* to that of R_{coil}. Thus when the temperature changes, the total resistance of the circuit is unaffected and the calibration of the meter remains accurate.

Figure 2.17*b* shows a static-pressure-probe design due to L. Prandtl. As the fluid flows over the surface of the probe, its velocity must increase since these streamlines are longer than those in the undisturbed flow. This velocity increase causes a drop in static pressure, so that a tap in the surface of the probe gives an incorrect reading. This underpressure error varies with the distance d_1 of the tap from the probe tip. Prandtl recognized that the probe support will have a stagnation point (line) along its front edge and that this overpressure will be felt upstream, the effect decreasing as the distance d_2 increases. By properly choosing the

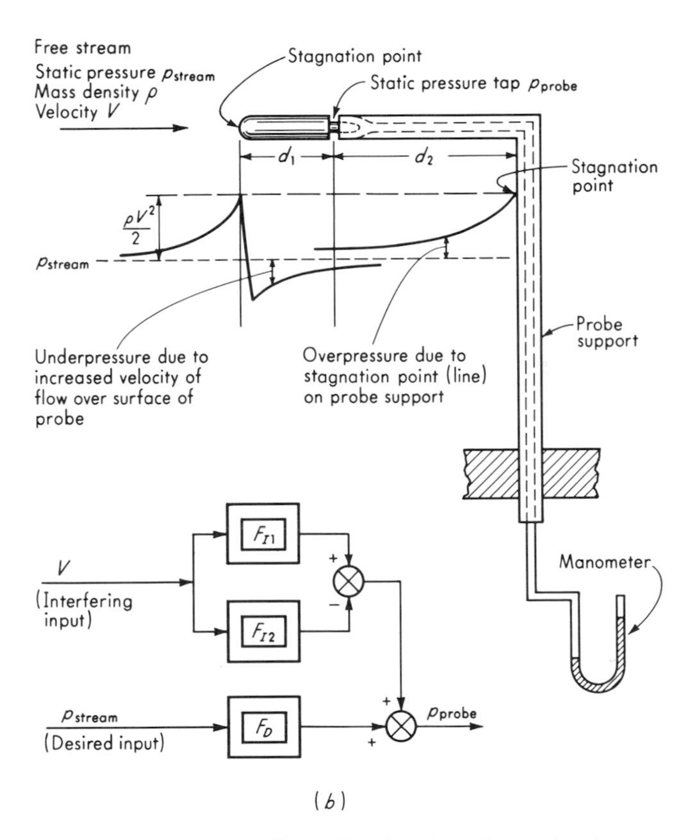

Fig. 2.17. *Examples of method of opposing inputs.*

distances d_1 and d_2 (by experimental test) these two effects can be made exactly to cancel each other, giving a true static-pressure value at the tap.

A device[1] for the measurement of the mass flow rate of gases is shown in Fig. 2.17*c*. The mass flow rate of gas through an orifice may be measured by measuring the pressure drop across the orifice, perhaps by means of a U-tube manometer. Unfortunately, the mass flow rate also depends on the density of the gas, which varies with pressure and

[1] National Instrument Laboratories, Inc., Washington, D.C.

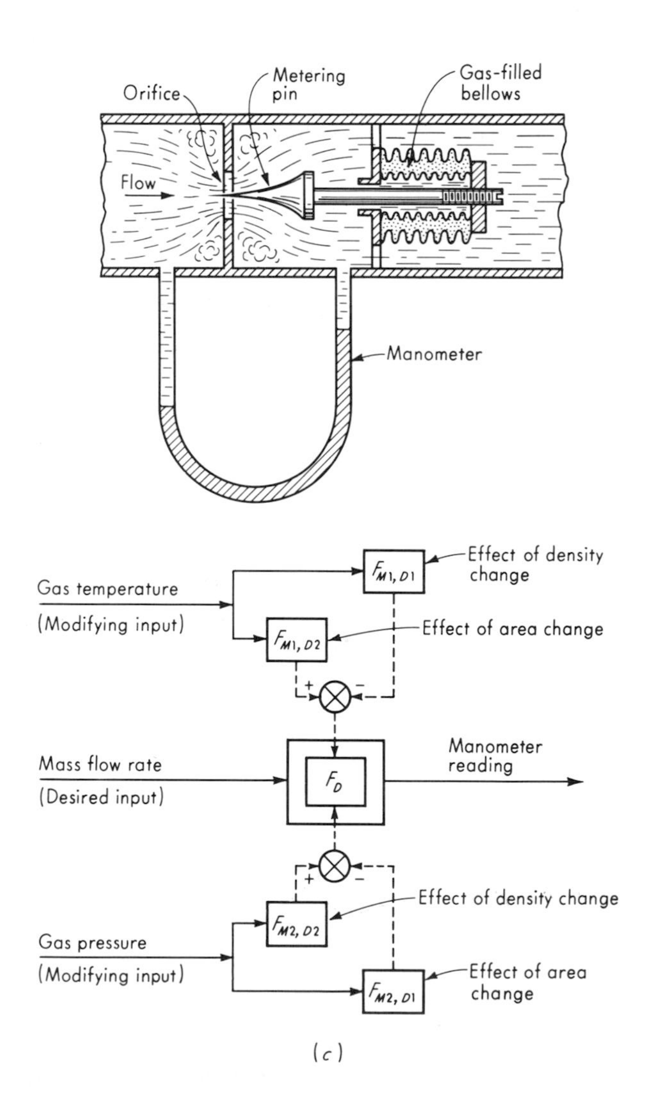

(*c*)

temperature. Thus the pressure-drop measuring device cannot usually be calibrated to give mass flow rate, since variations in gas temperature and pressure will give different mass flow rates for the same orifice pressure drop. The instrument of Fig. 2.17*c* overcomes this problem in an ingenious fashion. The flow rate through the orifice also depends on its flow area. Thus if the flow area could be varied in just the right way, this variation could compensate for pressure and temperature changes so that a given orifice pressure drop would *always* correspond to the same mass flow rate. This is accomplished by attaching the specially shaped metering pin to a gas-filled bellows as shown. When the temperature drops (causing an increase in density and therefore in mass flow rate) the gas in the bellows contracts, moving the metering pin into the orifice and thereby reducing the flow area. This returns the mass flow rate to its proper value. Similarly, should the pressure of the flowing gas

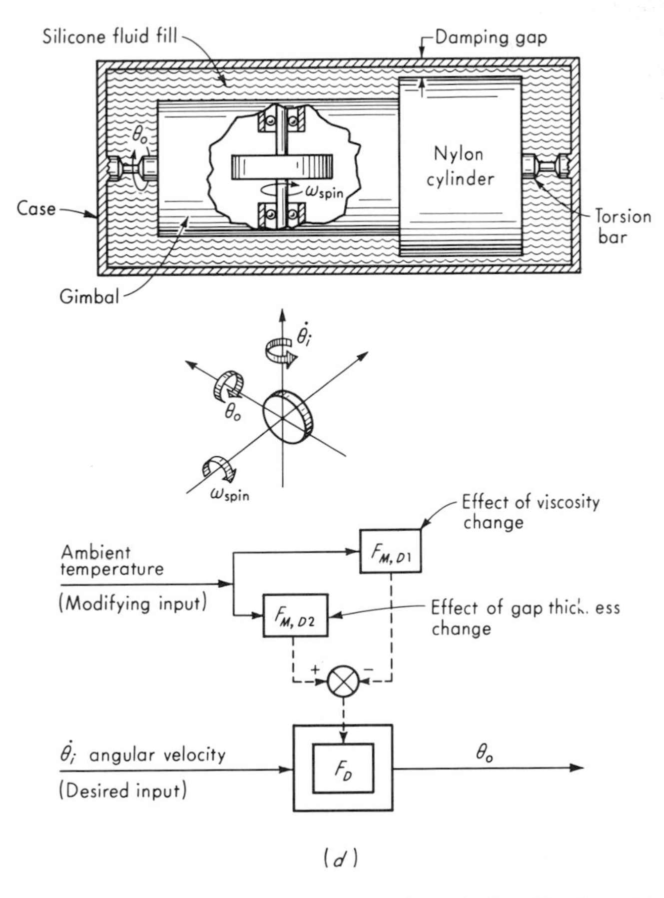

Fig. 2.17. (*Continued.*)

increase, causing an increase in density and mass flow rate, the gas-filled bellows would again be compressed, reducing the flow area and correcting the mass flow rate. The proper shape for the metering pin is revealed by a detailed analysis of the system.

A final example of the method of opposing inputs is given by the rate gyroscope of Fig. 2.17*d*. Such devices are widely used in aerospace vehicles for the generation of stabilization signals in the control system. The action of the device is that a vehicle rotation at angular velocity $\dot{\theta}_i$ causes a proportional rotation θ_o of the gimbal relative to the case. This rotation θ_o is measured by some motion pickup (not shown in Fig. 2.17*d*). A signal proportional to vehicle angular velocity is thus available, and this is useful in stabilizing the vehicle. When the vehicle undergoes rapid motion changes, however, the angle θ_o tends to oscillate, giving an incorrect angular-velocity signal. To control these oscillations, the gimbal rotation θ_o is damped by the shearing action of a viscous silicone fluid in a narrow damping gap. The damping effect varies with the viscosity of the fluid and the thickness of the damping gap. Although the viscosity of the silicone fluid is fairly constant, it does vary with ambient temperature, causing an undesirable change in damping characteristics. To compensate for this, a nylon cylinder is used in the gyro of Fig. 2.17*d*. When the temperature increases, viscosity drops, causing a loss of damping. Simultaneously, however, the nylon cylinder expands, narrowing the damping gap, and thus restoring the damping to its proper value. By proper choice of materials and geometry, the two effects may be made very nearly to cancel each other over the operating temperature range of the equipment.

2.6 Conclusion

This chapter has attempted to develop useful generalizations with regard to the functional elements and the input-output configurations of measuring instruments and systems. In the analysis of a given instrument or in the design of a new one, the starting point is the separation of the overall operation into its functional elements. Here one must take a broad view of *what* must be done but not be concerned with *how* it is actually to be accomplished. Once the general functional concepts have been clarified, the details of operation may fruitfully be considered. The ideas of active and passive transducers, analog and digital modes of operation, and null versus deflection methods give a systematic approach for either analysis or design.

Finally, compensation of spurious inputs and detailed evaluation of performance are facilitated by application of input-output block diagrams. These configuration diagrams show clearly which physical analyses must be made to evaluate performance with respect to accurate measurement of the desired inputs and rejection of spurious inputs. The evaluation of the relative quality of different instruments (or the same instrument with different numerical parameter values) requires the definition of performance criteria against which competitive designs may be compared. This is the subject of Chap. 3.

Problems

2.1 Make block diagrams such as Fig. 2.1, showing the functional elements of the instruments depicted in the following:
- *a.* Fig. 2.7.
- *b.* Fig. 2.8.
- *c.* Fig. 2.10*a*.
- *d.* Fig. 2.11. Take F as input and e_o as output.
- *e.* Fig. 2.14*g*. Take T_s as input and e_o as output.
- *f.* Fig. 2.17*b*. Take V as input and manometer Δh as output.
- *g.* Fig. 2.17*d*. Take θ_i as input and θ_o as output.

2.2 Identify the active transducers, if any, in the instruments of the following:
- *a.* Fig. 2.8.
- *b.* Fig. 2.10*a*.
- *c.* Fig. 2.11.
- *d.* Fig. 2.17*b*.
- *e.* Fig. 2.17*c*.

2.3 Consider a man driving a car along a road when he sees the opportunity to pass and decides to accelerate.
- *a.* If the light waves entering his eyes are considered input and accelerator-pedal travel as output, is the man functioning as an active or a passive transducer?
- *b.* If accelerator-pedal travel is considered input and car velocity as output, is the automobile engine an active or passive transducer?

2.4 Give an example of a null method of force measurement.

2.5 Give an example of a null method of voltage measurement.

2.6 Sketch and explain two possible modifications of the system of Fig. 2.4 which will allow measurement to $\frac{1}{10}$ revolution.

2.7 Identify desired, interfering, and modifying inputs for the systems of the following:
- *a.* Fig. 2.2.
- *b.* Fig. 2.3.
- *c.* Fig. 2.4.
- *d.* Fig. 2.5.

2.8 Why is tilt angle in Fig. 2.10*c* a modifying input?

2.9 Suppose in Eq. (2.7) that $K_{MO} = K_{SP} = e_i = 1.0$. Now let K_{MO} change by 10 percent to 1.1. What is the change in x_o? In Eq. (2.10) let $K_{MO} = K_{SP} = K_{FB} = e_i = 1.0$; $K_{AM} = 100$. Now let K_{MO} change by 10 percent to 1.1. What is the change in x_o? Investigate the effect of similar changes in K_{AM}, K_{SP}, and K_{FB}.

2.10 The natural frequency of oscillation of the balance wheel in a watch depends on the moment of inertia of the wheel and the spring constant of the (torsional) hairspring. A temperature rise results in a reduced spring constant, lowering the oscillation frequency. Propose a compensating means for this effect. Non-temperature-sensitive hairspring material is not an acceptable solution.

3
Generalized performance characteristics of instruments

3.1
Introduction If one is trying to choose, from commercially available instruments, the one most suitable for a proposed measurement, or, alternatively, if one is engaged in the design of instruments for specific measuring tasks, the subject of performance criteria assumes major proportions. That is, to make intelligent decisions, there must be some quantitative bases for comparing one instrument (or proposed design) with the possible alternatives. Chapter 2 has served as a useful preliminary to these considerations since there we developed systematic methods for breaking down the overall problem into its component parts. We

now propose to study in considerable detail the performance of measuring instruments and systems with regard to how well they measure the desired inputs and how thoroughly they reject the spurious inputs.

The treatment of instrument performance characteristics has generally been broken down into the subareas of *static characteristics* and *dynamic characteristics*, and this plan will be followed here. The reasons for such a classification are several. First, some applications involve the measurement of quantities that are constant or vary only quite slowly. Under these conditions it is possible to define a set of performance criteria that give a meaningful description of the quality of measurement without becoming concerned with dynamic descriptions involving differential equations. These criteria are called the static characteristics. Many other measurement problems are concerned with rapidly varying quantities; here the dynamic relations between the instrument input and output must be examined, generally by the use of differential equations. Performance criteria based on these dynamic relations constitute the dynamic characteristics.

Actually, static characteristics also influence the quality of measurement under dynamic conditions, but the static characteristics generally show up as nonlinear or statistical effects in the otherwise linear differential equations giving the dynamic characteristics. These effects would make the differential equations unmanageable, and so the conventional approach is to treat the two aspects of the problem separately. Thus the differential equations of dynamic performance generally neglect the effects of dry friction, backlash, hysteresis, statistical scatter, etc., even though they affect the dynamic behavior. These phenomena are more conveniently studied as static characteristics, and the overall performance of an instrument is then judged by a semiquantitative superposition of the static and dynamic characteristics. This approach is, of course, approximate but is a necessary expedient.

3.2 Static Characteristics

We begin our study of static performance characteristics by considering the meaning of the term static calibration.

The meaning of static calibration. All the static performance characteristics are obtained by one form or another of a process called static calibration. It is therefore appropriate to develop at this point a clear concept of what is meant by this term.

In general, static calibration refers to a situation where all inputs (desired, interfering, modifying) except one are kept at some constant values. The one input under study is then varied over some range of constant values, causing the output(s) to vary over some range of constant values. The input-output relations developed in this way comprise a static calibration *valid under the stated constant conditions of all the other inputs.* This procedure may be repeated, varying in turn each input considered to be of interest and thus developing a family of static input-output relations. One might then hope to describe the overall instrument static behavior by means of some suitable form of superposition of these individual effects. In some cases, if overall rather than individual effects are desired, the calibration procedure would specify the variation of several inputs simultaneously. It should also be understood that if one examines any practical instrument critically he will find many modifying and/or interfering inputs each of which might have quite small effects and which would be impractical to control. Thus the statement that *all* other inputs are held constant refers to an ideal situation which can only be approached, but never reached, in practice. The term *measurement method* has been used to describe the ideal situation while the term *measurement process* describes the (imperfect) physical realization of the measurement method.

The statement that one input is varied and all others held constant implies that all these inputs are determined (measured) independently of the instrument being calibrated. For interfering or modifying inputs (whose effects on the output should be relatively small in a good instrument), the measurement of these inputs usually need not be at an extremely high accuracy level. For example, suppose a pressure gage has temperature as an interfering input to the extent that a temperature change of 100°F causes a pressure error of 0.100 percent. Now, if we had measured the 100°F interfering input with a thermometer which itself had an error of 2.0 percent, the pressure error could actually have been 0.102 percent. It should be clear that the difference between an error of 0.100 and 0.102 percent is entirely negligible in most engineering situations. However, when calibrating the response of the instrument to its *desired* inputs one must exercise considerable care in choosing the means of determining the numerical values of these inputs. That is, if a pressure gage is inherently capable of an accuracy of 0.1 percent one must certainly be able to determine its input pressure during calibration with an accuracy somewhat greater than this. In other words, it is impossible to calibrate an instrument to an accuracy greater than that of the standard with which it is compared. A rule often followed is that the calibration standard should be at least about 10 times as accurate as the instrument being calibrated.

While we shall not go into a detailed discussion of standards at this point, it is of utmost importance that the person performing the calibration be able to answer the question: How do I know that this standard is capable of its stated accuracy? The ability to trace the accuracy of a standard back to its ultimate source in the fundamental standards of the National Bureau of Standards is termed *traceability*.

In performing a calibration the following steps are thus necessary:

1. Examine the construction of the instrument and identify and list all the possible inputs.
2. Decide, as best you can, which of the inputs will be significant in the application for which the instrument is to be calibrated.
3. Procure apparatus that will allow you to vary all significant inputs over the ranges considered necessary.
4. By holding some inputs constant, varying others, and recording the output(s), develop the desired static input-output relations.

We are now ready for a more detailed discussion of specific static characteristics. These characteristics may first be classified as either general or special. General static characteristics are of interest in *every* instrument. Special static characteristics are of interest only in a particular instrument. We shall concentrate mainly on general characteristics, leaving the treatment of special characteristics to later sections of the text in which specific instruments are discussed.

Accuracy, precision, and bias. When one makes a measurement of some physical quantity with an instrument and obtains a numerical value, he is usually concerned with how close this value may be to the "true" value. It is first necessary to understand that this so-called true value is, in general, unknown and unknowable, since perfectly exact definitions of the physical quantities that are to be measured are impossible. This can be illustrated by specific example, for instance, the length of a cylindrical rod. When we ask ourselves what we *really* mean by the length of this rod, we must consider such questions as these:

1. Are the two ends of the rod planes?
2. If they are planes, are they parallel?
3. If they are not planes, what sort of surfaces are they?
4. What about surface roughness?

We see that complex problems are introduced when we deal with a real object rather than an abstract geometrical solid. The term *true value*, then, refers to a value that would be obtained if the quantity under consideration were measured by an *exemplar method*,[1] that is, a method agreed

[1] Churchill Eisenhart, Realistic Evaluation of the Precision and Accuracy of Instrument Calibration Systems, *J. Res. Natl. Bur. Std., C*, vol. 67C, no. 2, April–June, 1963.

upon by experts as being sufficiently accurate for the purposes to which the data will ultimately be put.

We must also be concerned over whether we are describing the characteristics of a single reading of an instrument or the characteristics of a measurement process. If we are speaking of a single measurement, the *error* is the difference between the measurement and the corresponding true value, taken positive if the measurement is greater than the true value. When using an instrument, however, we are concerned with the characteristics of the measurement process associated with that instrument. That is, we may take a single reading, but this is a sample from a statistical population generated by the measurement process. If we know the characteristics of the process, we can *put bounds on* the error of the single measurement, although we cannot tell what the error itself is, since this would imply that we knew the true value. We are thus interested in being able to make statements about the accuracy (lack of error) of our readings. This can be done in terms of the concepts of *precision* and *bias* of the measurement process.

The measurement process consists of actually carrying out, as well as possible, the instructions for performing the measurement, which are the measurement method. (Since calibration is essentially a refined form of measurement, these remarks apply equally to the process of calibration.) If this process is repeated over and over again under *assumed* identical conditions, we get a large number of readings from the instrument. Usually these readings will not all be the same, and so we note immediately that we may *try* to assure identical conditions for each trial but it is never exactly possible. The data generated in this fashion may be used to describe the measurement process so that, if it is used in the future, we may be able to attach some numerical estimates of error to its outputs.

If the output data are to give a meaningful description of the measurement process, the data must form what is called a *random sequence*. Another way of saying this is that the process must be in a state of *statistical control.*[1] The concept of the state of statistical control is not a particularly simple one but we shall try to explain its essence briefly. We first note that it is meaningless to speak of the accuracy of an instrument as an isolated device; one must always consider the instrument plus its environment and method of use, that is, the instrument plus its inputs. This aggregate constitutes the measurement process. Every instrument has an infinite number of inputs; that is, the causes that can conceivably affect the output, if only very slightly, are limitless. Such effects as atmospheric pressure, temperature, and humidity are among the more

[1] *Ibid.*

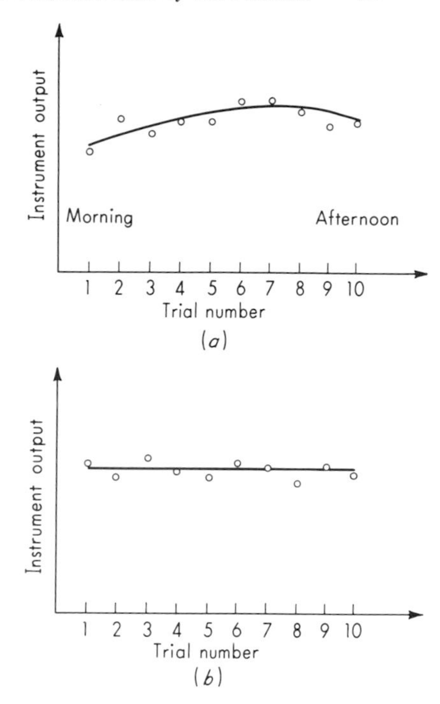

Fig. 3.1. *Effect of uncontrolled input on calibration.*

obvious but if one is willing to "split hairs" he can uncover a multitude of other physical causes which could affect the instrument with varying degrees of severity. In defining a calibration procedure for a specific instrument, one specifies that certain inputs be held "constant" within certain limits. These inputs, it is hoped, are the ones that contribute the largest components to the overall error of the instrument. The remaining infinite number of inputs is left uncontrolled, and it is hoped that each of these individually contributes only a very small effect and that in the aggregate their effect on the instrument output will be of a random nature. If this is indeed the case, the process is said to be in statistical control. Experimental proof that a process is in statistical control is not easy to come by; in fact, *strict* statistical control is unlikely of practical achievement. Thus one can only approximate this situation.

Lack of control is sometimes obvious, however, if one repeats a measurement and plots the result (output) versus the trial number. Figure 3.1*a* shows such a graph for the calibration of a particular instrument. In this instance it was ascertained after some study that the instrument was actually much more sensitive to temperature than had been thought. The original calibration was carried out in a room without temperature control; thus the room temperature varied from a low in the morning to a peak in the early afternoon and then dropped again in the

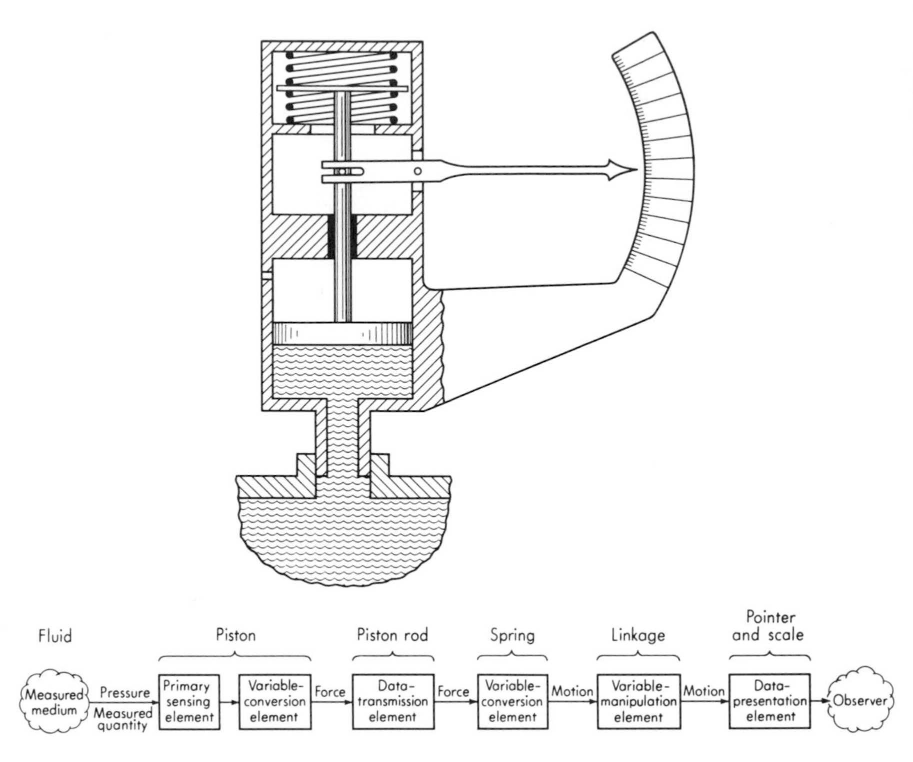

Fig. 3.2. *Pressure gage.*

late afternoon. Since the 10 trials covered a period of about one day, the trend of the curve is understandable. By performing the calibration in a temperature-controlled room, the graph of Fig. 3.1*b* was obtained. For the detection of more subtle deviations from statistical control, the methods of statistical-quality-control charts are useful.[1]

If the measurement process is in reasonably good statistical control and if one repeats a given measurement (or calibration point) over and over, he will generate a set of data exhibiting random scatter. As an example, consider the pressure gage of Fig. 3.2. Suppose we wish to determine the relationship between the desired input (pressure) and the output (scale reading). Other inputs which could be significant and which might have to be controlled during the pressure calibration include temperature, acceleration, and vibration. Temperature can cause expansion and contraction of instrument parts in such a way that the scale reading will change even though the pressure has remained constant.

[1] E. B. Wilson, Jr., "An Introduction to Scientific Research," chap. 9, McGraw-Hill Book Company, New York, 1952.

True pressure $= 10.000 \pm .001$ *psig*
Acceleration $= 0$
Vibration level $= 0$
Ambient temperature $= 70 \pm 1°F$

Trial number	*Scale reading, psig*
1	10.02
2	10.20
3	10.26
4	10.20
5	10.22
6	10.13
7	9.97
8	10.12
9	10.09
10	9.90
11	10.05
12	10.17
13	10.42
14	10.21
15	10.23
16	10.11
17	9.98
18	10.10
19	10.04
20	9.81

Fig. 3.3. *Pressure-gage calibration data.*

An acceleration of the instrument, which has a component along the axis of the piston rod, will cause a scale reading even though pressure has again remained unchanged. This input is significant if the pressure gage is to be used aboard a vehicle of some kind. A small amount of vibration may actually be helpful to the operation of an instrument since it may reduce the effects of static friction. Thus if the pressure gage is to be used by attaching it to a reciprocating air compressor (which always has some vibration) it may be more accurate under these conditions than it would be under calibration conditions where no vibration was provided. These examples should illustrate the general importance of carefully considering the relationship between the calibration conditions and the actual application conditions.

Suppose now that we have procured a sufficiently accurate pressure standard and have arranged to maintain the other inputs reasonably close to the actual application conditions. Repeated calibration at a given pressure (say 10 psig) might give the data of Fig. 3.3. Suppose we now

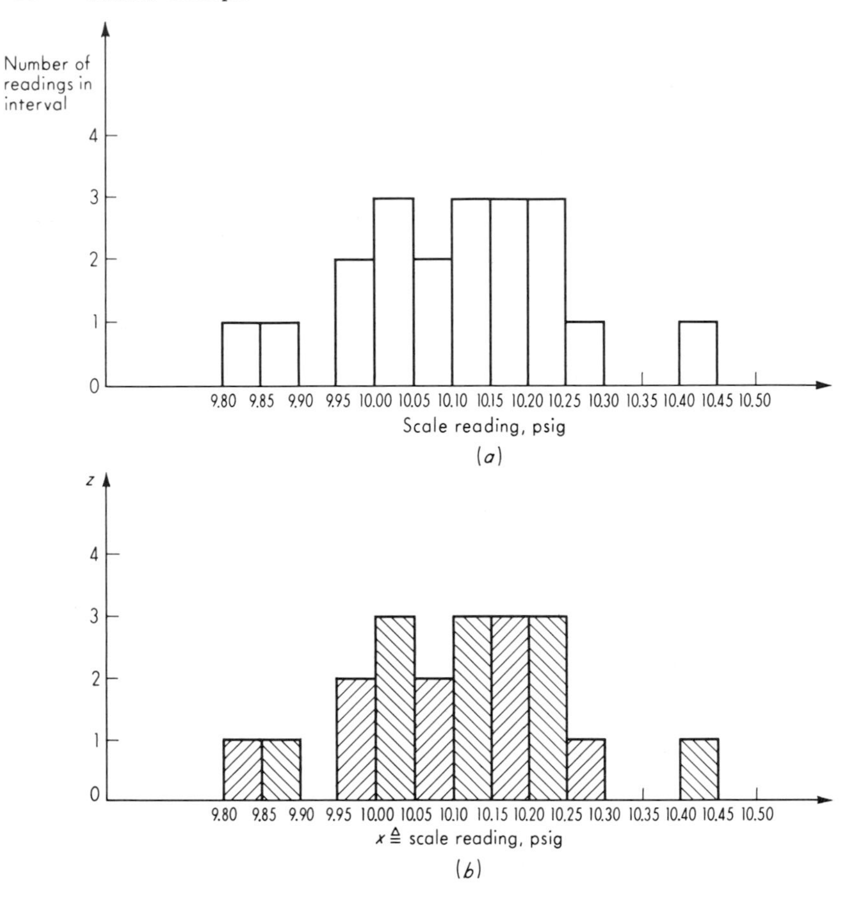

Fig. 3.4. Distribution of data.

order the readings from the lowest (9.81) to the highest (10.42) and see how many readings fall in each interval of, say, 0.05 psig, starting at 9.80. The result can be represented graphically as in Fig. 3.4*a*. Suppose we now define the quantity Z by

$$Z \triangleq \frac{(\text{number of readings in an interval})/(\text{total number of readings})}{\text{width of interval}} \tag{3.1}$$

and plot a "bar graph" with height Z for each interval. Such a "histogram" is shown in Fig. 3.4*b*. It should be clear from Eq. (3.1) that the area of a particular "bar" is numerically equal to the probability that a particular reading will fall in the associated interval. The area of the entire histogram must then be 1.0 (100 percent = 1.0), since there is

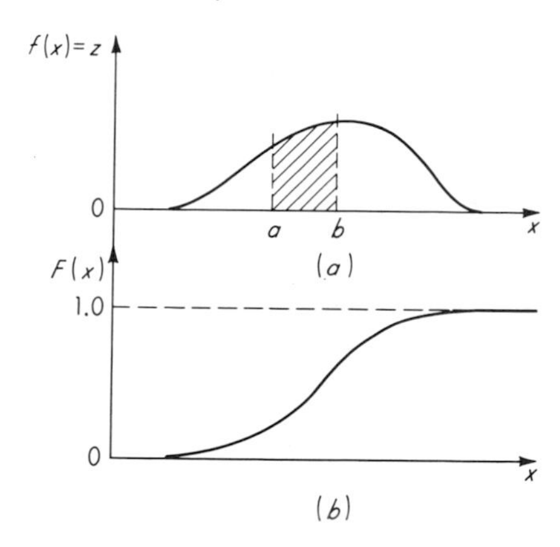

Fig. 3.5. *Probability distribution functions.*

100 percent probability that the reading will fall somewhere between the lowest and highest value, at least based on the data available. If it were now possible to take an infinite number of readings, each with an infinite number of significant digits, we could make the chosen intervals as small as we pleased and still have each interval contain a finite number of readings. Thus the steps in the graph of Fig. 3.4*b* would become smaller and smaller, the graph approaching in the limit a smooth curve. If we take this limiting abstract case as a mathematical model for the real physical situation, the function $Z = f(x)$ is called the *probability density function* for the mathematical model of the real physical process (see Fig. 3.5*a*). From the basic definition of Z, it should be clear that

$$\text{Probability of reading lying between } a \text{ and } b \triangleq P(a < x < b) = \int_a^b f(x)\, dx \tag{3.2}$$

From the infinite number of forms possible for probability density functions, a relatively small number have been found useful mathematical models for practical applications; in fact, *one* particular form is quite dominant. The probability information is sometimes given in terms of the *cumulative distribution function* $F(x)$, which is defined by

$$F(x) \triangleq \text{probability that reading is less than any chosen value of } x = \int_{-\infty}^{x} f(x)\, dx \tag{3.3}$$

and is shown in Fig. 3.5*b*.

The most useful density function or distribution is the normal or *gaussian* function, which is given by

$$f(x) = \frac{1}{\sqrt{2\pi}\,\sigma} e^{-(x-\mu)^2/2\sigma^2} \qquad -\infty < x < +\infty \tag{3.4}$$

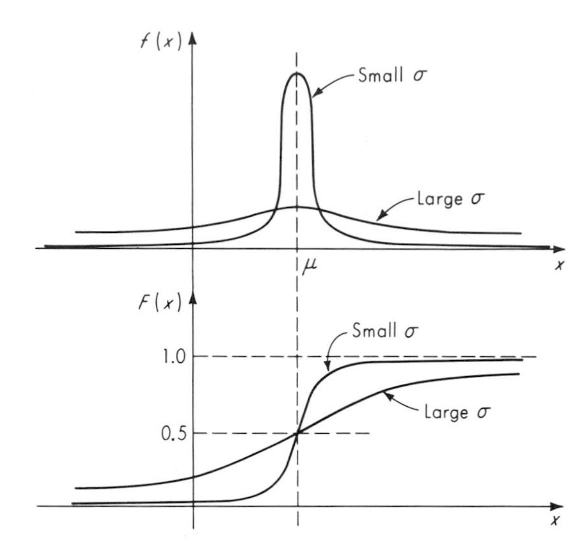

Fig. 3.6. Gaussian distribution.

Equation (3.4) defines a whole family of curves depending on the particular numerical values of μ (the mean value) and σ (the standard deviation). The shape of the curve is determined entirely by σ, μ serving only to locate its position along the x axis. The cumulative distribution function $F(x)$ cannot be written explicitly in this case because the integral of Eq: (3.3) cannot be carried out; however, the function has been tabulated by performing the integration by numerical means. Figure 3.6 shows that a small value of σ indicates a high probability that a "reading" will be found close to μ. Equation (3.4) also shows that there is a small probability that very large ($\rightarrow \pm\infty$) readings will occur. This is one of the reasons why a true gaussian distribution can never occur in the real world; physical variables are always limited to finite values. There is *zero* probability, for example, that the pointer on a pressure gage will read 100 psig when the range of the gage is only 20 psig. Real distributions must thus, in general, have their "tails" cut off, as in Fig. 3.7.

Although actual data may not conform *exactly* to the gaussian distribution, they very often are sufficiently close to allow use of the gaussian

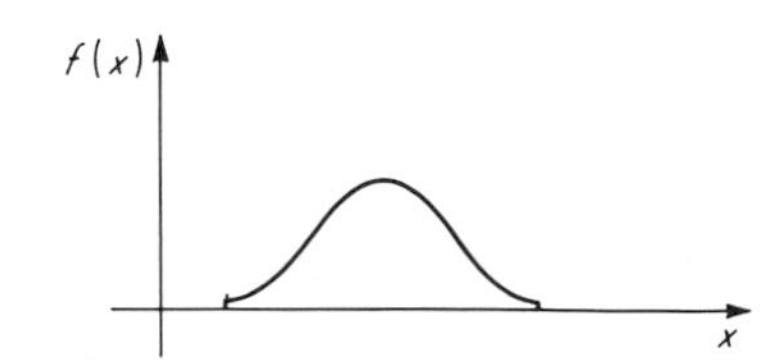

Fig. 3.7. Nongaussian distribution.

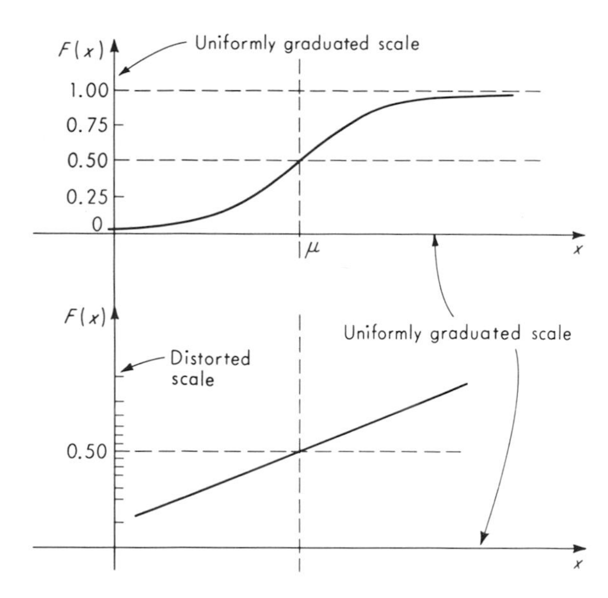

Fig. 3.8. *Rectification of gaussian curve.*

model in engineering work. It would be desirable to have available tests that would indicate whether the data are "reasonably" close to gaussian, and two such procedures will be explained briefly. It must be admitted, however, that in much practical work the time and effort necessary for such tests cannot be justified and the gaussian model is simply *assumed* until troubles arise which justify a closer study of the particular situation.

The first method of testing for an approximate gaussian distribution involves the use of probability graph paper. If one takes the cumulative distribution function for a gaussian distribution and suitably distorts the vertical scale of the graph, the curve can be made to plot as a straight line, as shown in Fig. 3.8. (This, of course, can be done with any curvilinear relation, not just probability curves.) Such graph paper is commercially available and may be used to give a rough qualitative test for conformity to the gaussian distribution. For example, consider the data of Fig. 3.3. These data may be plotted on gaussian probability graph paper as follows: First lay out on the uniformly graduated horizontal axis a numerical scale that includes all the pressure readings. Now the probability graph paper represents the cumulative distribution, so that the ordinate of any point represents the probability that a reading will be less than the abscissa of that particular point. This probability, in terms of the sample of data available, is simply the percentage (in decimal form) of points that fell at or below that particular value. Figure 3.9 shows

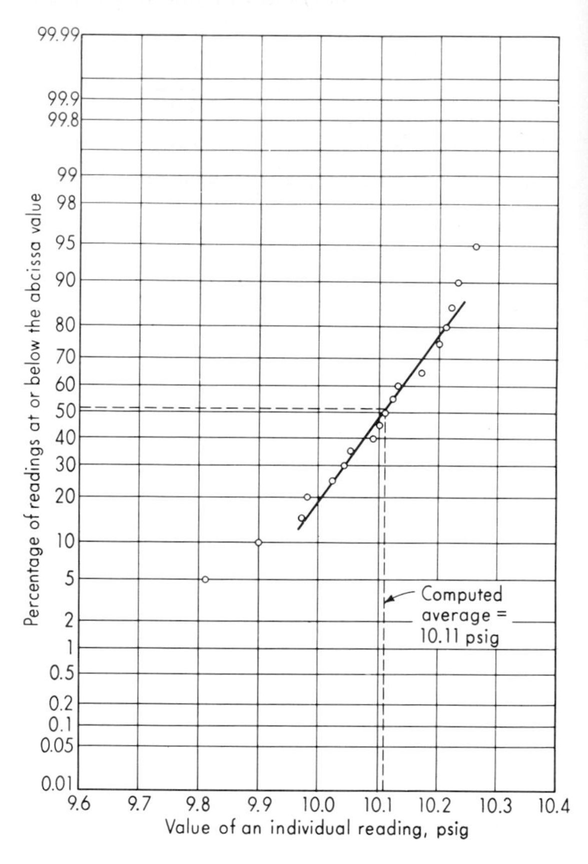

Fig. 3.9. *Graphical check of gaussian distribution.*

the resulting plot. Note that it is not possible to plot the highest point (10.42) since the 100 percent point cannot appear on the ordinate scale. For the data to be perfectly gaussian, the following requirements must be met:

1. All the points must fall exactly on one straight line.
2. The average of all the data points must intersect this line at an ordinate of exactly 0.50 (50 percent).

The procedure is thus to fit a straight line to the data points (usually by eye) and then see how well the above two requirements are met. It is advisable not to give much weight to the points at the low and high ends of the plot since these represent the "tails" of the distribution which are not accurately defined unless a very large number of readings have been made.

The results of Fig. 3.9 would be considered by most people to indicate a reasonable approximation to a gaussian distribution. However, this test is obviously only qualitative, and its main usefulness is perhaps in indicating *gross* departures from the theoretical distribution when they exist. Such deviations would lead one to examine the instrument and measurement process more closely before attempting to make any statistical statements about accuracy.

Another method of testing for "normality" (gaussian distribution) involves use of the chi-squared (χ^2) statistical test. This method puts the conclusions on a somewhat more quantitative basis but there is still an element of uncertainty involved, as there must be, since *perfect* gaussian distributions simply do not exist in nature. To use the χ^2 test, one must first develop some additional information. The exact gaussian distribution which is taken as a mathematical model of the real distribution has the parameters μ (mean or average value) and σ (standard deviation). The true values of these numbers cannot be exactly known; they can merely be estimated from the data taken on the real distribution. The estimate of μ is given the symbol $\bar{x}$ and is the ordinary average value

$$\bar{x} \triangleq \frac{\sum_{i=1}^{N} x_i}{N} \tag{3.5}$$

where $x_i \triangleq$ individual reading (3.6)

$N \triangleq$ total number of readings (3.7)

The estimate of σ is given the symbol s and the name sample standard deviation and is computed from

$$s \triangleq \sqrt{\frac{\sum_{i=1}^{N} (x_i - \bar{x})^2}{N-1}} \tag{3.8}$$

For the data of Fig. 3.3 we have $\bar{x} = 10.11$ psig and $s = 0.14$ psig. For a perfect gaussian distribution, it can be shown that

68 percent of the readings lie within $\pm\sigma$ of μ
95 percent of the readings lie within $\pm 2\sigma$ of μ (3.9)
99.7 percent of the readings lie within $\pm 3\sigma$ of μ

Thus, if we assume our real distribution is nearly gaussian, we might predict, for instance, that if more readings were taken 99.7 percent would fall within ± 0.42 psig of 10.11 psig. These estimates $\bar{x}$ and s of μ and σ can be improved by taking more readings. That is, suppose we took 20 more readings and computed $\bar{x}$ from these 20 readings. We would probably get a value different from 10.11. Thus $\bar{x}$ itself exhibits scatter.

We can describe this scatter by determining a standard deviation of the mean, $s_{\bar{x}}$, which would then allow statements such as (3.9) to be made about the closeness of $\bar{x}$ to μ. It can be shown[1] that

$$s_{\bar{x}} = \frac{s}{\sqrt{N-1}} \tag{3.10}$$

where s and N are as in Eqs. (3.7) and (3.8). This shows that we can get a better and better estimate of μ by using larger and larger samples. For our original sample of 20 readings we could make the statement: "There is a probability of 95 percent that $\bar{x}$ does not differ from μ by more than $\pm 0.28/\sqrt{19} = \pm 0.064$ psig." If we had 80 readings, this could be reduced to ± 0.032 psig (assuming the value of s stays about at 0.14).

Let us return now to the chi-squared test for estimating the approach to normality of a set of data. The first step is to order the data from the lowest to the highest and then group them so that no group has less than four or five members. For the test to be carried out at all, there must be at least four groups, and so we see that a sample of at least about 16 readings should be taken. The larger the sample, the more significant the test will be. The quantity χ^2 is defined as follows:

$$\chi^2 \triangleq \sum_{i=1}^{n} \frac{(n_0 - n_e)^2}{n_e} \tag{3.11}$$

where $n_0 \triangleq$ number of readings actually observed in given range (group)
$n_e \triangleq$ number of readings that would be observed in same range if distribution were normal, i.e. with $\mu = \bar{x}$ and $\sigma = s$
$n \triangleq$ number of groups

It is necessary to explain how the number n_e is calculated. We mentioned earlier that tables of the cumulative gaussian distribution are available.[2] These tables allow one to calculate the number n_e as follows: Entries in the table are values of $F(w)$, where

$$F(w) \triangleq \text{probability that reading falls in range from } -\infty \text{ to } w \tag{3.12}$$

$$w \triangleq \frac{x - \mu}{\sigma} \tag{3.13}$$

Definition (3.13) puts the tables on a nondimensional basis so that one table serves for all possible values of μ and σ. As an example, suppose we wish to calculate n_e for the first range, $-\infty$ to 10.00. This range of x

[1] A. M. Mood, "Introduction to the Theory of Statistics," pp. 133, 159, McGraw-Hill Book Company, New York, 1950.

[2] R. S. Burington, "Handbook of Mathematical Tables and Formulas," 2d ed., p. 257, McGraw-Hill Book Company, New York, 1940.

Group number	*Range of x*	*Range of w*	n_0	n_e	$\frac{(n_0 - n_e)^2}{n_e}$
1	$-\infty$ to 10.00	$-\infty$ to -0.79	4	4.296	0.020
2	10.00 to 10.095	-0.79 to -0.107	4	4.864	0.153
3	10.095 to 10.15	-0.107 to 0.286	4	3.080	0.274
4	10.15 to 10.215	0.286 to 0.75	4	3.220	0.189
5	10.215 to ∞	0.75 to ∞	4	4.532	0.062
					$\chi^2 = 0.698$

Fig. 3.10. *Tabulation for chi-squared test.*

corresponds to a range on w of $-\infty$ to -0.79 since we use Eq. (3.13) with $\mu = \bar{x} = 10.11$ and $\sigma = s = 0.14$ to calculate w. Now the probability that a reading falls in the range $-\infty$ to -0.79 is the same as the probability that it falls in $+0.79$ to $+\infty$ since the gaussian curve is perfectly symmetrical about $w = 0$. Entering the table for $w = 0.79$, we find $P(-\infty < w < 0.79)$ is 0.7852. Thus, $P(0.79 < w < \infty)$ is $(1.0000 - 0.7852) = 0.2148$, and we would expect in a sample of 20 trials that $(20)(0.2148) = 4.296$ readings would fall in the range $-\infty < x < 10.00$. Actually, we found that exactly four readings fell in this range. All other entries in Fig. 3.10 are found in the same manner.

To make the final interpretation of this test, one must know the number of "degrees of freedom." This is numerically equal to the number of groups minus 3, and so in the present example we have 2 degrees of freedom. The significance of the numerical value of χ^2 is given in Fig. 3.11, which is interpreted as follows: If we had a perfectly gaussian distribution with $\mu = 10.11$ and $\sigma = 0.14$ from which we drew a sample of 20 readings, we would *not*, in general, get a χ^2 value of zero. That is, because of the random nature of the readings, any finite sample will not exhibit the same properties as its "parent" population, and the smaller the sample the more likely it is that, just by chance, there is taken a sample that *appears* to be nongaussian. In Fig. 3.11 there is a probability of 5 percent that χ^2 would fall above the upper curve if the distribution were actually gaussian with $\mu = 10.11$ and $\sigma = 0.14$ and similarly a probability of 95 percent that χ^2 would fall above the lower curve. Thus if we compute a value of χ^2 that falls in either crosshatched region it is highly unlikely (though not impossible) that the sample came from the assumed gaussian distribution. (The 5 and 95 percent values were chosen somewhat arbitrarily. Tables[1] for other chosen percentages are available.)

[1] D. V. Huntsberger, "Elements of Statistical Inference," pp. 177, 259, Allyn and Bacon, Inc., Boston, 1961.

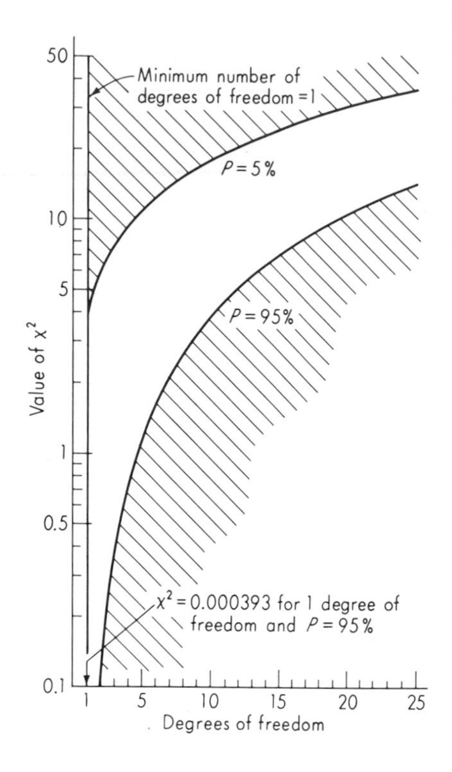

Fig. 3.11. *Chi-squared-test graph.*

If the point does not fall in the crosshatched areas, we can only say that there is no strong evidence of nongaussian behavior. It should be noted that, for small numbers of degrees of freedom, χ^2 can vary greatly (0.000393 to 3.84, almost 4 orders of magnitude for 1 degree of freedom) and still fall between the 5 and 95 percent lines. This simply means that for small samples (which give a low number of degrees of freedom) one cannot get a very sensitive test of normality. That is, even if the distribution *is* gaussian, small samples of it can very likely have a wide variation in the n_e's and thus in χ^2. However, large samples reduce the allowable range of χ^2 considerably (from 14.6 to 37.7 for 25 degrees of freedom). For our present example, $\chi^2 = 0.698$ for 2 degrees of freedom; thus normality is not disproved.

Most statistics texts recommend that if any of the n_e's are much less than 5 the groups be redefined to prevent this. Since Fig. 3.10 exhibits this problem, it might be in order to make some changes. The new groups and the results are shown in Fig. 3.12. The point ($\chi^2 = 0.3$, degrees of freedom = 1) again falls between the 5 and 95 percent lines and so our conclusions are unchanged.

Having considered the problem of determining the normality of scattered data, we now return to the main business of this section, that is,

Group number	*Range of x*	*Range of w*	n_0	n_e	$\frac{(n_0 - n_e)^2}{n_e}$
1	$-\infty$ to 10.03	$-\infty$ to -0.572	5	5.66	0.077
2	10.03 to 10.115	-0.572 to 0.0357	5	4.62	0.031
3	10.115 to 10.215	0.0357 to 0.75	6	5.18	0.130
4	10.215 to ∞	0.75 to ∞	4	4.532	0.062
					$\chi^2 = 0.300$

Fig. 3.12. *Tabulation for chi-squared test.*

definition of the terms accuracy, precision, and bias. Up to now we have been examining the situation wherein a single true value is applied repeatedly and the resulting measured values are recorded and analyzed. In an actual instrument calibration the true value is varied, in increments, over some range, causing the measured value also to vary over a range. Very often there is no multiple repetition of a given true value, the procedure being merely to cover the desired range in both the increasing and the decreasing direction. Thus a given true value is applied, at most, twice if one chooses to use the same set of true values for both increasing and decreasing readings.

As an example, suppose we wish to calibrate the pressure gage of Fig. 3.2 for the relation between the desired input (pressure) and the output (scale reading). Figure 3.13 gives the data for such a calibration over the range 0 to 10 psig. In this instrument (as in most but not all) the input-output relation is ideally a straight line. The *average calibration curve* for such an instrument is generally taken as a straight line which fits the scattered data points best as defined by some chosen criterion. The most common is the least-squares criterion which minimizes the sum of the squares of the vertical deviations of the data points from the fitted line. (The least-squares procedure can also be used to fit curves other than straight lines to scattered data if this should be desired.) The equation for the straight line is taken as

$$q_o = mq_i + b \tag{3.14}$$

where $q_o \triangleq$ output quantity (dependent variable) (3.15)
$q_i \triangleq$ input quantity (independent variable) (3.16)
$m \triangleq$ slope of line (3.17)
$b \triangleq$ intercept of line on vertical axis (3.18)

The equations for calculating m and b may be found in several

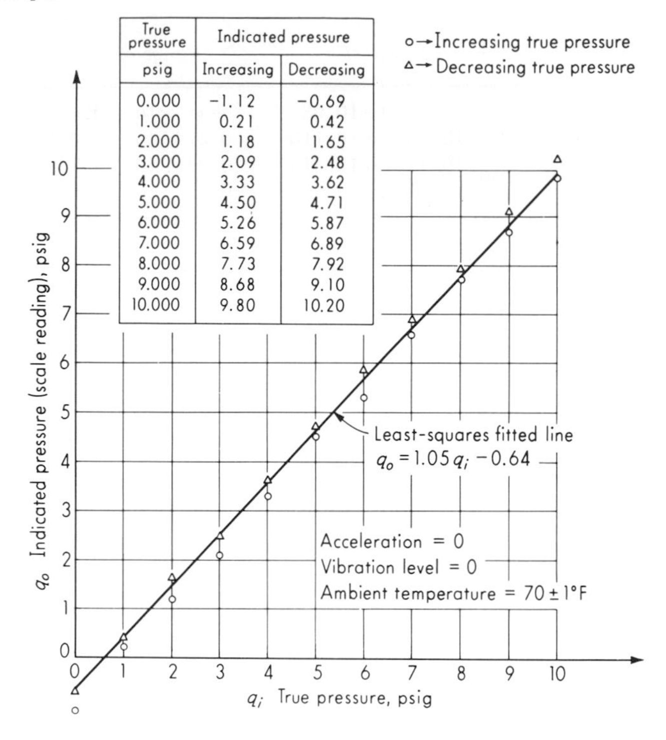

True pressure	Indicated pressure	
psig	Increasing	Decreasing
0.000	−1.12	−0.69
1.000	0.21	0.42
2.000	1.18	1.65
3.000	2.09	2.48
4.000	3.33	3.62
5.000	4.50	4.71
6.000	5.26	5.87
7.000	6.59	6.89
8.000	7.73	7.92
9.000	8.68	9.10
10.000	9.80	10.20

Fig. 3.13. *Pressure-gage calibration.*

references[1]:

$$m = \frac{N\Sigma q_i q_o - (\Sigma q_i)(\Sigma q_o)}{N\Sigma q_i^2 - (\Sigma q_i)^2} \tag{3.19}$$

$$b = \frac{(\Sigma q_o)(\Sigma q_i^2) - (\Sigma q_i q_o)(\Sigma q_i)}{N\Sigma q_i^2 - (\Sigma q_i)^2} \tag{3.20}$$

$$\text{where} \quad N \triangleq \text{total number of data points} \tag{3.21}$$

In the present example, calculation gives $m = 1.05$ and $b = -0.64$ psig. Since these values are derived from scattered data, it would be useful to have some idea of their possible variation. The standard deviations of m and b may be found from

$$s_m^2 = \frac{N s_{q_o}^2}{N\Sigma q_i^2 - (\Sigma q_i)^2} \tag{3.22}$$

$$s_b^2 = \frac{s_{q_o}^2 \Sigma q_i^2}{N\Sigma q_i^2 - (\Sigma q_i)^2} \tag{3.23}$$

$$\text{where} \quad s_{q_o}^2 = \frac{1}{N} \sum (mq_i + b - q_o)^2 \tag{3.24}$$

[1] H. D. Young, "Statistical Treatment of Experimental Data," p. 121, McGraw-Hill Book Company, New York, 1962.

The symbol s_{q_o} represents the standard deviation of q_o. That is, if q_i were fixed and then repeated over and over, q_o would give scattered values, the amount of scatter being indicated by s_{q_o}. If we assume that this s_{q_o} would be the same for *any* value of q_i, we can calculate s_{q_o} using *all* the data points of Fig. 3.13 and *without* having to repeat any one q_i many times. For the present example, calculation gives $s_{q_o} = 0.23$ psig. Then $s_m = 0.0154$ and $s_b = 0.091$ psig. Assuming a gaussian distribution and the 99.7 percent limits ($\pm 3s$), we could then give m as 1.05 ± 0.05 and b as -0.64 ± 0.27 psig.

In *using* the calibration results, the situation is one where q_o (the indicated pressure) is known and one wishes to make a statement about q_i (the true pressure). The least-squares line gives

$$q_i = \frac{q_o + 0.64}{1.05} \qquad (3.25)$$

However, the q_i value computed in this way must have some $\pm$ error limits put on it. These can be obtained since s_{q_i} can be computed from

$$s_{q_i}{}^2 = \frac{1}{N} \sum \left(\frac{q_o - b}{m} - q_i \right)^2 = \frac{s_{q_o}{}^2}{m^2} \qquad (3.26)$$

which in this example gives $s_{q_i} = 0.22$ psig. Thus if one were using this gage to measure an unknown pressure and got a reading of 4.32 psig, his estimate of the true pressure would be 4.72 ± 0.66 psig if he wished to use the $\pm 3s$ limits.

Another common method of giving bounds on the error uses the *probable error*, e_p. This is defined by

$$e_p \triangleq 0.674s \qquad (3.27)$$

A range of $\pm e_p$ includes the true value 50 percent of the time. In this case the above value would be quoted as 4.72 ± 0.15 psig. It should be clear that when one is using statements of this kind it is extremely important to state whether probable errors or $\pm 3s$ limits are used.

We should note that in computing s_{q_o} either of two approaches could be used. One might use data such as that in Fig. 3.13 and apply Eq. (3.24) or, alternatively, repeat a given q_i many times and compute s_{q_o} from Eq. (3.8). If s_{q_o} is actually the same for all values of q_i (as assumed above), these two methods should give the same answer for large samples. In computing s_{q_i}, however, the second method is not feasible because one cannot, in general, fix q_o in a calibration and then repeat that point over and over to get scattered values of q_i. This is because q_i is truly an independent variable (subject to one's choice) whereas q_o is dependent (not subject to choice). Thus, in computing s_{q_i}, an approach such as Eq. (3.26) is necessary.

A calibration such as that of Fig. 3.13 allows decomposition of the total error of a measurement process into two parts, the *bias* and the *imprecision*. That is, if one gets a reading of 4.32 psig, the true value is given as 4.72 ± 0.66 psig ($3s$ limits), the bias would be −0.40 psig, and the imprecision ±0.66 psig ($3s$ limits). Of course, once the instrument has been calibrated, the bias can be removed, and the only remaining error is that due to imprecision. The bias is also called the *systematic error* (since it is the same for each reading and can thus be removed by calibration). The error due to imprecision is called the *random error* or *nonrepeatability* since it is, in general, different for every reading and one can only put bounds on it but cannot remove it. Calibration is thus the process of removing bias and defining imprecision numerically. The *total inaccuracy* of the process is defined by the combination of bias and imprecision. If the bias is known, the total inaccuracy is entirely due to imprecision and can be specified by a single number such as s_{q_i}.

In actual engineering practice the accuracy of an instrument is usually given by a single numerical value; very often it is not made clear just what the precise meaning of this number is meant to be. Often, even though a calibration, as in Fig. 3.13, has been carried out, s_{q_i} is not calculated. The error is taken as the largest horizontal deviation of any data point from the fitted line. In Fig. 3.13 this occurs at $q_i = 0$ and amounts to 0.48 psig. The inaccuracy in this case might thus be quoted as ±4.8 percent of full scale. Note that this corresponds to about $\pm 2s_{q_i}$ in this case. This practice is no doubt due to the practical viewpoint that when one takes a measurement all he really wants is to say that it cannot be incorrect by more than some specific value; thus the "easy way out" is simply to give a single number. This would be legitimate if the bias were known to be zero (removed by calibration) and if the ± limit given were specified as being $\pm s$, $\pm 2s$, $\pm 3s$, or $\pm e_p$, since all these terms recognize the random nature of the error. However, if the bias is unknown (and not zero), the quotation of a single number for the total inaccuracy is somewhat unsatisfactory although it may be a necessary expedient.

One reason for this is that if one is trying to estimate the overall inaccuracy of a measurement system made up of a number of components, each of which has a known inaccuracy, the method of combining the individual inaccuracies is different for systematic errors (biases) than for random errors (imprecisions). Thus, if the number given for the total inaccuracy of a given component contains both bias and imprecision in unknown proportions, the calculation of overall system inaccuracy is confused. However, in many cases there is no alternative, and by calculation from theory, past experience, and/or judgment the experimenter must arrive at the best available estimate of the total inaccuracy, or *uncertainty* (as it is sometimes called), to be attached to the reading. In such cases a useful viewpoint is that one is willing to bet with certain odds (say 19 to 1) that the error falls within the given limits. Such limits may then be combined as if they were imprecisions in calculations of overall system error.[1,2]

Irrespective of the precise *meaning* to be attached to accuracy figures provided, say, by instrument manufacturers, the *form* of such specifications is fairly uniform. More often than not, accuracy is quoted as a percentage figure based on the full-scale reading of the instrument. Thus if a pressure gage has a range from 0 to 10 psig and a quoted inaccuracy of ± 1.0 percent of full scale, this is to be interpreted as meaning that no error greater than ± 0.1 psig can be expected for any reading that might be taken on this gage, provided it is "properly" used. The manufacturer may or may not be explicit about the conditions required for "proper use." Note that for an actual reading of 1 psig a 0.1 psig error is 10 *percent of the reading.*

Another method sometimes used gives the error as a percentage of the particular reading with a qualifying statement to apply to the low end of the scale. For example, a spring scale might be described as having an inaccuracy of ± 0.5 percent of reading or ± 0.1 lb_f, whichever is greater. Thus for readings less than 20 lb_f the error is constant at ± 0.1 lb_f, while for larger readings the error is proportional to the reading.

Combination of component errors in overall system-accuracy calculations. A measurement system is often made up of a chain of components each of which is subject to individual inaccuracy. If the individual inaccuracies are known, how is the overall inaccuracy computed? A similar problem occurs in experiments that use the results (measurements) from several different instruments to compute some quantity. If the inaccuracy of each instrument is known, how is the inaccuracy of the computed result estimated? Or, inversely, if there must be a certain accuracy in a computed result, what errors are allowable in the individual instruments?

To answer the above questions, consider the problem of computing a quantity N, where N is a known function of the n *independent* variables, $u_1, u_2, u_3, \ldots, u_n$. That is,

$$N = f(u_1, u_2, u_3, \ldots, u_n) \qquad (3.28)$$

The u's are the measured quantities (instrument or component outputs) and are in error by $\pm\Delta u_1$, $\pm\Delta u_2$, $\pm\Delta u_3$, $\ldots$, $\pm\Delta u_n$, respectively. These errors will cause an error ΔN in the computed result N. The Δu's

[1] S. J. Kline and F. A. McClintock, Describing Uncertainties in Single Sample Experiments, *Mech. Eng.*, vol. 75, p. 3, January, 1953.

[2] L. W. Thrasher and R. C. Binder, A Practical Application of Uncertainty Calculations to Measured Data, *Trans. ASME*, p. 373, February, 1957.

may be considered as absolute limits on the errors, as statistical bounds such as e_p's or $3s$ limits, or as uncertainties on which we are willing to give certain odds as including the actual error. However, the method of computing ΔN and the interpretation of its meaning are different for the first case as compared with the second and third. If the Δu's are considered as absolute limits on the individual errors and we wish to calculate similar absolute limits on the error in N, we could calculate

$$N \pm \Delta N = f(u_1 \pm \Delta u_1, u_2 \pm \Delta u_2, u_3 \pm \Delta u_3, \ldots, u_n \pm \Delta u_n) \tag{3.29}$$

and by subtracting N in Eq. (3.28) from $N \pm \Delta N$ in Eq. (3.29) finally obtain $\pm\Delta N$. This procedure is needlessly time-consuming, however, and an approximate solution valid for engineering purposes may be obtained by application of the Taylor series. Expanding the function f in a Taylor series, we get

$$\begin{aligned} f(u_1 \pm \Delta u_1, u_2 \pm \Delta u_2, \ldots, u_n \pm \Delta u_n) &= f(u_1, u_2, \ldots, u_n) \\ &+ \Delta u_1 \frac{\partial f}{\partial u_1} + \Delta u_2 \frac{\partial f}{\partial u_2} + \cdots + \Delta u_n \frac{\partial f}{\partial u_n} \\ &+ \frac{1}{2}\left[(\Delta u_1)^2 \frac{\partial^2 f_2}{\partial u_1} + \cdots\right] + \cdots \end{aligned} \tag{3.30}$$

where all the partial derivatives are to be evaluated at the known values of $u_1, u_2, \ldots, u_n$. That is, if the measurements have been made, the u's are all known as numbers and may be plugged into the expressions for the partial derivatives to give other numbers. In actual practice, the Δu's will all be small quantities and thus terms such as $(\Delta u)^2$ will be negligible. Equation (3.30) may then be given approximately as

$$\begin{aligned} f(u_1 + \Delta u_1, u_2 + \Delta u_2, \ldots, u_n + \Delta u_n) &= f(u_1, u_2, \ldots, u_n) \\ &+ \Delta u_1 \frac{\partial f}{\partial u_1} + \Delta u_2 \frac{\partial f}{\partial u_2} + \cdots + \Delta u_n \frac{\partial f}{\partial u_n} \end{aligned} \tag{3.31}$$

The absolute error E_a is then given by

$$E_a = \Delta N = \left|\Delta u_1 \frac{\partial f}{\partial u_1}\right| + \left|\Delta u_2 \frac{\partial f}{\partial u_2}\right| + \cdots + \left|\Delta u_n \frac{\partial f}{\partial u_n}\right| \tag{3.32}$$

The absolute-value signs are used because some of the partial derivatives might be negative, and for a positive Δu such a term would *reduce* the total error. Since an error Δu is, in general, just as likely to be positive as negative, to estimate the maximum possible error the absolute-value signs must be used as in Eq. (3.32). The form of Eq. (3.32) is very useful since it shows which variables (u's) exert the strongest influence on the accuracy of the overall result. That is, if, say, $\partial f/\partial u_3$ is a large number compared with the other partial derivatives this means that a small Δu_3 can have a large effect on the total error E_a. If the relative or percentage error E_r is desired, it is clearly given by

$$E_r = \frac{\Delta N}{N} \times 100 = \frac{100E_a}{N} \qquad (3.33)$$

The computed result may thus be expressed as either $N \pm E_a$ or $N \pm E_r$ percent, and the interpretation is that one is *certain* this error will not be exceeded since this is the way the Δu's were defined.

In carrying out the above computations, questions of significant figures and rounding off will occur. We here briefly review these matters for those not familiar with such procedures. A significant figure is any one of the digits 1, 2, 3, 4, 5, 6, 7, 8, 9; zero is a significant figure except when used to fix the decimal point or to fill the places of unknown or discarded digits. Thus in the number 0.000532 the significant figures are 5, 3, and 2, while in the number 2,076 *all* the digits, including the zero, are significant. For a number such as 2,300 the zeros may or may not be significant. To convey which figures are significant, one should write this as 2.3×10^3 if two significant figures are intended, 2.30×10^3 if three, 2.300×10^3 if four, and so forth.

In computations one often deals with numbers having unequal numbers of significant figures. For example, it may be necessary to multiply 4.62×0.317856. The first number is assumed good to three significant figures while the second is good to six. It can be shown that the product will be good only to three significant figures; therefore to save work, the six-figure number should be rounded off before multiplication. A number of rules have been proposed for this rounding procedure, and we now state one that is widely used:

> To round a number to n significant figures, discard all digits to the right of the nth place. If the discarded number is less than one-half a unit in the nth place, leave the nth digit unchanged. If the discarded number is greater than one-half a unit in the nth place, increase the nth digit by 1. If the discarded number is exactly one-half a unit in the nth place, leave the nth digit unchanged if it is an even number and add 1 to it if it is odd.

To determine to what extent numbers should be rounded, the following rules may be applied.

Addition. For addition, retain one more decimal digit in the more-accurate numbers than is contained in the least-accurate number. (The more-accurate numbers are those with the most significant figures.)

Then round off the result to the same decimal place as the least-accurate number.

Example.

$$\left.\begin{array}{r} 2.635 \\ 0.9 \\ 1.52 \\ 0.7345 \end{array}\right\} \longrightarrow \begin{array}{r} 2.64 \\ 0.9 \\ 1.52 \\ 0.73 \\ \hline 5.79 \end{array} \longrightarrow 5.8$$

Subtraction. For subtraction, round off the more-accurate number to the same number of decimal places as the less accurate before subtracting. Give the result to the same number of decimal places as the less-accurate figure.

Example.

$$\left.\begin{array}{r} 7.6345 \\ -\ 0.031 \\ \hline \end{array}\right\} \longrightarrow \begin{array}{r} 7.634 \\ -\ 0.031 \\ \hline 7.603 \end{array} \longrightarrow 7.603$$

Multiplication and division. For multiplication and division, round off the more-accurate numbers to one more significant figure than the least accurate before computing. Round the result to the same number of significant figures as the least-accurate number.

Example.

$$\frac{(1.2)(6.335)(0.0072)}{3.14159} \rightarrow \frac{(1.2)(6.34)(0.0072)}{3.14} \rightarrow 0.0174 \rightarrow 0.017$$

We can now give a step-by-step procedure for computing the overall error:

1. Tabulate all data, each with its $\pm$ error attached. All errors should be expressed to two significant figures. (Actually, one significant figure is often adequate.) The reason for this is that the errors themselves are not generally known very accurately and so it is foolish to carry many significant figures.
2. If the quantity to be computed is N, where $N = f(u_1,u_2, \ldots ,u_n)$, compute the partial derivatives $\partial f/\partial u_1$, $\partial f/\partial u_2$, $\ldots$, $\partial f/\partial u_n$ and evaluate each to slide-rule accuracy (three significant figures) by substituting in the basic data u_1, u_2, $\ldots$, u_n.
3. Using Eq. (3.32), compute E_a and round to two significant figures.
4. Compute N from Eq. (3.28) to one more decimal place than the rounded E_a of step 3. Thus if $E_a = \pm 0.062$, N should be computed as, say, 7.0516. This value is then rounded to the same number of decimal places as E_a, in this case to 7.052. In computing N, treat u_1, u_2, etc., as exact numbers; that is, they each have an infinite

number of significant figures. This viewpoint is necessary because one must be able to compute N to as many significant figures as required by E_a according to the above rule.

5. The result may then be quoted as

$$7.052 \pm 0.062 \qquad \text{absolute terms}$$

or

$$7.052 \pm 0.88 \text{ percent} \qquad \text{relative terms}$$

When the problem is one in which a certain overall accuracy is known to be required and one wishes to know what component accuracies are needed, the following method may be employed. It should be apparent that this problem is mathematically indeterminate since there are an infinite number of combinations of individual accuracies that could result in the same overall accuracy. The means of resolving this difficulty are to be found in the "method of equal effects." This principle merely assumes that each source of error will contribute an equal amount to the total error. Mathematically, if

$$\Delta N = \left|\frac{\partial f}{\partial u_1}\Delta u_1\right| + \left|\frac{\partial f}{\partial u_2}\Delta u_2\right| + \cdots + \left|\frac{\partial f}{\partial u_n}\Delta u_n\right|$$

then, if each term is assumed to be equal, we may write

$$\left|\frac{\partial f}{\partial u_1}\Delta u_1\right| = \left|\frac{\partial f}{\partial u_2}\Delta u_2\right| = \cdots = \left|\frac{\partial f}{\partial u_n}\Delta u_n\right| = \frac{\Delta N}{n} \tag{3.34}$$

Now the allowable overall error ΔN is known and so are n and u_1, u_2, . . . , u_n. Thus

$$\begin{aligned} \frac{\partial f}{\partial u_i}\Delta u_i &= \frac{\Delta N}{n} \\ \Delta u_i &= \frac{\Delta N}{n(\partial f/\partial u_i)} \end{aligned} \qquad i = 1, 2, 3, \ldots, n \tag{3.35}$$

and the allowable error Δu_i in each measurement may be calculated. (The partial derivatives are evaluated at the known values of u_i, u_2, . . . , u_n.) If a particular Δu_i turns out to be smaller than what can possibly be achieved by the instruments available it may be possible to relax this requirement if some *other* Δu_i can be made smaller than the value given by Eq. (3.35). That is, some instruments may give better accuracy than required by Eq. (3.35) while others may be unable to meet the requirements of Eq. (3.35). In such cases it may still be possible to meet the overall accuracy requirement; this may be checked by the formulas given.

When the Δu's are not considered as absolute limits of error but rather as statistical bounds such as $\pm 3s$ limits, probable errors, or uncertainties, the formulas for computing overall errors must be modified.

It can be shown[1] that the proper method of combining such errors is according to the root-sum square (rss) formula

$$E_{a_{rss}} = \sqrt{\left(\Delta u_1 \frac{\partial f}{\partial u_1}\right)^2 + \left(\Delta u_2 \frac{\partial f}{\partial u_2}\right)^2 + \cdots + \left(\Delta u_n \frac{\partial f}{\partial u_n}\right)^2} \tag{3.36}$$

The overall error $E_{a_{rss}}$ then has the same meaning as the individual errors. That is, if Δu_i represents a $\pm 3s$ limit on u_i, then $E_{a_{rss}}$ represents a $\pm 3s$ limit on N, and 99.7 percent of the values of N can be expected to fall within these limits. Equation (3.36) always gives a smaller value of error than does Eq. (3.32). Equation (3.35) must also be modified when this viewpoint is taken:

$$\Delta u_i = \frac{\Delta N}{\sqrt{n}\,(\partial f/\partial u_i)} \tag{3.37}$$

As an example of the above procedures, consider an experiment for measuring by means of a dynamometer the average power transmitted by a rotating shaft. The formula for horsepower can be written as

$$\text{hp} = \frac{2\pi RFL}{550t} \tag{3.38}$$

$$\begin{aligned} \text{where}\quad R &\triangleq \text{revolutions of shaft during time } t \\ F &\triangleq \text{force at end of torque arm, lb}_f \\ L &\triangleq \text{length of torque arm, ft} \\ t &\triangleq \text{time length of run, sec} \end{aligned} \tag{3.39}$$

A sketch of the experimental setup is shown in Fig. 3.14. The revolution counter is of the type shown in Fig. 2.4 and can be turned on and off with an electric switch. The instants of turning on and off are recorded by a stopwatch. If it is assumed the counter does not miss any counts, the maximum error in R is ± 1 revolution, because of the digital nature of the device (see Fig. 3.15).

There is a related error, however, in determining the time t since perfect synchronization of the starting and stopping of watch and counter is not possible. The stopwatch might be known to be a quite accurate time-measuring instrument but this does not guarantee that it will always measure the time interval intended. In assigning an error to t, then, we are not helped much by the watch manufacturer's guarantee of 0.10 percent inaccuracy if our synchronization error is much larger than this. This synchronization error is certainly not precisely known since it involves human factors. An experiment to determine its statistical characteristics would be a more expensive and involved undertaking than

[1] J. B. Scarborough, "Numerical Mathematical Analysis," 3d ed., p. 429, The Johns Hopkins Press, Baltimore, 1955.

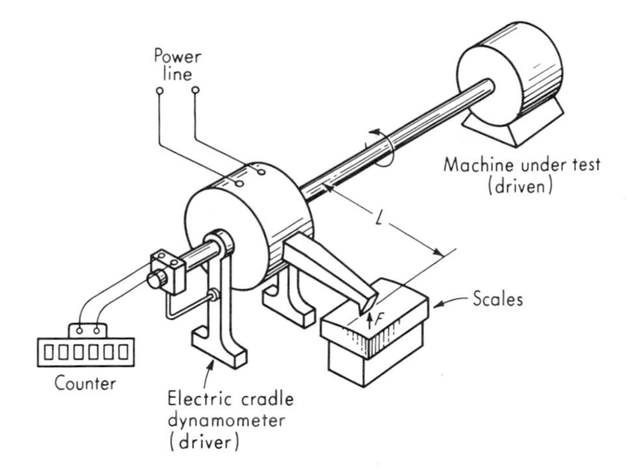

Fig. 3.14. *Dynamometer test setup.*

the power measurement of which it is a part. We are thus in the rather common position of having to rely on experience and judgment in arriving at an estimate of the proper numerical value, and we begin to appreciate that some of the statistical niceties and fine points of theory considered earlier may appear somewhat academic in such a situation. They are always useful in terms of the understanding of basic concepts that they develop; however, they cannot be relied upon to give clear-cut answers in situations where the basic data are ill-defined. In the present case, suppose it is decided that a total starting and stopping error will be taken as ± 0.50 sec. Whether this is to be considered as an absolute limit or as a $\pm 3s$ limit is somewhat meaningless when the basic number is arrived at in such an arbitrary fashion.

The measurement of the torque arm length L is also subject to

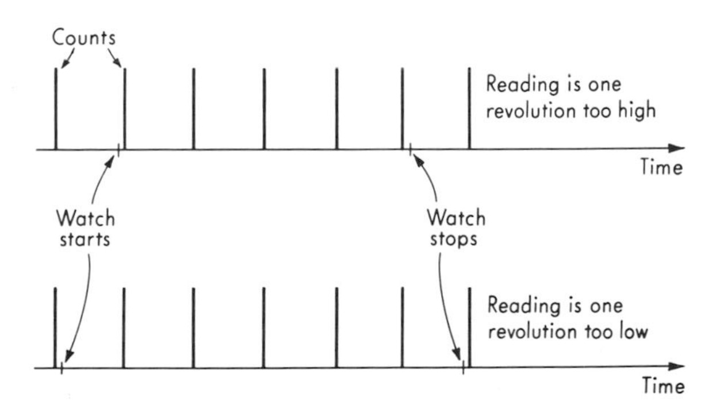

Fig. 3.15. *Revolution-counting error.*

similar vagaries, depending on the care taken in this particular measurement. Suppose we use a fairly rough procedure and decide an on error of ± 0.05 in.

The scales used to measure the force F can be statically calibrated with dead weights, giving a set of data analogous to that of Fig. 3.13. Suppose this is done and an s_{qi} of 0.0133 lb_f is obtained. The $\pm 3s$ limits would then be ± 0.040 lb_f. Again, however, the situation is not this simple. When actually used, the scales will be subject to vibration, which may reduce frictional effects and increase precision. At the same time, the pointer on the scale will not stand perfectly still when the dynamometer is running; thus in reading the scale we must perform a mental averaging process which may introduce more error. Such effects are clearly difficult to quantify, and we must again make a decision based partly on experience and judgment. Suppose we assume the two mentioned effects cancel each other and thus take ± 0.040 lb_f as the force measurement error.

If for a specific run the data are

$$\begin{aligned} R &= 1{,}202 \pm 1.0 \text{ revolutions} \\ F &= 10.12 \pm 0.040 \text{ lb}_f \\ L &= 15.63 \pm 0.050 \text{ in.} \\ t &= 60.0 \pm 0.50 \text{ sec} \end{aligned} \tag{3.40}$$

the calculation proceeds as follows: In terms of inch units, we have

$$\text{hp} = \frac{2\pi}{(550)(12)} \frac{FLR}{t} = K \frac{FLR}{t} \tag{3.41}$$

Then, computing the various partial derivatives to slide-rule accuracy gives

$$\frac{\partial(\text{hp})}{\partial F} = \frac{KLR}{t} = \frac{(0.000952)(15.63)(1{,}202)}{60} = 0.298 \text{ hp/lb}_f \tag{3.42}$$

$$\frac{\partial(\text{hp})}{\partial R} = \frac{KFL}{t} = \frac{(0.000952)(10.12)(15.63)}{60} = 0.00251 \text{ hp/revolution} \tag{3.43}$$

$$\frac{\partial(\text{hp})}{\partial L} = \frac{KFR}{t} = \frac{(0.000952)(10.12)(1{,}202)}{60} = 0.193 \text{ hp/in.} \tag{3.44}$$

$$\frac{\partial(\text{hp})}{\partial t} = -\frac{KFLR}{t^2} = -\frac{(0.000952)(10.12)(15.63)(1{,}202)}{3{,}600} = -0.0500 \text{ hp/sec} \tag{3.45}$$

If we now choose to consider the component errors as absolute limits and wish to compute the absolute limits on the overall error, we use Eq. (3.32) and get

$$E_a = (0.298)(0.040) + (0.00251)(1.0) + (0.193)(0.050) + (0.05)(0.50) \tag{3.46}$$

$$E_a = 0.0119 + 0.00251 + 0.00965 + 0.025 = 0.049 \text{ hp} \tag{3.47}$$

We now compute a rough value of hp by slide rule as

$$\text{hp} = \frac{(0.000952)(10.12)(15.63)(1{,}202)}{60} = 3.02 \qquad (3.48)$$

If we now follow the rule saying that hp should be computed to one more decimal place than E_a, we must calculate hp to four decimal places, which in this case means five significant figures. Thus

$$\text{hp} = \frac{(2.0000)(3.1416)(10.120)(15.630)(1{,}202.0)}{(550.00)(12.000)(60.000)} = 3.0167 \qquad (3.49)$$

which we round off to 3.017. The result may then be quoted as hp = 3.017 ± 0.049 hp or hp = 3.017 ± 1.6 percent. If the component errors were considered as having only one significant figure (which might well be the case here where they are in considerable doubt), the above computations can be simplified since fewer significant figures need be carried. The final result thus might be given more realistically as hp = 3.02 ± 0.05.

If the individual errors are thought of as $\pm 3s$ limits, then Eq. (3.36) should be used to compute $\pm 3s$ limits on hp. Let us carry this out to see the numerical significance.

$$E_{a_{rss}} = \sqrt{(0.0119)^2 + (0.00251)^2 + (0.00965)^2 + (0.025)^2} = 0.029 \text{ hp} \qquad (3.50)$$

We see that $E_{a_{rss}}$ is significantly smaller than E_a. We might say that the error is *possibly* as large as 0.049 hp but *probably* not larger than 0.029 hp. If the individual errors had been accurately known to be $\pm 3s$ limits, the word "probably" would have precise statistical meaning, otherwise not.

Finally, suppose we wish to measure hp to 0.5 percent accuracy in the previous example. What accuracies are needed in the individual measurements? We can use either Eq. (3.35) (if we wish to be conservative) or Eq. (3.37) (if we wish to give ourselves every chance of showing the measurement to be possible). Using Eq. (3.37), we get

$$\begin{aligned}
\Delta F &= \frac{(3.02)(0.005)}{\sqrt{4}\,(0.298)} = 0.025 \text{ lb}_f \\
\Delta R &= \frac{(3.02)(0.005)}{\sqrt{4}\,(0.0025)} = 3.0 \text{ revolutions} \\
\Delta L &= \frac{(3.02)(0.005)}{\sqrt{4}\,(0.193)} = 0.039 \text{ in.} \\
\Delta t &= \frac{(3.02)(0.005)}{\sqrt{4}\,(0.05)} = 0.15 \text{ sec}
\end{aligned} \qquad (3.51)$$

[If we use Eq. (3.35), all these allowable errors are cut in half.] If it

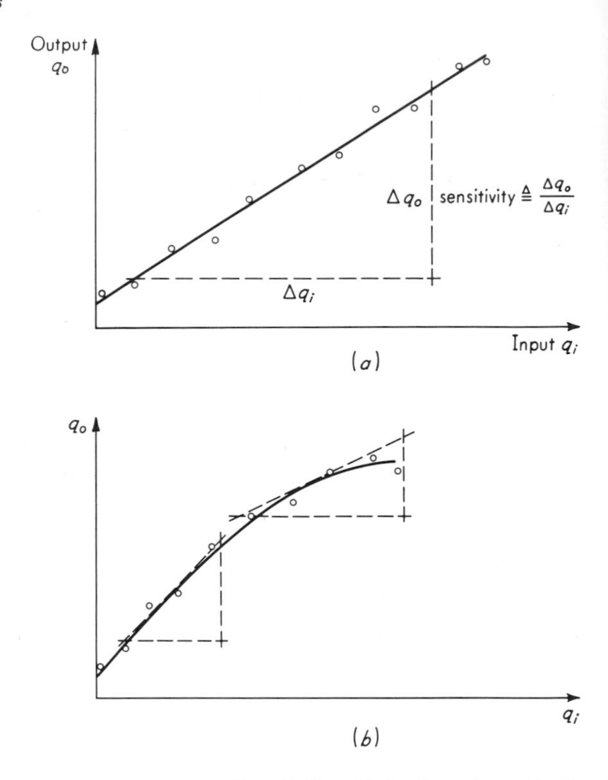

Fig. 3.16. Definition of sensitivity.

is found that the best instrument and technique available for measuring, say, F are good only to 0.04 lb_f rather than the 0.025 lb_f called for by Eq. (3.51), this does not necessarily mean that hp cannot be measured to 0.5 percent. However, it does mean that one or more of the other quantities R, L, and t *must* be measured *more* accurately than required by Eq. (3.51). Making one or more of these measurements more accurately may offset the excessive error in the F measurement. The given formulas allow calculation of whether this will be true or not.

Static sensitivity. When an input-output calibration such as that of Fig. 3.13 has been performed, the *static sensitivity* of the instrument can be defined as the slope of the calibration curve. If the curve is not nominally a straight line, the sensitivity will vary with the input value, as shown in Fig. 3.16*b*. To get a meaningful definition of sensitivity, the output quantity must be taken as the actual physical output, not the meaning attached to the scale numbers. That is, in Fig. 3.13 the output quantity was plotted as pounds per square inch gage; however, the actual physical output is an angular rotation of the pointer. Thus to define sensitivity properly, one must know the angular spacing of the pounds-per-square-inch-gage marks on the scale of the pressure gage. Suppose this is 5 angular degrees/psig. Since we have already calculated the slope in psig/psig as 1.05 in Fig. 3.13, we get the instrument static sensitivity as $(5)(1.05) = 5.25$ angular degrees/psig. In this form the sensitivity allows comparison of this pressure gage with others as regards its ability to detect pressure changes.

While the instrument's sensitivity to its desired input is of primary concern, its sensitivity to interfering and/or modifying inputs may also be of interest. As an example, consider temperature as an input to the pressure gage mentioned above. Temperature can cause a relative expansion and contraction that will result in a change in output reading even though the pressure has not changed. In this sense, it is an interfering input. Also, temperature can change the modulus of elasticity of the pressure-gage spring, thereby giving a change in the pressure sensitivity. In this sense, it is a modifying input. The first effect is often called a *zero drift* while the second is a *sensitivity drift* or *scale-factor drift.* These effects can be numerically evaluated by running suitable calibration tests. To evaluate zero drift the pressure is held at zero while the temperature is varied over a range and the output reading recorded. For reasonably small temperature ranges the effect is often nearly linear, and one can then quote the zero drift as, say, 0.01 angular degree/F°. Sensitivity drift may be found by fixing the temperature and running a pressure calibration to determine pressure sensitivity. Repeating this for various temperatures should show the effect of temperature on pressure sensitivity. Again, if this is nearly linear, one can specify sensitivity drift as, say, 0.0005 (angular degree/psig)/F°.

Figure 3.17 shows how the superposition of these two effects determines the total error due to temperature. If the instrument is used for measurement only and the temperature is known, numerical knowledge of zero drift and sensitivity drift allows correction of the readings. If the instrument is part of a large data-collection system or control system, such corrections may not be feasible; then knowledge of the drifts is used mainly to estimate overall system errors due to temperature.

The generalized input-output configuration of Fig. 2.9 can, for the present example, be made specific, as in Fig. 3.18. New symbology introduced here includes the *transfer function* and the *variable multiplier*. The output of the variable-multiplier symbol is taken as the product of the two inputs. The output of the transfer-function symbol is taken as the input times the function inside the box. In the present case, all the "functions" are merely constants. However, later we shall generalize this concept to include dynamic relations derived from differential equations relating input and output.

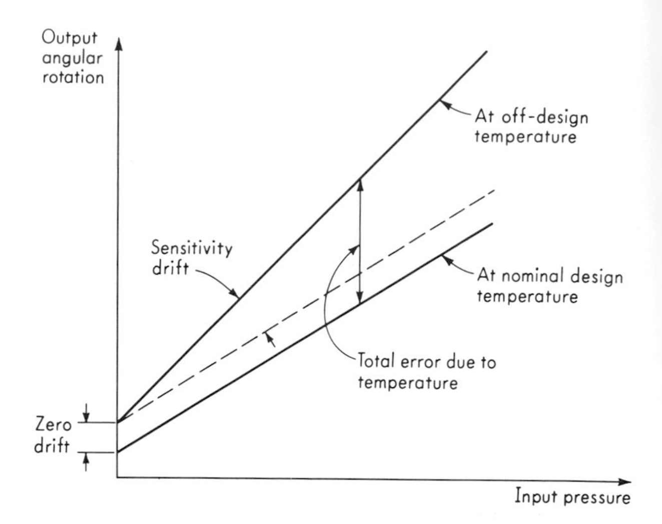

Fig. 3.17. *Zero and sensitivity drift.*

Linearity. If an instrument's calibration curve for desired input is not a straight line, the instrument may still be highly accurate. There are many applications, however, where linear behavior is most desirable. The conversion from a scale reading to the corresponding measured value of input quantity is most convenient if one merely has to multiply by a fixed constant rather than consult a nonlinear calibration curve or compute from a nonlinear calibration equation. Also, when the instrument is part of a larger data or control system, linear behavior of the parts often simplifies design and analysis of the whole. Thus specifications relating to the degree of conformity to straight-line behavior are common.

Several definitions[1] of linearity are possible. However, the so-called *independent linearity* seems to be preferable in many cases. Here the reference straight line is the least-squares fit, as in Fig. 3.13. The linearity is then simply a measure of the maximum deviation of any calibration points from this straight line. This may be expressed as a percent of the actual reading, a percent of full-scale reading, or a combination of the two. The last method is probably the most realistic and leads to the following type of specification:

$$\text{Independent nonlinearity} = \pm A \text{ percent of reading or } \pm B \text{ percent of full scale, whichever is greater} \tag{3.52}$$

The first part ($\pm A$ percent of reading) of the specification recognizes the

[1] L. P. Entin, Instrument Uncertainties: I and II, *Control Eng.*, December, 1959; February, 1960.

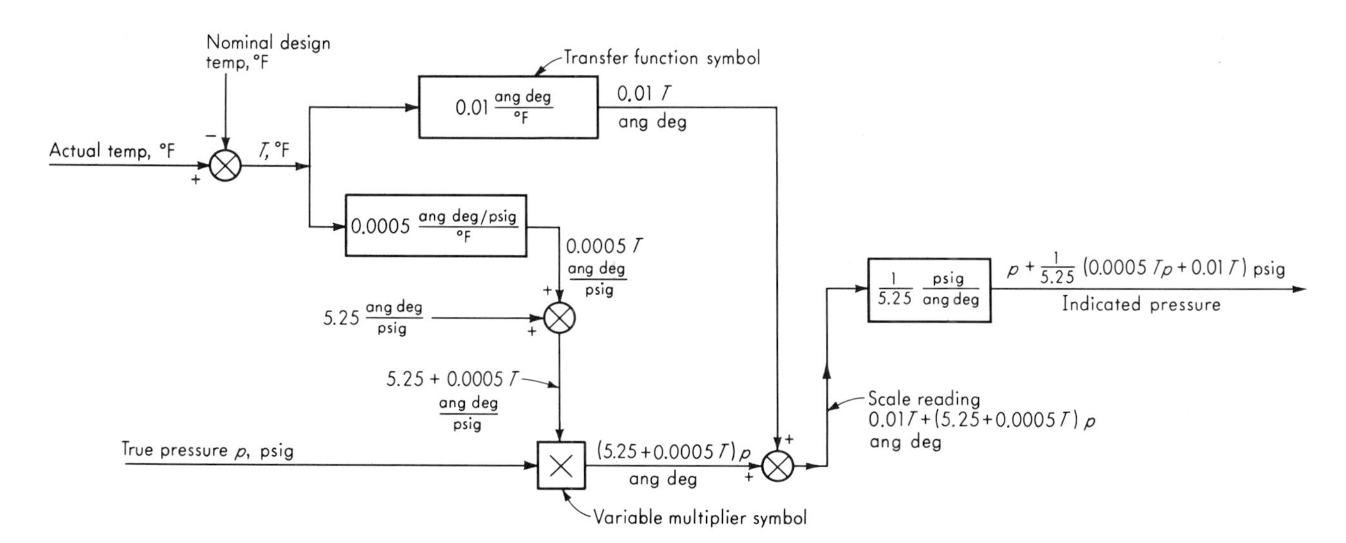

(The −0.64 psig zero bias of Fig. 3.13 is assumed to have been removed by gage adjustment)

Fig. 3.18. *Block diagram of pressure gage.*

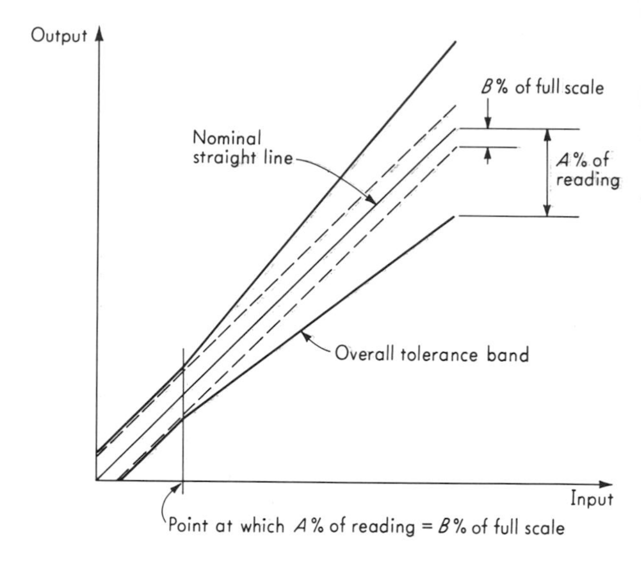

Fig. 3.19. Linearity specification.

desirability of a constant-percentage nonlinearity, while the second ($\pm B$ percent of full scale) recognizes the impossibility of testing for extremely small deviations near zero. That is, if a fixed percentage of reading is specified, the absolute deviations approach zero as the readings approach zero. Since the test equipment should be about 10 times as accurate as the instrument under test, this leads to impossible requirements on the test equipment. Figure 3.19 shows the type of tolerance band allowed by specifications of the form (3.52).

It should be pointed out that in instruments that are considered essentially linear the specification of nonlinearity is equivalent to a specification of overall inaccuracy when the common (nonstatistical) definition of inaccuracy is used. Thus in many commercial linear instruments only a linearity specification (and not an accuracy specification) may be given. The reverse (an accuracy specification but not a linearity specification) may be the case if nominally linear behavior is implied by the quotation of a fixed sensitivity figure.

In addition to overall accuracy requirements, linearity specifications are often useful in dividing the total error into its component parts. Such a division is sometimes advantageous in choosing and/or applying measuring systems for a particular application in which, perhaps, one type of error is more important than another. In such cases, different definitions of linearity may be especially suitable for certain types of systems. The Scientific Apparatus Makers Association standard load-cell (force-measuring device) terminology,[1] for instance, defines linearity as follows: "The maximum deviation of the calibration curve from a straight line drawn between no-load and full-scale load outputs, expressed as a percentage of the full-scale output and measured on increasing load only." The breakdown of total inaccuracy into its component parts will be carried further in the next few sections where hysteresis, resolution, etc., are considered.

Threshold, resolution, hysteresis, and dead space. Consider a situation wherein the pressure gage of Fig. 3.2 has the input pressure slowly and smoothly varied from zero to full scale and then back to zero. If there were no friction due to sliding of moving parts, the input-output graph might appear as in Fig. 3.20*a*. The noncoincidence of loading and unloading curves is due to the internal friction or hysteretic damping of the stressed parts (mainly the spring). That is, all the energy put into the stressed parts upon loading is not recoverable upon unloading, because of the second law of thermodynamics, which rules out perfectly reversible processes in the real world. Certain materials exhibit a minimum of internal friction, and they should be given consideration in designing highly stressed instrument parts, provided that their other properties are also suitable for the specific application. For instruments with a usable range on both sides of zero, the behavior is as shown in Fig. 3.20*b*.

If it were possible to reduce internal friction to zero but external sliding friction were still present, the results might be as in Fig. 3.20*c* and *d*, where a constant coulomb (dry) friction force is assumed. If there is any free play or looseness in the mechanism of an instrument, a curve of similar shape will result.

Hysteresis effects also show up in electrical phenomena. One example is found in the relation between output voltage and input field current in a d-c generator, which is similar in shape to Fig. 3.20*b*. This effect is due to the magnetic hysteresis of the iron in the field coils.

In a given instrument a number of causes such as those just mentioned may combine to give an overall hysteresis effect which might result in an input-output relation as in Fig. 3.20*e*. The numerical value of hysteresis can be specified in terms of either input or output and is usually given as a percentage of full scale. When the total hysteresis has a large component of internal friction, time effects during hysteresis testing may confuse matters since sometimes significant relaxation and recovery effects are present. Thus in going from one point to another in Fig. 3.20*e* one may get a different output reading immediately after

[1] Standard Load Cell Terminology and Definitions: II, Scientific Apparatus Makers Association, Chicago, Jan. 11, 1962.

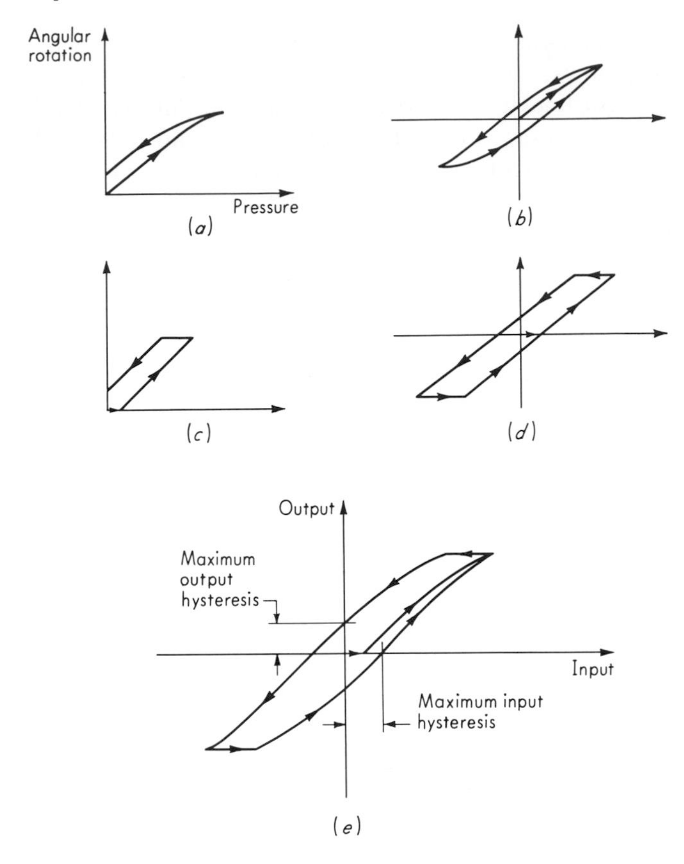

Fig. 3.20. Hysteresis effects.

changing the input than if some time elapses before the reading is taken. If this is the case, the time sequence of the test must be clearly specified if reproducible results are to be obtained.

If the instrument input is very gradually increased from zero there will be some minimum value below which no output change can be detected. This minimum value defines the *threshold* of the instrument. In specifying threshold, the first detectable output change is often described as being any "noticeable" or "measurable" change. Since these terms are somewhat vague, to improve reproducibility of threshold data it may be preferred to state a definite numerical value for output change for which the corresponding input is to be called the threshold.

If the input is slowly increased from some arbitrary (nonzero) input value, it will again be found that the output does not change at all until a certain input increment is exceeded. This increment is called the *resolution;* again, to reduce ambiguity, it is defined as the input increment that gives some small but definite numerical change in the output. Thus resolution defines the smallest measurable input *change* while threshold defines the smallest measurable *input*. Both threshold and resolution may be given either in absolute terms or as a percentage of full-scale reading. An instrument with large hysteresis does not necessarily have poor resolution; internal friction in a spring can give a large hysteresis, but even small changes in input (force) cause corresponding changes in deflection, giving high resolution.

The terms dead space, dead band, and dead zone are sometimes used interchangeably with the term hysteresis. However, they may be defined as the total range of input values possible for a given output and may thus be numerically twice the hysteresis as defined in Fig. 3.20*e*. Since none of these terms is completely standardized, one should always be sure which definition is meant.

Scale readability. Since the majority of instruments that have analog (rather than digital) output are read by a human observer noting the position of a "pointer" on a calibrated scale, it is usually desirable for the data taker to state his opinion as to how closely he believes he can read this scale. This characteristic, *which depends on both the instrument and the observer*, is called the scale readability. While this characteristic should logically be *implied* by the number of significant figures recorded in the data, it is probably good practice for the observer to stop and think about this before taking data and then *record* the scale readability he decides upon on the data sheet. It may also be appropriate at this point to suggest that all data, including scale readabilities, be given in decimal rather than fractional form. Since some instrument scales are calibrated in $\frac{1}{4}$'s, $\frac{1}{2}$'s, etc., this requires the data taker to convert to decimal form before recording data. This procedure is considered preferable to recording a piece of data as, say, $21\frac{1}{4}$ and then *later* trying to decide whether 21.250 or 21.3 was meant.

Span. The range of measured variable that an instrument is designed to measure is sometimes called the span. Equivalent terminology also in use states the "low operating limit" and "high operating limit." For essentially linear instruments, the term "linear operating range" is also common. A related term, which, however, implies dynamic fidelity also, is the *dynamic range*. This is the ratio of the largest to the smallest dynamic input that the instrument will faithfully measure. The number representing the dynamic range is often given in decibels, where the decibel (db) value of a number N is defined as db $\triangleq$ 20 $\log_{10} N$. Thus a dynamic range of 60 db indicates the instrument can handle a range of input sizes of 1,000 to 1.

Generalized static stiffness and input impedance. It has been mentioned before that the introduction of any measuring instrument into a measured medium always results in the extraction of some energy from the medium, thereby changing the value of the measured quantity from its undisturbed state and thus making perfect measurements theoretically impossible. Since the instrument designer wishes to approach perfection as nearly as practicable, some numerical means of characterizing this "loading" effect of the instrument on the measured medium would be helpful in comparing competitive instrument designs. The concepts of *stiffness* and *input impedance*[1] are intended to serve such a function. While both these terms are useful for both static and dynamic conditions, we here introduce them by considering their static aspects only.

In Fig. 2.1 and subsequent schematic and block diagrams the connection of functional elements by single lines perhaps gives the impression that the transfer of information and energy is described by a single variable only. Closer examination reveals that energy transfers require the specification of two variable quantities for their description. The definitions of stiffness and generalized input impedance are in terms of two such variables. At the input of each component in a measuring system there exists a variable q_{i1} with which one is primarily concerned, insofar as information transmission is concerned. At the same point, however, there is associated with q_{i1} another variable q_{i2} such that the product $q_{i1}q_{i2}$ has the dimensions of power and represents an instantaneous rate of energy withdrawal from the preceding element. When these two signals are identified, one can define the generalized input impedance Z_{gi} by

$$Z_{gi} \triangleq \frac{q_{i1}}{q_{i2}} \qquad (3.53)$$

if q_{i1} is an "effort variable." (The definition of an "effort variable" will be given shortly.) [At this point we consider only systems where (3.53) is an ordinary algebraic equation. However, the concept of impedance can easily be extended to dynamic situations and (3.53) then must be given a more general interpretation.] Using (3.53), we see that the power drain $P = q_{i1}^2/Z_{gi}$ and that *a large input impedance is needed to keep the power drain small.* The concept of generalized impedance (and of course the terminology itself) is a generalization of electrical impedance, and we first give some examples from this perhaps somewhat more familiar field.

Consider a voltmeter of the common type shown in Fig. 2.17. Suppose this meter is to be applied to a circuit in order to measure an unknown

[1] R. G. Boiten, The Mechanics of Instrumentation, *Proc. Inst. Mech. Engrs. (London)*, vol. 177, no. 10, 1963.

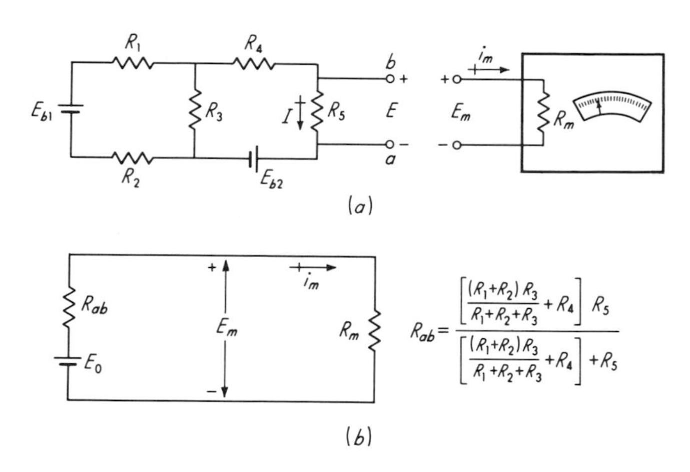

Fig. 3.21. *Voltmeter loading effect.*

voltage E as in Fig. 3.21*a*. As soon as the meter is attached to the terminals a, b, the circuit is changed and the value of E is no longer the same. For the meter alone, the input variable of direct interest (q_{i1}) is the terminal voltage E_m. If we look for an associated variable (q_{i2}) which when multiplied by q_{i1} gives the power withdrawal we find the meter current i_m meets these requirements. In this example, then, $Z_{gi} = E_m/i_m = R_m$, the meter resistance.

Further to illustrate the significance of input impedance, let us determine just how much error is caused when the meter is connected to the circuit. To facilitate this, we first give without proof a very useful network theorem called Thévenin's theorem.[1] Consider any network made up of linear, bilateral impedances and generators (or batteries). A linear impedance is one whose elements (R,L,C) do not change value with the magnitude of the current or voltage. Most resistances, capacitances, and air-core inductances are linear; iron-core inductances are nonlinear. A bilateral impedance is one that transmits energy equally well in either direction. Resistances, capacitances, and inductances are essentially bilateral. Vacuum tubes are unilateral since they effectively transmit energy only in one direction (from grid circuit to plate circuit, *not* the reverse).

A linear bilateral network is shown in Fig. 3.22*a* as a "black box" with terminals A, B. A load of impedance Z_l may be connected across the terminals A, B. When the load Z_l is *not* connected, a voltage will, in general, exist at terminals A, B. This is called E_o, the open-circuit

[1] K. Y. Tang, "Alternating Current Circuits," 2d ed., p. 202, International Textbook Company, Scranton, Pa., 1952.

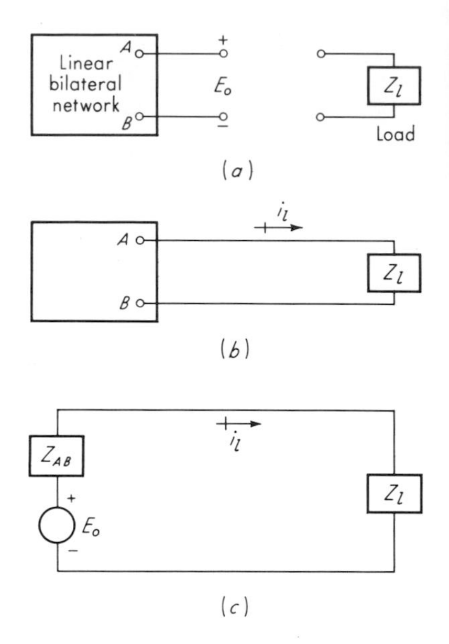

Fig. 3.22. *Thévenin's theorem.*

output voltage of the network. Also with Z_l *not* connected, it is possible to determine the impedance Z_{AB} between terminals A and B. When this is done, any batteries or generators in the network are to be replaced by their internal impedance. If their internal impedance is assumed to be zero, they are replaced by a short circuit (just a wire with no resistance). Thévenin's theorem then states: If the load Z_l is connected as shown in Fig. 3.22*b* a current i_l will flow. This current will be the *same* as the current that flows in the fictitious equivalent circuit of Fig. 3.22*c*. Thus the network, no matter how complex, may be replaced by a single impedance (the output impedance) Z_{AB} in series with a single voltage source E_o.

Applying Thévenin's theorem to Fig. 3.21*a*, we get Fig. 3.21*b*. We see here that the value E_m indicated by the meter is *not* the true value E but rather

$$E_m = \frac{R_m}{R_{ab} + R_m} E \tag{3.54}$$

and that if E_m is to approach E we must have $R_m \gg R_{ab}$. Thus our earlier statement about the desirability of high input impedance can now be made more specific. *The input impedance must be high relative to the output impedance of the system to which the load is connected.* Assuming that it is possible to define generalized input and output impedances Z_{gi} and Z_{go} in nonelectrical as well as electrical systems, we may generalize

Eq. (3.54) to

$$q_{i1m} = \frac{Z_{gi}}{Z_{go} + Z_{gi}} q_{i1u} = \frac{1}{Z_{go}/Z_{gi} + 1} q_{i1u} \qquad (3.55)$$

where $q_{i1m} \triangleq$ measured value of effort variable
$q_{i1u} \triangleq$ undisturbed value of effort variable

Of course, if we knew both Z_{gi} and Z_{go}, we could correct q_{i1m} by means of Eq. (3.55). However, this would be inconvenient; also Z_{go} is not always known, especially in nonelectrical systems, where definition of *both* Z_{gi} and Z_{go} is not always straightforward. Thus a high value of Z_{gi} is desirable since then corrections are unnecessary and the actual values of either Z_{gi} or Z_{go} need not be known.

To achieve a high value of input impedance for any instrument, not just voltmeters, a number of paths are open to the designer. We shall now describe three of them, using the voltmeter as a specific example. The most obvious approach is to leave the configuration of the instrument unchanged but to change the numerical values of physical parameters so that the input impedance is increased. In the voltmeter of Fig. 3.21 this is simply accomplished by winding the coil in such a way (higher resistance material and/or more turns) that R_m is increased. While this accomplishes the desired result, certain undesirable effects also appear. Since this type of voltmeter is basically a *current*-sensitive rather than a voltage-sensitive device, an increase in R_m will *reduce* the magnetic torque available from a given impressed voltage. Thus if the spring constant of the restraining springs is not changed, the angular deflection for a given voltage (the sensitivity) is reduced. To bring the sensitivity back to its former value, we must reduce the spring constant. Also, because of lower torque levels, pivot bearings with less friction must be employed. These design changes generally result in a less rugged and reliable instrument so that this method of increasing input impedance is limited in the degree of improvement possible before other performance features are compromised. This situation will be found to occur in most instruments, not just in this specific example.

If input impedance is to be increased without compromising other characteristics, different approaches are needed. One of general usefulness employs a change of configuration of the instrument so as to include an auxiliary power source. The concept is that a rugged instrument requires a fair amount of power to actuate its output elements but that this power need *not* necessarily be taken from the measured medium. Rather, the low power signal from the primary sensing element may *control* the output of the auxiliary power source so as to realize a power-amplifying effect.

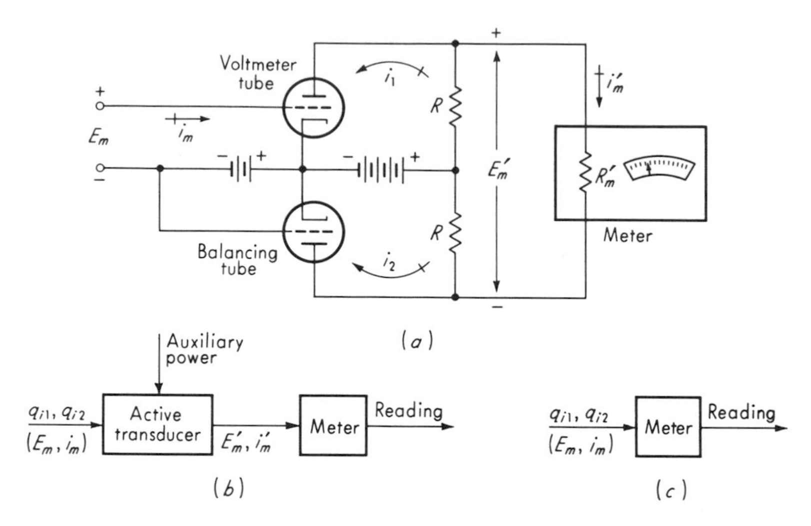

Fig. 3.23. *Vacuum-tube voltmeter.*

Continuing our voltmeter example, this approach is exemplified by the vacuum-tube voltmeter (VTVM). Such a device is shown in rudimentary form in Fig. 3.23*a*. When the input voltage is zero (short circuit across the E_m terminals), if the two tubes are perfectly identical and the two resistances R are exactly equal, the meter voltage E'_m must be zero, because of symmetry. If an input E_m is applied, the grid bias on each tube is no longer the same, the currents i_1 and i_2 will no longer be equal, and a meter voltage E'_m will exist. While the meter current i'_m may still be as large as in a conventional meter, the current that determines the input impedance is i_m, the grid current, which can be extremely small. Thus a very high input impedance can be realized while still employing a rugged meter element. A block diagram for the VTVM is given in Fig. 3.23*b* which may be compared with that for an ordinary meter in Fig. 3.23*c*.

Still another approach to the problem of increasing input impedance uses the principle of feedback or null balance. For the specific area of voltage measurements, this technique is exemplified by the potentiometer. The most simple form of this instrument is shown in Fig. 3.24. It should be clear that in Fig. 3.24*a* each position of the sliding contact corresponds to a definite voltage between the terminals *a* and *b*. Thus the scale can be calibrated and any voltage between zero and the battery voltage obtained by properly positioning the slider. If one now connects in an unknown voltage E_m and a galvanometer (current detector) as in Fig. 3.24*b*, if E_m is less than the battery voltage there will be some point on the slider scale at which the voltage picked off the slide-wire just equals

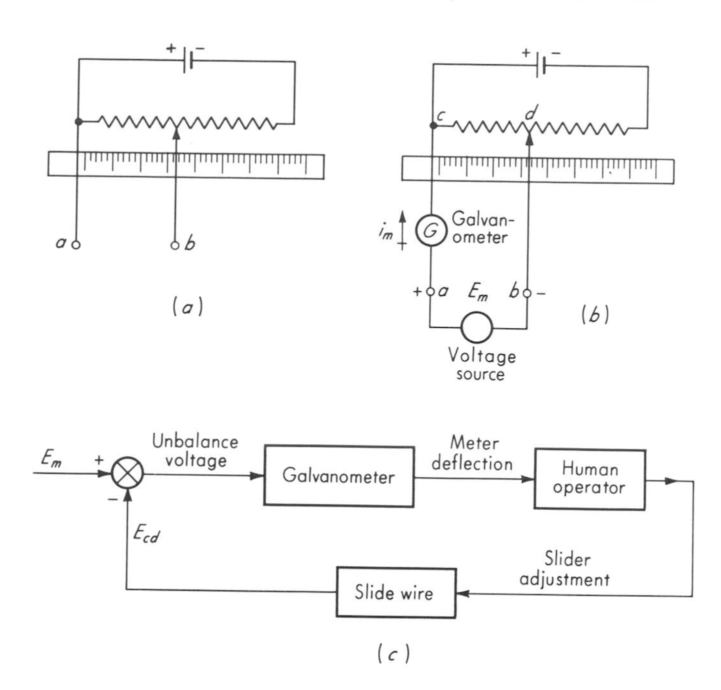

Fig. 3.24. *Potentiometer voltage measurement.*

the unknown E_m. This point of null balance can be detected by a zero deflection of the galvanometer since the net loop voltage a, c, d, b, a will be zero. The unknown voltage can then be read from the calibrated scale.

We should note that under conditions of perfect balance the current drawn from the unknown voltage source is exactly zero, thus giving an infinite input impedance. In actual practice, there must always remain some unknown unbalance current since a galvanometer always has a threshold below which currents cannot be detected. The interpretation of potentiometric voltage measurement as a feedback scheme is given in Fig. 3.24*c*. While the manual balancing described above is adequate when the unknown voltage is relatively constant, the procedure may be made automatic by use of the instrument servomechanism (in this case called a self-balancing potentiometer) as shown in Fig. 2.7. Then, by providing a pen and recording chart, varying voltages may be accurately measured.

In generalizing the concepts of input impedance, a reasonable starting point might be a listing of q_{i1} and corresponding q_{i2} variables for some common measurement situations. In general, the quantity q_{i1}, which is of primary concern, may be either a *flow variable* or an *effort variable.*

The concepts of flow and effort variables are discussed by Paynter.[1] Briefly, energy transfer across the boundaries of a system may be defined in terms of two variables, the product of which gives the instantaneous power. One of these variables, the flow variable, is an *extensive* variable, in the sense that its magnitude depends on the extent of the system taking part in the energy exchange. The other variable, the effort variable, is an *intensive* variable, whose magnitude is independent of the amount of material being considered. In the literature, flow variables are also called "through" variables, and effort variables are called "across" variables. When q_{i1} is an effort variable, Eq. (3.53) and subsequent developments apply. However, if q_{i1} is a flow variable, the situation is somewhat different. It is then appropriate to define a *generalized input admittance* Y_{gi} as

$$Y_{gi} \triangleq \frac{\text{flow variable}}{\text{effort variable}} = \frac{q_{i1}}{q_{i2}} \tag{3.56}$$

rather than a generalized input impedance Z_{gi},

$$Z_{gi} = \frac{\text{effort variable}}{\text{flow variable}} \tag{3.57}$$

We can then write the power drain of the instrument from the measured medium in terms of the measured variable q_{i1} as

$$P = q_{i1}q_{i2} = \frac{q_{i1}^2}{Y_{gi}} \tag{3.58}$$

and we note that now a large value of input admittance Y_{gi} is required in order to minimize the power drain. A familiar electrical example of this situation is the ammeter. In Fig. 3.25*a* we are interested in measuring the current by means of an ammeter inserted into the circuit as shown. Applying Thévenin's theorem, we can reduce Fig. 3.25*a* to Fig. 3.25*b*. We see that the measured value of the current is given by

$$I_m = \frac{E_{ab}}{R_{ab} + R_m} = \frac{E_{ab}}{1/Y_{ab} + 1/Y_m} \tag{3.59}$$

whereas the true (undisturbed) value of the current would be

$$I_u = \frac{E_{ab}}{R_{ab}} = \frac{E_{ab}}{1/Y_{ab}} \tag{3.60}$$

It is now clear that if I_m is to approach I_u we must use an ammeter with $Y_m \gg Y_{ab}$; that is, the meter resistance must be sufficiently *low*, just the *opposite* of that desired in a voltmeter. This result can be generalized

[1] H. M. Paynter, "Analysis and Design of Engineering Systems," p. 18, The M.I.T. Press, Cambridge, Mass., 1960.

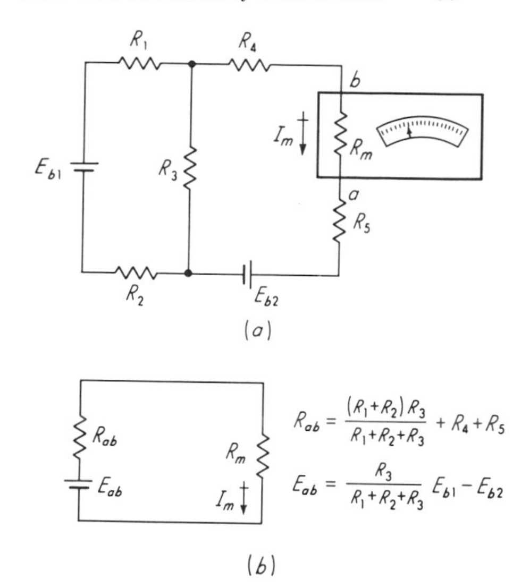

Fig. 3.25. *Ammeter loading effect.*

to apply to all effort variables (such as voltage) and all flow variables (such as current). Equation (3.55) is applicable to those cases in which the measured variable q_{i1} is an effort variable, and Eq. (3.61) gives the corresponding relationship when q_{i1} is a flow variable.

$$q_{i1m} = \frac{1}{Y_{go}/Y_{gi} + 1} q_{i1u} \qquad (3.61)$$

where $Y_{go} \triangleq$ generalized output admittance of preceding element
$Y_{gi} \triangleq$ generalized input admittance of instrument
$q_{i1m} \triangleq$ measured value of flow variable
$q_{i1u} \triangleq$ undisturbed value of flow variable

For some instruments, in the case of a static input, the *power* drain from the preceding element is zero in the steady state, although some total *energy* is removed in going from one steady state to another. In such an instance the concepts of impedance and admittance are not as directly useful as one would like, and it is appropriate to consider the concepts of *static stiffness* and *static compliance*. These make it possible to characterize the *energy* drain (in the same way that impedance and admittance serve to define the *power* drain) in those situations where impedance or admittance becomes infinite and thus not directly meaningful. The terms stiffness and compliance come from the terminology of mechanical systems which afford some of the best examples of the application of these concepts. However, we shall generalize their definitions to include all types of physical systems, just as we did with impedance and admittance.

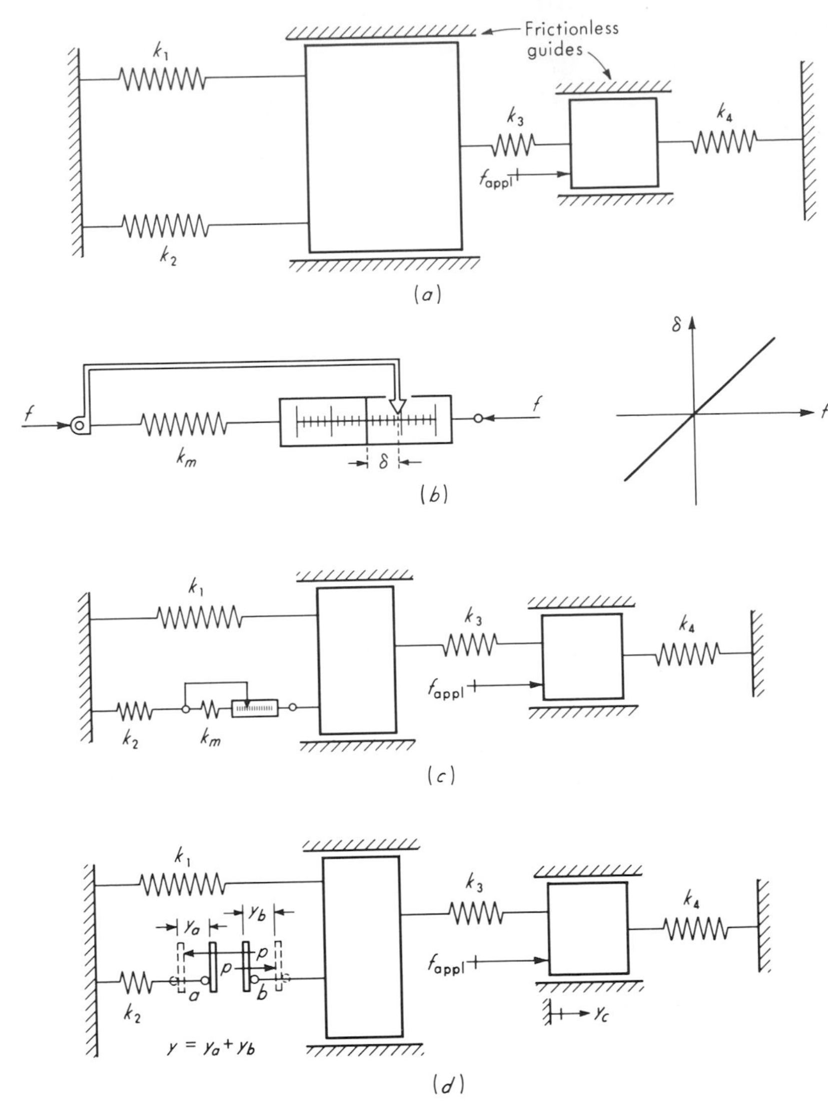

Fig. 3.26. *Force-gage loading effect.*

Consider the system of Fig. 3.26*a* as an idealized model of some elastic structure under applied load f_{appl}. This load will cause forces in the various structural members. Suppose we wish to measure the force in the member represented by the spring k_2. A common method of force measurement employs a calibrated elastic link whose deflection is proportional to force; thus a deflection measurement allows a force measurement. Such a device, with spring constant k_m, is shown in Fig. 3.26*b*. To measure the force in link k_2 we insert the force-measuring device "in series" with the link k_2 as shown in Fig. 3.26*c*. The usual difficulty is encountered here in that the insertion of the measuring instrument alters the condition of the measured system and thus changes the measured variable from its undisturbed value. We wish to assess the nature and amount of this error and shall use this example to introduce the concept of static stiffness.

The measured variable, force, is an effort variable. Thus if we try to use the impedance concept, we must find an associated flow variable whose product with force will give power. Since mechanical power has dimensions ft-lb_f/sec, we have

$$\text{Flow variable} = \frac{\text{power}}{\text{effort variable}} = \frac{\text{ft lb}_f/\text{sec}}{\text{lb}_f} = \frac{\text{ft}}{\text{sec}} = \text{velocity} \qquad (3.62)$$

Mechanical impedance is thus given by

$$\text{Mechanical impedance} = \frac{\text{effort variable}}{\text{flow variable}} = \frac{\text{force}}{\text{velocity}} \qquad (3.63)$$

If we now calculate the static mechanical impedance of an elastic system by applying a constant force and noting the resulting velocity, we get

$$\text{Static mechanical impedance} = \frac{\text{force}}{0} = \infty \qquad (3.64)$$

This difficulty may be overcome by using energy rather than power in the definition of the variable associated with the measured variable. If this is done, a new term for the ratio of the two variables must be introduced, since the use of mechanical impedance as the ratio of force to velocity is well established. We thus define

$$\text{Mechanical static stiffness} \triangleq \frac{\text{force}}{\text{displacement}} = \frac{\text{force}}{\int(\text{velocity})\,dt} \qquad (3.65)$$

$$\text{since} \quad \text{Energy} = (\text{force})(\text{displacement}) \qquad (3.66)$$

Thus, in general, whenever the measured variable is an effort variable and the static impedance is infinite, instead of using impedance one uses a generalized static stiffness S_g defined by

$$S_g \triangleq \frac{\text{effort variable}}{\int(\text{flow variable})\,dt} \qquad (3.67)$$

If this is done, it can be shown that the same formulas can be used for calculating the error due to inserting the measuring instrument as were used for impedance, except S is used instead of Z. Thus, Eq. (3.55) becomes

$$q_{i1m} = \frac{S_{gi}}{S_{go} + S_{gi}} q_{i1u} = \frac{1}{S_{go}/S_{gi} + 1} q_{i1u} \tag{3.68}$$

where $q_{i1m} \triangleq$ measured value of effort variable
$q_{i1u} \triangleq$ undisturbed value of effort variable
$S_{gi} \triangleq$ generalized static input stiffness of measuring instrument
$S_{go} \triangleq$ generalized static output stiffness of measured system

Let us now apply these general concepts to the specific case at hand. The output stiffness of the system of Fig. 3.26*a* at the point of insertion of the measuring device is simply the ratio of force p to deflection y at the terminals a, b in Fig. 3.26*d*. This stiffness can be found theoretically by applying a fictitious load p and calculating the resulting y, or if the structure (or a scale model) has been constructed one can obtain the stiffness experimentally by applying known loads and measuring the resulting deflections. A theoretical analysis might proceed as below:

$$\Sigma \text{ forces} = 0$$

$$\begin{cases} p - y_b(k_1) + k_3(y_c - y_b) = 0 & (3.69) \\ f_{\text{appl}} - k_3(y_c - y_b) - k_4 y_c = 0 & (3.70) \end{cases}$$

$$\begin{cases} (-k_1 - k_3)y_b + (k_3)y_c = -p & (3.71) \\ (k_3)y_b + (-k_3 - k_4)y_c = -f_{\text{appl}} & (3.72) \end{cases}$$

using determinants,

$$y_b = \frac{\begin{vmatrix} -p & k_3 \\ -f_{\text{appl}} & -(k_3 + k_4) \end{vmatrix}}{\begin{vmatrix} -(k_1 + k_3) & k_3 \\ k_3 & -(k_3 + k_4) \end{vmatrix}} = \frac{p(k_3 + k_4) + f_{\text{appl}}(k_3)}{(k_3 + k_4)(k_1 + k_3) - k_3^2} \tag{3.73}$$

The output stiffness is now obtained from Eq. (3.73) by letting f_{appl} be zero:

$$S_{go} = \frac{p}{y} = \frac{p}{y_a + y_b} = \frac{p}{p/k_2 + p(k_3 + k_4)/[(k_3 + k_4)(k_1 + k_3) - k_3^2]} \tag{3.74}$$

$$S_{go} = \frac{1}{1/k_2 + (k_3 + k_4)/[(k_3 + k_4)(k_1 + k_3) - k_3^2]} \tag{3.75}$$

The input stiffness of the measuring instrument is given by

$$S_{gi} = \frac{\text{force}}{\text{displacement}} = k_m \tag{3.76}$$

We may now apply Eq. (3.68) to get

$$\frac{\text{Measured value of force}}{\text{True value of force}} = \frac{k_m}{\dfrac{1}{1/k_2 + (k_3 + k_4)/(k_1k_3 + k_1k_4 + k_3k_4)} + k_m} \tag{3.77}$$

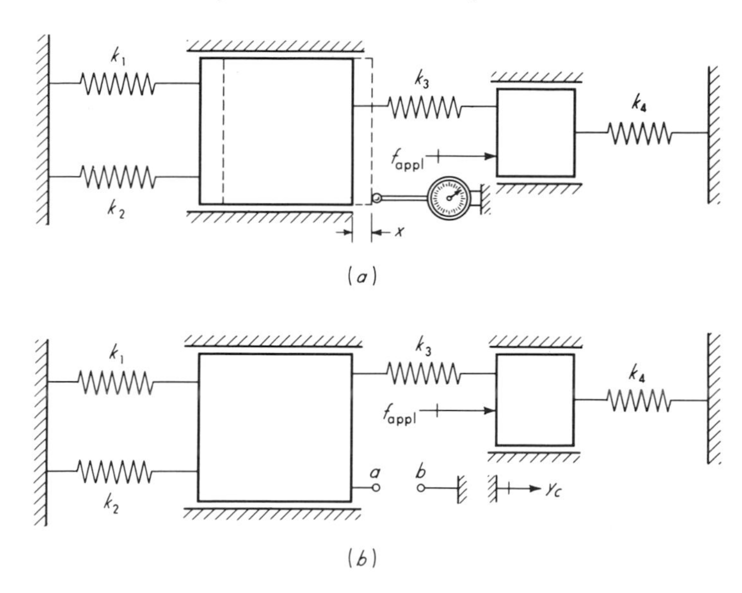

Fig. 3.27. *Displacement-gage loading effect.*

From Eq. (3.68) it is apparent that in general we should like to have $S_{gi} \gg S_{go}$ in order to have the measured value close to the true value. In this example, this requirement corresponds to

$$k_m \gg \frac{1}{1/k_2 + (k_3 + k_4)/(k_1k_3 + k_1k_4 + k_3k_4)} \qquad (3.78)$$

Thus the measuring device must have a sufficiently stiff spring.

We saw earlier that when the measured variable is not an effort variable, admittance rather than impedance is a more convenient tool. Again, however, under static conditions it may happen that admittance is infinite; thus a concept analogous to stiffness is needed to facilitate the treatment of such situations. For such cases the generalized compliance C_g is defined by

$$C_g \triangleq \frac{\text{flow variable}}{\int(\text{effort variable})\, dt} \qquad (3.79)$$

As a mechanical example, suppose we wish to measure the displacement x in Fig. 3.27*a* by means of a dial indicator. Such indicators generally have a spring load to ensure positive contact with the body whose motion is being measured. This spring load adds a force to the measured system, thereby causing error in the motion measurement. It is clear that the indicator spring load should be as light as possible, but we wish to make more quantitative statements about measurement accuracy.

Again, it is possible to show the applicability of the admittance relations [Eq. (3.61)] if we replace admittance by compliance:

$$q_{i1m} = \frac{1}{C_{go}/C_{gi} + 1} q_{i1u} \tag{3.80}$$

where $q_{i1m} \triangleq$ measured value of flow variable
$q_{i1u} \triangleq$ undisturbed value of flow variable
$C_{gi} \triangleq$ generalized static input compliance of measuring instrument
$C_{go} \triangleq$ generalized static output compliance of measured system

We note that for accurate measurement we require $C_{gi} \gg C_{go}$.

In our example the measured variable is displacement, which is a flow variable. If we try to use admittance concepts, we can find the associated effort variable in the usual way:

$$\text{Power} = (\text{effort variable})(\text{flow variable}) \tag{3.81}$$

$$\frac{\text{ft-lb}_f}{\text{sec}} = (\text{effort variable})(\text{displacement, ft}) \tag{3.82}$$

$$\text{Effort variable} = \frac{\text{lb}_f}{\text{sec}} = \text{rate of change of force} \tag{3.83}$$

The admittance is then

$$Y = \frac{\text{flow variable}}{\text{effort variable}} = \frac{\text{displacement}}{\text{rate of change of force}} \tag{3.84}$$

and if we apply this definition to the case of a static load on a spring, we get $Y = \infty$. In this case, however, the compliance would be

$$C_g = \frac{\text{flow variable}}{\int(\text{effort variable})\,dt} = \frac{\text{displacement}}{\text{force}} = \frac{1}{\text{spring constant}} \tag{3.85}$$

In our example the output compliance of the measured system is the ratio of displacement to force for the terminals a, b of Fig. 3.27b. If we apply a fictitious force p between the terminals a, b a displacement y will occur; it may be computed as follows:

$$\Sigma \text{ forces} = 0$$

$$\begin{cases} p - y(k_1 + k_2) + k_3(y_c - y) = 0 & (3.86) \\ f_{\text{appl}} - k_3(y_c - y) - k_4 y_c = 0 & (3.87) \end{cases}$$

$$\begin{cases} (-k_1 - k_2 - k_3)y + (k_3)y_c = -p & (3.88) \\ (k_3)y + (-k_3 - k_4)y_c = -f_{\text{appl}} & (3.89) \end{cases}$$

$$y = \frac{\begin{vmatrix} -p & k_3 \\ -f_{\text{appl}} & (-k_3 - k_4) \end{vmatrix}}{\begin{vmatrix} (-k_1 - k_2 - k_3) & k_3 \\ k_3 & (-k_3 - k_4) \end{vmatrix}}$$

$$= \frac{p(k_3 + k_4) + f_{\text{appl}}(k_3)}{(k_3 + k_4)(k_1 + k_2 + k_3) - k_3^2} = \frac{p(k_3 + k_4) + f_{\text{appl}}(k_3)}{(k_3 + k_4)(k_1 + k_2) + k_3 k_4} \tag{3.90}$$

We can now get C_{go} from Eq. (3.90) by letting $f_{\text{appl}} = 0$:

$$C_{go} = \frac{y}{p} = \frac{k_3 + k_4}{(k_3 + k_4)(k_1 + k_2) + k_3 k_4} = \frac{1}{k_1 + k_2 + k_3 k_4/(k_3 + k_4)} \tag{3.91}$$

If the spring constant of the dial indicator is k_m, the input compliance of the measuring instrument is given by

$$C_{gi} = \frac{1}{k_m} \tag{3.92}$$

We then have

$$\frac{\text{Measured value of deflection}}{\text{True value of deflection}} = \frac{1}{\dfrac{k_m}{k_1 + k_2 + k_3 k_4/(k_3 + k_4)} + 1} \tag{3.93}$$

and k_m must be sufficiently small in order to get accurate displacement measurement.

To illustrate the general applicability of the concepts of impedance, admittance, stiffness, and compliance to measurement problems, Fig. 3.28 has been compiled. In the first column are listed some of the physical quantities commonly measured, each one identified as a flow variable or an effort variable. The appropriate associated variables which give either power or energy when multiplied with the measured variable are then listed. The last four columns indicate the dimensions of the appropriate loading criteria for that particular measurement. For effort variables, both impedance and stiffness are given; which one to use depends on the nature of the specific instrument. The fact that admittance and compliance are *not* given for effort variables does not mean that they could not be defined but merely that there is no need to consider them when the methods explained earlier in this section are used. Similar statements apply to those measured variables that are flow variables.

The basic formulas (3.55), (3.61), (3.68), and (3.80) were given without a detailed proof. At this point we wish to show the justification for these results and also to make clearer the physical meaning of impedance, admittance, stiffness, and compliance for physical systems in general. This discussion will also indicate more clearly how one calculates theoretically or measures experimentally these important system characteristics.

Consider first Fig. 3.29, showing two separate elements which we shall subsequently wish to interconnect in the order shown. Our objective is to determine the characteristics of the *individual* devices that must

Measured variable	*Associated variable*					
	Power based	*Energy based*	*Impedance*	*Admittance*	*Stiffness*	*Compliance*
Voltage (effort)	Current	Charge	$\frac{\text{volts}}{\text{amp}}$		$\frac{\text{volts}}{\text{coul}}$	
Current (flow)	Voltage	$\int$(voltage) dt		$\frac{\text{amp}}{\text{volt}}$		$\frac{\text{amp}}{\text{volt-sec}}$
Force (effort)	Translational velocity	Translational displacement	$\frac{\text{lb}_f}{\text{fps}}$		$\frac{\text{lb}_f}{\text{ft}}$	
Translational displacement (flow)	d/dt(force)	Force		$\frac{\text{in.}}{\text{lb}_f\text{/sec}}$		$\frac{\text{in.}}{\text{lb}_f}$
Torque (effort)	Rotational velocity	Rotational displacement	$\frac{\text{ft-lb}_f}{\text{rad/sec}}$		$\frac{\text{ft-lb}_f}{\text{rad}}$	
Rotational displacement (flow)	d/dt(torque)	Torque		$\frac{\text{rad}}{\text{ft-lb}_f\text{/sec}}$		$\frac{\text{rad}}{\text{ft-lb}_f}$
Translational velocity (flow)	Force	$\int$(force) dt		$\frac{\text{fps}}{\text{lb}_f}$		$\frac{\text{fps}}{\text{lb}_f\text{-sec}}$
Rotational velocity (flow)	Torque	$\int$(torque) dt		$\frac{\text{rad/sec}}{\text{ft-lb}_f}$		$\frac{\text{rad/sec}}{\text{ft-lb}_f\text{-sec}}$

Measured variable	*Associated variable*		*Impedance*	*Admittance*	*Stiffness*	*Compliance*
	Power based	*Energy based*				
Translational acceleration (flow)	$\int(\text{force})\,dt$	$\int[\int(\text{force})\,dt]\,dt$		$\frac{\text{ft/sec}^2}{\text{lb}_f\text{-sec}}$		$\frac{\text{ft/sec}^2}{\text{lb}_f\text{-sec}^2}$
Rotational acceleration (flow)	$\int(\text{torque})\,dt$	$\int[\int(\text{torque})\,dt]\,dt$		$\frac{\text{rad/sec}^2}{\text{ft-lb}_f\text{-sec}}$		$\frac{\text{rad/sec}^2}{\text{ft-lb}_f\text{-sec}^2}$
Fluid pressure (effort)	Volume flow rate	Volume flow	$\frac{\text{lb}_f\text{/ft}}{\text{ft}^3\text{/sec}}$		$\frac{\text{lb}_f\text{/ft}}{\text{ft}^3}$	
Volume flow rate (flow)	Fluid pressure	$\int(\text{fluid pressure})\,dt$		$\frac{\text{ft}^3\text{/sec}}{\text{lb}_f\text{/ft}^2}$		$\frac{\text{ft}^3\text{/sec}}{(\text{lb}_f\text{/ft}^2)\text{-sec}}$
Temperature (effort)	Heat-transfer rate per unit temperature difference	Total heat transfer per unit temperature difference	$\frac{\text{F}^\circ}{\text{Btu/(sec-F}^\circ)}$		$\frac{\text{F}^\circ}{\text{Btu/F}^\circ}$	

Fig. 3.28. *Loading parameters for common variables.*

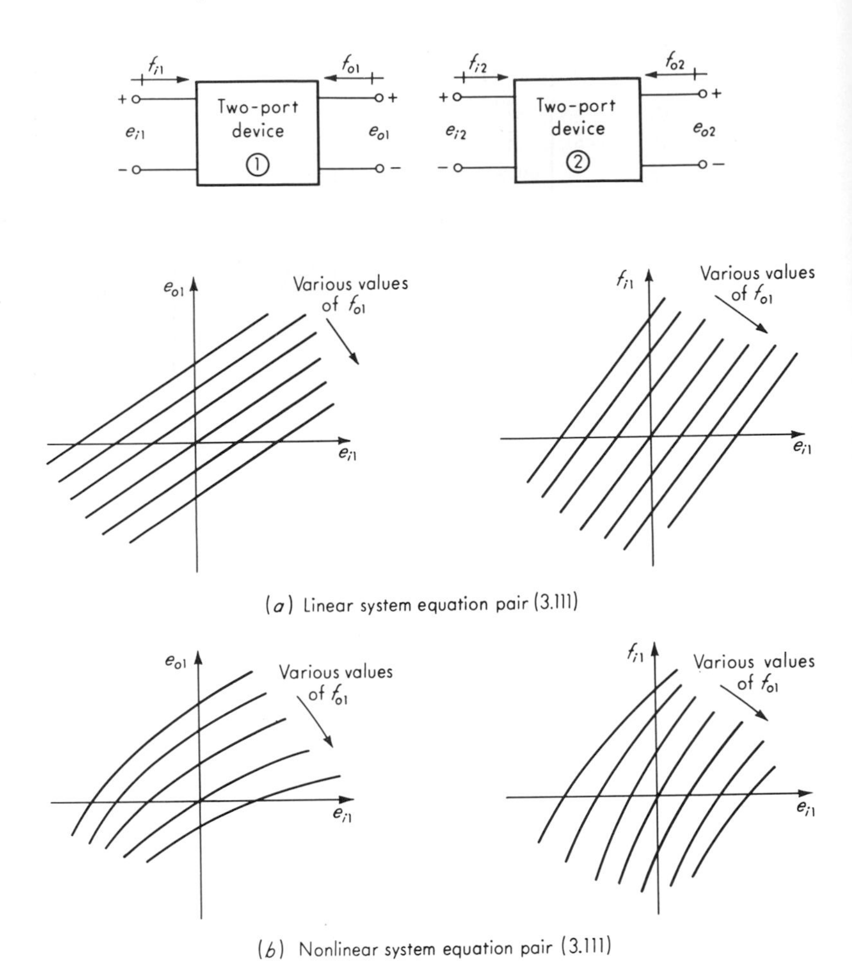

(a) Linear system equation pair (3.111)

(b) Nonlinear system equation pair (3.111)

Fig. 3.29. *Generalized loading configuration.*

be known in order to predict their overall operation when connected together. We first assume that each element may be characterized as a "two-port."[1] This means essentially that the only significant energy exchanges that take place between the device and others that might be connected to it occur at two places ("ports"), which we denote as the input and the output. This does not necessarily mean that a third (or fourth) flow of energy through the "boundaries" of the device might not exist but merely that such additional energy flows are assumed constant and unaffected by changes in the two main energy flows. For example, an electronic amplifier has mainly an input voltage (and current) and an

[1] Paynter, *op. cit.*, p. 50.

output voltage (and current). However, it actually also has a third "port," the connection to its power supply. However, if the power supply is assumed *constant*, it can be included within the boundaries of the amplifier itself, allowing characterization of such amplifiers as two-port.

In Fig. 3.29, the symbols e represent effort variables whereas f's denote flow variables. Any one of the four variables associated with each of the devices may be considered as being determined by the values of the other variables. That is, we may consider any one variable as a *dependent* variable whose value is determined by the values of several independent variables. It is important to note that the quantities in question (the e's and f's) are all available at the "terminals" of the device and may be measured experimentally *without any knowledge of the internal details of the device. Thus complex devices for which no adequate theory exists may be studied experimentally by such methods.*

Suppose we choose to consider first the quantity e_{o1} as a dependent variable. Then we might at first think that the independent variables would be e_{i1}, f_{i1}, and f_{o1}, that is,

$$e_{o1} = e_{o1}(e_{i1}, f_{i1}, f_{o1}) \tag{3.94}$$

This, however, is not the case, since e_{i1}, f_{i1}, and f_{o1} are not all independent. If any two of these three are assigned values, the value of the third is determined by the system and is not open to independent choice. The truth of this statement may be demonstrated as follows: Similarly to Eq. (3.94), we could write

$$f_{o1} = f_{o1}(e_{i1}, f_{i1}, e_{o1}) \tag{3.95}$$

$$e_{i1} = e_{i1}(f_{i1}, f_{o1}, e_{o1}) \tag{3.96}$$

$$f_{i1} = f_{i1}(e_{i1}, e_{o1}, f_{o1}) \tag{3.97}$$

We note now, however, from (3.95) that f_{o1} depends on e_{i1} and f_{i1}. Thus it would not be correct to have e_{i1}, f_{i1}, *and* f_{o1} as independent variables in (3.94). Similar inconsistencies can be found in all four formulas. It is thus clear that three independent variables are too many. If we try two instead, we can write

$$e_{o1} = e_{o1}(e_{i1}, f_{i1}) \tag{3.98}$$

$$\text{or} \quad e_{o1} = e_{o1}(e_{i1}, f_{o1}) \tag{3.99}$$

$$\text{or} \quad e_{o1} = e_{o1}(f_{i1}, f_{o1}) \tag{3.100}$$

$$\text{Also} \quad f_{o1} = f_{o1}(e_{i1}, e_{o1}) \tag{3.101}$$

$$\text{or} \quad f_{o1} = f_{o1}(e_{i1}, f_{i1}) \tag{3.102}$$

$$\text{or} \quad f_{o1} = f_{o1}(e_{o1}, f_{i1}) \tag{3.103}$$

$$\text{Also} \quad e_{i1} = e_{i1}(e_{o1}, f_{i1}) \tag{3.104}$$

$$\text{or} \quad e_{i1} = e_{i1}(e_{o1}, f_{o1}) \tag{3.105}$$

$$\text{or} \quad e_{i1} = e_{i1}(f_{i1}, f_{o1}) \tag{3.106}$$

$$\text{Also} \quad f_{i1} = f_{i1}(e_{i1}, e_{o1}) \tag{3.107}$$

$$\text{or} \quad f_{i1} = f_{i1}(e_{i1}, f_{o1}) \tag{3.108}$$

$$\text{or} \quad f_{i1} = f_{i1}(e_{o1}, f_{o1}) \tag{3.109}$$

We can now choose any one of the above equations and immediately find its companion (another equation with the *same* independent variables) and thus define the system in terms of two "input" quantities (the independent variables) and two "output" quantities (the dependent variables). Enumerating all possibilities, we get

$$\begin{aligned} e_{o1} &= e_{o1}(e_{i1},f_{i1}) \\ f_{o1} &= f_{o1}(e_{i1},f_{i1}) \end{aligned} \tag{3.110}$$

$$\begin{aligned} e_{o1} &= e_{o1}(e_{i1},f_{o1}) \\ f_{i1} &= f_{i1}(e_{i1},f_{o1}) \end{aligned} \tag{3.111}$$

$$\begin{aligned} e_{o1} &= e_{o1}(f_{i1},f_{o1}) \\ e_{i1} &= e_{i1}(f_{i1},f_{o1}) \end{aligned} \tag{3.112}$$

$$\begin{aligned} f_{o1} &= f_{o1}(e_{i1},e_{o1}) \\ f_{i1} &= f_{i1}(e_{i1},e_{o1}) \end{aligned} \tag{3.113}$$

$$\begin{aligned} f_{o1} &= f_{o1}(e_{o1},f_{i1}) \\ e_{i1} &= e_{i1}(e_{o1},f_{i1}) \end{aligned} \tag{3.114}$$

$$\begin{aligned} e_{i1} &= e_{i1}(e_{o1},f_{o1}) \\ f_{i1} &= f_{i1}(e_{o1},f_{o1}) \end{aligned} \tag{3.115}$$

Thus we see that, in general, one can choose values for any two of the four quantities and then the values of the other two are determined. If the physical system is strictly linear, its static input-output characteristics can be displayed graphically for any one of the equation pairs (3.110) to (3.115) in a fashion similar to Fig. 3.29*a*. If it is nonlinear in a continuous (smooth) fashion, the curves might be as in Fig. 3.29*b*.

We are now in a position to derive Eq. (3.55). Taking the first equation of pair (3.111), we may write

$$de_{o1} = \left.\frac{\partial e_{o1}}{\partial e_{i1}}\right|_{f_{o1}=\text{const}} de_{i1} + \left.\frac{\partial e_{o1}}{\partial f_{o1}}\right|_{e_{i1}=\text{const}} df_{o1} \tag{3.116}$$

We now define

$$\left.\frac{\partial e_{o1}}{\partial e_{i1}}\right|_{f_{o1}=\text{const}} \triangleq \text{no-load static transfer function} \triangleq K \tag{3.117}$$

$$\left.\frac{\partial e_{o1}}{\partial f_{o1}}\right|_{e_{i1}=\text{const}} \triangleq \text{generalized output impedance } Z_{go} \tag{3.118}$$

The physical interpretation here is that we consider the system originally in equilibrium with e_{i1}, f_{i1}, e_{o1}, and f_{o1} all at some constant values. Now, if e_{i1} changes by de_{i1} and if f_{o1} should stay constant [the simplest case is where f_{o1} is constant *at zero* because device 1 is open circuit (unloaded) at its output], the change in output is given by

$$de_{o1} = K\, de_{i1} \tag{3.119}$$

If, however, the output of device 1 *is* connected to the input of device 2, then f_{o1} will not be constant and de_{o1} will be different from $K\,de_{i1}$. The amount of "loading error" depends on df_{o1}; to find it, we must know the input impedance of device 2. To define this, we use the second equation of pair (3.112) and apply it to device 2.

$$de_{i2} = \left.\frac{\partial e_{i2}}{\partial f_{i2}}\right|_{f_{o2}=\text{const}} df_{i2} + \left.\frac{\partial e_{i2}}{\partial f_{o2}}\right|_{f_{i2}=\text{const}} df_{o2} \tag{3.120}$$

If device 2 has no third device connected to its output, we may take $df_{o2} = 0$ and get

$$de_{i2} = \left.\frac{\partial e_{i2}}{\partial f_{i2}}\right|_{f_{o2}=\text{const}} df_{i2} \tag{3.121}$$

We now define

$$\left.\frac{\partial e_{i2}}{\partial f_{i2}}\right|_{f_{o2}=\text{const}} \triangleq \text{generalized input impedance } Z_{gi} \tag{3.122}$$

From Fig. 3.29, we note that, when devices 1 and 2 are connected, $e_{o1} = e_{i2}$ and $f_{o1} = -f_{i2}$. We thus get

$$de_{o1} = K\,de_{i1} + Z_{go}\,df_{o1} = K\,de_{i1} - \frac{Z_{go}}{Z_{gi}}\,de_{o1} \tag{3.123}$$

$$\left(1 + \frac{Z_{go}}{Z_{gi}}\right) de_{o1} = K\,de_{i1} \tag{3.124}$$

Now, $K\,de_{i1}$ is the value de_{o1} would have if device 2 were not connected to device 1. This corresponds to q_{i1u} of Eq. (3.55). We can thus write

$$de_{o1} = \frac{1}{Z_{go}/Z_{gi} + 1}\,K\,de_{i1} \tag{3.125}$$

thereby proving Eq. (3.55). Equations (3.61), (3.68), and (3.80) can all be established in similar fashion.

When the physical devices are strictly linear, the small changes de_{o1}, de_{i1}, etc., can be replaced by the actual quantities e_{o1}, e_{i1}, etc., and the partial derivatives are constant for all values of the other independent variable. That is, in Eq. (3.116), for example, the term

$$\left.\frac{\partial e_{o1}}{\partial e_{i1}}\right|_{f_{o1}=\text{const}}$$

would be numerically the same for any and all values of f_{o1} that one might choose. One generally chooses the simplest value of f_{o1} with which to work theoretically or experimentally. This is $f_{o1} \equiv 0$. The other partial derivatives are similarly handled. Thus, in linear systems, the various impedances, admittances, transfer functions, etc., are all constant and *independent of the size* of the various signals e_{i1}, e_{o1}, etc., in the devices.

In nonlinear devices, on the other hand, the system terminal

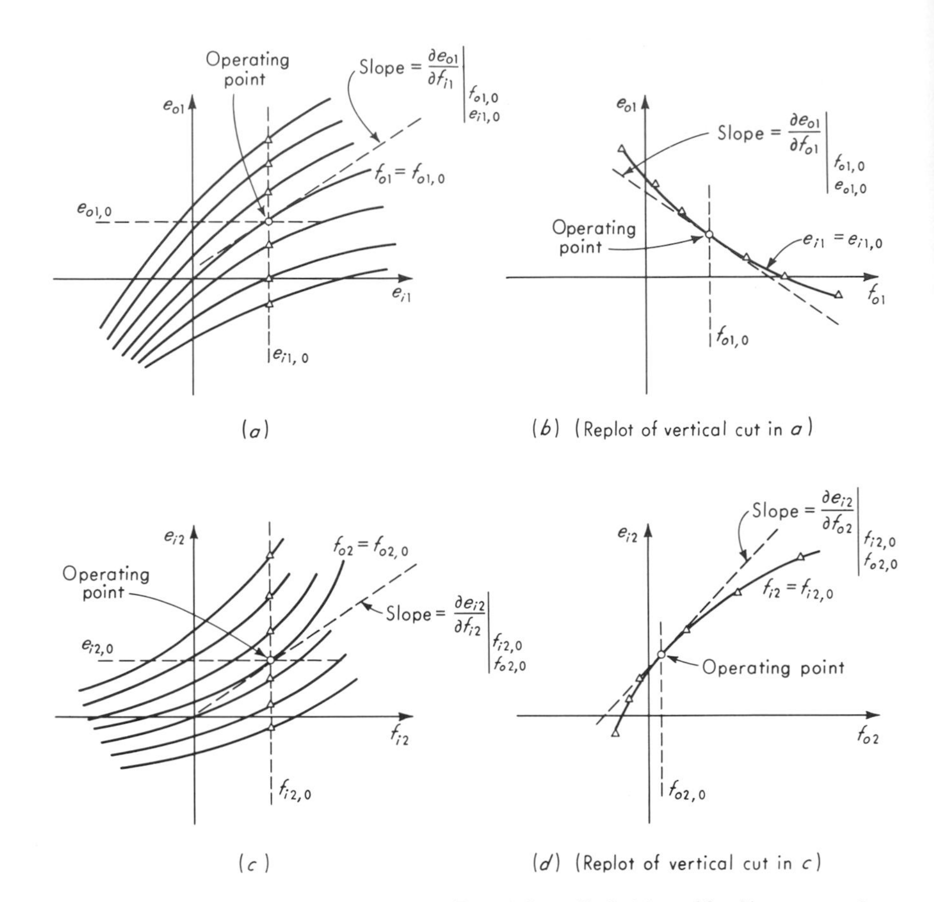

Fig. 3.30. *Definition of loading parameters.*

characteristics vary with the size of the signal and give accurate results only if the signal variations are limited to *small* excursions from some operating point. That is, we assume, in Fig. 3.29*b*, for example, that the independent variables f_{o1} and e_{i1} are set at some fixed values $f_{o1,0}$ and $e_{i1,0}$, as shown in Fig. 3.30*a*. The changes in the output quantities caused by small changes in the input quantities can then be predicted by Eq. (3.116), using the curve slopes of Fig. 3.30*a* and *b* as the numerical values of the partial derivatives. A similar interpretation of Eq. (3.120) for the second device leads to Fig. 3.30*c* and *d*. The "operating points" cannot be chosen arbitrarily. They must correspond to a stable equilibrium condition when the two devices are interconnected, since then $e_{o1,0}$ must equal $e_{i2,0}$ and $f_{o1,0}$ must equal $-f_{i2,0}$.

As an example of the above procedures, consider a situation where one wishes to measure the volume flow rate of a motor-driven pump by means of a flowmeter connected at its discharge line. With no flowmeter attached, the pump is assumed to discharge to atmospheric pressure (0 psig) at a certain flow rate. When the flowmeter is attached, it will cause a pressure drop across itself; thus, if the flowmeter discharges to atmosphere, the pump discharge is now above atmosphere. Depending on the type of pump, this increase in discharge pressure will cause a greater or lesser change in flow rate; thus the flowmeter will not be measuring the no-load flow accurately. A generalization of this problem might be stated as follows: Suppose the pump (with flowmeter attached) is driven at a speed ω_0 with a shaft torque T_0. This will result in pump discharge pressure $p_{p,0}$ and a volume flow rate Q_0. If the pump speed is changed by an amount $d\omega$, the flow rate will change by an amount dQ. How much different will this dQ be from the dQ that would occur for the same $d\omega$ if the flowmeter were not present?

Figure 3.31 illustrates this situation and its analysis by means of admittance techniques. Since the measured variable Q is a flow variable, we search the equation list (3.110) to (3.115) for equations that will allow us to define the output admittance Y_{go} of the pump and the input admittance Y_{gi} of the flowmeter. Also, since we are concerned with a change in speed ω (rather than torque T), we desire an equation containing the flow variable (rather than the effort variable) as one of the independent variables. For the pump, the first equation of pair (3.114) gives

$$dQ = \left.\frac{\partial Q}{\partial \omega}\right|_{p_p = p_{p,0}} d\omega + \left.\frac{\partial Q}{\partial p_p}\right|_{\omega = \omega_0} dp_p \qquad (3.126)$$

We define

$$\left.\frac{\partial Q}{\partial \omega}\right|_{p_p = p_{p,0}} \triangleq \text{no-load static transfer function} \triangleq K_{Q,\omega} \qquad (3.127)$$

$$\left.\frac{\partial Q}{\partial p_p}\right|_{\omega = \omega_0} \triangleq \text{output admittance} \triangleq Y_{go} \qquad (3.128)$$

The numerical values of these parameters may be established by experiments, which give the graphical results of Fig. 3.31*c*. Turning now to the flowmeter, we use the second equation of pair (3.113) to get

$$-dQ = \left.\frac{\partial Q}{\partial p_p}\right|_{p_d = p_{d,0}} dp_p + \left.\frac{\partial Q}{\partial p_d}\right|_{p_p = p_{p,0}} dp_d \qquad (3.129)$$

We define

$$\left.\frac{\partial Q}{\partial p_p}\right|_{p_d = p_{d,0}} \triangleq \text{input admittance} \triangleq Y_{gi} \qquad (3.130)$$

$$\left.\frac{\partial Q}{\partial p_d}\right|_{p_p = p_{p,0}} \triangleq \text{flowmeter output admittance} \triangleq Y_{go.f} \qquad (3.131)$$

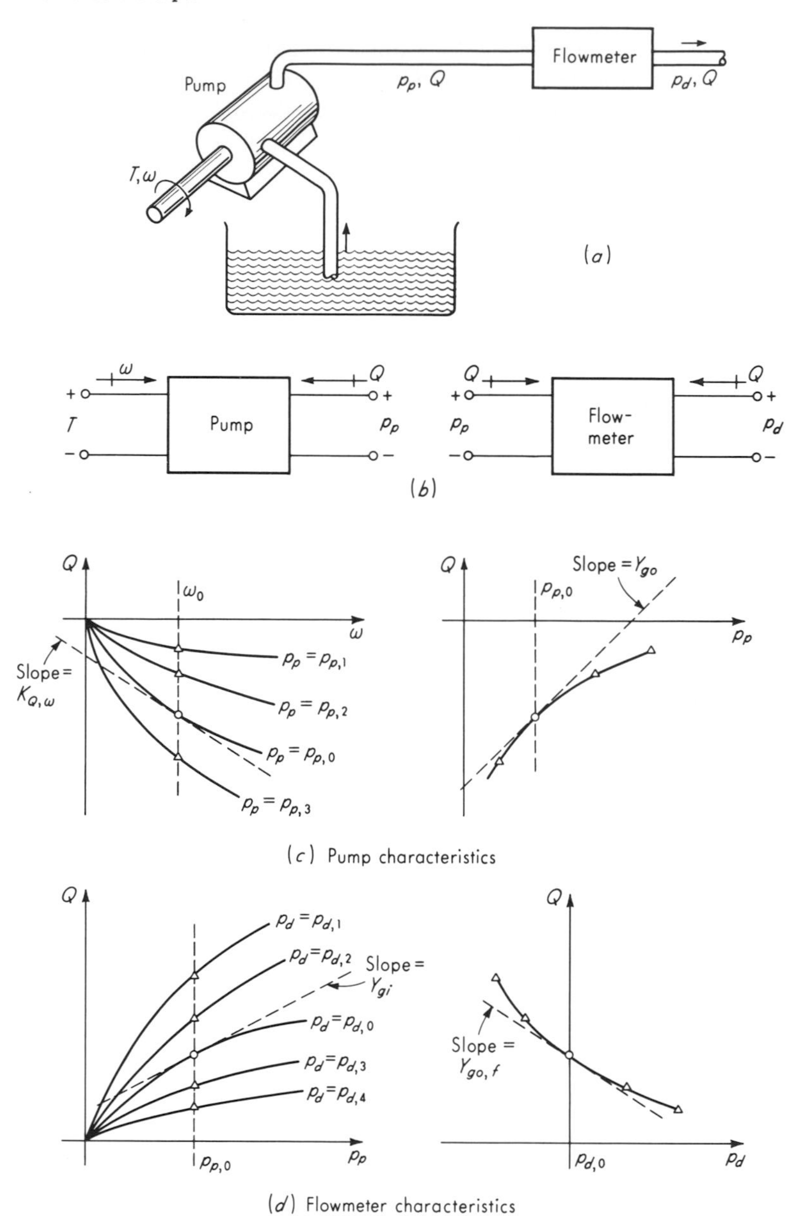

Fig. 3.31. Example of loading analysis.

Figure 3.31*d* illustrates the method of obtaining these parameters from experimental tests on the flowmeter.

If we assume that the flowmeter output is not attached to the input of some other device but discharges directly to atmosphere, then $p_d =$ const and $dp_d = 0$. Combining Eqs. (3.126) and (3.129), we get

$$dQ = K_{Q,\omega}\,d\omega + Y_{go}\,dp_p = K_{Q,\omega}\,d\omega - \frac{Y_{go}}{Y_{gi}}\,dQ \tag{3.132}$$

$$\left(1 + \frac{Y_{go}}{Y_{gi}}\right) dQ = K_{Q,\omega}\,d\omega \tag{3.133}$$

$$dQ = \frac{1}{1 + Y_{go}/Y_{gi}}\,K_{Q,\omega}\,d\omega \tag{3.134}$$

If the flowmeter were not present, dp_p in Eq. (3.126) would be zero, giving

$$dQ = K_{Q,\omega}\,d\omega \tag{3.135}$$

Thus the flowmeter causes an "error" (flow change) that depends on the numerical value of Y_{go}/Y_{gi}. If Y_{go}/Y_{gi} is very small compared with 1, the error becomes negligible.

In concluding this section on loading effects, the following comments are appropriate. In every measurement, the instrument input causes a load on the measured-medium output. If the instrument system consists of several interconnected stages (the "elements" of Fig. 2.1), there may, furthermore, be significant loading effects between stages. This is often the case when the "elements" are general-purpose devices such as amplifiers and recorders which are connected together in different ways at different times to create a measurement system suited to a particular problem. In using such a "building block" approach, loading problems must be carefully considered, and methods such as have been outlined above must be used if satisfactory and predictable results are to be obtained. If, on the other hand, the elements are merely functional components (permanently connected together) of a specific instrument, loading between elements may be a necessary consideration for the instrument *designer* but not for the user. The user need concern himself only with the loading situation at the measured-medium–primary-sensing-element interface.

While the impedance-type concepts discussed in this section are extremely useful in studying the disturbing effects of measuring instruments on measured media, some such effects do not lend themselves to this type of approach. For example, if a pitot tube is inserted into a flow field to measure flow velocity, the presence of the tube distorts the velocity field without necessarily extracting any energy or power. Thus, our impedance-type concepts are not directly applicable to such situations.

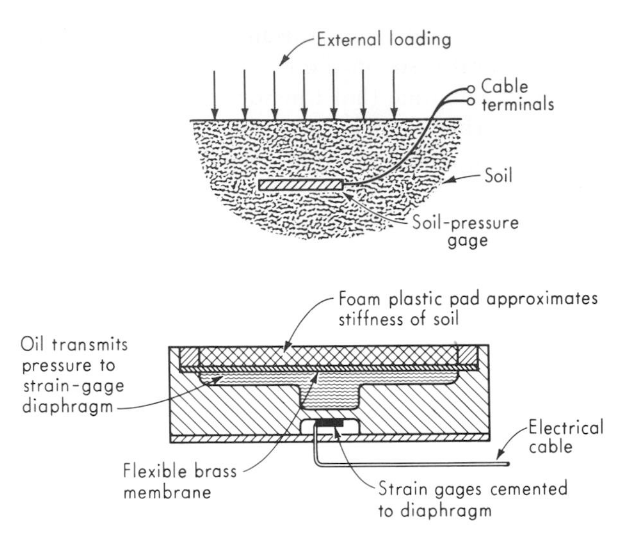

Fig. 3.32. *Soil-pressure gage.*

A related example, in which impedance-type concepts are useful (though not in the way described in this section) is given by Boiten.[1] An instrument for measuring subsoil pressures due to surface loading is useful in soil mechanics studies related to the construction of roads, airfields, etc. Figure 3.32 shows the essential features of such a device. While it is usually desirable that a force-measuring device have the highest possible stiffness, this criterion is *not* correct for this particular application. The reason is that the presence of the pressure gage in the soil causes a distortion of the pressure field if the stiffness of the gage is either higher or lower than the stiffness of the surrounding soil. That is, a very stiff gage will read too high when the external loading is applied, since the soil around it will compress more than the gage; thus the gage carries a disproportionate share of the total load. Similarly, an excessively compliant gage will read too low a pressure. The best gage is one whose stiffness just *matches* that of the soil in which it is to be used and thus does not distort the pressure field in the soil.

The feature distinguishing this example from those treated earlier in this section is that it deals with a distributed-parameter- ("field-") type of model for the physical system whereas our earlier discussions were all in terms of lumped-parameter ("network") models. The pitot-tube example also falls in the distributed-parameter category. While

[1] R. G. Boiten, The Mechanics of Instrumentation, *Proc. Inst. Mech. Engrs. (London)*, vol. 177, no. 10, 1963.

impedance concepts have been applied to distributed-parameter models, we shall not pursue this subject further in this text.

Concluding remarks on static characteristics. In this section we have attempted to present the most significant static characteristics commonly used in describing instrument performance. It is not possible to list *all* the specific static characteristics that might be pertinent in a particular instrument; rather, those of general interest were considered. It should also be emphasized that the terminology of the measurement field has not been thoroughly standardized. Therefore one should be careful to determine the precise definition intended when an apparently familiar term is encountered.

Also, errors not intrinsically associated with the instrument itself, but due to human factors involved in taking readings or incurred by incorrect installation of the instrument, were not considered in detail. However, such errors could certainly be included in our general framework merely by extending the concept of "inputs" to include human factors and installation effects.

3.3

Dynamic Characteristics

We begin our study of the dynamic characteristics of measurement systems by postulating a generalized mathematical model embodying the features pertinent to the characterization of the dynamic relation between any particular input and output.

Generalized mathematical model of measurement system. As in so many other areas of engineering application (vibration theory, circuit theory, automatic-control theory, aircraft stability and control theory, etc.), the most widely useful mathematical model for the study of measurement-system dynamic response is the ordinary linear differential equation with constant coefficients. We assume that the relation between any particular input (desired, interfering, or modifying) and the output can, by application of suitable simplifying assumptions, be put in the form

$$a_n \frac{d^n q_o}{dt^n} + a_{n-1} \frac{d^{n-1} q_o}{dt^{n-1}} + \cdots + a_1 \frac{dq_o}{dt} + a_0 q_o = b_m \frac{d^m q_i}{dt^m} + b_{m-1} \frac{d^{m-1} q_i}{dt^{m-1}} + \cdots + b_1 \frac{dq_i}{dt} + b_0 q_i \qquad (3.136)$$

where $q_o \triangleq$ output quantity
$q_i \triangleq$ input quantity
$t \triangleq$ time
a's, b's $\triangleq$ combinations of system physical parameters, assumed constant

If we define the differential operator $D \triangleq d/dt$, Eq. (3.136) can be written as

$$(a_n D^n + a_{n-1} D^{n-1} + \cdots + a_1 D + a_0) q_o = (b_m D^m + b_{m-1} D^{m-1} + \cdots + b_1 D + b_0) q_i \tag{3.137}$$

The solution of equations of this type has been put on a systematic basis using either the "classical" method of D operators or the Laplace-transform method. With the D-operator method, the complete solution q_o is obtained in two separate parts as

$$q_o = q_{ocf} + q_{opi} \tag{3.138}$$

where $q_{ocf} \triangleq$ complementary-function part of solution
$q_{opi} \triangleq$ particular-integral part of solution

The solution q_{ocf} has n arbitrary constants; q_{opi} has none. These n arbitrary constants may be evaluated numerically by imposing n initial conditions on Eq. (3.138). The solution q_{ocf} is obtained by calculating the n roots of the algebraic *characteristic equation*

$$a_n D^n + a_{n-1} D^{n-1} + \cdots + a_1 D + a_0 = 0 \tag{3.139}$$

Once these roots $r_1, r_2, \ldots, r_n$ have been found, the complementary-function solution is immediately written by following the rules stated below.

1. Real roots, unrepeated. For each real unrepeated root r one term of the solution is written as Ce^{rt}, where C is an arbitrary constant. Thus, for example, roots -1.7, $+3.2$, and 0 give a solution $C_1 e^{-1.7t} + C_2 e^{3.2t} + C_3$.
2. Real roots, repeated. For each root r which appears p times, the solution is written as $(C_0 + C_1 t + C_2 t^2 + \cdots + C_{p-1} t^{p-1}) e^r$. Thus, if there are roots $-1, -1, +2, +2, +2, 0, 0$ the solution is written as $(C_0 + C_1 t)e^{-t} + (C_2 + C_3 t + C_4 t^2)e^{2t} + (C_5 + C_6 t)$.
3. Complex roots, unrepeated. A complex root has the general form $a + ib$. It can be shown that if the a's of Eq. (3.139) are themselves real numbers (which they generally will be since they are physical quantities such as mass, spring, rate, etc.) then if any complex roots occur they will always occur in pairs of the form $a \pm ib$. For each such root pair, the corresponding solution is $Ce^{at} \sin (bt + \phi)$, where C and ϕ are the two arbitrary constants. Thus roots $-3 \pm i4$, $2 \pm i5$, and $0 \pm i7$ give a solution $C_0 e^{-3t} \sin (4t + \phi_0) + C_1 e^{2t} \sin (5t + \phi_1) + C_2 \sin (7t + \phi_2)$.
4. Complex roots, repeated. For each pair of complex roots $a \pm ib$ which appears p times the solution is $C_0 e^{at} \sin (bt + \phi_0) + C_1 t e^{at}$

$\sin(bt + \phi_1) + \cdots + C_{p-1}t^{p-1}e^{at}\sin(bt + \phi_{p-1})$. Roots $-3 \pm i2$, $-3 \pm i2$, and $-3 \pm i2$ thus give a solution $C_0e^{-3t}\sin(2t + \phi_0) + C_1te^{-3t}\sin(2t + \phi_1) + C_2t^2e^{-3t}\sin(2t + \phi_2)$.

The complete complementary-function solution is simply the algebraic sum of the individual parts found from the four rules. Whereas the above method for finding q_{ocf} *always* works, no universal method for finding the particular solution q_{opi} exists. This is because q_{opi} depends on the form of q_i, and one can always define a sufficiently "pathological" form of q_i to prevent solution for q_{opi}. However, if q_i is restricted to functions of prime engineering interest a relatively simple method for finding q_{opi} is available. This is the *method of undetermined coefficients*, which will now be briefly reviewed. Since the method does not work for all q_i, the first question to be answered is whether it will work for the q_i of interest. For a given q_i, the right side of Eq. (3.136) is some known function of time $f(t)$. To test whether this method can be applied, we repeatedly differentiate $f(t)$ and then examine the functions created by these differentiations. There are three possibilities:

1. After a certain-order derivative, all higher derivatives are zero.
2. After a certain-order derivative, all higher derivatives have the same functional form as some lower-order derivative.
3. Upon repeated differentiation, new functional forms continue to arise.

If case 1 or 2 occurs, the method will work. If case 3 occurs, this method will not work, and others must be tried. If the method is applicable, the solution q_{opi} is immediately written as

$$q_{opi} = Af(t) + Bf'(t) + Cf''(t) + \cdots \tag{3.140}$$

where the right-hand side includes one term for each functionally different form found by examining $f(t)$ and all its derivatives. The constants A, B, C, etc., can be found *immediately* (they do *not* depend on the initial conditions) by substituting q_{opi} [as given in Eq. (3.140)] into Eq. (3.136) and requiring (3.136) to be an *identity*. This procedure always generates as many simultaneous algebraic equations in the unknowns A, B, C, etc., as there are unknowns; thus the equations can be solved for A, B, C, etc.

The operational transfer function. In the analysis, design, and application of measurement systems the concept of the operational transfer function is very useful. The operational transfer function relating

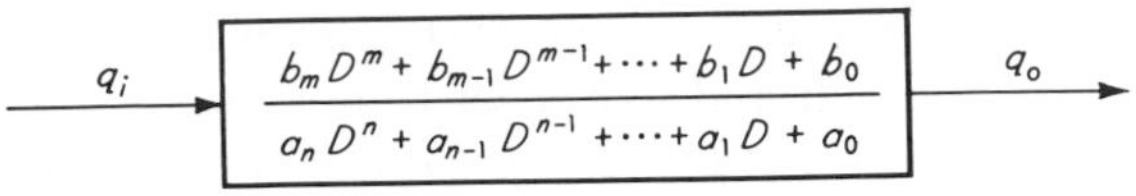

$$q_i \rightarrow \boxed{\frac{b_m D^m + b_{m-1} D^{m-1} + \cdots + b_1 D + b_0}{a_n D^n + a_{n-1} D^{n-1} + \cdots + a_1 D + a_0}} \rightarrow q_o$$

Fig. 3.33. General operational transfer function.

output q_o to input q_i is defined by treating Eq. (3.137) as if it were an algebraic relation and forming the ratio output/input:

$$\text{Operational transfer function} \triangleq \frac{q_o}{q_i}(D) \triangleq \frac{b_m D^m + b_{m-1} D^{m-1} + \cdots + b_1 D + b_0}{a_n D^n + a_{n-1} D^{n-1} + \cdots + a_1 D + a_0} \tag{3.141}$$

In writing transfer functions, one always writes $(q_o/q_i)(D)$, not just q_o/q_i, to emphasize that the transfer function is a *general* relation between q_o and q_i and very definitely *not* the instantaneous ratio of the time-varying quantities q_o and q_i.

One of the several useful features of transfer functions is their utility for graphic symbolic depiction of system dynamic characteristics by means of block diagrams. That is, if we wish to depict graphically a device with transfer function (3.141) we can draw a block diagram as in Fig. 3.33. Furthermore, the transfer function is helpful in determining the overall characteristics of a system made up of components whose individual transfer functions are known. This combination is most simply achieved when there is negligible loading (the input impedance of the second device much higher than the output impedance of the first, etc.) between the connected devices. For this case, the overall transfer function is simply the product of the individual ones since the output of the preceding device becomes the input of the following one. Figure 3.34 illustrates this procedure. When significant loading *is* present, one may apply the impedance concepts of Sec. 3.2 (extended to the dynamic

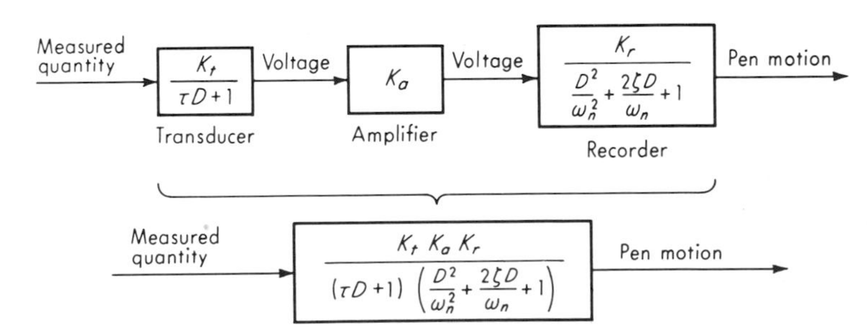

Fig. 3.34. Combination of individual transfer functions.

case) or simply analyze the complete system "from scratch" without using the individual transfer functions.

In the technical literature, the Laplace-transform method is in common use for the study of linear systems. When such methods are employed, the *Laplace transfer function* is defined as the ratio of the Laplace transform of the output quantity to the Laplace transform of the input quantity when all initial conditions are zero. Thus, analogous to Eq. (3.141), the Laplace transfer function would be written as

$$\frac{q_o(s)}{q_i(s)} \triangleq \frac{q_o}{q_i}(s) \triangleq \frac{b_m s^m + b_{m-1}s^{m-1} + \cdots + b_1 s + b_0}{a_n s^n + a_{n-1}s^{n-1} + \cdots + a_1 s + a_0} \tag{3.142}$$

where $s \triangleq \sigma + i\omega$ is the complex variable of the Laplace transform. We note that, so far as the *form* of the transfer function is concerned, one can shift from the Laplace form to the D-operator form (or vice versa) simply by interchanging s and D. Thus, if one encounters a block diagram using the Laplace notation, he can *always* convert to the D notation by a simple substitution. All the methods we shall subsequently develop may then be applied to the operational transfer function.

The sinusoidal transfer function. In studying the quality of measurement under dynamic conditions, we shall be analyzing the response of measurement systems to certain "standard" inputs. One of the most important of such responses is the steady-state response to a sinusoidal input. Here the input q_i is of the form $A_i \sin \omega t$. If one waits for all transient effects to die out (the complementary-function solution of a stable linear system always eventually dies out) it will be seen that the output quantity q_o will be a sine wave of exactly the same frequency (ω) as the input. However, the amplitude of the output may differ from that of the input, and a phase shift may be present. These results are easily shown by obtaining the particular (steady-state) solution by means of the method of undetermined coefficients. Since the frequency is the same, the relation between the input and output sine waves is completely specified by giving their amplitude ratio and phase shift. Both these quantities, in general, change when the driving frequency ω changes. Thus the *frequency response* of a system consists of curves of amplitude ratio and phase shift as a function of frequency. Figure 3.35 illustrates these concepts.

While the frequency response of any linear system may be obtained by getting the particular solution of its differential equation with

$$q_i = A_i \sin \omega t$$

much quicker and easier methods are available. These methods depend on the concept of the sinusoidal transfer function. The sinusoidal

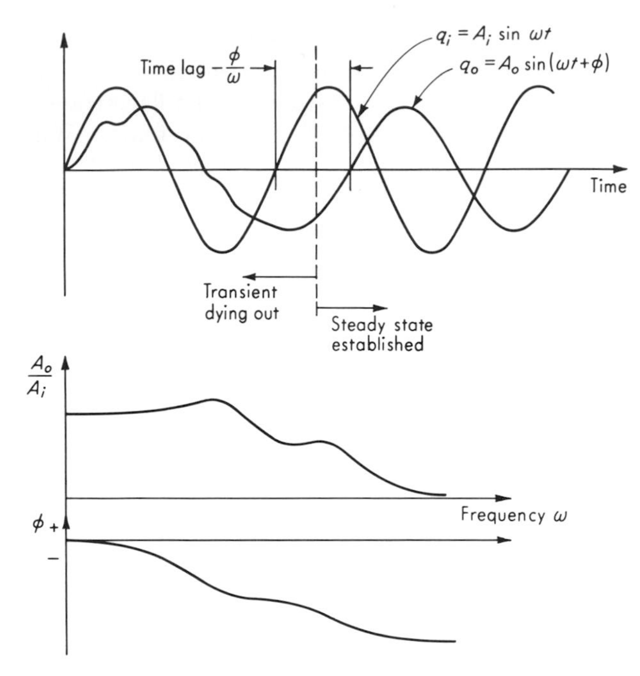

Fig. 3.35. Frequency-response terminology.

transfer function of a system is obtained by substituting $i\omega$ for D wherever it appears in the operational transfer function:

$$\text{Sinusoidal transfer function} \triangleq \frac{q_o}{q_i}(i\omega)$$

$$\triangleq \frac{b_m(i\omega)^m + b_{m-1}(i\omega)^{m-1} + \cdots + b_1(i\omega) + b_0}{a_n(i\omega)^n + a_{n-1}(i\omega)^{n-1} + \cdots + a_1(i\omega) + a_0} \tag{3.143}$$

where $i \triangleq \sqrt{-1}$

$\omega \triangleq$ frequency, rad/time

For any given frequency ω, Eq. (3.143) shows that $(q_o/q_i)(i\omega)$ is a complex number, which can always be put in the polar form $M\underline{/\phi}$. *We shall prove that the magnitude M of the complex number is the amplitude ratio A_o/A_i while the angle ϕ is the phase angle by which the output q_o leads the input q_i.* (If the output *lags* the input, ϕ is negative.)

The proof of the above statement is most readily demonstrated by means of the rotating-vector or phasor method of representing sinusoidal quantities. By a well-known trigonometric identity, we may write, in general,

$$Ae^{i\theta} = A(\cos\theta + i\sin\theta) = A\cos\theta + iA\sin\theta \tag{3.144}$$

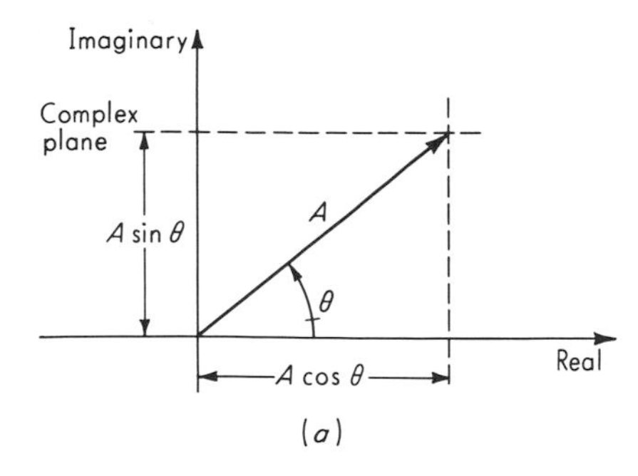

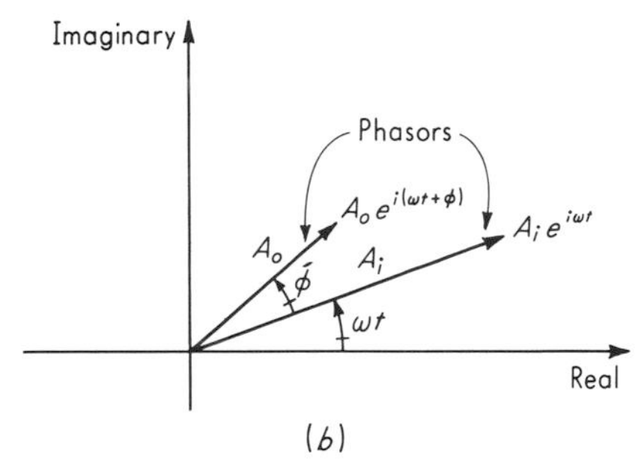

Fig. 3.36. *Phasor representation of sine waves.*

The complex number represented by the right-hand side can be exhibited graphically as in Fig. 3.36*a*. If we now apply this general result to the specific problem of representing q_o and q_i, we get

$$\text{For input } q_i \begin{cases} \text{let } A = A_i \text{ and } \theta = \omega t \\ A_i e^{i\omega t} = A_i \cos \omega t + iA_i \sin \omega t \end{cases} \tag{3.145}$$

$$\text{For output } q_o \begin{cases} \text{let } A = A_o \text{ and } \theta = \omega t + \phi \\ A_o e^{i(\omega t+\phi)} = A_o \cos (\omega t + \phi) + iA_o \sin (\omega t + \phi) \end{cases} \tag{3.146}$$

We note that the frequency ω of sinusoidal oscillation is also the angular velocity of rotation of the phasors of Fig. 3.36*b*. The phasors both rotate at the same angular velocity ω but maintain a fixed angle ϕ between them.

In carrying out our proof, we shall need to be able to differentiate phasor quantities. Since the amplitude A and the quantity i are constants, we have, in general,

$$\frac{d}{dt}(Ae^{i\theta}) = \left(i\frac{d\theta}{dt}\right)Ae^{i\theta} \tag{3.147}$$

and in particular, if $A = A_i$ and $\theta = \omega t$,

$$\frac{d}{dt}(A_i e^{i\omega t}) = (i\omega)A_i e^{i\omega t} \tag{3.148}$$

or, if $A = A_o$ and $\theta = \omega t + \phi$,

$$\frac{d}{dt}[A_o e^{i(\omega t+\phi)}] = (i\omega)A_o e^{i(\omega t+\phi)} \tag{3.149}$$

Clearly, for any higher derivative, we would get

$$\frac{d^n}{dt^n}(A_i e^{i\omega t}) = (i\omega)^n(A_i e^{i\omega t}) \tag{3.150}$$

and
$$\frac{d^n}{dt^n}[A_o e^{i(\omega t+\phi)}] = (i\omega)^n[A_o e^{i(\omega t+\phi)}] \tag{3.151}$$

Thus, differentiating a phasor quantity n times with respect to time t may be achieved simply by multiplying it by $(i\omega)^n$.

Suppose we now consider Eq. (3.137) for the sinusoidal steady-state case. Then *every* term on each side of the equation will be a sinusoidally varying quantity since repeated differentiation of sine waves gives only more sine waves (or cosines, which can be replaced by sines with a phase angle). We now convert the differential equation (3.137) into a complex algebraic equation by replacing each sinusoidal term by its phasor representation. This is *not* a matter of simple substitution since the sinusoidal terms are not *equal* to the phasor quantities; rather they are *represented* by the phasor quantities. We must thus be careful to show that, when the new phasor (complex-number) equation is satisfied, we are guaranteed that the original system differential equation is also satisfied. We can then perform any desired manipulations on the complex-number equation with assurance that correct results will be obtained. This is done by first replacing the sinusoidal terms by their phasor representations:

$$\begin{aligned} a_n(i\omega)^n A_o e^{i(\omega t+\phi)} + a_{n-1}(i\omega)^{n-1}A_o e^{i(\omega t+\phi)} + \cdots \\ + a_1(i\omega)A_o e^{i(\omega t+\phi)} + a_0 A_o e^{i(\omega t+\phi)} = b_m(i\omega)^m A_i e^{i\omega t} \\ + b_{m-1}(i\omega)^{m-1}A_i e^{i\omega t} + \cdots + b_1(i\omega)A_i e^{i\omega t} + b_0 A_i e^{i\omega t} \end{aligned} \tag{3.152}$$

This complex-number equation can be satisfied only if the real parts on the left equal the real parts on the right and similarly for the imaginary parts. Thus, if Eq. (3.152) is enforced, we are guaranteed that the equation given by the imaginary parts will also be satisfied. If we obtain the first few terms in this equation, the pattern should be obvious. We have

$$\operatorname{Im}[a_0 A_o e^{i(\omega t+\phi)}] = a_0 A_o \sin(\omega t+\phi) = a_0 q_o$$
$$\operatorname{Im}[b_0 A_i e^{i\omega t}] = b_0 A_i \sin\omega t = b_0 q_i$$

lowest-order terms in the original differential equation

$$\operatorname{Im}[a_1(i\omega)A_o e^{i(\omega t+\phi)}] = \operatorname{Im}\{a_1(i\omega)A_o[\cos(\omega t+\phi) + i\sin(\omega t+\phi)]\} = a_1\omega A_o\cos(\omega t+\phi) = a_1 Dq_o$$
$$\operatorname{Im}[b_1(i\omega)A_i e^{i\omega t}] = \operatorname{Im}[b_1(i\omega)A_i(\cos\omega t + i\sin\omega t)] = b_1\omega A_i\cos\omega t = b_1 Dq_i$$

next terms in original differential equation

It should be clear now that requiring Eq. (3.152) to hold is *equivalent* to requiring (3.137) to hold, even though they are *not* the same equation.

We now manipulate Eq. (3.152) as follows to prove our final result:

$$[a_n(i\omega)^n + a_{n-1}(i\omega)^{n-1} + \cdots + a_1(i\omega) + a_0]A_o e^{i(\omega t+\phi)} = [b_m(i\omega)^m + b_{m-1}(i\omega)^{m-1} + \cdots + b_1(i\omega) + b_0]A_i e^{i\omega t} \tag{3.153}$$

$$\frac{A_o e^{i(\omega t+\phi)}}{A_i e^{i\omega t}} = \frac{b_m(i\omega)^m + b_{m-1}(i\omega)^{m-1} + \cdots + b_1(i\omega) + b_0}{a_n(i\omega)^n + a_{n-1}(i\omega)^{n-1} + \cdots + a_1(i\omega) + a_0} \triangleq \frac{q_o}{q_i}(i\omega) \tag{3.154}$$

$$\text{Now} \quad \frac{A_o e^{i(\omega t+\phi)}}{A_i e^{i\omega t}} = \frac{A_o}{A_i}e^{i\phi} = \frac{A_o}{A_i}(\cos\phi + i\sin\phi) \tag{3.155}$$

$$\cos\phi + i\sin\phi = \sqrt{\cos^2\phi + \sin^2\phi}\,\underline{/\phi} = 1\,\underline{/\phi}$$

$$\text{and thus} \quad \frac{q_o}{q_i}(i\omega) = \frac{A_o}{A_i}\underline{/\phi} = M\underline{/\phi} \tag{3.156}$$

Equation (3.156) states that at any chosen frequency ω the magnitude of the complex number $(q_o/q_i)(i\omega)$ is numerically the amplitude ratio A_o/A_i while the angle of the complex number is the angle by which the output leads the input. Therefore our desired result is proved.

The zero-order instrument. While the general mathematical model of Eq. (3.136) is adequate for handling any linear measurement system, certain special cases occur so frequently in practice that they warrant separate consideration. Furthermore, more complicated systems can profitably be studied as combinations of these simple special cases.

The simplest possible special case of Eq. (3.136) occurs when all the a's and b's other than a_0 and b_0 are assumed to be zero. The differential equation then degenerates into the simple algebraic equation

$$a_0 q_o = b_0 q_i \tag{3.157}$$

Any instrument or system that closely obeys Eq. (3.157) over its intended range of operating conditions is defined to be a *zero-order instrument*. Actually, two constants a_0 and b_0 are not necessary, and so we define the static sensitivity (or steady-state "gain") as follows:

$$q_o = \frac{b_0}{a_0}q_i = Kq_i \tag{3.158}$$

$$K \triangleq \frac{b_0}{a_0} \triangleq \text{static sensitivity} \tag{3.159}$$

Since the equation $q_o = Kq_i$ is an algebraic equation, it is clear that, no matter how q_i might vary with time, the instrument output (reading) follows it *perfectly* with no distortion or time lag of any sort. Thus, the zero-order instrument represents ideal or perfect dynamic performance.

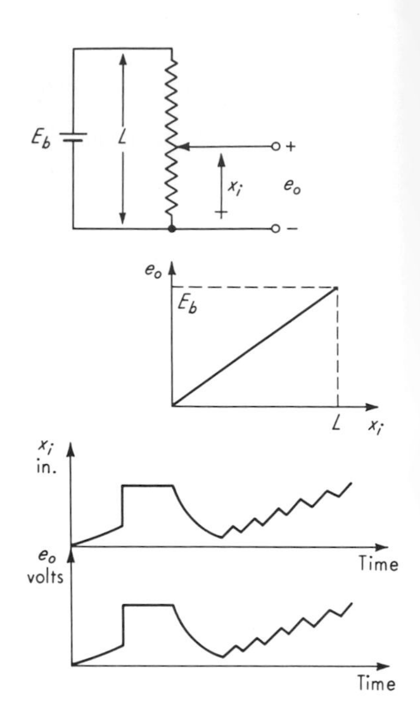

Fig. 3.37. *Zero-order instrument.*

A practical example of a zero-order instrument is the displacement-measuring potentiometer. Here (see Fig. 3.37) a strip of resistance material is excited with a voltage and provided with a sliding contact. If the resistance is linearly distributed along the length L, we may write

$$e_o = \frac{x_i}{L} E_b = K x_i \qquad (3.160)$$

$$\text{where} \quad K \triangleq \frac{E_b}{L} \qquad \text{volts/in.}$$

If one examines this measuring device more critically, he will find that it is not *exactly* a zero-order instrument. This is simply a manifestation of the *universal* rule that no mathematical model can *exactly* represent *any* physical system. In our present example, we would find that, if we wish to *use* a potentiometer for motion measurements, we must attach to the output terminals some voltage-measuring device (such as an oscilloscope). Such a device will always draw some current (however small) from the potentiometer. Thus, when x_i changes, the potentiometer winding current will also change. This in itself would cause no dynamic distortion or lag *if the potentiometer were a pure resistance.* However, the idea of a pure resistance is a *mathematical model,* not a real system; thus the potentiometer will have some (however small) inductance and capacitance. If x_i is varied relatively slowly, these parasitic inductance and capacitance effects will not be apparent. However, for sufficiently fast variation of x_i these effects are no longer negligible and cause dynamic errors between x_i and e_o. The reasons a potentiometer is normally called a zero-order instrument are as follows:

1. The parasitic inductance and capacitance can be made very small by design.
2. The speeds ("frequencies") of motion to be measured are not high enough to make the inductive or capacitive effects noticeable.

Another aspect of nonideal behavior in a real potentiometer comes to light when we realize that the sliding contact must be attached to the body whose motion is to be measured. Thus, there is a mechanical loading effect, due to the inertia of the sliding contact and its friction, which will cause the measured motion x_i to be different from that which would occur were the potentiometer not present. This effect is thus different *in kind* from the inductive and capacitive phenomena mentioned earlier since they affected the relation [Eq. (3.160)] between e_o and x_i whereas the mechanical loading has no effect on this relation but, rather, makes x_i different from the undisturbed case.

The first-order instrument. If in Eq. (3.136) all a's and b's other than a_1, a_0, and b_0 are taken as zero, we get

$$a_1 \frac{dq_o}{dt} + a_0 q_o = b_0 q_i \tag{3.161}$$

Any instrument that follows this equation is by definition a first-order instrument. There may be some conflict here between mathematical terminology and common engineering usage. In mathematics, a first-order *equation* has the general form

$$a_1 \frac{dq_o}{dt} + a_0 q_o = (b_m D^m + b_{m-1} D^{m-1} + \cdots + b_1 D + b_0) q_i \tag{3.162}$$

where m could have any numerical value. However, through long usage, in engineering one commonly understands a first-order *system* to be defined by Eq. (3.161). Since in technical presentations both words *and equations* are generally employed, confusion on this point is rarely a problem.

While Eq. (3.161) has three parameters a_1, a_0, and b_0, only two are really essential since the whole equation could always be divided through by either a_1, a_0, or b_0, thus making the coefficient of one of the terms numerically equal to 1. The most useful procedure is to divide through

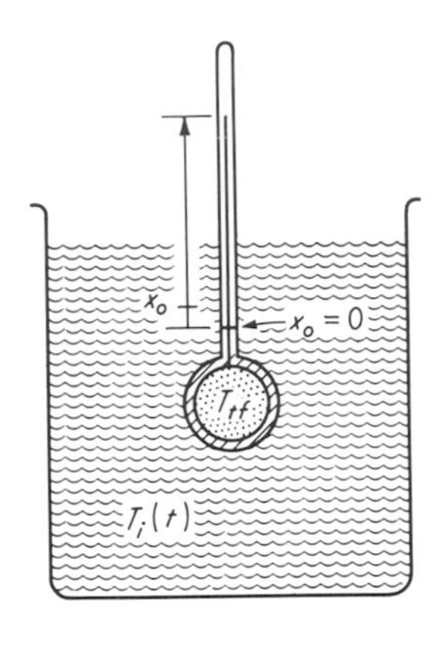

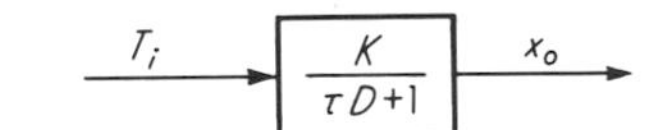

Fig. 3.38. *First-order instrument.*

by a_0, giving

$$\frac{a_1}{a_0}\frac{dq_o}{dt} + q_o = \frac{b_0}{a_0} q_i \tag{3.163}$$

which becomes

$$(\tau D + 1)q_o = Kq_i \tag{3.164}$$

when we define

$$K \triangleq \frac{b_0}{a_0} \triangleq \text{static sensitivity} \tag{3.165}$$

$$\tau \triangleq \frac{a_1}{a_0} \triangleq \text{time constant} \tag{3.166}$$

The time constant τ always has the dimensions of time, while the static sensitivity K has the dimensions output/input. For *any*-order instrument, K is always defined as b_0/a_0 and always has the same physical meaning, that is, the amount of output per unit input when the input is static (constant), because under such conditions all the derivative terms in the differential equation are zero. The operational transfer function of any first-order instrument is

$$\frac{q_o}{q_i}(D) = \frac{K}{\tau D + 1} \tag{3.167}$$

As an example of a first-order instrument, let us consider the liquid-in-glass thermometer of Fig. 3.38. The input (measured) quantity here is the temperature $T_i(t)$ of the fluid surrounding the bulb of the thermometer, and the output is the displacement x_o of the thermometer fluid in the capillary tube. We assume the temperature $T_i(t)$ is uniform throughout the fluid at any given time but may vary with time in an arbitrary fashion. The principle of operation of such a thermometer is the thermal expansion of the filling fluid which drives the liquid column up or down in response to temperature changes. Since this liquid column has inertia, mechanical lags will be involved in moving the liquid from one level to another. However, we shall assume this lag is negligible compared with the thermal lag involved in transferring heat from the surrounding fluid through the bulb wall and into the thermometer fluid. This assumption rests (as all such assumptions necessarily must) on experience, judgment, order-of-magnitude calculations, and, ultimately, experimental verification (or refutation) of the results predicted by the analysis. Assumption of negligible mechanical lag allows us to relate the temperature of the fluid in the bulb to the reading x_o by the instantaneous (algebraic) equation

$$x_o = \frac{K_{ex} V_b}{A_c} T_{tf} \qquad (3.168)$$

where $x_o \triangleq$ displacement from reference mark, in.
$T_{tf} \triangleq$ temperature of fluid in bulb (assumed uniform throughout bulb volume), $T_{tf} = 0$ when $x_o = 0$, °F
$K_{ex} \triangleq$ differential expansion coefficient of thermometer fluid and bulb glass, $\text{in.}^3/(\text{in.}^3\text{-F}°)$
$V_b \triangleq$ volume of bulb, in.^3
$A_c \triangleq$ cross-sectional area of capillary tube, in.^2

To get a differential equation relating input and output in this thermometer, we consider conservation of energy over an infinitesimal length of time dt for the thermometer bulb:

$$\text{(Heat in)} - \text{(heat out)} = \text{energy stored}$$

$$UA_b(T_i - T_{tf})\,dt - 0\text{(assume no heat loss)} = V_b \rho C\, dT_{tf} \qquad (3.169)$$

where $U \triangleq$ overall heat-transfer coefficient across bulb wall, $\text{Btu}/(\text{in.}^2\text{-F}°\text{-sec})$
$A_b \triangleq$ heat-transfer area of bulb wall, in.^2
$\rho \triangleq$ mass density of thermometer fluid, $\text{lb}_m/\text{in.}^3$
$C \triangleq$ specific heat of thermometer fluid, $\text{Btu}/(\text{lb}_m\text{-F}°)$

Equation (3.169) involves many assumptions:

1. The bulb wall and fluid films on each side are pure resistance to heat transfer with no heat-storage capacity. This will be a good assumption if the heat-storage capacity (mass) (specific heat) of the bulb wall and fluid films is small *compared* with $C\rho V_b$ for the bulb.
2. The overall coefficient U is constant. Actually, film coefficients and bulb-wall conductivity all change with temperature, but these changes are quite small so long as the temperature does not vary over wide ranges.

3. The heat-transfer area A_b is constant. Actually, expansion and contraction would cause this to vary, but this effect should be quite small.
4. No heat is lost from the thermometer bulb by conduction up the stem. Heat loss will be small if the stem is of small diameter, made of a poor conductor, and is immersed in the fluid over a great length and if the exposed end is subjected to an air temperature not much different from T_i and T_{tf}.
5. The mass of fluid in the bulb is constant. Actually mass must enter or leave the bulb whenever the level in the capillary tube changes. For a fine capillary and a large bulb this effect should be small.
6. The specific heat C is constant. Again, this fluid property varies with temperature but the variation is slight except for large temperature changes.

The above list of assumptions is not complete but should give some appreciation of the discrepancies between a mathematical model and the real system it represents. Many of these assumptions could be relaxed to get a more accurate model but one would pay a heavy price in increased mathematical complexity. The choice of assumptions that are *just good enough* for the needs of the job at hand is one of the most difficult and important tasks of the engineer.

Returning to Eq. (3.169), we may write it as

$$V_b \rho C \frac{dT_{tf}}{dt} + UA_b T_{tf} = UA_b T_i \qquad (3.170)$$

Using Eq. (3.168), we get

$$\frac{\rho C A_c}{K_{ex}} \frac{dx_o}{dt} + \frac{UA_b A_c}{K_{ex} V_b} x_o = UA_b T_i \qquad (3.171)$$

which we recognize to be the form of Eq. (3.163), and so we immediately define

$$K \triangleq \frac{K_{ex} V_b}{A_c} \qquad \text{in./F}^\circ \qquad (3.172)$$

$$\tau \triangleq \frac{\rho C V_b}{UA_b} \qquad \text{sec} \qquad (3.173)$$

Having shown a concrete example of a first-order instrument, let us now return to the problem of examining the dynamic response of first-order instruments in general. Once one has obtained the differential equation relating the input and output of an instrument he can study its dynamic performance by taking the input (quantity to be measured) to be some known function of time and then solving the differential equation for the output as a function of time. If the output is closely proportional to the input at all times, the dynamic accuracy is good. The fundamental

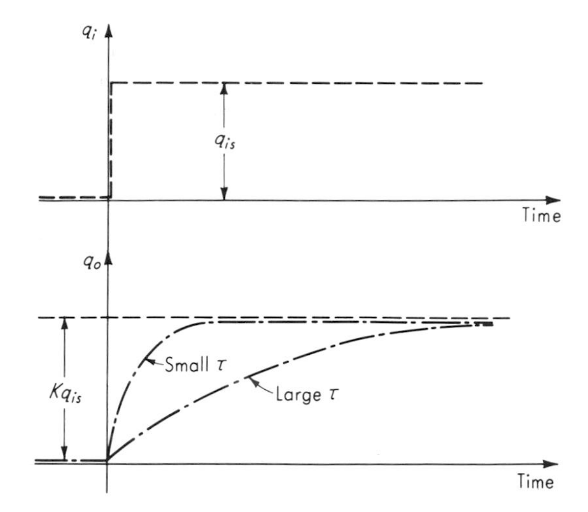

Fig. 3.39. *Step-function response of first-order instrument.*

difficulty in this approach lies in the fact that, in actual practice, the quantities to be measured usually do not follow some simple mathematical function but rather are of a random nature. Fortunately, however, much can be learned about instrument performance by examining the response to certain rather simple "standard" input functions. That is, just as one is not able to analyze the real *system* but rather an idealized model of it, so also one cannot work with the real *inputs* to a system but rather with simplified representations of them. This simplification of inputs (just as that of systems) can be carried out at several different levels, leading to either simple, rather inaccurate input functions that are readily handled mathematically or, alternatively, complex, more accurate representations that lead to mathematical difficulties.

We commence our study by considering several quite simple standard inputs that are in wide use. Although these inputs are, in general, only crude approximations to the actual inputs, they are extremely useful for studying the effects of parameter changes in a given instrument or comparing the *relative* performance of two competitive measurement systems.

Step response of first-order instruments. To apply a step input to a system, we assume that initially it is in equilibrium, with $q_i = q_o = 0$, when at time $t = 0$ the input quantity increases instantly an amount q_{is} (see Fig. 3.39). For $t > 0$, Eq. (3.164) becomes

$$(\tau D + 1)q_o = Kq_{is} \qquad (3.174)$$

It can be shown generally (by mathematical reasoning) or in any specific physical problem, such as the thermometer (by physical reasoning), that the initial condition for this situation is $q_o = 0$ for $t = 0^+$ ($t = 0^+$ means an infinitesimal time after $t = 0$). The complementary-function solution is

$$q_{ocf} = Ce^{(-t)/\tau} \qquad (3.175)$$

while the particular solution is

$$q_{opi} = Kq_{is} \qquad (3.176)$$

giving the complete solution as

$$q_o = Ce^{(-t)/\tau} + Kq_{is} \qquad (3.177)$$

Applying the initial condition,

$$0 = C + Kq_{is}$$
$$C = -Kq_{is}$$

giving finally

$$q_o = Kq_{is}[1 - e^{(-t)/\tau}] \qquad (3.178)$$

Examination of Eq. (3.178) shows that the speed of response depends *only* on the value of τ and is faster if τ is smaller. Thus in first-order instruments one strives to minimize τ for faithful dynamic measurements.

These results may be nondimensionalized by writing

$$\frac{q_o}{Kq_{is}} = 1 - e^{(-t)/\tau} \qquad (3.179)$$

and then plotting q_o/Kq_{is} versus t/τ as in Fig. 3.40*a*. This curve is then universal for any value of K, q_{is}, or τ that might be encountered. We could also define the measurement error e_m as

$$e_m \triangleq q_i - \frac{q_o}{K} \qquad (3.180)$$

$$e_m = q_{is} - q_{is}[1 - e^{(-t)/\tau}]$$

and nondimensionalize for plotting in Fig. 3.40*b* as

$$\frac{e_m}{q_{is}} = e^{(-t)/\tau} \qquad (3.181)$$

A dynamic characteristic useful in characterizing the speed of response of any instrument is the *settling time*. This is the time (after application of a step input) for the instrument to reach and stay within a stated plus-and-minus tolerance band around its final value. A small settling time is thus indicative of fast response. It is obvious that the numerical value of a settling time depends on the percentage tolerance band used; one must always state this. Thus one speaks of, say, a 5 percent settling time. For a first-order instrument a 5 percent settling time is equal to three time constants (see Fig. 3.41). Other percentages may be and are used in actual practice.

Knowing now that fast response requires a small value of τ, we can examine any specific first-order instrument to see what physical changes

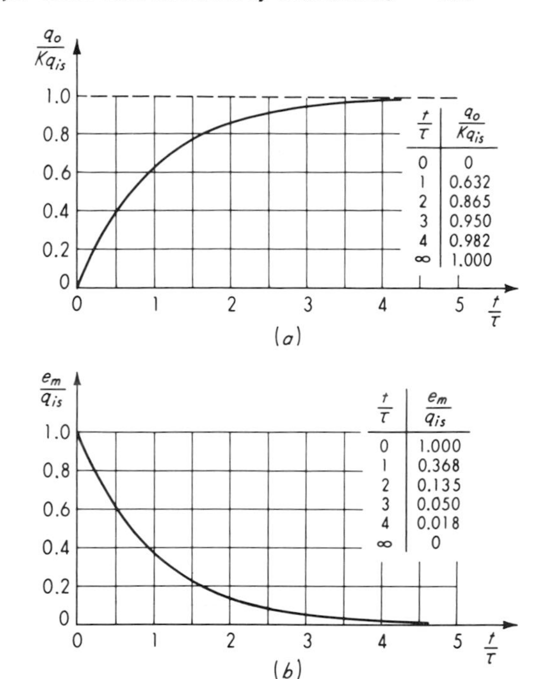

Fig. 3.40. *Nondimensional step-function response.*

would be needed to reduce τ. If we use our thermometer example, Eq. (3.173) shows that τ may be reduced by the following:

1. Reducing ρ, C, and V_b
2. Increasing U and A_b

Since ρ and C are properties of the fluid filling the thermometer they cannot be varied independently of one another, and so for small τ we

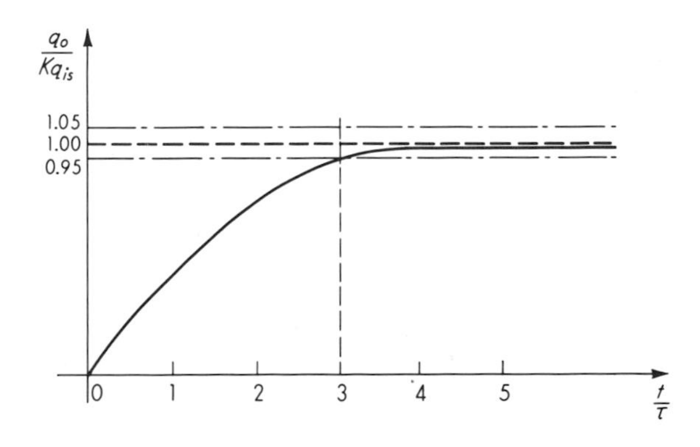

Fig. 3.41. *Settling-time definition.*

search for fluids with a small ρC product. The bulb volume V_b may be reduced but this will also reduce A_b unless some extended-surface heat-transfer augmentation (such as fins on the bulb) is introduced. Even more significant is the effect of reduced V_b on the static sensitivity K, as given by Eq. (3.172). We see that attempts to reduce τ by reducing V_b will result in reductions in K. Thus increased speed of response is traded off for lower sensitivity. This tradeoff is not unusual and will be observed in many other instruments.

The fact that τ depends on U means that one cannot state that a certain *thermometer* has a certain time constant but only that a certain thermometer *used in a certain fluid under certain heat-transfer conditions* (say free or forced convection) has a certain time constant. This is because U depends partly on the value of the film coefficient of heat transfer at the outside of the bulb, which varies greatly with changes in fluid (liquid or gas), flow velocity, etc. For example, a thermometer in stirred oil might have a time constant of 5 sec while the same thermometer in stagnant air would have a τ of perhaps 100 sec. Thus one must always be careful in giving (or using) performance data to be sure that the conditions of use correspond to those in force during calibration or that proper corrections are applied.

Ramp response of first-order instruments. To apply a ramp input to a system, we assume that initially the system is in equilibrium, with $q_i = q_o = 0$, when at $t = 0$ the input q_i suddenly starts to change at a constant rate $\dot{q}_{is}$. We thus have

$$q_i = q_o = 0 \qquad t \leq 0$$
$$q_i = \dot{q}_{is}t \qquad t \geq 0$$
$$\text{and therefore} \quad (\tau D + 1)q_o = K\dot{q}_{is}t \qquad (3.182)$$

The necessary initial condition can again be shown to be $q_o = 0$ for $t = 0^+$. Solution of Eq. (3.182) gives

$$q_{ocf} = Ce^{(-t)/\tau}$$
$$q_{opi} = K\dot{q}_{is}(t - \tau)$$
$$q_o = Ce^{(-t)/\tau} + K\dot{q}_{is}(t - \tau)$$

and applying the initial condition gives

$$q_o = K\dot{q}_{is}[\tau e^{(-t)/\tau} + t - \tau] \qquad (3.183)$$

We again define measurement error e_m by

$$e_m \triangleq q_i - \frac{q_o}{K} = \dot{q}_{is}t - \dot{q}_{is}\tau e^{(-t)/\tau} - \dot{q}_{is}t + \dot{q}_{is}\tau \qquad (3.184)$$

$$e_m = \underbrace{-\dot{q}_{is}\tau e^{(-t)/\tau}}_{\substack{\text{transient error} \\ e_{m,t}}} + \underbrace{\dot{q}_{is}\tau}_{\substack{\text{steady-} \\ \text{state} \\ \text{error} \\ e_{m,ss}}} \qquad (3.185)$$

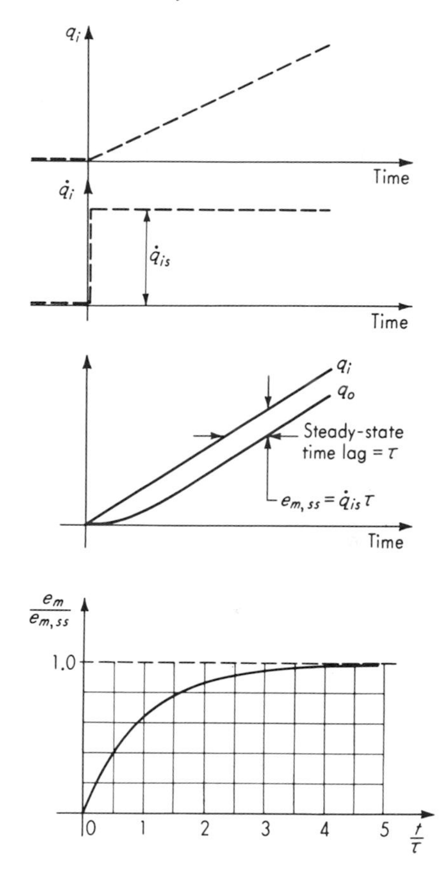

Fig. 3.42. Ramp response of first-order instrument.

We note that the first term of e_m will gradually disappear as time goes by, and so it is called the *transient error*. The second term, however, persists forever and is thus called the *steady-state error*. The transient error disappears more quickly if τ is small. The steady-state error is directly proportional to τ; thus small τ is desirable here also. Steady-state error also increases directly with $\dot{q}_{is}$, the rate of change of the measured quantity. In steady state, the horizontal (time) displacement between input and output curves is seen to be τ, and so one may make the interpretation that the instrument is reading what the input *was* τ sec ago. The above results, together with a nondimensionalized representation, are given graphically in Fig. 3.42.

Frequency response of first-order instruments. Equation (3.143) may be applied directly to the problem of finding the response of first-order systems to sinusoidal inputs. We have

$$\frac{q_o}{q_i}(i\omega) = \frac{K}{i\omega\tau + 1} = \frac{K}{\sqrt{\omega^2\tau^2 + 1}} \underline{/\tan^{-1} - \omega\tau} \qquad (3.186)$$

Thus the amplitude ratio is

$$\frac{A_o}{A_i} = \left| \frac{q_o}{q_i}(i\omega) \right| = \frac{K}{\sqrt{\omega^2\tau^2 + 1}} \qquad (3.187)$$

and the phase angle

$$\phi = \angle \frac{q_o}{q_i}(i\omega) = \tan^{-1} - \omega\tau \qquad (3.188)$$

The ideal frequency response (zero-order instrument) would have

$$\frac{q_o}{q_i}(i\omega) = K\angle 0° \qquad (3.189)$$

Thus a first-order instrument approaches perfection if Eq. (3.186) approaches Eq. (3.189). We see this occur if the product $\omega\tau$ is sufficiently small. Thus for *any* τ there will be some frequency of input ω below which measurement is accurate, or, alternatively, if a q_i of high frequency ω must be measured, the instrument used must have a sufficiently small τ. Again we see that accurate dynamic measurement requires a small time constant.

If one were concerned with the measurement of *pure* sine waves only, the above considerations would not be very pertinent since if one knew the frequency and τ he could easily correct for amplitude attenuation and phase shift by simple calculations. In actual practice, however, q_i is often a combination of several sine waves of different frequencies. An example will show the importance of adequate frequency response under such conditions. Suppose we must measure a q_i given by

$$q_i = 1 \sin 2t + 0.3 \sin 20t \qquad (3.190)$$

(where t is in seconds) with a first-order instrument whose τ is 0.2 sec. Since this is a linear system we may use the superposition principle to find q_o. We first evaluate the sinusoidal transfer function at the two frequencies of interest.

$$\left.\frac{q_o}{q_i}(i\omega)\right|_{\omega=2} = \frac{K}{\sqrt{0.16 + 1}} \angle -21.8° = 0.93K\angle -21.8° \qquad (3.191)$$

$$\left.\frac{q_o}{q_i}(i\omega)\right|_{\omega=20} = \frac{K}{\sqrt{16 + 1}} \angle -76° = 0.24K\angle -76° \qquad (3.192)$$

We can then write q_o as

$$q_o = (1)(0.93K) \sin (2t - 21.8°) + (0.3)(0.24K) \sin (20t - 76°) \qquad (3.193)$$

$$\frac{q_o}{K} = 0.93 \sin (2t - 21.8°) + 0.072 \sin (20t - 76°) \qquad (3.194)$$

Since ideally $q_o/K = q_i$, comparison of Eq. (3.194) with (3.190) shows the presence of considerable measurement error. A graph of these two

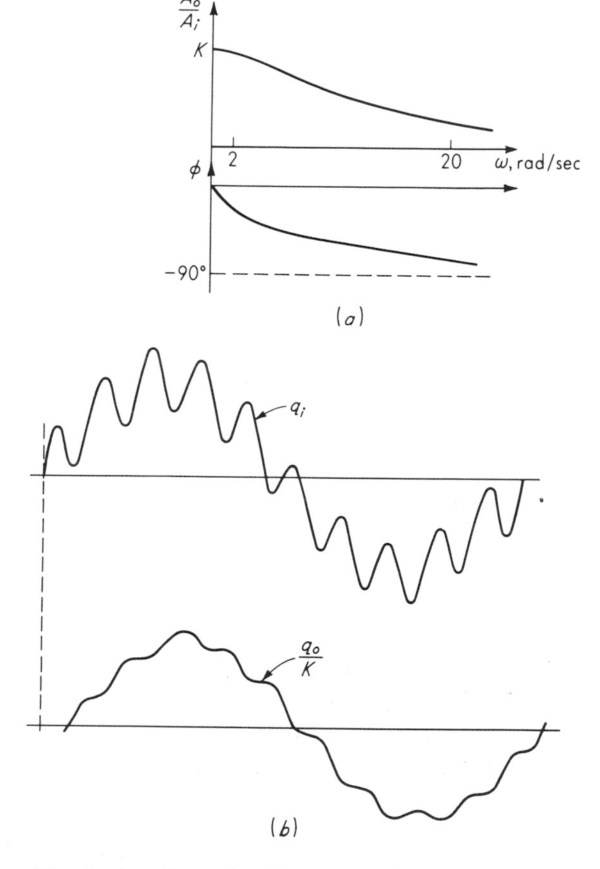

Fig. 3.43. *Example of inadequate frequency response.*

equations in Fig. 3.43*b* shows that the instrument gives a severely distorted measurement of the input. Furthermore, the high-frequency (20 rad/sec) component present in the instrument output is now so small relative to the low-frequency component that any attempts at correction are not only inconvenient but also inaccurate.

Suppose we now consider use of an instrument with $\tau = 0.002$ sec. We then have

$$\frac{q_o}{q_i}(i\omega)\Big|_{\omega=2} = \frac{K}{\sqrt{1.6 \times 10^{-5} + 1}}\underline{/-0.23^\circ} = 1.00K\underline{/-0.23^\circ} \qquad (3.195)$$

$$\frac{q_o}{q_i}(i\omega)\Big|_{\omega=20} = \frac{K}{\sqrt{1.6 \times 10^{-3} + 1}}\underline{/-2.3^\circ} = 1.00K\underline{/-2.3^\circ} \qquad (3.196)$$

$$\text{giving} \quad \frac{q_o}{K} = 1.00 \sin(2t - 0.23^\circ) + 0.3 \sin(20t - 2.3^\circ) \qquad (3.197)$$

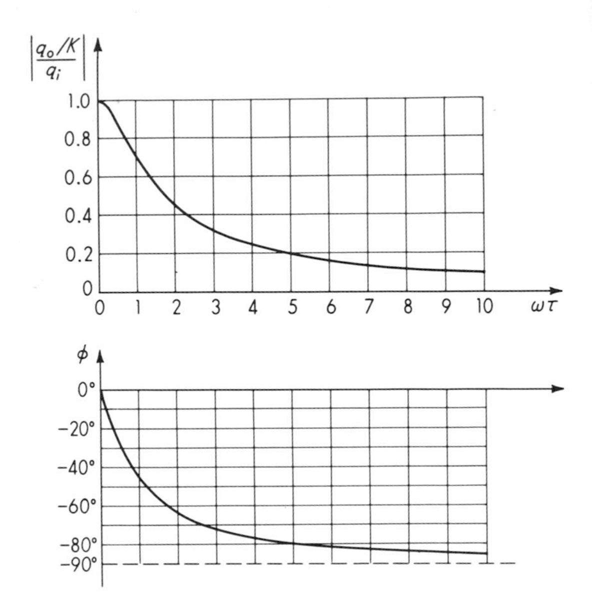

Fig. 3.44. Frequency response of first-order system.

Comparison of Eq. (3.190) and (3.197) shows clearly that this instrument faithfully measures the given q_i.

A nondimensional representation of the frequency response of any first-order system may be obtained by writing Eq. (3.186) as

$$\frac{q_o/K}{q_i}(i\omega) = \frac{1}{\sqrt{(\omega\tau)^2 + 1}} \underline{/\tan^{-1} - \omega\tau} \tag{3.198}$$

and plotting as in Fig. 3.44.

Impulse response of first-order instruments. The final standard input we shall consider is the *impulse function.* Consider the pulse function $p(t)$ defined graphically in Fig. 3.45*a*. The impulse function of "strength" (area) A is defined by the limiting process

$$\text{Impulse function of strength } A \triangleq \lim_{T\to 0} p(t) \tag{3.199}$$

We see that this "function" has rather peculiar properties. Its time duration is infinitesimal, its peak is infinitely high, and its area is A. If A is taken as 1, it is called the *unit* impulse function, $u_1(t)$. Thus an impulse function of any strength A may be written as $Au_1(t)$. This rather peculiar function plays an important role in system dynamic analysis, as we shall see in more detail later.

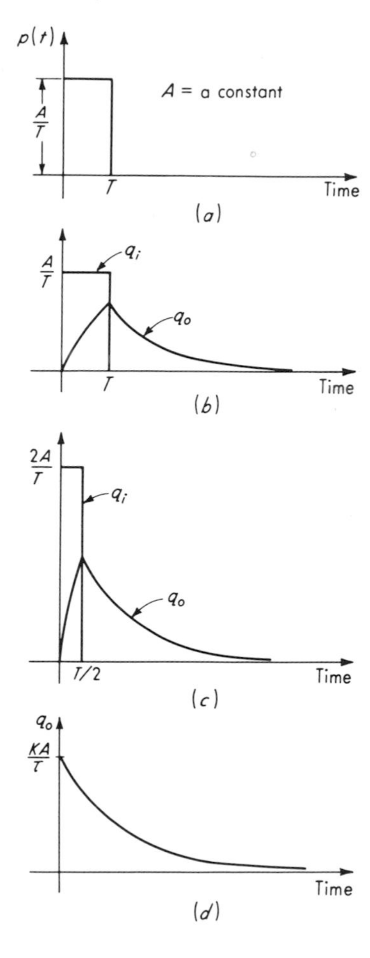

Fig. 3.45. *Impulse response of first-order system.*

We shall now find the response of a first-order instrument to an impulse input. We do this by finding the response to the pulse $p(t)$ and then applying the limiting process to the result. For $0 < t < T$ we have

$$(\tau D + 1)q_o = Kq_i = \frac{KA}{T} \qquad (3.200)$$

Since, up until time T, this is no different from a *step* input of size A/T, our initial condition is $q_o = 0$ at $t = 0^+$, and the complete solution is

$$q_o = \frac{KA}{T}[1 - e^{(-t)/\tau}] \qquad (3.201)$$

However, this solution is valid only up to time T. At this time we have

$$q_o \bigg|_{t=T} = \frac{KA}{T}[1 - e^{(-T)/\tau}] \qquad (3.202)$$

Now for $t > T$, our differential equation is

$$(\tau D + 1)q_o = Kq_i = 0 \qquad (3.203)$$

$$\text{giving} \quad q_o = Ce^{(-t)/\tau} \qquad (3.204)$$

The constant C is found by imposing initial condition (3.202),

$$\frac{KA}{T}[1 - e^{(-T)/\tau}] = Ce^{(-T)/\tau} \qquad (3.205)$$

$$C = \frac{KA[1 - e^{(-T)/\tau}]}{Te^{(-T)/\tau}} \qquad (3.206)$$

giving finally

$$q_o = \frac{KA[1 - e^{(-T)/\tau}]e^{(-t)/\tau}}{Te^{(-T)/\tau}} \qquad (3.207)$$

Figure 3.45*b* shows a typical response, and Fig. 3.45*c* shows the effect of cutting T in half. As T is made shorter and shorter, the first part ($t < T$) of the response becomes of negligible consequence so that we can get an expression for q_o by taking the limit of Eq. (3.207) as $T \to 0$.

$$\lim_{T\to 0} \left\{\frac{KA[1 - e^{(-T)/\tau}]}{Te^{(-T)/\tau}}\right\} e^{(-t)/\tau} = KAe^{(-t)/\tau} \lim_{T\to 0} \frac{1 - e^{(-T)/\tau}}{Te^{(-T)/\tau}} \qquad (3.208)$$

$$\lim_{T\to 0} \frac{1 - e^{(-T)/\tau}}{T} = \frac{0}{0} \qquad \text{an indeterminate form}$$

Applying L'Hospital's rule,

$$\lim_{T\to 0} \frac{1 - e^{(-T)/\tau}}{T} = \lim_{T\to 0} \frac{(1/\tau)e^{(-T)/\tau}}{1} = \frac{1}{\tau} \qquad (3.209)$$

Thus we have finally for the impulse response of a first-order instrument

$$q_o = \frac{KA}{\tau} e^{(-T)/\tau} \qquad (3.210)$$

which is plotted in Fig. 3.45*d*.

We note that the output q_o is also "peculiar" in that it has an infinite (vertical) slope at $t = 0$ and thus goes from zero to a finite value in infinitesimal time. Such behavior is clearly impossible for a physical system since it requires energy transfer at an infinite rate. In our thermometer example, for instance, to cause the temperature of the fluid in the bulb *suddenly* to rise a finite amount requires an infinite rate of heat transfer. Mathematically, this infinite rate of heat transfer is provided by having the input $T_i(t)$ be infinite, i.e., an impulse function. In actuality, of course, T_i cannot go to infinity; however, if it is large enough and of sufficiently short duration (relative to the response speed of the system) the system may respond very nearly as it would for a perfect impulse.

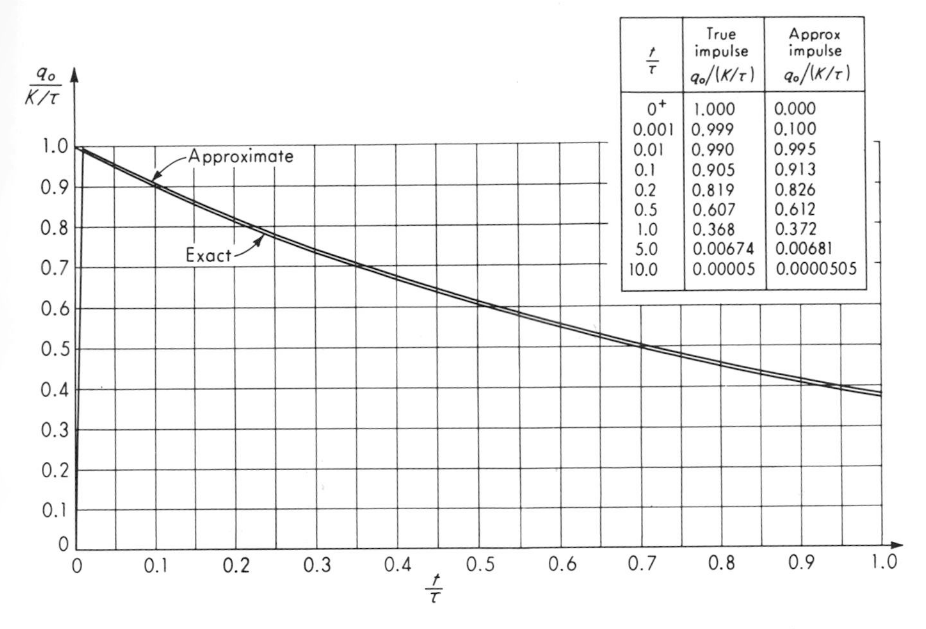

$\frac{t}{\tau}$	True impulse $q_o/(K/\tau)$	Approx impulse $q_o/(K/\tau)$
0^+	1.000	0.000
0.001	0.999	0.100
0.01	0.990	0.995
0.1	0.905	0.913
0.2	0.819	0.826
0.5	0.607	0.612
1.0	0.368	0.372
5.0	0.00674	0.00681
10.0	0.00005	0.0000505

Fig. 3.46. *Exact and approximate impulse response.*

To illustrate this, suppose in Fig. 3.45a we take $A = 1$ and $T = 0.01\tau$. The response to this approximate unit impulse is

$$q_o = \frac{100K}{\tau}[1 - e^{(-t)/\tau}] \qquad 0 \le t \le T \qquad (3.211)$$

$$q_o = \frac{100K(1 - e^{-0.01})e^{(-t)/\tau}}{\tau e^{-0.01}} \qquad T \le t \le \infty \qquad (3.212)$$

Figure 3.46 gives a tabular and graphical comparison of the exact and approximate response, showing excellent agreement. The agreement is quite acceptable in most cases if T/τ is even as large as 0.1. It can also be shown that *the shape of the pulse is immaterial;* as long as its duration is sufficiently short, only its *area* matters. The plausibility of this statement may be shown by integrating the terms in the differential equation as follows:

$$\tau \frac{dq_o}{dt} + q_o = Kq_i \qquad (3.213)$$

$$\int_0^{0^+} \tau \, dq_o + \int_0^{0^+} q_o \, dt = \int_0^{0^+} Kq_i \, dt \qquad (3.214)$$

$$\tau \left(q_o \Big|_{0^+} - q_o \Big|_0 \right) + 0$$

$$= K \text{ (area under } q_i \text{ curve from } t = 0 \text{ to } t = 0^+) \qquad (3.215)$$

$$q_o \Big|_{0^+} = \frac{K}{\tau} \text{ (area of impulse)} \qquad (3.216)$$

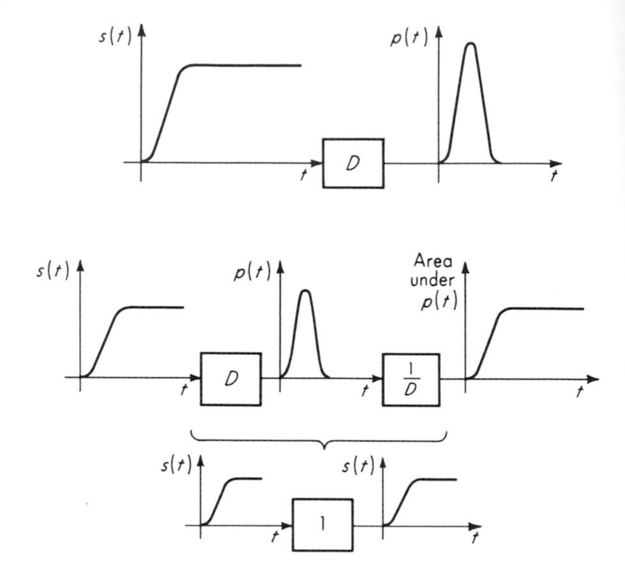

Fig. 3.47. Approximate step and impulse functions.

This analysis holds strictly for an exact impulse and is a good approximation for a pulse of arbitrary shape if its duration is sufficiently short. It should be noted that, since the right side of the differential equation (3.213) is zero for $t > 0^+$, an impulse (or short pulse) is equivalent to a zero forcing function and a nonzero initial ($t = 0^+$) condition. That is, the solution of

$$(\tau D + 1)q_o = 0$$

$$q_o = \frac{K}{\tau} \qquad \text{at } t = 0^+ \tag{3.217}$$

is exactly the same as the impulse response.

Another interesting aspect of the impulse function is its relation to the step function. Since a perfect step function is also physically unrealizable because it changes from one level to another in infinitesimal time, consider an approximation such as in Fig. 3.47. If this approximate step function is fed into a differentiating device the output will be a pulse-type function. As the approximate step function is made to approach the mathematical ideal more and more closely the output of the differentiating device will approach a perfect impulse function. *In this sense, the impulse function may be thought of as the derivative of the step function,* even though the discontinuities in the step function preclude the rigorous application of the basic definition of the derivative. In Fig. 3.47 the truth of these assertions is demonstrated by passing the output of the differentiating device through an integrating device ($1/D$).

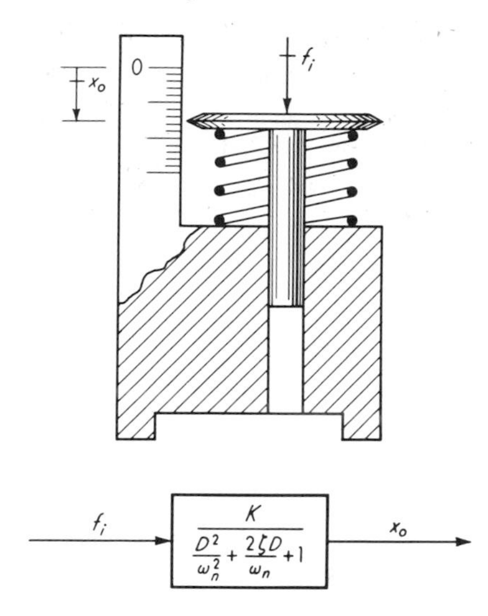

Fig. 3.48. *Second-order instrument.*

The second-order instrument. A second-order instrument is defined as one that follows the equation

$$a_2 \frac{d^2 q_o}{dt^2} + a_1 \frac{dq_o}{dt} + a_0 q_o = b_0 q_i \qquad (3.218)$$

Again, a second-order *equation* could have more terms on the right-hand side, but in common engineering usage, Eq. (3.218) is generally accepted as defining a second-order *system*.

The essential parameters in Eq. (3.218) can be reduced to three:

$$K \triangleq \frac{b_0}{a_0} \triangleq \text{static sensitivity} \qquad (3.219)$$

$$\omega_n \triangleq \sqrt{\frac{a_0}{a_2}} \triangleq \text{undamped natural frequency, rad/time} \qquad (3.220)$$

$$\zeta \triangleq \frac{a_1}{2\sqrt{a_0 a_2}} \triangleq \text{damping ratio, dimensionless} \qquad (3.221)$$

giving

$$\left(\frac{D^2}{\omega_n{}^2} + \frac{2\zeta D}{\omega_n} + 1\right) q_o = K q_i \qquad (3.222)$$

The operational transfer function is thus

$$\frac{q_o}{q_i}(D) = \frac{K}{D^2/\omega_n{}^2 + 2\zeta D/\omega_n + 1} \qquad (3.223)$$

A good example of a second-order instrument is the force measuring spring scale of Fig. 3.48. We assume the applied force f_i has frequency components only well below the natural frequency of the spring itself. Then the main dynamic effect of the spring may be taken into account by adding one-third of the spring's mass to the main moving mass. This total mass we call M. The spring is assumed linear with spring constant K_s lb$_f$/in. Although in a real scale there might be considerable dry friction, we assume perfect film lubrication and therefore a viscous damping effect with constant B lb$_f$/(in./sec).

The scale can be adjusted so that $x_o = 0$ when $f_i = 0$ (gravity force will then drop out of the equation), giving

$$\Sigma \text{ forces} = (\text{mass})(\text{acceleration})$$

$$f_i - B\frac{dx_o}{dt} - K_s x_o = M\frac{d^2x_o}{dt^2} \tag{3.224}$$

$$(MD^2 + BD + K_s)x_o = f_i \tag{3.225}$$

Noting this to fit the second-order model, we immediately define

$$K \triangleq \frac{1}{K_s} \quad \text{in./lb}_f \tag{3.226}$$

$$\omega_n \triangleq \sqrt{\frac{K_s}{M}} \quad \text{rad/sec} \tag{3.227}$$

$$\zeta \triangleq \frac{B}{2\sqrt{K_s M}} \tag{3.228}$$

Step response of second-order instruments. For a step input of size q_{is} we get

$$\left(\frac{D^2}{\omega_n^2} + \frac{2\zeta D}{\omega_n} + 1\right) q_o = Kq_{is} \tag{3.229}$$

with initial conditions

$$q_o = 0 \quad \text{at } t = 0^+$$
$$\frac{dq_o}{dt} = 0 \quad \text{at } t = 0^+ \tag{3.230}$$

The particular solution of Eq. (3.229) is clearly $q_{opi} = Kq_{is}$. The complementary-function solution takes on one of three possible forms, depending on whether the roots of the characteristic equation are real and unrepeated (overdamped case), real and repeated (critically damped case), or complex (underdamped case). The complete solutions of Eq. (3.229) with initial conditions (3.230) are, in nondimensional form,

$$\frac{q_o}{Kq_{is}} = -\frac{\zeta + \sqrt{\zeta^2 - 1}}{2\sqrt{\zeta^2 - 1}} e^{(-\zeta + \sqrt{\zeta^2 - 1})\omega_n t} + \frac{\zeta - \sqrt{\zeta^2 - 1}}{2\sqrt{\zeta^2 - 1}} e^{(-\zeta - \sqrt{\zeta^2 - 1})\omega_n t} + 1 \quad \text{overdamped} \tag{3.231}$$

$$\frac{q_o}{Kq_{is}} = -(1 + \omega_n t)e^{-\omega_n t} + 1 \quad \text{critically damped} \tag{3.232}$$

$$\frac{q_o}{Kq_{is}} = -\frac{e^{-\zeta\omega_n t}}{\sqrt{1 - \zeta^2}} \sin\left(\sqrt{1 - \zeta^2}\,\omega_n t + \phi\right) + 1 \quad \text{underdamped} \tag{3.233}$$

$$\phi \triangleq \sin^{-1}\sqrt{1 - \zeta^2}$$

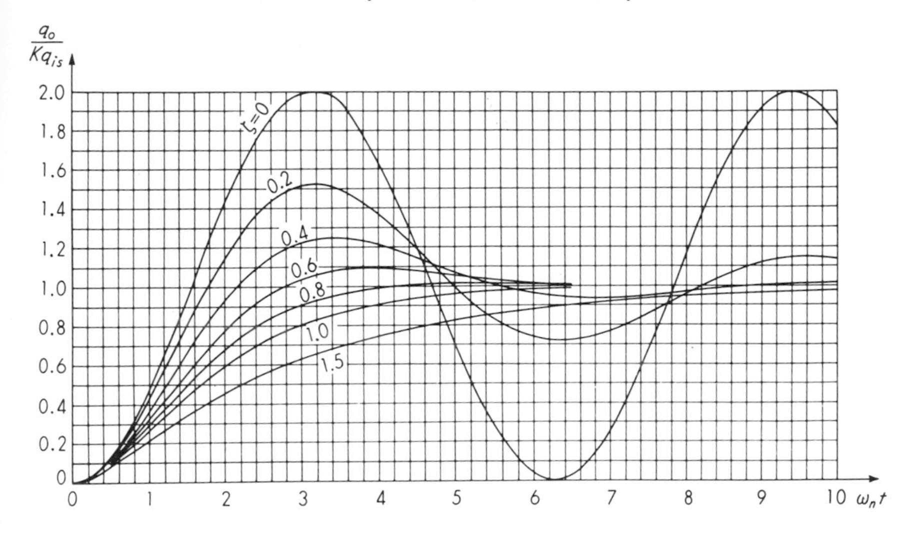

Fig. 3.49. *Nondimensional step-function response of second-order instruments.*

Since t and ω_n always appear as the product $\omega_n t$, the curves of q_o/Kq_{is} may be plotted against $\omega_n t$, making them universal for any ω_n, as in Fig. 3.49. This fact also shows that ω_n *is a direct indication of speed of response.* For a given ζ, doubling ω_n will halve the response time since $\omega_n t$ (and thus q_o/Kq_{is}) achieves the same value at one-half the time. The effect of ζ is not clearly perceived from the equations but is evident from the graphs. An increase in ζ reduces oscillation but also slows the response in the sense that the first crossing of the final value is retarded. A settling time may actually be a better indication of response speed; however, the optimum value of ζ will then vary with the chosen tolerance band. For example, if we choose a 10 percent settling time, the curve for $\zeta = 0.6$ gives a settling time of about $0.24/\omega_n$, and this is optimum since ζ either larger or smaller gives a longer settling time. However, if we had chosen a 5 percent settling time, a ζ between 0.7 and 0.8 gives the shortest value. In choosing a proper ζ value for a practical application, the situation is further complicated by the fact that the real inputs will not be step functions and their *actual* form influences what will be the best ζ value. If the actual inputs are quite variable in form, some compromise must be struck. It will be found that many commercial instruments use $\zeta = 0.6$ to 0.7. We shall show shortly that this range of ζ gives good frequency response over the widest frequency range.

Terminated-ramp response of second-order instruments. Under certain circumstances the response of second order instruments to perfect step inputs is misleading. The best example of this is perhaps found in

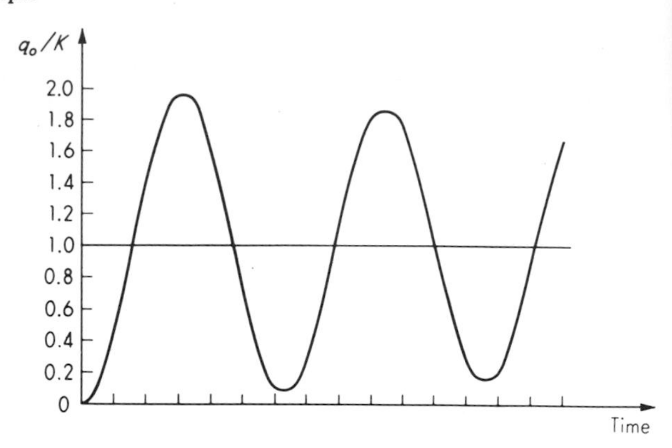

Fig. 3.50. *Step response of poorly damped system.*

piezoelectric pressure pickups, accelerometers, etc. While these devices will be discussed in detail later, for the present it is sufficient to state that they usually have an extremely high natural frequency and very little damping ($\zeta < 0.01$, often). Based on a perfect step input, such an instrument appears highly undesirable because of its large overshoot and strong oscillation (Fig. 3.50). Actually, these instruments may give excellent response. The explanation of this apparent inconsistency lies in the fact that *perfect* step inputs do not occur in nature, since a macroscopic quantity cannot change a finite amount in an infinitesimal time. Thus a more realistic input than the step is the *terminated-ramp input,* defined in Fig. 3.51. This input has a *finite* slope equal to $1/T$, whereas a step input has an infinite slope. By letting T get smaller and smaller, one can approach the perfect step input. For a second-order system we would have mathematically

$$\left(\frac{D^2}{\omega_n^2} + \frac{2\zeta D}{\omega_n} + 1\right) q_o = Kq_i \qquad (3.234)$$

$$q_i = \begin{cases} \dfrac{t}{T} & 0 \le t \le T \\ 1.0 & T \le t < \infty \end{cases} \qquad (3.235)$$

$$q_o = \frac{dq_o}{dt} = 0 \qquad \text{at } t = 0^+ \qquad (3.236)$$

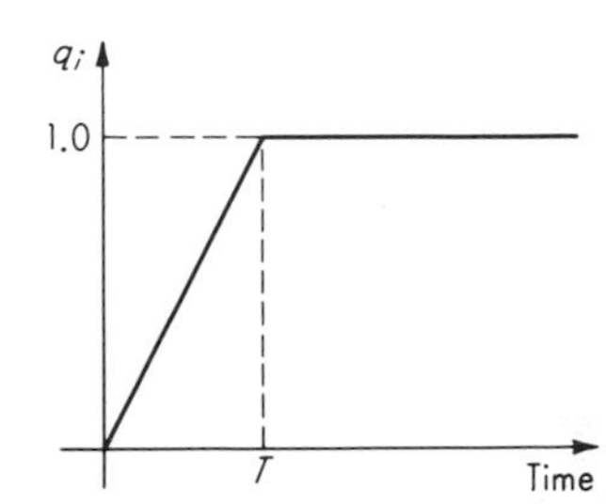

Fig. 3.51. *Terminated-ramp input.*

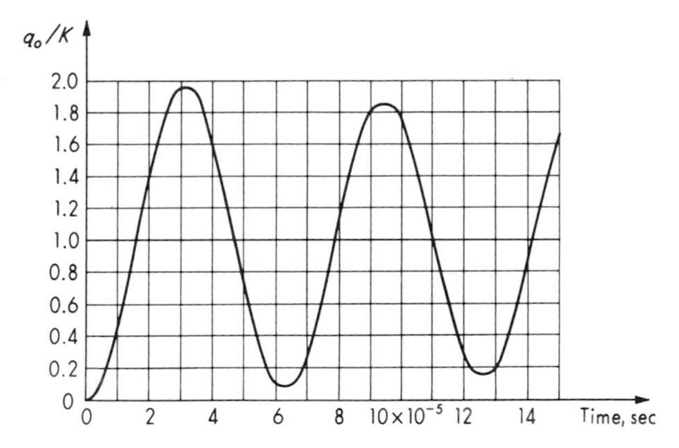

Fig. 3.52. *Step response of poorly damped system.*

Since we are concerned here with lightly damped systems, we obtain the solution only for the underdamped case as

$$\frac{q_o}{K} = \frac{t}{T} - \frac{2\zeta}{\omega_n T} + \frac{1}{\omega_n T\sqrt{1-\zeta^2}} e^{-\zeta\omega_n t} \sin\left(\sqrt{1-\zeta^2}\,\omega_n t + \phi\right) \qquad 0 \le t \le T \tag{3.237}$$

$$\frac{q_o}{K} = \left[\frac{t}{T} - \frac{2\zeta}{\omega_n T} + \frac{1}{\omega_n T\sqrt{1-\zeta^2}} e^{-\zeta\omega_n t} \sin\left(\sqrt{1-\zeta^2}\,\omega_n t + \phi\right)\right] - \left[\frac{t}{T} - 1 - \frac{2\zeta}{\omega_n T} + \frac{1}{\omega_n T\sqrt{1-\zeta^2}} e^{-\zeta\omega_n (t-T)} \sin\left(\sqrt{1-\zeta^2}\,\omega_n (t-T) + \phi\right)\right] \qquad T \le t < \infty \tag{3.238}$$

$$\phi \triangleq 2\tan^{-1}\frac{\sqrt{1-\zeta^2}}{\zeta} \tag{3.239}$$

From Eq. (3.237) we note immediately that, for $0 \le t \le T$, the following is true:

1. There is a steady-state error of size $2\zeta/\omega_n T$.
2. The transient error can be no larger than $1/(\omega_n T\sqrt{1-\zeta^2})$.

Thus if $\zeta = 0$ (no damping) the steady-state error is zero and the "transient" error is a sustained sine wave of amplitude $1/\omega_n T$. *Therefore if ω_n is sufficiently large relative to $1/T$ the transient error can be made very small even if the damping is practically nonexistent.* This result is based on Eq. (3.237) but similar results are obtained from (3.238) for $T \le t \le \infty$,

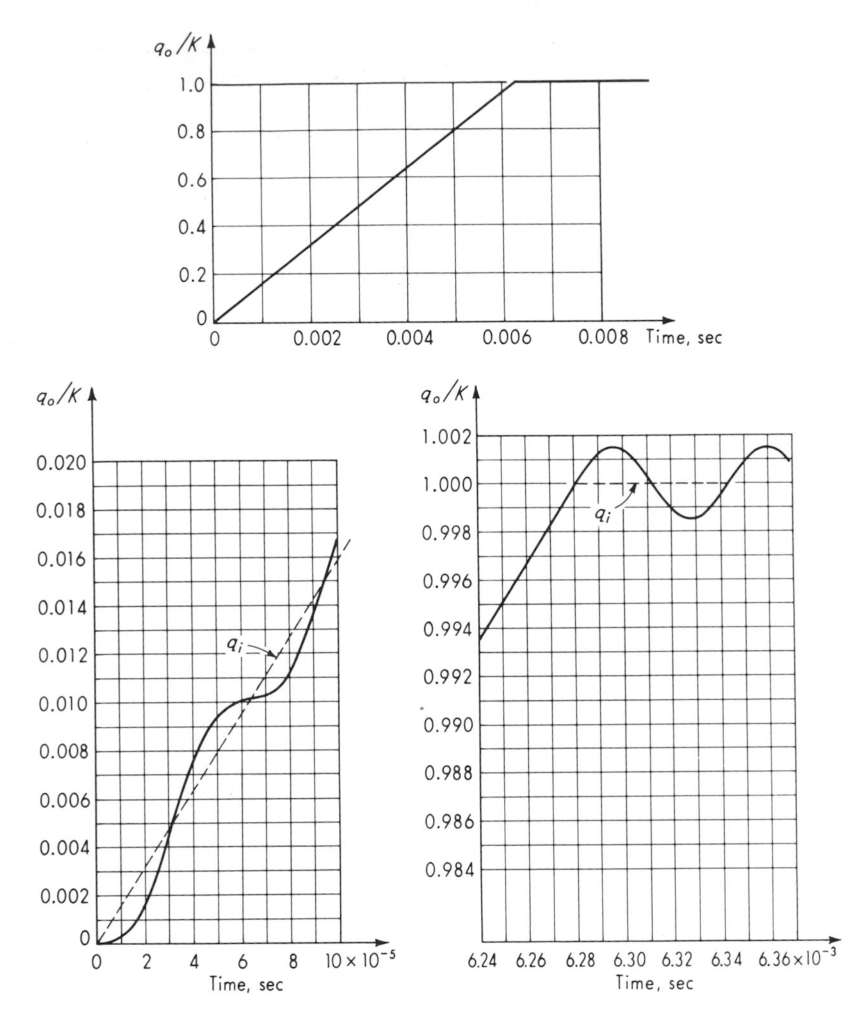

Fig. 3.53. *Terminated-ramp response of poorly damped system.*

since the transient induced at $t = 0$ by the increasing ramp is essentially the same as that induced at $t = T$ by a decreasing ramp. That is, the q_i of Fig. 3.51 is really a superposition of an increasing ramp starting at $t = 0$ and a decreasing ramp starting at $t = T$.

As a numerical example, suppose a pressure pickup with $\zeta = 0.01$ and $\omega_n = 100{,}000$ rad/sec is subjected to terminated-ramp-type inputs with $T = 0.00628$ sec. The step response of such an instrument is shown in Fig. 3.52 and indicates the severe overshooting and oscillation which could lead one to reject the instrument. Figure 3.53, however,

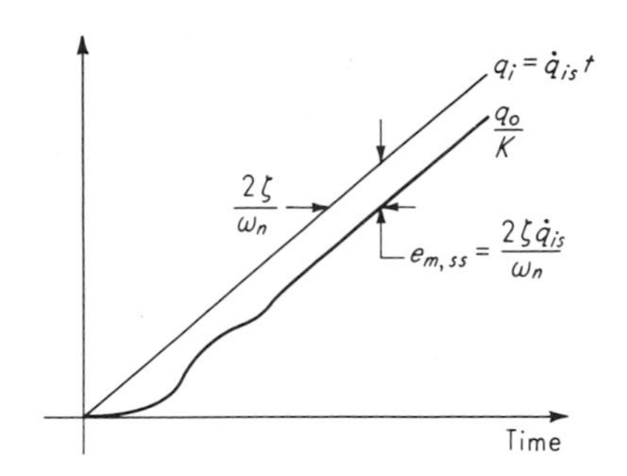

Fig. 3.54. *Ramp response of second-order instrument.*

shows the terminated-ramp response corresponding to the *actual* input. It is clear that the response is almost perfect. The conclusion, then, is that if the instrument is not subjected to any inputs more rapid in variation than that stated above it should prove quite acceptable.

Ramp response of second-order instruments. The differential equation here is

$$\left(\frac{D^2}{\omega_n{}^2} + \frac{2\zeta D}{\omega_n} + 1\right) q_o = K\dot{q}_{is} t \qquad (3.240)$$

$$q_o = \frac{dq_o}{dt} = 0 \qquad \text{at } t = 0^+$$

The solutions are found to be

$$\frac{q_o}{K} = \dot{q}_{is} t - \frac{2\zeta \dot{q}_{is}}{\omega_n}\left[1 + \frac{2\zeta^2 - 1 - 2\zeta\sqrt{\zeta^2 - 1}}{4\zeta\sqrt{\zeta^2 - 1}} e^{(-\zeta + \sqrt{\zeta^2 - 1})\omega_n t} + \frac{-2\zeta^2 + 1 - 2\zeta\sqrt{\zeta^2 - 1}}{4\zeta\sqrt{\zeta^2 - 1}} e^{(-\zeta - \sqrt{\zeta^2 - 1})\omega_n t}\right] \qquad \text{overdamped} \qquad (3.241)$$

$$\frac{q_o}{K} = \dot{q}_{is} t - \frac{2\dot{q}_{is}}{\omega_n}\left[1 - e^{-\omega_n t}\left(1 + \frac{\omega_n t}{2}\right)\right] \qquad \text{critically damped} \qquad (3.242)$$

$$\frac{q_o}{K} = \dot{q}_{is} t - \frac{2\zeta \dot{q}_{is}}{\omega_n}\left[1 - \frac{e^{-\zeta\omega_n t}}{2\zeta\sqrt{1 - \zeta^2}} \sin\left(\sqrt{1 - \zeta^2}\, \omega_n t + \phi\right)\right] \qquad (3.243)$$

$$\tan \phi = \frac{2\zeta\sqrt{1 - \zeta^2}}{2\zeta^2 - 1} \qquad \text{underdamped}$$

Figure 3.54 shows the general character of the response. There is a steady-state error $2\zeta\dot{q}_{is}/\omega_n$. Since the value of q_{is} is set by the measured quantity, the steady-state error can be reduced only by reducing ζ and increasing ω_n. For a given ω_n, reduction in ζ results in larger oscillations. There is also a steady-state time lag $2\zeta/\omega_n$. Figure 3.55 gives a set of nondimensionalized curves summarizing system behavior.

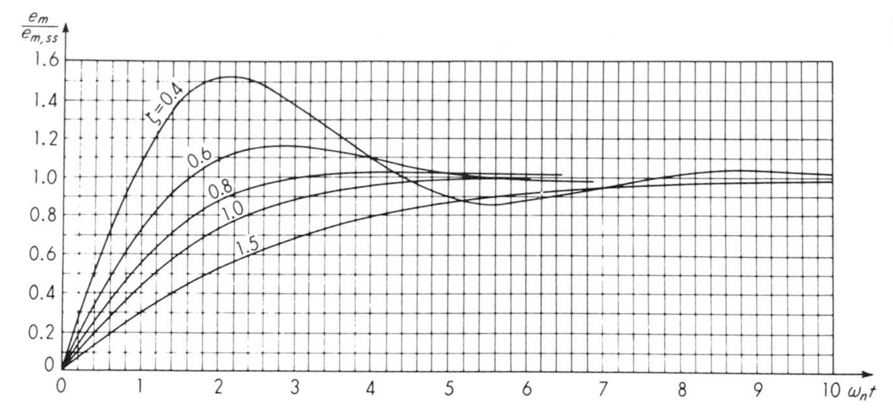

Fig. 3.55. Nondimensional ramp response.

Frequency response of second-order instruments. The sinusoidal transfer function is

$$\frac{q_o}{q_i}(i\omega) = \frac{K}{(i\omega/\omega_n)^2 + (2\zeta i\omega/\omega_n) + 1} \tag{3.244}$$

which can be put in the form

$$\frac{q_o/K}{q_i}(i\omega) = \frac{1}{\sqrt{[1 - (\omega/\omega_n)^2]^2 + 4\zeta^2\omega^2/\omega_n^2}} \underline{/\phi} \tag{3.245}$$

$$\phi \triangleq \tan^{-1} \frac{2\zeta}{\omega/\omega_n - \omega_n/\omega} \tag{3.246}$$

Figure 3.56 gives the nondimensionalized frequency-response curves. Clearly, increasing ω_n will increase the range of frequencies for which the amplitude-ratio curve is relatively flat; thus a high ω_n is needed to measure accurately high-frequency q_i's. An optimum range of values for ζ is indicated by both the amplitude-ratio and phase-angle curves. The widest flat amplitude ratio exists for ζ of about 0.6 to 0.7. While zero phase angle would be ideal it is rarely possible to realize this even approximately. Actually, if the main interest is in q_o reproducing the correct *shape* of q_i and if a time delay is acceptable, we shall show shortly that ϕ need not be zero; rather it should vary *linearly* with frequency ω. Examining the phase curves of Fig. 3.56, we note that the curves for $\zeta = 0.6$ to 0.7 are nearly straight for the widest frequency range. These considerations lead to the widely accepted choice of $\zeta = 0.6$ to 0.7 as the optimum value of damping for second-order instruments. There are exceptions, however, as noted in the section on terminated-ramp response.

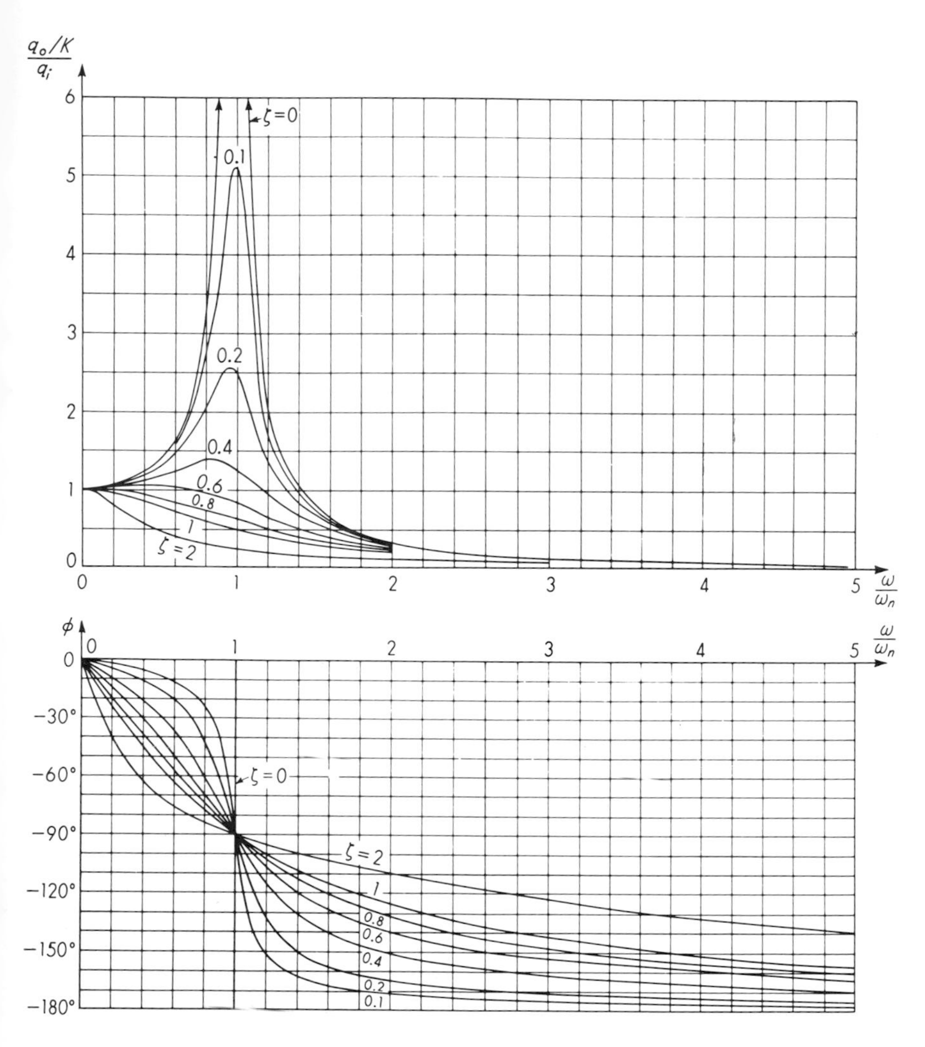

Fig. 3.56. Frequency response of second-order system.

Impulse response of second-order instruments. In the section on first-order instruments we showed that the impulse response is equivalent to the free (unforced) response if the initial ($t = 0^+$) conditions produced by the impulse are taken into account. To find the initial conditions produced by applying an impulse of area A to a second-order instrument redraw the block diagram of Fig. 3.57*a* as in Fig. 3.57*b*. (The equivalence of the two diagrams is easily demonstrated by tracing through the signals in Fig. 3.57*b* to get the differential equation relating q_o to q_i.) In Fig. 3.57*c*

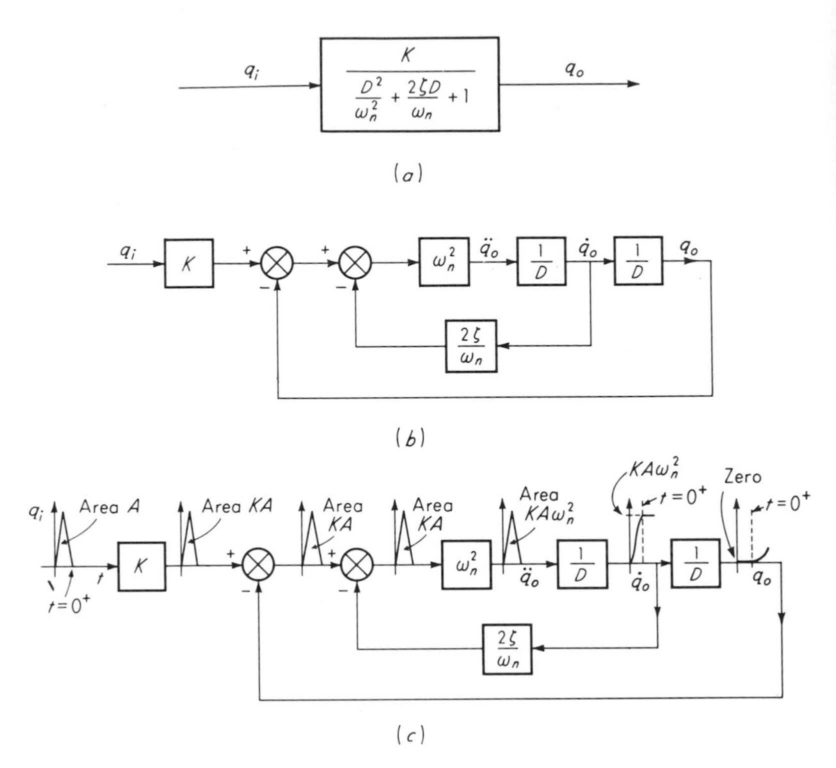

Fig. 3.57. Block-diagram analysis of impulse response.

the impulse is applied at q_i, and the "propagation" of this input signal is traced through the rest of the diagram. This analysis shows that at $t = 0^+$ we have $q_o = 0$ and $\dot{q}_o = KA\omega_n^2$. The differential equation to be solved is then

$$\left(\frac{D^2}{\omega_n^2} + \frac{2\zeta D}{\omega_n} + 1\right) q_o = 0$$

$$q_o = 0 \qquad \frac{dq_o}{dt} = KA\omega_n^2 \qquad \text{at } t = 0^+ \tag{3.247}$$

The solutions are found to be

$$\frac{q_o}{KA\omega_n} = \frac{1}{2\sqrt{\zeta^2 - 1}}\left[e^{(-\zeta+\sqrt{\zeta^2-1})\omega_n t} - e^{(-\zeta-\sqrt{\zeta^2-1})\omega_n t}\right] \qquad \text{overdamped} \tag{3.248}$$

$$\frac{q_o}{KA\omega_n} = \omega_n t e^{-\omega_n t} \qquad \text{critically damped} \tag{3.249}$$

$$\frac{q_o}{KA\omega_n} = \frac{1}{\sqrt{1-\zeta^2}} e^{-\zeta\omega_n t} \sin\left(\sqrt{1-\zeta^2}\,\omega_n t\right) \qquad \text{underdamped} \tag{3.250}$$

Figure 3.58 displays these results graphically.

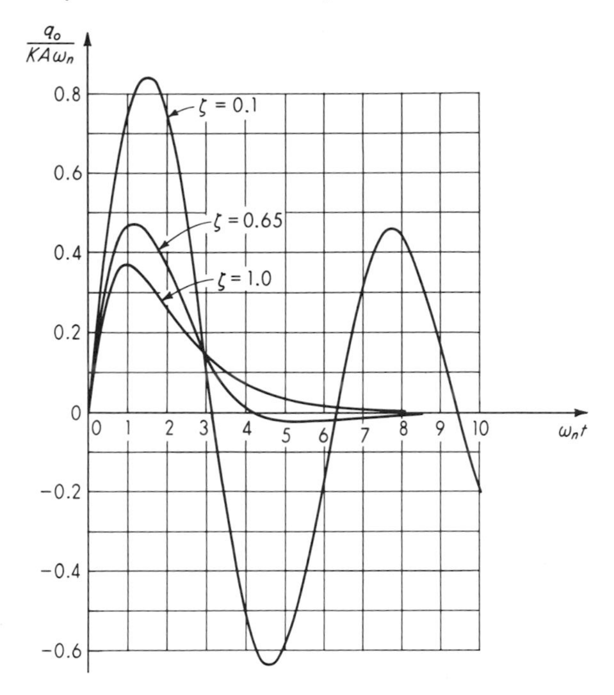

Fig. 3.58. Nondimensional impulse response of second-order system.

Dead-time elements. Some components of measuring systems are adequately represented as dead-time elements. A dead-time element is defined as a system in which the output is exactly the same form as the input but occurs τ_{dt} sec (the dead time) later. Mathematically,

$$q_o(t) = Kq_i(t - \tau_{dt}) \qquad t \geq \tau_{dt} \tag{3.251}$$

This type of element is also called a pure delay or a transport lag. An example of such an effect is found in pneumatic signal-transmission systems. A pressure signal at one end of a length of pneumatic tubing will cause no response at all at the other end until the pressure wave has had time to propagate the distance between them. Because this speed of propagation is the same as the speed of sound, a 1,000-ft length of tubing will have a dead time of about 1 sec, since the speed of sound in standard air is about 1,000 fps.

The response of dead-time elements to the standard inputs is easily found. For steps, ramps, and impulses the results are given in Fig. 3.59. For sinusoidal input we have

$$q_i = A_i \sin \omega t$$

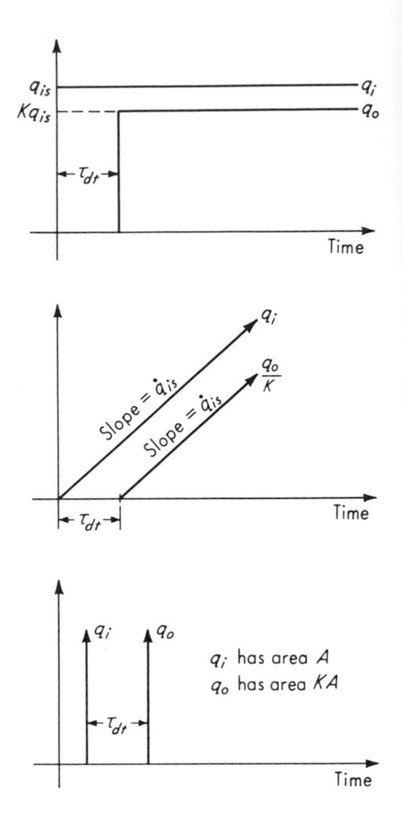

Fig. 3.59. Dead-time responses.

and from Eq. (3.251)

$$q_o = KA_i \sin \omega(t - \tau_{dt}) \tag{3.252}$$

$$q_o = KA_i \sin (\omega t - \omega\tau_{dt}) = KA_i \sin (\omega t + \phi) \tag{3.253}$$

$$\phi \triangleq -\omega\tau_{dt}$$

$$\text{Thus} \quad \frac{q_o/K}{q_i}(i\omega) = 1\underline{/\phi} = e^{-i\omega\tau_{dt}} \tag{3.254}$$

The frequency-response curves for a dead-time element are shown in Fig. 3.60.

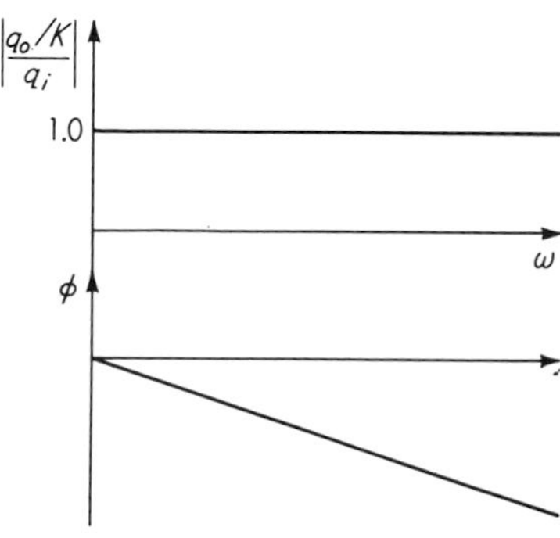

Fig. 3.60. Dead-time frequency response.

Logarithmic plotting of frequency-response curves. We shall find the frequency response of measurement systems extremely useful, so that rapid methods for getting the amplitude-ratio and phase-angle curves would be helpful. Certain logarithmic methods are in wide use and will now be explained.

The sinusoidal transfer function of a measurement system can generally be put in the form

$$\frac{q_o}{q_i}(i\omega) = \frac{\left\{\begin{array}{l} K(i\omega)^n(i\omega\tau_1 + 1) \cdots (i\omega\tau_m + 1)[(i\omega/\omega_{n1})^2 + 2\zeta_1 i\omega/\omega_{n1} + 1] \\ \quad \cdots [(i\omega/\omega_{nr})^2 + 2\zeta_r i\omega/\omega_{nr} + 1](e^{-i\omega\tau_{dt1}}) \ldots (e^{-i\omega\tau_{dtp}}) \end{array}\right\}}{\left\{\begin{array}{l} (i\omega\tau_I + 1) \cdots (i\omega\tau_M + 1)[(i\omega/\omega_{NI})^2 + 2\zeta_I i\omega/\omega_{NI} + 1] \\ \quad \cdots [(i\omega/\omega_{nR})^2 + 2\zeta_R i\omega/\omega_{nR} + 1] \end{array}\right\}} \tag{3.255}$$

This follows from the fact that the polynomials in the numerator and denominator of Eq. (3.141) can, in general, be *factored* into terms of the form $(D)^n$, $\tau D + 1$, and $D^2/\omega_n^2 + 2\zeta D/\omega_n + 1$. Replacing D by $i\omega$ then gives Eq. (3.255) when dead-time elements are also included.

Since Eq. (3.255) is in the form of a *product* of complex numbers, the use of logarithms suggests itself as a means of replacing multiplication by addition. That is,

$$\frac{q_o}{q_i}(i\omega) = G_1(i\omega)G_2(i\omega) \cdots G_u(i\omega) \tag{3.256}$$

where the $G(i\omega)$ functions represent the various terms of Eq. (3.255). The amplitude ratio would be given by

$$\left|\frac{q_o}{q_i}(i\omega)\right| = |G_1(i\omega)|\,|G_2(i\omega)| \cdots |G_u(i\omega)| \tag{3.257}$$

A widely used logarithmic method uses the *decibel notation* to express amplitude ratios. An amplitude ratio A is given in decibels (db) by

$$\text{Decibel value} \triangleq \text{db} \triangleq 20 \log_{10} A \tag{3.258}$$

Then

$$20 \log_{10}\left|\frac{q_o}{q_i}(i\omega)\right| = 20 \log_{10} [|G_1(i\omega)|\,|G_2(i\omega)| \cdots |G_u(i\omega)|] \tag{3.259}$$

$$20 \log_{10}\left|\frac{q_o}{q_i}(i\omega)\right| = 20 \log_{10} |G_1(i\omega)| + 20 \log_{10} |G_2(i\omega)| + \cdots + 20 \log_{10} |G_u(i\omega)| \tag{3.260}$$

Thus, if one gets the amplitude-ratio curves for the individual terms in

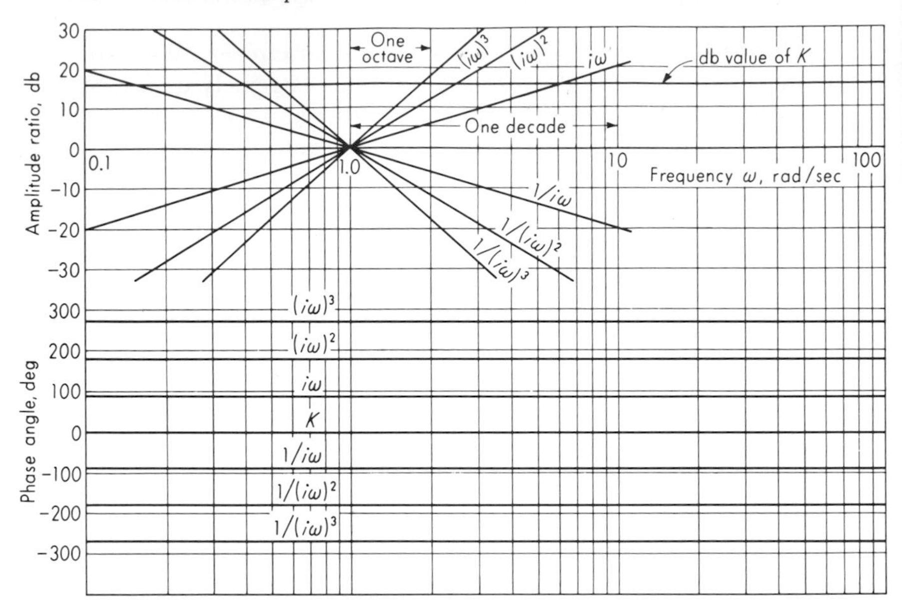

Fig. 3.61. *Integrator and differentiator frequency response.*

Eq. (3.255) in the decibel form, the overall decibel curve is obtained by simple graphical addition. The phase-angle curves are also obtained by simple addition since one adds phase angles when multiplying complex numbers.

We now show how to obtain with a minimum of effort the amplitude-ratio (db) and phase-angle curves for each of the types of terms in Eq. (3.255). To put the curves in their simplest form, we plot against the logarithm of frequency rather than frequency itself. The simplest term is the sensitivity K which is a real number $K\underline{/0°}$; thus its decibel curve is a straight horizontal line through the decibel value of K while its phase-angle curve is a straight horizontal line through zero degrees (see Fig. 3.61). Figure 3.62 gives a convenient table for interconverting decibel and ordinary-number values.

The next type of term is $(i\omega)^n$, where $n = \pm 1, \pm 2, \ldots$. The phase angle of such terms is constant with frequency and is given by $90n°$. The amplitude ratio is ω^n, so that the decibel value is

$$20 \log_{10} \omega^n = 20n \log_{10} \omega$$

Since we plot against $\log_{10} \omega$, the decibel curves become straight lines of slope $20n$ db/decade (see Fig. 3.61). A *decade* is defined as any 10-to-1 frequency range; an *octave* is any 2-to-1 range.

Number A	0	1	2	3	4	5	6	7	8	9
0		−40.00	−33.98	−30.46	−27.96	−26.02	−24.44	−23.10	−21.94	−20.92
0.1	−20.00	−19.17	−18.42	−17.72	−17.08	−16.48	−15.92	−15.39	−14.89	−14.42
0.2	−13.98	−13.56	−13.15	−12.77	−12.40	−12.04	−11.70	−11.37	−11.06	−10.76
0.3	−10.46	−10.16	−9.90	−9.63	−9.37	−9.12	−8.87	−8.64	−8.40	−8.18
0.4	−7.96	−7.74	−7.54	−7.33	−7.13	−6.94	−6.74	−6.56	−6.38	−6.20
0.5	−6.02	−5.85	−5.68	−5.51	−5.35	−5.19	−5.04	−4.88	−4.73	−4.58
0.6	−4.44	−4.29	−4.15	−4.01	−3.88	−3.74	−3.61	−3.48	−3.35	−3.22
0.7	−3.10	−2.97	−2.85	−2.73	−2.62	−2.50	−2.38	−2.27	−2.16	−2.05
0.8	−1.94	−1.83	−1.72	−1.62	−1.51	−1.41	−1.31	−1.21	−1.11	−1.01
0.9	−0.92	−0.82	−0.72	−0.63	−0.54	−0.45	−0.35	−0.26	−0.18	−0.09
1.0	0.00	0.09	0.17	0.26	0.34	0.42	0.51	0.59	0.67	0.75
1.1	0.83	0.91	0.98	1.06	1.14	1.21	1.29	1.36	1.44	1.51
1.2	1.58	1.66	1.73	1.80	1.87	1.94	2.01	2.08	2.14	2.21
1.3	2.28	2.35	2.41	2.48	2.54	2.61	2.67	2.73	2.80	2.86
1.4	2.92	2.98	3.05	3.11	3.17	3.23	3.29	3.35	3.41	3.46
1.5	3.52	3.58	3.64	3.69	3.75	3.81	3.86	3.92	3.97	4.03
1.6	4.08	4.14	4.19	4.24	4.30	4.35	4.40	4.45	4.51	4.56
1.7	4.61	4.66	4.71	4.76	4.81	4.86	4.91	4.96	5.01	5.06
1.8	5.11	5.15	5.20	5.25	5.30	5.34	5.39	5.44	5.48	5.53
1.9	5.58	5.62	5.67	5.71	5.76	5.80	5.85	5.89	5.93	5.98
2.0	6.02	6.44	6.85	7.23	7.60	7.96	8.30	8.63	8.94	9.25
3.0	9.54	9.83	10.10	10.37	10.63	10.88	11.13	11.36	11.60	11.82
4.0	12.04	12.26	12.46	12.67	12.87	13.06	13.26	13.44	13.62	13.80
5.0	13.98	14.15	14.32	14.49	14.65	14.81	14.96	15.12	15.27	15.42
6.0	15.56	15.71	15.85	15.99	16.12	16.26	16.39	16.52	16.65	16.78
7.0	16.90	17.03	17.15	17.27	17.38	17.50	17.62	17.73	17.84	17.95
8.0	18.06	18.17	18.28	18.38	18.49	18.59	18.69	18.79	18.89	18.99
9.0	19.08	19.18	19.28	19.37	19.46	19.55	19.65	19.74	19.82	19.91

db value corresponding to A

Fig. 3.62. *Decibel conversion table.*

Terms of the form $(i\omega\tau + 1)$ or $1/(i\omega\tau + 1)$ give, respectively,

$$\text{db} = 20 \log_{10} \sqrt{(\omega\tau)^2 + 1} \tag{3.261}$$

$$\text{and} \quad \text{db} = -20 \log_{10} \sqrt{(\omega\tau)^2 + 1} \tag{3.262}$$

When $\omega\tau \gg 1$, these become

$$\text{db} \approx 20 \log_{10} \omega\tau = 20 \log_{10} \tau + 20 \log_{10} \omega \tag{3.263}$$

$$\text{and} \quad \text{db} \approx -20 \log_{10} \omega\tau = -20 \log_{10} \tau - 20 \log_{10} \omega \tag{3.264}$$

We see that both of these represent straight lines of slope ± 20 db/decade, and these straight lines will be the high-frequency asymptotes of the actual amplitude-ratio curves. Similarly, for $\omega\tau \ll 1$,

$$\text{db} \approx 20 \log_{10} 1 = 0 \tag{3.265}$$

$$\text{and} \quad \text{db} \approx -20 \log_{10} 1 = 0 \tag{3.266}$$

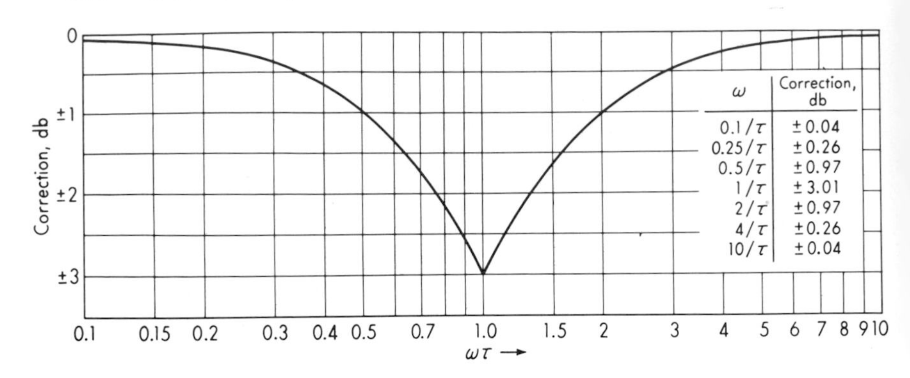

ω	Correction, db
0.1/τ	±0.04
0.25/τ	±0.26
0.5/τ	±0.97
1/τ	±3.01
2/τ	±0.97
4/τ	±0.26
10/τ	±0.04

Fig. 3.63. *First-order-system amplitude-ratio corrections.*

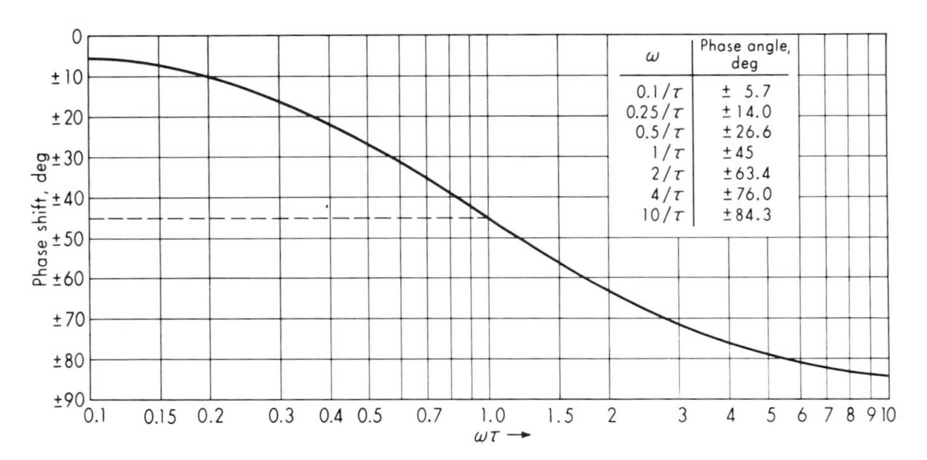

ω	Phase angle, deg
0.1/τ	± 5.7
0.25/τ	±14.0
0.5/τ	±26.6
1/τ	±45
2/τ	±63.4
4/τ	±76.0
10/τ	±84.3

Fig. 3.64. *First-order-system phase angle.*

so that the low-frequency asymptote is simply the 0-db line. The two straight-line asymptotes will meet at $\omega\tau = 1$ because this is where (3.263) and (3.264) are zero. The point $\omega = 1/\tau$ is called the "breakpoint" or "corner frequency." In plotting curves for such terms, one first locates the breakpoint and then draws the two asymptotes. The true curve is obtained by correcting the straight-line asymptotes at several points, using the data of Fig. 3.63. The phase-angle curves may be quickly plotted, using the data of Fig. 3.64. A numerical example illustrating these methods is given in Fig. 3.65.

Terms of the form $[(i\omega/\omega_n)^2 + 2\zeta i\omega/\omega_n + 1]^{\pm 1}$ have low-frequency asymptotes of 0 db and high-frequency asymptotes of slope ± 40 db/decade.

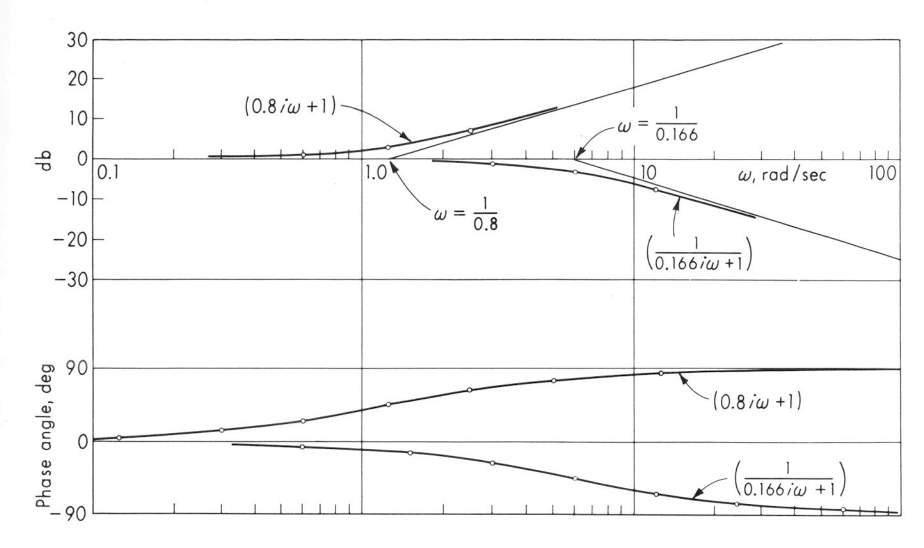

Fig. 3.65. *Example of first-order terms.*

They intersect at $\omega = \omega_n$. The exact curves for a given ζ are obtained by applying the corrections of Fig. 3.66. The phase-angle curves are obtained from Fig. 3.67. Figure 3.68 gives numerical examples.

The final type of term to be considered is the dead-time term $e^{-i\omega\tau_{dt}}$. Since the amplitude ratio is 1.0 for all frequencies, the decibel curve is simply the 0-db line. The phase-angle curve is easily plotted from $\phi = -\omega\tau_{dt}$ for any given dead time.

To illustrate the procedure for combining the individual terms to obtain the overall frequency-response curves, we consider the following example:

$$\frac{q_o}{q_i}(i\omega) = \frac{4.4(i\omega)}{(i\omega + 1)(0.2i\omega + 1)} \tag{3.267}$$

Figure 3.69 shows the procedure and results.

Response of a general form of instrument to a periodic input. Our approach to the dynamic response of measurement systems has, to this point, been limited in two ways. First, we considered only rather simple types of instruments (zero-order, first-order, and second-order) and, secondly, we subjected these instruments only to rather simple inputs (steps, ramps, sine waves, and impulses). At this point, by applying more advanced mathematical tools, we begin to remove both these limitations. We shall see that the concept of frequency response plays a central role

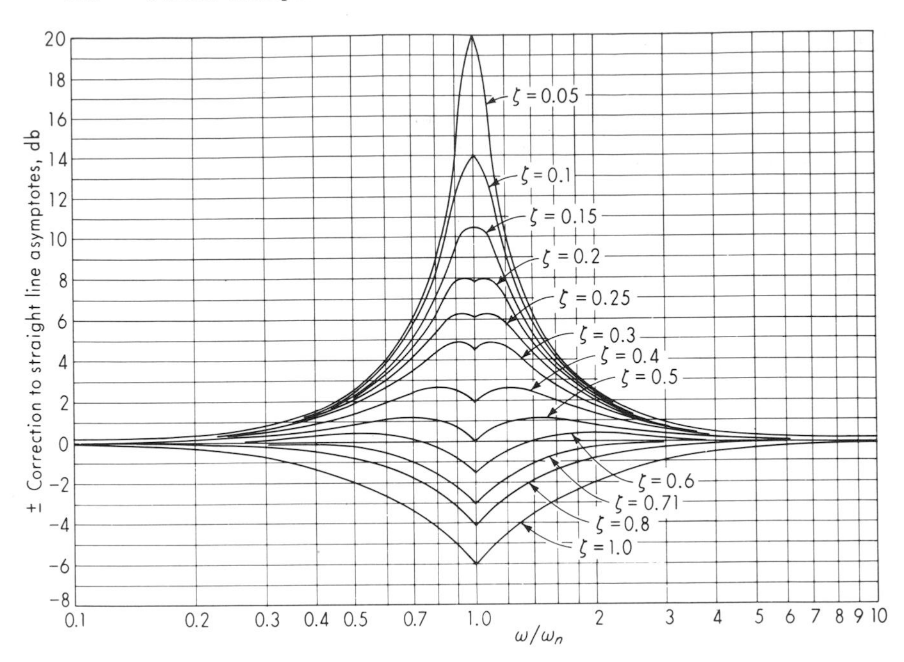

Fig. 3.66. *Second-order-system amplitude-ratio corrections.*

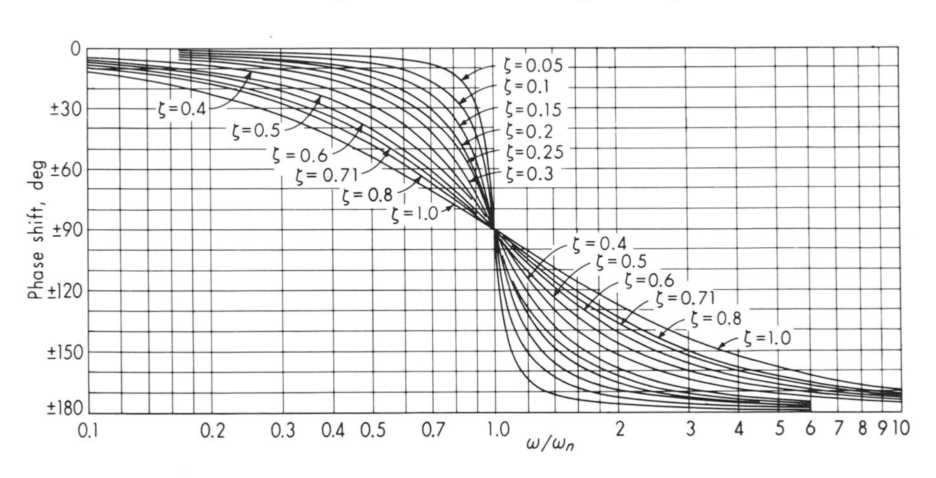

Fig. 3.67. *Second-order-system phase angle.*

in these developments. The first step involves the study of the response of a general (linear, time-invariant) instrument to periodic inputs.

By a periodic function we mean one that repeats itself cyclically over and over again, as in Fig. 3.70. If this function meets the *Dirichlet conditions* (it must be single-valued, finite, and have a finite number of

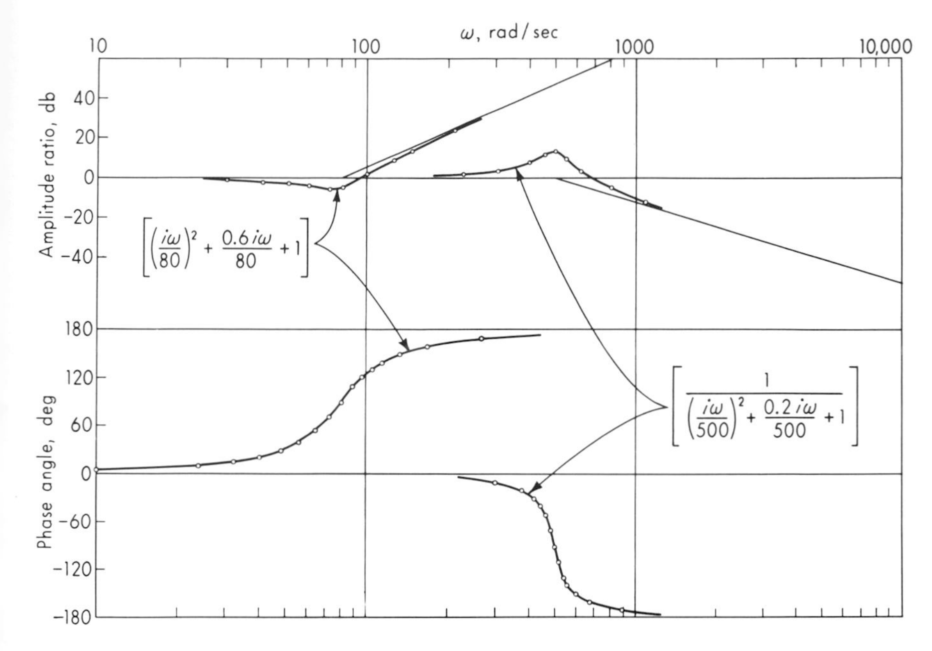

Fig. 3.68. Example of second-order terms.

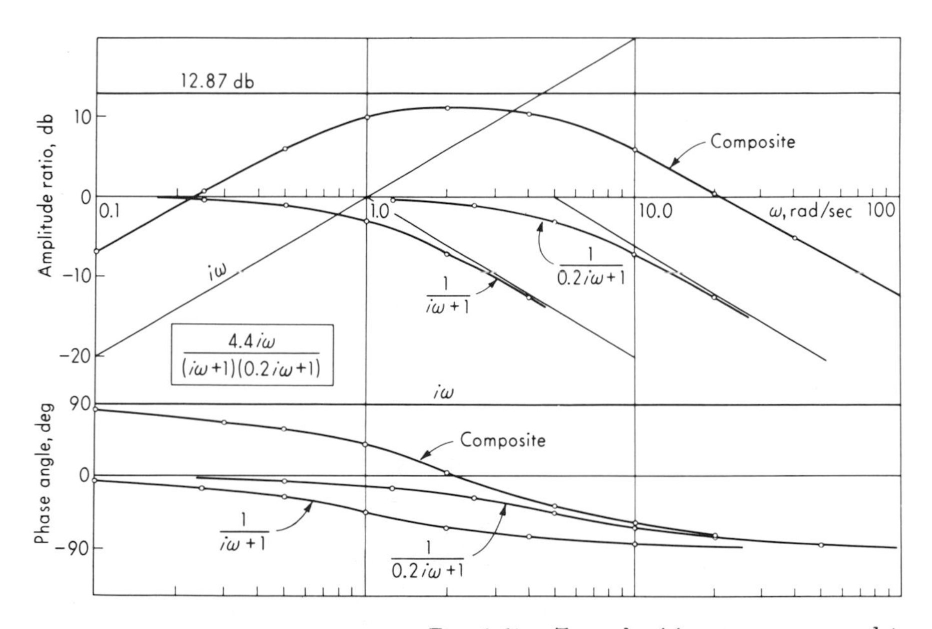

Fig. 3.69. Example of frequency-response plot.

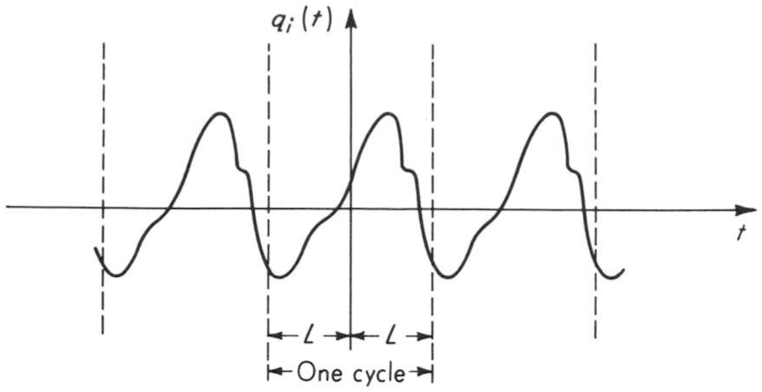

Fig. 3.70. General periodic function.

discontinuities and maxima and minima in one cycle), it may be represented by a *Fourier series.*[1] That is,

$$q_i(t) = q_{i,\mathrm{av}} + \frac{1}{L}\left(\sum_{n=1}^{\infty} a_n \cos\frac{n\pi t}{L} + \sum_{n=1}^{\infty} b_n \sin\frac{n\pi t}{L}\right) \tag{3.268}$$

$$\text{where} \quad q_{i,\mathrm{av}} \triangleq \text{average value of } q_i = \frac{1}{2L}\int_{-L}^{L} q_i(t)\,dt \tag{3.269}$$

$$a_n \triangleq \int_{-L}^{L} q_i(t)\cos\frac{n\pi t}{L}\,dt \tag{3.270}$$

$$b_n \triangleq \int_{-L}^{L} q_i(t)\sin\frac{n\pi t}{L}\,dt \tag{3.271}$$

(The origin of the t coordinate may be chosen wherever most convenient.) We see then that any periodic function satisfying the Dirichlet conditions can be replaced by a sum of terms consisting of a constant and sine and cosine waves of various frequencies. If one wishes, any *pair* of sine and cosine terms of the *same* frequency can be replaced by a *single* sine wave of the same frequency at some phase angle since

$$A\cos\omega t + B\sin\omega t = C\sin(\omega t + \alpha) \tag{3.272}$$

$$\text{where} \quad C \triangleq \sqrt{A^2 + B^2} \tag{3.273}$$

$$\alpha \triangleq \tan^{-1}\frac{A}{B} \tag{3.274}$$

The Fourier series will usually be an *infinite* series, and to get a *perfect* reconstruction of $q_i(t)$ from the series an infinite number of terms would have to be added. Fortunately, perfect reproduction is not required in engineering applications; thus generally $q_i(t)$ is *approximated* by a truncated (cut off after a certain number of terms) series. Just how many terms

[1] B. J. Ley, S. G. Lutz, and C. F. Rehberg, "Linear Circuit Analysis," chap. 6, McGraw-Hill Book Company, New York, 1959.

to use depends on the form of $q_i(t)$ (if it has very sharp changes more terms are required) and also on the use to which the information is to be put. Often, less than 10 "harmonics" (the first 10 different frequencies) are adequate.

The method of obtaining the desired terms in the Fourier series depends on the nature of $q_i(t)$. If $q_i(t)$ is given as a known mathematical formula, Eqs. (3.268) to (3.271) may be employed. If the required integrations cannot be performed [because of the complexity of $q_i(t)$], one can use one of several approximate numerical integration schemes or he can *plot* the functions to be integrated [such as $q_i(t) \cos (n\pi t/L)$] and use a planimeter to perform the integrations. If this plotting and planimetering (area measuring) are accurately done, the exact values of the desired coefficients will be obtained. This method is also feasible if $q_i(t)$ is not given by a formula but rather by a table of experimental data or a chart from a recording device. An alternative scheme is to choose a definite number of terms (such as 12) in the series and then adjust the coefficients of these terms so that $q_i(t)$ is fitted *exactly* at 12 points but an error exists at all other points. This method of obtaining the approximate Fourier series is in error in two ways:

1. Only a finite number of terms is used.
2. The terms that *are* used are not exact.

Both these types of errors can be reduced by using more terms. Such methods have been reduced to rather simple computation schemes in books on numerical analysis[1] and are thus quite popular. We now give two of these without proof. The *method of* 12 *ordinates* is given in Fig. 3.71 and the method of 24 ordinates in Fig. 3.72.

There have also been developed various mechanical, electrical, and optical instruments for obtaining the Fourier coefficients from a graphical or electrical record of $q_i(t)$. Some of these will be touched on in a later chapter under the topic Signal Analyzers.

Once the Fourier series for a particular $q_i(t)$ has been found, the *steady-state* response of any instrument to this input may be found by use of frequency-response techniques and the principle of superposition. That is, the response for each individual sinusoidal term is found and then they are added algebraically to get the total response. By use of Eqs. (3.272) to (3.274) all terms in the Fourier series can be put in the form

$$A_{ik} \sin (\omega_k t + \alpha_k)$$

[1] J. B. Scarborough, "Numerical Mathematical Analysis," 3d ed., chap. 17, The Johns Hopkins Press, Baltimore, 1958.

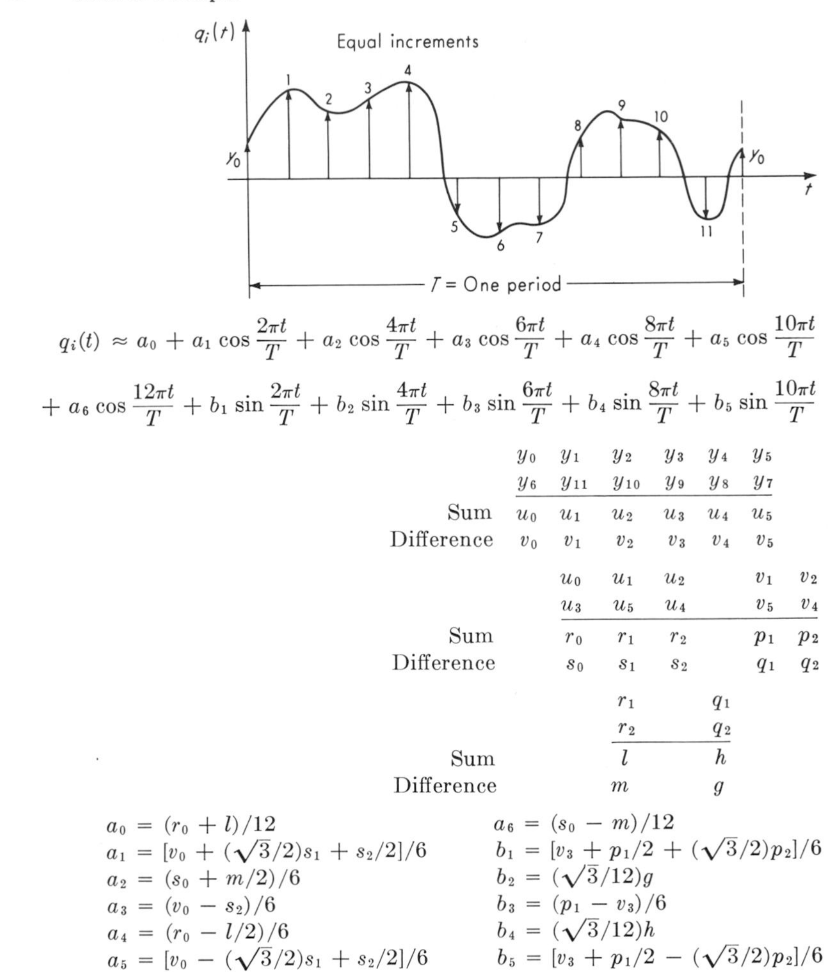

Fig. 3.71. *Method of 12 ordinates.*

We now define the complex number $Q_i(i\omega_k)$ by

$$Q_i(i\omega_k) \triangleq A_{ik}\underline{/\alpha_k} \tag{3.275}$$

For example, the constant term -7.2 becomes $7.2\underline{/180°}$, and the term $9.3 \sin (20t + 37°)$ becomes $9.3\underline{/37°}$. When the Fourier series representing $q_i(t)$ is expressed in this form, it is called $Q_i(i\omega)$, the *input-frequency spectrum*. Thus, if

$$q_i(t) = A_{i0} + A_{i1} \sin (\omega_1 t + \alpha_1) + A_{i2} \sin (\omega_2 t + \alpha_2) + \cdots \tag{3.276}$$

then

$$Q_i(i\omega) = |A_{i0}|\underline{/0° \text{ or } 180°} + A_{i1}\underline{/\alpha_1} + A_{i2}\underline{/\alpha_2} + \cdots \tag{3.277}$$

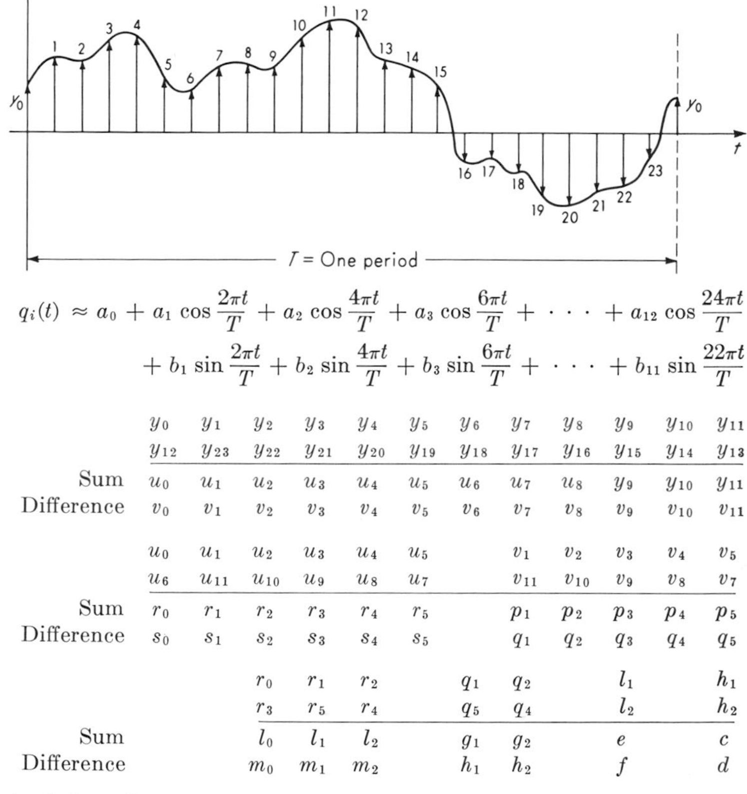

$$q_i(t) \approx a_0 + a_1 \cos\frac{2\pi t}{T} + a_2 \cos\frac{4\pi t}{T} + a_3 \cos\frac{6\pi t}{T} + \cdots + a_{12} \cos\frac{24\pi t}{T}$$
$$+ b_1 \sin\frac{2\pi t}{T} + b_2 \sin\frac{4\pi t}{T} + b_3 \sin\frac{6\pi t}{T} + \cdots + b_{11} \sin\frac{22\pi t}{T}$$

	y_0	y_1	y_2	y_3	y_4	y_5	y_6	y_7	y_8	y_9	y_{10}	y_{11}
	y_{12}	y_{23}	y_{22}	y_{21}	y_{20}	y_{19}	y_{18}	y_{17}	y_{16}	y_{15}	y_{14}	y_{13}
Sum	u_0	u_1	u_2	u_3	u_4	u_5	u_6	u_7	u_8	y_9	y_{10}	y_{11}
Difference	v_0	v_1	v_2	v_3	v_4	v_5	v_6	v_7	v_8	v_9	v_{10}	v_{11}

	u_0	u_1	u_2	u_3	u_4	u_5		v_1	v_2	v_3	v_4	v_5
	u_6	u_{11}	u_{10}	u_9	u_8	u_7		v_{11}	v_{10}	v_9	v_8	v_7
Sum	r_0	r_1	r_2	r_3	r_4	r_5		p_1	p_2	p_3	p_4	p_5
Difference	s_0	s_1	s_2	s_3	s_4	s_5		q_1	q_2	q_3	q_4	q_5

			r_0	r_1	r_2		q_1	q_2		l_1		h_1
			r_3	r_5	r_4		q_5	q_4		l_2		h_2
Sum			l_0	l_1	l_2		g_1	g_2		e		c
Difference			m_0	m_1	m_2		h_1	h_2		f		d

In the formulas below, $C = \cos 15° = 0.9659258$, $S = \sin 15° = 0.2588190$.

$$
\begin{aligned}
a_0 &= (l_0 + e)/24 \\
a_1 &= [v_0 + Cs_1 + (\sqrt{3}/2)s_2 + (1/\sqrt{2})s_3 + (1/2)s_4 + Ss_5]/12 \\
a_2 &= [s_0 + (\sqrt{3}/2)m_1 + (1/2)m_2]/12 \\
a_3 &= [v_0 + (1/\sqrt{2})(s_1 - s_3 - s_5) - s_4]/12 \\
a_4 &= [m_0 + (1/2)f]/12 \\
a_5 &= [v_0 + Ss_1 - (\sqrt{3}/2)s_2 - (1/\sqrt{2})s_3 + (1/2)s_4 + Cs_5]/12 \\
a_6 &= (s_0 - m_2)/12 \\
a_7 &= [v_0 - Ss_1 - (\sqrt{3}/2)s_2 + (1/\sqrt{2})s_3 + (1/2)s_4 - Cs_5]/12 \\
a_8 &= [l_0 - (1/2)e]/12 \\
a_9 &= [v_0 - (1/\sqrt{2})(s_1 - s_3 - s_5) - s_4]/12 \\
a_{10} &= [s_0 - (\sqrt{3}/2)m_1 + (1/2)m_2]/12 \\
a_{11} &= [v_0 - Cs_1 + (\sqrt{3}/2)s_2 - (1/\sqrt{2})s_3 + (1/2)s_4 - Ss_5]/12 \\
a_{12} &= (m_0 - f)/24
\end{aligned}
$$

$$
\begin{aligned}
b_1 &= [Sp_1 + (1/2)p_2 + (1/\sqrt{2})p_3 + (\sqrt{3}/2)p_4 + Cp_5 + v_6]/12 \\
b_2 &= [(1/2)g_1 + (\sqrt{3}/2)g_2 + q_3]/12 \\
b_3 &= [p_2 - v_6 + (1/\sqrt{2})(p_1 + p_3 - p_5)]/12 \\
b_4 &= (\sqrt{3}/24)c \\
b_5 &= [Cp_1 + (1/2)p_2 - (1/\sqrt{2})p_3 - (\sqrt{3}/2)p_4 + Sp_5 + v_6]/12 \\
b_6 &= (g_1 - q_3)/12 \\
b_7 &= [Cp_1 - (1/2)p_2 - (1/\sqrt{2})p_3 + (\sqrt{3}/2)p_4 + Sp_5 - v_6]/12 \\
b_8 &= (\sqrt{3}/24)d \\
b_9 &= [v_6 - p_2 + (1/\sqrt{2})(p_1 + p_3 - p_5)]/12 \\
b_{10} &= [(1/2)g_1 - (\sqrt{3}/2)g_2 + q_3]/12 \\
b_{11} &= [Sp_1 - (1/2)p_2 + (1/\sqrt{2})p_3 - (\sqrt{3}/2)p_4 + Cp_5 - v_6]/12
\end{aligned}
$$

Fig. 3.72. *Method of 24 ordinates.*

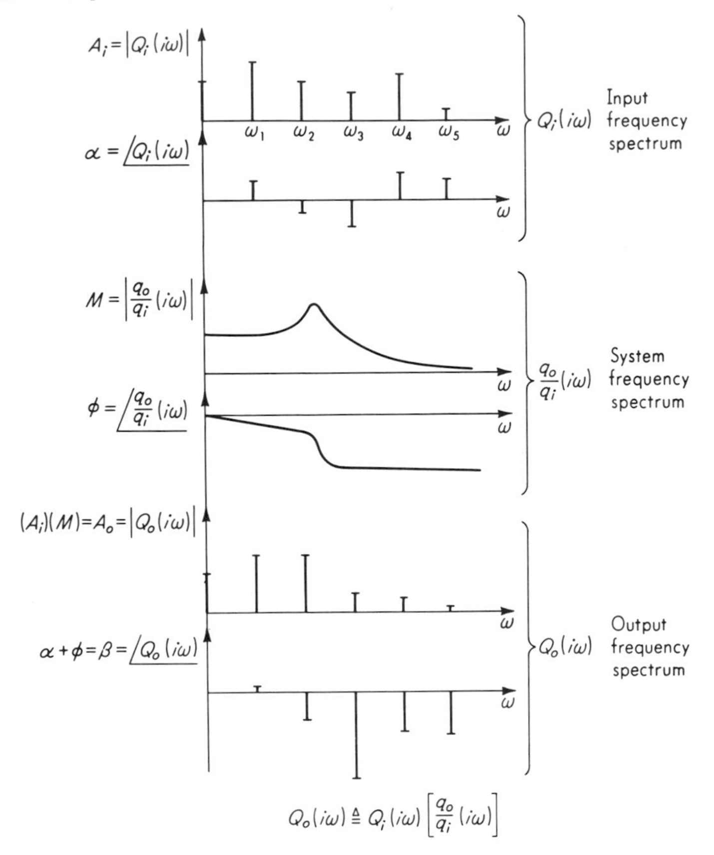

Fig. 3.73. *System response to periodic input.*

Such a spectrum, which exists only at isolated frequencies, is called a *discrete spectrum.*

Now, we recall that the sinusoidal transfer function of the system, $(q_o/q_i)(i\omega)$, is also a complex number for any given frequency. An alternative name for the sinusoidal transfer function is the *system frequency spectrum.* Such a spectrum, which exists at *all* frequencies, is called a *continuous spectrum.* If we pick any frequency ω_k and multiply the complex numbers $Q_i(i\omega_k)$ and $(q_o/q_i)(i\omega_k)$ we get another complex number which we define as $Q_o(i\omega_k)$. If we do this for *all* frequencies and add all the $Q_o(i\omega_k)$, the sum is called $Q_o(i\omega)$, the *output frequency spectrum,* which has the form

$$Q_o(i\omega) = A_{o0}\underline{/0^\circ \text{ or } 180^\circ} + A_{o1}\underline{/\beta_1} + A_{o2}\underline{/\beta_2} + \cdots \tag{3.278}$$

This is now interpreted as in Eqs. (3.276) and (3.277) to give

$$q_o(t) = \pm A_{o0} + A_{o1}\sin(\omega_1 t + \beta_1) + A_{o2}\sin(\omega_2 t + \beta_2) + \cdots \tag{3.279}$$

Since $q_i(t)$ is periodic, the frequencies ω_2, ω_3, etc., are all integer multiples of ω_1 and thus $q_o(t)$ will also be a periodic function. For accurate measurement, $q_i(t)$ and $q_o(t)$ must, of course, have nearly identical wave forms. The validity of the statement

$$Q_o(i\omega) = Q_i(i\omega)\left[\frac{q_o}{q_i}(i\omega)\right] \tag{3.280}$$

used in the above manipulations follows easily from the basic definition of the sinusoidal transfer function, the superposition theorem, and the rules for multiplying complex numbers. Figure 3.73 illustrates the method graphically.

Response of a general form of instrument to a transient input. By a transient input we mean a $q_i(t)$ that is identically zero for all values of time greater than some finite value t_o, that is, an input that eventually dies out. For transient inputs of specific mathematical form we can usually solve the differential equation and get $q_o(t)$ directly. For q_i's given by experimental data or, more importantly, if we wish to bring out certain important results of a *general* (not restricted to a specific type of q_i) nature, the methods of Fourier transforms[1,2,3] or Laplace transforms[1,2,3] are useful. We now present the methods of applying these techniques, without proof of their validity.

The *direct Fourier transform* $Q_i(i\omega)$ (or the Laplace transform with $s = i\omega$) of the transient input $q_i(t)$ which is zero for $t < 0$ is given by

$$Q_i(i\omega) \triangleq \int_0^\infty q_i(t)\cos\omega t\,dt - i\int_0^\infty q_i(t)\sin\omega t\,dt \tag{3.281}$$

where ω can take all values from $-\infty$ to $+\infty$. Equation (3.281) is said to transform the input function from the time domain $[q_i(t)]$ to the frequency domain $[Q_i(i\omega)]$. The function $Q_i(i\omega)$ is also called the frequency spectrum of the input and plays the same role for transient inputs as Eq. (3.277) does for periodic inputs. However, whereas $Q_i(i\omega)$ is a *discrete* spectrum for $q_i(t)$ periodic, it is a *continuous* spectrum for $q_i(t)$ transient. That is, if one carries out Eq. (3.281) for a given $q_i(t)$, he will find $Q_i(i\omega)$ to be a complex number which varies with (is a function of) frequency ω and exists for *all* ω, not just at isolated points. As an example, consider the transient input of Fig. 3.74*a*. Applying Eq.

[1] Ley, Lutz, and Rehberg, *op. cit.*

[2] M. F. Gardner and J. L. Barnes, "Transients in Linear Systems," John Wiley & Sons, Inc., New York, 1942.

[3] A. Papoulis, "The Fourier Integral and Its Applications," McGraw-Hill Book Company, New York, 1962.

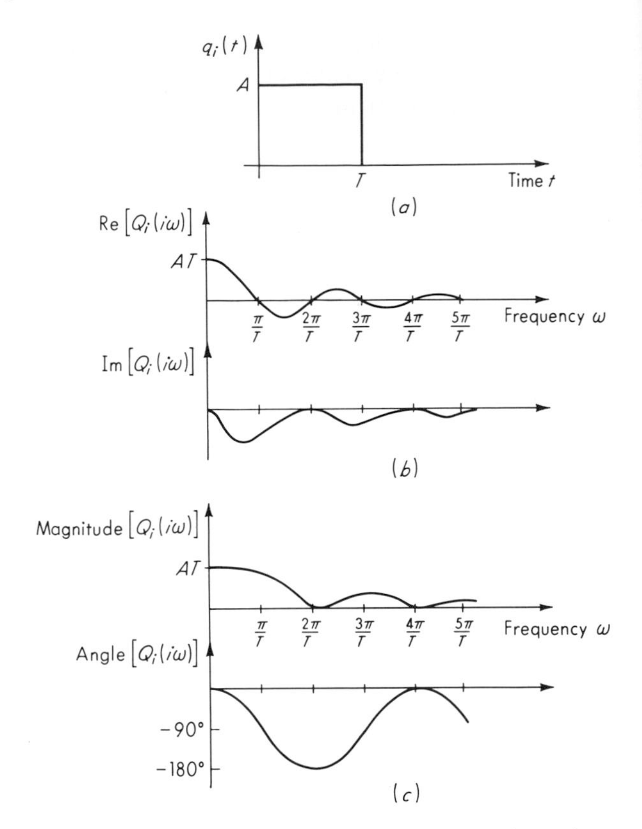

Fig. 3.74. *Frequency spectrum of transient.*

(3.281) we get

$$Q_i(i\omega) = \int_0^T A \cos \omega t \, dt - i \int_0^T A \sin \omega t \, dt \tag{3.282}$$

$$Q_i(i\omega) = \underbrace{\frac{A \sin \omega T}{\omega}}_{\text{real part}} + \underbrace{i \frac{A}{\omega} (-1 + \cos \omega T)}_{\text{imaginary part}} \tag{3.283}$$

or, alternatively,

$$Q_i(i\omega) = \underbrace{\frac{\sqrt{2}\, A}{\omega} \sqrt{1 - \cos \omega T}}_{\text{magnitude}} \; \underbrace{/\alpha}_{\text{angle}} \tag{3.284}$$

where

$$\tan \alpha = \frac{\cos \omega T - 1}{\sin \omega T} \tag{3.285}$$

Plots of this frequency spectrum are given in Fig. 3.74*b* and *c*. [While $Q_i(i\omega)$ exists for both positive and negative values of ω, because of symmetry, we can often get the desired results by considering only the range $0 \leq \omega < +\infty$. The stated symmetry consists of the following:

1. Re $Q_i(-i\omega)$ = Re $Q_i(+i\omega)$
2. Im $Q_i(-i\omega)$ = $-$Im $Q_i(+i\omega)$
3. Magnitude $Q_i(-i\omega)$ = magnitude $Q_i(+i\omega)$
4. Angle $Q_i(-i\omega)$ = $-$angle $Q_i(+i\omega)$

Most of our graphs and calculation methods will employ the range $0 \leq \omega < +\infty$, but $Q_i(-i\omega)$ always exists and can be found from a given $Q_i(+i\omega)$ by application of the above symmetry rules.] These graphs indicate the "frequency content" of the transient input just as the Fourier series indicates the frequency content of a periodic input. Thus we see that if T is small large values of $Q_i(i\omega)$ persist out to higher frequencies than if T is large. Therefore a short-duration pulse is said to have more high-frequency content than a long one. It is important to point out that the concept of frequency content for transients is not as clear-cut as for periodic functions; $q_i(t)$ can *not*, for a transient, be built up by simply adding distinct sine waves because $Q_i(i\omega)$ is now a *continuous function and no distinct frequencies exist.*

A further illustration of this distinction may be found by examining the dimensions of $Q_i(i\omega)$ in both cases. As an example, consider a $q_i(t)$ which is a pressure, $\text{lb}_f/\text{in.}^2$. If this pressure is periodic, Eqs. (3.268), etc., show that $Q_i(i\omega)$ has the same dimensions as $q_i(t)$, that is, $\text{lb}_f/\text{in.}^2$. Now, however, if the pressure is a transient, Eq. (3.281) gives

$$Q_i(i\omega) = \int_0^\infty (\text{lb}_f/\text{in.}^2) \underbrace{\cos \omega t}_{\text{dimensionless}} (\text{sec}) - i\,(\text{same dimensions}) \qquad (3.286)$$

We see that $Q_i(i\omega)$ now has dimensions of $\text{lb}_f\text{-sec/in.}^2$ or, reinterpreting this, $(\text{lb}_f/\text{in.}^2)/(\text{rad/sec})$. That is, $Q_i(i\omega)$ is thought of as the amount of signal *per unit frequency increment* rather than the actual amount of signal at a discrete frequency. This is analogous to the concept of distributed (rather than concentrated) loads in strength of materials. When a beam has sand (or water, etc.) piled on it, the applied load at any particular *point* is zero, but over an *area* the load is the force density times the area. Similarly a transient signal has no discrete frequencies but does contain a certain amount of signal within any frequency *band*. Thus, for a transient, $Q_i(i\omega)$ may be thought of as the *density* of signal per frequency bandwidth rather than as the signal itself.

The main purpose of using Eq. (3.281) is to convert functions from the time domain to the frequency domain, perform certain desired operations (which are *easier* or more *revealing* in the frequency domain than in the time domain), and then convert the information back to the time domain since this is the more familiar and directly applicable (in an engineering sense) form. The conversion from frequency domain to time domain is

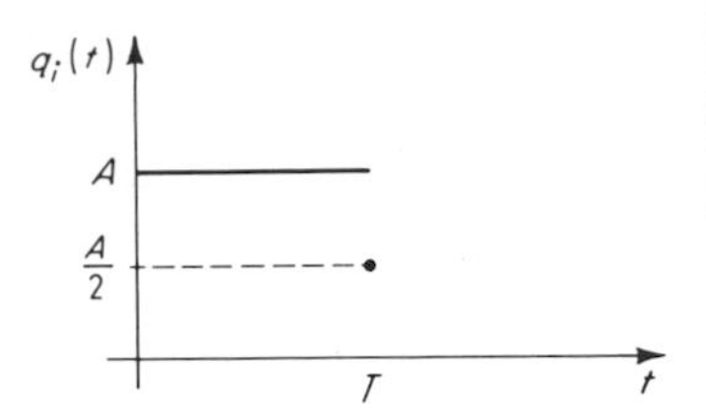

Fig. 3.75. *Single-valued definition of transient.*

accomplished by the *inverse-Fourier-* (or Laplace-) transform formula given by

$$\begin{aligned} q_i(t) &= \frac{2}{\pi}\int_0^\infty \operatorname{Re}\,[Q_i(i\omega)]\cos\omega t\,d\omega \qquad t > 0 \\ q_i(t) &\equiv 0 \qquad t < 0 \end{aligned} \tag{3.287}$$

Since these transformations are unique, if a $Q_i(i\omega)$ for a given $q_i(t)$ is found from Eq. (3.281), it should be possible to reconstruct the original $q_i(t)$ from the $Q_i(i\omega)$ by Eq. (3.287). Carrying this out for our example, we get

$$\begin{aligned} q_i(t) &= \frac{2}{\pi}\int_0^\infty \frac{A\sin\omega T}{\omega}\cos\omega t\,d\omega \qquad t > 0 \\ q_i(t) &\equiv 0 \qquad t < 0 \end{aligned} \tag{3.288}$$

After some transformations, this can be put in a standard form found in integral tables and gives

$$q_i(t) = \begin{cases} 0 & t < 0 \\ A & 0 < t < T \\ A/2 & t = T \\ 0 & T < t \end{cases} \tag{3.289}$$

This function is shown in Fig. 3.75, and we see that it is practically identical to Fig. 3.74*a*. Actually, in Fig. 3.74*a*, we were not mathematically precise in defining $q_i(t)$ at $t = T$ since the graph shows it taking on *all* values between 0 and A. The usual practice for such discontinuities is to define the function as single-valued and equal to the midpoint. The Fourier transform is set up on this basis and thus always gives results similar to (3.289). The Fourier *series* for a periodic function with step discontinuities also behaves in this fashion; that is, it converges to the midpoint. Thus in using numerical schemes such as those of Figs. 3.71 and 3.72, if an ordinate falls right on a discontinuity, the midpoint should be used as the numerical entry in the computation schedule.

To get a better feeling for the above methods and also to show how they are graphically or numerically applied to functions (data) for which mathematical formulas are not available, let us consider the following development. Figure 3.76*a* shows a typical transient $q_i(t)$ as might be

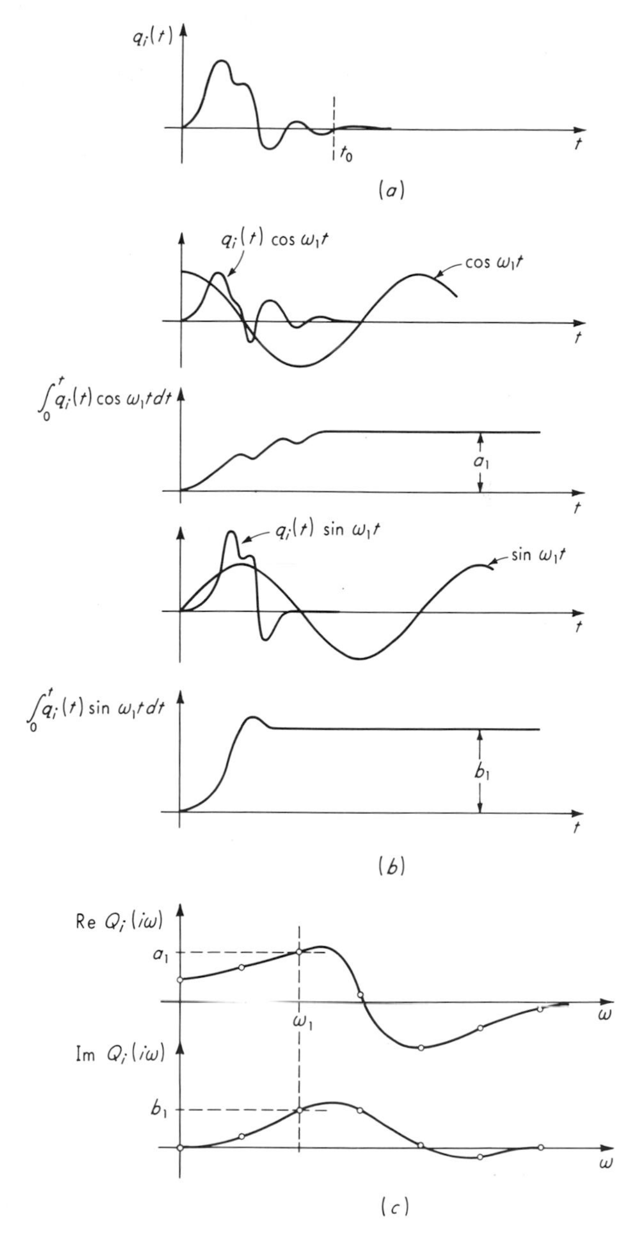

Fig. 3.76. *Graphical interpretation of direct Fourier transform.*

experimentally recorded in, say, a shock test. The first thing we note is that, for all practical purposes, the transient is ended in a finite length of time t_0. Thus, since $q_i(t)$ is a multiplying factor in the integrand and becomes zero for $t > t_0$, we may write Eq. (3.281) as

$$Q_i(i\omega) = \int_0^{t_0} q_i(t) \cos \omega t \, dt - i \int_0^{t_0} q_i(t) \sin \omega t \, dt \qquad (3.290)$$

We obtain $Q_i(i\omega)$ one point (frequency) at a time as follows:

1. Choose a numerical value of ω, say ω_1.
2. Now, $\cos \omega_1 t$ is a perfectly definite curve and may be plotted against t.
3. Multiply $q_i(t)$ and $\cos \omega_1 t$ point by point to get the curve $q_i(t) \cos \omega_1 t$.
4. Integrate, by any suitable numerical, graphical, or machine means, the curve $q_i(t) \cos \omega_1 t$ from $t = 0$ to $t = t_0$. Call the integral (area under curve) a_1.
5. Repeat the above procedure for $q_i(t) \sin \omega_1 t$ and call the integral b_1.
6. $Q_i(i\omega_1)$ is then $a_1 + ib_1$.
7. Repeat for as many ω's as desired to generate the curves for $Q_i(i\omega)$ versus ω.

Figure 3.76 illustrates these procedures.

To appreciate the difference in "frequency content" between a "slow" transient and a "fast" one, we consider Fig. 3.77. Here, $Q_i(i\omega)$ is found (for a high value of ω) for both a slow transient (Fig. 3.77*a*) and a fast one (Fig. 3.77*b*). It is clear that $Q_i(i\omega)$ will be nearly zero for ω's at or above the chosen ω_1 for the slow transient; thus its frequency content is limited to lower frequencies. The fast transient, on the other hand, has a nonzero value for $Q_i(i\omega_1)$ and thus "contains" frequencies at and somewhat above this value. For any real-world transient, one can always find *some* ω_1 high enough to make $Q_i(i\omega_1) \approx 0$; that is, all *real* transients are limited in frequency content at the high end. An "unreal" (mathematically possible only) transient is the impulse function which we can easily show to contain *all* frequencies from 0 to ∞ and all in equal "strength." For an impulse of area A

$$Q_i(i\omega) = \int_0^{\infty} Au_1(t) \cos \omega t \, dt - i \int_0^{\infty} Au_1(t) \sin \omega t \, dt \qquad (3.291)$$

$$Q_i(i\omega) = A - i0 = A \qquad \text{for any finite } \omega \qquad (3.292)$$

Thus Fig. 3.78 shows the frequency content of an impulse. This property of an impulse makes it most useful as a "test signal" for investigating unknown systems, since all frequencies will be excited equally and the

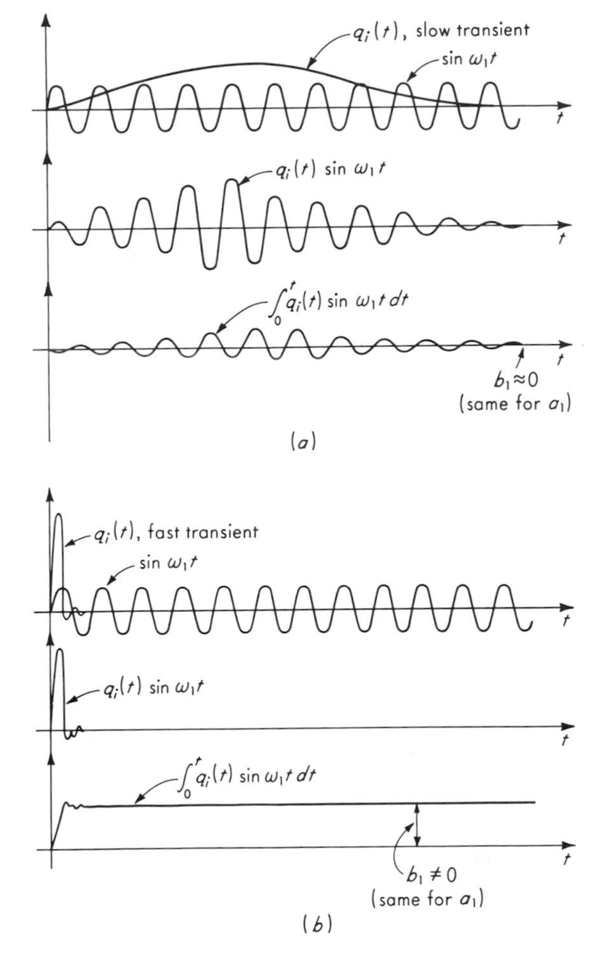

Fig. 3.77. *Frequency content of fast and slow transients.*

true nature of the system will be revealed by its response. We shall develop this important concept in more detail later.

The process of *inverse* transformation can also be interpreted graphically. The defining equation (3.287) may, in actual practice, be written as

$$q_i(t) = \frac{2}{\pi}\int_0^{\omega_0} \mathrm{Re}\,[Q_i(i\omega)]\cos\omega t\,d\omega \qquad t > 0$$
$$q_i(t) \equiv 0 \qquad t < 0 \tag{3.293}$$

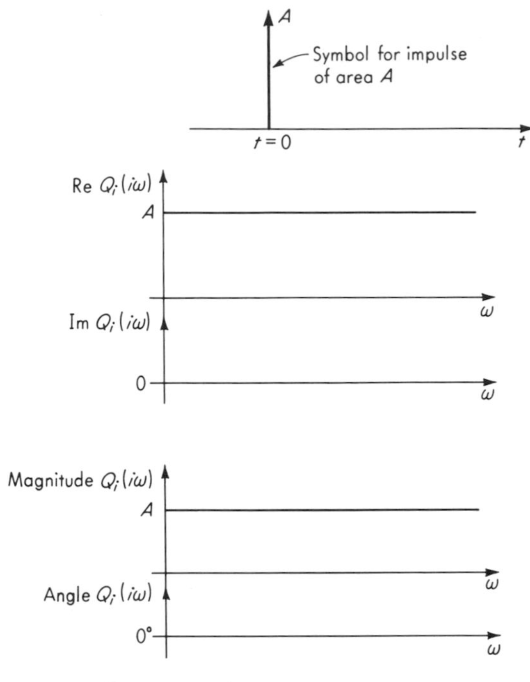

Fig. 3.78. Frequency spectrum of impulse.

since all $Q_i(i\omega)$ representing physical quantities become approximately equal to zero for ω greater than some finite value ω_0. This follows directly from the fact that the $Q_i(i\omega)$ come from the $q_i(t)$ and they cannot contain infinitely high frequencies. A step-by-step procedure for finding $q_i(t)$ one point at a time from a given $Q_i(i\omega)$ is as follows:

1. Choose a numerical value of t, say t_1.
2. Now, $\cos \omega t_1$ is a perfectly definite curve and may be plotted against ω.
3. Multiply $\text{Re}\,[Q_i(i\omega)]$ and $\cos \omega t_1$ point by point to get the curve $\text{Re}\,[Q_i(i\omega)] \cos \omega t_1$.
4. Integrate, by any suitable numerical, graphical, or machine means, the curve $\text{Re}\,[Q_i(i\omega)] \cos \omega t_1$ from $\omega = 0$ to $\omega = \omega_0$. The integral (area under curve) is $(\pi/2)q_i(t_1)$, from Eq. (3.293). Plot $q_i(t_1)$ versus t.
5. Repeat for as many t's as desired to generate the curve $q_i(t)$ versus t.

Figure 3.79 illustrates this procedure.

Since the direct and inverse transformations described above are widely used, methods for speeding up the calculations have been devel-

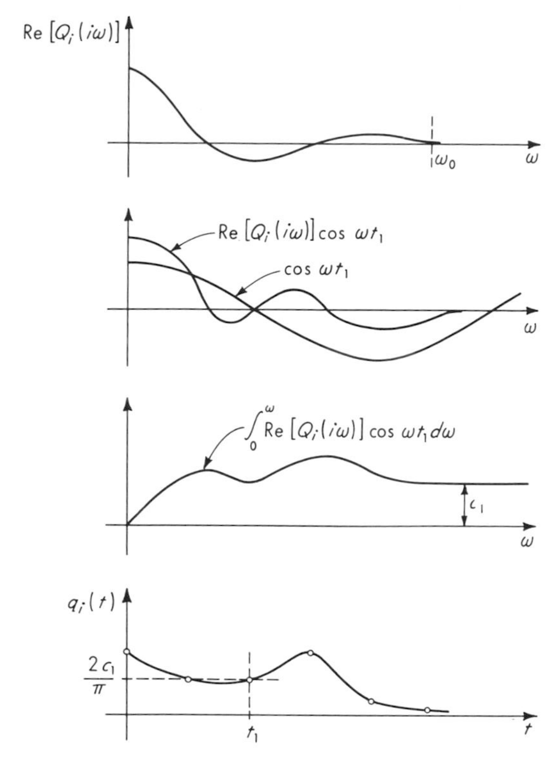

Fig. 3.79. *Graphical interpretation of inverse Fourier transform.*

oped.[1,2,3] Further treatment of these methods can be found in the bibliography entries at the end of this chapter.

The main usefulness of the above transform methods is based on the important result[4] relating the Fourier transform of the input signal $Q_i(i\omega)$, the system frequency response $(q_o/q_i)(i\omega)$, and the Fourier transform of

[1] C. R. Huss and J. J. Donegan, Method and Tables for Determining the Time Response to a Unit Impulse from Frequency-response Data and for Determining the Fourier Transform of a Function of Time, *NACA, Tech. Note* 3598, January, 1956.

[2] C. R. Huss and J. J. Donegan, Tables for the Numerical Determination of the Fourier Transform of a Function of Time and the Inverse Fourier Transform of a Function of Frequency, with Some Applications to Operational Calculus Methods, *NACA, Tech. Note* 4073, October, 1957.

[3] NACA is National Advisory Committee on Aeronautics; name changed in 1958 to National Aeronautics and Space Administration (NASA).

[4] J. A. Aseltine, "Transform Method in Linear System Analysis," McGraw-Hill Book Company, New York, 1958.

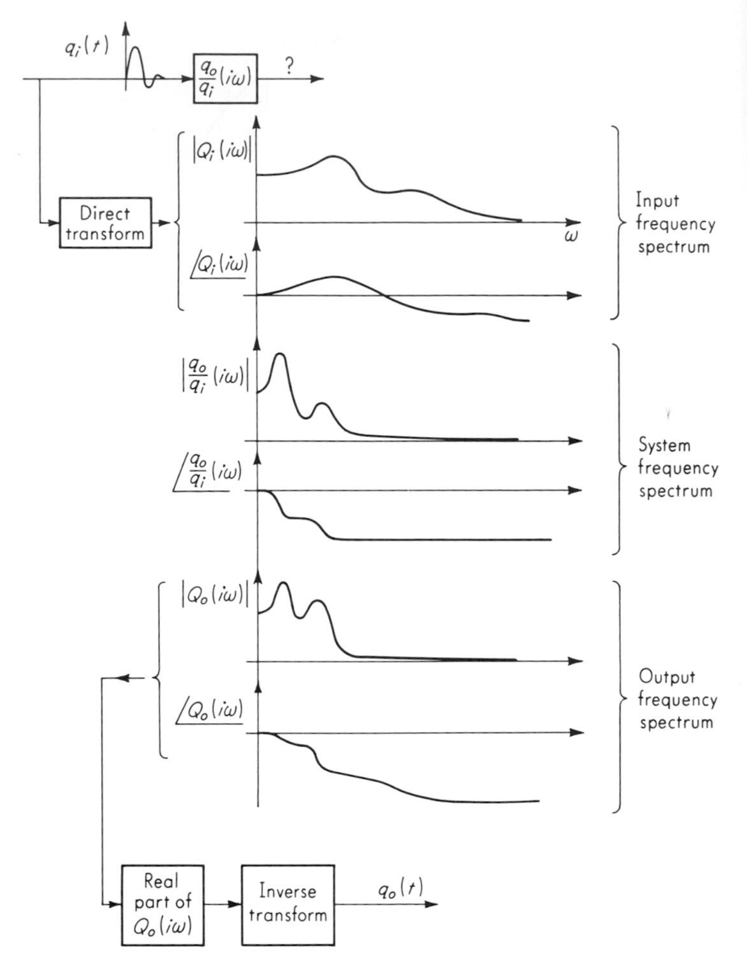

Fig. 3.80. *System response to transient input.*

the output signal $Q_o(i\omega)$:

$$Q_o(i\omega) = Q_i(i\omega)\left[\frac{q_o}{q_i}(i\omega)\right] \tag{3.294}$$

The restriction on this result is that the system and the input must be such that all derivatives (other than the highest derivative) and integrals of $q_o(t)$ that appear in the differential equation relating $q_o(t)$ and $q_i(t)$ must be zero at $t = 0^+$ (an infinitesimal time after the input is applied). That is, the initial conditions must be zero; the system starts from rest. Many important practical problems meet this requirement. (Also, for nonzero initial conditions, the method may still be applied if the input is suitably modified.[1])

The meaning of Eq. (3.294) is that when the frequency-response curves for a system are known, if a transient input $q_i(t)$ is applied, one can transform $q_i(t)$ to $Q_i(i\omega)$, multiply $Q_i(i\omega)$ and $(q_o/q_i)(i\omega)$ point by point to get $Q_o(i\omega)$, and then inverse-transform $Q_o(i\omega)$ to $q_o(t)$ to get the output or response of the system. Figure 3.80 illustrates the procedure.

From a measurement-system point of view, Eq. (3.294) has the following important interpretation. For accurate measurement, $q_o(t) \approx Kq_i(t)$, and since the Fourier transforms are unique [only one possible $F(i\omega)$ for each $f(t)$ and vice versa], this requires $Q_o(i\omega) \approx KQ_i(i\omega)$. Since $Q_o(i\omega)$ is obtained by multiplying $Q_i(i\omega)$ by $(q_o/q_i)(i\omega)$, this means $(q_o/q_i)(i\omega)$ must be $K\underline{/0°}$ over the entire range of frequencies for which $Q_i(i\omega)$ is not practically zero *but can be anything elsewhere.* The requirement that $(q_o/q_i)(i\omega)$ be $K\underline{/0°}$ for *all* frequencies for perfect measurement is obvious without the use of transform methods. The condition that this need be so only for a definite, finite *range* of frequencies (corresponding to the frequency content of the input) is the contribution of the transform methods and is of great practical significance since it puts much more realistic demands on the measurement system. A further relaxation of these requirements (which allows phase shift) will be developed later in this chapter.

Frequency spectrum of amplitude-modulated signals. Interest in amplitude-modulated signals stems mainly from two considerations:

1. Physical data that are to be measured and interpreted sometimes are amplitude-modulated.
2. Certain types of measurement systems intentionally introduce amplitude modulation for one or more benefits this process may supply.

While, in general, the signal that modulates the amplitude of a carrier wave may be of any form (single sine wave, general periodic function, random wave, transient, etc.) and the carrier may also be given different forms (sine wave, square wave, etc.), the process is perhaps most easily understood for a single sine wave modulating a sinusoidal carrier. The modulation process is basically one of multiplying the signal carrying the information by a carrier wave of constant frequency and amplitude; see

[1] Aseltine, *op. cit.*

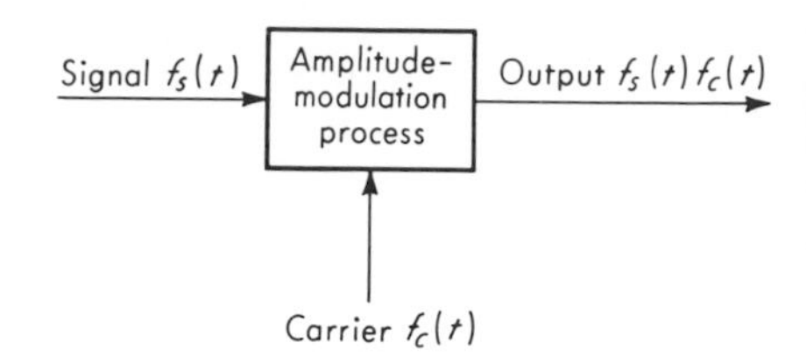

Fig. 3.81. *Amplitude modulation.*

Fig. 3.81. For our simple example we have

$$\text{Output} = (A_s \sin \omega_s t)(A_c \sin \omega_c t) \tag{3.295}$$

where $A_s \triangleq$ amplitude of signal
$\omega_s \triangleq$ frequency of signal
$A_c \triangleq$ amplitude of carrier
$\omega_c \triangleq$ frequency of carrier

The frequency ω_c is greater (usually considerably greater) than ω_s. For such a situation the output has the shape shown in Fig. 3.82*a*. The frequency spectrum of such a signal is easily obtained from the following trigonometric identity:

$$\sin \alpha \sin \beta \equiv \tfrac{1}{2} \cos (\alpha - \beta) - \tfrac{1}{2} \cos (\alpha + \beta) \tag{3.296}$$

Applying this to Eq. (3.295), we get

$$\text{Output} = \frac{A_s A_c}{2} [\cos (\omega_c - \omega_s) t - \cos (\omega_c + \omega_s) t] \tag{3.297}$$

$$\text{Output} = \frac{A_s A_c}{2} \sin [(\omega_c - \omega_s) t + 90°] + \frac{A_s A_c}{2} \sin [(\omega_c + \omega_s) t - 90°] \tag{3.298}$$

We see that the frequency spectrum of this signal is a discrete spectrum existing only at the frequencies $\omega_c - \omega_s$ and $\omega_c + \omega_s$, the so-called *side frequencies*. If such a signal is the input $q_i(t)$ to a measurement system, one can find the steady-state output easily by the methods of Fig. 3.73.

Some applications of these concepts may be appropriate at this point. In a first, rough consideration of the vibration and noise of shafts with gears, one would perhaps expect the important frequencies to be those corresponding to the rotational speeds of the shafts and those corresponding to the tooth-meshing frequencies. For example, a shaft with a 20-tooth gear running at 200 rps would be expected to generate noise at 200 and 4,000 cps. However, actual noise measurements in such situations may show the peak noise to occur at frequencies different from those

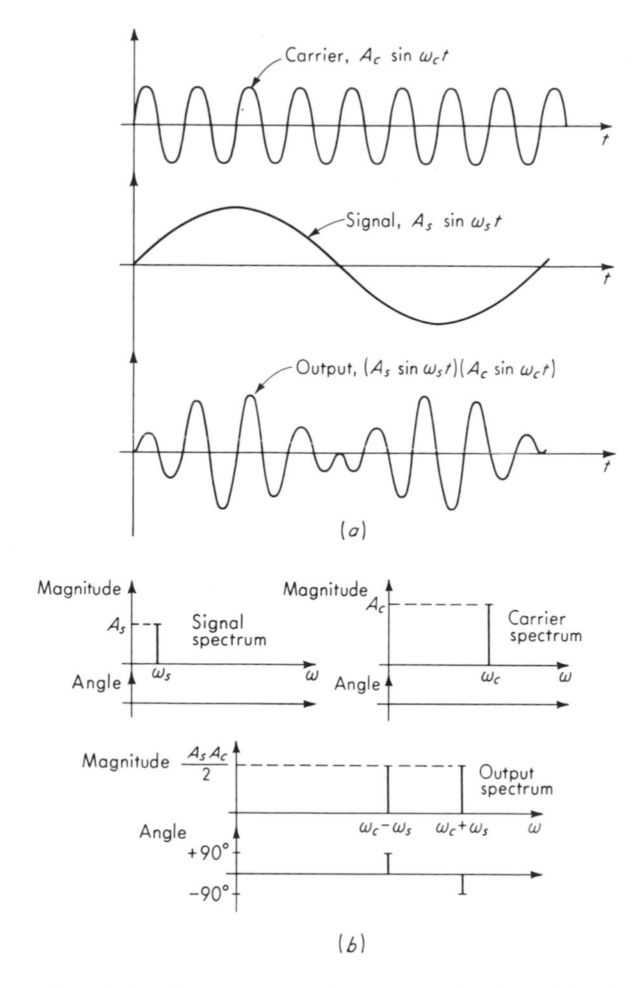

Fig. 3.82. *Frequency spectrum of amplitude-modulated signals.*

expected from the above crude analysis.[1] These discrepancies may often be resolved by the application of amplitude-modulation concepts as follows:

For a pair of absolutely true-running gears, the tooth forces (which cause vibration and, thereby, noise) would have a fundamental frequency equal to the tooth-meshing frequency. These forces would not be pure

[1] P. K. Stein, "Measurement Engineering," vol. I, sec. 17, Stein Engineering Services, Inc., Phoenix, Ariz., 1962.

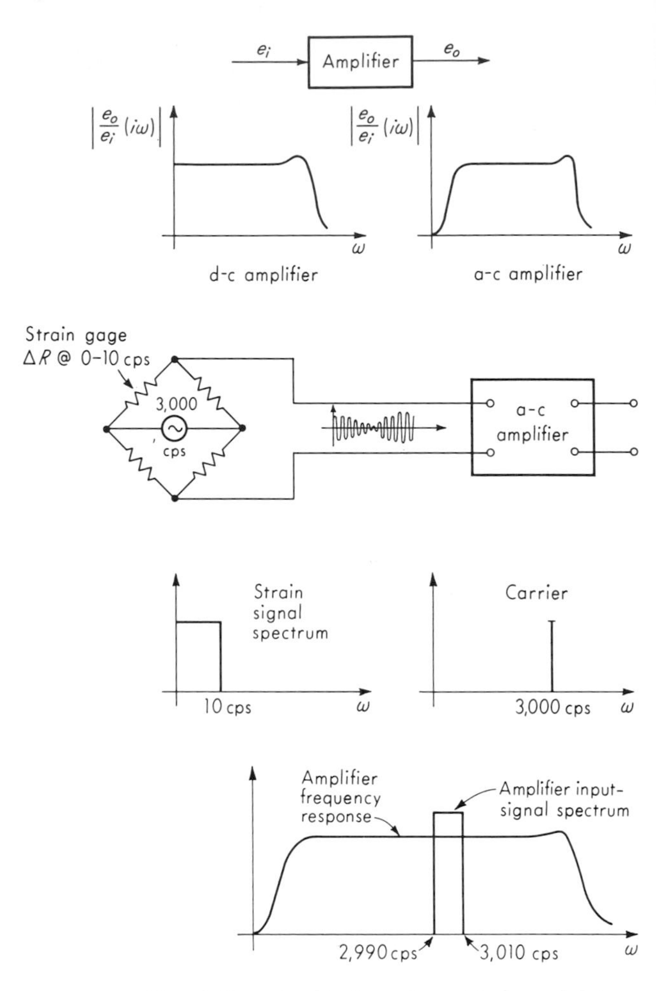

Fig. 3.83. Application of amplitude modulation.

sine waves but *would* be periodic. Thus one could get a Fourier series for them. For simplicity, let us assume these forces to be pure sine waves of fixed amplitude. In an actual set of gears there is always some eccentricity or "runout"; that is, the gears are closer together at some points in their rotation than at others. It is postulated that this runout leads to a force amplitude that *varies* as the gear rotates; that is, the tooth-force amplitude is *modulated* as a function of rotational position. If this is so, the frequencies of generated noise (corresponding to tooth-force frequencies) would be expected to be the side frequencies generated by modulating the 4,000-cps tooth-meshing frequency with the 200-cps (once per rotation) runout frequency. These frequencies (3,800 and

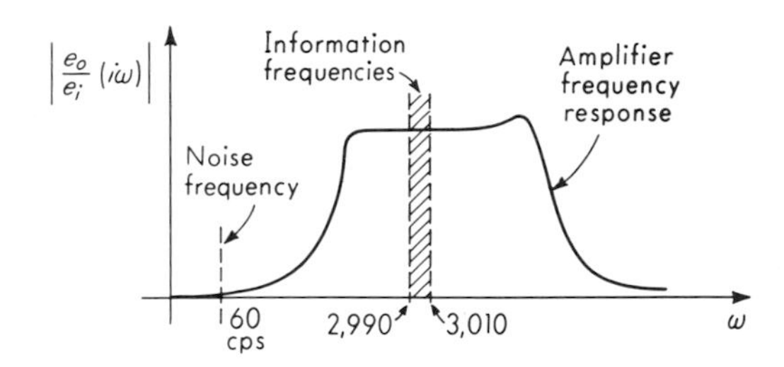

Fig. 3.84. *Noise rejection through amplitude modulation.*

4,200 cps in our example) have actually been measured, confirming the original conjecture. Without the amplitude modulation concepts, the engineer would have been hard pressed to explain these frequencies in the measured data.

Another interesting example is found in the carrier amplifier. To measure easily and record very small voltages coming from transducers (such as strain gages) requires a very-high-gain amplifier. Because of drift problems, a high-gain amplifier is easier to build as an a-c, rather than a d-c, unit. An a-c amplifier, however, does not amplify constant or slowly varying voltages and so would appear to be unsuitable for measuring static strains. This problem is overcome by exciting the strain-gage bridge with alternating voltage (say 5 volts at 3,000 cps) rather than direct current. Thus when the bridge is unbalanced by strain-induced resistance changes, the output voltage will be a 3,000-cps a-c voltage whose amplitude will be modulated by the strain changes. Thus if we are measuring strains which vary from, say, 0 cps (static) to 10 cps, the amplifier will have an input-frequency spectrum bounded by 2,990 and 3,010 cps. This range of frequencies is easily handled by an a-c amplifier. Figure 3.83 illustrates these concepts. This "shifting" of the information frequencies from one part of the frequency range to another is the basis of many useful applications of amplitude modulation.

As a final example, suppose that the wires leading from the bridge to the amplifier in Fig. 3.83 are subjected to a stray 60-cps field from surrounding a-c machinery and a 60-cps noise, or "hum," is superimposed (additively) on the desired signals. This 60-cps noise could easily be larger than the desired strain signals. With a carrier system, however, this noise may be easily eliminated merely by designing the a-c amplifier so that it does not respond to 60 cps. Since the desired band of frequencies is 2,990 to 3,010 cps, making the low-frequency cutoff of the amplifier greater than 60 cps is not difficult. Figure 3.84 illustrates this situation.

We shall now extend the amplitude-modulation concept to signals other than just a single sine wave. If the modulating signal is a periodic function $f_i(t)$ it may be expanded in a Fourier series to get the output of the modulator as [see Eq. (3.268)]

$$\text{Output} = \left[f_{i,\text{av}} + \frac{1}{L}\left(\sum_{n=1}^{\infty} a_n \cos\frac{n\pi t}{L} + \sum_{n=1}^{\infty} b_n \sin\frac{n\pi t}{L}\right)\right] A_c \sin \omega_c t \tag{3.299}$$

which can be written as

$$\begin{aligned}\text{Output} &= A_0A_c \sin \omega_c t + (A_1A_c \cos \omega_1 t \sin \omega_c t \\ &+ A_2A_c \cos \omega_2 t \sin \omega_c t + \cdots) + (B_1A_c \sin \omega_1 t \sin \omega_c t \\ &+ B_2A_c \sin \omega_2 t \sin \omega_c t + \cdots)\end{aligned} \tag{3.300}$$

Now, $\sin \alpha \sin \beta \equiv \frac{1}{2} \cos (\alpha - \beta) - \frac{1}{2} \cos (\alpha + \beta)$

and $\sin \alpha \cos \beta \equiv \frac{1}{2} \sin (\alpha + \beta) + \frac{1}{2} \sin (\alpha - \beta)$

and so

$$\begin{aligned}\text{Output} = A_0A_c \sin \omega_c t + C_1\{&\sin [(\omega_c + \omega_1)t - \alpha_1] \\ &+ \sin [(\omega_c - \omega_1)t + \alpha_1]\} + \cdots\end{aligned} \tag{3.301}$$

where

$$C_1 \triangleq \frac{A_c}{2}\sqrt{A_1^2 + B_1^2} \qquad \alpha_1 \triangleq \tan^{-1}\frac{B_1}{A_1} \tag{3.302}$$

We see that the spectrum of the output signal is a discrete spectrum containing the frequencies ω_c, $\omega_c \pm \omega_1$, $\omega_c \pm \omega_2$, $\omega_c \pm \omega_3$, etc. That is, each frequency component of the modulating signal produces one pair of side frequencies (see Fig. 3.85). If the output of the modulator is applied to the input of a system with known frequency response the methods of Fig. 3.73 can again be used to find the steady-state output.

If the modulating signal is a transient, the spectrum of the modulator output may be obtained with the help of the modulation theorem[1,2] for Fourier (or Laplace) transforms. If the modulating signal is a transient $f_i(t)$ it will have a Fourier transform $F_i(i\omega)$ which can be obtained in the usual ways. The modulation theorem leads to the following result if $f_i(t)$ is multiplied by the carrier $A_c \sin \omega_c t$ to produce the modulated output:

$$\text{Fourier transform of modulated output} \triangleq |Q_i(i\omega)|\underline{/Q_i(i\omega)} \tag{3.303}$$

where

$$|Q_i(i\omega)| = \frac{A_c}{2}\text{ magnitude }\{F_i[i(\omega - \omega_c)]\} \tag{3.304}$$

$$\underline{/Q_i(i\omega)} = \text{angle }\{F_i[i(\omega - \omega_c)]\} - 90° \tag{3.305}$$

[1] G. A. Korn and T. M. Korn, "Mathematical Handbook for Scientists and Engineers," p. 219, McGraw-Hill Book Company, New York, 1961.

[2] Papoulis, *op. cit.*, p. 15.

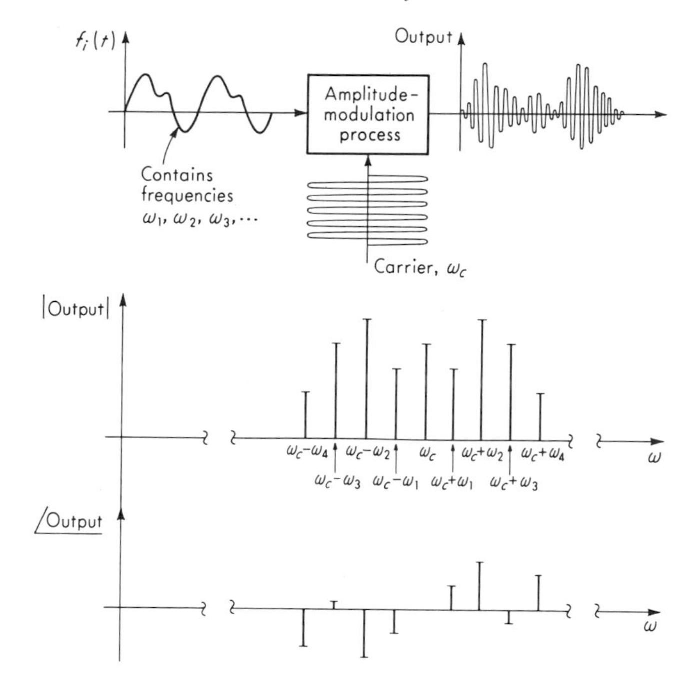

Fig. 3.85. *Frequency spectrum when modulating signal is periodic.*

and $0 \leq \omega < \infty$. Note that the argument (independent variable) of the F_i function is now $i(\omega - \omega_c)$, so that if one wants to find $Q_i(i\,4{,}000)$ and $\omega_c = 3{,}000$, he must evaluate $F_i(i\,1{,}000)$. Also note that, to get $Q_i(i\omega)$ for $0 < \omega < \infty$, one must know $F_i(i\omega)$ for *negative* ω's, since any $\omega < \omega_c$ gives $F_i i(\omega - \omega_c)$ a negative argument. While we have generally worked with positive ω's, the transform for negative ω's always exists and is easily found from the previously given symmetry rules. The spectrum given by Eq. (3.303) will be a continuous one and if the modulated output is applied as an input to some system with known frequency response the corresponding output can be obtained by the methods of Fig. 3.80.

For measurement systems in which amplitude modulation is intentionally introduced to allow the use of carrier-amplifier techniques, the carrier frequency must be considerably greater (usually 5 to 10 times) than any significant frequencies present in the modulating signal. For such a situation the pertinent frequency spectra are as shown in Fig. 3.86. We see again that the amplitude-modulation process shifts the frequency spectrum by the amount ω_c.

When amplitude modulation is intentionally introduced to facilitate data handling in one way or another, it generally plays the role of an

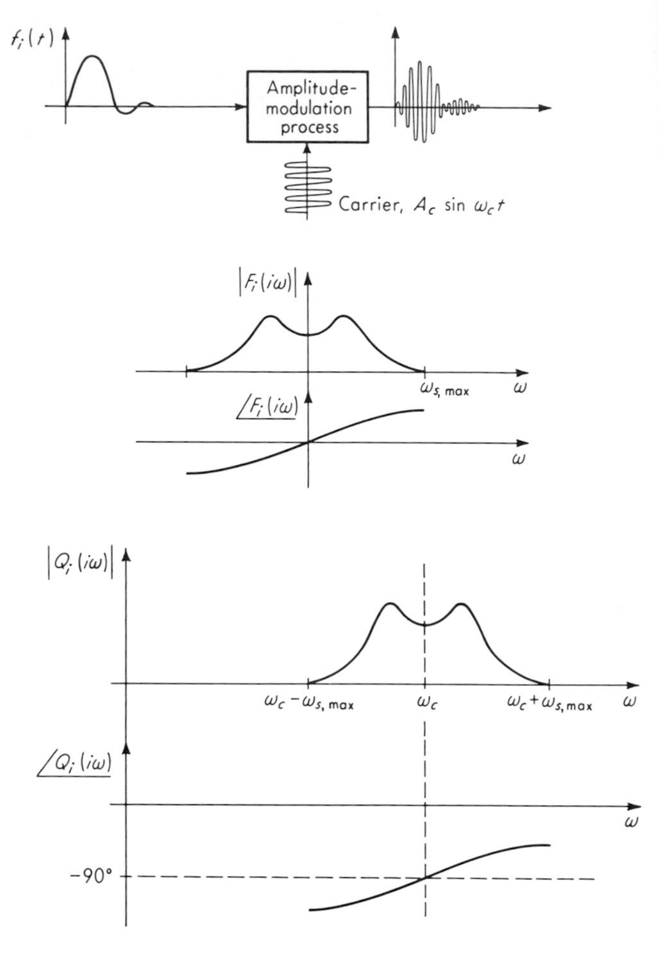

Fig. 3.86. *Frequency spectrum when modulating signal is a transient.*

intermediate step, and the amplitude-modulated signal is not usually considered a suitable final readout. Rather, the original form of the modulating signal (the basic measured data from, say, a transducer) should be recovered. The process for accomplishing this involves *demodulation* (or *detection*, as it is sometimes called) and filtering. Demodulation may be full-wave, half-wave, phase-sensitive, or non-phase-sensitive (Fig. 3.87). We here treat the form giving the best reproduction of the original data, full-wave phase-sensitive demodulation, and consider only the process, not the hardware for accomplishing it. Again it is necessary to consider whether the form of the original signal was single

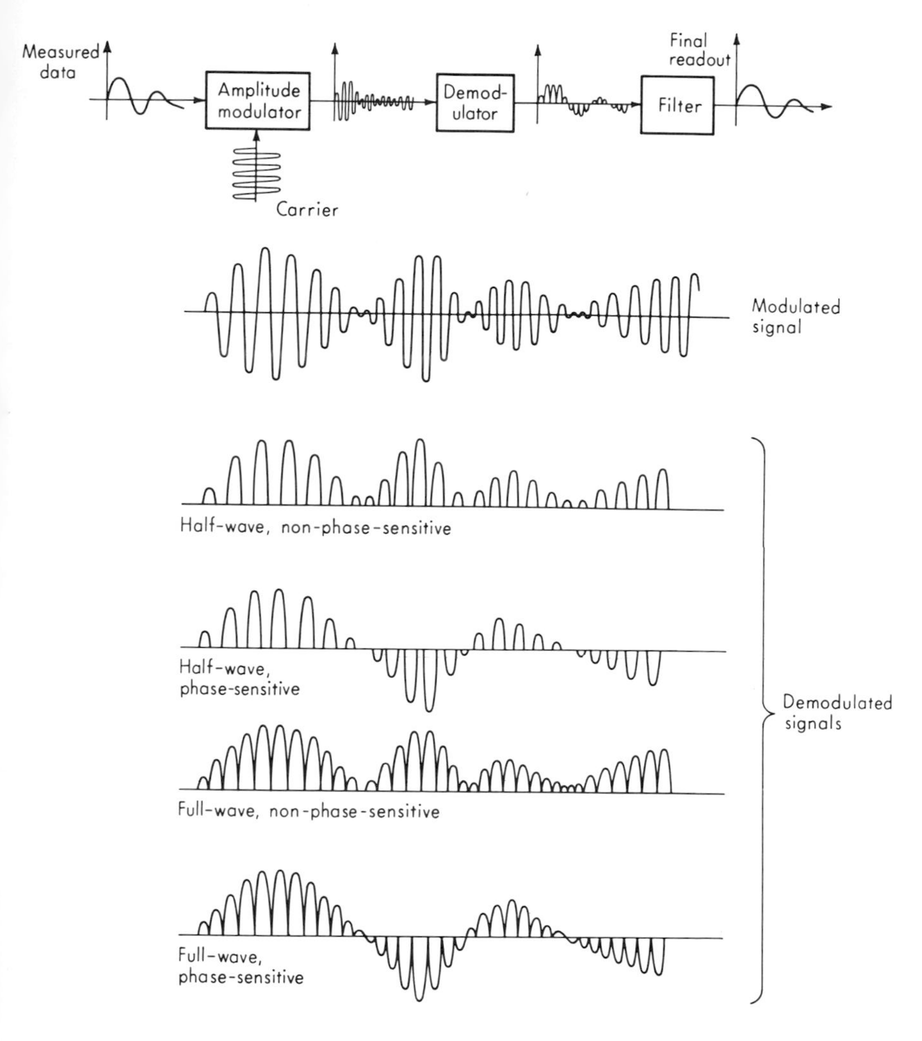

Fig. 3.87. *Types of demodulation.*

sine wave, periodic function, or transient. For a single sine wave $A_s \sin \omega_s t$ which is modulating a carrier $A_c \sin \omega_c t$ the expression for the full-wave phase-sensitive demodulated signal is

$$\text{Demodulator output} = (A_s \sin \omega_s t)|A_c \sin \omega_c t| \qquad (3.306)$$

as seen from Fig. 3.88*a*. Now $|A_c \sin \omega_c t|$ is a periodic function and may be expanded in a Fourier series by application of Eq. (3.268). The results

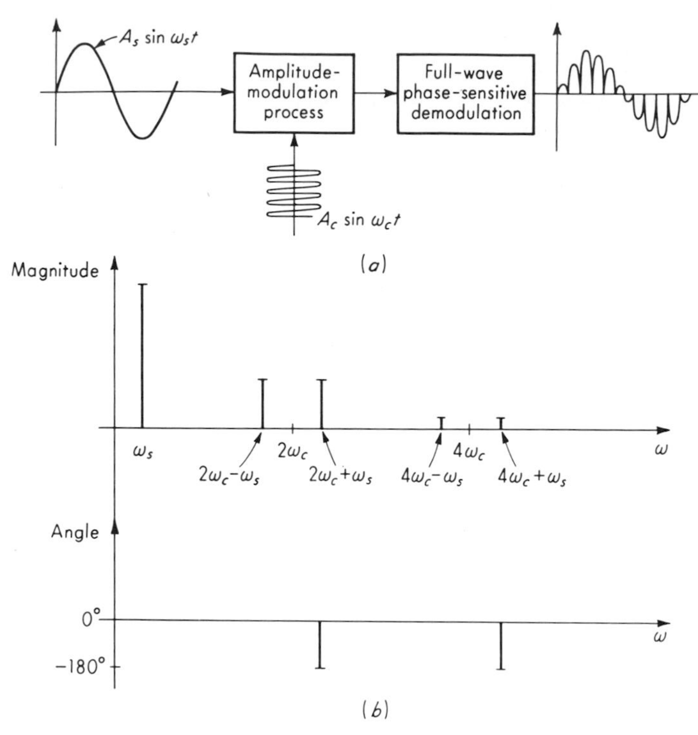

Fig. 3.88. *Frequency spectrum of full-wave phase-sensitive demodulation.*

are

$$|A_c \sin \omega_c t| = \frac{2A_c}{\pi}\left(1 - \tfrac{2}{3}\cos 2\omega_c t - \tfrac{2}{15}\cos 4\omega_c t + \cdots + \frac{2}{1-4n^2}\cos 2n\omega_c t + \cdots\right) \qquad n = 3, 4, 5, \ldots \tag{3.307}$$

Equation (3.306) can then be written as

$$\text{Demodulator output} = (A_s \sin \omega_s t)\left[\frac{2A_c}{\pi}\left(1 - \tfrac{2}{3}\cos 2\omega_c t - \tfrac{2}{15}\cos 4\omega_c t + \cdots\right)\right] \tag{3.308}$$

which when multiplied gives

$$\text{Demodulator output} = \frac{2A_cA_s}{\pi}\sin \omega_s t - \frac{4A_cA_s}{3\pi}\sin \omega_s t \cos 2\omega_c t - \frac{4A_cA_s}{15\pi}\sin \omega_s t \cos 4\omega_c t + \cdots \tag{3.309}$$

Now, terms of the form $(\sin \omega_s t)(\cos 2n\omega_c t)$ can, by a trigonometric identity, be written as $[\sin (2n\omega_c + \omega_s)t - \sin (2n\omega_c - \omega_s)t]/2$. We can thus write Eq. (3.309) as

$$\text{Demodulator output} = \frac{2A_cA_s}{\pi} \sin \omega_s t - \frac{2A_cA_s}{3\pi} [\sin (2\omega_c + \omega_s)t - \sin (2\omega_c - \omega_s)t] - \frac{2A_cA_s}{15\pi} [\sin (4\omega_c + \omega_s)t - \sin (4\omega_c - \omega_s)t] + \cdots \tag{3.310}$$

From this we see that the frequency spectrum of the demodulator output signal is a discrete spectrum with frequency content at ω_s, $2\omega_c \pm \omega_s$, $4\omega_c \pm \omega_s$, etc., as shown in Fig. 3.88*b*. If this signal were an input to a system of known frequency response (such as the filter of Fig. 3.87) the output of this system may be found by the methods of Fig. 3.73. If the output of the filter is to look like the original data, the filter must be designed to *reject* the frequencies $2\omega_c \pm \omega_s$, $4\omega_c \pm \omega_s$, etc., while *passing* with a minimum of distortion the signal frequency ω_s. The design of such a low-pass filter is made simpler if the passband and the rejection band are more widely separated. This is the basis of our earlier statement that carrier frequencies are usually chosen to be 5 to 10 times the highest expected signal frequency.

When the modulating signal is a periodic wave rather than a single sine wave a procedure similar to that just used is employed, except now the modulating signal is *also* expressed as a Fourier series of the form

$$\text{Modulating signal} = A_{s0} + A_{s1} \sin (\omega_s t + \alpha_1) + A_{s2} \sin (2\omega_s t + \alpha_2) + \cdots \tag{3.311}$$

When this is multiplied by $|A_c \sin \omega_c t|$ as given by Eq. (3.307), we find exactly the same situation as for a single sine wave, but it must be applied for *each* signal frequency ($0, \omega_s, 2\omega_s, 3\omega_s$, etc.). The frequency spectrum of the demodulated signal will thus be a discrete spectrum with frequency content at $\omega = 0$, (ω_s, $2\omega_c \pm \omega_s$, $4\omega_c \pm \omega_s$, $6\omega_c \pm \omega_s$, etc.), ($2\omega_s$, $2\omega_c \pm 2\omega_s$, $4\omega_c \pm 2\omega_s$, $6\omega_c \pm 2\omega_s$, etc.), etc. Figure 3.89 illustrates these concepts. Again, if such a signal is applied to a system of known frequency response the methods of Fig. 3.73 allow calculation of the output.

When the modulating signal is a transient $f_i(t)$ the demodulated signal will be $f_i(t)|A_c \sin \omega_c t|$, which can be written as

$$\text{Demodulated signal} = f_i(t) \left[\frac{2A_c}{\pi} \left(1 - \tfrac{2}{3} \cos 2\omega_c t - \tfrac{2}{15} \cos 4\omega_c t + \cdots \right) \right] \tag{3.312}$$

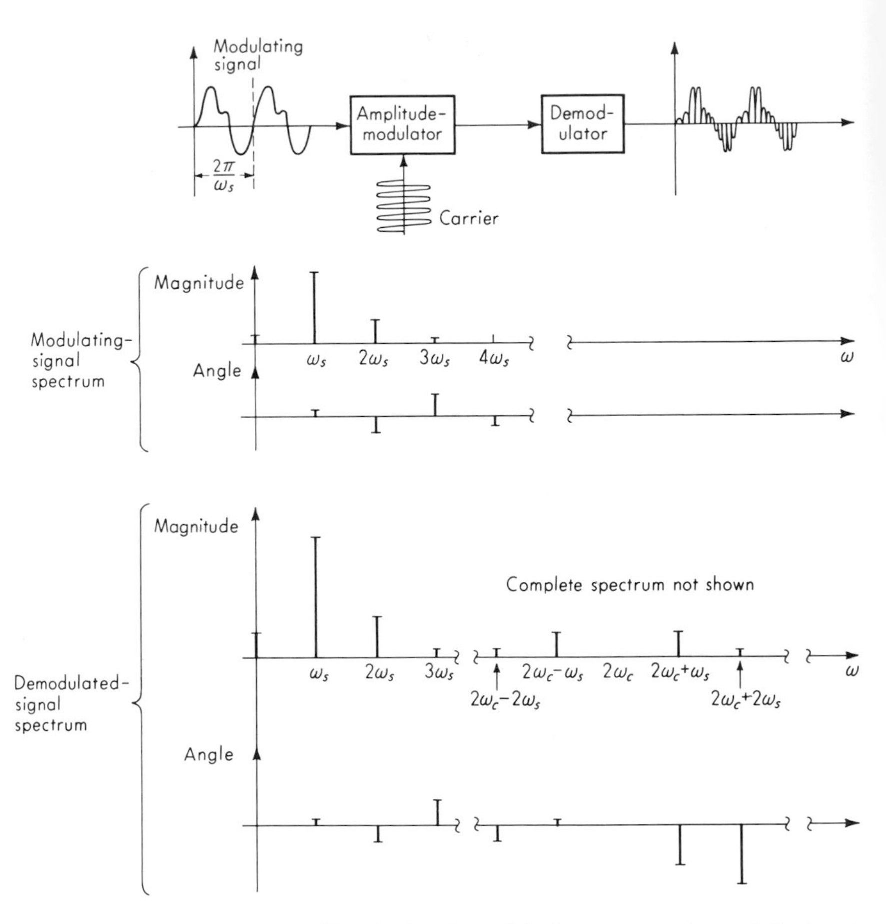

Fig. 3.89. Demodulation spectrum for periodic input.

or, multiplying,

$$\text{Demodulated signal} = \frac{2A_c}{\pi} f_i(t) - \frac{4A_c}{3\pi} f_i(t) \cos 2\omega_c t - \frac{4A_c}{15\pi} f_i(t) \cos 4\omega_c t + \cdots \tag{3.313}$$

Application of the modulation theorem to each of the modulated terms of Eq. (3.313) leads to the result

$$\text{Fourier transform of demodulated signal} \triangleq |Q_i(i\omega)| \underline{/Q_i(i\omega)}$$

where

$$|Q_i(i\omega)| = \frac{2A_c}{\pi} |F_i(i\omega)| + \frac{2A_c}{3\pi} |F_i[i(\omega - 2\omega_c)]| + \frac{2A_c}{15\pi} |F_i[i(\omega - 4\omega_c)]| + \cdots \tag{3.314}$$

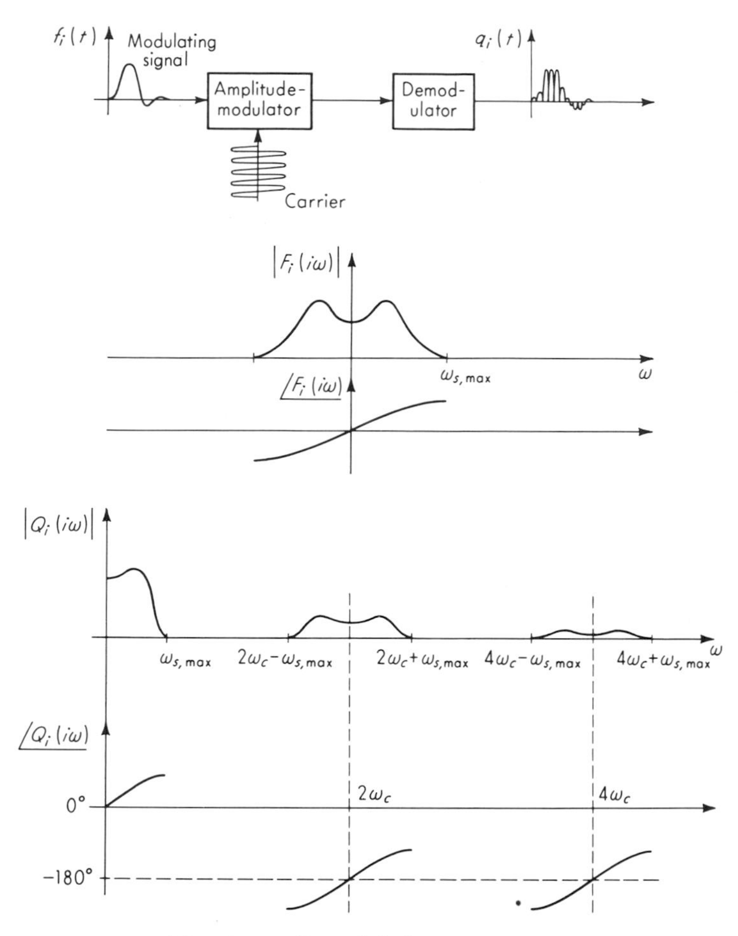

Fig. 3.90. *Demodulation spectrum for transient input.*

and $0 \leq \omega \leq \infty$. The expression for $\angle Q_i(i\omega)$ is not easily given since in Fourier-transforming Eq. (3.313) (each term may be treated separately since superposition is allowed) we find $Q_i(i\omega)$ as a *sum* of complex numbers rather than a product. To find the overall angle of a sum of complex numbers one must express *each* number as $a_i + ib_i$ and then add the a's and b's to get the overall $a + ib$. The angle is then $\tan^{-1}(b/a)$. The angle due to transforming $f_i(t)$ is $\angle F_i(i\omega)$, the angle due to transforming $-f_i(t) \cos 2\omega_c t$ is $-180° + \angle F_i[i(\omega - 2\omega_c)]$, the angle due to transforming $-f_i(t) \cos 4\omega_c t$ is $-180° + \angle F_i[i(\omega - 4\omega_c)]$, etc. These results allow calculation of $\angle Q_i(i\omega)$. The frequency spectrum of such a signal is shown in Fig. 3.90. Since it is a continuous spectrum, the response of a system to such an input may be found by the methods of Fig. 3.80. In Fig. 3.90

it is assumed that the $F_i(i\omega)$ is practically zero for $\omega > \omega_{s,\max}$ and that $\omega_c > \omega_{s,\max}$. For such a situation the phase-angle calculation is greatly simplified, since the individual terms of Eq. (3.314) do not coexist over any frequency range. That is, for $0 < \omega < \omega_{s,\max}$, only the first term is nonzero; for $(2\omega_c - \omega_{s,\max}) < \omega < (2\omega_c + \omega_{s,\max})$, only the second term is nonzero, etc. Thus the overall phase angle is determined by one term only within each of the specified frequency bands.

This concludes our treatment of amplitude-modulated and -demodulated signals. The methods developed can be readily applied to the other common variations such as half-wave demodulation, non-phase-sensitive demodulation, non-sinusoidal-carrier wave forms (square wave, for instance), etc. Non-phase-sensitive demodulation cannot detect a sign change in the modulating signal. Half-wave systems shift the side frequencies of demodulated signals to $\omega_c \pm \omega_s$ rather than $2\omega_c \pm \omega_s$, thus making the filtering problem more difficult. They also have less amplitude than full-wave systems. Nonsinusoidal carriers may be useful in reducing heating effects in resistive transducers such as strain gages. The carrier wave form is such as to give a high ratio of peak value to rms (effective heating) value. The output signal is related to peak value while the power dissipated in the strain gage is related to rms value; thus a high peak/rms ratio increases output for a given allowable heating level. A carrier in the form of a train of high, narrow pulses satisfies this sort of criterion.

Characteristics of random signals. The final class of signals we consider is the so-called random or stochastic type. Such signals are of increasing importance since they serve as more realistic mathematical models of many physical processes than do deterministic signals. By a random signal we shall mean one that can be described only statistically before it actually occurs; that is, it cannot be described by a specific function of time prior to its occurrence. Of course, *all* signals in the real world have some degree of randomness, so that we should be clear that we are *always* dealing with random signals even though we may take a specific time function (periodic, transient, etc.) as a *model* of what is really going on to simplify analysis in some types of problems.

Figure 3.91*a* shows a time record of a typical random signal such as might be measured by a vibration pickup mounted on a booster-rocket structure subjected to acoustic-pressure forces generated by the rocket exhaust. These pressures are strongly random in that no specific frequencies are apparent in a time record. The stresses caused by such pressures can lead to fatigue failure of the structure. Thus engineers are concerned with means to analyze the effects of such random forcing functions on structures and machinery. Since experimental methods are

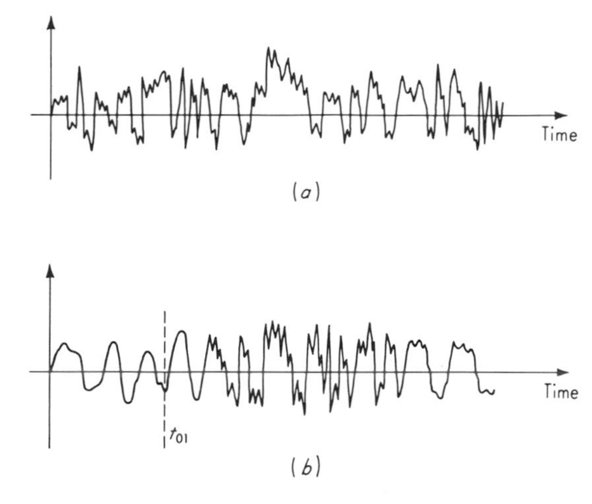

Fig. 3.91. *Random signal.*

needed in much of this work, accurate measurement of random signals is important. We shall see that frequency-response techniques are again most useful for this type of problem.

First, we should note that, as far as deciding on the requirements for a measuring system is concerned, *after* a random process has occurred and there is a time record of it, one can define the function as zero for some time $t > t_0$, treat it as a transient, and calculate its Fourier transform to determine its frequency content and thus the required frequency response of the instrumentation. The only problem here is the selection of the cutoff time t_0 so as to have an adequate statistical "sample" of the random function. In Fig. 3.91*b*, for example, if we based our calculation on the record for $0 < t < t_{01}$ we should miss completely the high-frequency character apparent in the record for $t > t_{01}$. The existence of some minimum valid cutoff time t_0 implies that the random process is *stationary;* that is, its statistical properties (such as average value, mean-square value, etc.) do not change with time. When this is true, there will exist some t_0 corresponding to a chosen level of confidence in the results.

We should be clear that, in dealing with random processes, theoretically an *infinite* record length is needed to give precise results, and results based on finite-length records must always be qualified by statistical statements referring to the *probability* of the result being correct within a certain percentage. This situation leads to an engineering trade-off since long records are desirable from accuracy considerations but are undesirable in terms of the cost involved in obtaining and analyzing them. Also, in many situations the maximum available length of record is limited by the lifetime of the device under study, which in the case of a missile or rocket may be quite short. If an adequately long record is available, the minimum allowable t_0 can be found by choosing a small t_0 and calculating the Fourier transform. Then a larger t_0 is chosen and the Fourier transform again evaluated. If the second transform differs significantly from the first, the first t_0 was not long enough. By choosing successively longer t_0's one will find some range of t_0 beyond which further increases in t_0 cause no significant change in the transform. Any t_0 beyond this range would thus be considered acceptably long. (In actual practice, more refined statistical procedures may be required, and are available,[1] to resolve questions of this sort.)

While the above concepts are adequate for understanding the requirements put on measurement systems for random variables, they do not cover the means available for statistically *describing* the signals. That is, while the exact form of a random function cannot be predicted ahead of time, certain of its statistical characteristics *can* be predicted; these may be useful in predicting (in a statistical way) the output of some physical system that has the random variable as an input. We now develop some of the more common methods of statistically describing random signals. These methods fall mainly into two groups: those concerned with describing the magnitude of the variable and those concerned with describing the rapidity of change (frequency content) of the variable.

If we call a random variable $q_i'(t)$, the *average* or *mean value* $\overline{q_i'(t)}$ is defined by

$$\overline{q_i'(t)} \triangleq \lim_{T \to \infty} \left(\frac{1}{T}\right) \int_0^T q_i(t)\, dt \tag{3.315}$$

Since this can be thought of as a constant component of the total signal, it is not random and is usually subtracted from the total signal to give a signal with zero mean value. Also, many real random processes inherently have zero mean value. For these reasons, from here on we consider only signals $q_i(t)$ with zero mean value. An indication of the magnitude of the random variable is the *mean-squared value* $\overline{q_i^2(t)}$ given by

$$\overline{q_i^2(t)} \triangleq \lim_{T \to \infty} \left(\frac{1}{T}\right) \int_0^T q_i^2(t)\, dt \tag{3.316}$$

The mean-squared value has dimensions of $[q_i(t)]^2$. Thus to get a measure of the size of $q_i(t)$ itself, the *root-mean-square* (rms) value $q_i(t)_{\text{rms}}$ is defined by

$$q_i(t)_{\text{rms}} \triangleq \sqrt{\overline{q_i^2(t)}} \tag{3.317}$$

[1] J. S. Bendat, L. D. Enochson, and A. G. Piersol, Analytical Study of Vibration Data Reduction Methods, *NASA*, *N*64-15529, 1963.

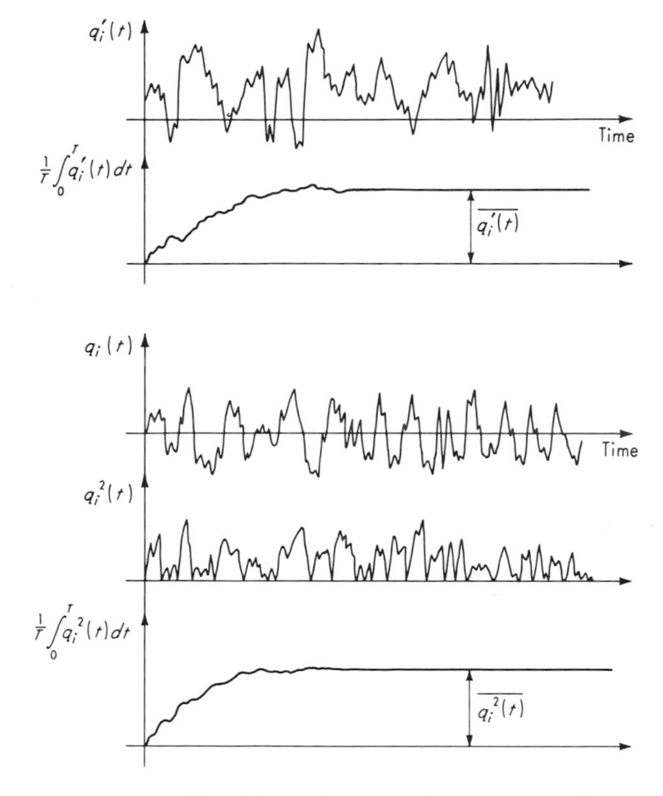

Fig. 3.92. *Average and mean-square values.*

The above definitions are illustrated in Fig. 3.92. The quantities $\overline{q_i^2(t)}$ and $q_i(t)_{\text{rms}}$ give an indication of the overall size of $q_i(t)$ but no clue as to the distribution of "amplitude," that is, the probability of occurrence of large or small values of $q_i(t)$. The specification of this important information is provided by the *amplitude-distribution function* (probability density function) $W_1(q_i)$. To define this function, we consider Fig. 3.93. We define the probability P that $q_i(t)$ will be found between some specific value q_{i1} and $q_{i1} + \Delta q_i$ by

$$\text{Probability } [q_{i1} < q_i < (q_{i1} + \Delta q_i)] \triangleq P[q_{i1}, (q_{i1} + \Delta q_i)] = \lim_{T\to\infty} \frac{\Sigma\, \Delta t_i}{T} \tag{3.318}$$

where $\Sigma\, \Delta t_i$ represents the total time spent by $q_i(t)$ within the band Δq_i during the time interval T. We now define $W_1(q_i)$ by

$$\text{Amplitude-distribution function} \triangleq W_1(q_i) \triangleq \lim_{\Delta q_i\to 0} \frac{P[q_i, (q_i + \Delta q_i)]}{\Delta q_i} \tag{3.319}$$

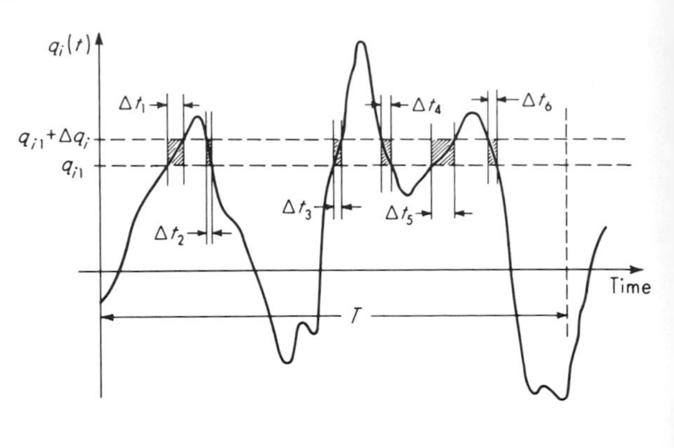

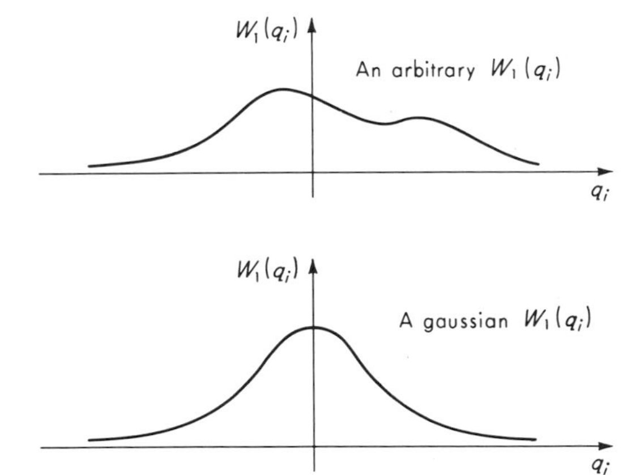

Fig. 3.93. Probability-density-function definition.

From this definition it should be clear that

$$W_1(q_i)\, dq_i = \text{probability that } q_i \text{ lies in } dq_i \tag{3.320}$$

and thus

$$\int_{q_{i1}}^{q_{i2}} W_1(q_i)\, dq_i = \text{probability } q_i \text{ lies between } q_{i1} \text{ and } q_{i2} \tag{3.321}$$

The function $W_1(q_i)$ can theoretically take on an infinite number of different forms; however, certain forms have been found to be adequate mathematical models for real physical processes. The most common of these is the gaussian or normal distribution given by

$$W_1(q_i) = \frac{1}{\sqrt{2\pi}\,\sigma} e^{-q_i^2/2\sigma^2} \tag{3.322}$$

where $\sigma \triangleq$ standard deviation. Whether a given random process closely approximates this form must usually be found experimentally by means of instruments based on Eqs. (3.318) and (3.319). Since the limiting processes in these equations can never be exactly realized in a physical instrument, we again must be satisfied with statements regarding the *probability* that a process is gaussian rather than the *certainty* that it is.

While the quantities $q_i(t)_{\text{rms}}$ and $W_1(q_i)$ are usually sufficient to describe the magnitude of a random variable, they give no indication as to the *rapidity* of variation in time. That is, two random processes could both be gaussian with the same numerical value of σ but one could be much more rapidly varying than the other. To describe the time aspect of random variables the concepts of *autocorrelation function* and *mean-square spectral density* (power spectral density) are employed.[1] The autocorrelation function $R(\tau)$ of a random variable $q_i(t)$ is given by

$$R(\tau) \triangleq \lim_{T\to\infty} \left(\frac{1}{T}\right) \int_0^T q_i(t)q_i(t+\tau)\,dt \qquad (3.323)$$

The function $q_i(t+\tau)$ is simply $q_i(t)$ shifted in time by τ sec. Thus, to find $R(\tau)$ one selects a value of τ, say 2 sec, plots against t the functions $q_i(t)$ and $q_i(t+2)$, multiplies them together point by point, integrates the product curve from 0 to T, and divides by T. This procedure is then repeated for other values of τ to generate the curve $R(\tau)$ versus τ. In actual practice, the shifting of $q_i(t)$ by τ sec is sometimes accomplished by writing $q_i(t)$ on magnetic tape at one point and reading it off at another. The time delay τ is then simply

$$\tau = \frac{\text{distance between read and write heads}}{\text{tape velocity}}$$

The multiplication of $q_i(t)$ and $q_i(t+\tau)$, the integration, and the division by T can all be accomplished by standard electronic-analog-computer components. [Equation (3.323) can also be implemented on a digital computer successfully.] Figure 3.94 illustrates these concepts. For $\tau = 0$, Eq. (3.323) gives $R(0) = \overline{q_i^2(t)}$; that is, the autocorrelation function is numerically equal to the mean-square value for $\tau = 0$.

To appreciate the relation between $R(\tau)$ and the rapidity of variation of $q_i(t)$, consider Fig. 3.95 where both a slowly varying and a rapidly varying $q_i(t)$ are shown. For *any* $q_i(t)$, fast or slow, when $\tau = 0$, $q_i(t)q_i(t+\tau)$ is *positive* for all t. Thus integration gives the largest possible value, $\overline{q_i^2(t)}$. Any shift ($\tau \neq 0$) will "misalign" the positive and negative parts of $q_i(t)$ and $q_i(t+\tau)$, causing the product curve to be sometimes positive, sometimes negative. Thus integration of this curve gives a smaller value than for $\tau = 0$. If $q_i(t)$ is rapidly varying, it takes

[1] S. H. Crandall, "Random Vibration," John Wiley & Sons, Inc., New York, 1958.

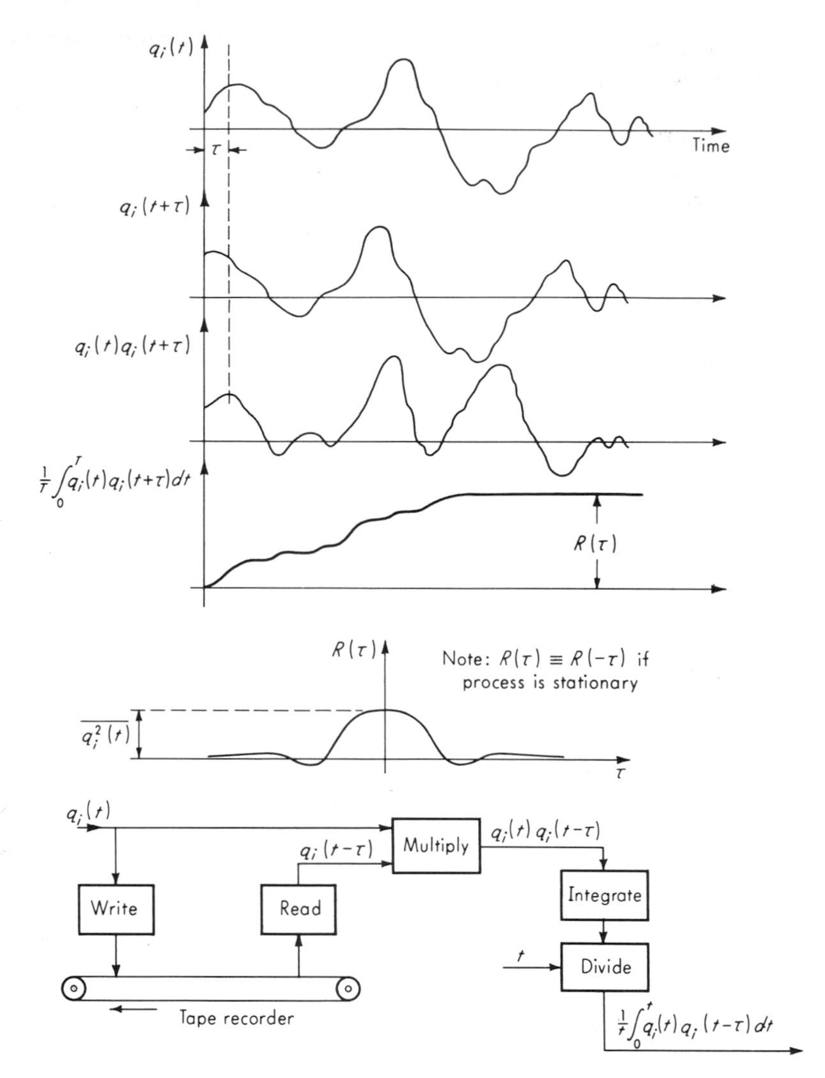

Fig. 3.94.. *Autocorrelation-function definition.*

only a small shift (small τ) to cause this misalignment, whereas a slowly varying $q_i(t)$ requires a larger shift before $R(\tau)$ drops off significantly. Thus a sharp peak in $R(\tau)$ at $\tau = 0$ indicates the presence of rapid variation (strong high-frequency content) in $q_i(t)$.

The mean-square spectral density (power spectral density) is another method of determining the frequency content of a random signal. The mean-square spectral density is proportional to the Fourier transform of $R(\tau)$ and conveys in the frequency domain exactly the same information

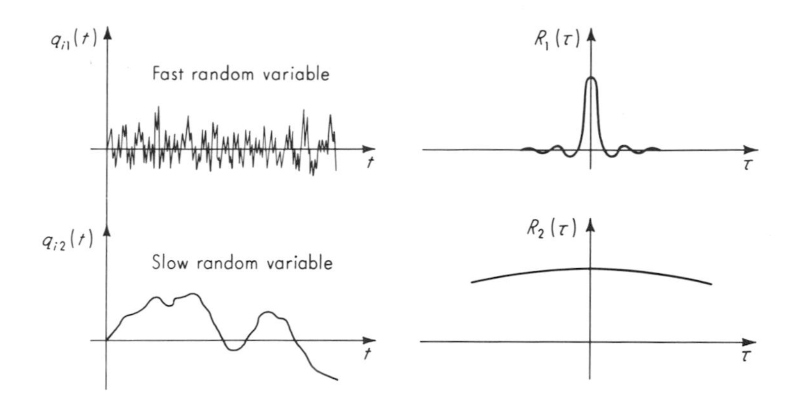

Fig. 3.95. *Frequency significance of autocorrelation function.*

that $R(\tau)$ conveys in the time domain. While they are mathematically related, in actual practice where they must be determined experimentally, one or the other may be preferable. It appears that in random-vibration work the mean-square-spectral-density approach is largely preferred. To develop this concept, let us consider first a periodic function expanded into a Fourier series to give

$$q_i(t) = A_{i1} \sin (\omega_1 t + \alpha_1) + A_{i2} \sin (2\omega_1 t + \alpha_2) + \cdots \tag{3.324}$$

It is easy to show that the total mean-square value of $q_i(t)$ is equal to the sum of the individual mean-square values for each of the harmonic terms:

$$\overline{q_i^2(t)} = \overline{q_{i1}^2(t)} + \overline{q_{i2}^2(t)} + \cdots \tag{3.325}$$

Thus the contribution of each frequency to the overall mean-square value is easily found.

We should like now to develop, for a *random* function, a related technique that will show how the total mean-square value is "distributed" over the frequency range. We first note that for a random function no isolated, discrete frequencies exist; thus the frequency spectrum is a continuous one. The concept of mean-square spectral density is perhaps most clearly visualized in terms of the instrumentation used to measure it experimentally. Figure 3.96*a* shows the arrangement necessary to measure the *overall* mean-square value of a signal $q_i(t)$. To find out how much each part of the frequency range contributes to this overall value we simply filter out (with a narrow-band-pass filter of bandwidth $\Delta\omega$) all (ideally) frequencies other than the narrow band of interest and then perform the squaring and averaging operations on what remains, as in

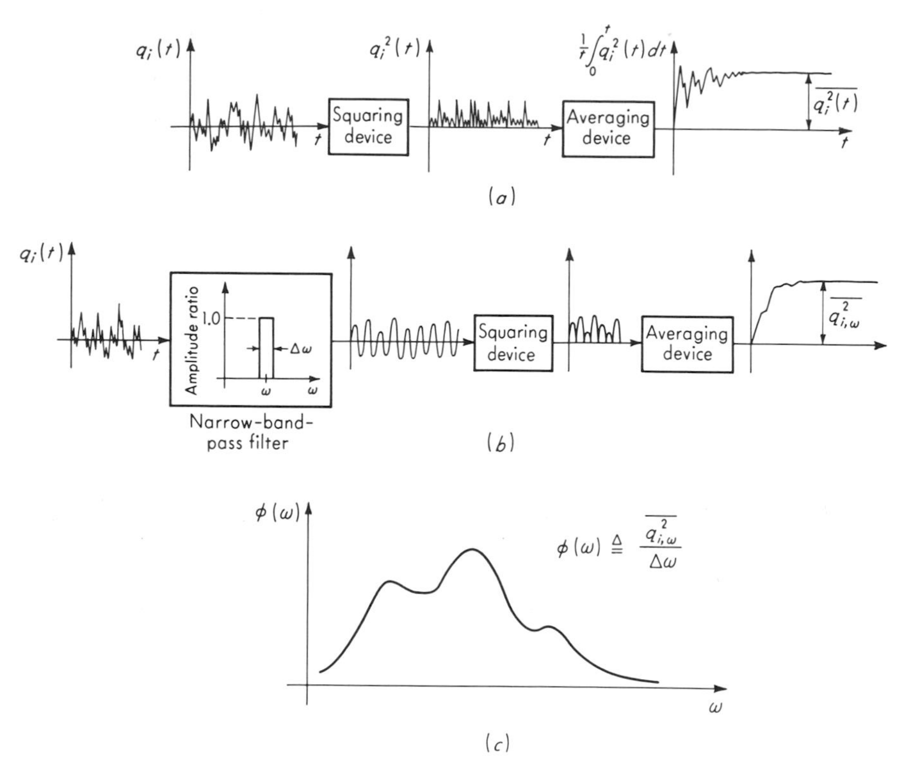

Fig. 3.96. *Mean-square spectral-density definition.*

Fig. 3.96*b*. The results of this operation are thus the mean-square value of that part of the signal $q_i(t)$ lying within the chosen frequency band. We call this value $\overline{q_{i,\omega}^2}$. The filter is adjustable in the sense that we can shift its passband anywhere along the frequency axis, and so we can obtain $\overline{q_{i,\omega}^2}$ for any chosen center frequency ω. A narrow passband is desirable for resolving closely spaced peaks in the spectrum but is undesirable in terms of the increased time required to cover a given frequency range in small steps rather than large. Thus a compromise is needed. The mean-square spectral density $\phi(\omega)$ is defined by

$$\phi(\omega) \triangleq \frac{\overline{q_{i,\omega}^2}}{\Delta\omega} \tag{3.326}$$

and represents the "density" (amount per unit frequency bandwidth) of the mean-square value since $\phi(\omega)\,\Delta\omega = \overline{q_{i,\omega}^2}$. If we evaluate $\phi(\omega)$ for a whole range of frequencies we can plot it as a curve versus ω as in Fig. 3.96*c*. Note that the dimensions of $\phi(\omega)$ are those of $q_i^2(t)/(\text{rad/sec})$.

The total area under the $\phi(\omega)$-versus-ω curve will be the total mean-square value $\overline{q_i^2(t)}$.

The reader will find that in the literature the term power spectral density is used almost exclusively for the quantity we have called mean-square spectral density, except that power spectral density is π times $\phi(\omega)$. This is a carry-over from communications engineering where the concept was originally developed and where the signal $q_i(t)$ is a voltage applied to a 1-ohm resistor. Under these conditions $\phi(\omega)$ would have dimensions of power/(rad/sec). Actually, however, the concept of $\phi(\omega)$ is a *mathematical* one related to the mean-square value and *not* a physical one related to electrical engineering. In physical applications the dimensions of $\phi(\omega)$ would be $(°F)^2$/(rad/sec) if $q_i(t)$ is temperature, g^2/(rad/sec) if $q_i(t)$ is acceleration, etc., which are in general *not* power/(rad/sec). However, the term power spectral density seems to be firmly entrenched and probably will continue to prevail. The author here merely suggests that mean-square spectral density might be more appropriate terminology. Thus when dealing with, say, random pressures, one would refer to the "mean-square spectral density of pressure."

Perhaps the main interest in the mean-square spectral density lies in the fact that if a $q_i(t)$ with a known $\phi_i(\omega)$ is applied as input to a linear system of known frequency response, the mean-square spectral density $\phi_o(\omega)$ of the output is easily computed from the relation[1]

$$\phi_o(\omega) = \phi_i(\omega) \left| \frac{q_o}{q_i}(i\omega) \right|^2 \qquad (3.327)$$

We can thus compute the $\phi_o(\omega)$ curve point by point; see Fig. 3.97. The area under this curve will be the total mean-square value of $q_o(t)$. Furthermore, if the input is gaussian the output will also be gaussian. It is then possible to make statements such as

$$\begin{aligned} |q_o(t)| &> \sqrt{\overline{q_o^2(t)}} && 31.7\% \text{ of the time} \\ |q_o(t)| &> 2\sqrt{\overline{q_o^2(t)}} && 4.6\% \text{ of the time} \\ |q_o(t)| &> 3\sqrt{\overline{q_o^2(t)}} && 0.3\% \text{ of the time} \end{aligned} \qquad (3.328)$$

Other useful results of this nature can be found in the literature.[2]

A particular form of $\phi(\omega)$ is of great usefulness in practice. This is the so-called *white noise*. For a mathematically perfect white noise $\phi(\omega)$ is equal to a constant for all frequencies; that is, it contains all frequencies in equal amounts (see Fig. 3.98). This makes it useful as a test signal, just as the impulse is useful as a transient test signal because of its

[1] Aseltine, *op. cit.*

[2] Bendat, Enochson, and Piersol, *op. cit.*

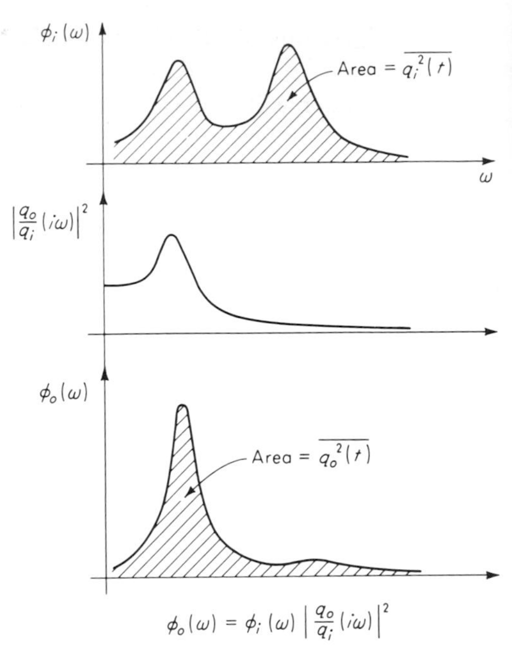

Fig. 3.97. *System response to random input.*

uniform frequency content. From Eq. (3.327), if $\phi_i(\omega) = C$

$$\phi_o(\omega) = C\left|\frac{q_o}{q_i}(i\omega)\right|^2 \tag{3.329}$$

and thus

$$\left|\frac{q_o}{q_i}(i\omega)\right| = \sqrt{\frac{\phi_o(\omega)}{C}} \tag{3.330}$$

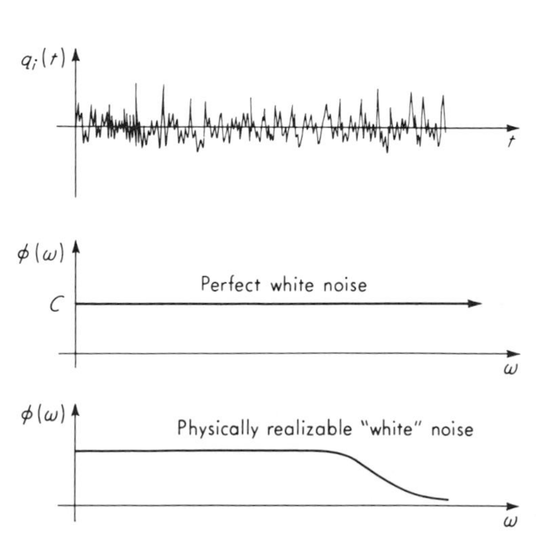

Fig. 3.98. *White noise.*

While it is not possible to build a generator of perfect white noise, the flat range of $\phi(\omega)$ of a practical generator can be quite large, and as long as its flat range extends beyond the frequency response of the system being tested, the "nonwhiteness" will not present any difficulty. If a white-noise generator is available, one can "construct" almost any $\phi(\omega)$ that he wishes simply by passing the white noise through a suitably designed filter according to Eq. (3.329). That is, $|(q_o/q_i)(i\omega)|^2$ for the filter must have the shape desired in $\phi(\omega)$. This technique is widely used in analog-computer simulation studies of aircraft gust response, control-system evaluation, etc., where the frequency characteristics of some random-input quantity are known and it is desired to study their effect on some system.

Most of the previous material has referred to statistical properties of a *single* random variable. Useful practical results may be derived by consideration of two random variables. The *joint amplitude-distribution function* (joint probability density function) of two random variables $q_1(t)$ and $q_2(t)$ is given by

$$W_1(q_1,q_2) = \lim_{T\to\infty} \lim_{\substack{\Delta q_1 \to 0 \\ \Delta q_2 \to 0}} \frac{1}{T\,\Delta q_1\,\Delta q_2} \Sigma\, \Delta t_i \qquad (3.331)$$

where $\Sigma\, \Delta t_i$ represents the total time (during the time T) that $q_1(t)$ and $q_2(t)$ spent *simultaneously* in the bands $q_1 + \Delta q_1$ and $q_2 + \Delta q_2$. Figure 3.99 illustrates these concepts. From the basic definition it should be clear that

$$\text{Probability } (q_{1a} < q_1 < q_{1b},\ q_{2a} < q_2 < q_{2b}) = \int_{q_{2a}}^{q_{2b}} \int_{q_{1a}}^{q_{1b}} W_1(q_1,q_2)\, dq_1\, dq_2 \qquad (3.332)$$

Again, $W_1(q_1,q_2)$ can take an infinite variety of forms. The most useful is probably the bivariate gaussian (normal) distribution.[1] The main purpose in experimentally measuring $W_1(q_1,q_2)$ is, just as for $W_1(q_i)$, to determine whether the physical data follow approximately some simple mathematical form such as the gaussian. If this can be proved, many useful theoretical results can be applied. Also, certain calculations can be made directly from $W_1(q_1,q_2)$. For example, if q_1 and q_2 represent the random vibratory motions of two adjacent machine parts, knowledge of $W_1(q_1,q_2)$ allows calculation of the probability that the two parts will strike each other. In actual practice, engineering applications of $W_1(q_1,q_2)$ are somewhat limited because of the difficulty of measuring this function.

[1] A. M. Mood, "Introduction to the Theory of Statistics," p. 165, McGraw-Hill Book Company, New York, 1950.

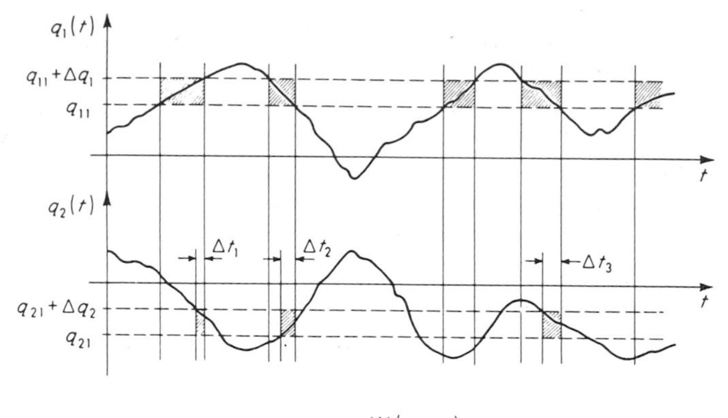

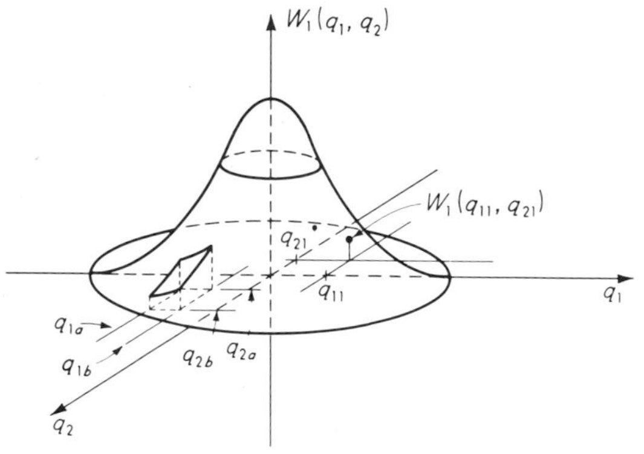

Fig. 3.99. *Bivariate probability density function.*

The *cross-correlation function* $R_{q_1q_2}(\tau)$ for two random variables $q_1(t)$ and $q_2(t)$ is defined by

$$R_{q_1q_2}(\tau) = \lim_{T\to\infty} \left(\frac{1}{T}\right) \int_0^T q_1(t)q_2(t+\tau)\,dt \qquad (3.333)$$

An example[1] of its application might be as follows: Suppose a source of vibratory motion $q_1(t)$ exists at one point in a structure and causes a vibratory response motion $q_2(t)$ at another point in the structure. Suppose also that the transmission of vibration from the first point to the second could occur by either (or both) of two mechanisms. The first mechanism is by acoustic (air-pressure) wave propagation through the air separating the two points; the second is by elastic wave propagation through the metallic structure connecting the two points. Since the propagation velocities of waves in air and metal are greatly different, one would expect that an input at q_1 would cause an output at q_2 that would be

[1] Bendat, Enochson, and Piersol, *op. cit.*

delayed by different times, depending on whether the transmission was mainly through the air or mainly through the metal. If one knows the transmission path length and the wave velocity, these delays can be calculated. Suppose the air-path delay is 0.01 sec and the structure-path delay is 0.002 sec. If one now experimentally measures $R_{q_1q_2}(\tau)$ and finds a large peak at $\tau = 0.01$ sec and a smaller one at $\tau = 0.002$ sec, he would conclude that the acoustic transmission is responsible for most of the vibration at q_2. Thus, since $R_{q_1q_2}$ is a measure of the correlation ("relatedness") between two signals delayed by various amounts, it can be used as a diagnostic tool for investigating the presence and/or nature of the relation, as in the above example.

Information equivalent to that contained in the cross-correlation function but in a different (and often more practically useful) form is found in the *cross mean-square spectral density* (cross-power spectral density) $\phi_{q_1q_2}(\omega)$ given by

$$\phi_{q_1q_2}(\omega) = C_{q_1q_2}(\omega) - iQ_{q_1q_2}(\omega) \tag{3.334}$$

where $C_{q_1q_2}(\omega) \triangleq$ cospectrum

$$\triangleq \lim_{T\to\infty} \lim_{\Delta\omega\to 0} \frac{1}{T\,\Delta\omega} \int_0^T (q_{1\Delta\omega})(q_{2\Delta\omega})\,dt \tag{3.335}$$

$Q_{q_1q_2}(\omega) \triangleq$ quad spectrum

$$\triangleq \lim_{T\to\infty} \lim_{\Delta\omega\to 0} \frac{1}{T\,\Delta\omega} \int_0^T (q_{1\Delta\omega})_{90^\circ}(q_{2\Delta\omega})\,dt \tag{3.336}$$

$q_{1\Delta\omega} \triangleq$ output of narrow- $(\Delta\omega)$ band filter whose input is $q_1(t)$ (3.337)

$q_{2\Delta\omega} \triangleq$ output of narrow- $(\Delta\omega)$ band filter whose input is $q_2(t)$

$(q_{1\Delta\omega})_{90^\circ} \triangleq$ signal $q_{1\Delta\omega}$ with phase shift of 90° (3.338)

A block diagram of instrumentation necessary to measure $\phi_{q_1q_2}(\omega)$ is given in Fig. 3.100. Note that $\phi_{q_1q_2}(\omega)$ is a *complex* quantity whereas $\phi(\omega)$ is real. Perhaps the main application of the cross mean-square spectral density is in the experimental determination of the sinusoidal transfer function $(q_o/q_i)(i\omega)$ of a linear system. From Eq. (3.327) we see that using the ordinary mean-square spectral density one can find only the magnitude (not the phase angle) of the transfer function if $\phi_i(\omega)$ is known and $\phi_o(\omega)$ is measured. The cross mean-square spectral density determines both magnitude and phase according to the following equation:

$$\frac{q_o}{q_i}(i\omega) = \frac{\phi_{q_iq_o}(\omega)}{\phi_{q_i}(\omega)} \tag{3.339}$$

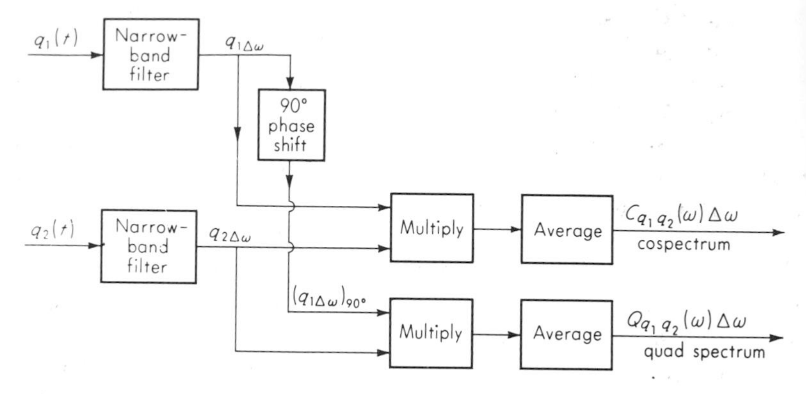

Fig. 3.100. *Cross mean-square spectral density.*

where $\phi_{q_iq_o}(\omega) \triangleq$ cross mean-square spectral density of q_i and q_o
$\phi_{q_i}(\omega) \triangleq$ mean-square spectral density of q_i

We see that it is necessary to measure one (ordinary) mean-square spectral density and one cross mean-square spectral density.

This concludes our treatment of random signals. The most important commonly useful results have been presented and the main terminology developed. Further theoretical and practical details may be found in the literature. The earlier-referenced work of Bendat, Enochson, and Piersol was found particularly useful from a practical-application viewpoint by the author.

Requirements on the instrument transfer function to ensure accurate measurement. Having up to this point expressed many different forms of signals in terms of the common denominator of frequency content, we can now give a general statement of the requirements that a measurement system must meet in order to measure accurately a given form of input.

1. For perfect shape reproduction with no time delay between q_i and q_o and with:
   a. Periodic inputs. $(q_o/q_i)(i\omega)$ must equal $K\underline{/0°}$ for all frequencies contained in q_i with significant amplitude.
   b. Transient inputs. $(q_o/q_i)(i\omega)$ must equal $K\underline{/0°}$ for the entire frequency range in which the Fourier transform of $q_i(t)$ has significant magnitude.
   c. Amplitude-modulated signals. Same criteria as in parts *a* and *b*, depending on whether the modulating signal is periodic or transient.

d. Demodulated signals. $(q_o/q_i)(i\omega)$ for everything following the demodulator should be $K\underline{/0°}$ for all significant frequency bands of the modulating signal and should be zero for all carrier and side frequency bands produced by the modulation process.

e. Random signals. $(q_o/q_i)(i\omega)$ must equal $K\underline{/0°}$ over the entire frequency band where $\phi_{q_i}(\omega)$ is significantly larger than zero.

If the output signal $q_o(t)$ is to be used strictly for measurement rather than as a signal in a feedback-control system, a time delay between q_o and q_i is usually not objectionable as long as the shape of q_i is properly reproduced at q_o. (In control systems, time delay is damaging to stability and therefore is objectionable.) Thus a more easily attained (and usually acceptable) criterion is as follows:

2. For perfect shape reproduction with time delay τ_{dt} between q_i and q_o and with:

a. Periodic inputs.
b. Transient inputs.
c. Amplitude-modulated signals.
d. Demodulated signals.
e. Random signals.

Same as 1 except that wherever $(q_o/q_i)(i\omega) = K\underline{/0°}$ is required now $(q_o/q_i)(i\omega) = K\underline{/-\omega\tau_{dt}}$ is required. *That is, amplitude ratio is constant but phase lag increases linearly with frequency* ω.

The validity of the above statements is readily perceived by consideration of Fig. 3.101. To get the output in the frequency domain, the frequency-domain input is always multiplied by the sinusoidal transfer function. If this transfer function is $K\underline{/-\omega\tau_{dt}}$, the output will have a magnitude equal to K times the input magnitude and an angle equal to the angle of the input minus $\omega\tau_{dt}$. This is exactly what we would get if we passed $q_i(t)$ through a pure gain K followed by a dead time τ_{dt}. Thus when we inverse-transform to get from $Q_o(i\omega)$ to $q_o(t)$, we are *bound* to get $Kq_i(t)$ delayed by τ_{dt} sec. Thus, while the *actual* instrument transfer function will be of the form of Eq. (3.255), over the pertinent range of frequencies it must effectively amount to $K\underline{/-\omega\tau_{dt}}$ if accurate wave-form reproduction is to be expected. This is the basis for the selection of $\zeta \approx 0.6$ to 0.7 in second-order instruments since this range of ζ values makes the amplitude ratio most nearly constant and the phase-angle-versus-frequency curve most nearly linear over the widest possible frequency range for a given ω_n.

It should be noted that the above accuracy criteria do not actually state any *numerical* results. That is, no statement of the form "If the amplitude ratio is flat within $\pm x$ percent and the phase angle linear within $\pm y$ percent over the range of frequencies in which $|Q_i(i\omega)|$ (Fourier-series term coefficient or Fourier-transform magnitude) is greater than

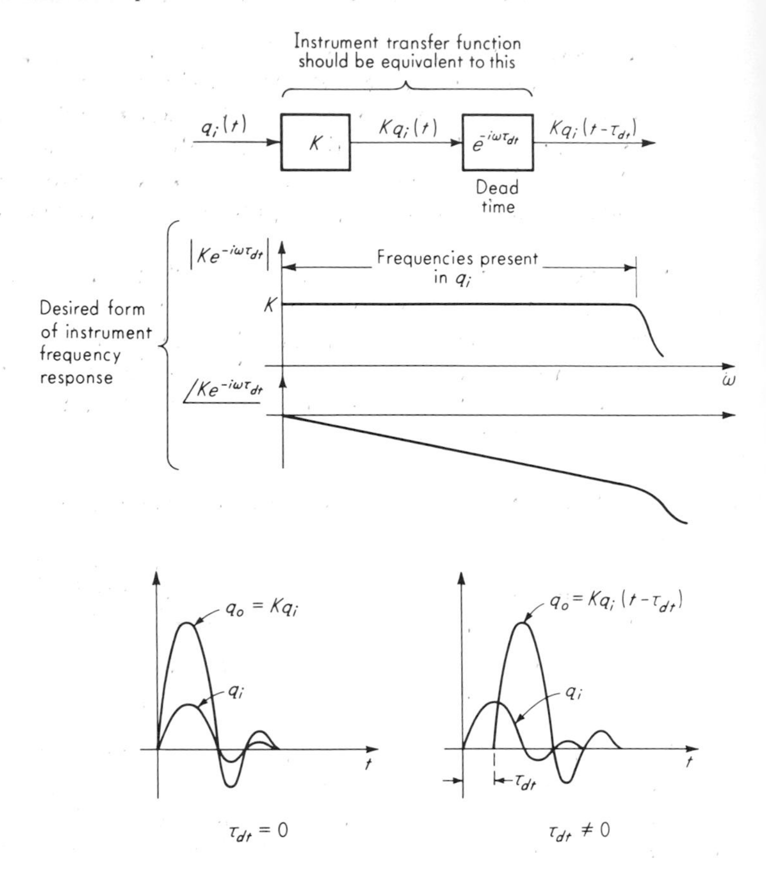

Fig. 3.101. *Requirements for accurate measurement.*

z percent of its maximum value, then $q_o(t)$ will be within $\pm w$ percent of $Kq_i(t - \tau_{dt})$ at all times" is or *can be* made. While this sort of statement would be exceedingly useful, unfortunately it cannot be made in any *general* sense. For specific forms of $q_i(t)$ and $(q_o/q_i)(i\omega)$ one could investigate mathematically the effect of variations in numerical values of system parameters on dynamic accuracy by means of transform methods or directly from the differential equations. The use of analog and/or digital computers may be most helpful in such studies.

As an example, suppose one has a measuring system whose form is known but some (or all) of its numerical values are open to choice. Suppose it is intended to measure a random variable $q_i(t)$ whose $\phi_{q_i}(\omega)$ is known and may be simulated electrically with a white-noise generator and suitable shaping filters. The arrangement of Fig. 3.102a might be used to study the effect of varying measurement-system parameters on the

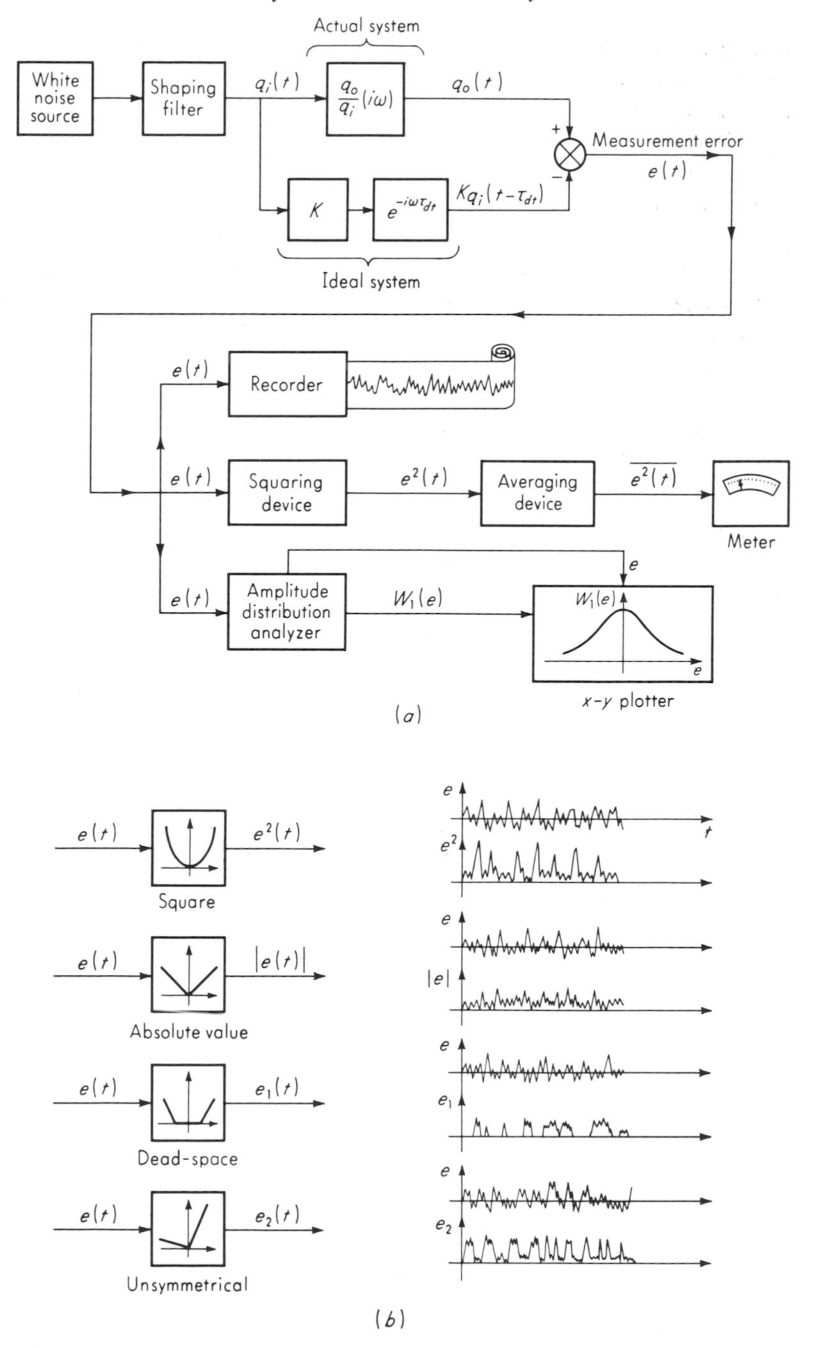

Fig. 3.102. *Experimental system-accuracy analysis.*

measurement error. Several measures of error are made available to the investigator in this setup. An overall measure is provided as a single number by the mean-squared error. A visual display of the error as a function of time is provided by the chart recorder. A statistical analysis of the error amplitude distribution is provided by a suitable computer. This gives $W_1(e)$ which allows prediction of the probability of errors of a particular magnitude. Altogether a rather complete picture of the nature of the error is provided. By changing the parameters of $(q_o/q_i)(i\omega)$ and noting the effect on the various aspects of the error, an optimum measuring system may be found for the particular $q_i(t)$. Similar arrangements could be used for transient or periodic inputs.

The possibility of defining measures of error that are particularly slanted at some specific application should be kept in mind. For example, the mean-squared error weights large errors out of proportion to small ones, because of the squaring operation $[2^2 \neq 2(1^2)]$. If a uniform weighting is desired, the absolute value of error rather than the square might be used. The absolute value of an electrical signal is easily obtained in an analog computer. If errors less than a certain threshold value are of *no* consequence, a dead-space element can be used to give an output only when $|e(t)| > e_{\text{threshold}}$. If negative errors are less important than positive errors, an unsymmetrical element can be used. Figure 3.102*b* illustrates these concepts. Because of the versatility of analog computers, the possibilities are endless.

Numerical correction of dynamic data. Theoretically, if $q_o(t)$ (the actual measured data) is known and if $(q_o/q_i)(i\omega)$ for the measurement system is known one can always reconstruct a *perfect* record of $q_i(t)$ by the following process:

1. Transform $q_o(t)$ to $Q_o(i\omega)$.
2. Apply the formula

$$Q_o(i\omega) = Q_i(i\omega) \frac{q_o}{q_i} (i\omega)$$

   in the inverse sense as

$$Q_i(i\omega) = \frac{Q_o(i\omega)}{(q_o/q_i)(i\omega)}$$

   to find $Q_i(i\omega)$.
3. Inverse-transform $Q_i(i\omega)$ to $q_i(t)$.

This procedure theoretically will give the exact $q_i(t)$ whether the measurement system meets the $K\underline{/-\omega\tau_{dt}}$ requirements or not. In actual practice, of course, while the measurement system does not have to meet $K/\underline{-\omega\tau_{dt}}$, it *does* have to respond fairly strongly to all frequencies present in q_i; otherwise some parts of the q_o frequency spectrum will be so small as to be submerged in the unavoidable "noise" present in all systems and thus be unrecoverable by the above mathematical process. This process is tedious and introduces its own errors; therefore one would not use this approach unless absolutely necessary. However, it is a usable alternative in those situations where measurement systems meeting $K/\underline{-\omega\tau_{dt}}$ cannot be constructed with the present state of the art.

An important variation of the above process has been successfully applied in cases where the primary sensor is inadequate but can be cascaded with frequency-sensitive elements whose transfer functions make up the deficiencies in the primary sensor. The above computations are then in a sense automatically and continuously carried out by the compensating equipment to reconstruct $q_i(t)$. This subject is discussed in detail later under the topic Dynamic Compensation.

Experimental determination of measurement-system parameters. While theoretical analysis of instruments is vital to reveal the basic relationships involved in the operation of a device, it is rarely accurate enough to provide usable numerical values for critical parameters such as sensitivity, time constant, natural frequency, etc. Thus calibration of instrument systems is a necessity. We have already discussed static calibration; we here concentrate on dynamic characteristics.

For zero-order instruments the response is instantaneous and so no dynamic characteristics exist. The only parameter to be determined is the static sensitivity K, which is found by static calibration.

For first-order instruments the static sensitivity K is also found by static calibration. There is only one parameter pertinent to dynamic response, the time constant τ, and this may be found by a variety of methods. One common method applies a step input and measures τ as the time to achieve 63.2 percent of the final value. This method is influenced by inaccuracies in the determination of the $t = 0$ point and also gives no check as to whether the instrument is really first-order. A preferred method uses the data from a step-function test replotted semilogarithmically to get a better estimate of τ and also to check conformity to true first-order response. This method goes as follows: From Eq. (3.178) we can write

$$\frac{q_o - Kq_{is}}{Kq_{is}} = -e^{-t/\tau} \qquad (3.340)$$

$$1 - \frac{q_o}{Kq_{is}} = e^{-t/\tau} \qquad (3.341)$$

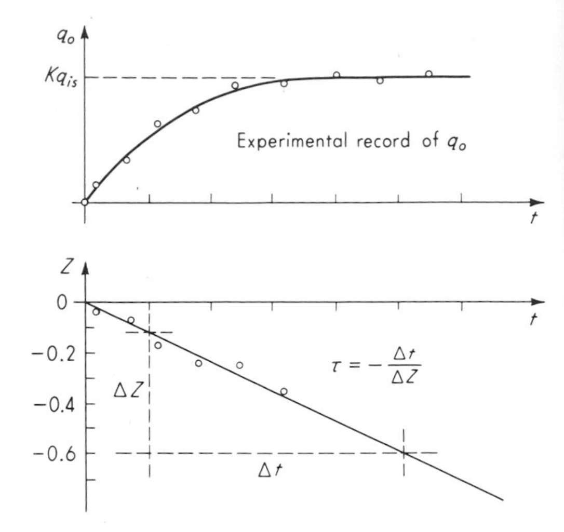

Fig. 3.103. *Step-function test of first-order system.*

Now we define

$$Z \triangleq \log_e \left(1 - \frac{q_o}{Kq_{is}}\right) \tag{3.342}$$

and then

$$Z = \frac{-t}{\tau} \qquad \frac{dZ}{dt} = \frac{-1}{\tau} \tag{3.343}$$

Thus if we plot Z versus t we get a straight line whose slope is numerically $-1/\tau$. Figure 3.103 illustrates the procedure. This gives a more accurate value of τ since the best line through *all* the data points is used rather than just two points, as in the 63.2 percent method. Furthermore, if the data points fall nearly on a straight line we are assured that the instrument is behaving as a first-order type. If the data deviate considerably from a straight line we know the instrument is not truly first-order and a τ value obtained by the 63.2 percent method would be quite misleading.

An even stronger verification (or refutation) of first-order dynamic characteristics is available from frequency-response testing, although at considerable cost of time and money if the system is not completely electrical, since nonelectrical sine-wave generators are neither common nor necessarily cheap. If the equipment is available, the system is subjected to sinusoidal inputs over a wide frequency range and the input and output recorded. Amplitude ratio and phase angle are plotted on the logarithmic scales. If the system is truly first-order the amplitude ratio follows the typical low- and high-frequency asymptotes (slope 0 and −20 db/decade) and the phase angle approaches −90° asymptotically. If these characteristics are present, the numerical value of τ is found by finding ω at the

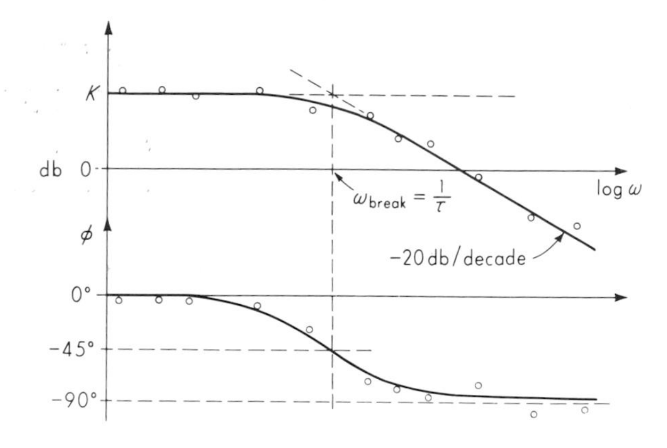

Fig. 3.104. *Frequency-response test of first-order system.*

breakpoint and using $\tau = 1/\omega_{break}$ (see Fig. 3.104). Deviations from the above amplitude and/or phase characteristics indicate non-first-order behavior.

For second-order systems, K is found from static calibration, and ζ and ω_n can be obtained in a number of ways from step or frequency-response tests. Figure 3.105a shows a typical step-function response for an underdamped second-order system. The values of ζ and ω_n may be found from the relations

$$\zeta = \sqrt{\frac{1}{\left(\frac{\pi}{\log_e (a/A)}\right)^2 + 1}} \tag{3.344}$$

$$\omega_n = \frac{2\pi}{T\sqrt{1 - \zeta^2}} \tag{3.345}$$

When a system is lightly damped, any fast transient input will produce a response similar to Fig. 3.105b. Then ζ can be closely approximated by

$$\zeta \approx \frac{\log_e (x_1/x_n)}{2\pi n} \tag{3.346}$$

This approximation assumes $\sqrt{1 - \zeta^2} \approx 1.0$, which is quite accurate when $\zeta < 0.1$, and ω_n can again be found from Eq. (3.345). In applying Eq. (3.345), if several cycles of oscillation appear in the record it is more accurate to determine the period T as the average of as many distinct cycles as are available rather than from a single cycle. If a system is strictly linear and second-order, the value of n in Eq. (3.346) is immaterial; the same value of ζ will be found for any number of cycles. Thus if ζ is calculated for, say, $n = 1, 2, 4$, and 6 and *different* numerical values of ζ are obtained, one knows the system is not following the postulated mathe-

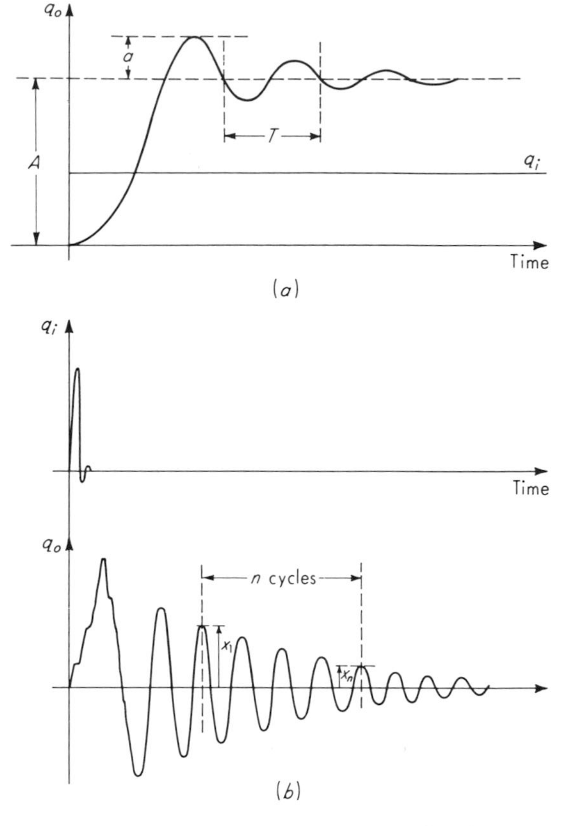

Fig. 3.105. Second-order-system step and pulse tests.

matical model. For overdamped systems ($\zeta > 1.0$) no oscillations exist, and the determination of ζ and ω_n becomes more difficult. Usually it is easier to express the system response in terms of two time constants τ_1 and τ_2 rather than ζ and ω_n. From Eq. (3.231) we can write

$$\frac{q_o}{Kq_{is}} = \frac{\tau_1}{\tau_2 - \tau_1} e^{-t/\tau_1} - \frac{\tau_2}{\tau_2 - \tau_1} e^{-t/\tau_2} + 1 \qquad (3.347)$$

where

$$\tau_1 \triangleq \frac{1}{(\zeta - \sqrt{\zeta^2 - 1})\omega_n} \qquad (3.348)$$

$$\tau_2 \triangleq \frac{1}{(\zeta + \sqrt{\zeta^2 - 1})\omega_n} \qquad (3.349)$$

To find τ_1 and τ_2 from a step-function-response curve we may proceed as follows[1]:

[1] N. A. Anderson, Step-analysis Method of Finding Time Constant, *Instr. Control Systems*, p. 130, November, 1963.

1. Define the "percent incomplete response" R_{pi} as

$$R_{pi} \triangleq [1 - q_o/Kq_{is}]100$$

2. Plot R_{pi} on a logarithmic scale versus time t on a linear scale. This curve will approach a straight line for large t if the system is second-order. Extend this line back to $t = 0$ and note the value P_1 where this line intersects the R_{pi} scale. Now, τ_1 is the time at which the straight-line asymptote has the value $0.368P_1$.
3. Now plot on the same graph a new curve which is the difference between the straight-line asymptote and R_{pi}. If this new curve is not a straight line, the system is not second-order. If it is a straight line, the time at which this line has the value $0.368(P_1 - 100)$ is numerically equal to τ_2.

Figure 3.106 illustrates this procedure. Once τ_1 and τ_2 are found, ζ and ω_n can be determined from Eqs. (3.348) and (3.349) if desired. Other methods[1] for finding τ_1 and τ_2 are available in the literature. Frequency-response methods may also be used to find ζ and ω_n or τ_1 and τ_2. Figure 3.107 shows the application of these techniques. The methods shown use the amplitude-ratio curve only. If phase-angle curves are available, they constitute a valuable check on conformance to the postulated model.

For measurement systems of arbitrary form (as contrasted to first- and second-order types), description of the dynamic behavior in terms of frequency response is usually desired. Ideally, if one has the amplitude-ratio and phase curves for a wide range of frequency, curve fitting (by cut-and-try or other methods) should give the sinusoidal transfer function in numerical form. The straight-line asymptotes of the decibel plot are particularly helpful in judging the location of time constants and natural frequencies when these are not closely spaced on the frequency axis. Actually, the frequency response in *graphical* form is all that is really needed since we have shown how one can get the response to any input from these graphs.

Since frequency-response testing is expensive and time-consuming, consideration should be given to impulse testing. A perfect impulse input is not needed; the only requirements are the following:

1. The approximate impulse must be of sufficiently short duration so that its Fourier-transform-magnitude curve is flat out to frequencies just beyond the point where the tested system's response practically cuts off.

[1] G. M. Hoerner, Second-order System Characteristics from Initial Step Response, *Control Eng.*, p. 93, December, 1962.

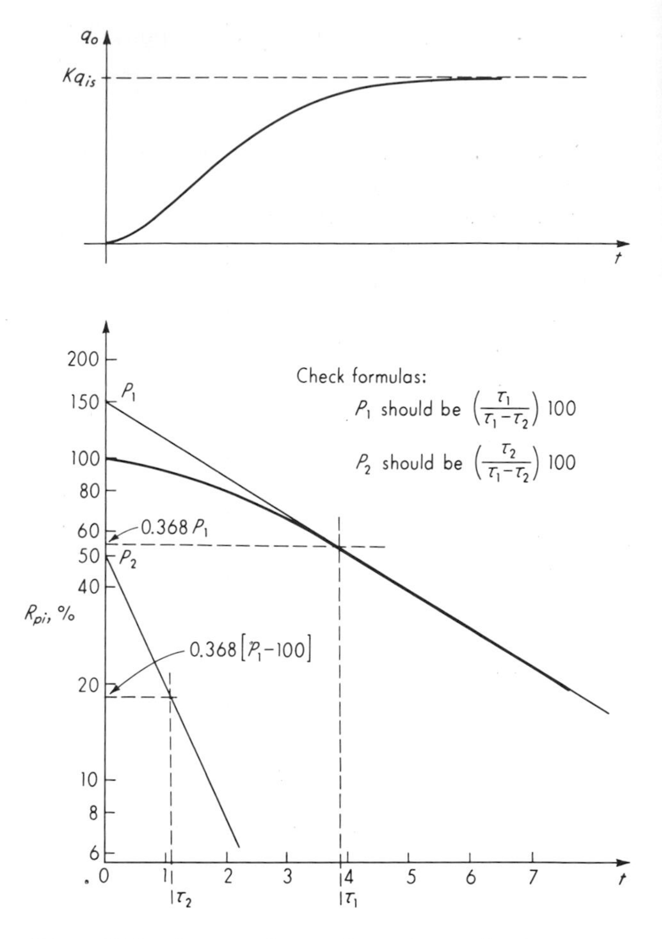

Fig. 3.106. Step test for overdamped second-order systems.

2. The approximate impulse must be strong enough to cause an output response that is large enough to be accurately measurable.

The first requirement implies some knowledge of the "unknown" system's characteristics prior to the test, but one can usually estimate the upper bounds of response easily from theory or experience. The second requirement conflicts with the first since the strength (area) of the pulse is proportional to the duration; thus short durations (dictated by fast-response systems) require very large peak values in order to maintain adequate area. If a suitable impulse test can be set up, one records $q_i(t)$ and $q_o(t)$ and then computes the system frequency response as follows:

$$Q_o(i\omega) = Q_i(i\omega)\,\frac{q_o}{q_i}\,(i\omega) \qquad (3.350)$$

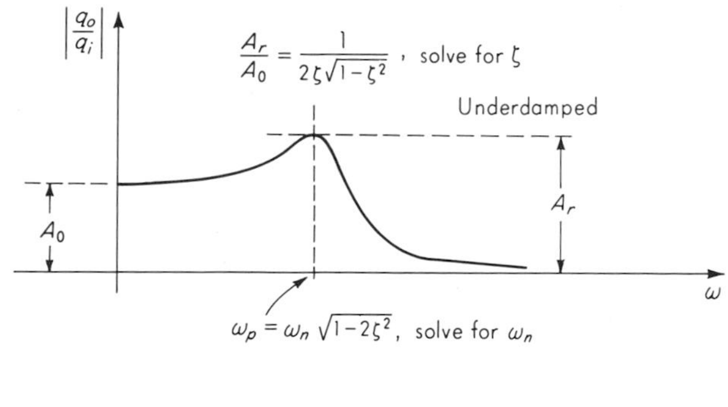

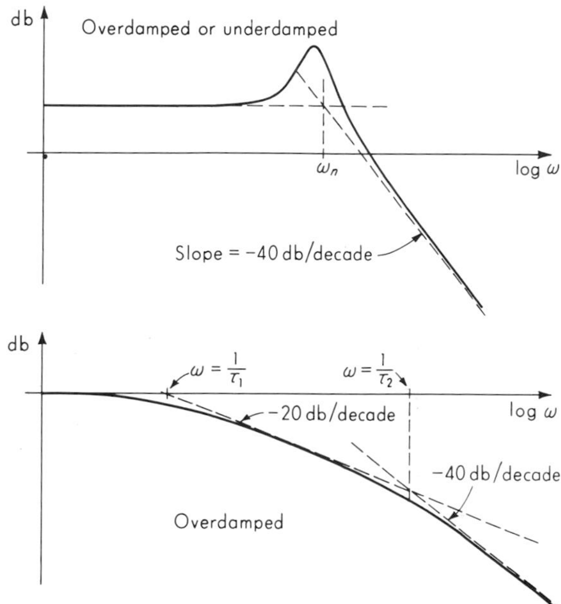

Fig. 3.107. *Frequency-response test of second-order system.*

But, for a "short enough" pulse,

$$Q_i(i\omega) = A_i\underline{/0°} \qquad A_i = \text{area of } q_i(t) \tag{3.351}$$

for all pertinent frequencies. Thus

$$Q_o(i\omega) = A_i \frac{q_o}{q_i}(i\omega) \tag{3.352}$$

$$\text{and} \quad \frac{q_o}{q_i}(i\omega) = \frac{Q_o(i\omega)}{A_i} \tag{3.353}$$

Since $q_o(t)$ is available, one can compute $Q_o(i\omega)$ in the usual way and thus get $(q_o/q_i)(i\omega)$. In fact, if one is satisfied that $q_i(t)$ is fast enough, and if one does not care what the static sensitivity of the system is (since that is easily obtained by static calibration), there is no need to measure $q_i(t)$ at all since A_i in Eq. (3.353) is merely a constant factor in $(q_o/q_i)(i\omega)$.

Loading effects under dynamic conditions. The treatment of loading effects by means of impedance, admittance, etc., was treated in Sec. 3.2 for static conditions. All these results can be immediately transferred to the case of dynamic operation by generalizing the definitions in terms of transfer functions. The basic equations relating the undisturbed value q_{i1u} and the actual measured value q_{i1m} at the input of a device are

$$q_{i1m} = \frac{1}{Z_{go}/Z_{gi} + 1} q_{i1u} \qquad (3.55)$$

$$q_{i1m} = \frac{1}{Y_{go}/Y_{gi} + 1} q_{i1u} \qquad (3.61)$$

$$q_{i1m} = \frac{1}{S_{go}/S_{gi} + 1} q_{i1u} \qquad (3.68)$$

$$q_{i1m} = \frac{1}{C_{go}/C_{gi} + 1} q_{i1u} \qquad (3.80)$$

The quantities Z, Y, S, and C were previously considered to be the ratios of small changes in two related system variables under stated conditions. To generalize these concepts, we now define the quantities Z, Y, S, and C as *transfer functions* relating the same two variables under the same conditions except now dynamic operation is to be considered. That is, we must get (theoretically or experimentally) $Z(D)$, $Y(D)$, $S(D)$, and $C(D)$ if we wish to use operational transfer functions and $Z(i\omega)$, $Y(i\omega)$, $S(i\omega)$, and $C(i\omega)$ if we wish to use frequency-response methods.

Usually the frequency-response form is most useful if these quantities must be found experimentally. This means, then, that in finding, say, $Z(i\omega)$, one of the two variables involved in the definition of Z plays the role of an "input" quantity which we vary sinusoidally at different frequencies. This causes a sinusoidal change in the other ("output") variable, and we can thus speak of an amplitude ratio and phase angle between these two quantities, making $Z(i\omega)$ now a complex number that varies with frequency. (If the system is somewhat nonlinear the effective approximate Z now becomes a function also of input amplitude. This situation was adequately described under static conditions in Sec. 3.2.) In Eq. (3.55), for example, both Z_{go} and Z_{gi} would now be complex numbers; if these were known, we could calculate the amplitude and phase of q_{i1m} if the amplitude, phase, and frequency of a sinusoidal q_{i1u} were given. The quantity q_{i1m} would then be the *actual* input (q_i) to the measuring device, and we could calculate q_o if the transfer function $(q_o/q_i)(i\omega)$ were known. That is,

$$Q_o(i\omega) = \frac{1}{Z_{go}(i\omega)/Z_{gi}(i\omega) + 1}\left[\frac{q_o}{q_i}(i\omega)\right] Q_{i1u}(i\omega) \tag{3.354}$$

One could thus define a *loaded transfer function* $(q_o/q_{i1u})(i\omega)$ as

$$\frac{q_o}{q_{i1u}}(i\omega) \triangleq \frac{1}{Z_{go}(i\omega)/Z_{gi}(i\omega) + 1}\frac{q_o}{q_i}(i\omega) \tag{3.355}$$

where $q_o \triangleq$ actual output of measuring device which has no load at *its* output

$q_{i1u} \triangleq$ measured variable value that would exist if measuring device caused *no* loading on measured medium

Equations (3.61), (3.68), and (3.80) may be modified in similar fashion. Also, if differential equations relating $q_o(t)$ and $q_{i1u}(t)$ are desired, we may write

$$\frac{q_o}{q_{i1u}}(D) = \frac{1}{Z_{go}(D)/Z_{gi}(D) + 1}\frac{q_o}{q_i}(D) \tag{3.356}$$

and then obtain the differential equation in the usual way by "cross-multiplying":

$$\begin{aligned}\{[Z_{go}(D) + Z_{gi}(D)](a_nD^n + a_{n-1}D^{n-1} + \cdots a_1D \\ + a_0)\}q_o = \{[Z_{gi}(D)](b_mD^m + b_{m-1}D^{m-1} \\ + \cdots + b_1D + b_0)\}q_{i1u}\end{aligned} \tag{3.357}$$

An example of the above methods will be helpful. Consider a device for measuring translational velocity as in Fig. 3.108*a*. The unloaded transfer function relating the output displacement x_o and the input (measured) velocity v_i is obtained as follows:

$$B_i(\dot{x}_i - \dot{x}_o) - K_{is}x_o = M_i\ddot{x}_o \tag{3.358}$$

$$\frac{x_o}{v_i}(D) = \frac{K_i}{D^2/\omega_{ni}^2 + 2\zeta_iD/\omega_{ni} + 1} \tag{3.359}$$

where $K_i \triangleq$ instrument static sensitivity $\triangleq \dfrac{B_i}{K_{is}}$ in./(in./sec) (3.360)

$\zeta_i \triangleq$ instrument damping ratio $\triangleq \dfrac{B_i}{2\sqrt{K_{is}M_i}}$ (3.361)

$\omega_{ni} \triangleq$ instrument undamped natural frequency $\triangleq \sqrt{\dfrac{K_{is}}{M_i}}$ rad/sec (3.362)

We see that the instrument is second-order and will thus measure v_i accurately for frequencies sufficiently low relative to ω_{ni}. Suppose we now

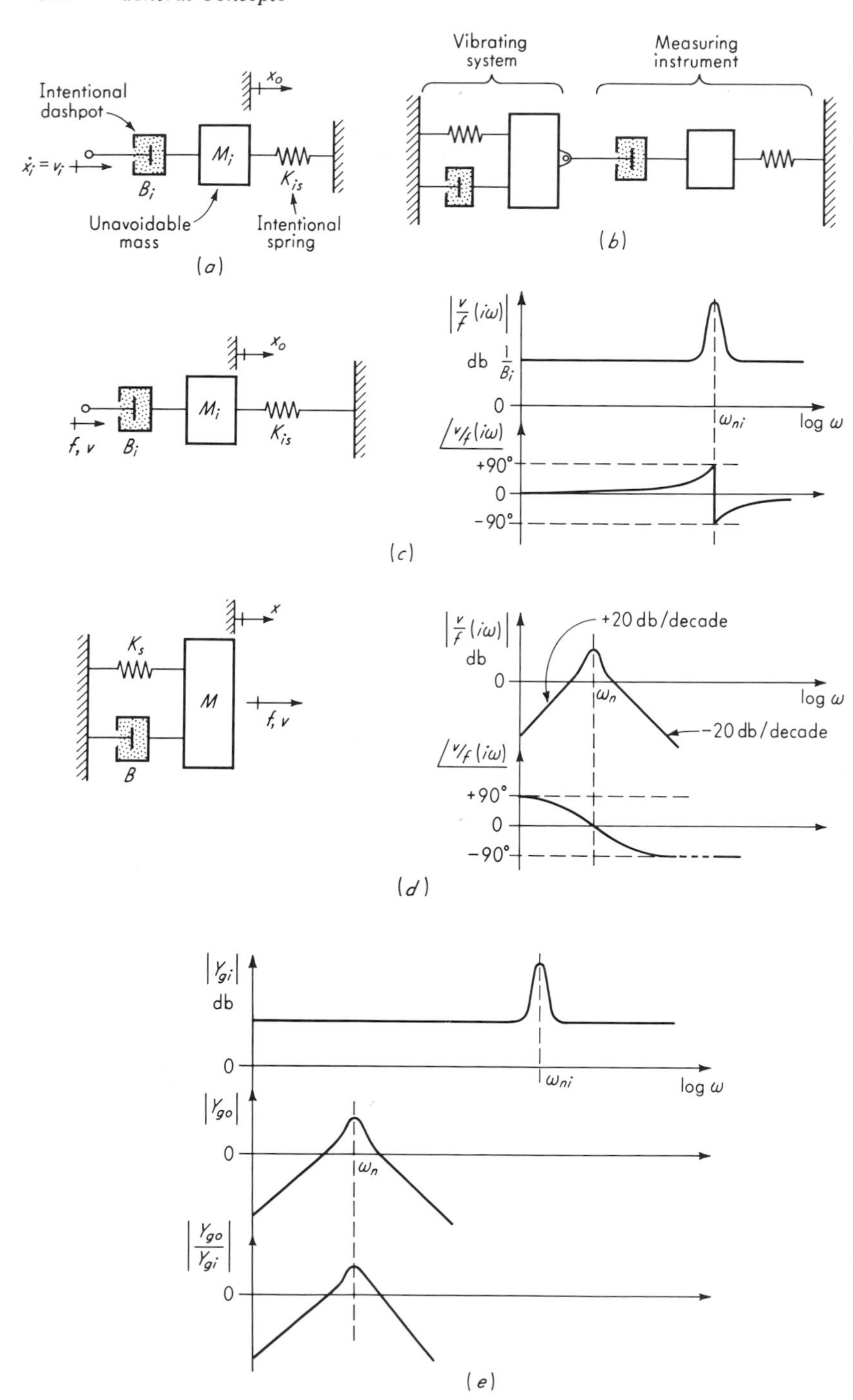

Fig. 3.108. Example of dynamic-loading analysis.

attach this instrument to a vibrating system whose velocity we wish to measure, as in Fig. 3.108*b*. The presence of the measuring instrument will distort the velocity we are trying to measure. The character of this distortion may be assessed by application of Eq. (3.61), since the measured quantity is velocity (a flow variable; see Fig. 3.28) and thus admittance is the appropriate quantity to use. We determine the input admittance $Y_{gi}(D) = (v/f)(D)$ from Fig. 3.108*c* as follows:

$$f - K_{is}x_o = M_i\ddot{x}_o \tag{3.363}$$

Also

$$f = B_i(v - \dot{x}_o) \tag{3.364}$$

and, eliminating x_o, we get

$$Y_{gi}(D) = \frac{v}{f}(D) = \frac{(1/B_i)(D^2/\omega_{ni}^2 + 2\zeta_i D/\omega_{ni} + 1)}{D^2/\omega_{ni}^2 + 1} \tag{3.365}$$

Figure 3.108*c* also shows the frequency characteristics of this input admittance. The output admittance $Y_{go}(D) = (v/f)(D)$ of the measured system is obtained from Fig. 3.108*d*:

$$f - B\dot{x} - K_s x = M\ddot{x} \tag{3.366}$$

$$Y_{go}(D) = \frac{v}{f}(D) = \frac{(1/K_s)D}{D^2/\omega_n^2 + 2\zeta D/\omega_n + 1} \tag{3.367}$$

The frequency characteristic of this output admittance is shown in Fig. 3.108*d*. We may now write

$$\frac{x_o}{v_{i1u}}(D) = \frac{1}{Y_{go}(D)/Y_{gi}(D) + 1}\,\frac{x_o}{v_i}(D) \tag{3.368}$$

$$\frac{x_o}{v_{i1u}}(D) = \frac{1}{\underbrace{\dfrac{(1/K_s)D}{D^2/\omega_n^2 + 2\zeta D/\omega_n + 1}\,\dfrac{(D^2/\omega_{ni}^2) + 1}{(1/B_i)(D^2/\omega_{ni}^2 + 2\zeta_i D/\omega_{ni} + 1)} + 1}_{\text{loading effect}}}\;\frac{K_i}{D^2/\omega_{ni}^2 + 2\zeta_i D/\omega_{ni} + 1} \tag{3.369}$$

where $x_o \triangleq$ actual output of measuring device
$v_{i1u} \triangleq$ velocity that would exist if measuring device caused no loading

Figure 3.108*e* shows that in this example the loading effect is most serious for frequencies near the natural frequency of the measured system but approaches zero for both very low and very high frequencies. Since the loading effects can be expressed in frequency terms, they can be handled for all kinds of inputs by using appropriate Fourier series, transform, or mean-square spectral density.

Problems

3.1 For the system of Fig. 2.3:

a. Explain how you would carry out a static calibration to determine the relation between the desired input and the output.

b. The temperature of the air surrounding the capillary tube is an interfering input. Explain how you would calibrate the relation between this input and the output.

c. The elevation difference between the Bourdon tube and the bulb is another interfering input. Discuss means for its calibration.

3.2 Does the system of Fig. 2.4 require calibration? Explain.

3.3 What fundamental difficulties arise in trying to define the true temperature of a physical body?

3.4 Slide a coin along a smooth surface, trying to make it come to rest at a drawn line. Measure the distance of the coin from the line. Repeat 100 times and check the resulting data for conformance to a gaussian distribution, using probability graph paper.

3.5 Using the data generated in Prob. 3.4, apply the chi-squared test for conformance to a gaussian distribution.

3.6 In Eq. (2.6), solve for the strain ϵ in terms of the other parameters; $\epsilon = f(GF, R_g, E_b, R_a, e_o)$. Then take the natural log of both sides; $\ln \epsilon = \ln f$. Now take the differential of both sides so that terms such as $d\epsilon/\epsilon$, de_o/e_o, dR_a/R_a, etc., are formed. This will give the percentage error $d\epsilon/\epsilon$ in ϵ as a function of the percentage errors in the other parameters. If GF, R_g, E_b, R_a, and e_o are all measured to ± 1 percent error, what is the possible error in the computed value of ϵ?

3.7 Is the logarithmic differentiation method of Prob. 3.6 applicable to all forms of functional relations? Explain. Hint: Apply it to the relation $w = \sin x + 5y^3 - 6e^z$.

3.8 The discharge coefficient C_q of an orifice can be found by collecting the water that flows through during a timed interval when it is under a constant head h. The formula is

$$C_q = \frac{W}{t\rho A\sqrt{2gh}}$$

Find C_q and its possible error if:

$W = 865 \pm 0.5\ \text{lb}_m$ $A = \pi d^2/4$ $d = 0.500 \pm 0.001$ in.

$t = 600.0 \pm 2$ sec $g = 32.17 \pm 0.1\%$ ft/sec^2

$\rho = 62.36 \pm 0.1\%\ \text{lb}_m/\text{ft}^3$ $h = 12.02 \pm 0.01$ ft

considering both the following:

a. The errors are the absolute limits.

b. The errors are $\pm 3s$ limits.

3.9 In Prob. 3.8 if C_q must be measured within ± 0.5 percent for the numerical mean values given, what errors are allowable in the measured data? Use the method of equal effects.

3.10 Static calibration of an instrument gives the data of Fig. P3.1. Calculate the following:

a. The best-fit straight line
b. s_m and s_b
c. s_{q_i}
d. q_i and its error limits if the instrument is used after calibration and reads $q_o = 5.72$

q_i	q_o *Increasing values*	q_o *Decreasing values*
0	−0.07	+0.01
5	1.08	1.16
10	2.05	2.10
15	3.27	3.29
20	4.28	4.36
25	5.41	5.45
30	6.43	6.53
35	7.57	7.61
40	8.66	8.75

Fig. P3.1

3.11 In Fig. 3.21, what percent error may be expected in measuring the voltage across R_5 if $R_1 = R_2 = R_3 = R_4 = R_5 = 100$ ohms and $R_m = 1{,}000$ ohms? If $R_m = 10{,}000$ ohms?

3.12 Repeat Prob. 3.11 except now the voltage across R_3 is to be measured.

3.13 In Fig. 3.25, what percent error may be expected in measuring the current through R_5 if $R_1 = R_2 = R_3 = R_4 = R_5 = 100$ ohms and $R_m = 10$ ohms? If $R_m = 1$ ohm?

3.14 Repeat Prob. 3.13 except now the current through R_3 is to be measured.

3.15 In Fig. 3.26, what percent error may be expected in measuring the force in k_2 if $k_1 = k_2 = k_3 = k_4 = 100$ lb$_f$/in. and $k_m = 1{,}000$ lb$_f$/in.? If $k_m = 10{,}000$ lb$_f$/in.?

3.16 Repeat Prob. 3.15 except now the force in k_3 is to be measured.

3.17 In Fig. 3.27, what percent error may be expected in measuring the deflection x if $k_1 = k_2 = k_3 = k_4 = 1$ lb$_f$/in. and $k_m = 0.1$ lb$_f$/in.? If $k_m = 0.01$ lb$_f$/in.?

3.18 Repeat Prob. 3.17 except now the motion of the right-hand block is to be measured.

3.19 Using methods similar to those used in proving Eq. (3.55), prove the following:
a. Eq. (3.61)
b. Eq. (3.68)
c. Eq. (3.80)

3.20 A mercury thermometer has a capillary tube of 0.010-in. diameter. If the bulb is made of a zero-expansion material, what volume must it have if a sensitivity of 0.10 in./F° is desired? Assume operation near 70°F. If the bulb is spherical and is immersed in stationary air, estimate the time constant.

3.21 A balloon carrying a first-order thermometer with a 15-sec time constant rises through the atmosphere at 20 fps. Assume temperature varies with altitude at 0.3 F°/100 ft. The balloon radios temperature and altitude readings back to the ground. At 10,000 ft the balloon says the temperature is 30°F. What is the true altitude at which 30°F occurs?

3.22 A first-order instrument must measure signals with frequency content up to 100 cps with an amplitude inaccuracy of 5 percent. What is the maximum allowable time constant? What will be the phase shift at 50 and 100 cps?

3.23 For the spring scale of Fig. 3.48, discuss the tradeoff between sensitivity and speed of response resulting from changes in K_s.

3.24 Derive Eqs. (3.237) to (3.239).

3.25 Find the transfer function of a spring scale (Fig. 3.48) whose mass is negligible. Show that the steady-state time lag for a ramp input is the same whether mass is zero or not.

3.26 Plot decibel and phase-angle curves for the following systems:

a. $$\frac{q_o}{q_i}(D) = \frac{10D}{10D + 1}$$

b. $$\frac{q_o}{q_i}(D) = \frac{D}{(D + 1)(5D + 1)}$$

c. $$\frac{q_o}{q_i}(D) = \frac{10e^{-2D}}{0.01D^2 + 0.1D + 1}$$

d. $$\frac{q_o}{q_i}(D) = \frac{100D^2}{(D + 1)^2(10D + 1)^2}$$

e. $$\frac{q_o}{q_i}(D) = \frac{5(0.25D^2 + 1)}{0.01D^2 + 0.1D + 1}$$

3.27 Find $Q_i(i\omega)$ for the $q_i(t)$ of Fig. P3.2 by the following methods:

a. The exact analytical method
b. The method of 12 ordinates
c. The method of 24 ordinates

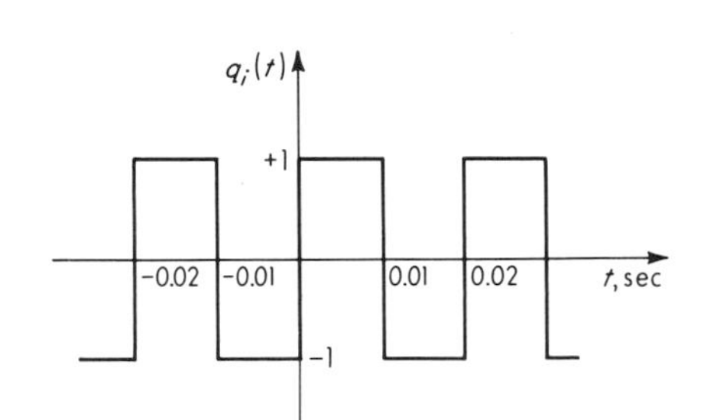

Fig. P3.2

3.28 Repeat Prob. 3.27 for Fig. P3.3.

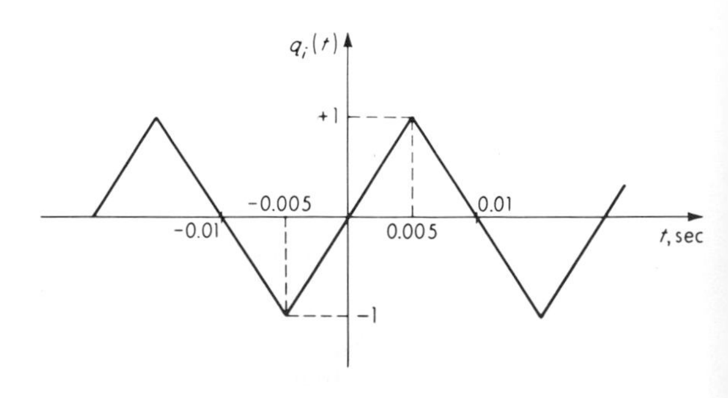

Fig. P3.3

3.29 If the $q_i(t)$ of Prob. 3.27 is the input to a first-order system with a gain of 1 and a time constant of 0.001 sec, find $Q_o(i\omega)$ and $q_o(t)$ for the steady state, using methods a, b, and c of Prob. 3.27.

3.30 Repeat Prob. 3.29 except use $q_i(t)$ from Prob. 3.28.

3.31 In Fig. 3.82 let the carrier be a square wave as in Fig. P3.4. Find the frequency spectrum of the output signal of the modulator.

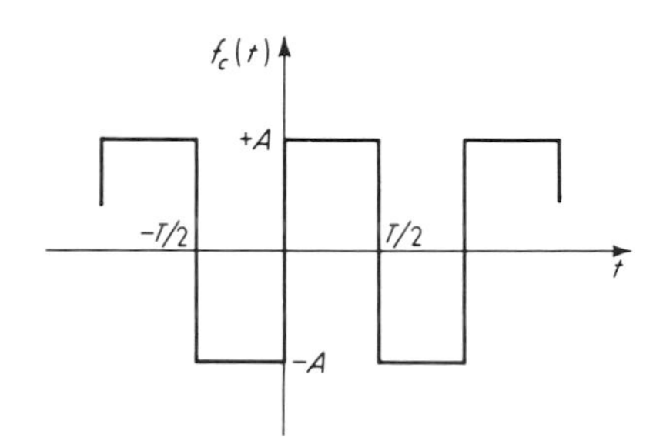

Fig. P3.4

3.32 Repeat Prob. 3.31 if the carrier is a square wave as in Fig. P3.5.

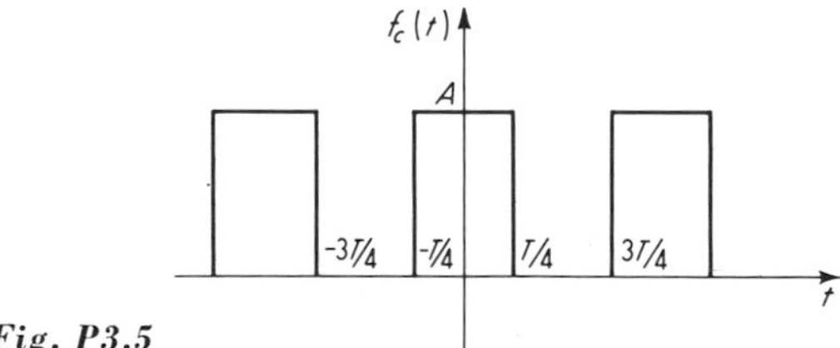

Fig. P3.5

3.33 In an analog-computer study it is desired to simulate a random atmospheric turbulence whose mean-square spectral density $\phi_t(\omega)$ is adequately represented as $10/(1 + 0.0001\omega^2)$, where ω is in radians per second. A white-noise generator having $\phi_{wn}(\omega) = 10$ is available. Select a suitable filter configuration and numerical values to follow the generator and produce the desired $\phi_t(\omega)$. The output of the noise generator should "see" a filter input resistance of 10,000 ohms.

3.34 Tests on a gyroscope show that it can withstand any random vibration along a given axis if the frequency content is between 0 and 1,000 rad/sec and the rms acceleration is less than 80 in./sec^2. This gyro is to be mounted in a rocket where it will be subjected to acoustic-pressure-induced vibration. The transfer function between pressure and acceleration and the mean-square spectral density of pressure are as given in Fig. P3.6. Will this gyro withstand the vibration?

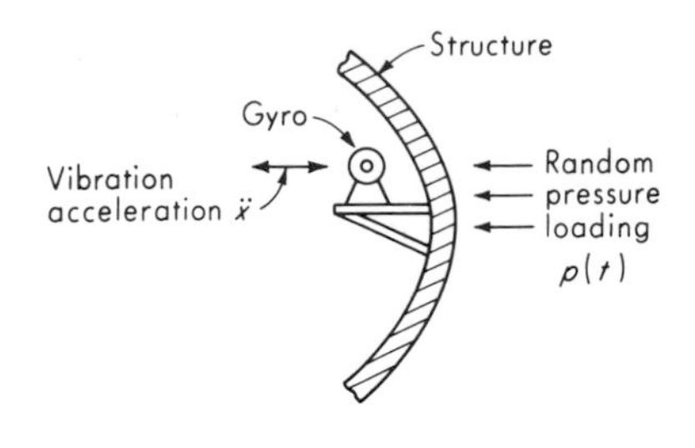

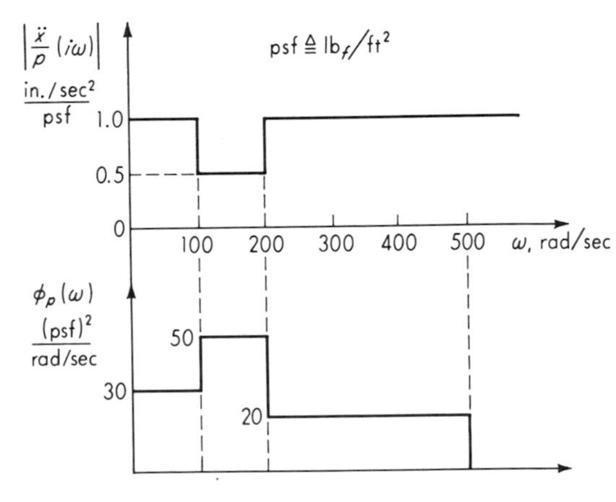

Fig. P3.6

3.35 Derive Eq. (3.344).

3.36 Derive Eq. (3.346).

3.37 Explain how the sinusoidal transfer function of a system may be obtained from measured records of $q_i(t)$ and $q_o(t)$ if q_i is a transient of any shape whatever.

3.38 Reanalyze the force-measuring problem of Fig. 3.26 for *dynamic* operation, assuming the two blocks have masses M_1 and M_2. That is, get the operational transfer function analogous to Eq. (3.77).

3.39 Reanalyze the displacement-measuring problem of Fig. 3.27 for *dynamic* operation, assuming the two blocks have masses M_1 and M_2. That is, get the operational transfer function analogous to Eq. (3.93).

3.40 Reanalyze the voltage-measuring problem of Fig. 3.21 for dynamic operation; i.e., replace the batteries with sources of time-varying voltage. Also, let the voltage-measuring device be an oscilloscope with R_m shunted by a capacitor C_m.

3.41 Reanalyze the current-measuring problem of Fig. 3.25 for dynamic operation; i.e., replace the batteries with sources of time-varying voltage. Also, let the current-measuring device be a galvanometer which has an inductance L_m in series with R_m.

Bibliography

1. H. E. Koenig and W. A. Blackwell: "Electromechanical System Theory," McGraw-Hill Book Company, New York, 1961.

2. C. L. Cuccia: "Harmonics, Sidebands, and Transients in Communication Engineering," McGraw-Hill Book Company, New York, 1952.
3. T. N. Whitehead: "The Design and Use of Instruments and Accurate Mechanisms," Dover Publications, Inc., New York, 1954.
4. V. L. Lebedev: Random Processes in Electrical and Mechanical Systems, *NASA, Tech. Transl., F*-61, 1961.
5. C. C. Perry: The Least Squares Method, *Machine Design*, p. 210, May 12, 1960.
6. V. R. Boulton: Economics of Instrumentation Precision, *Aerospace Eng.*, p. 30, March, 1961.
7. C. T. Morrow: Averaging Time and Data-reduction Time for Random Vibration Spectra, *J. Acoust. Soc. Am.*, vol. 30, no. 6, p. 572, June, 1958.
8. N. R. Goodman et al.: Frequency Response from Stationary Noise: Two Case Histories, *Technometrics*, p. 245, May, 1961.
9. R. L. Hammon: An Application of Random Process Theory to Gyro Drift Analysis, *IRE Trans. PGANE*, vol. ANE-7, no. 3, September, 1960.
10. D. E. Cartwright et al.: Digital Techniques for the Study of Sea Waves, Ship Motion and Allied Processes, *Trans. Soc. Instr. Tech. (London)*, p. 1, March, 1962.
11. J. T. Broch: Automatic Recording of Amplitude Density Curves, *B & K Tech. Rev.*, B & K Instruments, Cleveland, Ohio, no. 4, 1959.
12. J. T. Broch: Recording of Narrow Band Noise, *B & K Tech. Rev.*, B & K Instruments, Cleveland, Ohio, no. 4, 1960.
13. K. R. Thorson and Q. R. Bohne: Application of Power Spectral Methods in Airplane and Missile Design, *J. Aero/Scope Sci.*, p. 107, February, 1960.
14. J. C. Laurence: Intensity, Scale and Spectra of Turbulence in Mixing Region of Free Subsonic Jet, *NACA, Tech. Notes* 3561, 1955.
15. J. R. Rice et al.: On the Prediction of Some Random Loading Characteristics Relevant to Fatigue, *NASA, CR*-56152, 1964.
16. W. A. Wildhack et al.: Accuracy in Measurements and Calibrations, *NBS, Tech. Notes* 262, 1965.

part II

Measuring Devices

4 Motion measurement

4.1 Introduction

We commence our study of specific measuring devices with motion measurement since it is based on two of the fundamental quantities in nature (length and time) and also because so many other quantities such as force, pressure, temperature, etc., are often measured by transducing them to motion and then measuring this resulting motion. As indicated in the chapter title, our main interest is in motion (a *changing* displacement). Thus we shall not go extensively into dimensional measurement or gaging of fixed lengths, angles, hole diameters, etc., except as this relates to standards or calibration of motion-measuring devices.

We are also mainly (though not exclusively) concerned with electromechanical transducers which convert motion quantities into electrical quantities. The intent is not to present a catalog listing of the myriad physical effects which have been, or might be, used as the basis of a motion transducer but rather to provide sufficient detail for practical application of the relatively small number of transducer types which form the basis of the majority of practical measurements. The above-mentioned catalog-listing type of information is extremely useful to one who has a measurement problem not solvable by one of the standard techniques and who must therefore invent and/or develop a new instrument. Material of this type is available in several references.[1,2]

4.2

Fundamental Standards The four fundamental quantities of the International Measuring System, for which independent standards have been defined, are length, time, mass, and temperature. Units and standards for all other quantities are *derived* from these. In motion measurement the fundamental quantities are length and time. Prior to 1960 the standard of length was the carefully preserved platinum-iridium International Meter Bar at Sèvres, France. In 1960 the meter was redefined in terms of the wavelength of a krypton-86 lamp as "the length equal to 1,650,763.73 wavelengths in vacuum corresponding to the transition between the energy levels $2p_{10}$ and $5d_5$ of the atom krypton 86."[3] This standard is believed[4] to be reproducible to about 2 parts in 10^8 and can be applied at this precision level to measurements of length in the range of about 10^{-8} to 40 in.[5]

The above National Prototype Standard is not available for routine calibration work. Rather, to protect such top-level standards from deterioration, the National Bureau of Standards has set up National Reference Standards and, below these, Working Standards. Further down the line in accuracy are the so-called Interlaboratory Standards, which are standards sent in to the National Bureau of Standards for

[1] K. S. Lion, "Instrumentation in Scientific Research," McGraw-Hill Book Company, New York, 1959.

[2] C. F. Hix, Jr., and R. P. Alley, "Physical Laws and Effects," John Wiley & Sons, Inc., New York, 1958.

[3] A. G. McNish, Fundamentals of Measurement, *Electro-Technol. (New York)*, p. 113, May, 1963.

[4] W. A. Wildhack, NBS—Source of American Standards, *ISA J.*, p. 45, February, 1961.

[5] L. B. Wilson and H. W. Martin, The Measurements Gap, *Space/Aeron.*, p. 84, March, 1964.

calibration and certification by factories and laboratories all over the country. These last-mentioned standards are the ones usually readily available to the working engineer for calibration of motion transducers.

The fundamental unit of time is the second, which was redefined for scientific use as 1/31,556,925.9747 of the tropical year at 12^h ephemeris time, 0 January 1900,[1] by the International Committee on Weights and Measures in 1956. A serious fault in this definition is that no one can measure an interval of time by direct comparison with the interval of time defining the second. Rather, lengthy astronomical measurements over several years are necessary to relate the current value of the mean solar second to the basic standard. These measurements and calculations result in an estimated probable error of about 1 part in 10^9, which is quite poor compared with the precision implied in the basic definition of the second. To remedy this difficulty, meteorologists in 1964 again redefined the second in terms of the frequencies of atomic resonators.[2] Now the second is defined as the interval of time corresponding to 9,192,631,770 cycles of the atomic resonant frequency of cesium 133. Already it is possible for independent laboratories to construct cesium-beam resonators which agree in frequency within a few parts in 10^{11}. The hydrogen maser gives promise of extending this to 10^{13}.

The above short discussion was concerned with the *fundamental* standards of length and time rather than the practical working standards with which most engineers will be concerned. These practical standards and associated calibration procedures will be discussed in each specific section, such as relative displacement, acceleration, etc.

4.3

Relative Displacement, Translational and Rotational

We consider here devices for measuring the translation along a line of one point relative to another and the plane rotation about a single axis of one line relative to another. Such displacement measurements are of great interest as such and also because they form the basis of many transducers for measuring pressure, force, acceleration, temperature, etc., as shown in Fig. 4.1.

Calibration. Static calibration of translational devices can often be satisfactorily accomplished using ordinary dial indicators or micrometers as the standard. When used directly to measure the displacement of the transducer, these devices usually are suitable to read to the nearest

[1] McNish, *loc. cit.*

[2] Time Standards, *Instr. Control Systems,* p. 87, October, 1965.

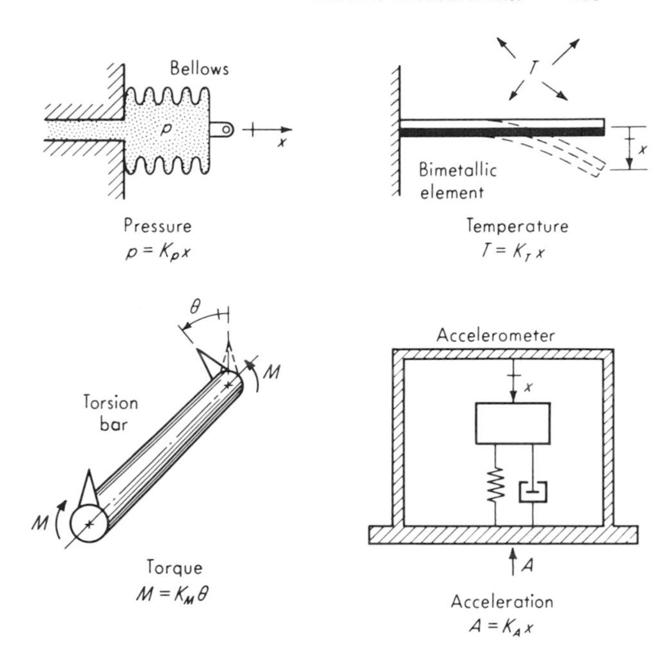

Fig. 4.1. *Applications of displacement measurement.*

0.0001 in. If smaller increments are necessary, lever arrangements (about a 10:1 ratio is fairly easy to achieve) or wedge-type mechanisms (about 100:1) can be employed for motion reduction.[1] The Mikrokator,[2] a unique mechanical gage of high sensitivity, may also be useful in measuring small motions down to a few millionths of an inch.

If accuracy to 0.0001 in. or better is required, such equipment should itself be calibrated against gage blocks, or (for maximum accuracy) gage blocks should be used *directly* to calibrate the transducer. *Gage blocks* are small blocks of hard, dimensionally stable steel or other material, made up in sets which can be stacked up to provide accurate dimensions over a wide range and in small steps. They are the basic working length standards of industry. As purchased from the manufacturer, their dimensions are accurate to ± 8 μin. for working grade blocks, ± 4 μin. for reference grade, and ± 2 μin. for all blocks up to 1 in. (± 2 μin./in. for blocks longer than 1 in.) for master blocks. If these tolerances are too large, the blocks can be sent to the National Bureau of Standards and calibrated against light wavelengths to the nearest 10^{-7} in. Some pre-

[1] H. C. Roberts, "Mechanical Measurements by Electrical Methods," chap. 13, The Instruments Publishing Co., Pittsburgh, Pa., 1951.

[2] C. E. Johansson Gage Co., Dearborn, Mich.

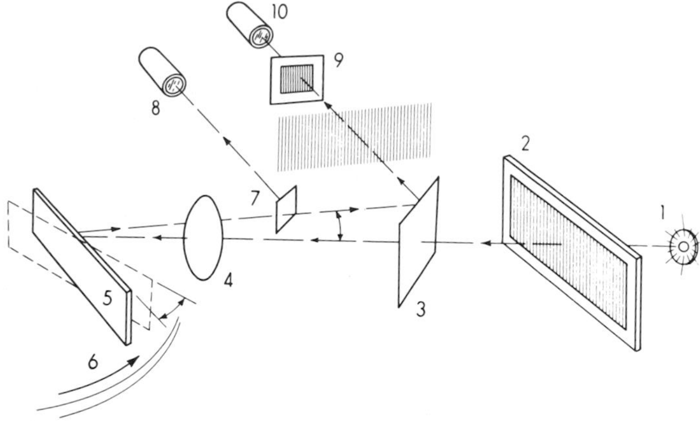

The Midarm system is comprised of an optical unit and an electronic unit. The simplified diagram shows how Midarm operates. Light from a monochromatic light point source (1) *passes through a grid* (2), *a beam splitter* (3), *a collimating lens* (4), *and strikes a mirror* (5) *mounted on the rotating specimen to be tested* (6). *The image is reflected back into the system where it is directed by a beam splitter* (7) *to the reference photosensor* (8). *The image is also reflected by the beam splitter* (3) *through a second grid* (9) *to the control photosensor* (10). *As the test specimen (mirror) rotates, the image of the first grid passes across the second grid which allows minimum and maximum amounts of light to reach the control photosensor. The output voltage of the photosensor has a period of* 12.8 *arc-sec, the angle subtended by the grid spaces. The Midarm has digital output pulses at* 12.8 *arc-sec and an analog voltage output of* 30 *volts/arc-sec.*

Fig. 4.2. *The Midarm system.*

cision-manufacturing operations currently require and use the latter calibration service. When calibrating transducers to very high accuracies it is extremely important to control all interfering and/or modifying inputs such as ambient temperature, electrical excitation to the transducer, etc.

Rotational or angular displacement is not itself a fundamental quantity since it is based on length, and so a fundamental standard is not necessary. However, reference and working standards for angles (and thus angular displacement) are desirable and available. The basic standards (against which other standards or instruments may be calibrated) are called *angle blocks.*[1] These are carefully made steel blocks about $\frac{5}{8}$ in. wide and 3 in. long, with a specified angle between the two contact surfaces. Just as for length gage blocks, these angle blocks can

[1] C. E. Haven and A. G. Strong, Assembled Polygon for the Calibration of Angle Blocks, *Natl. Bur. Std. (U.S.), Handbook* 77, vol. 3, p. 318, 1961.

be stacked to "build up" any desired angle accurately and in small increments. The blocks can be calibrated to an accuracy of 0.1 second of arc by the National Bureau of Standards.[1]

Recent developments in inertial guidance systems have required angle and angular rate measurements on rotating components to an accuracy approaching or exceeding the capability of National Bureau of Standards calibration. Combinations of optical and electronic principles have led to the development of instruments such as the Midarm[2] system to meet these requirements. This instrument will, with relative convenience, measure angular displacement with an accuracy of 0.05 second of arc and a repeatability of 0.02 second of arc. Figure 4.2 shows a simplified diagram of this instrument.

Rotational transducers rarely require such accuracy for calibration nor can the laborious and expensive techniques necessary to realize these limits be economically justified. Thus most static calibration of angular-displacement transducers can adequately be carried out using more convenient and readily available equipment. Examples[3] of such equipment which should be available in a precision machine shop are the circular division tester (range 360°, microscope reads to 0.1 minute of arc, precision of scale disk ± 20 seconds of arc), the optical dividing head (range 360°, scale reads to 1.0 minute of arc, working accuracy ± 20 seconds of arc), and the division tester with telescope and collimator (accuracy ± 2 seconds of arc). In some applications even cruder devices such as ordinary machine-tool index heads, calibrated dials, etc., may be perfectly adequate.

Resistive potentiometers. Basically, a resistive potentiometer consists of a resistance element provided with a movable contact. The contact motion can be translation, rotation, or a combination of the two (helical motion in a multiturn rotational device), thus allowing measurement of rotary and translatory displacements. Translatory devices have strokes from about 0.1 to 20 in., and rotational ones range from about 10° to as much as 60 full turns. The resistance element is excited with either d-c or a-c voltage, and the output voltage is (ideally) a linear function of the input displacement. Resistance elements in common use may be classified as wire-wound, carbon-film, or conducting-plastic.

If the distribution of resistance with respect to translational or angular travel of the wiper (moving contact) is linear, the output voltage e_o will faithfully duplicate the input motion x_i or θ_i if the terminals at e_o

[1] Independent Standards Laboratory, *Instr. Control Systems,* p. 478, March, 1961.

[2] Razdow Laboratories, Inc., Newark, N.J.

[3] Carl Zeiss, Inc., New York.

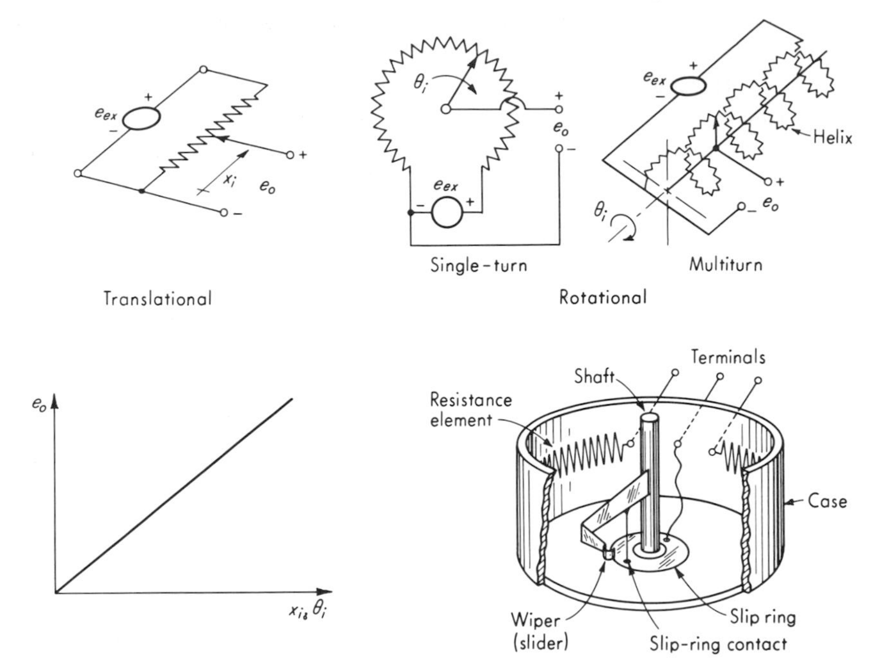

Fig. 4.3. *Potentiometer displacement transducer.*

are open circuit (no current drawn at the output). (For a-c excitation, x_i or θ_i amplitude-modulate e_{ex}, and e_o does not look like the input motion.) The usual situation, however, is one in which the potentiometer output voltage is the input to a meter or recorder that draws some current from the potentiometer. Thus a more realistic circuit is as shown in Fig. 4.4. Analysis of this circuit gives

$$\frac{e_o}{e_{ex}} = \frac{1}{1/(x_i/x_t) + (R_p/R_m)[1 - (x_i/x_t)]} \tag{4.1}$$

which becomes for ideal ($R_p/R_m = 0$ for an open circuit) conditions

$$\frac{e_o}{e_{ex}} = \frac{x_i}{x_t} \tag{4.2}$$

Thus for no "loading" the input-output curve is a straight line. In actual practice, $R_m \neq \infty$ and Eq. (4.1) shows a nonlinear relation between e_o and x_i. This deviation from linearity is shown in Fig. 4.4. The maximum error is about 12 percent of full scale if $R_p/R_m = 1.0$ and drops to about 1.5 percent when $R_p/R_m = 0.1$. For values of $R_p/R_m < 0.1$ the position of maximum error occurs in the neighborhood of $x_i/x_t = 0.67$, and the maximum error is approximately $15(R_p/R_m)$ percent of full scale.

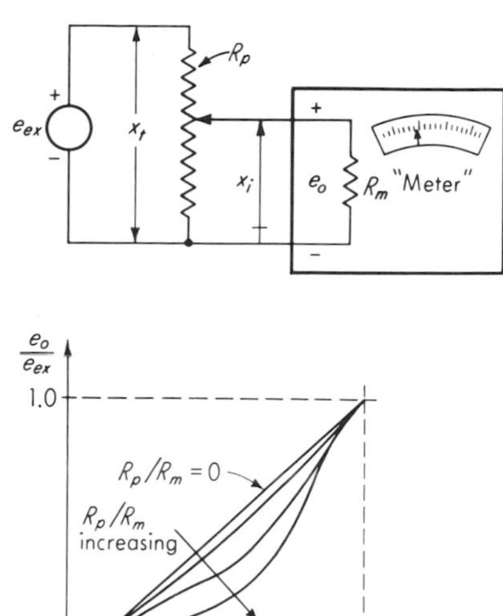

Fig. 4.4. *Potentiometer loading effect.*

We see that to achieve good linearity, for a "meter" of a given resistance R_m, one should choose a potentiometer of sufficiently *low* resistance relative to R_m. This requirement conflicts with the desire for high sensitivity. Since e_o is directly proportional to e_{ex}, it would seem possible to get any sensitivity desired simply by increasing e_{ex}. This is not actually the case, however, since potentiometers have definite power ratings related to their heat-dissipating capacity. Thus a manufacturer may design a series of potentiometers, say single-turn 2-in.-diameter, with a wide range (perhaps 100 to 100,000 ohms) of total resistance R_p but all these will be essentially the same size and mechanical configuration, giving the same heat-transfer capability and thus the same power rating, say about 5 watts at 70°F ambient. If the heat dissipation is limited to P watts, the maximum allowable excitation voltage is given by

$$\max e_{ex} = \sqrt{PR_p} \tag{4.3}$$

Thus a low value of R_p allows only a small e_{ex} and therefore a small sensitivity. Choice of R_p must thus be influenced by a tradeoff between loading and sensitivity considerations. The maximum available sensitivity of potentiometers varies considerably from type to type and also with size in a given type. It can be calculated from the manufacturer's data on maximum allowable voltage, current, or power and the maximum stroke. The shorter-stroke devices generally have higher sensitivity. *Extreme* values are of the order of 15 volts/deg for short-stroke rotational types ("sector" potentiometers) and 300 volts/in. for short-stroke (about

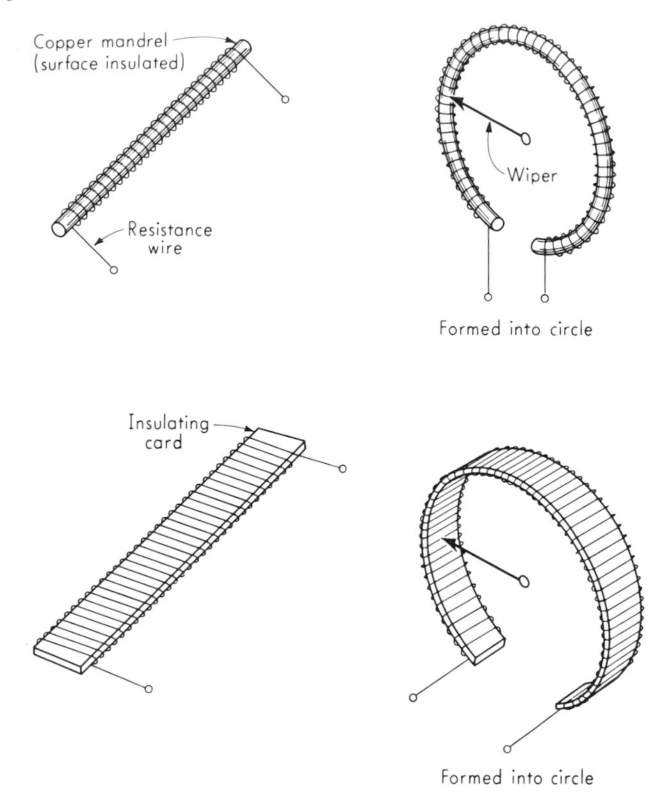

Fig. 4.5. *Construction of resistance elements.*

$\frac{1}{4}$ in.) translational pots. It must be emphasized that these are maximum values and that the usual application involves a much smaller (10 to 100 times smaller) sensitivity.

The resolution of potentiometers is strongly influenced by the construction of the resistance element. An obvious approach is to use a single slide-wire as the resistance, giving an essentially continuous stepless resistance variation as the wiper travels over it. Such potentiometers are available but are limited to rather small resistance values since the length of wire is limited by the desired stroke in a translational device and by space restrictions (diameter) in a rotational one. Resistance of a given length of wire can be increased by decreasing the diameter but this is limited by strength and wear considerations.

To get sufficiently high resistance values in small space the wire-wound resistance element is widely used. The resistance wire is wound on a mandrel or card which is then formed into a circle or helix if a rotational device is desired (see Fig. 4.5). With such a construction the

Wiper
Resistance wire
x

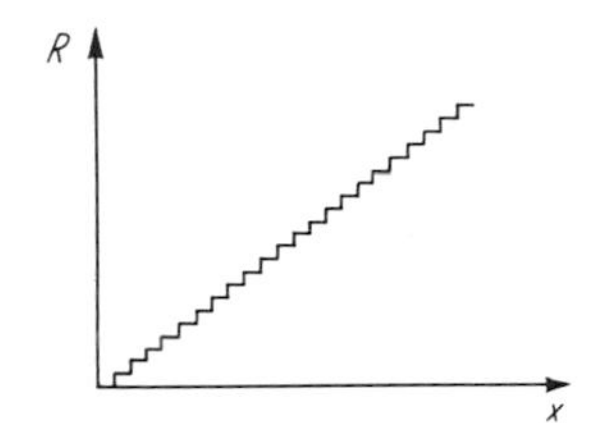

Fig. 4.6. *Resolution of wire-wound potentiometers.*

variation of resistance is not a linear continuous change but actually proceeds in small steps as the wiper moves from one turn of wire to the next (see Fig. 4.6). This phenomenon results in a fundamental limitation on the resolution in terms of resistance-wire size. For instance, if a translational device has 500 turns of resistance wire on a card 1 in. long, motion changes smaller than 0.002 in. cannot be detected. (This is slightly conservative since the wiper, in going from one turn to the next, goes through an intermediate position in which it is touching *both* turns at once.[1,2] The resolution thus actually varies from one position to another, the *worst* value being that given by a simple counting of turns per inch.) The actual practical limit for wire spacing according to current practice is between 500 and 1,000 turns per inch.[3] For translational devices, resolution is thus limited to 0.001 to 0.002 in. while single-turn rotational devices can trade off increased diameter D for increased angular resolution according to the relation

$$\text{Best angular resolution} = \frac{0.12 \text{ to } 0.24}{D} \quad \text{degrees} \qquad D \text{ in inches} \tag{4.4}$$

It should be noted that resolution is intimately related to total resistance since the fine wire required to get close wire spacing will naturally have a high resistance. Thus one cannot choose total resistance and resolution independently. If extremely fine resolution and high resistance are required a carbon-film or conductive-plastic resistance ele-

[1] H. Gray, How to Specify Resolution for Potentiometer Servos, *Control Eng.*, p. 129, November, 1959.

[2] C. A. Mounteer, The Effective Resolution of Wire-wound Potentiometers, *Giannini Tech. Notes*, G. M. Giannini Co., Pasadena, Calif., January–February, 1959.

[3] S. A. Davis and B. K. Ledgerwood, "Electromechanical Components for Servomechanisms," p. 53, McGraw-Hill Book Company, New York, 1961.

ment may be indicated. Carbon-film elements may have a resolution as good as 5×10^{-6} in. (the fact that it is not infinitesimally small is due to *granularity* in the surface) but the overall resolution of the potentiometer is somewhat poorer because of mechanical defects in bearings and wiper springs. In applying carbon-film devices it is important to keep in mind that the wiper is a relatively high-resistance contact (wire-wound devices have very low contact resistance). Thus the amount of current drawn from the potentiometer must be kept quite small; otherwise the iR drop across the wiper will cause errors in the output voltage.

Another approach to increased resolution involves the use of multi-turn potentiometers. The resistance element is in the form of a helix, and the wiper travels along a "lead screw." The number of wires per inch of element is still limited, as mentioned above, but an increase in resolution can be obtained by introducing gearing between the shaft whose motion is to be measured and the potentiometer shaft. For example, one rotation of the measured shaft could cause 10 rotations of the potentiometer shaft; thus the resolution of measured-shaft motion is increased by a factor of 10. Multiturn potentiometers are available up to about 60 turns. For translational devices various motion-amplifying mechanisms could be used in similar fashion.

Most potentiometers used for motion measurement are intended to give a linear input-output relation and are used as purchased, without calibration. Thus a specification of linearity is essentially equivalent to one of accuracy. Potentiometers are available in a wide range of linearities and corresponding prices. Linearity depends greatly on the uniformity of the resistance winding but errors in this can be corrected by adding fixed resistances in series and/or parallel at proper locations on the winding. This procedure also can correct loading errors so as to give a linear relation for a heavily loaded potentiometer.[1] The best nonlinearities commercially available range from 1 percent of full scale for $\frac{1}{2}$-in.-diameter single-turn pots through 0.02 percent for a 2-in.-diameter multiturn to 0.002 percent for a 10-in.-diameter multiturn. The best nonlinearities of translational pots are about 0.05 to 0.10 percent of full scale. It should be noted that accuracy can be no better (and is generally worse) than one-half the resolution; thus resolution places a limit on accuracy.

Noise in potentiometers refers to spurious-output-voltage fluctuations occurring during motion of the slider and includes the effects of resolution. In addition, various mechanical and electrical defects produce noise. In a wire-wound pot, motion of the slider over the resistance wires may cause bouncing of the contact at certain speeds, thus causing

[1] Davis and Ledgerwood, *op. cit.*, p. 59.

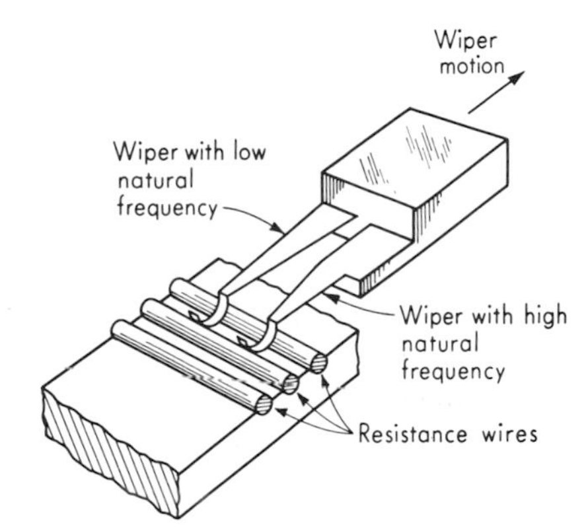

Fig. 4.7. *Antivibration wiper construction.*

intermittent contact. This phenomenon becomes particularly significant if the speed and wire spacing are such as to produce forces of frequency near the resonant frequency of the spring-loaded contact. Contacts are sometimes made in the configuration of Fig. 4.7 to overcome this problem. Here the resonant frequency of each section of the wiper is different. Thus if one section is resonating at a certain speed the other will be off resonance and making continuous contact. Another possibility lies in filling the interior of the potentiometer with a damping fluid to limit resonant amplitudes. This also generally increases the shock and vibration tolerance of the unit. Another source of noise is found in dirt and wear products which come between the contact surface and the winding. Even if no dirt or wear products are present, the contact resistance of a moving contact varies during motion, and if any load current is flowing through this contact a spurious iR voltage appears in the output. This effect also occurs at the slip-ring contact. Numerical values of noise voltage quoted in specifications generally include all sources of noise and correspond to a definite speed and current.[1]

The dynamic characteristic of potentiometers (considering displacement as input and voltage as output) is essentially that of a zero-order instrument since the impedance of the winding is almost purely resistive at the motion frequencies for which the device is usable. However, the mechanical loading imposed on the measured motion by the inertia and friction of the potentiometer's moving parts should be carefully considered. The friction is usually mostly dry friction, and the manufacturer generally supplies numerical values of the starting and running friction force or torque. These values vary over a wide range, depending on the construction of the potentiometer. Special low-friction rotary pots have starting torques as small as 0.003 oz-in. More conventional instruments

[1] PPMA Conference Report, *Electromech. Design*, p. 8, April, 1964.

may have 0.1 to 0.5 oz-in. or more. Translational pots have friction values from less than 1 oz to over 1 lb. Inertia values for both rotary and translatory pots vary widely with size. A typical $\frac{7}{8}$-in.-diameter single-turn pot has a moment of inertia of 0.12 g-cm^2 while a 2-in.-diameter 10-turn pot has about 18 g-cm^2. Moving masses of translatory pots have weights ranging from fractions of an ounce to several ounces.

Since the measured variable in a pot is displacement (a flow variable) the pertinent loading quantity is the generalized input admittance or compliance. The presence of dry friction (a discontinuous nonlinear effect) prevents one from defining an exact input admittance or compliance. If friction is small, it may be neglected and the system analyzed as if it contained inertia only. We may choose to work with either admittance or compliance. If we choose compliance, the input compliance of a pure mass is given by

$$C_{gi}(D) \triangleq \frac{\text{displacement}}{\text{force}}(D) = \frac{x}{f}(D) \tag{4.5}$$

and since

$$f = M\ddot{x} = MD^2x \tag{4.6}$$

we get

$$C_{gi}(D) = \frac{1}{MD^2} \tag{4.7}$$

or, in frequency-response terms,

$$C_{gi}(i\omega) = \frac{1}{(i\omega)^2 M} \tag{4.8}$$

For negligible loading effect the term C_{go}/C_{gi}, where C_{go} is the output compliance of the measured system, must be small compared with 1.0. Since $C_{gi}(i\omega)$ approaches infinity at low frequencies, loading is negligible at low frequencies so long as $C_{go}(i\omega)$ does not also approach infinity as $\omega \to 0$. The range of low frequencies for which $C_{gi}(i\omega)$ is large can be extended by decreasing the moving mass M.

Finally, selection of potentiometers should take into account various environmental factors such as high or low temperatures, shock and vibration, humidity, and altitude. These may act as modifying and/or interfering inputs so as seriously to degrade instrument performance. Under good environmental conditions the life of a potentiometer may be more than 20 million full strokes or rotations.

Resistance strain gages. Consider a conductor of uniform cross-sectional area A and length L, made of a material with resistivity ρ. The resistance R of such a conductor is given by

$$R = \frac{\rho L}{A} \tag{4.9}$$

If this conductor is now stretched or compressed, its resistance will change because of dimensional changes (length and cross-sectional area) and also because of a fundamental property of materials called *piezoresistance*[1] (pronounced pī-ēzō-resistance) which indicates a dependence of resistivity ρ on the mechanical strain. To find how a change dR in R depends on the basic parameters, we differentiate Eq. (4.9) to get

$$dR = \frac{A(\rho\, dL + L\, d\rho) - \rho L\, dA}{A^2} \tag{4.10}$$

Since volume $V = AL$, $dV = A\, dL + L\, dA$. Also

$$dV = L(1 + \epsilon)A(1 - \epsilon\nu)^2 - AL \tag{4.11}$$

where $\epsilon \triangleq$ unit strain
$\nu \triangleq$ Poisson's ratio

Since ϵ is small, $(1 - \nu\epsilon)^2 \approx 1 - 2\nu\epsilon$ and Eq. (4.11) becomes

$$dV = AL\epsilon(1 - 2\nu) = A\, dL + L\, dA \tag{4.12}$$

and since $\epsilon \triangleq dL/L$

$$A\, dL(1 - 2\nu) = A\, dL + L\, dA \tag{4.13}$$

$$-2\nu A\, dL = L\, dA \tag{4.14}$$

Substituting in Eq. (4.10)

$$dR = \frac{\rho A\, dL + LA\, d\rho + 2\nu\rho A\, dL}{A^2} \tag{4.15}$$

and thus
$$dR = \frac{\rho\, dL(1 + 2\nu)}{A} + \frac{L\, d\rho}{A} \tag{4.16}$$

Dividing by Eq. (4.9) gives

$$\frac{dR}{R} = \frac{dL}{L}(1 + 2\nu) + \frac{d\rho}{\rho} \tag{4.17}$$

and finally

$$\text{Gage factor} \triangleq \frac{dR/R}{dL/L} = \underbrace{1}_{\substack{\text{resistance}\\ \text{change due}\\ \text{to length}\\ \text{change}}} + \underbrace{2\nu}_{\substack{\text{resistance}\\ \text{change due to}\\ \text{area change}}} + \underbrace{\frac{d\rho/\rho}{dL/L}}_{\substack{\text{resistance change due}\\ \text{to piezoresistance}\\ \text{effect}}} \tag{4.18}$$

Thus if the gage factor is known, measurement of dR/R allows measurement of the strain $dL/L = \epsilon$. This is the principle of the resistance strain gage. The term $(d\rho/\rho)/(dL/L)$ can also be expressed as $\pi_1 E$, where

$\pi_1 \triangleq$ longitudinal piezoresistance coefficient
$E \triangleq$ modulus of elasticity

The material property π_1 can be either positive or negative. Poisson's

[1] C. M. Harris and C. E. Crede (eds.), "Shock and Vibration Handbook," vol. 1, p. 16-35, McGraw-Hill Book Company, New York, 1961.

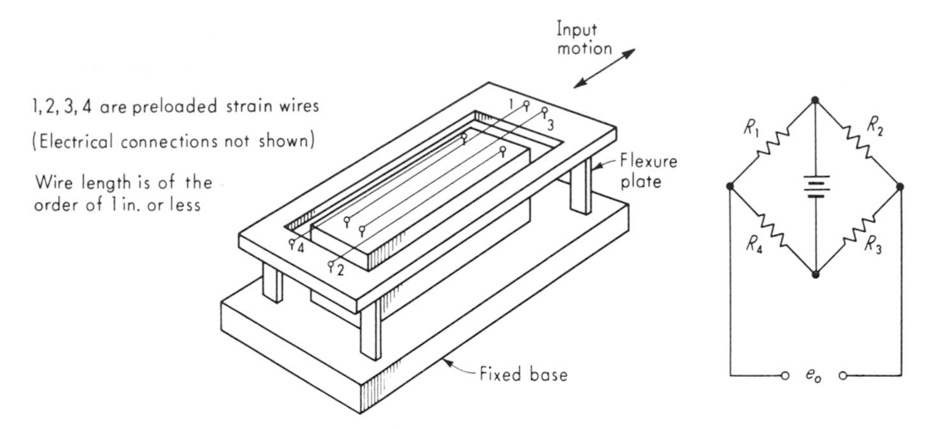

Fig. 4.8. *Unbonded strain gage.*

ratio is always between 0 and 0.5 for all materials. The most common type of strain gage uses one of the two alloys Advance (55 percent copper, 45 percent nickel) or Iso Elastic (36 percent nickel, 8 percent chromium, 4 percent manganese, silicon, and molybdenum; remainder iron). Advance gives a gage factor of about 2 and Iso Elastic about 3.5.

About 1960, strain gages based on semiconductor materials rather than metals began to become commercially available. While these are somewhat more expensive and more difficult to apply than metallic gages, their outstanding virtue is a very high gage factor of about 130. From Eq. (4.18) we can see that in the common metallic gages most of the resistance change comes from dimensional changes whereas in semiconductor gages most of it comes from piezoresistance effects. Ideally the gage factor would be a constant, and for metallic gages it can generally be treated as such. In semiconductor gages, however, π_1 varies somewhat with strain so that a nonlinear strain/resistance relationship exists. This tends to complicate the interpretation of readings from such gages. Intensive development is rapidly overcoming the disadvantages of semiconductor gages and they are already used in considerable quantities, especially in load cells, accelerometers, and other transducer applications.

For metallic gages two different methods of utilizing the above basic principle are in common use. These correspond to the *unbonded* and *bonded* strain gage. In the unbonded gage (shown in simplified form in Fig. 4.8) the resistance wires (about 0.001-in. diameter) are stretched between two frames which can move relative to each other as guided by flexure plates. Since the wires would buckle if compressive forces were applied, an internal preload greater than any expected external compressive load is employed. Under these conditions, applied motion to the right increasingly stretches wires 1 and 3 and reduces the tension in wires 2 and 4. A motion to the left does just the reverse, and so motions in both directions can be measured so long as the preload is not overcome. The resistance wires are generally connected in a bridge circuit (shown in its simplest form in Fig. 4.8). With the preload present but no external load applied, the bridge is balanced if $R_1/R_4 = R_2/R_3$. Adjustable resistors are generally provided in the bridge to accomplish this. An external load will then cause variation in resistance of the wires, unbalancing the bridge and causing an output voltage e_o in proportion to the motion. The motions directly measurable by gages of this type are very small, of the order of 0.0015 in. full scale.

Unbonded gages are used mainly as elements of force and pressure transducers and accelerometers rather than directly as displacement pickups. In these applications the strain wires often serve as the necessary spring element in transducing force to deflection, in addition to being the displacement sensor. The allowable force on the wires is very small, about 0.15 oz for the maximum deflection of 0.0015 in. Variation in the number, size, and length of wires allows design for a range of force and deflection values. The resolution of such devices is infinitesimally small since the resistance variation is a smooth change. Inaccuracy is of the order of 0.15 percent of full scale for a typical[1] unit. The sensitivity for the recommended 5-volt bridge excitation is 40 mv full-scale output for 0.0024-in. full-scale displacement (30 g full-scale force), that is, 16.7 volts/in. or 0.60 volt/lb$_f$. Thermal-sensitivity shift is 0.01 percent/F° between −65 and +250°F while thermal zero shift is 0.01 percent of full scale/F° between −65 and +250°F. The resistance of the bridge arms is nominally 350 ohms. With displacement considered as the input and bridge voltage e_o as output, the response of such instruments is essentially instantaneous (zero-order), if wave propagation is neglected. When measuring the displacement of a system, mechanical loading effects are again determined by the ratio of measured-system-output compliance to instrument-input compliance. Owing to the use of flexures rather than bearings, friction is negligible, and the instrument input is characterized by a moving mass M and a spring constant K_s determined by the resistance wires and the flexures. For such a situation, input compliance is given by

$$C_{gi}(D) \triangleq \frac{\text{displacement}}{\text{force}}(D) = \frac{x}{f}(D)$$

where

$$f - K_s x \doteq M\ddot{x} = MD^2x \tag{4.19}$$

$$C_{gi}(D) = \frac{1}{MD^2 + K_s} \tag{4.20}$$

$$C_{gi}(i\omega) = \frac{1}{(i\omega)^2 M + K_s} = \frac{1}{K_s - M\omega^2} \tag{4.21}$$

[1] Universal transducing cell, Statham Instruments Inc., Los Angeles, Calif.

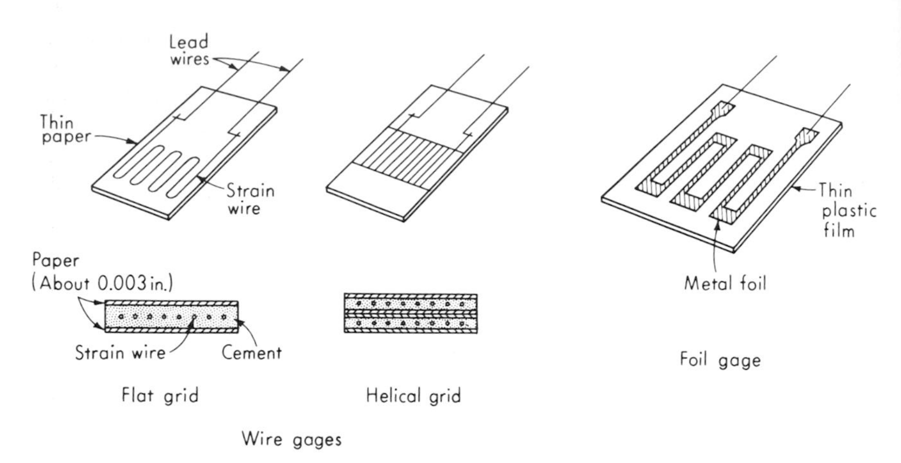

Fig. 4.9. *Bonded strain gages.*

As $\omega \to 0$, the static case is approached, and $C_{gi} \to 1/K_s$. Thus there will be a loading error for static operation unless the measured system has $C_{go}(i\omega) = 0$ for $\omega = 0$ (an infinitely stiff system). Since large compliance C_{gi} is desired to minimize loading, K_s should be small. Also note that C_{gi} increases with increasing frequency, reaching an infinite peak at $\omega^2 = K_s/M$ and then decreasing for higher frequencies.

Bonded metallic strain gages[1] use elements of wire in a flat grid or (flattened) helical construction or a thin metal foil printed and etched to give a grid-type pattern (see Fig. 4.9). Gages are available in sizes from about 6 in. in length to about $\frac{1}{64}$ in. These gages must be cemented to the surface whose strain is to be measured with a cement suitable to the environmental conditions. Once cemented down, the gages cannot be removed and reused. For wire gages the wire size is about 0.001-in. diameter. Foil gages can be made somewhat thinner; using a foil of about 0.00015-in. thickness and plastic film of 0.001 in. gives an overall thickness of about 0.001 in., somewhat thinner than wire gages.

When the gages are properly cemented down, they effectively become a part of the surface to which they are fastened and undergo essentially the same strain as that surface. They work equally well in both tension and compression since the matrix of cement surrounding the wire or foil completely prevents buckling. With suitable auxiliary electronic equipment, strains down to about 10^{-7} in./in. can be detected. While the useful upper strain limit (about 0.01 in./in.) for most gages is set by the elastic limit of the strain wires, special "post-yield" gages can

[1] C. C. Perry and H. R. Lissner, "The Strain Gage Primer," McGraw-Hill Book Company, New York, 1955.

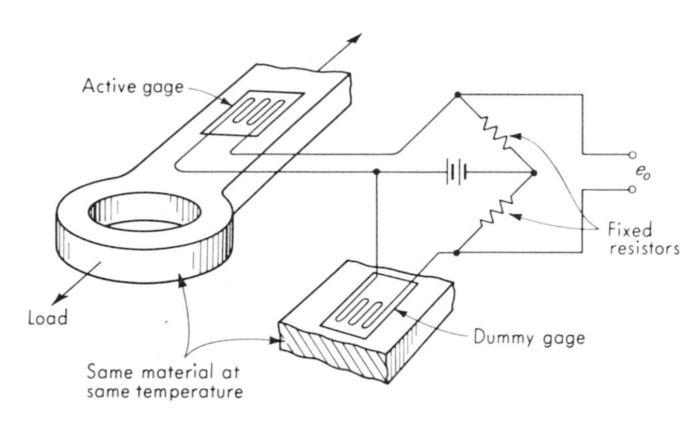

Fig. 4.10. *Strain-gage temperature compensation.*

be used to measure strains as great as 0.1 in./in. The gages are mainly sensitive to the component of strain along their longitudinal axis; however, there is some small transverse sensitivity because of the loops at the end of each turn of wire. This effect is usually less than 1 percent and is reduced even more in foil-type gages by making the end "loops" of greater cross section than the main portion. The total resistance of individual gages ranges from about 40 to 2,000 ohms, with 120, 350, and 1,000 ohms being common standard values. Heating of the gages limits the maximum allowable gage current. This varies with the type of gage and the heat-transfer conditions but is of the order of 0.030 amp.

Temperature is an important interfering input for strain gages since resistance changes with *both* strain and temperature. Since strain-induced resistance changes are quite small, the temperature effect can assume major proportions. Another aspect of temperature sensitivity is found in the possible differential thermal expansion of the gage and the underlying material. This can cause a strain and resistance change in the gage even though the material is not subjected to an external load. These temperature effects can be compensated in various ways. In Fig. 4.10 a "dummy" gage (identical to the active gage) is cemented to a piece of the same material as is the active gage and placed so as to assume the same temperature. The dummy and active gages are placed in adjacent legs of a Wheatstone bridge; thus resistance changes due to the temperature coefficient of resistance and differential thermal expansion will have no effect on the bridge output voltage whereas resistance changes due to an applied load will unbalance the bridge in the usual way (see text on bridge circuits). Another approach to this problem involves special inherently temperature-compensated gages. These gages are designed to be used on a specific material and have expansion and resistance properties

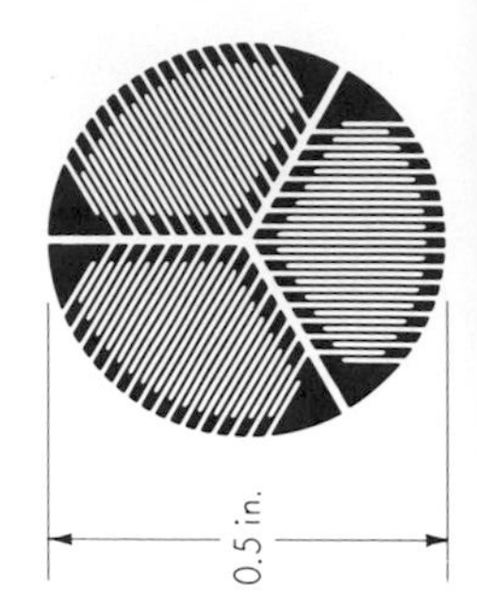

Fig. 4.11. *Foil rosette of Baldwin-Lima-Hamilton Corp.*

such that the two effects very nearly cancel each other and no dummy gage is required.

Temperature also can act as a modifying input in that it may change the gage factor. With metallic gages this effect is usually quite small except at extremely low or high temperatures. Semiconductor gages are more seriously affected in this way; however compensation is possible. Although the above temperature problems must be carefully considered in each application, strain gages have been successfully employed from liquid-helium temperature (7°R) to the order of 2000°F. However, these extreme (especially the high temperature) applications require special techniques and yield results of lower accuracy than is obtained in routine room-temperature situations.

Whereas the strain in a body is defined at a *point*, the element of a resistance strain gage extends over a finite *area*. Thus what strain is actually measured by such an instrument? Obviously it is some sort of average strain over the sensitive area of the gage. If the strain is uniaxial, if the gage is aligned with this axis, and if the strain gradient (rate of change of strain with distance along the surface) is constant, the average strain indicated by the gage will numerically equal the "point" strain at the midsection of the gage. The midpoint of the gage length is generally marked on the gage by the manufacturer. If the strain gradient is not constant and its form is unknown, the strain value read by the gage cannot be associated with any specific point. For such situations the smallest practical gage should be used in order to reduce this uncertainty. When the direction and magnitude of the maximum strain at a point are completely unknown, it can be shown that strain measurements in three different directions are sufficient to calculate the strain in any direction and thus its maximum value. To facilitate such measurements, strain-gage "rosettes" which combine the necessary three gages into one easily applied assembly have been developed. They are available in both wire and foil types, Fig. 4.11 showing a foil rosette.

To use a strain gage for strain measurement, one must know its resistance and gage factor. The resistance of a gage can be measured by standard techniques. However, the gage factor of an individual bonded gage cannot be found without cementing it to a simple member (for which strain can be accurately calculated from theory or measured by some independent means), applying known loads or deflections, and measuring the resulting resistance changes. The difficulty is that this calibrated gage *cannot* now be removed from the calibration member and recemented to a member whose unknown strain is to be measured. The manufacturer supplies along with a gage both its resistance and its gage factor but the gage factor of that particular gage has *not* itself been determined. Rather, periodic samples of the gage production are taken and calibrated, and the manufacturer relies on careful statistical quality control to ensure that the gage factors of gages sold do not deviate more than a specified percentage from the quoted figures. Thus a typical purchased gage might have a resistance of 120 ± 0.4 ohms and a gage factor of 2.14 ± 1 percent stated on the package. The resistance value can be measured more accurately by the user but the typical ± 1 percent uncertainty in the gage factor represents a basic limitation on gage accuracy for strain measurements.

Used directly, the bonded strain gage is useful for measuring only very small displacements (strains). However, larger displacements may be measured by bonding the gage to a flexible element such as a thin cantilever beam and applying the unknown displacement to the end of the beam, as in Fig. 4.12. For such an application the gage factor need not be accurately known since the overall system can be calibrated by applying known displacements to the end of the beam and measuring the resulting bridge output voltage. The configuration shown is temperature-compensated without the need for dummy gages and has four times the sensitivity of a single gage because of judicious application of bridge-circuit properties. Such transducers may be accurate to 0.1 percent of full scale.

The dynamic response of bonded strain gages with respect to faithfully reproducing as a resistance variation the strain variation of the underlying surface is very good. The dynamic effects of wave propagation in the cement and strain wires seem to be negligible for frequencies up to at least 50,000 cps, and so a zero-order dynamic model is generally adequate. The loading effect of the cement and strain wires on the underlying structure is generally negligible except for very thin members, in which the stiffening effect of the strain gage may reduce the measured strain to a value considerably lower than that present without the gage. Compliance techniques can be applied to study this effect.

The voltage output from metallic strain-gage circuits is quite small (a few microvolts to a few millivolts), and so amplification is generally needed. As an example, consider the measurement of a stress level of

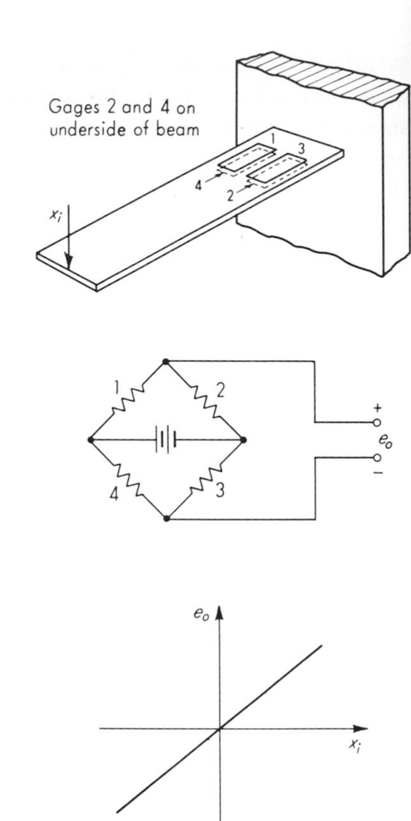

Fig. 4.12. *Beam displacement transducer.*

1,000 psi in steel with a single active gage of 120 ohms resistance and a gage factor of 2.0. If a bridge circuit of all equal arms is used, the maximum allowable bridge voltage for 30-ma gage current is

$$e_{ex} = (240)(0.030) = 7.2 \text{ volts} \tag{4.22}$$

The strain ϵ is $1{,}000/30 \times 10^6 = 3.33 \times 10^{-5}$ in./in., so that

$$\Delta R = (\text{gage factor})(\epsilon)(R) = (2)(3.33 \times 10^{-5})(120) = 7.99 \times 10^{-3} \text{ ohm} \tag{4.23}$$

For the given bridge arrangement,

$$e_o = e_{ex}\left(\frac{1}{4R}\right)\Delta R = \frac{(7.2)(7.99 \times 10^{-3})}{480} = 0.12 \text{ mv} \tag{4.24}$$

Based on limitations of the gage alone, the smallest detectable strain depends on the thermal or Johnson-noise[1] voltage generated in every

[1] E. B. Wilson, Jr., "An Introduction to Scientific Research," p. 116, McGraw-Hill Book Company, New York, 1952.

resistance because of the random motion of its electrons. This random voltage is essentially a white noise of spectral density $4kTR$ volts²/cps, where

$$k \triangleq \text{Boltzmann's constant} = 1.38 \times 10^{-23}\ \text{joule/K°} \qquad (4.25)$$
$$T \triangleq \text{absolute temperature of resistor, °K} \qquad (4.26)$$
$$R \triangleq \text{resistance, ohms} \qquad (4.27)$$

Thus if this voltage were measured by a hypothetical noise-free oscilloscope with a bandwidth of Δf cps, the measured rms voltage would be

$$E_{\text{noise,rms}} = \sqrt{4kTR\,\Delta f} \qquad \text{volts} \qquad (4.28)$$

As an example, a strain gage of $R = 120$ ohms at 300°K over a bandwidth of 100,000 cps would put out an rms noise voltage of 0.45 μv. Comparing this with our earlier calculation of the signal due to 1,000-psi stress, we see that the signal/noise ratio would be $120/0.45 = 267:1$. Suppose, however, that we wish to measure 1-psi stress rather than 1,000. The signal is then 0.12 μv, which is less than the noise; therefore the signal would be lost in the noise. Amplification under these conditions is of no use since the signal and noise are both amplified. This simple example does not cover other methods that have been developed to reduce this limitation but it should be understood that the limitation is a fundamental one and can be reduced but not overcome. Similar random fluctuations limit the measurable threshold of all physical variables.[1] In practical strain-gage measurement systems, Johnson noise of resistances other than the strain gage and other sources of noise in tubes, etc., actually limit the system resolution.

The useful operating life of bonded and unbonded strain gages is heavily influenced by environmental conditions. Unlike potentiometers, which *wear* out as a result of friction at the sliding contact, failure of strain gages, when it occurs, is often chargeable to fatigue of the metal wires or solder joints because of cyclic stressing. Since the metals used for strain wires are generally nonferrous, no endurance limit (stress below which failure *never* occurs) exists, and any stress, if repeated often enough, will ultimately cause failure. Under normal conditions this may take many millions of cycles, and so a long useful life is generally to be expected. Gradual or sudden failure of the cement is also responsible for some failures. For long-term applications a protective covering of wax or other material to prevent entrance of humidity may be indicated.

Differential transformers. Figure 4.13 shows schematic and circuit diagrams for translational and rotational linear variable-differential-transformer (LVDT) displacement pickups. The excitation of such devices is

[1] *Ibid.*

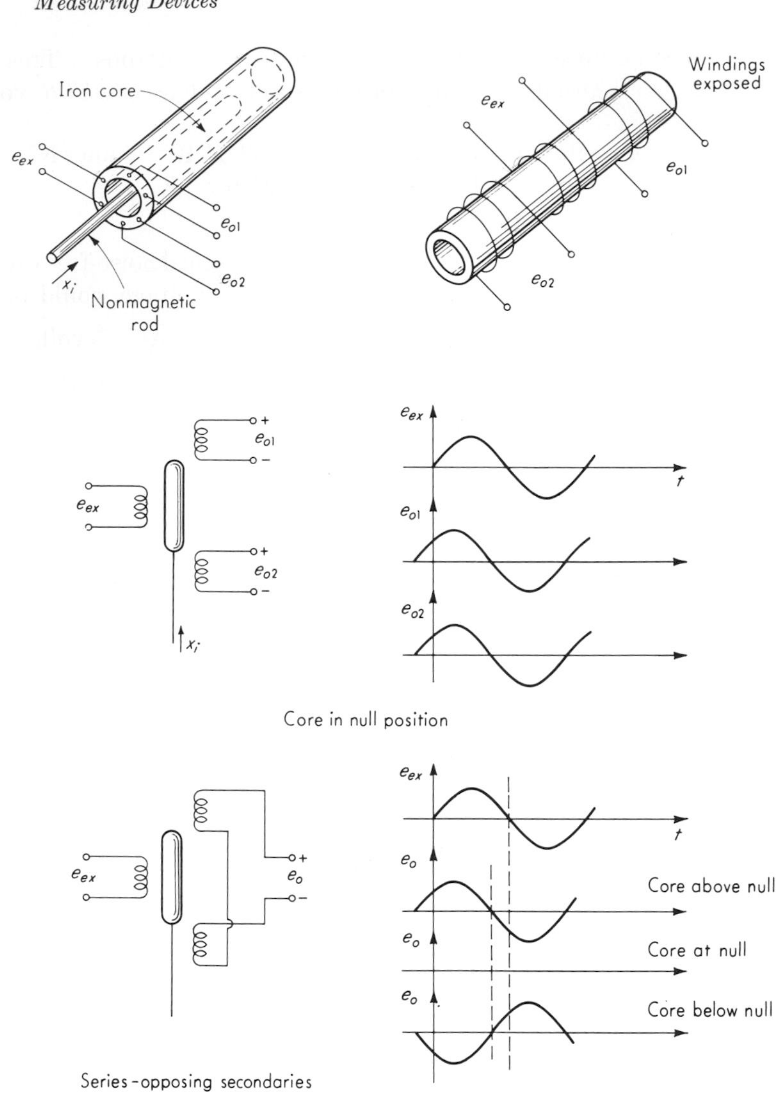
Iron core
e_{ex}
e_{o1}
e_{o2}
x_i
Nonmagnetic rod
Windings exposed
e_{ex}
e_{o1}
e_{o2}
e_{o1}
e_{ex}
e_{o2}
x_i
t
Core in null position
e_{ex}
e_o
t
Core above null
Core at null
Core below null
Series-opposing secondaries

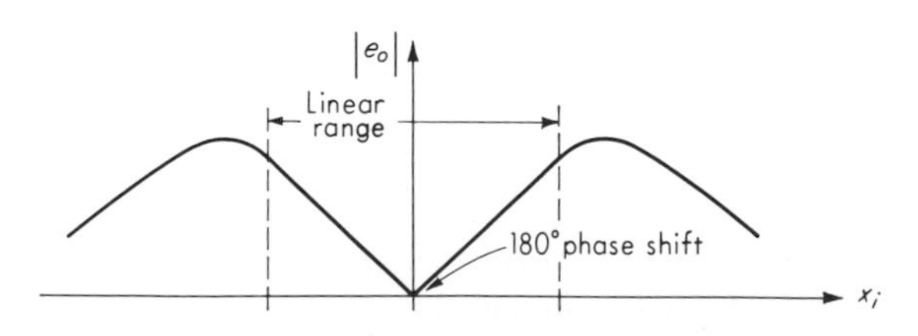
$|e_o|$
Linear range
180° phase shift
x_i

normally a sinusoidal voltage of 3 to 15 volts rms amplitude and frequency of 60 to 20,000 cps. The two identical secondary coils have induced in them sinusoidal voltages of the same frequency as the excitation; however, the amplitude varies with the position of the iron core. When the secondaries are connected in series opposition, a null position exists ($x_i \triangleq 0$) at which the net output e_o is essentially zero. Motion of the core from null then causes a larger mutual inductance (coupling) for one coil and a smaller mutual inductance for the other, and the amplitude of e_o becomes a nearly linear function of core position for a considerable range either side of null. The voltage e_o undergoes a 180° phase shift in going through null. The output e_o is generally out of phase with the excitation e_{ex}; however, this varies with the frequency of e_{ex}, and for each differential transformer there exists a particular frequency (numerical value supplied by the manufacturer) at which this phase shift is zero. If the differential transformer is used with some readout system that requires a small phase shift between e_o and e_{ex} (some carrier-amplifier systems require this), excitation at the correct frequency can solve this problem. If the output voltage is applied directly to an a-c meter or an oscilloscope this phase shift is not a problem.

The origin of this phase shift can be seen from analysis of Fig. 4.14. Applying Kirchhoff's voltage-loop law, we get

$$i_pR_p + L_p\left(\frac{di_p}{dt}\right) - e_{ex} = 0 \qquad (4.29)$$

Now the voltage induced in the secondary coils is given by

$$e_{s1} = M_1\frac{di_p}{dt} \qquad (4.30)$$

$$e_{s2} = M_2\frac{di_p}{dt} \qquad (4.31)$$

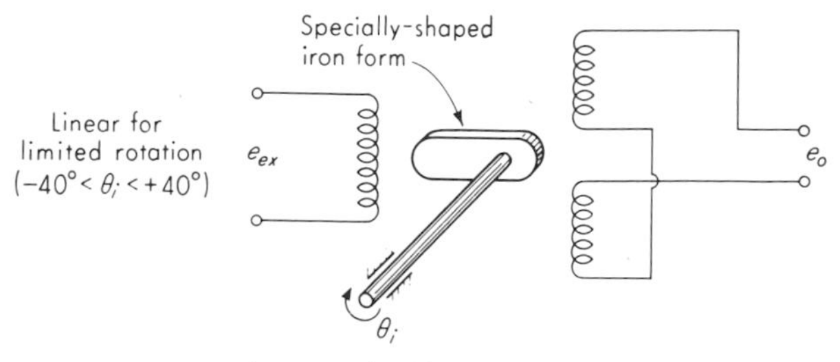

Fig. 4.13. *Differential transformer.*

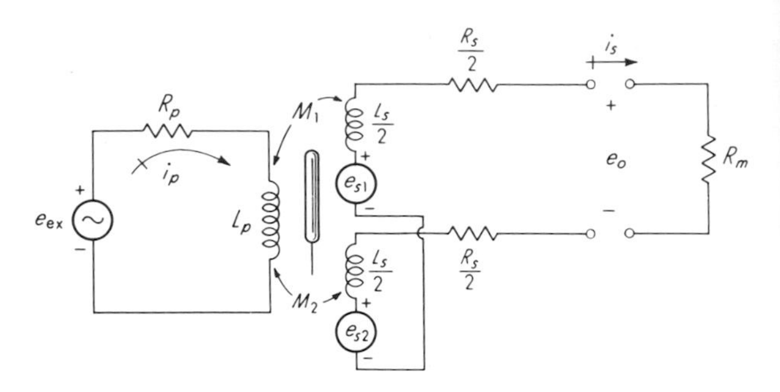

Fig. 4.14. *Circuit analysis.*

where M_1 and M_2 are the respective mutual inductances. The net secondary voltage e_s is then given by

$$e_s = e_{s1} - e_{s2} = (M_1 - M_2)\frac{di_p}{dt} \tag{4.32}$$

The net mutual inductance $M_1 - M_2$ is the quantity that varies linearly with core motion. If the output is open circuit (no voltage-measuring device attached), we have for a fixed core position

$$e_o = e_s = (M_1 - M_2)\frac{D}{L_pD + R_p}e_{ex} \tag{4.33}$$

and thus

$$\frac{e_o}{e_{ex}}(D) = \frac{[(M_1 - M_2)/R_p]D}{\tau_pD + 1} \qquad \tau_p \triangleq \frac{L_p}{R_p} \tag{4.34}$$

In terms of frequency response,

$$\frac{e_o}{e_{ex}}(i\omega) = \frac{(M_1 - M_2)/R_p}{\sqrt{(\omega\tau_p)^2 + 1}}\underline{/\phi} \qquad \phi = 90° - \tan^{-1}\omega\tau_p \tag{4.35}$$

thus demonstrating the phase shift between e_o and e_{ex}. If a voltage-measuring device of input resistance R_m is attached to the output terminals, a current i_s will flow, and we can write

$$-\frac{(M_1 - M_2)D}{L_pD + R_p}e_{ex} + (R_s + R_m)i_s + L_s\frac{di_s}{dt} = 0 \tag{4.36}$$

Then

$$i_s = \frac{M_1 - M_2}{R_p(R_s + R_m)}\frac{De_{ex}}{(\tau_pD + 1)(\tau_sD + 1)} \qquad \tau_s \triangleq \frac{L_s}{R_s + R_m} \tag{4.37}$$

and since $e_o = i_sR_m$,

$$\frac{e_o}{e_{ex}}(D) = \frac{(M_1 - M_2)R_m}{R_p(R_s + R_m)}\frac{D}{(\tau_pD + 1)(\tau_sD + 1)} \tag{4.38}$$

Since the frequency response of $(e_o/e_{ex})(i\omega)$ has a phase angle of $+90°$ at low frequencies and $-90°$ at high, somewhere in between it will be zero,

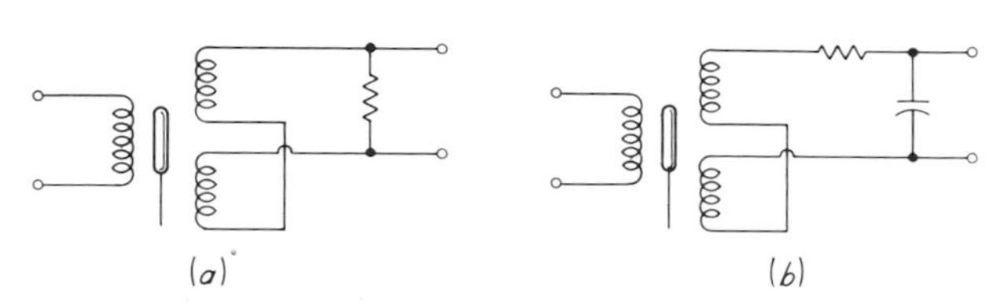

Two possible methods for retarding a leading phase angle

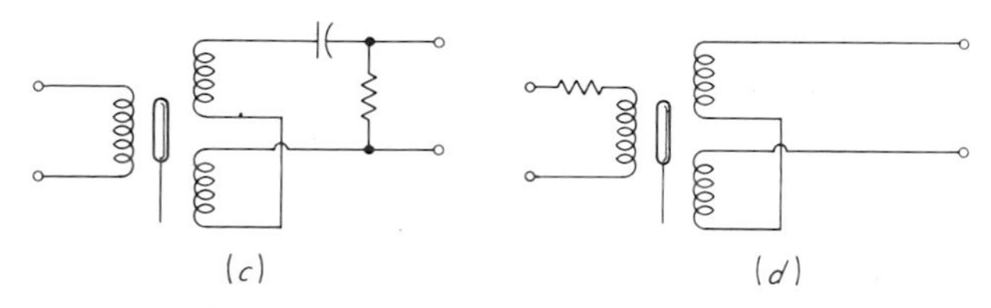

Two possible methods for advancing a lagging phase angle

Fig. 4.15. *Phase-angle-adjustment circuits.*

as mentioned earlier. If, for some reason, the excitation frequency cannot be adjusted to this value, the same effect may be achieved for a given frequency by one of the methods[1,2] shown in Fig. 4.15.

While the output voltage at the null position is ideally zero, harmonics in the excitation voltage and stray capacitance coupling between the primary and secondary usually result in a small but nonzero null voltage. Under usual conditions this is less than 1 percent of the full-scale output voltage and may be quite acceptable. Methods of reducing this null when it is objectionable are available. First, the preferred connection shown in Fig. 4.16*a* should be used if a balanced (center-tapped) excitation-voltage source is available. The grounding shown tends to reduce capacitance-coupling effects. If a center-tapped voltage source is not available, the arrangement of Fig. 4.16*b* can be used. With the core at the null position and the output-measuring device connected, the potentiometer is adjusted until the minimum null reading is obtained. The values of R and R_p are not critical but should be as low as possible without loading (drawing excessive current from) the excitation source.

The output of a differential transformer is a sine wave whose amplitude is proportional to the core motion. If this output is applied to an a-c voltmeter the meter reading can be directly calibrated in motion units. This arrangement is perfectly satisfactory for measurement of static or very slowly varying displacements except that the meter will give exactly

[1] A. Miller, Differential Transformers, *The Right Angle,* The Sanborn Co., Waltham, Mass., August, 1956; November, 1956.

[2] Schaevitz Engineering, Camden, N.J., *Bull.* AA-1A.

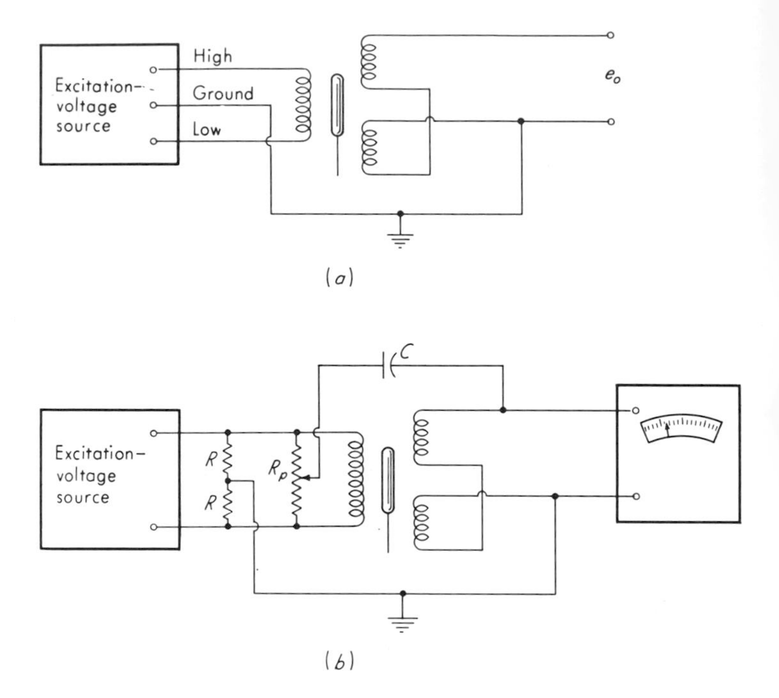

Fig. 4.16. *Methods for null reduction.*

the same reading for displacements of equal amount on *either* side of the null since the meter is not sensitive to the 180° phase change at null. Thus one cannot tell to which side of null the reading applies without some independent check. Furthermore, if rapid core motions are to be measured, the meter cannot follow or record them, and an oscillograph or oscilloscope must be used as a readout device. These instruments record the actual wave form of the output as an amplitude-modulated sine wave, which is usually undesirable; what is desired is an output-voltage record that looks like the mechanical motion being measured. To achieve the desired results, demodulation and filtering must be performed; if it is necessary to detect unambiguously the motions on both sides of null, the demodulation must be phase-sensitive. Many different circuits are available for performing these operations. We show here only one arrangement, which is quite simple. To use this approach, all four output leads of the LVDT must be accessible (some have the series opposition connection *internal* to the case and would thus not be applicable to the following discussion).

Figure 4.17*c* shows the circuit arrangement for phase-sensitive demodulation using semiconductor diodes. Ideally these pass current

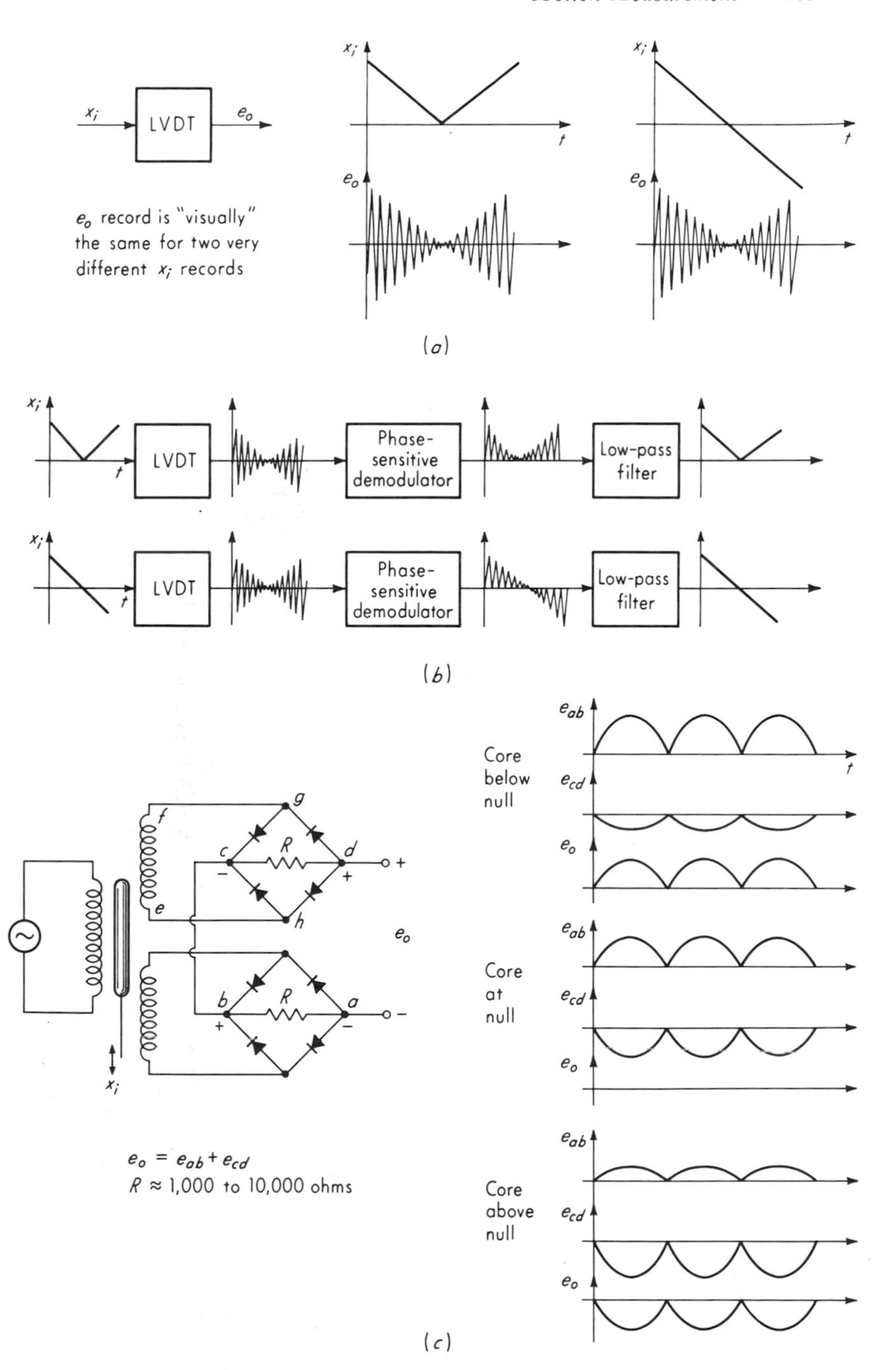

Fig. 4.17. *Demodulation and filtering.*

only in one direction; thus when f is positive and e is negative the current path is *efgcdhe*, while when f is negative and e positive the path is *ehcdgfe*. The current through R is therefore always from c to d. A similar situation exists in the lower diode bridge. For static or very slowly varying core displacements the voltage e_o may be applied directly to a d-c voltmeter. The meter will act as an electromechanical low-pass filter, the needle assuming a position corresponding to the average value of the rectified sine wave e_o. If motions both sides of null are to be measured, a meter with zero in the center of the scale will eliminate the need for switching lead wires when e_o goes negative. When rapid core motions are to be measured, this d-c meter arrangement is useless since the meter movement cannot follow variations more rapid than about 1 cps. It is then necessary to connect e_o of Fig. 4.17*c* to the input of a low-pass filter which will pass the frequencies present in x_i but reject all those (higher) frequencies produced by the modulation process. The design of such a filter is eased by making the LVDT excitation frequency much higher than the x_i frequencies.

If a frequency ratio of 10:1 or more is feasible, a simple RC filter as in Fig. 4.18*a* may be adequate. The output of this filter then becomes the input to an oscillograph or oscilloscope. For example, suppose we wish to measure a transient x_i whose Fourier transform has dropped to insignificant magnitude for all frequencies higher than 1,000 cps. Suppose also an LVDT system with an excitation frequency of 10,000 cps is available. The frequencies produced by the modulation process will thus lie in the band 19,000 to 21,000 cps. Suppose that we desire the "ripple" due to frequencies at 19,000 cps and higher to be no more than 5 percent. The filter time constant $\tau_f = R_f C_f$ can then be calculated as

$$0.05 = \frac{1}{\sqrt{[(19{,}000)(6.28)\tau_f]^2 + 1}} \tag{4.39}$$

$$\tau_f = 0.00017 \text{ sec} \tag{4.40}$$

At the highest motion frequency (1,000 cps) this filter has an amplitude ratio of 0.68 and a phase shift of $-47°$; thus it will distort the high-frequency portion of the x_i transient considerably. A more selective (sharper cutoff) filter would help this situation. Consider the double RC filter of Fig. 4.18*b*. The value of τ_f for a 5 percent ripple is now obtained from

$$0.05 = \frac{1}{[(19{,}000)(6.28)\tau_f]^2 + 1} \tag{4.41}$$

$$\tau_f = 0.000037 \text{ sec} \tag{4.42}$$

Now, at 1,000 cps the amplitude ratio is 0.98 and the phase angle is $-13°$. Since the phase angle of this filter from $\omega = 0$ to $\omega = 6{,}280$ rad/sec is

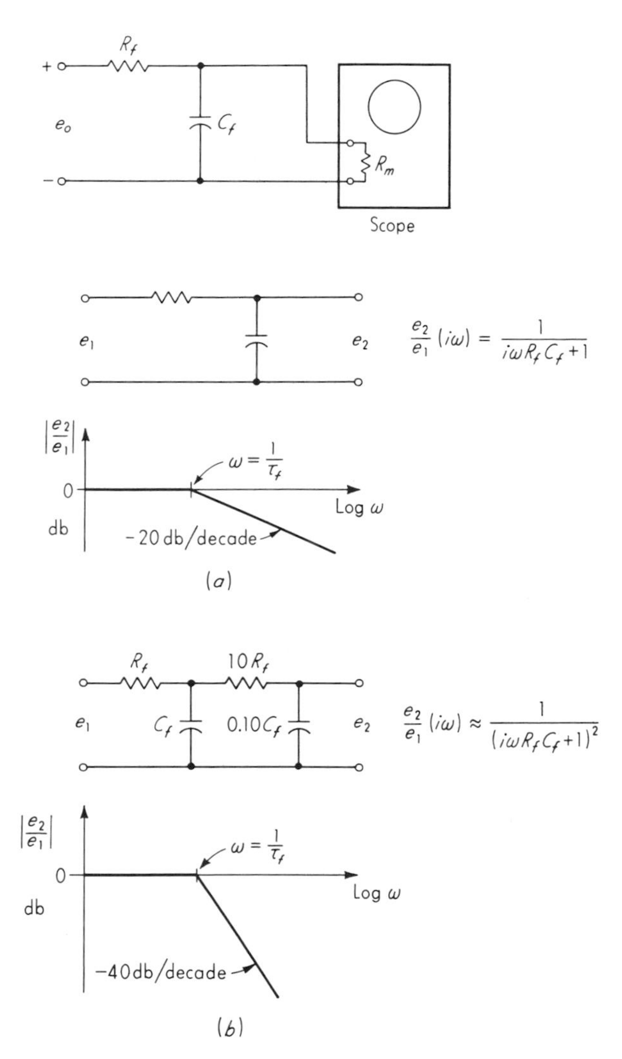

Fig. 4.18. *Filter frequency response.*

nearly linear and the amplitude ratio is nearly flat (1.0 to 0.98), the wave form of the transient will be faithfully reproduced but with a delay (dead time) of about $13/[(57.3)(6{,}280)] = 36\ \mu\text{sec}$. The above calculations give the desired value of τ_f but not R_f and C_f directly. In going from the demodulator circuit to the filter and then to the oscilloscope, transfer functions of the individual elements can be multiplied together only if no significant loading of the successive stages is present. As a rule of thumb the impedance level should go up about 10 to 1 for each successive stage. A typical oscilloscope input resistance is about 10^6 ohms. This suggests that in Fig. 4.18*b* $10R_f$ could be about 10^5 ohms and R_f thus 10^4 ohms. In Fig. 4.17*c* the demodulator R can be of the order of 10^3 ohms, and so the overall chain should not show much loading effect and our above calculations should be fairly accurate. In any case, experimental checks of the system should be performed to verify the final design. If R_f of 10^4 ohms is satisfactory, C_f will then be $(37 \times 10^{-6})/10^4 = 0.0037\ \mu f$.

The full-range stroke of commercially available translational LVDT's ranges from about ± 0.005 to about ± 3 in., with other sizes available as specials. The nonlinearity of standard units is of the order of 0.5 percent of full scale, with 0.1 percent possible by selection. Sensitivity with normal excitation voltage of 3 to 6 volts is of the order of 0.6 to 30 mv per 0.001 in., depending on frequency of excitation (higher frequency gives more sensitivity) and stroke (smaller strokes usually have higher sensitivity). Some special units have sensitivity as high as 1 to 1.5 volts per 0.001 in. Since the coupling variation due to core motion is a continuous phenomenon, the resolution of LVDT's is infinitesimal. Amplification of the output voltage allows detection of motions down to a few microinches. There is no physical contact between the core and the coil form; thus there is no friction or wear. There are, however, small radial and longitudinal magnetic forces on the core if it is not centered radially and at the null position. These are in the nature of magnetic "spring" forces in that they increase with motion from the equilibrium point. They are rarely more than 0.1 to 0.3 g and are thus often negligible. Rotary LVDT's have a nonlinearity of about ± 1 percent of full scale for travel of $\pm 40°$ and ± 3 percent for $\pm 60°$. The sensitivity is of the order of 10 to 20 mv/deg. The moving mass (core) of LVDT's is quite small, ranging from less than 0.1 g in small units to 5 g or more in larger ones. There is a small radial clearance (air gap) between the core and the hole in which it moves. Motion in the radial direction produces a small output signal but this undesirable transverse sensitivity is usually less than 1 percent of the longitudinal sensitivity.

The dynamic response of LVDT's is limited mainly by the excitation frequency, since it must be much higher than the core-motion frequencies so as to be able to distinguish between them in the amplitude-modulated output signal. For adequate demodulation and filtering, a frequency ratio much less than 10:1 presents problems. Since few differential transformers are designed to be excited by more than 20,000 cps, the useful range of motion frequencies is limited to about 2,000 cps. This is adequate for many applications. The mechanical loading effect on the measured system is mainly mass; thus the compliance analysis of Eq. (4.5), etc., is applicable. If the small "magnetic spring" force is not negligible, Eq. (4.19), etc., should be used.

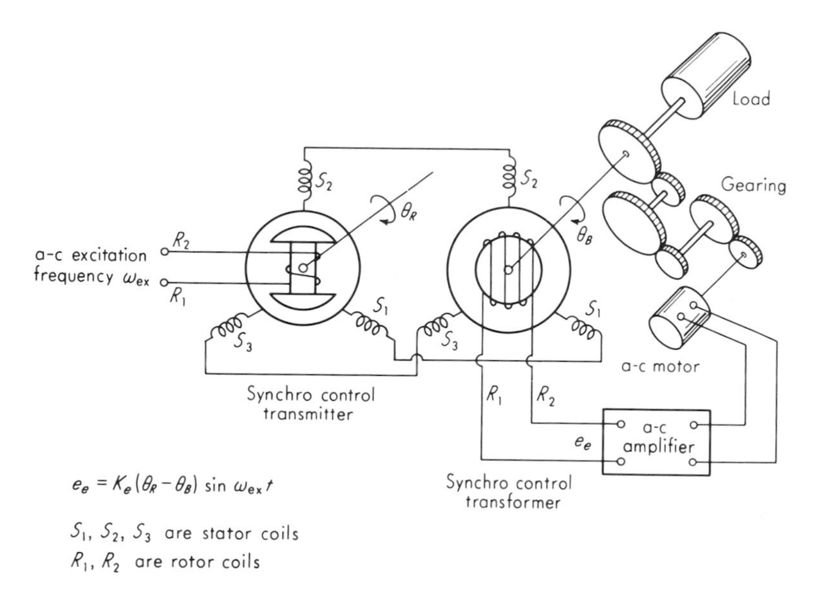

Fig. 4.19. *A-C servomechanism.*

Synchros and induction potentiometers. The term synchro is applied to a family of a-c electromechanical devices which, in various forms, perform the functions of angle measurement, voltage and/or angle addition and subtraction, remote angle transmission, and computation of rectangular components of vectors. In this section we are concerned only with the angle-measuring function; equipment for performing the other functions is covered in later appropriate chapters.

Synchros for angle measurement are most utilized as components of servomechanisms (automatic motion-control feedback systems) where they are used to measure and compare the actual rotational position of a load with its commanded position, as in Fig. 4.19. To perform this function two different types of synchros, the control transmitter and the control transformer, are used. The error voltage signal e_e is an a-c voltage of the same frequency as the excitation and of amplitude proportional (for small error angles) to the error angle $\theta_R - \theta_B$. Its phase changes by 180° at the null point; thus the direction of the error is detected. When $\theta_R = \theta_B$, the error voltage (and thus the amplifier output and motor input) is zero and the system stays at rest. If a command rotation θ_R is now put in, $e_e \neq 0$ and the motor will rotate so as to return θ_B to correspondence with θ_R.

The physical construction of the control transmitter and control transformer is identical except that the transmitter has a salient-pole

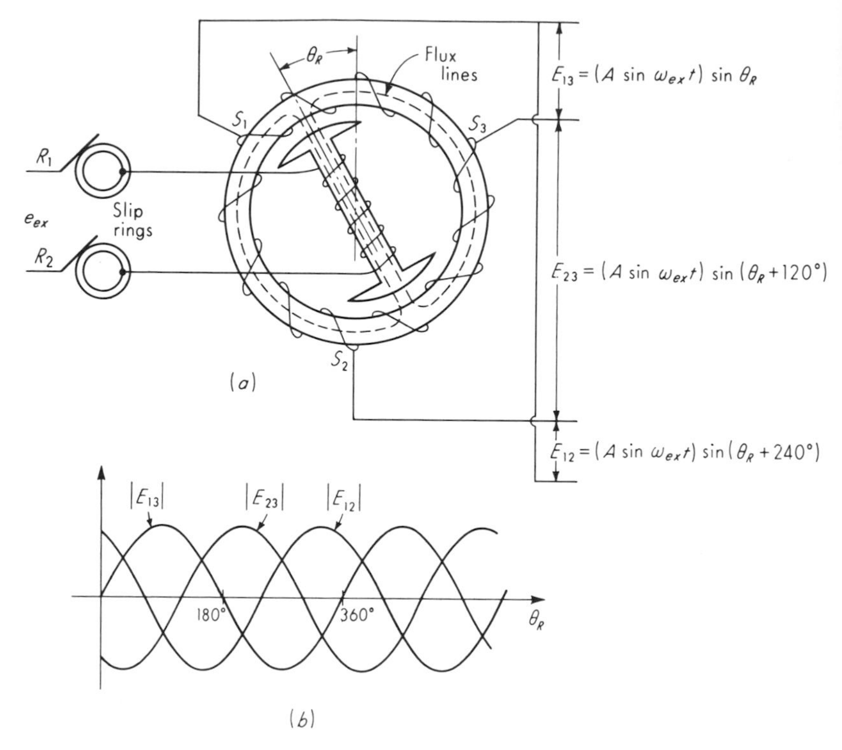

Fig. 4.20. *Synchro.*

("dumbbell") rotor while the transformer has a cylindrical rotor. The construction is similar to that of a wound-rotor induction motor. Figure 4.20*a* shows the coil arrangement of the transmitter alone. Basically, rotation of the rotor changes the mutual inductance (coupling) between the rotor coil and the stator coils. For a given stator coil the open-circuit output voltage is sinusoidal in time and varies in amplitude with rotor position, also sinusoidally, as shown in Fig. 4.20*b*. The three voltage signals from the stator coils uniquely define the angular position of the rotor. When these three voltages are applied to the stator coils of a control transformer, they produce a resultant magnetomotive force aligned in the same direction as that of the transmitter rotor. The rotor of the transformer acts as a "search coil" in detecting the direction of its stator field. If the axis of this coil is aligned with the field, the maximum voltage is induced into the transformer rotor coil. If the axis is perpendicular to the field, zero voltage is induced, giving the null position mentioned above. The output-voltage amplitude actually varies sinusoidally with the misalignment angle, but for small angles the sine and the angle are nearly equal, giving a linear output.

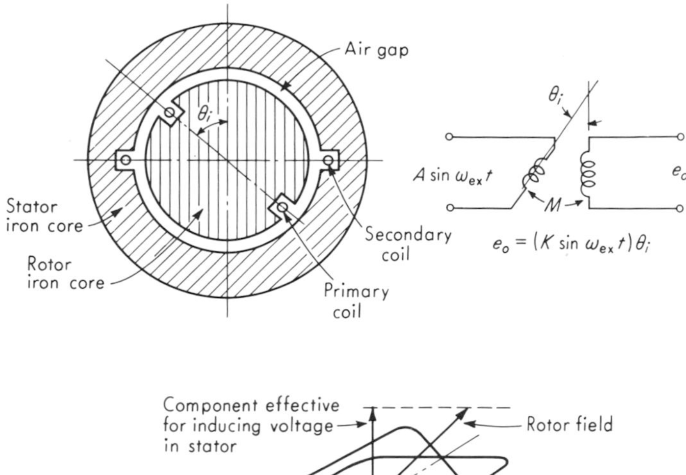

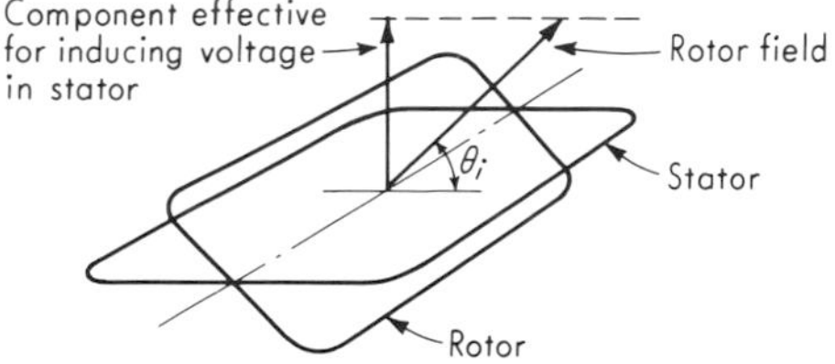

Fig. 4.21. *Induction potentiometer.*

In an induction potentiometer there is one winding on the rotor and one on the stator. (Additional dummy windings are sometimes used to improve accuracy, however.) Both these windings are concentrated; thus for simplicity we show them as single-turn coils in Fig. 4.21. The primary winding (rotor) is excited with alternating current. This induces a voltage into the secondary (stator). The amplitude of this output voltage varies with the mutual inductance (coupling) between the two coils, and this varies with the angle of rotation. For single-turn (concentrated) coils the variation with angle would be sinusoidal and only a small linear range around null would be obtained. By carefully *distributing* the rotor and stator windings a linear relation for up to $\pm 90°$ rotation may be obtained.

While synchros and induction pots could be designed to work at a variety of excitation frequencies, standard commercial units are generally available only for 60 or 400 cps. The physical size ranges from about $\frac{1}{2}$- to 3-in. diameter. Sensitivities of both synchros and induction pots are of the order of 1 volt/degree rotation while the residual voltage at null is of the order of 10 to 100 mv. For a standard synchro transmitter-transformer pair the misalignment of the two shafts when rotated from an originally established electrical null to any other null position within a complete rotation is of the order of 10 angular minutes. This type of error puts a basic limit on the positioning accuracy of servo systems using synchros. Synchro pairs and induction pots are capable of continuous rotation although the linear range of induction pots is limited to about ± 60 to $\pm 90°$. Within this range the nonlinearity is of the order of 0.25 percent.

Just as in LVDT's, synchros and induction pots require some sort of phase-sensitive demodulation to obtain a signal of the same form as the mechanical-motion input. When used in a-c servomechanisms, the conventional two-phase a-c servomotor itself accomplishes this function without any additional equipment. In a strictly measurement (as opposed to control) application, some sort of phase-sensitive demodulator is needed if an electrical output signal of the same form as the mechanical input is required. The dynamic response is limited by the excitation frequency and demodulator filtering requirements, just as in LVDT's. The mechanical loading of these rotary components on the measured system is mainly the inertia of the rotor.

Variable-inductance and variable-reluctance pickups. Closely related to LVDT's and synchros but in practice distinguished from them by name is a family of motion pickups variously called variable-inductance, variable-reluctance, or variable-permeance (permeance is the reciprocal of reluctance) pickups or transducers. The terminology used for these pickups is not uniform nor necessarily descriptive of their basic principles of operation. We are here concerned mainly with describing some common examples rather than trying to develop a systematic nomenclature.

Figure 4.22*a* shows the arrangement of a typical translational variable-inductance transducer. Outwardly the physical size and shape are very similar to an LVDT. Again there is a movable iron core which provides the mechanical input. However, only two inductance coils are present; they generally form two legs of a bridge which is excited with alternating current of 5 to 30 volts at 60 to 5,000 cps. With the core at the null position, the inductance of the two coils is equal, the bridge is balanced, and e_o is zero. A core motion from null causes a change in the reluctance of the magnetic paths for each of the coils, increasing one and decreasing the other. This reluctance change causes a proportional change in inductance for each coil, a bridge unbalance, and thus an output voltage e_o. By careful construction, e_o can be made a nearly linear function of x_i over the rated displacement range.

Two alternative methods of forming the bridge circuit are shown in Fig. 4.22*b*. The total transducer impedance (Z_1 plus Z_2) at the excitation frequency is of the order of 100 to 1,000 ohms. The resistors R are usually about the same value as Z_1 and Z_2, and the input impedance of

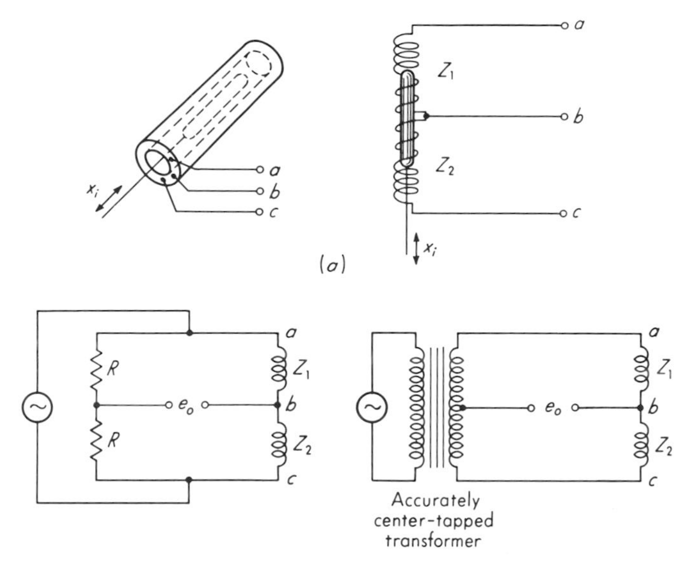

Fig. 4.22. *Variable-inductance pickup.*

the voltage-measuring device at e_o should be at least $10R$. If the bridge output must be worked into a low-impedance load, R must be quite small. To get high sensitivity, high excitation voltage is needed; this causes a high power loss (heating) in the resistors R. To solve this problem, a center-tapped transformer circuit may be used. Here the bridge is mainly inductive, and less power is consumed with corresponding less heating.

Such variable-inductance transducers are available in strokes of about 0.1 in. to as much as 200 in. The resolution is infinitesimal, and the nonlinearity ranges from about 1 percent of full scale for standard units to 0.02 percent for special units of rather long stroke. Sensitivity is of the order of 5 to 40 volts/in. Rotary versions using specially shaped rotating cores have a nonlinearity of the order of 0.5 to 1 percent of full scale over a $\pm 45°$ range. The sensitivity is about 0.1 volt/degree.

Figure 4.23 shows another common version of the variable-reluctance principle. This particular application is an accelerometer for measurement of accelerations in the range $\pm 4g$. Since the force required to accelerate a mass is proportional to the acceleration, the springs supporting the mass in Fig. 4.23 deflect in proportion to the acceleration; thus a displacement measurement allows an acceleration measurement. The

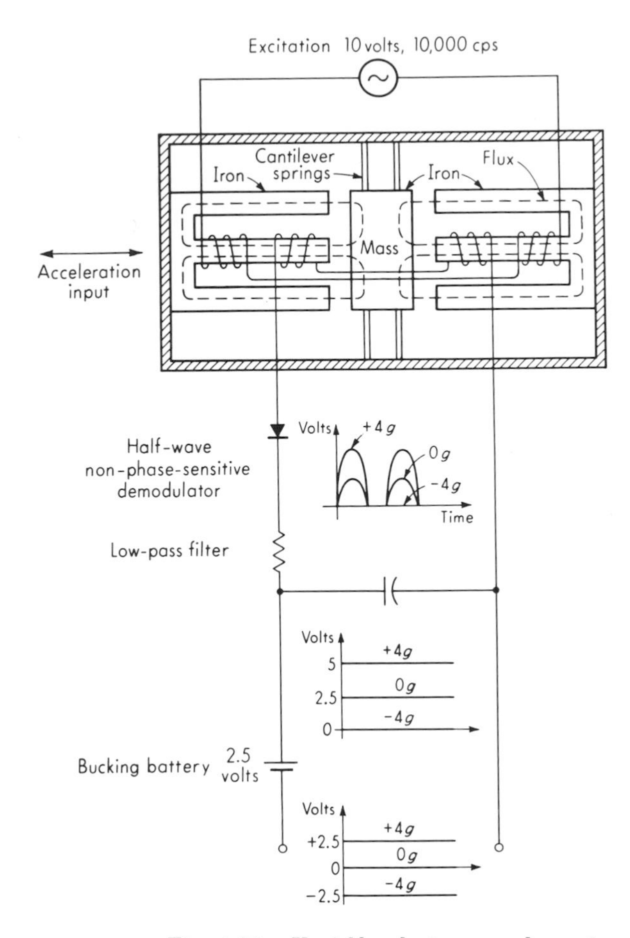

Fig. 4.23. *Variable-reluctance accelerometer.*

mass is of iron and thus serves as both an inertial element for transducing acceleration to force and also as a magnetic circuit element for transducing motion to reluctance.

We shall consider the complete instrument here since it has several features of general interest with regard to displacement measurement. Ordinarily, such an instrument would be constructed so that the iron core would be halfway between the two E frames when the acceleration was zero, thus giving zero output voltage for zero acceleration. However, to detect motion on both sizes of zero (corresponding to plus or minus accelerations) a fairly involved phase-sensitive demodulator would be required. It was desired to save the cost, weight, and space of this demodulator, and so another solution (which can also be used with LVDT's and similar devices) was proposed. With zero-acceleration input the iron core and springs were adjusted so that the core was offset to one side by an amount equal to the spring deflection corresponding to $4g$ acceleration. Thus, with no acceleration applied, the output voltage was not zero but some specific value (2.5 volts in this particular case). Then, when $+4g$ of acceleration was applied the output went to 5.0 volts, and when $-4g$ was applied the output went to zero. In this way a relatively simple demodulator and filter circuit can be used to provide direction-sensitive motion measurement. The main drawback of this scheme (which argues against its use except when necessary) is the loss of linearity. This is because the greatest linearity is found around the null position. Thus for a given total stroke it is better to put one-half of it on each side of null rather than all on one side, as in the above scheme.

Returning to the basic motion-measuring principle of Fig. 4.23, the primary coils set up a flux dependent on the reluctance of the magnetic path. The main reluctance is the air gap. When the core is in the neutral position, the flux is the same for both halves of the secondary coil, and since they are connected in series opposition the net output voltage is zero. A motion of the core increases the reluctance (air gap) on one side and decreases it on the other, causing more voltage to be induced into one half of the secondary coil than the other and thus a net output voltage. Motion in the other direction causes the reverse action, with a 180° phase shift occurring at null. The output voltage is half-wave, non-phase-sensitive rectified (demodulated) and filtered to produce an output of the same form as the acceleration input. If the 2.5-volt output for zero-acceleration input is objectionable, it can be bucked out with a 2.5-volt battery of opposite polarity connected externally to the accelerometer. The actual full-scale motion of the mass in this particular instrument is just a few thousandths of an inch, giving a displacement sensitivity for the variable-reluctance element of almost 1,000 volts/in.

The final variable-reluctance element we shall consider is the Microsyn,[1] a rotary component shown in Fig. 4.24 and widely used in sensitive gyroscopic instruments. The sketch shows the instrument in the null position where the voltages induced in coils 1 and 3 (which aid each other) are just balanced by those of coils 2 and 4 (which also aid each other but oppose 1 and 3). Motion of the input shaft from the null (say clockwise) increases the reluctance (decreases the induced voltage) of coils 1 and 3 and decreases the reluctance (increases the voltage) of coils 2 and 4, thus giving a net output voltage e_o. Motion in the opposite direction causes a similar effect except the output voltage has a 180° phase

[1] P. H. Savet (ed.), "Gyroscopes: Theory and Design," p. 332, McGraw-Hill Book Company, New York, 1961.

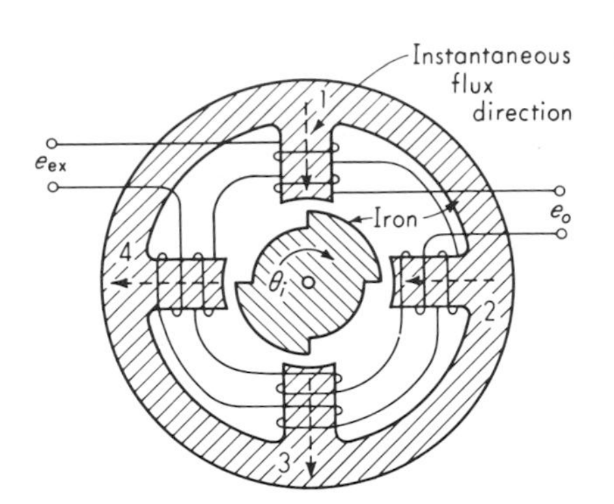

Fig. 4.24. Microsyn.

shift. If a direction-sensitive d-c output is required, a phase-sensitive demodulator is necessary.

The excitation voltage is 5 to 50 volts at 60 to 5,000 cps. Sensitivity is of the order of 0.2 to 5 volts/degree rotation. Nonlinearity is about 0.5 percent of full scale for $\pm 7°$ rotation and 1.0 percent for $\pm 10°$. The null voltage is extremely small, being less than the output signal generated by 0.01° of rotation; thus very small motions can be detected. The magnetic-reaction torque is also extremely small. Since there are no coils on the rotor, no slip rings (with their attendant friction) are needed.

Capacitance pickups. A rotational or translatory motion may be used in many ways to change the capacitance of a variable capacitor.[1,2,3] The resulting capacitance change can be converted to a usable electrical signal by means of a variety of circuitry.[1,2,3] We here consider only a few typical applications. Capacitance-type pickups tend to be special-purpose devices developed for a particular problem rather than ready-made commercial devices. The associated electronics is somewhat more complex than for the more common types of transducers. However, their mechanical simplicity, very small mechanical loading effects, and potential high sensitivity make them attractive in a number of applications, and some commercial general-purpose devices now available are quite convenient to use.

The most common form of variable capacitor used in motion transducers is the parallel-plate capacitor with a variable air gap. Theory

[1] C. H. Harris and C. E. Crede (eds.), "Shock and Vibration Handbook," vol. 1, p. 14-1, McGraw-Hill Book Company, New York, 1961.

[2] H. C. Roberts, "Mechanical Measurements by Electrical Methods," chaps. 3 and 9, The Instruments Publishing Co., Pittsburgh, Pa., 1951.

[3] R. R. Batcher and W. Moulic, "The Electronic Control Handbook," chaps. 2 and 3, The Instruments Publishing Co., Pittsburgh, Pa., 1946.

gives the capacitance of such an arrangement as

$$C = \frac{0.225A}{x} \tag{4.43}$$

where $C \triangleq$ capacitance, pf
$A \triangleq$ plate area, in.^{2}
$x \triangleq$ plate separation, in.

For example, the capacitance of an air capacitor with 1-in.^{2} plates separated by 0.01 in. is 22.5 pf. The impedance $1/i\omega C$ of this capacitor at a frequency of, say, 10,000 cps has a magnitude of 708,000 ohms. This very high impedance level of capacitance gages is responsible for some of their main problems with respect to spurious noise voltages, sensitivity to length and position of connecting cables, and requirement for high-input-impedance electronics. From Eq. (4.43) we also note that the variation of capacitance with plate separation x is nonlinear (hyperbolic); thus the percentage change in x from a chosen "neutral" position must be small if good linearity is to be achieved. The sensitivity of capacitance to changes in plate separation may be computed from Eq. (4.43):

$$\frac{dC}{dx} = -\frac{0.225A}{x^2} \tag{4.44}$$

We note that the sensitivity increases as x decreases. However, the *percentage* change in C is equal to the *percentage* change in x for small changes about *any* neutral position, as shown by

$$\frac{dC}{dx} = -\frac{C}{x} \tag{4.45}$$

$$\frac{dC}{C} = -\frac{dx}{x} \tag{4.46}$$

Perhaps the simplest useful circuit is that employed with capacitor microphones (see Sec. 6.10 on microphones for a detailed analysis). Figure 4.25 shows the arrangement. When the capacitor plates are stationary with a separation x_0, no current flows and $e_o = E_b$. If there is then a relative displacement x_1 from the x_0 position, a voltage $e_1 \triangleq e_o - E_b$ is produced and is related to x_1 by

$$\frac{e_1}{x_1}(D) = \frac{K\tau D}{\tau D + 1} \tag{4.47}$$

where

$$K \triangleq \frac{E_b}{x_0}, \text{ volts/in.} \tag{4.48}$$

$$\tau \triangleq 0.225 \times 10^{-12} \frac{AR}{x_0} \quad \text{sec} \tag{4.49}$$

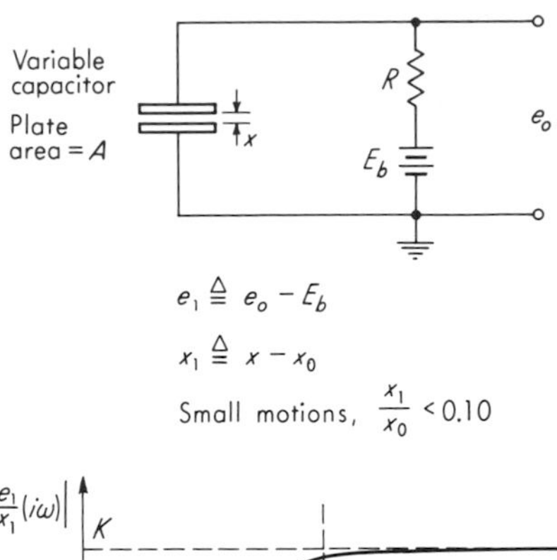

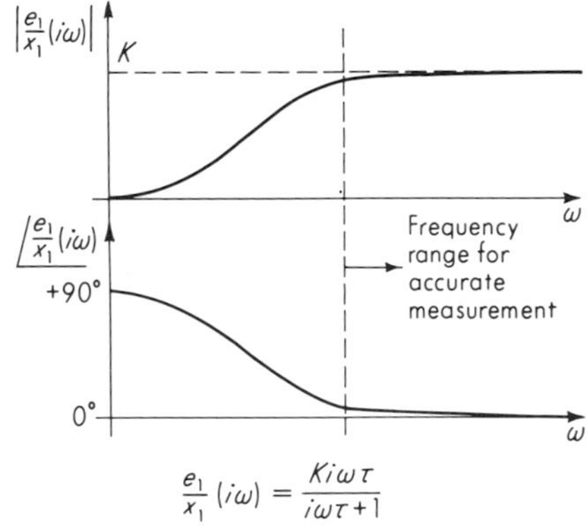

Fig. 4.25. Capacitive transducer.

Equation (4.47) shows that this arrangement does not allow measurement of static displacements since e_1 is zero in steady state for any value of x_1. For sufficiently rapid variations in x_1, however, the signal e_1 will faithfully measure the motion. This is most easily seen from the frequency response

$$\frac{e_1}{x_1}(i\omega) = \frac{K\tau i\omega}{i\omega\tau + 1} \tag{4.50}$$

which becomes for $\omega\tau \gg 1$

$$\frac{e_1}{x_1}(i\omega) \approx K \tag{4.51}$$

Thus e_1 follows x_1 accurately under these conditions. A microphone usually need not measure sound pressures slower than about 20 cps, and so the above arrangement is perfectly satisfactory. To make $\omega\tau \gg 1$ for low frequencies requires a large τ. For a given capacitor and x_0, the value of τ can be increased only by increasing R. Typically, R will be 10^6 ohms or more. Thus to prevent loading of the capacitance transducer circuit the readout device connected to the e_o terminals must have a

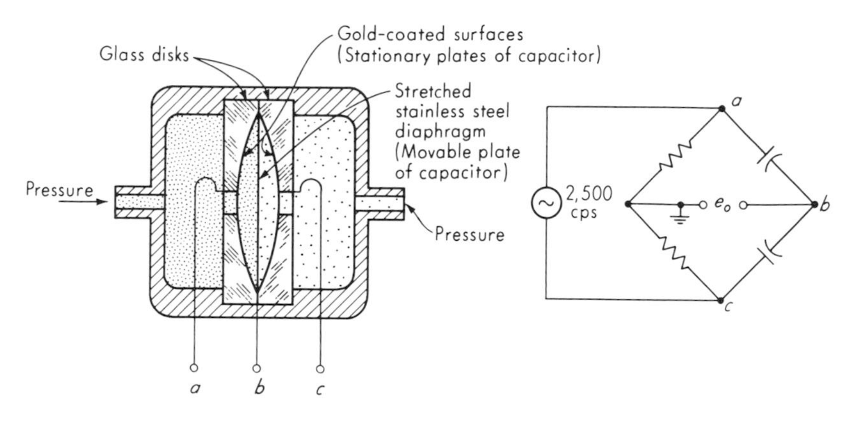

Fig. 4.26. *Differential-capacitor pressure pickup.*

high (10^7 ohms or more) input impedance. This usually requires use of a cathode-follower type of isolation amplifier.

The use of a variable differential (three-terminal) capacitor with a bridge circuit is shown in the Equibar[1] differential pressure transducer of Fig. 4.26. Spherical depressions of a depth of about 0.001 in. are ground into the glass disks; then these depressions are gold-coated to form the fixed plates of a differential capacitor. A thin stainless-steel diaphragm is clamped between the disks and serves as the movable plate. With equal pressures applied to both ports, the diaphragm is in a neutral position, the bridge is balanced, and e_o is zero. If one pressure is greater than the other, the diaphragm deflects in proportion, giving an output at e_o in proportion to the differential pressure. For the opposite pressure difference, e_o exhibits a 180° phase change. The high impedance level again requires a cathode-follower amplifier at e_o. A direction-sensitive d-c output can be obtained by conventional phase-sensitive demodulation and filtering. Balance resistors necessary for initially nulling the bridge are not shown in Fig. 4.26. This method (as opposed to that of Fig. 4.25) allows measurement of static deflections. Such differential-capacitor arrangements also exhibit considerably greater linearity than do single capacitor types.[2]

An ingenious method[3] of circumventing the nonlinear relationship [Eq. (4.43)] between x and C is shown in Fig. 4.27. This technique employs a high-gain feedback amplifier using an approach common in

[1] Trans-Sonics, Inc., Burlington, Mass.

[2] N. H. Cook and E. Rabinowicz, "Physical Measurement and Analysis," p. 142, Addison-Wesley Publishing Company, Inc., Reading, Mass., 1963.

[3] Wayne Kerr Corp., Philadelphia, Pa.

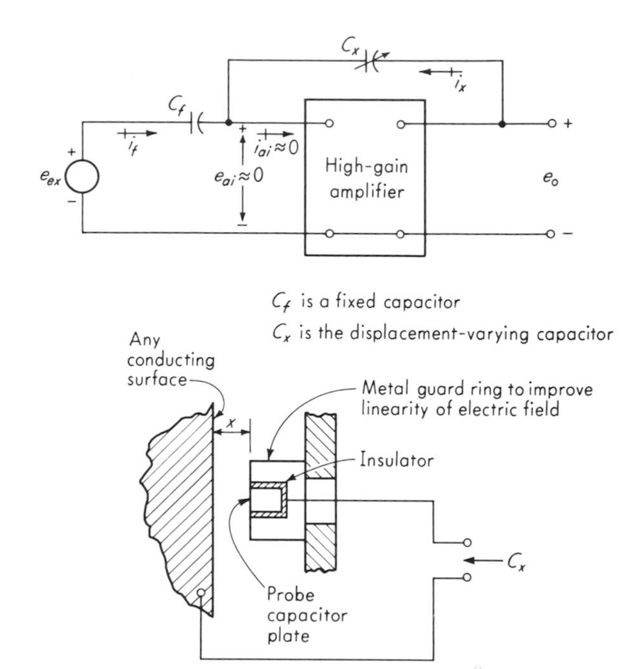

Fig. 4.27. *Feedback-type capacitive pickup.*

electronic analog computers. The assumptions necessary to an analysis of this circuit depend on the following characteristics of so-called "operational" amplifiers:

1. The input impedance is so high that the amplifier input current may be taken as zero relative to other currents.
2. The gain is so high that if the output voltage of the amplifier is not saturated the input voltage is extremely small and may be taken as zero relative to other voltages. For example, a typical amplifier has linear output for the range ± 100 volts and a gain of 10^8 volts/volt. Thus the maximum input for linear operation is 10^{-6} volt.

Using these assumptions in Fig. 4.27, we can write

$$\frac{1}{C_f}\int i_f\,dt = e_{ex} - e_{ai} = e_{ex} \qquad (4.52)$$

$$\frac{1}{C_x}\int i_x\,dt = e_o - e_{ai} = e_o \qquad (4.53)$$

$$i_f + i_x - i_{ai} = 0 = i_f + i_x \qquad (4.54)$$

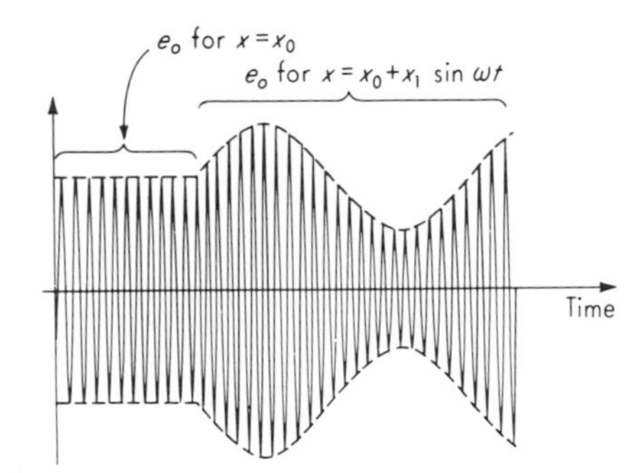

Fig. 4.28. *Wave form for sinusoidal displacement.*

Manipulation then gives

$$e_o = \frac{1}{C_x}\int i_x\,dt = -\frac{1}{C_x}\int i_f\,dt = -\frac{C_f}{C_x}e_{ex} \tag{4.55}$$

$$e_o = -\frac{C_f x e_{ex}}{0.225A} = Kx \tag{4.56}$$

Equation (4.56) shows that the output voltage is now *directly* proportional to the plate separation x; linearity is thus achieved for both large and small motions. In the commercial instrument described above, e_{ex} is a 50-kc sine wave of fixed amplitude. The output e_o is also a 50-kc sine wave which is rectified and applied to a d-c voltmeter calibrated directly in distance units.

For vibratory displacements, e_o will be an amplitude-modulated wave as in Fig. 4.28. The average value of this wave after rectification is still the mean separation of the plates and can still be read by the same meter as used for static displacements. The vibration amplitude around this mean position is extracted by applying the e_o signal also to a demodulator and a low-pass filter with cutoff at 10 kc. The output of this filter is applied to a peak-to-peak voltmeter directly calibrated in vibration amplitude and is also available at a jack for connection to an oscilloscope for viewing of the vibration wave form. The instrument is provided with six different probes ranging from 0.0447 to 1.0 in. in diameter of the capacitance plate and covering the full-scale displacement ranges of 0.001, 0.005, 0.01, and 0.5 in., respectively. The overall system accuracy is of the order of 2 percent of full scale. Any flat conductive surface may serve as the second plate of the variable capacitor. Thus in vibrating machine parts the parts themselves may often perform this function. The resolution of 0.5 percent of full scale indicates that (with the 0.001-in. full-scale probe) it is possible to detect motion as small as 5 μin.

The final capacitance-type device we consider here involves the

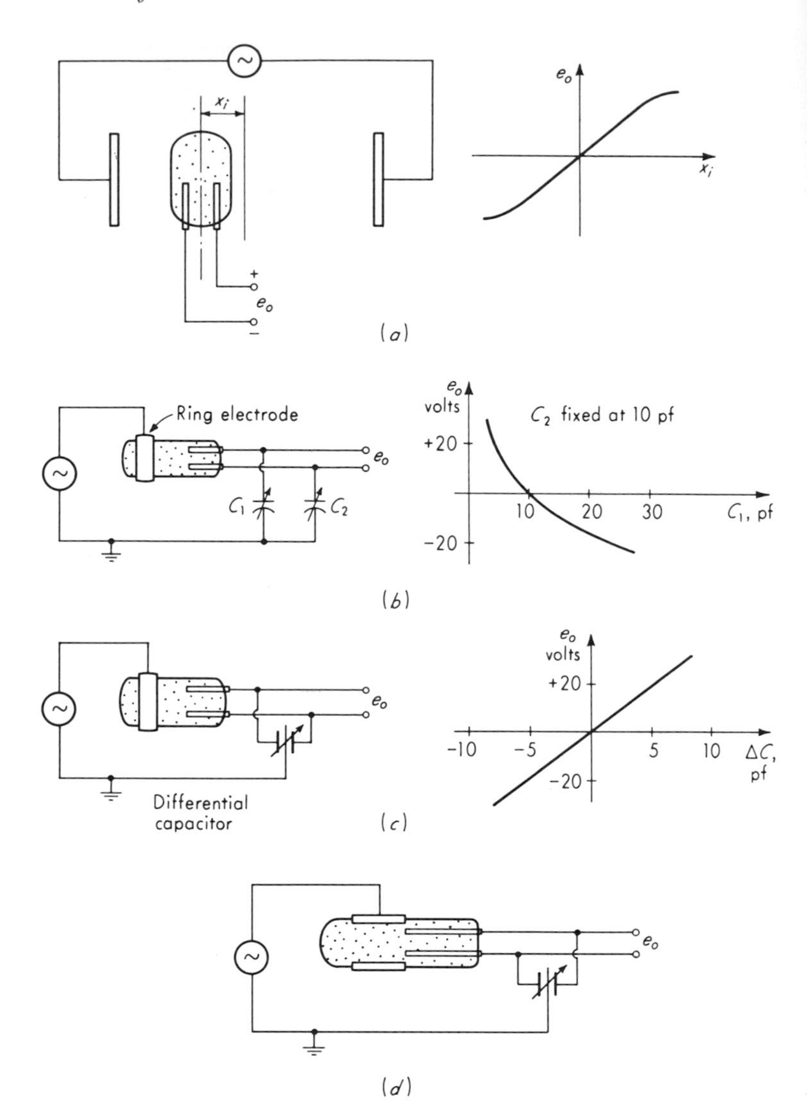

Fig. 4.29. *Ionization transducer.*

use of the T-42 ionization transducer.[1,2] While this transducer system involves other fundamental physical effects in addition to capacitance, we include it here because the variation of capacitance is used in many

[1] K. S. Lion, Mechanic-Electric Transducer, *Rev. Sci. Instr.*, vol. 27, no. 4, pp. 222–225, April, 1956.

[2] The Decker Corp., Bala-Cynwyd, Pa.

of its practical applications. The basic physical effect employed in the tube shown in Fig. 4.29*a* is the development of a d-c voltage across the internal electrodes of a glass tube, which contains gas under reduced pressure, when the tube is exposed to an electric field created by the external parallel-plate electrodes connected to a radio-frequency (250 kc in the commercial version) voltage source. This d-c voltage varies with the position x_i, being zero at the neutral position and varying linearly (with a polarity change at null) over a small range on either side. The sensitivity of this arrangement can be as high as several thousand volts per millimeter.

While the basic arrangement of Fig. 4.29*a* has been employed in several practical transducers, a modified form of the tube which employs capacitance-variation techniques has been found useful for general-purpose motion measurement. Figure 4.29*b* and *c* shows these modifications and typical performance. A further refinement leading to increased tube life is pictured in Fig. 4.29*d*. This arrangement has a sensitivity of about 2 volts/pf capacitance change for an initial capacitance of 10 pf. The output impedance of all these transducer systems is high (of the order of 1 megohm); thus cathode-follower-type amplifiers are needed at the output. Commercial instruments based on Fig. 4.29*d* plus associated electronics give about 0.2 volt/percent change in capacitance in the preferred range of 5 to 20 pf. Frequency response is from zero to about 1,000 cps. Nonlinearity depends on the capacitor arrangement used and is typically about 1 percent of full scale.

Piezoelectric transducers. When certain solid materials are deformed, they generate within them an electric charge. This effect is reversible in that if a charge is applied the material will mechanically deform in response. These actions are given the name piezoelectric (pī-ēzō-electric) effect.[1] This electromechanical energy-conversion principle is usefully applied in both directions. The mechanical-input/electrical-output direction is the basis of many instruments for measuring acceleration, force, and pressure. It also can be used as a means of generating high-voltage low-current electrical power such as is used in spark-ignition engines and electrostatic dust filters. The electrical-input/mechanical-output direction is applied in small vibration shakers, sonar systems for acoustic detection and location of underwater objects, and industrial ultrasonic nondestructive test equipment.

The materials that exhibit a significant and useful piezoelectric effect fall into two main groups: natural (quartz, rochelle salt) and synthetic (lithium sulfate, ammonium dihydrogen phosphate) crystals

[1] Harris and Crede, *op. cit.*, chap. 16, pt. II.

and polarized ferroelectric ceramics (barium titanate). Because of their natural asymmetrical structure, the crystal materials exhibit the effect without further processing. The ferroelectric ceramics must be artificially polarized by applying a strong electric field to the material (while it is heated to a temperature above the Curie point of that material) and then slowly cooling with the field still applied. (The Curie temperature is the temperature above which a material loses its ferroelectric properties; thus it limits the highest temperature at which such materials may be used.) When the external field is removed from the cooled material a remanent polarization is retained and the material will now exhibit the piezoelectric effect.

The piezoelectric effect can be made to respond to (or cause) mechanical deformations of the material in many different modes, such as thickness expansion, transverse expansion, thickness shear, and face shear. The mode of motion effected depends on the shape and orientation of the body relative to the crystal axes and the location of the electrodes. Metal electrodes are plated onto selected faces of the piezoelectric material so that lead wires can be attached for bringing in or leading out the electrical charge. Since the piezoelectric materials are insulators, the electrodes also become the plates of a capacitor. A piezoelectric element used for converting mechanical motion to electrical signals may thus be thought of as a charge generator and a capacitor. Mechanical deformation generates a charge; this charge then results in a definite voltage appearing between the electrodes according to the usual law for capacitors, $E = Q/C$. The piezoelectric effect is direction-sensitive in that tension produces a definite voltage polarity while compression produces the opposite.

We shall illustrate the main characteristics of piezoelectric motion-to-voltage transducers by considering only one common mode of deformation, thickness expansion. For this mode the physical arrangement is as in Fig. 4.30*b*. Various double-subscripted physical constants are used to describe numerically the phenomena occurring. The convention is that the first subscript refers to the direction of the electrical effect and the second to that of the mechanical effect, using the axis-numbering system of Fig. 4.30*a*.

Two main families of constants, the g constants and the d constants, will be considered. For a barium titanate thickness-expansion device the pertinent g constant is g_{33}, which is defined as

$$g_{33} \triangleq \frac{\text{field produced in direction 3}}{\text{stress applied in direction 3}} = \frac{e_o/t}{f_i/(wl)} \tag{4.57}$$

Thus if one knows g for a given material and the dimension t he can calculate the output voltage per unit applied stress. Typical g values are

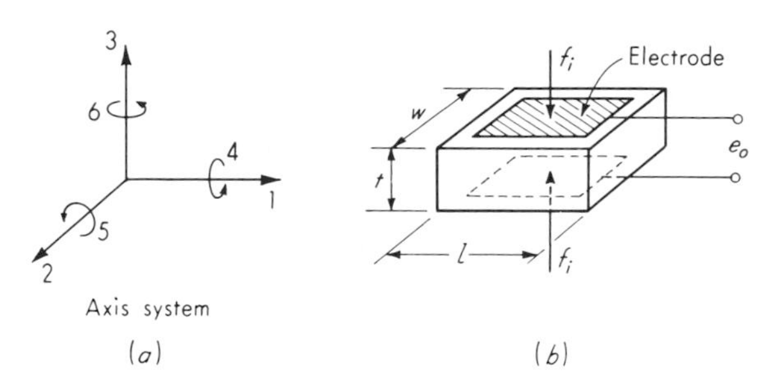

Fig. 4.30. *Piezoelectric transducer.*

12×10^{-3} (volt/m)/(newtons/m^2) for barium titanate and 50×10^{-3} for quartz. Thus, for example, a quartz crystal 0.1 in. thick would have a sensitivity of 0.88 volt /psi, illustrating the large voltage output for small stress typical of piezoelectric devices.

To relate applied force to generated charge the d constants can be defined as

$$d_{33} \triangleq \frac{\text{charge generated in direction 3}}{\text{force applied in direction 3}} = \frac{Q}{f_i} \qquad (4.58)$$

Actually, d_{33} can be calculated from g_{33} if the dielectric constant ϵ of the material is known, since

$$C = \frac{\epsilon wl}{t} \qquad (4.59)$$

$$g_{33} \triangleq \frac{\text{field}}{\text{stress}} = \frac{e_o wl}{tf_i} = \frac{e_o C}{\epsilon f_i} = \frac{Q}{\epsilon f_i} = \frac{d_{33}}{\epsilon} \qquad (4.60)$$

$$d_{33} = \epsilon g_{33} \qquad (4.61)$$

The dielectric constant of quartz is about 4.06×10^{-11} farad/m while for barium titanate it is $1{,}250 \times 10^{-11}$. For quartz, then,

$$d_{11} = \epsilon g_{11} = (4.06 \times 10^{-11})(50 \times 10^{-3}) = 2.03 \text{ pcoul/newton} \qquad (4.62)$$

(The subscripts 11 are used because in quartz the thickness-expansion mode is along the crystallographic axis conventionally called axis 1.) Sometimes it is desired to express the output charge or voltage in terms of deflection (rather than stress or force) of the crystal since it is really the *deformation* that causes the charge generation. To do this, one must know the modulus of elasticity, which is 8.6×10^{10} newtons/m^2 for quartz and 12×10^{10} for barium titanate.

With the above brief introduction as background, we now proceed to consider piezoelectric elements as displacement transducers. The ultimate purpose is generally force, pressure, or acceleration measurement but we shall consider only the conversion from displacement to voltage. For analysis purposes it is necessary to consider the transducer, connecting cable, and associated amplifier as a unit. The transducer impedance is generally very high; thus the amplifier is usually a high-impedance cathode follower used for isolation purposes rather than voltage gain. The cable capacitance can be significant, especially for long cables. For the transducer alone, if a static deflection x_i is applied and maintained, a transducer terminal voltage will be developed but the charge will slowly leak off through the leakage resistance of the transducer. Since R_{leak} is generally very large (the order of 10^{11} ohms) this decay would be very slow, perhaps allowing at least a quasi-static response. However, when an external voltage-measuring device of low input impedance is connected to the transducer the charge leaks off very rapidly, preventing the measurement of static displacements. Even relatively high-impedance cathode followers do not generally allow static measurements. Some commercially available[1] systems using quartz transducers (very high leakage resistance) and electrometer input amplifiers (very high input impedance) achieve an effective total resistance of 10^{14} ohms which gives a sufficiently slow leakage to allow static measurements.

To put the above discussion on a quantitative basis, we consider Fig. 4.31. The charge generated by the crystal can be expressed as

$$q = K_q x_i \tag{4.63}$$

where

$$K_q \triangleq \text{coul/in.} \tag{4.64}$$

$$x_i \triangleq \text{deflection, in.} \tag{4.65}$$

The resistances and capacitances of Fig. 4.31*b* can be combined as in 4.31*c*. We also convert the charge generator to a more familiar current generator according to

$$i_{cr} = \frac{dq}{dt} = K_q \left(\frac{dx_i}{dt}\right) \tag{4.66}$$

We may then write

$$i_{cr} = i_C + i_R \tag{4.67}$$

$$e_o = e_C = \frac{\int i_C\, dt}{C} = \frac{\int (i_{cr} - i_R)\, dt}{C} \tag{4.68}$$

$$C\left(\frac{de_o}{dt}\right) = i_{cr} - i_R = K_q\left(\frac{dx_i}{dt}\right) - \frac{e_o}{R} \tag{4.69}$$

$$\frac{e_o}{x_i}(D) = \frac{K\tau D}{\tau D + 1} \tag{4.70}$$

where

$$K \triangleq \text{sensitivity} \triangleq \frac{K_q}{C}\ \text{volts/in.} \tag{4.71}$$

$$\tau \triangleq \text{time constant} \triangleq RC,\ \text{sec} \tag{4.72}$$

[1] Kistler Instrument Corp., North Tonawanda, N.Y.

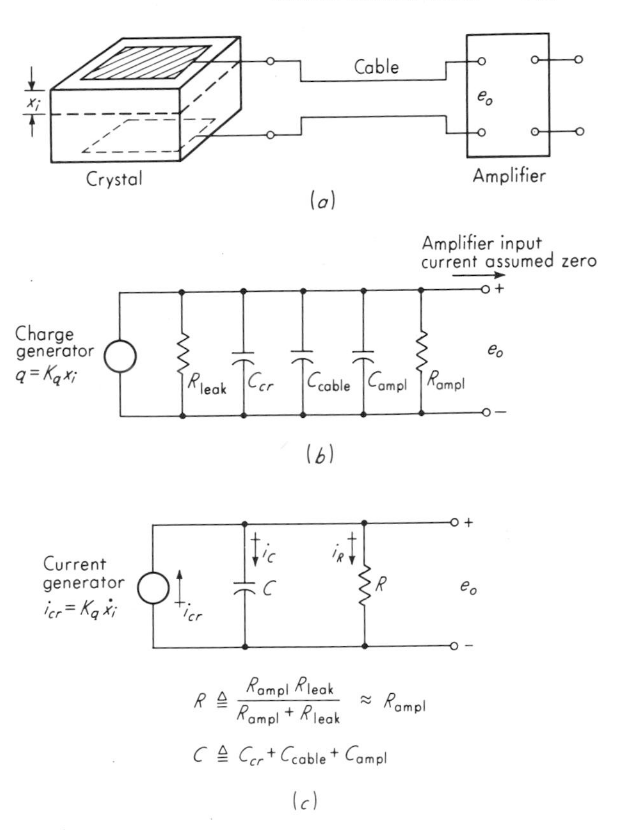

Fig. 4.31. *Equivalent circuit for piezoelectric transducer.*

We see that, just as in the capacitance pickup of Eq. (4.25), the steady-state response to a constant x_i is zero; thus we cannot measure static displacements. For a flat amplitude response within, say, 5 percent the frequency must exceed ω_1, where

$$(0.95)^2 = \frac{(\omega_1\tau)^2}{(\omega_1\tau)^2 + 1} \tag{4.73}$$

$$\omega_1 = \frac{3.02}{\tau} \tag{4.74}$$

Thus a large τ gives an accurate response at lower frequencies.

The response of these transducers is further illuminated by considering the displacement input of Fig. 4.32. The differential equation is

$$(\tau D + 1)e_o = (K\tau D)x_i \tag{4.75}$$

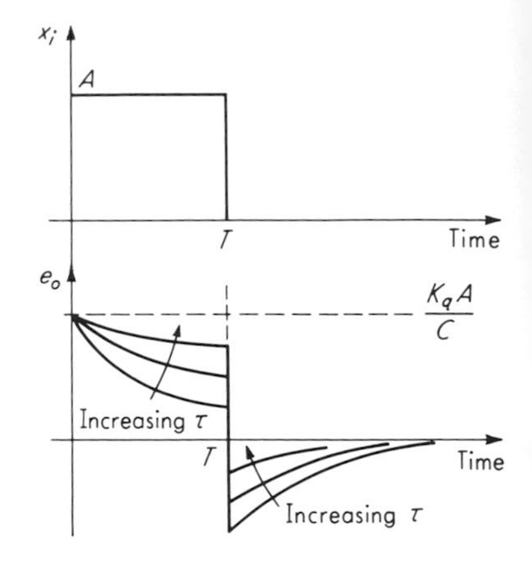

Fig. 4.32. *Pulse response of piezoelectric transducer.*

Since $x_i = A$ for $0 < t < T$, this becomes

$$(\tau D + 1)e_o = 0 \qquad (4.76)$$

Now at $t = 0^+$ the displacement x_i is A and so the charge *suddenly* increases to K_qA, and the crystal capacitor voltage and therefore e_o *suddenly* increases to K_qA/C. Thus our initial condition is

$$e_o = \frac{K_qA}{C} \qquad \text{at } t = 0^+ \qquad (4.77)$$

Solving Eq. (4.76) with initial condition (4.77) gives

$$e_o = \frac{K_qA}{C} e^{-t/\tau} \qquad 0 < t < T \qquad (4.78)$$

Equation (4.78) holds until $t = T$. At this instant we must stop using it because of the change in x_i. For $T < t < \infty$ the differential equation is

$$(\tau D + 1)e_o = 0 \qquad (4.79)$$

At $t = T^-$ Eq. (4.78) is still valid and

$$e_o = \frac{K_qA}{C} e^{-T/\tau} \qquad (4.80)$$

Now, at $t = T$, x_i suddenly drops an amount A, causing a sudden decrease in charge of K_qA and a sudden decrease in e_o of K_qA/C from its value at $t = T^-$. Thus at $t = T^+$, e_o is given by

$$e_o = \frac{K_qA}{C} (e^{-T/\tau} - 1) \qquad (4.81)$$

which becomes the initial condition for Eq. (4.79). The solution then becomes

$$e_o = \frac{K_qA}{C} (e^{-T/\tau} - 1)e^{-(t-T)/\tau} \qquad T < t < \infty \qquad (4.82)$$

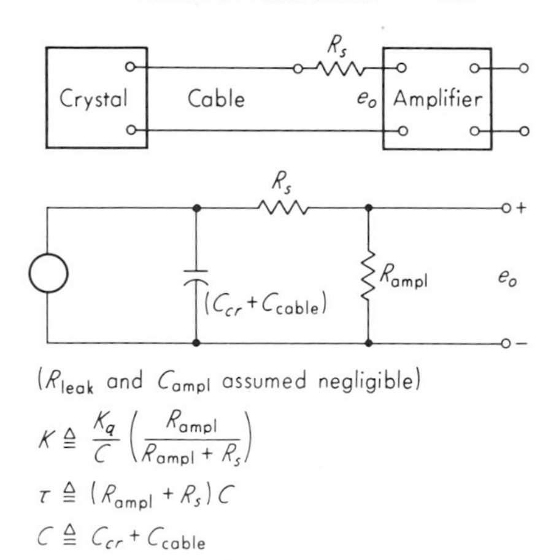

Fig. 4.33. *Use of series resistor to increase time constant.*

Figure 4.32 shows the complete process for three different values of τ. It is clear that a large τ is desirable for faithful reproduction of x_i. If the decay and "undershoot" at $t = T$ is to be kept within, say, 5 percent of the true value, τ must be at least $20T$. If an increase of τ is required in a specific application, it may be achieved by increasing either or both R and C. An increase in C is easily obtained by connecting an external shunt capacitor across the transducer terminals, since shunt capacitors add directly. The price paid for this increase in τ is a loss of sensitivity according to $K = K_q/C$. This may often be tolerated because of the initial high sensitivity of piezoelectric devices. An increase in R generally requires an amplifier of greater input resistance. If sensitivity can be sacrificed, a series resistor connected external to the amplifier, as shown in Fig. 4.33, will increase τ without the need of obtaining a different amplifier.

Detailed data on static and dynamic performance characteristics of piezoelectric transducers will be deferred to the respective sections on force, pressure, and acceleration measurement where these data will be more meaningful.

Electro-optical devices. The combination of classical optical principles with modern electronic developments has led to a variety of useful electro-optical measuring instruments. The Midarm system of Fig. 4.2 is one example of this class. Here we consider another, the Optron[1] displacement follower, shown in Fig. 4.34.

Light emanating from the phosphorescent spot on the surface of a special cathode-ray tube is sharply focused onto the plane of the edge of

[1] Optron Corp., Santa Barbara, Calif.

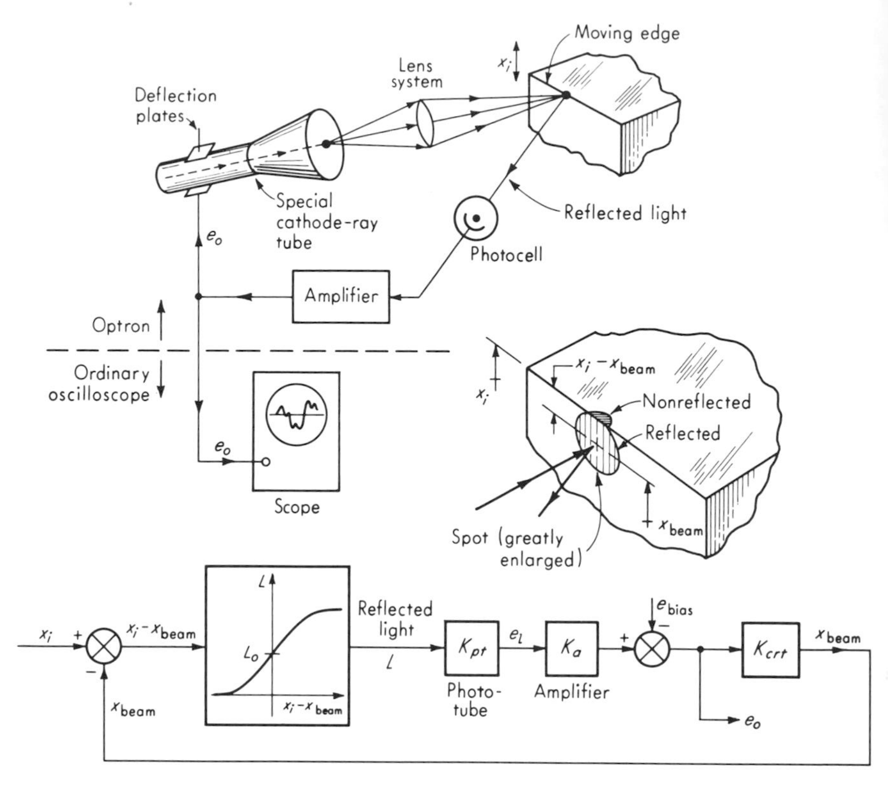

Fig. 4.34. Optron displacement pickup.

the body whose motion x_i is to be measured. If the spot is completely above this edge, no light will be reflected back to the photocell, and its output voltage e_l will be zero. The voltage applied to the vertical deflection plates of the special cathode-ray tube is then equal to $-e_{\text{bias}}$, and the spot motion x_{beam} will be $-K_{crt}e_{\text{bias}}$, driving the beam downward. The voltage e_{bias} is sufficient to drive the beam down past the edge. As it comes onto the edge from the top, first a little light and then more and more is reflected back to the photocell. If the spot is completely below the edge, further downward motion produces no more reflected light, saturating this effect. This relation between $x_i - x_{\text{beam}}$ and reflected light is somewhat nonlinear, as shown in the block diagram of Fig. 4.34. As the spot first comes onto the edge from above, the photocell puts out voltage e_l. This is amplified an amount K_a, which is very large, so that the spot moves only a small distance onto the edge before e_lK_a is larger than e_{bias}, and x_{beam} is now positive, driving the beam upward. If it is driven too far up, e_l again becomes smaller than e_{bias} and the spot is again driven down. Thus this feedback system tends at all times to "slave" or "lock on" the point of light to the moving edge. Under these conditions, $x_{\text{beam}} \approx x_i$; since the voltage $e_o = x_{\text{beam}}/K_{crt}$ is also applied to the measuring oscilloscope, its trace is an accurate record of x_i.

The block diagram of Fig. 4.34 shows all effects as occurring instantaneously; thus it is adequate only for an analysis of static behavior. However, this analysis is quite revealing and so we carry it through. The nonlinear relation between $x_i - x_{\text{beam}}$ and reflected light L may be linearized for small excursions about an operating point L_0 as follows:

$$L \approx L_0 + \frac{dL}{d(x_i - x_{\text{beam}})}\bigg|_{(x_i - x_{\text{beam}})=0} (x_i - x_{\text{beam}})$$
$$= L_0 + K_L(x_i - x_{\text{beam}}) \tag{4.83}$$

The phototube, amplifier, and cathode-ray tube are assumed linear although it will be seen that only the cathode-ray tube need be strictly linear. From the block diagram we can write

$$\{[L_0 + K_L(x_i - x_{\text{beam}})]K_{pt}K_a - e_{\text{bias}}\}K_{crt} = x_{\text{beam}} \tag{4.84}$$

It is possible to choose $L_0K_{pt}K_a = e_{\text{bias}}$, thus giving

$$(K_LK_{pt}K_aK_{crt} + 1)x_{\text{beam}} = (K_LK_{pt}K_aK_{crt})x_i \tag{4.85}$$

$$x_{\text{beam}} = \frac{K_LK_{pt}K_aK_{crt}}{1 + K_LK_{pt}K_aK_{crt}} x_i \tag{4.86}$$

We now by design make $K_LK_{pt}K_aK_{crt} \ggg 1$ (usually by making K_a very large) to get

$$x_{\text{beam}} \approx x_i \tag{4.87}$$

thus showing that x_{beam} "tracks" x_i. Equation (4.87) holds, whether K_L, K_{pt}, K_a, or K_{crt} is constant or variable (due to time drift and/or nonlinearity). The *only* requirement is that their product at all times should be much greater than 1.0. This is a result of the feedback principle employed. This feedback unfortunately also puts an upper allowable value on $K_LK_{pt}K_aK_{crt}$ (the loop gain) because of stability problems which would become apparent if dynamic effects were taken into account. However, proper design allows a sufficiently high value of loop gain to achieve high accuracy without instability. Since the output of the instrument is the voltage $e_o = x_{\text{beam}}/K_{crt}$, we see that K_{crt} *must* be a constant if e_o is to be in proportion to x_{beam} and thus to x_i. The special cathode-ray tube must therefore have a linear voltage/(spot deflection) characteristic.

The Optron described above must be used in low ambient-light levels, the equipment being covered with a blackout cloth where convenient. An adjustable optical system gives four ranges as follows:

Range	*Full scale* x_i, *in.*	*Resolution, in.*	*Working distance, in.*
1	0.1	0.0001	0.5
2	0.25	0.0002	2.0
3	1.0	0.001	3.5
4	4.0	0.004	9.0

Nonlinearity is 0.2 percent of full scale over the central 75 percent of range. Nonrepeatability is 0.1 percent of full scale over the same range. The frequency response is from 0 to 5,000 cps with the amplitude ratio down 3 db at 5,000 cps. The output signal is 40 volts full scale.

A more recent version of the Optron uses a lens or telescope to form an optical image of a black-and-white target fastened to the object under study on the photocathode of an image-dissector tube. This tube projects an "electron image" of the target onto an aperture behind which is located a photomultiplier tube. The output of the photomultiplier tube is amplified and drives the deflection coil of the image-dissector tube. The deflection coil positions the electron image on the aperture so that the black-white boundary always splits the aperture. Motion of the target causes motion of the photocathode optical image and the corresponding electron image at the aperture. However, as soon as the image at the aperture starts to deviate from the neutral position the photomultiplier output tends to drive it back by means of the deflection coil.

Although the hardware is different, the basic concept, feedback analysis, and performance are quite analogous to those of the system of Fig. 4.34. An advantage of this system (see Fig. 4.35) over the earlier version is that it is usable in ordinary room light (40 ft-c) and is "passive" in the sense that the Optron system does *not* provide the light source illuminating the target, as did the special cathode-ray tube of Fig. 4.34. The working distance can thus be greatly extended to hundreds of feet or even miles and the instrument applied to such problems as missile tracking. At a working distance of 200 ft the full-scale range of displacement is 10 in. with a resolution of 0.050 in. With special lenses, a working distance of 0.65 in., a full-scale range of 0.050 in., and a resolution of 12 μin. are claimed. The frequency response is flat from 0 to 5,000 cps.

Photographic techniques. The application of still and motion-picture photography often allows qualitative and quantitative analysis of com-

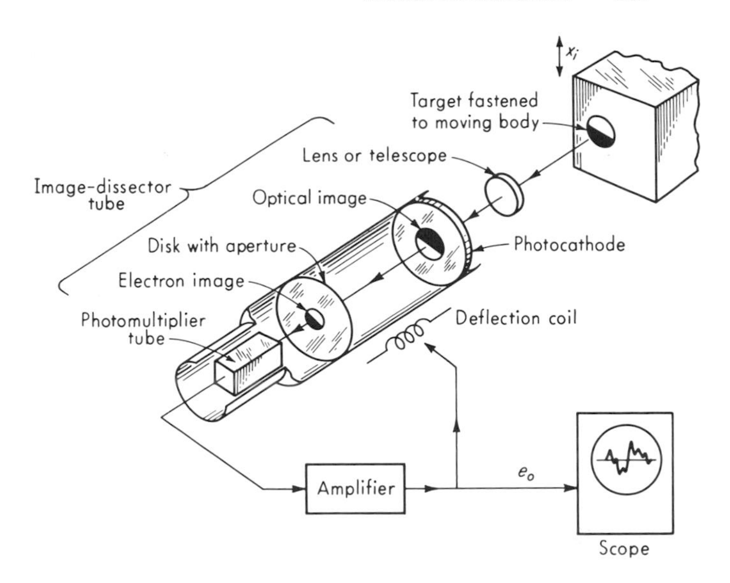

Fig. 4.35. *Optron displacement pickup.*

plex motions that would be difficult by other methods.[1] We here touch briefly on some of the more common applications.

Perhaps the simplest application of still photography is the single-flash "stop-action" technique.[2] The objective is to "freeze" a motion at a particular phase of its occurrence to allow detailed visual study of some physical phenomenon. The equipment usually employed consists of a still camera, a stroboscopic light source, and some means of triggering a single flash of the strobe light at the desired instant. If the experiment can be performed in a darkened room, the procedure consists of manually opening the camera shutter, allowing the phenomenon to occur, triggering the light at the desired instant, and then manually closing the shutter. If triggering, focus, and exposure are correct, a photo of the phenomenon, "frozen" at the instant of the light flash, will be obtained. Such photos can be most helpful in understanding complex physical processes in fluid motion or moving machine parts. Of course, the effective freezing of the motion depends on the flash duration being sufficiently short compared with the velocity of the motion. Flash durations of the order of 1 to 3 μsec are readily available. Thus, for example, a velocity of 1,000 fps will cause a "blurring" of 0.012 to 0.036 in. at the object. The actual blurring

[1] W. G. Hyzer, "Engineering and Scientific High-speed Photography," The Macmillan Company, New York, 1962.

[2] "Handbook of High-speed Photography," General Radio Co., West Concord, Mass.

on the film will be this value times the image/object ratio of the camera setup. If a 1-in. object shows up on the film as 0.1 in., for example, the above blurring would amount to 0.0012 to 0.0036 in. on the film, which would generally be considered acceptable. If the experiment cannot be carried out in a darkened room, the opening and closing of the shutter must be synchronized with the flash and the open time of the shutter must be short enough so as not to overexpose the film from the room light.

If a displacement-time record is desired, a multiple-flash still-camera technique may be employed. The setup is essentially the same as above except the strobe light flashes repetitively at a known rate. The result is a multiple-exposure photo showing the moving object in successive positions which are separated by known increments of time. By including a calibrated length scale in the photo (preferably in the plane of the motion) numerical values of displacement at specific time intervals may be measured.

High-speed motion-picture photography is used to study motions that occur too rapidly for the eye to analyze properly. This is accomplished by taking the pictures at a high camera picture frequency (frames per second) and then projecting the film at a low projector picture frequency. The lowest usable projector frequency is about 16 frames per second since lower frequencies result in flickering because the human eye's persistence of vision is about 0.06 sec. The highest usable camera picture frequency depends on the construction of the camera. Relatively small, portable cameras are available with picture frequencies up to about 20,000 frames per second, and these are in relatively wide use in industry. Larger, more complex, and expensive cameras are available where the application dictates their higher speed. Their picture frequencies are up to several million per second. The time magnification is defined as the ratio of the camera frequency to the projector frequency; thus if 16 is the projector frequency, magnifications of several hundred thousand to one are achievable while 1,500:1 is not unusual in common industrial practice. Aside from picture frequency, the shutter speed of the camera must be sufficiently high to prevent blurring of the individual frames, just as in still photography. For 16-mm film a blur of 0.002 in. is considered acceptable. The shutter-speed requirement can be greatly relaxed by the use of a synchronized short-duration electronic flash as the light source, since the flash duration then controls blur no matter what the shutter speed. Selection of a proper camera picture rate may be judged roughly by the rule that the projection of the complete motion to be visually analyzed should take about 2 to 10 sec. Thus a motion occurring in 0.001 sec requires a time magnification of 2,000 to 10,000. For vibratory motions the camera picture rate must be several (preferably 5 to 10) times the highest vibration frequency.

The main features of all photographic motion-analysis techniques include their noncontacting nature, the ability to see the overall motion of a body or fluid rather than just that of isolated points, and the capacity to measure quantitatively a general plane motion (any combination of translation and rotation) rather than just a pure rotation or translation, as for potentiometers, differential transformers, etc.

Photoelastic and brittle-coating stress-analysis techniques. Since both these methods are really strain- rather than stress-sensitive and since strain is a small displacement, we include a brief treatment in this section on displacement measurement.

Photoelastic methods[1] depend on the property of birefringence under load exhibited by certain natural or synthetic transparent materials. Birefringence (double refraction) under load refers to the phenomenon wherein light travels at different speeds in a transparent material, depending on the direction of travel relative to the directions of the principle stresses and also depending on the magnitude of the difference between the principal stresses, for a two-dimensional stress field. By constructing models (from suitable transparent materials) of the same shape as the part to be stress-analyzed and shining suitably polarized light through them while they are subjected to loads proportional to those expected in actual service, a pattern of light and dark fringes appears which shows the stress distribution throughout the piece and allows numerical calculation of stresses at any chosen point.

By use of the "frozen stress" technique, the method can be extended to three-dimensional problems. A three-dimensional plastic model is subjected to simultaneous load and high temperature. The load is maintained while the specimen is slowly cooled to room temperature whereupon the load is released. It will be found that a residual stress pattern identical to that produced by the load will be "frozen" into the specimen. Furthermore, the model may now be carefully sliced in various directions to produce flat (two-dimensional) slabs which may be photoelastically analyzed to determine three-dimensional stresses. By combining photoelastic and high-speed photographic techniques, the method may be extended to dynamic studies such as the propagation of shock waves through solid bodies.

A recent extension of photoelastic techniques, reflective photoelasticity, does not require construction of a plastic model. Rather, the metal part itself (with its surface polished or aluminum-painted for reflectivity) is coated with a liquid photoelastic material which hardens

[1] M. M. Frocht, "Photoelasticity," vols. I and II, John Wiley & Sons, Inc., New York, 1941, 1948.

and bonds to the surface. Polarized light is directed onto the part and reflects from the shiny undersurface. Dark and light patterns again reveal the stress distribution when a load is applied.

Photoelastic methods, as compared with bonded resistance strain gages, give an overall picture of the stress distribution in a part. This is very helpful in locating and numerically evaluating stress concentrations and in redesigning the part for optimum material use. The method also does not disturb the local stress field as a strain gage might.

In the brittle-coating stress-analysis technique[1] a special lacquerlike material is sprayed on the actual part to be analyzed and the coating allowed to dry. Application of load causes visible cracking of the brittle coating. The direction of the cracking shows the direction of maximum stress while the spacing of the cracks indicates magnitude. Under favorable conditions, numerical values of stress can be calculated from measured crack spacing to about 10 percent accuracy. The main features of the method are its simplicity, low cost, and speed in giving an overall picture of stress distribution. It is often used in conjunction with electric resistance strain gages, the brittle coating locating the points of maximum stress and its direction so that strain gages can be applied at the proper places and in the proper orientation for accurate strain measurement.

Displacement-to-pressure transducer (nozzle-flapper). The nozzle-flapper-transducer principle is widely used in precision gaging equipment and also as a basic component of pneumatic and hydraulic measurement and control apparatus. Figure 4.36 shows the general arrangement. Fluid at a regulated pressure is supplied to a fixed flow restriction and a variable flow restriction connected in series. The variable flow restriction is varied by moving the "flapper" to change the distance x_i. This causes a change in output pressure p_o which, for a limited range of motion, is nearly proportional to x_i and extremely sensitive to it. A pressure-measuring device connected to p_o can thus be calibrated to read x_i. Ideally (pressure-containing chambers rigid; fluid incompressible) a sudden change in x_i would cause an instantaneous change in p_o. Actually, the dynamics are approximately those of a linear, first-order system for small changes in x_i. The time constant is determined for gases by the compressibility of the gas; for liquids the elastic deformation of the pressure-sensing device often controls.

We shall analyze the system of Fig. 4.36*a* for the case of a gaseous medium since a majority of practical applications utilize low-pressure ($p_s \approx 20$ to 30 psig) air as the working fluid. The principle of conserva-

[1] C. C. Perry and H. R. Lissner, "The Strain Gage Primer," chap. 13, McGraw-Hill Book Company, New York, 1955.

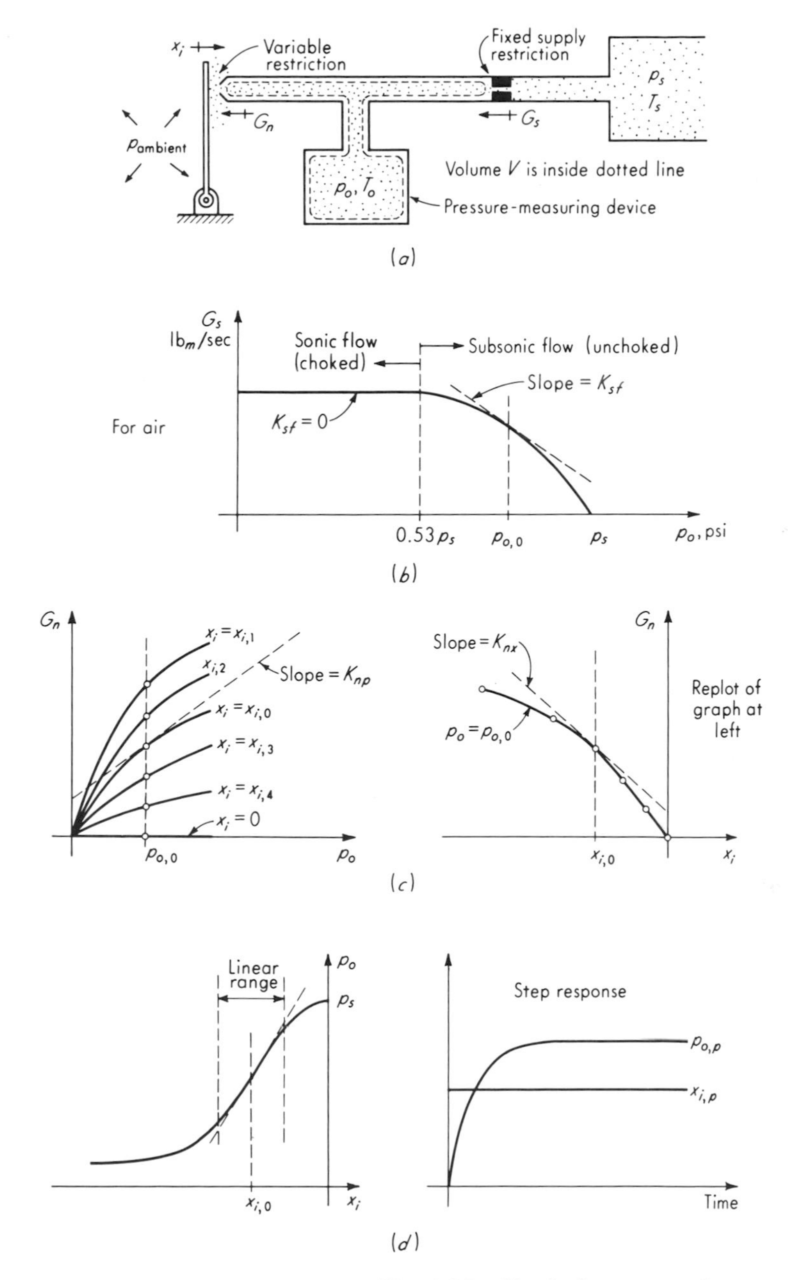

Fig. 4.36. *Nozzle-flapper transducer.*

tion of mass is applied to the volume V by stating that during a time interval dt the difference between entering mass and leaving mass must show up as an additional mass storage in V.

It is necessary to obtain expressions for the mass flow rates G_s and G_n. We assume that supply pressure p_s and temperature T_s are constant. Then G_s depends on p_o only; however, the dependence is nonlinear, and so we employ a linearized (perturbation) analysis which will be valid for small changes from an operating point. We can write

$$G_s = G_s(p_o) \approx G_{s,0} + \left.\frac{dG_s}{dp_o}\right|_{p_o = p_{o,0}} (p_o - p_{o,0}) = G_{s,0} + K_{sf}p_{o,p} \tag{4.88}$$

where

$$G_{s,0} \triangleq \text{value of } G_s \text{ at equilibrium operating point} \tag{4.89}$$
$$p_{o,0} \triangleq \text{value of } p_o \text{ at equilibrium operating point} \tag{4.90}$$
$$p_{o,p} \triangleq \text{small change (perturbation) in } p_o \text{ from } p_{o,0} \tag{4.91}$$
$$K_{sf} \triangleq \text{value of } dG_s/dp_o \text{ at } p_{o,0} \text{ (a constant)} \tag{4.92}$$

The function $G_s(p_o)$ can be found theoretically from fluid mechanics and thermodynamics (with the help of an experimental orifice-discharge coefficient) or entirely by experiment for a given orifice. Its general shape is given in Fig. 4.36*b*.

In finding the nozzle mass flow rate G_n, we assume that the process from p_s, T_s to p_o, T_o is a perfect-gas, work-free, adiabatic process. Also, the velocity of the gas at pressure p_o in volume V is assumed zero. Thus the gas at p_o is essentially in a stagnation state, and since the stagnation enthalpy for a perfect-gas, work-free, adiabatic process is constant, the temperature T_o is nearly the same as T_s and remains nearly constant. If T_o may be assumed constant, the nozzle mass flow rate depends only on p_o and x_i. The relationship between G_n, p_o, and x_i can be found from theory (with experimental corrections) or from experiment alone for a specific device. The relationship is again nonlinear, and so a perturbation analysis is in order.

$$G_n(p_o,x_i) \approx G_{n,0} + \left.\frac{\partial G_n}{\partial p_o}\right|_{\substack{x_{i,0}\\ p_{o,0}}} (p_o - p_{o,0}) + \left.\frac{\partial G_n}{\partial x_i}\right|_{\substack{x_{i,0}\\ p_{o,0}}} (x_i - x_{i,0}) \tag{4.93}$$

$$G_n \approx G_{n,0} + K_{np}p_{o,p} + K_{nx}x_{i,p} \tag{4.94}$$

Figure 4.36*c* shows how K_{np} and K_{nx} could be found from experimental data.

The mass storage in volume V can be treated by using the perfect-gas law $p_oV = MRT_o$. We assume V, R, and T_o to be constant. Then,

$$p_o = \frac{RT_o}{V} M \tag{4.95}$$

$$p_o + p_{o,p} = \frac{RT_o}{V} (M_0 + M_p) \tag{4.96}$$

$$\frac{dp_{o,p}}{dt} = \frac{RT_o}{V}\frac{dM_p}{dt} \tag{4.97}$$

By conservation of mass during a time interval dt,

$$(\text{Mass in}) - (\text{mass out}) = \text{additional mass stored}$$

$$(G_{s,0} + K_{sf}p_{o,p})\,dt - (G_{n,0} + K_{np}p_{o,p} + K_{nx}x_{i,p})\,dt = dM_p = \frac{V}{RT_o}\,dp_{o,p} \tag{4.98}$$

If the operating point $p_{o,0}$, $x_{i,0}$ is an equilibrium condition, $G_{s,0} = G_{n,0}$. We then have

$$\frac{V}{RT_o}\frac{dp_{o,p}}{dt} + (K_{np} - K_{sf})p_{o,p} = (-K_{nx})x_{i,p} \tag{4.99}$$

This is clearly a first-order system, and so we define

$$K \triangleq \frac{-K_{nx}}{K_{np} - K_{sf}} \qquad \text{psi/in.} \tag{4.100}$$

$$\tau \triangleq \frac{V}{RT_o(K_{np} - K_{sf})} \qquad \text{sec} \tag{4.101}$$

to give

$$(\tau D + 1)p_{o,p} = Kx_{i,p} \tag{4.102}$$

and

$$\frac{p_{o,p}}{x_{i,p}}(D) = \frac{K}{\tau D + 1} \tag{4.103}$$

To improve speed of response (decrease τ) the volume V should be minimized. Since T_o is usually the ambient temperature, R and T_o are not available for adjustment. An increase in $K_{np} - K_{sf}$ will decrease τ but at the expense of sensitivity, as shown by Eq. (4.100).

A relatively crude device of this type made up for student laboratory use had a nozzle diameter of $\frac{1}{32}$ in. and a volume V of the order of 1 in.^{3}. For a supply-orifice diameter of $\frac{1}{32}$ in. and a supply pressure of 25 psig this device had a K (at the most sensitive part of its range) of about 2,000 psi/in. and a τ of about 0.12 sec. It was quite linear over a range of about ± 0.002 in. around $x_{i,0} = 0.004$ in. By changing only the supply-orifice diameter to $\frac{1}{64}$ in. the sensitivity was raised to about 8,000 psi/in. while τ increased to 0.24 sec. The linear range was now about ± 0.0005 in. around $x_{i,0} = 0.0015$ in. Since 1 psi = 27.7 in. of water, a water manometer used to read p_o gives a (easily readable) 0.1-in. change for an x_i change of only 0.45×10^{-6} in. when the sensitivity is 8,000 psi/in. This illustrates the great sensitivity of this transducer.

A useful approximate expression for the static sensitivity may be easily obtained by assuming incompressible flow. The results are accurate for liquids and a good estimate for gases if pressure changes are not large. The mass flow through the supply orifice is now

$$G_s = \frac{C_d \pi d_s^2}{4}\sqrt{2\rho(p_s - p_o)} \tag{4.104}$$

where $C_d \triangleq$ discharge coefficient
$d_s \triangleq$ supply-orifice diameter
$\rho \triangleq$ fluid mass density

and we are neglecting the velocity of approach since the orifice area is very small compared with the upstream passage. The flow area for the nozzle is taken as the peripheral area of a cylinder of height x_i and diameter d_n, the nozzle diameter. This is true only for small values of x_i. The discharge coefficient for this configuration may be different from that for the supply orifice and may vary somewhat with x_i but here we take it to be the same as C_d in Eq. (4.104). We have then

$$G_n = C_d \pi d_n x_i \sqrt{2\rho(p_o - p_{\text{ambient}})} \qquad (4.105)$$

For steady state, $G_n = G_s$ so that (taking $p_{\text{ambient}} = 0$ psig) we get

$$p_o = \frac{p_s}{1 + 16(d_n^2 x_i^2 / d_s^4)} \qquad (4.106)$$

The sensitivity dp_o/dx_i varies with x_i and is found to have its maximum value at $x_i = 0.14 d_s^2/d_n$. This maximum value is

$$K_{\max} = \frac{2.6 d_n p_s}{(d_s)^2} \quad \text{psi/in.} \qquad (4.107)$$

Thus we see that large d_n, large p_s, and small d_s lead to high sensitivity.

Digital displacement transducers. More and more, measuring instruments are being required to communicate with digital computers. The amount of raw data generated by large-scale test programs is so great that automated computer reduction of these data to meaningful form is a necessity. Also, feedback-control systems for complex processes are increasingly dependent on digital computers for partial or complete control-action generation. Thus, measuring devices that form a basic part of these overall systems must be compatible with the digital nature of the computer.

There is some question whether any "true" digital transducers exist, since a clear-cut definition is not at present available. Rather, many people prefer to think of all transducers as analog with an analog-to-digital converter "built in" to a greater or lesser extent. We shall not try to resolve this semantic problem here but rather shall simply state that in practice two main methods are used to realize digital signals. The first involves converting the analog variable to a shaft rotation (or translation) and then using one of the many types of *shaft-angle encoder* to generate digital voltage signals. The other approach is to convert the analog variable to an analog voltage which is then converted to a digital voltage by one of the many types of *voltage-to-digital* converters. The shaft-angle encoder and the voltage-to-digital converter are perhaps the closest approach to true digital transducers, the first for motion, the second for voltage. In this section we consider the shaft-angle encoder[1] for transducing analog motion to digital voltage.

While analog voltages can be transmitted with a single pair of wires, digital voltages for input to a computer require a pair of wires for each digit of the signal. That is, a voltage of 564 volts is communicated to a digital computer in three separate pieces. The computer is told that the units digit is 4, the tens digit 6, and the hundreds digit 5. (Actually the computer works on the binary rather than the decimal system, but the concept is the same.) If the computer worked on the decimal system, it would require components that could recognize 10 different states (0 through 9) for each digit. The most simple and reliable arrangement is to use components that recognize only two states. If only two states are needed, one can be "on" and the other "off." Thus the basic components can be essentially "switches" that are either open or closed. This type of hardware leads naturally to representing numbers in the binary (rather than decimal) system since each binary digit (called a *bit*) requires only two states to specify it completely.

Any number can be expressed in binary form by breaking it down into a sum of various powers of 2. For example,

$$10 = 1(2^3) + 0(2^2) + 1(2^1) + 0(2^0) \qquad (4.108)$$

The coefficients 1, 0, 1, 0 are actually sufficient to specify the number completely if it is known that the binary system is being used. Thus the notation used to give the number 10 in binary form is simply 1010. It takes four bits to give this number; thus four "switches" in the computer are needed. To communicate the number 10 to the computer, the first and third switches would be closed (a closed switch corresponds to 1) while the second and fourth would be open (an open switch corresponds to zero). With four switches, the computer could handle only numbers in the range 0 to 15 in steps of 1; thus its "resolution" would be only $100/15 = 6.7$ percent. To get more accuracy (resolution), more switches (bits) must be included in the computer. To handle noninteger numbers, negative powers of 2 are used. Thus

$$\begin{aligned} 9.72 = {} & \underbrace{1(2^3) + 0(2^2) + 0(2^1) + 1(2^0) + 1(2^{-1}) + 0(2^{-2}) + 1(2^{-3}) + 1(2^{-4}) + 1(2^{-5})}_{9.71875} + \cdots \end{aligned} \qquad (4.109)$$

We see from this example that a decimal number may not have an exact binary equivalent but that one can come as close as one wishes by adding

[1] P. Barr, Shaft Position Encoders, *Electromech. Design*, p. 165, January, 1964.

more bits. When noninteger numbers are expressed in binary, a *binary point* is placed just to the right of the 2^0 position; thus 9.72 expressed as closely as possible in nine-bit binary is 1001.10111.

While the computer itself generally works in the pure binary system, other *codes* are sometimes useful where communication between a man (who is most familiar with the decimal system) and the machine is necessary. The binary-coded decimal (BCD) code is widely used for such purposes. Here each decimal digit is given by its binary equivalent. Since the numbers 0 through 9 must be representable, this requires four bits (2^3, 2^2, 2^1, 2^0) for each decimal digit. Thus the decimal number 872.5 would be written as 1000 0111 0010. 0101 in the binary-coded decimal system, requiring a total of 16 bits. In a pure binary system this number would be 1101100000.1, requiring only 11 bits; thus the binary-coded decimal code trades off ease of conversion for efficiency. The code given above should not be called *the* binary-coded decimal code since other ways of coding decimal into binary are also in use.

With the above brief introduction as background, we may now consider the operation of a typical shaft-angle encoder. The principle is most easily visualized for a translational rather than a rotary motion. Figure 4.37 shows such a device in schematic form. The encoder shown has four tracks (bits) and is divided into conducting and insulating portions, with the smallest increment being 0.01 in. As the scale moves under the brushes, the respective lamp circuits are made or broken so that the sum of the numbers shown on the readout lamps is at every instant equal to the displacement in hundredths of an inch. With four bits the maximum travel is 0.15 in.; longer travel requires more tracks (bits). If resolution finer than 0.01 in. is needed, a more closely spaced black-white pattern can be produced or some motion-amplifying mechanism (gearing, etc.) employed. If the displacement information is to be used in a digital computer rather than just visually displayed, the brushes are connected to the computer input "switches" rather than to the lamps.

For rotational motions, the pattern of Fig. 4.37 is simply deformed so that the length of the scale becomes the circumference of a circle on a flat disk. The brushes are then disposed along a radial line on the disk. Figure 4.38 shows a typical disk. Many detail variations on the basic principle are in use. To eliminate wear and friction of the brushes and to improve resolution, optical readout using clear and opaque segments, light sources, and photocells is used. Magnetic readout without contact is also possible. A typical unit with optical readout has 13 tracks (giving a resolution of $1/2^{13} = 1/8{,}192$) in a case only 4 in. in diameter and 7 in. long.

One important defect in the arrangement of Fig. 4.37 is that, if the brushes ("reading line") and the grid patterns are not perfectly aligned,

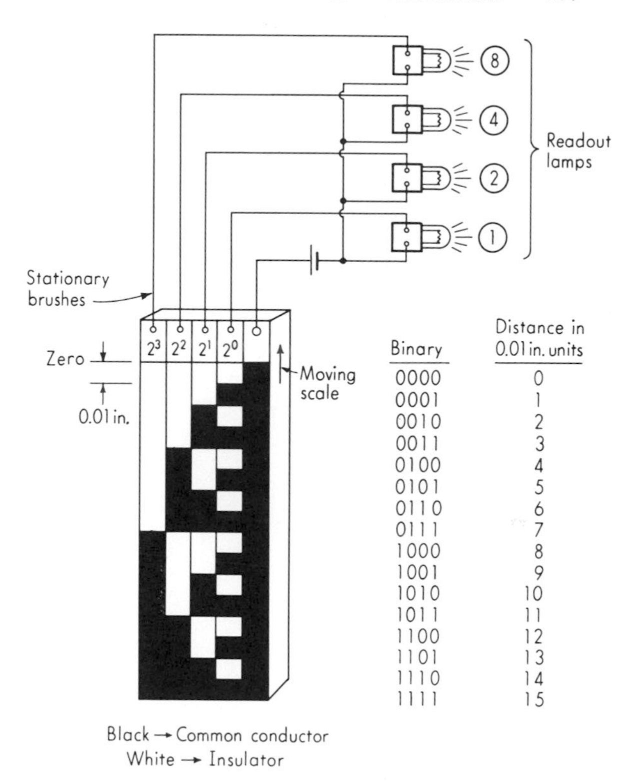

Binary	Distance in 0.01 in. units
0000	0
0001	1
0010	2
0011	3
0100	4
0101	5
0110	6
0111	7
1000	8
1001	9
1010	10
1011	11
1100	12
1101	13
1110	14
1111	15

Fig. 4.37. *Digital transducer.*

in moving from one position to the next an error of more than one unit can occur. For example, suppose the reading is 9 (1001 binary) and in going to 10, because of misalignment, the 2^1 digit changes before the 2^0 digit does, giving binary 1011. Now 1011 binary is 11 decimal, and so the reading jumps from 9 to 11. This problem is overcome in practical

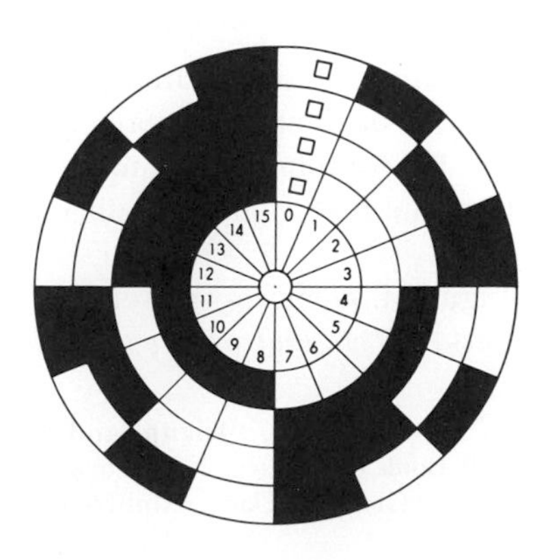

Fig. 4.38. *Code disk for shaft encoder.*

Ordinary decimal	Cyclic decimal
1	1
2	2
3	3
4	4
5	5
6	6
7	7
8	8
9	9
10	19
11	18
12	17
13	16
14	15
15	14
16	13
17	12
18	11
19	10
20	20
21	21
22	22
23	23
24	24
25	25

Fig. 4.39. *Cyclic decimal system.*

encoders either by special brush arrangements or by changing the code to one that changes only one digit in going from one number to the next. The Gray[1] (cyclic binary) code is one such in common use.

The Giannini Datex[2] code combines the cyclic binary code with a binary-coded decimal code as follows: First the ordinary decimal number is converted to "cyclic" decimal according to the rules:

1. An ordinary decimal digit that follows an even digit is not changed.
2. An ordinary decimal digit that follows an odd digit is changed to its 9's complement (9 minus the digit).
3. The first (most significant) digit is not changed.

Figure 4.39 shows the relation between ordinary and cyclic decimal numbers. *Note that any two adjacent numbers differ in only one decimal digit*

[1] J. T. Tou, "Digital and Sampled-data Control Systems," p. 372, McGraw-Hill Book Company, New York, 1959.

[2] Giannini Corp., Monrovia, Calif., *Bull.* 001, 1955.

Ordinary decimal	*Cyclic decimal*	*Brush number*	*Binary-coded decimal (Datex)* Hundreds digit 9 10 11 12	Tens digit 5 6 7 8	Units digit 1 2 3 4
000	000		1 0 0 0	1 0 0 0	1 0 0 0
001	001		1 0 0 0	1 0 0 0	1 1 0 0
009	009		1 0 0 0	1 0 0 0	1 0 0 1
010	019		1 0 0 0	1 1 0 0	1 0 0 1
011	018		1 0 0 0	1 1 0 0	1 1 0 1
099	090		1 0 0 0	1 0 0 1	1 0 0 0
100	190		1 1 0 0	1 0 0 1	1 0 0 0
199	100		1 1 0 0	1 0 0 0	1 0 0 0
200	200		0 1 0 0	1 0 0 0	1 0 0 0

Fig. 4.40. *Binary-coded decimal system.*

for the cyclic decimal system. The next step is to convert the cyclic decimal to binary, using a special binary-coded decimal code. Recall that each decimal digit requires four bits; however, with four bits one can write 16 decimal numbers (0 through 15) while only 10 (0 through 9) are necessary. Thus one can *choose* which of the 16 to use. Based on the desire for error checking, reducing power-supply requirements, and ease of obtaining the 9's complement of any number, it can be shown that the best choice for a binary-coded decimal code here is as follows:

Decimal	*Binary-coded decimal (Datex)*
0	1000
1	1100
2	0100
3	0110
4	0010
5	0011
6	0111
7	0101
8	1101
9	1001

An encoder based on this system requires 4 brushes for each decimal digit. Thus to handle three-digit decimal numbers, 12 brushes are needed. Figure 4.40 shows some typical conversions. Note that for a unit change in the decimal number only one contact change in the encoder is ever needed, thus preventing the large errors possible in a straight binary system.

The encoders discussed above are all of the so-called "absolute"-position type in that the contact pattern uniquely identifies any given position. A different method, the "incremental" system, merely gives a pulse each time an increment of motion occurs. These pulses are put into a counter which adds pulses for forward motion and subtracts them for reverse motion. The "contents" of the counter are thus at any time an indication of the position. This method uses only one "track" to generate the pulses, and its output is not in the binary form desired for direct computer input; thus a separate pulse-count-to-binary converter is required. The main problem with such systems is that if an error occurs it persists for all later motions.

4.4

Relative Velocity, Translational and Rotational We consider here devices for measuring the velocity of translation, along a line, of one point relative to another and the plane rotational velocity about a single axis of one line relative to another.

Calibration. The measurement of rotational (angular) velocity is probably more common than that of translational velocity. Since translation can generally be obtained from rotation by suitable gearing or mechanisms, we consider mainly the calibration of rotational devices.

Perhaps the most difficult area of angular rate measurements is the extremely slow rotations associated with inertial guidance equipment. Angular velocities of 1 revolution/day (earth's rate) and less are of interest. Electro-optical devices such as the Midarm system (discussed under displacement calibration) enable measurement (and thus calibration of other less accurate transducers) of these low angular velocities with an accuracy of the order of 0.0002 degree/hr, using a measuring time of 1 min.

For higher angular and linear velocities perhaps the most convenient calibration scheme uses a combination of a toothed wheel, a simple magnetic proximity pickup, and an electronic EPUT (events per unit time) meter (see Fig. 4.41). The angular rotation is provided by some adjustable-speed drive of adequate stability. The toothed iron wheel passing under the proximity pickup produces an electrical pulse each time one tooth passes. These pulses are fed to the EPUT meter which counts them over an accurate time period (say 1.00000 sec), displays the result visually for a few seconds to enable reading, and then repeats the process

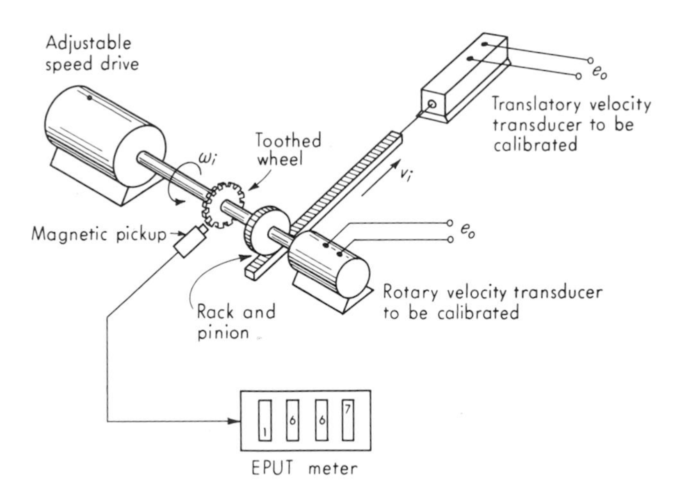

Fig. 4.41. *Velocity-calibration setup.*

over and over. The stability of the rotational drive is easily checked by observing the variation of the EPUT meter readings from one sample to another. The inaccuracy of pulse counting is ± 1 pulse plus the error in the counter time base, which is of the order of 1 ppm. The overall accuracy achieved depends on the stability of the motion source, the angular velocity being measured, and the number of teeth on the wheel. If the motion source were *absolutely* stable (no change in velocity whatever), very accurate measurement could be achieved simply by counting pulses over a long period of time, since then the average velocity and the instantaneous velocity would be identical. If the motion source has some drift, however, the time sample must be fairly short. For example, a shaft rotating at 1,000 rpm with a 100-tooth wheel produces 1,667 pulses in a 1-sec sample period. The inaccuracy here would be 1 part in 1,667 (the 1-ppm time-base error is totally negligible) or 0.06 percent. If the shaft rotated at 10 rpm the error would be 6 percent. Slow rotations can be accurately measured by such means if the toothed wheel is placed on a shaft which is sufficiently geared up from the shaft driving the transducer being calibrated.

The above procedure uses relatively simple equipment and generally provides entirely adequate accuracy. Other simpler and less accurate procedures can be used if they are adequate for their intended purpose. These usually consist of simply comparing the reading of a velocity transducer known to be accurate with the reading of the transducer to be calibrated when they are both experiencing the same velocity input.

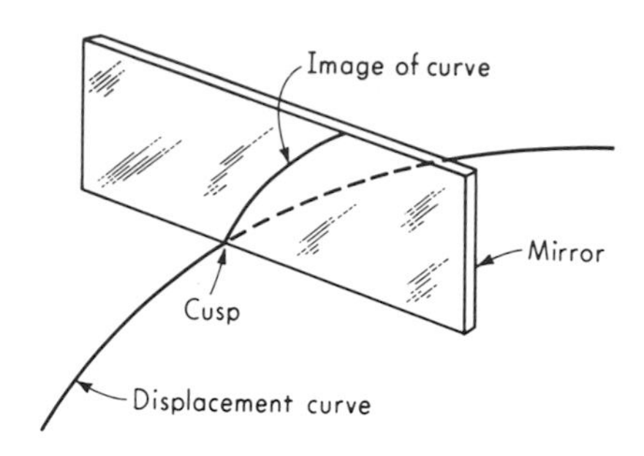

Adjust mirror until image and curve have smooth juncture (no cusp). Mirror will then lie on normal to curve.

Fig. 4.42. Graphical differentiation.

Velocity by graphical differentiation of displacement records. If one has a graphical record of displacement versus time it is possible to obtain velocity versus time by graphical measurement[1] of instantaneous slope. This procedure can be fairly accurate for smooth displacement curves but becomes increasingly less reliable for rapidly changing displacements. A mirror used as in Fig. 4.42 to find the normal to the curve is helpful in establishing the tangent.

Velocity by electrical differentiation of displacement voltage signals. The output voltage of any displacement transducer may be applied to the input of a suitable differentiating circuit to obtain a voltage proportional to velocity (see Sec. 10.4 on differentiating circuits). The main problem is that differentiation accentuates any low-amplitude, high-frequency noise present in the displacement signal. Thus a carbon-film potentiometer would be preferable to the wire-wound type, and demodulated and filtered signals from a-c transducers may cause trouble because of the remaining ripple at carrier frequency. Workable systems using electrical differentiation are possible, however, with adequate attention to details.

Average velocity from measured Δx and Δt. Often a value of average velocity over a short distance or time interval is adequate, and a continuous velocity/time record is not required. A useful basic method is somehow (optically, magnetically, etc.) to generate a pulse when the moving object passes two locations whose spacing is accurately known. If the velocity were constant, any spacing could be used, large spacing of course leading to greater accuracy. If the velocity is varying, the

[1] F. A. Willers, "Practical Analysis," p. 158, Dover Publications, Inc., New York, 1947.

spacing Δx should be small enough so that the average velocity over Δx is not very different from the velocity at either end of Δx. The same technique is applicable to rotational motion.

Figure 4.43*a* shows the application of a variable-reluctance proximity pickup such as was used in Fig. 4.41. When magnetic material passes close in front of the face of the pickup the reluctance of the magnetic path changes with time, generating a voltage in the coil. These pickups are simple and cheap and give a large output voltage (often several volts) under typical operating conditions. The output voltage increases with velocity and closeness of the external moving iron to the pickup. Display of the two pulses on a single sweep of an oscilloscope with a calibrated time base allows measurement of the average velocity. Greater accuracy may be achieved by applying the voltage pulses to an electronic time-interval meter. Figure 4.43*b* shows a similar arrangement using electro-optical techniques. Such an arrangement could replace the magnetic pickup in Fig. 4.41 also.

Photography of the motion, using a stroboscopic lamp flashing at a known rate, also provides velocity data of this type.

Mechanical flyball angular-velocity sensor. A classical rotary speed-measuring device still in wide use today, especially as a measuring element of industrial speed-control systems for engines, turbines, etc., is the flyball. Figure 4.44 shows the general arrangement schematically. Since the centrifugal force varies as the square of input velocity ω_i, the output x_o will not vary linearly with speed if an ordinary linear spring is used. For *small* changes in ω_i a linearized model may be used to show that the transfer function between ω_i and x_o is essentially of the form

$$\frac{x_o}{\omega_i}(D) = \frac{K}{D^2/\omega_n^2 + 2\zeta D/\omega_n + 1} \qquad (4.110)$$

The nonlinear static relation between ω_i and x_o for large speed changes may be acceptable in some systems. Where it is not, a nonlinear spring with $F_{\text{spring}} = K_s x_o^2$ can be used to get a linear overall characteristic since, at balance,

$$\text{Centrifugal force} = \text{spring force}$$

$$F_c = K_c\omega_i^2 = F_{\text{spring}} = K_s x_o^2 \qquad (4.111)$$

and thus $x_o = \sqrt{K_c/K_s}\,\omega_i$, a linear relationship.

A variation[1] on this principle uses a pneumatic force-balance system to replace the spring and produces a standard 3- to 15-psig air-pressure signal proportional to ω_i^2. Since this is the standard pressure range of

[1] G. C. Carroll, "Industrial Process Measuring Instruments," p. 187, McGraw-Hill Book Company, New York, 1962.

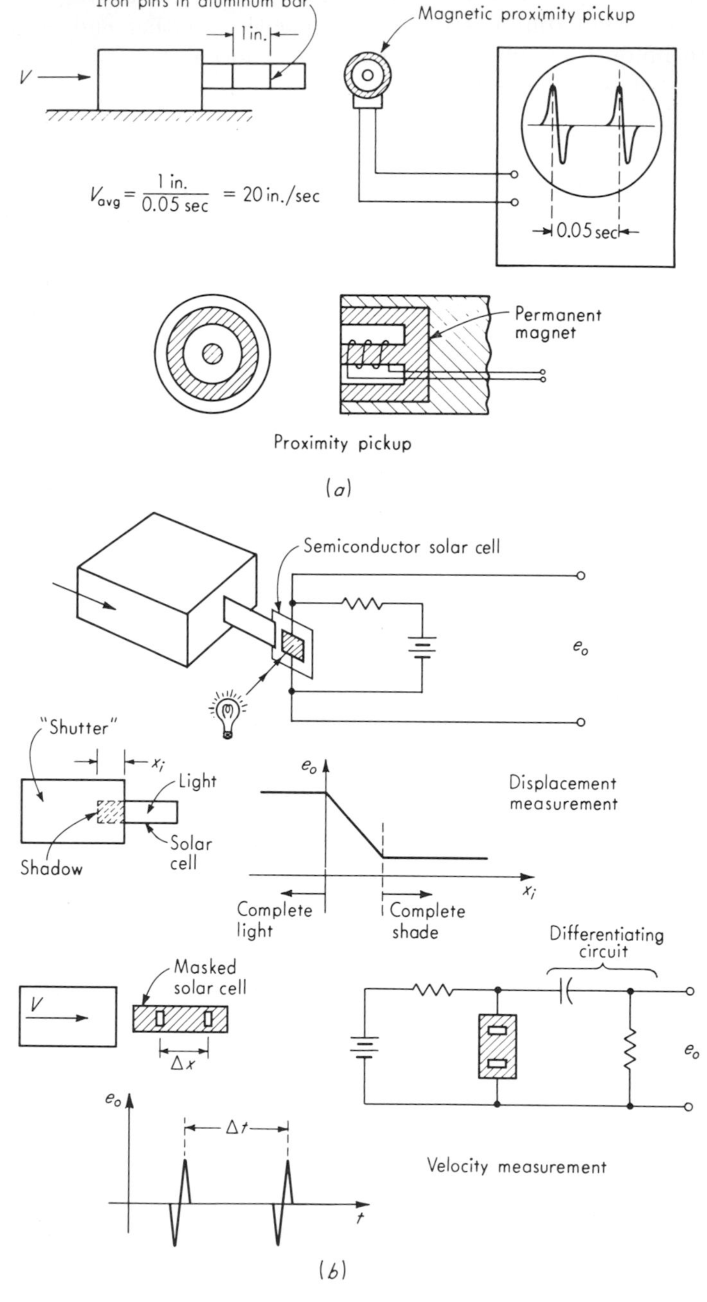

Fig. 4.43. *Velocity measurement as* $\Delta x/\Delta t$.

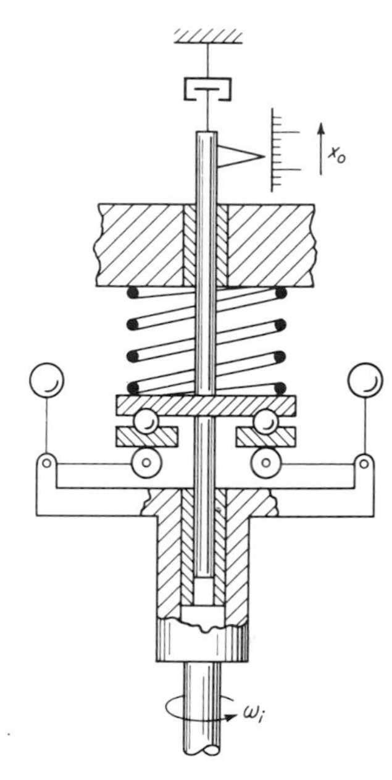

Fig. 4.44. *Flyball velocity pickup.*

industrial-process-control systems, this speed transducer can be directly incorporated into such systems.

Mechanical revolution counters and timers. When continuous reading and an electrical output signal are not required, a variety of mechanical revolution counters (with or without built-in timers) are available. They are generally supplied with a variety of rubber-tipped wheels which transmit by friction the motion to be measured to the counter input shaft.

Magnetic and photoelectric pulse-counting methods. The arrangement of Fig. 4.41, using magnetic pickups (or photocells and light sources with slotted wheels or black-and-white targets) and discussed under Calibration is also often used for measurement since the equipment needed is quite widely available in industry today. If an analog signal (varying d-c voltage) proportional to speed is desired, electronic devices called frequency-to-voltage converters can be connected to the pickup output terminals. However, this arrangement will not generally give as high accuracy as the pulse-counting technique.

Stroboscopic methods. Rotational velocity may be conveniently measured by using electronic stroboscopic lamps which flash at a known and adjustable rate. The light is directed onto the rotating member which itself usually has spokes, gear teeth, or some other feature enabling

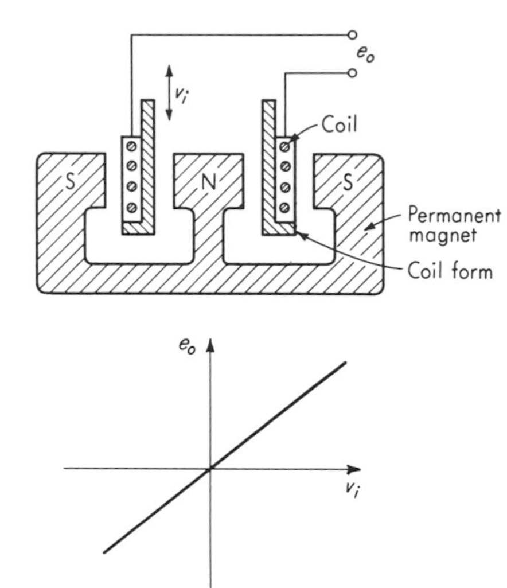

Fig. 4.45. Moving-coil velocity pickup.

"lock on." If not, a simple black-and white paper target can be attached. The frequency of lamp flashing is adjusted until the "target" appears motionless. At this setting, the lamp frequency and motion frequency are identical, and the numerical value can be read from the lamp's calibrated dial to an inaccuracy of about ± 1 percent of the reading. The range of lamp frequency of a typical unit[1] is 110 to 25,000 flashes per minute. Speeds greater than 25,000 rpm can be measured by the following technique. Synchronism can be achieved at any flashing rate r that is an integral submultiple of the speed to be measured, n. The flashing rate is adjusted until synchronism is achieved at the largest possible flashing rate, say r_1. The flashing rate is then slowly decreased until synchronism is again achieved at a rate r_2. The unknown speed n is then given by

$$n = \frac{r_1 r_2}{r_1 - r_2} \tag{4.112}$$

For very high speeds, r_1 and r_2 are close together, giving poor accuracy. Accuracy can be improved by reducing the flashing rate below r_2 until synchronism is again achieved. This procedure can be continued, obtaining synchronism N times $(r_1, r_2, r_3, \ldots, r_N)$. The speed n is then given by

$$n = \frac{r_1 r_N (N - 1)}{r_1 - r_N} \tag{4.113}$$

This procedure can extend the upper range to about 250,000 rpm.

Translational-velocity transducers (moving-coil and moving-magnet pickups). The moving-coil pickup of Fig. 4.45 is based on the

[1] Strobotac, General Radio Co., West Concord, Mass.

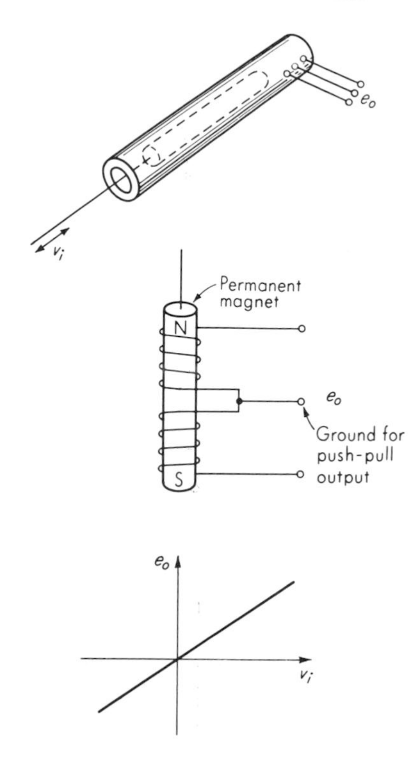

Fig. 4.46. *Velocity pickup.*

law of induced voltage

$$e_o = (Blv_i)10^{-8} \qquad (4.114)$$

where e_o = terminal voltage, volts
B = flux density, gauss
l = length of coil, cm
v_i = relative velocity of coil and magnet, cm/sec

Since B and l are constant, the output voltage follows the input velocity linearly and reverses polarity when the velocity changes sign. Such pickups are widely used for the measurement of vibratory velocities. Since the flux density available from permanent magnets is limited to the order of 10,000 gauss, increase in sensitivity can be achieved only by increase in the length of wire in the coil. To keep the coil small, this requires fine wire and thus high resistance. High-resistance coils require a high-resistance voltage-measuring device at e_o to prevent loading. A typical pickup of about 500 ohms resistance has a sensitivity of 0.15 volt/(in./sec) and a full-scale displacement of 0.15 in. with a nonlinearity of ± 1 percent. A more sensitive coil used in a seismometer (instrument to measure earth shocks) has 500,000 ohms resistance and a sensitivity of 115 volts/(in./sec).

The Sanborn[1] LVsyn shown in Fig. 4.46 uses a permanent-magnet

[1] Sanborn Co., Waltham, Mass.

core moving inside a form wound with two coils connected as shown. Units are available in full-range strokes from about 0.5 to 9.0 in. Sensitivity varies from about 0.1 to 0.65 volt/(in./sec), and coil resistance from 2,000 to 32,000 ohms. The nonlinearity is about 1 percent while core weights range from 3.5 to 69 g.

D-C tachometer generators for rotary-velocity measurement. An ordinary d-c generator (using either a permanent magnet or separately excited field) produces an output voltage roughly proportional to speed. By emphasizing certain aspects of design, such a device can be made an accurate instrument for measuring speed rather than a machine for producing power. The basic principle is again Eq. (4.114), which when applied to the rotational configuration of a d-c generator becomes

$$e_o = \frac{n_p n_c \phi N}{60 n_{pp}} 10^{-8} \qquad (4.115)$$

where
$e_o \triangleq$ average output voltage, volts
$n_p \triangleq$ number of poles
$n_c \triangleq$ number of conductors in armature
$\phi \triangleq$ flux per pole, lines
$N \triangleq$ speed, rpm
$n_{pp} \triangleq$ number of parallel paths between positive and negative brushes

The voltage e_o is a d-c voltage proportional to speed which reverses polarity when the angular velocity reverses. A small superimposed ripple voltage is present because of the finite number of conductors. A typical high-accuracy unit[1] (permanent magnet) has a sensitivity of 7 volts/1,000 rpm, a rated speed of 5,000 rpm, nonlinearity of 0.07 percent over a range 0 to 3,600 rpm, ripple voltage 2 percent of average voltage for speeds above 100 rpm, friction torque of 0.2 in.-oz, rotor inertia of 7 g-cm^2, output impedance of 2,800 ohms, and a total weight of 3 oz.

A special d-c tachometer[2] of unique design for use where a limited ($\pm 15°$) angular travel is acceptable exhibits a very high sensitivity. A 1-in.-diameter model gives 500 volts/1,000 rpm while a 3-in.-diameter gives 30,000 volts/1,000 rpm. The nonlinearity is ± 9 percent for $\pm 15°$ travel, and the operating torque is 500 g-cm. In this generator the permanent magnet rotates while the coil is stationary, and no commutator is needed because of the limited travel.

[1] General Precision Inc., Kearfott Div., Little Falls, N.J.
[2] Armstrong Whitworth Equipment, Hucclecote, Gloucester, England.

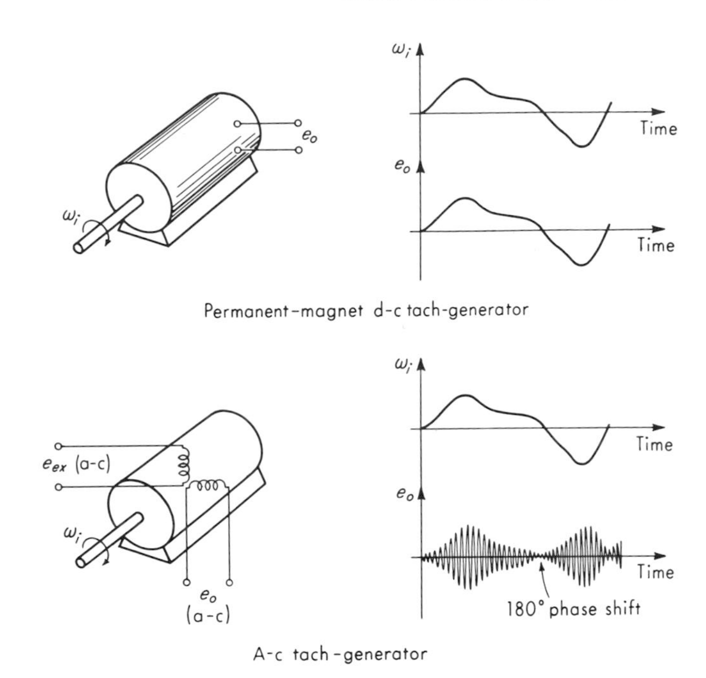

Fig. 4.47. *Tachometer generators.*

A-C tachometer generators for rotary-velocity measurement. An a-c two-phase squirrel-cage induction motor can be used as a tachometer by exciting one phase with its usual a-c voltage and taking the voltage appearing at the second phase as output. With the rotor stationary, the output voltage is essentially zero. Rotation in one direction causes at the output an a-c voltage of the same frequency as the excitation and of an amplitude proportional to the instantaneous speed. This output voltage is in phase with the excitation. Reversal of rotation causes the same action except the phase of the output shifts 180°. While squirrel-cage rotors are sometimes used, the most accurate units employ a drag-cup rotor. This does not change the basic operating characteristics.

A typical high-accuracy unit[1] is excited by 115-volt/400-cps voltage, has a sensitivity of 2.8 volts/1,000 rpm, nonlinearity of 0.05 percent from 0 to 3,600 rpm, negligible rotor friction, rotor inertia of 7 g-cm^2, and a total weight of 6.7 oz. Most commercial a-c tachometers are designed to be used in a-c servomechanisms which conventionally operate on either 60 or 400 cps, and so they are generally designed for operation at these frequencies. For general-purpose motion measurement, the frequency

[1] General Precision Inc., Kearfott Div., Little Falls, N.J.

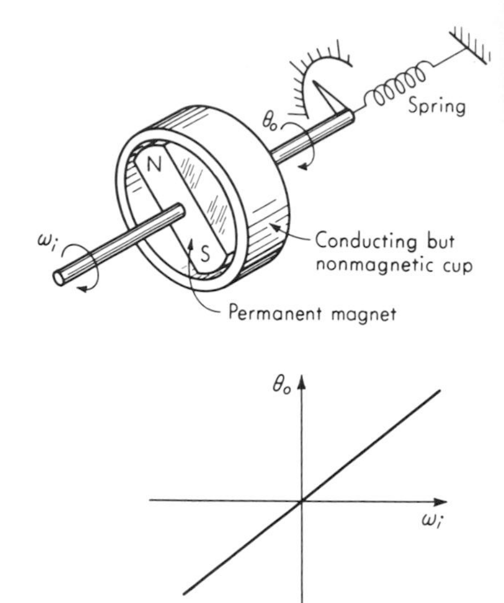

Fig. 4.48. Drag-cup velocity pickup.

response of such units is limited (as are all a-c or "carrier"-type devices) by the carrier frequency to about one-tenth to one-fifth of the carrier frequency. It is usually possible, however, to excite a tachometer designed for 400 cps at considerably higher frequencies if necessary, although some or all of the performance characteristics may change value.

Eddy-current drag-cup tachometer. Figure 4.48 shows schematically an eddy-current tachometer. Rotation of the magnet induces voltages into the cup which thereby produce circulating eddy currents in the cup material. These eddy currents interact with the magnet field to produce a torque on the cup in proportion to the relative velocity of magnet and cup. This causes the cup to turn through an angle θ_o until the linear spring torque just balances the magnetic torque. Thus in steady state the angle θ_o is directly proportional to ω_i, the input velocity. If an electrical output signal is desired, any low-torque displacement transducer can be used to measure θ_o. Dynamic operation is governed by the rotary inertia of parts moving with θ_o, the spring stiffness, and the viscous damping effect of the eddy-current coupling between magnet and cup, leading to a second-order response of the form

$$\frac{\theta_o}{\omega_i}(D) = \frac{K}{D^2/\omega_n^2 + 2\zeta D/\omega_n + 1} \qquad (4.116)$$

Nonlinearity of the order of 0.3 percent can be achieved in such units.

4.5

Relative-acceleration Measurements Transducers directly sensitive to the relative acceleration of two bodies are not generally commercially available. This is due to the scarcity of physical effects directly producing an electrical signal proportional to relative acceleration and also to the wide availability of ("seismic") transducers for measuring *absolute* acceleration. One possibility (though not widely used) is the a-c tachometer generator described above. If the excitation winding is supplied with direct current rather than alternating current, the output voltage will be a varying direct current proportional to the relative angular acceleration of the rotor and stator. The author does not know of any commercial devices of this kind that are specifically designed and marketed for acceleration measurement. One would thus have to experiment with a given a-c tachometer to see whether it was satisfactory for the particular acceleration measurement.

Other possibilities for relative-acceleration measurement include graphical or electrical differentiation of displacement or velocity signals or records. The double differentiation required of displacement records or voltages can rarely be accurately performed except with very smooth signals. Single differentiation of velocity signals may be practical in some instances.

4.6

Seismic- (Absolute-) Displacement Pickups Figure 4.49 shows the general construction of a seismic-displacement pickup for translatory or rotary motions. These devices are used almost exclusively for measurement of vibratory displacements in those (many) cases where a fixed reference for relative-displacement measurement is not available. That is, the vibration of a body can be measured with any of the relative-motion transducers discussed earlier in this chapter, but only if one end of the transducer can be attached to a stationary reference. For measurements on moving vehicles, such references are not generally available, and in many other situations measurement of absolute motion is easier and more desirable. The basic principle of seismic- (absolute-) displacement pickups is simply to measure (with any convenient relative-motion transducer) the relative displacement of a mass connected by a soft spring to the vibrating body. For frequencies above the natural frequency, this relative displacement is also very nearly the absolute displacement since the mass tends to stand still.

To obtain a quantitative measure of performance for such systems we analyze the configuration of Fig. 4.49*a*. The rotational configuration

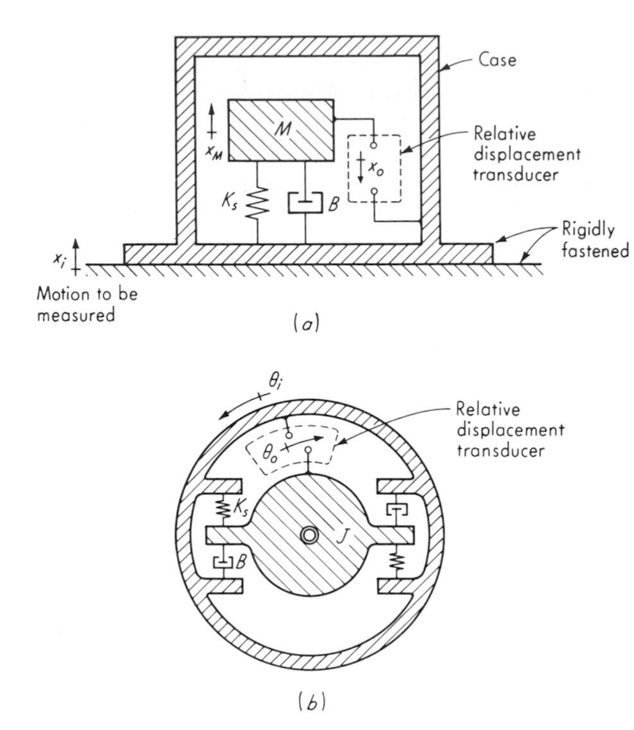

Fig. 4.49. *Translational and rotational seismic pickups.*

is completely analogous. Newton's law may be applied to the mass M as follows:

$$K_s x_o + B\dot{x}_o = M\ddot{x}_M = M(\ddot{x}_i - \ddot{x}_o) \qquad (4.117)$$

where x_i and x_M are the absolute displacements and we have chosen our reference for x_o such that x_o is zero when the gravity force (weight of M) is acting along the x axis statically. Manipulation gives

$$\frac{x_o}{x_i}(D) = \frac{D^2/\omega_n^2}{D^2/\omega_n^2 + 2\zeta D/\omega_n + 1} \qquad (4.118)$$

$$\text{where} \quad \omega_n \triangleq \sqrt{\frac{K_s}{M}}$$

$$\zeta \triangleq \frac{B}{2\sqrt{K_s M}}$$

Since the pickup is intended mainly as a vibration sensor, the frequency response is of prime interest.

$$\frac{x_o}{x_i}(i\omega) = \frac{(i\omega)^2/\omega_n^2}{(i\omega/\omega_n)^2 + 2\zeta i\omega/\omega_n + 1} \qquad (4.119)$$

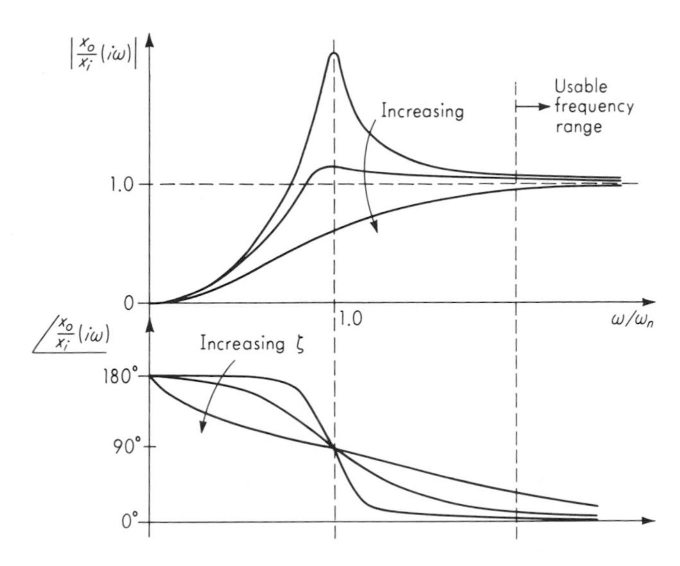

Fig. 4.50. *Seismic-displacement-pickup frequency response.*

This is graphed in Fig. 4.50. We note that there is no response to static displacement inputs and that ω_n should be much less than the lowest vibration frequency ω for accurate displacement measurement. For frequencies much above ω_n, $(x_o/x_i)(i\omega) \rightarrow 1\underline{/0°}$, indicating perfect measurement. The characteristics of the relative-displacement transducer in converting x_o to a voltage e_o must also be considered. Since the force in spring K_s is directly proportional to x_o, if strain gages are used they can be applied directly to this spring, which may be in the form of a cantilever beam. Since a low ω_n is desired, either a large mass or soft spring (or both) is necessary. To keep size (and thereby loading on the measured system) to a minimum, soft springs are preferred to large masses. Intentional damping in the range $\zeta = 0.6$ to 0.7 is often employed to minimize resonant response to slow transients.

4.7

Seismic- (Absolute-) Velocity Pickups The application here is again limited to vibratory velocities, and the basic configuration is exactly the same as in Fig. 4.49. To measure velocity $\dot{x}_i$ rather than displacement x_i, three possibilities will be considered. First, a voltage signal from a displacement pickup may be sent to an electrical differentiation circuit. Second (and this is the most practical), the relative-displacement transducer of Fig. 4.49 is replaced by a relative-velocity transducer

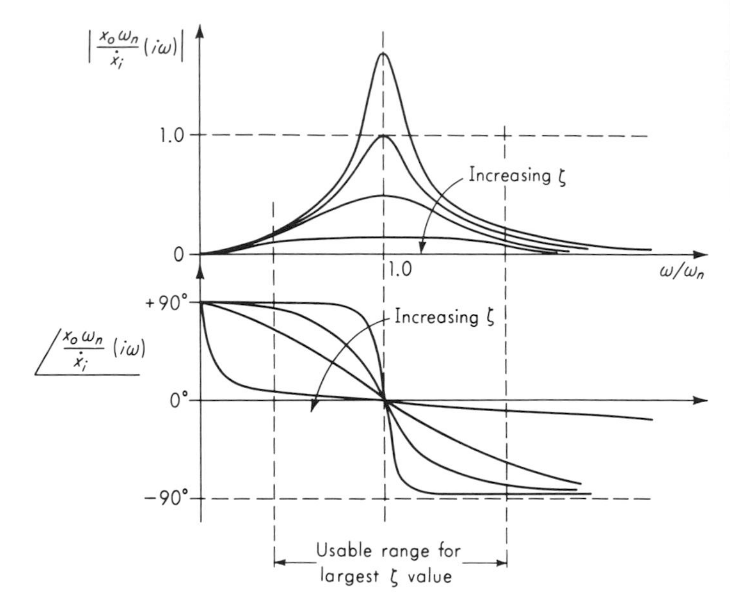

Fig. 4.51. *Seismic-velocity-pickup frequency response.*

(usually a moving-coil pickup). Then, since $e_o = K_e\dot{x}_o$, we have

$$\frac{e_o}{\dot{x}_i}(D) = \frac{K_e D^2/\omega_n^2}{D^2/\omega_n^2 + 2\zeta D/\omega_n + 1} \tag{4.120}$$

and accurate velocity measurement is possible if $\omega \gg \omega_n$. Signals from such pickups may be readily integrated electrically to get displacement information. The third possibility is revealed by rewriting Eq. (4.118) as

$$\frac{x_o}{Dx_i}(D) = \frac{D}{D^2 + 2\zeta\omega_n D + \omega_n^2} \tag{4.121}$$

$$\frac{x_o}{\dot{x}_i}(i\omega) = \frac{1}{2\zeta\omega_n - i[(\omega_n^2 - \omega^2)/\omega]} \tag{4.122}$$

Now if we wish x_o to be a measure of $\dot{x}_i$, then $(x_o/\dot{x}_i)(i\omega) \approx$ const. From Eq. (4.122) we see that this will be the case if $(\omega_n^2 - \omega^2)/\omega \approx 0$, since then

$$\frac{x_o}{\dot{x}_i}(i\omega) \approx \frac{1}{2\zeta\omega_n} \tag{4.123}$$

Now $(\omega_n^2 - \omega^2)/\omega \approx 0$ if $\omega \approx \omega_n$. This would allow measurement only at frequency ω_n. However, if ζ is made very large, the range of frequencies around ω_n for which $(\omega_n^2 - \omega^2)/\omega$ is negligible compared with $2\zeta\omega_n$ is fairly broad. Figure 4.51 shows this graphically. While a possibility, this approach is rarely employed in practice. The required large value of ζ reduces the sensitivity [see Eq. (4.123)], and *both* very low and very high frequencies are not accurately measured.

4.8

Seismic- (Absolute-) Acceleration Pickups (Accelerometers)

The most important pickup for vibration, shock, and general-purpose absolute-motion measurement is the accelerometer. This instrument is commercially available in a wide variety of types and ranges to meet correspondingly diverse application requirements. The basis for this popularity lies in the following features:

1. Frequency response is from zero to some high limiting value. Steady accelerations can (except in piezoelectric types) be measured.
2. Displacement and velocity can be easily obtained by electrical integration, which is much preferred to differentiation.
3. Measurement of transient (shock) motions is more readily achieved than with displacement or velocity pickups.
4. Destructive forces in machinery, etc., are often more closely related to acceleration than to velocity or displacement.

The basic accelerometer configuration is again that of Fig. 4.49. The operating principle is as follows: Suppose the acceleration $\ddot{x}_i$ to be measured is constant. Then, in steady state, the mass M will be at rest relative to the case and thus its absolute acceleration will also be $\ddot{x}_i$. If mass M is accelerating at $\ddot{x}_i$ there must be some force to cause this acceleration, and if M is not moving relative to the case, this force can come only from the spring. Since spring deflection x_o is proportional to force, which in turn is proportional to acceleration, x_o is a measure of acceleration $\ddot{x}_i$. Thus absolute-acceleration measurement is reduced to the measurement of the force required to accelerate a known mass (sometimes called the "proof" mass). This dependence on mass leads to problems (mainly in inertial guidance systems, not in vibration measurement) since a mass also experiences forces due to gravitational fields. Thus an accelerometer cannot distinguish between a force due to acceleration and a force due to gravity.

The majority of accelerometers may be classified as either deflection type or null-balance type. Those used for vibration and shock measurement are generally the deflection type whereas those used for measurement of gross motions of vehicles (submarines, aircraft, spacecraft, etc.) may be either type, with the null-balance being used when extreme accuracy is needed.

Deflection-type accelerometers. A large number of practical accelerometers have the configuration of Fig. 4.49 and differ only in details, such as the spring element used, relative-motion transducer used, and type of damping provided. Since the desired input is now $\ddot{x}_i$, we can rewrite Eq. (4.118) as

$$\frac{x_o}{D^2 x_i}(D) = \frac{x_o}{\ddot{x}_i}(D) = \frac{K}{D^2/\omega_n^2 + 2\zeta D/\omega_n + 1} \tag{4.124}$$

where

$$K \triangleq \frac{1}{\omega_n^2} \quad \text{in./(in./sec}^2\text{)} \tag{4.125}$$

Since output voltage $e_o = K_e x_o$ for many motion transducers, Eq. (4.124) has the correct form for the acceleration-to-voltage transfer function also. We see that the accelerometer is an ordinary second-order instrument; thus all our previous work on this type is immediately applicable. The frequency response extends from 0 to some fraction of ω_n, depending on the accuracy required and the damping. Because sensitivity $K = 1/\omega_n^2$, high-frequency response must be traded for sensitivity. Since the dynamic characteristics of second-order instruments have already been thoroughly discussed, here we shall discuss mainly the specific characteristics of commercially available instruments.

Accelerometers using resistive potentiometers as their motion pickup are intended mainly for slowly varying accelerations and low-frequency vibration. A typical family[1] of such instruments offers nine models covering the range of $\pm 1g$ full scale to $\pm 50g$ full scale. The natural frequencies range from 12 to 86 cps, and ζ is 0.5 to 0.8 over the temperature range -65 to $+165$°F, using a temperature-compensated liquid damping arrangement. Potentiometer resistance may be selected in the range 1,000 to 10,000 ohms, with corresponding resolution of 0.45 to 0.25 percent of full scale. The potentiometer power rating is 0.5 watt at $+165$°F. The sensitivity to acceleration at right angles to the desired axis (cross-axis sensitivity) is less than ± 1 percent of the sensitivity to the desired direction. The operating life is 2,000,000 cycles. Overall inaccuracy is ± 1 percent of full scale or less at room temperature. This increases to ± 1.8 percent if the temperature is allowed to vary over the design range of -65 to $+165$°F. Size is about a 2-in. cube; weight is about 1 lb.

Unbonded-strain-gage accelerometers use the strain wires as the spring element and also as the motion transducer. They are useful for general-purpose motion measurement and also for vibration up to relatively high frequencies. They are available in a wide range of characteristics, typical[2] values including ± 0.5 to $\pm 200g$ full scale, natural

[1] Bourns, Inc., Riverside, Calif.
[2] Statham Instruments Inc., Los Angeles, Calif.

frequency 17 to 800 cps, excitation voltage about 10 volts alternating or direct current, full-scale output ± 20 to ± 50 mv, resolution less than 0.1 percent, inaccuracy 1 percent of full scale or less, cross-axis sensitivity less than 2 percent, and damping ratio (using silicone-oil damping) of 0.6 to 0.8 at room temperature. (Temperature-compensated models are also available.) These instruments can be made quite small and light, a typical size being $\frac{1}{2}$ by $\frac{1}{2}$ by 2 in., with a weight of 26 g.

Bonded-strain-gage accelerometers generally use a mass supported by thin flexure beams, with strain gages cemented to the beam so as to achieve maximum sensitivity, temperature compensation, and insensitivity to cross-axis and angular acceleration. Their characteristics are similar to those of unbonded-gage accelerometers except that size and weight tend to be greater. Silicone-oil damping is again widely used.

A recently developed strain-gage accelerometer[1] using semiconductor materials exhibits many desirable properties. Full-scale range is $\pm 250g$ with full-scale output of ± 250 mv (10 volts d-c excitation). The natural frequency is greater than 10,000 cps with a damping ratio of about 0.06. This low damping is due to material hysteresis and air drag; no intentional damping device is employed. The usable frequency range (amplitude ratio flat to 5 percent) is from 0 to 2,000 cps. In this range, phase shift is very small because of the light damping. The light damping causes no "ringing" problems as long as shock inputs do not contain much frequency content near the 10-kc resonant frequency (see text on terminated-ramp inputs). The cross-axis sensitivity is 1 to 3 percent while nonlinearity and hysteresis are 1 percent of full scale. Temperature compensation is provided to give an operating range of -65 to $+250$°F. The thermal zero shift is 0.02 percent of full scale/F° over this range, and the maximum sensitivity shift is about -7 percent at the extremes of this range. The size is about $\frac{5}{8}$-in. diameter and 1 in. high, with a weight of about 1 oz.

A family of liquid-damped differential-transformer accelerometers[2] exhibits the following characteristics: full-scale ranges from ± 2 to $\pm 700g$, natural frequency from 35 to 620 cps, nonlinearity 1 percent of full scale, full-scale output about 1 volt with excitation of 10 volts at 2,000 cps, damping ratio 0.6 to 0.7 at 70°F, residual voltage at null less than 1 percent of full scale, and hysteresis less than 1 percent of full scale. The size is about a 2-in. cube, with a weight of 4 oz.

A variable-reluctance accelerometer[3] using eddy-current damping has full-scale ranges of ± 1 to $\pm 40g$, natural frequency 16 to 100 cps, damping

[1] Endevco Corp., Pasadena, Calif.

[2] Schaevitz Engineering, Camden, N.J.

[3] Honeywell Inc., Boston, Mass.

ratio 0.6 ±0.2 from −65 to +250°F, 25 volts full-scale output with 26-volt/400-cps excitation, hysteresis 0.15 percent of full scale, cross-axis sensitivity 0.5 percent, and nonlinearity of ±0.25 percent for one-half range and ±1.6 percent full range. The threshold and resolution are each $0.0001g$. The size is about a 2-in. cube, and the weight is about 1 lb.

Piezoelectric accelerometers are in wide use for shock and vibration measurements. In general they do not give an output for constant acceleration because of the basic characteristics of piezoelectric motion transducers. They do, however, have large output-voltage signals and can have very high natural frequencies (higher than any other type) which are necessary for accurate shock measurements. In general no intentional damping is provided, material hysteresis being the only source of energy loss. This results in a very low (about 0.01) damping ratio but this is acceptable because of the very high natural frequency. The transfer function is a combination of Eqs. (4.70) and (4.124):

$$\frac{e_o}{\ddot{x}_i}(D) = \frac{(K_q/C\omega_n^2)\tau D}{(\tau D + 1)(D^2/\omega_n^2 + 2\zeta D/\omega_n + 1)} \tag{4.126}$$

The low-frequency response is limited by the piezoelectric characteristic $\tau D/(\tau D + 1)$ while the high-frequency response is limited by mechanical resonance. The damping ratio ζ of piezoelectric accelerometers is not usually quoted by the manufacturer but can be taken as zero for most practical purposes. The accurate (5 percent high at the high-frequency end and 5 percent low at the low-frequency end) frequency range of such an accelerometer is $3/\tau < \omega < 0.2\omega_n$. Accurate low-frequency response requires large τ, which is usually achieved by use of high-impedance (cathode-follower) amplifiers or charge amplifiers. Some quartz accelerometers and electrometer amplifiers have large enough τ to allow measurement of constant accelerations.

Typical construction of a piezoelectric accelerometer is shown in Fig. 4.52. The "crystal" is preloaded to about 10,000-psi stress by screwing down the cap on the hemispherical spring. This prestressing puts the piezoelectric material at a more linear part of its stress-charge curve. It also allows measurement of acceleration in both directions without the crystal going into tension. When the preload is applied, a voltage of a certain polarity is developed but this soon leaks off to zero. Any further deflection (due to acceleration forces) gives a plus or minus charge, depending on the direction of the motion. Figure 4.53 shows some other constructions designed to minimize cross-axis sensitivity.

Piezoelectric accelerometers are available in a wide range of characteristics; we quote only a few typical examples.[1] A single low-g instru-

[1] Endevco Corp., Pasadena, Calif.

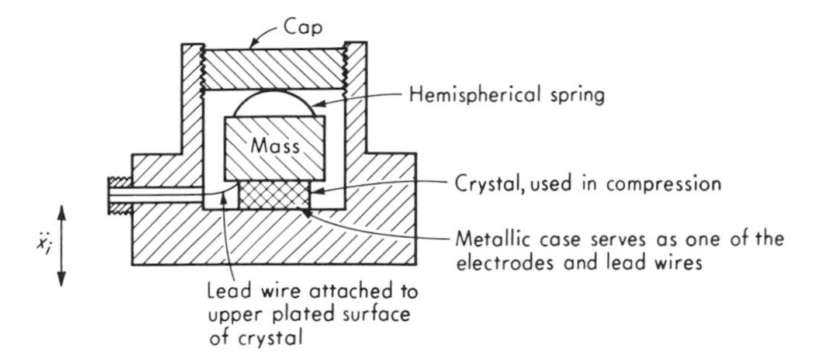

Fig. 4.52. *Piezoelectric-accelerometer construction.*

ment has a sensitivity of 50 mv/g, will measure accelerations from 0.03 to 1,000g with a nonlinearity of 1 percent of full scale, and has a natural frequency of 20 kc, flat frequency response ± 5 percent from 20 to 4,000 cps when used with 100-megohm input impedance, capacitance of 600 pf with 3 ft of cable, cross-axis sensitivity of 5 percent, sensitivity drift of ± 10 percent from -30 to $+230$°F, size about a 1-in. cube, and a weight of 2 oz.

A shock accelerometer has a sensitivity of 5 mv/g, range 0 to 10,000g with 1 percent nonlinearity, natural frequency of 35 kc, flat frequency response ± 5 percent from 0.1 to 7,000 cps with charge amplifier, pulse response ± 5 percent for pulses shorter than 0.66 sec (see Fig. 4.32), no ringing for pulses longer than 0.15 msec, capacitance of 100 pf, cross-axis sensitivity of 5 percent, sensitivity drift ± 1 percent from -100 to ± 350°F, size about a 0.7-in. cube, and a weight of 1 oz.

Some special piezoelectric-accelerometer characteristics available in specific models include water-cooled units usable at 2200°F, triaxial units combining three mutually perpendicular elements in a 1-in.-cube case, high-capacitance (7,000 pf) units insensitive to cable capacitance and giving good low-frequency response with relatively low-input-impedance measuring equipment, and miniature units weighing the order of 1 g for small-object testing.

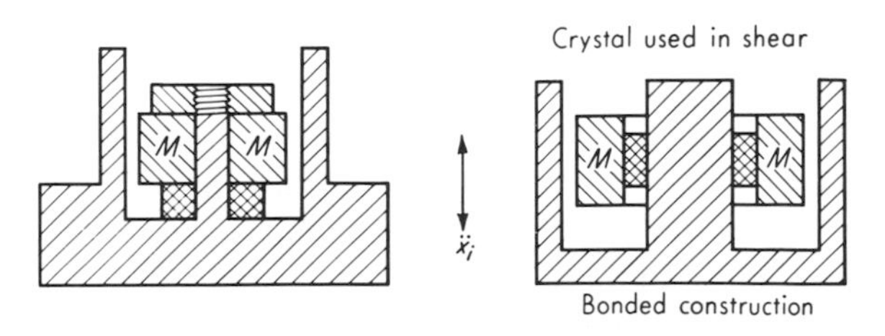

Fig. 4.53. *Piezoelectric-accelerometer construction.*

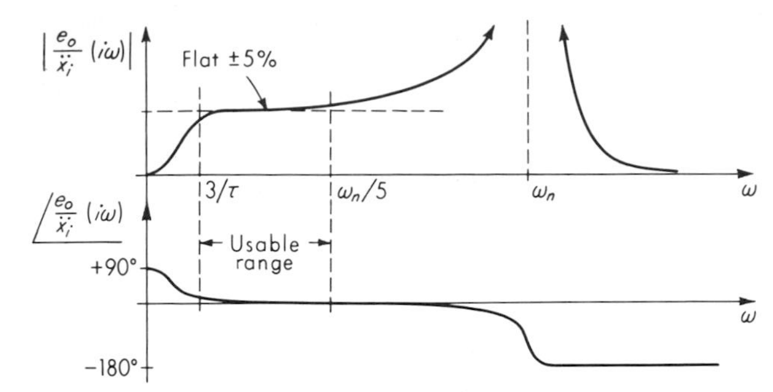

Piezoelectric accelerometer frequency response

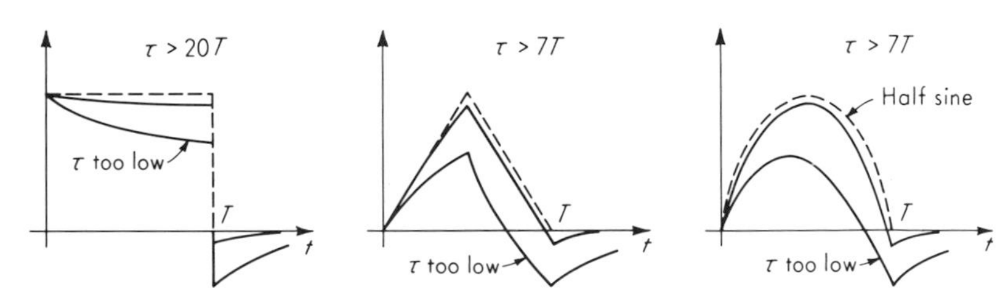

Requirements for accurate peak measurements ±5%

Low-frequency response problems

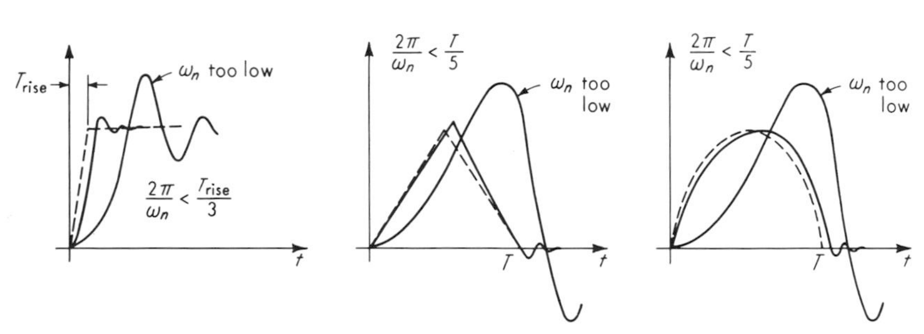

Requirements for accurate peak measurements ±10%

High-frequency response problems

Fig. 4.54. *Piezoelectric-accelerometer response.*

The use of charge amplifiers rather than voltage amplifiers for piezoelectric devices is becoming quite common because of their advantages of insensitivity to cable capacitance and better low-frequency-response characteristics. These instruments are discussed later in the text in the section on amplifiers.

While the general dynamic-response techniques of Chap. 3 are applicable to accelerometer problems, Fig. 4.54 summarizes some results useful in choosing piezoelectric accelerometers for practical applications. The low-frequency problems are peculiar to piezoelectric types whereas the high-frequency problems are common to all accelometers.

Angular accelerometers based on the configuration of Fig. 4.49*b* can be constructed by using various pickups, just as in translational accelerometers. An interesting variation on the basic principle is found in the Statham liquid-rotor angular accelerometer.[1] The inertial mass is liquid contained in a circular case having a flexure-mounted paddle. Rotational acceleration of the case causes a fluid pressure on the paddle whose resulting deflection is measured with a suitable motion pickup. A family of such devices consisting of 11 models exhibits the following characteristics: Full-scale range is from ± 1.5 to $\pm 3{,}000$ rad/sec^2, natural frequency 3 to 150 cps, damping ratio 0.7 ± 0.1 at room temperature, cross-axis (angular) sensitivity 2 percent, nonlinearity and hysteresis ± 2 percent of full scale, sensitivity to linear acceleration from 0.3 percent of full scale/g for the lowest range to 0.02 percent for the highest, and sensitivity to angular velocity about axes other than the instrument axis of about 0.05 rad/sec^2 for an angular velocity of 5 rad/sec. The pickup is either an unbonded strain gage (10 volts excitation, ± 25-mv full-scale output) or a two-arm inductive bridge (10 volts at 3,000 cps excitation, ± 0.75-volt full-scale output). The size and weight range from 9-in. diameter by 3 in. high (8 lb) for the lowest-range instrument to 3-in. diameter by 3 in. high ($1\frac{1}{2}$) lb) for the highest range.

Null-balance- (servo-) type accelerometers. So-called servo accelerometers using the principle of feedback have been developed for applications requiring greater accuracy than is generally achieved with instruments using mechanical springs as the force-to-displacement transducer. In these null-balance instruments the acceleration-sensitive mass is kept very close to the zero-displacement position by sensing this displacement and generating a magnetic force which is proportional to this displacement and which always opposes motion of the mass from neutral. This restoring force plays the same role as the mechanical spring force in a conventional accelerometer. Thus one may consider the mechanical spring to have been replaced by an electrical "spring." The advantages derived from this approach are the greater linearity and lack of hysteresis of the electrical spring as compared with the mechanical one. Also, in some cases, electrical damping (which can often be made less temperature-sensitive than mechanical damping) may be employed. There is also the possibility of testing the static and dynamic performance of the device just

[1] G. N. Rosa, Some Design Considerations for Liquid Rotor Angular Accelerometers, *Statham Instr. Notes* 26, Statham Instruments Inc., Los Angeles, Calif.

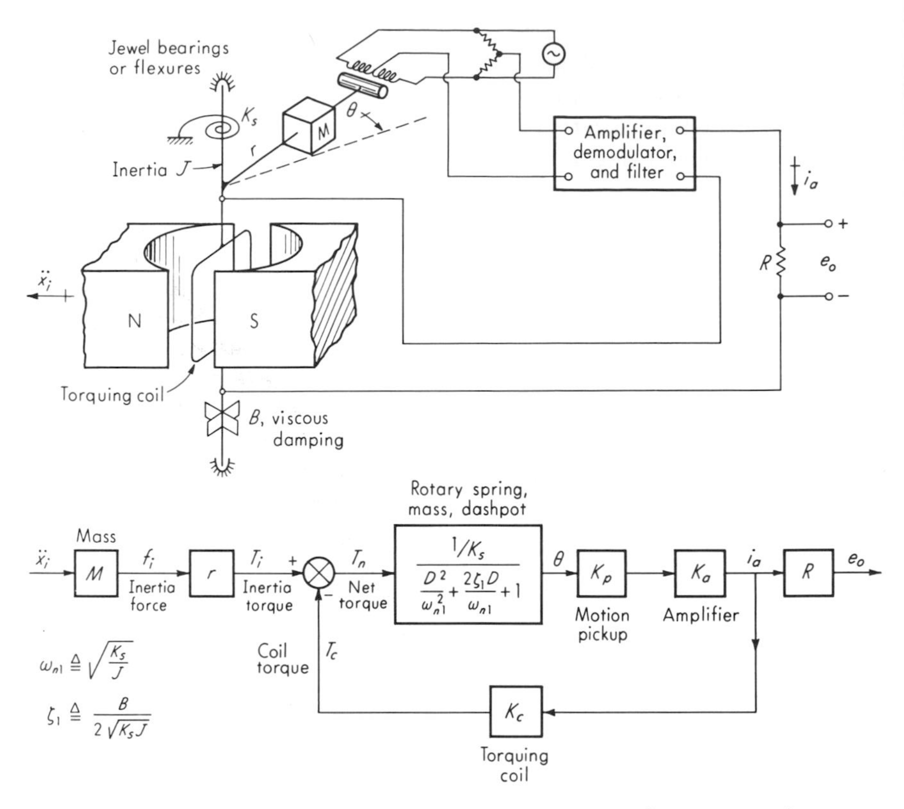

Fig. 4.55. *Servo-type accelerometer.*

prior to a test run by introducing electrically excited test forces into the system. This convenient and rapid remote self-checking feature can be quite important in complex and expensive tests where it is extremely important that all systems operate correctly before the test is commenced. These servo accelerometers are usually used for general-purpose motion measurement and low-frequency vibration. They are also particularly useful in acceleration-control systems since the desired value of acceleration can be put into the system by introducing a proportional current i_a from some external source.

Figure 4.55 illustrates in simplified fashion the operation of a typical instrument designed to measure a translational acceleration $\ddot{x}_i$. (Angular acceleration can also be measured by these techniques by using an obvious mechanical modification.) The acceleration $\ddot{x}_i$ of the instrument case causes an inertia force f_i on the sensitive mass M, tending to make it pivot in its bearings or flexure mount. The rotation θ from neutral is sensed by an inductive pickup and is amplified, demodulated, and filtered to produce a current i_a directly proportional to the motion from null. This current is passed through a precision stable resistor R to produce the output-voltage signal and also is applied to a coil suspended in a magnetic field. The current through the coil produces a magnetic torque on the coil (and the attached mass M) which acts to return the mass to neutral. The current required to produce a coil magnetic torque that just balances the inertia torque due to $\ddot{x}_i$ is directly proportional to $\ddot{x}_i$; thus e_o is a measure of $\ddot{x}_i$. Since a nonzero displacement θ is necessary to produce a current i_a, the mass is not returned exactly to null but comes very close because a high-gain amplifier is used. Analysis of the block diagram reveals the details of performance as follows:

$$\left(Mr\ddot{x}_i - \frac{e_o K_c}{R}\right)\frac{K_p K_a/K_s}{D^2/\omega_{n1}^2 + 2\zeta_1 D/\omega_{n1} + 1} = \frac{e_o}{R} \tag{4.127}$$

$$\left[\frac{D^2}{\omega_{n1}^2} + \frac{2\zeta_1 D}{\omega_{n1}} + \left(1 + \frac{K_c K_p K_a}{K_s}\right)\right] e_o = \frac{MrRK_pK_a}{K_s}\ddot{x}_i \tag{4.128}$$

Now, by design, the amplifier gain K_a is made large enough so that $K_cK_pK_a/K_s \gg 1.0$, so that then

$$\frac{e_o}{\ddot{x}_i}(D) = \frac{K}{D^2/\omega_n^2 + 2\zeta D/\omega_n + 1} \tag{4.129}$$

where

$$K \triangleq \frac{MrR}{K_c} \qquad \text{volts/(in./sec}^2) \tag{4.130}$$

$$\omega_n \triangleq \omega_{n1}\sqrt{\frac{K_pK_aK_c}{K_s}} \qquad \text{rad/sec} \tag{4.131}$$

$$\zeta \triangleq \frac{\zeta_1}{\sqrt{K_pK_aK_c/K_s}} \tag{4.132}$$

Equation (4.130) shows that the sensitivity now depends only on M, r, R, and K_c, all of which can be made constant to a high degree. This demonstrates again the usefulness of high-gain feedback in shifting the requirements for accuracy and stability from many components to a few chosen ones where the requirements can be met. As in all feedback systems, the gain cannot be made arbitrarily high because of dynamic instability; however, a sufficiently high gain can be achieved to give excellent performance. Turning to Eq. (4.131), we see that ω_n is increased from the basic spring-mass frequency ω_{n1} by the factor $\sqrt{K_pK_aK_c/K_s}$, another benefit of high-gain feedback. However, ζ is decreased by the same factor, and so ζ_1 must be made sufficiently high to compensate for this.

A typical accelerometer[1] of this kind, using a flexure pivot, is available in full-scale ranges of ± 10 to $\pm 100g$, natural frequency of 100 to

[1] Systron-Donner Corp., Concord, Calif.

250 cps, damping ratio of from 0.3 to 5, cross-axis sensitivity of 0.1 percent, resolution better than 0.0001 percent of full scale, nonlinearity and hysteresis each better than 0.005 percent of full scale, and has a full-scale output of ± 7.5 volts with considerable current capacity (± 1.2 to ± 12 ma).

Accelerometers for inertial navigation. Inertial navigation is accomplished in principle by measuring the absolute acceleration (usually in terms of three mutually perpendicular components of the total-acceleration vector) of the vehicle and then integrating these acceleration signals twice to obtain displacement from an initial known starting location. Thus instantaneous position is always known without the need for any communication with the world outside the vehicle. To keep the accelerometers' sensitive axes always oriented parallel to their original starting positions, elaborate stable platforms using gyroscopic references and feedback systems are necessary. Since the accelerometers are also sensitive to gravitational force, this force must be computed and corrections applied continuously. Since the inertial navigation system measures absolute motion, systems for navigation over the earth's surface (such as for submarines) must include means for compensating for the earth's own motions.

While accelerometers for such navigation systems must operate on essentially the same basic principles that we have considered above, their extreme performance requirements and the desire to obtain integrals of the acceleration rather than acceleration itself lead to special techniques and configurations. The desire for compatibility with the required data-processing computers (often digital) also influences the designer's choice of alternatives. The details of these applications are beyond the scope of this book but may be found in numerous references.[1,2,3,4,5,6]

Mechanical loading of accelerometers on the test object. The attachment of an accelerometer to a vibrating system results in a change in the motion measured as compared with the undisturbed case. We can apply general impedance principles to this problem to calculate the significance of this effect in any particular instance. In doing so, a useful

[1] H. B. Sabin, 17 Ways to Measure Acceleration, *Control Eng.*, p. 106, February, 1961.

[2] J. M. Slater and D. E. Wilcox, How Precise Are Inertial Components?, *Control Eng.*, p. 86, July, 1958.

[3] *Sperry Engineering Review*, spring, 1964.

[4] J. M. Slater, Inertial Guidance Notes, North American Aviation Corp., Autonetics Div.

[5] C. F. Savant et al., "Principles of Inertial Navigation," McGraw-Hill Book Company, New York, 1961.

[6] P. H. Savet (ed.), "Gyroscopes: Theory and Design," McGraw-Hill Book Company, New York, 1961.

simplification, which is adequate in most cases, is to regard the entire accelerometer as one rigid mass equal to the total mass of the instrument. This approximation generally holds since accelerometers are used below their natural frequency and there is thus little relative motion of the proof mass and the instrument case.

4.9

Calibration of Vibration Pickups While the response of vibration pickups to interfering and modifying inputs such as temperature, acoustic noise, and magnetic fields is often of interest, we are here concerned with the response to the desired input of displacement, velocity, or acceleration. An excellent reference giving a more complete treatment of this subject is the publication "American Standard Methods for the Calibration of Shock and Vibration Pickups 52.2-1959."[1] We shall here briefly touch on some of the main points only.

The calibration methods in wide use may be classified into three broad types: constant acceleration, sinusoidal motion, and transient motion. Constant-acceleration methods (which are suitable only for calibrating accelerometers) include the tilting-support method and the centrifuge. The tilting-support method utilizes the accelerometer's inherent sensitivity to gravity. Static "accelerations" over the range $\pm 1g$ may be accurately applied by fastening the accelerometer to a tilting support whose tilt angle from vertical is accurately measured. This method requires that the accelerometer respond to static accelerations; therefore most piezoelectric devices cannot be calibrated in this way. The accuracy of the method depends on the accuracy of angle measurement and the knowledge of the local gravity. The accuracy is of the order of $\pm 0.0003g$. In the centrifuge method the sensitive axis of the accelerometer is radially disposed on a rotating horizontal disk so that it experiences the normal acceleration of uniform circular motion. Static accelerations in the range 0 to $60{,}000g$ are achievable with an accuracy of ± 1 percent. The allowable weight of the pickup varies from 100 lb at $100g$ to 1 lb at $60{,}000g$.

The sinusoidal-motion method is exemplified by the calibration facility of the National Bureau of Standards.[2] This consists of a modified electrodynamic vibration shaker which has been carefully designed to provide uniaxial pure sinusoidal motion and which is equipped with an

[1] Available from the American Standards Association, Inc., 10 E. 40th St., New York, N.Y., 10016.

[2] R. R. Bouche, Improved Standard for the Calibration of Vibration Pickups, *Exptl. Mechanics*, April, 1961.

accurately calibrated moving-coil velocity pickup to measure its table motion. If a motion is known to be purely sinusoidal, knowledge of its velocity and frequency allows accurate calculation of the displacement and acceleration. (The motion frequency is easily obtained with high accuracy by electronic counters.) This technique is thus useful for displacement, velocity, or acceleration pickups. The particular equipment referred to above can calibrate pickups (obtaining both amplitude ratio and phase angle) of a weight up to 2 lb over the frequency range 8 to 2,000 cps. The acceleration range available is 0 to $25g$, velocity range is 0 to 50 in./sec, and displacement range is 0 to 0.5 in. Accuracy is ± 1 percent from 8 to 900 cps and ± 2 percent from 900 to 2,000 cps.

At frequencies up to about 200 cps the displacement of a vibrating shake table can be quite accurately measured by viewing a suitable target (a point-light-source-illuminated piece of 320-grit emery cloth cemented to the table is convenient) through a measuring microscope. The illuminated grit crystals generate short straight lines of length equal to the peak-to-peak displacement amplitude in the microscope viewing field. The length of the lines can be measured with an inaccuracy of the order of 0.0001 in. This inaccuracy limits the usable frequency range of the method since the displacement amplitude corresponding to the acceleration levels of practical interest becomes very small as the frequency increases. For example, consider a sinusoidal vibration of $10g$ peak acceleration at 1,000 cps. The peak-to-peak displacement would be 0.000196 in., and an inaccuracy of 0.0001 in. would represent a 50 percent error. Thus, at high frequencies, more accurate displacement-measuring techniques are needed. The optical interferometer fringe-disappearance method[1] is used at amplitudes of the order of 4×10^{-6} in. and allows calibration at frequencies exceeding 10,000 cps.

Transient-motion calibration methods include the physical pendulum,[2] ballistic pendulum,[2] and drop-test[2] techniques. The latter two methods (used for accelerometers) are of greatest interest and will be briefly discussed. Both employ the concept that the velocity change Δv during a time interval $\Delta t = t_2 - t_1$ is given by

$$\Delta v = \int_{t_1}^{t_2} a\, dt \qquad (4.133)$$

where $a \triangleq$ acceleration. The procedure involves measurement, by some independent means, of the velocity change Δv of a rigid body to which the pickup is attached and simultaneous recording of the output voltage e_o of

[1] J. Johansson, Accelerometer Calibration, *Instr. Control Systems*, p. 79, December, 1963.

[2] C. M. Harris and C. E. Crede (eds.), "Shock and Vibration Handbook," vol. I, chap. 18, McGraw-Hill Book Company, New York, 1961.

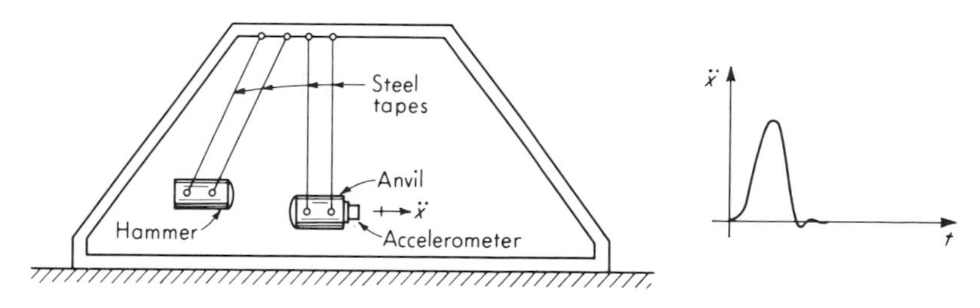

Fig. 4.56. *Ballistic pendulum.*

the pickup. This output voltage is given by

$$e_o = Ka \qquad (4.134)$$

where K is the unknown sensitivity [volts/(in./sec²)] of the pickup. Thus we may write

$$\Delta v = \frac{1}{K}\int_{t_1}^{t_2} e_o\, dt \qquad (4.135)$$

and thus

$$K = \frac{\int_{t_1}^{t_2} e_o\, dt}{\Delta v} \qquad (4.136)$$

The integral of e_o can be obtained by connecting the output of the accelerometer to an integrating circuit or by recording e_o on paper or film and then measuring the area under the curve with a planimeter.

The physical layout of the ballistic pendulum is shown in Fig. 4.56. The hammer is raised on its suspension tapes and allowed to impact the anvil and attached accelerometer. The velocity change can be measured by means of any suitable relative-velocity transducer, or, alternatively, it may be calculated from the maximum height of rise of the anvil after the impact. The pulse duration can be varied by changing the shape and/or material of the impacting surfaces, typical values being of the order of 0.001 sec.

A drop-test calibrator is shown in Fig. 4.57. The velocity change

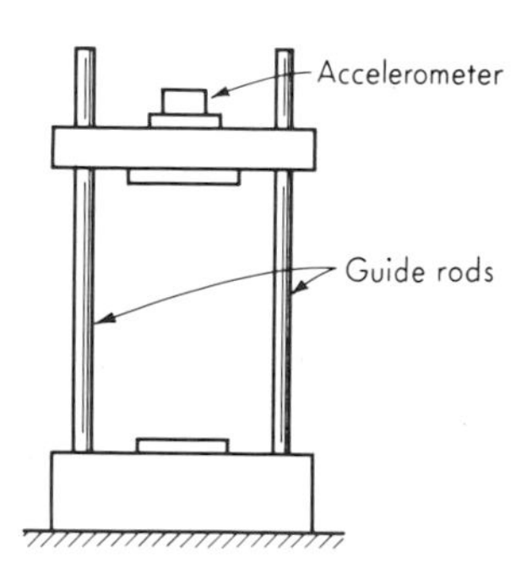

Fig. 4.57. *Drop-test apparatus.*

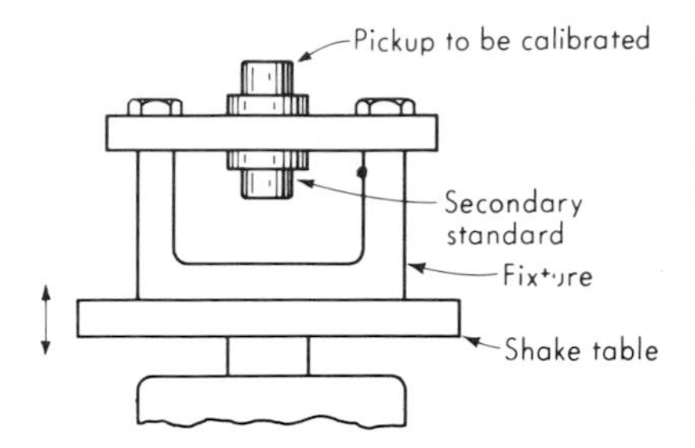

Fig. 4.58. Calibration fixture.

is again measured by means of conventional transducers or may be calculated (assuming low friction) from the height of fall h_1 and the height of rebound h_2 as

$$\Delta v = \sqrt{2gh_1} + \sqrt{2gh_2} \qquad (4.137)$$

Perhaps the main usefulness of the above transient techniques lies in their ability to provide high acceleration values (up to several thousand g's) with large high-frequency content (short pulse duration). Such tests, while in general not as accurate for determining numerical values of instrument characteristics as the frequency-response tests, provide a very severe test of the accelerometer's freedom from internal resonances and should thus be included in the calibration programs for any accelerometers to be used for shock work.

As a final comment on calibration methods, it should be noted that once a pickup has been accurately calibrated (say by the National Bureau of Standards) it becomes a secondary standard against which other pickups may be calibrated by direct comparison. This can readily be done by fastening both pickups to a common rigid fixture mounted on a vibration shake table and then applying the frequencies and amplitudes desired, as in Fig. 4.58.

4.10

Jerk Pickups In some measurement and control applications the rate of change of acceleration, or jerk, d^3x/dt^3, must be measured. An obvious approach is to apply the electrical output from an accelerometer to a differentiating circuit. A more subtle technique which avoids the noise-accentuating problems of differentiating circuits is applied in the Donner Jerkmeter.[1] By ingenious use of feedback principles, this null-balance instrument provides both acceleration and jerk signals of good quality. The physical configuration is essentially that of Fig. 4.55 with

[1] Systron Donner Corp., Concord, Calif.

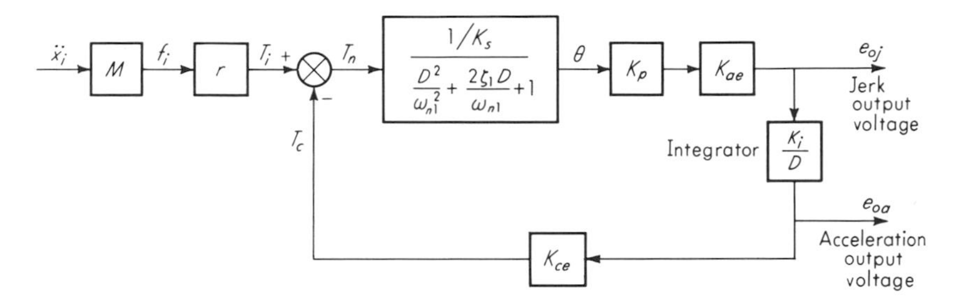

Fig. 4.59. *Jerkmeter block diagram.*

the addition of an electronic integrator. The resulting block diagram is shown in Fig. 4.59. Analysis gives

$$\left(MrD^2x_i - \frac{K_iK_{ce}e_{oj}}{D}\right)\frac{K_pK_{ae}/K_s}{D^2/\omega_{n1}^2 + 2\zeta_1 D/\omega_{n1} + 1} = e_{oj} \tag{4.138}$$

leading to the differential equation

$$\left(\frac{K_s}{K_iK_pK_{ae}K_{ce}\omega_{n1}^2}D^3 + \frac{2\zeta_1 K_s}{K_iK_pK_{ae}K_{ce}\omega_{n1}}D^2 + \frac{K_s}{K_iK_pK_{ae}K_{ce}}D + 1\right)e_{oj} = \frac{Mr}{K_iK_{ce}}D^3x_i \tag{4.139}$$

We note that the relationship between jerk input D^3x_i and voltage output e_{oj} is that of a third-order differential equation. The static sensitivity is easily seen to be Mr/K_iK_{ce} volts/(in./sec^3); however, the dynamic behavior is not obvious since we have not previously considered third-order instruments. For any given set of numerical values, the cubic characteristic equation of Eq. (4.139) will have three numerical roots. These will be either three real roots or one real root plus a pair of complex conjugates. In an actual design the latter situation often prevails. This means that the transfer function of Eq. (4.139) will generally be of the form

$$\frac{e_{oj}}{D^3x_i}(D) = \frac{K}{(\tau D + 1)(D^2/\omega_n^2 + 2\zeta D/\omega_n + 1)} \tag{4.140}$$

where

$$K \triangleq \frac{Mr}{K_iK_{ce}}$$

and τ, ζ, and ω_n can be found if numerical values for all the constants are given. The frequency response for Eq. (4.140) is easily plotted by using logarithmic methods. Actually, the frequency response for Eq. (4.139) is quite revealing. It is clear that, for sufficiently low frequencies, $e_{oj} \approx (Mr/K_iK_{ce})\dddot{x}_i$ for *any* values of the system constants since the first three terms involve ω^3, ω^2, and ω as a factor and thus go to zero as $\omega \to 0$. To increase the usable frequency range, the *coefficients* of the ω^3, ω^2, and ω terms must be made small. This can be done by making $K_s/K_iK_pK_{ae}K_{ce}$ small, which corresponds to making the gain of the feedback loop large. A limit is placed on the gain, however, since excessive gain will cause dynamic instability and resultant destruction of the instrument.

The Routh stability criterion[1] shows that for a cubic characteristic equation of the form $a_3D^3 + a_2D^2 + a_1D + a_0 = 0$ stability is assured if all the a's are positive and $a_0a_3 < a_1a_2$. In our case this becomes

$$\frac{K_s}{K_iK_pK_{ae}K_{ce}\omega_{n1}^2} < \frac{2\zeta_1K_s^2}{(K_iK_pK_{ae}K_{ce})^2\omega_{n1}} \tag{4.141}$$

or

$$\zeta_1 > \frac{K_iK_pK_{ae}K_{ce}}{2K_s\omega_{n1}} \tag{4.142}$$

This shows that if $K_s/K_iK_pK_{ae}K_{ce}$ is made small to increase the usable frequency range, the damping ζ_1 must be correspondingly increased to retain adequate stability. This required damping effect can also be obtained by adding proper electrical compensating networks to the circuit. This approach is often used. No matter how the damping is achieved, however, a tradeoff will always be necessary between frequency range and stability.

The resolution, nonlinearity, and hysteresis of a commercially available Jerkmeter are each better than 0.1 percent of full scale. Instruments with full-scale jerk values ranging from ± 0.5 to $\pm 20g$/sec and full-scale acceleration values of ± 1 to $\pm 30g$ are available. The full-scale acceleration and jerk output voltages are both ± 7.5 volts, while the instrument weight is about 8 oz and its size 3 in. long by 1.5 in. square.

4.11
Pendulous (Gravity-referenced) Angular-displacement Sensors

In a number of applications the measurement of angular displacement relative to the local vertical (gravity vector) is a useful technique. Examples include sensing elements for control systems of road-paving and scraping machines; drainage-tile-laying machines; alignment of construction forms, piles, and bridges; and attitude control of vehicles (such as submarines) and torpedoes or missiles. These relatively simple instruments (basically plumb bobs with electrical output) can sometimes replace more complex and expensive gyroscopic instruments which perform similar functions. Their main disadvantages relative to gyros are their

[1] E. O. Doebelin, "Dynamic Analysis and Feedback Control," p. 175, McGraw-Hill Book Company, New York, 1962.

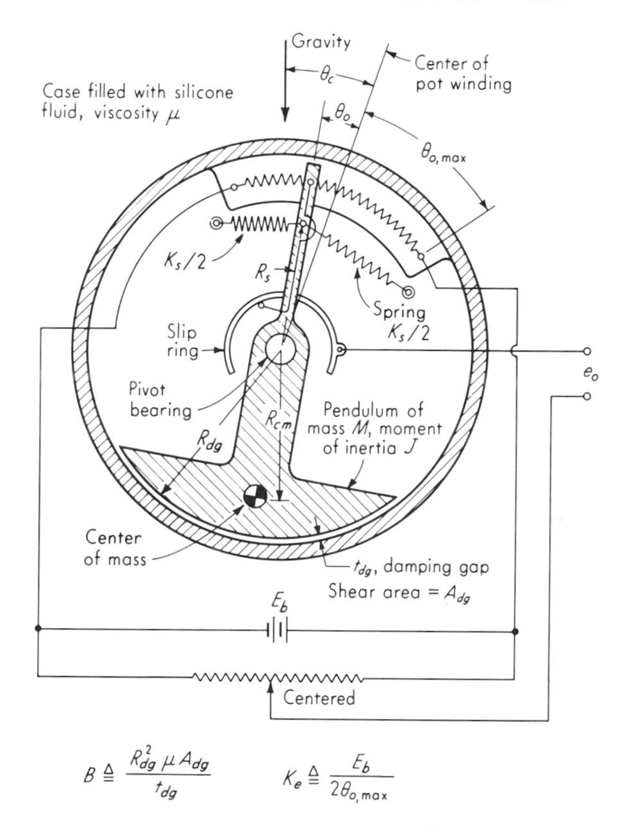

Fig. 4.60. *Pendulum displacement sensor.*

sensitivity to interfering translatory acceleration inputs and their dependence on a gravity field. (They will not work in essentially gravity-free space.)

Figure 4.60 shows a typical configuration of a single-axis pendulum-type sensor. The desired input to be measured is the case rotation angle θ_c. Most commercial sensors do not include the springs K_s; we include them here because their presence makes possible interesting and potentially useful dynamic behavior. For the usual case of no springs, K_s is simply set equal to zero. The damping effect is not essential to the theoretical operation of the device but is included in most practical instruments to reduce oscillations at pendulum frequency caused by transient interfering inputs. A variety of electrical displacement transducers may be employed, depending on the required characteristics; a potentiometer is shown for simplicity. The following assumptions are justifiable for most purposes in simplifying the analysis:

1. Angles are small enough so that the sine and the angle are nearly equal and the cosine is nearly 1.
2. The inertia effect of the fluid on the pendulum motion is negligible.
3. The damping effect of the fluid is limited to the damping gap.
4. All dry-friction effects in pot wipers, bearings, and slip rings may be neglected for dynamic analysis.
5. The buoyant force on the pendulum is negligible.
6. The springs provide a linear restoring torque.

The analysis is left for the problems at the end of this chapter. However the results are as follows: When the springs are present ($K_s \neq 0$) we have

$$\frac{e_o}{\theta_c}(D) = \frac{K(D^2/\omega_{n1}^2 + 1)}{D^2/\omega_{n2}^2 + 2\zeta D/\omega_{n2} + 1} \tag{4.143}$$

where

$$K \triangleq \frac{MgR_{cm}K_e}{R_s^2K_s + MgR_{cm}} \tag{4.144}$$

$$\omega_{n1} \triangleq \sqrt{\frac{MgR_{cm}}{J}} \tag{4.145}$$

$$\omega_{n2} \triangleq \sqrt{\frac{R_s^2K_s + MgR_{cm}}{J}} \tag{4.146}$$

$$\zeta \triangleq \frac{B}{2\sqrt{J(R_s^2K_s + MgR_{cm})}} \tag{4.147}$$

The frequency response of this system is shown in Fig. 4.61*a*. Note the "notch-filter" effect at ω_{n1} followed by a resonant peak near ω_{n2}. If the springs are removed (the usual case) we get $\omega_{n1} = \omega_{n2} = \omega_n$ and $K_e = K$, giving

$$\frac{e_o}{\theta_c}(D) = \frac{K(D^2/\omega_n^2 + 1)}{D^2/\omega_n^2 + 2\zeta D/\omega_n + 1} \tag{4.148}$$

and the frequency response of Fig. 4.61*b*.

The pendulum sensor is unfortunately sensitive to horizontal accelerations, and so its application is ruled out where such accelerations are large enough to cause a significant output signal. A simple analysis shows that a steady horizontal acceleration A_x will cause an output voltage K_eA_x/g for an instrument with no springs.

A commercial pendulum[1] using a potentiometer motion pickup has a full-scale range of $\pm 8°$, natural frequency of 2 cps, damping ratio of 0.6, and resolution of 0.1° (0.05° if vibration is present). Pendulums are also available for measuring rotation about two mutually perpendicular axes.

[1] Honeywell Inc., Minneapolis, Minn.

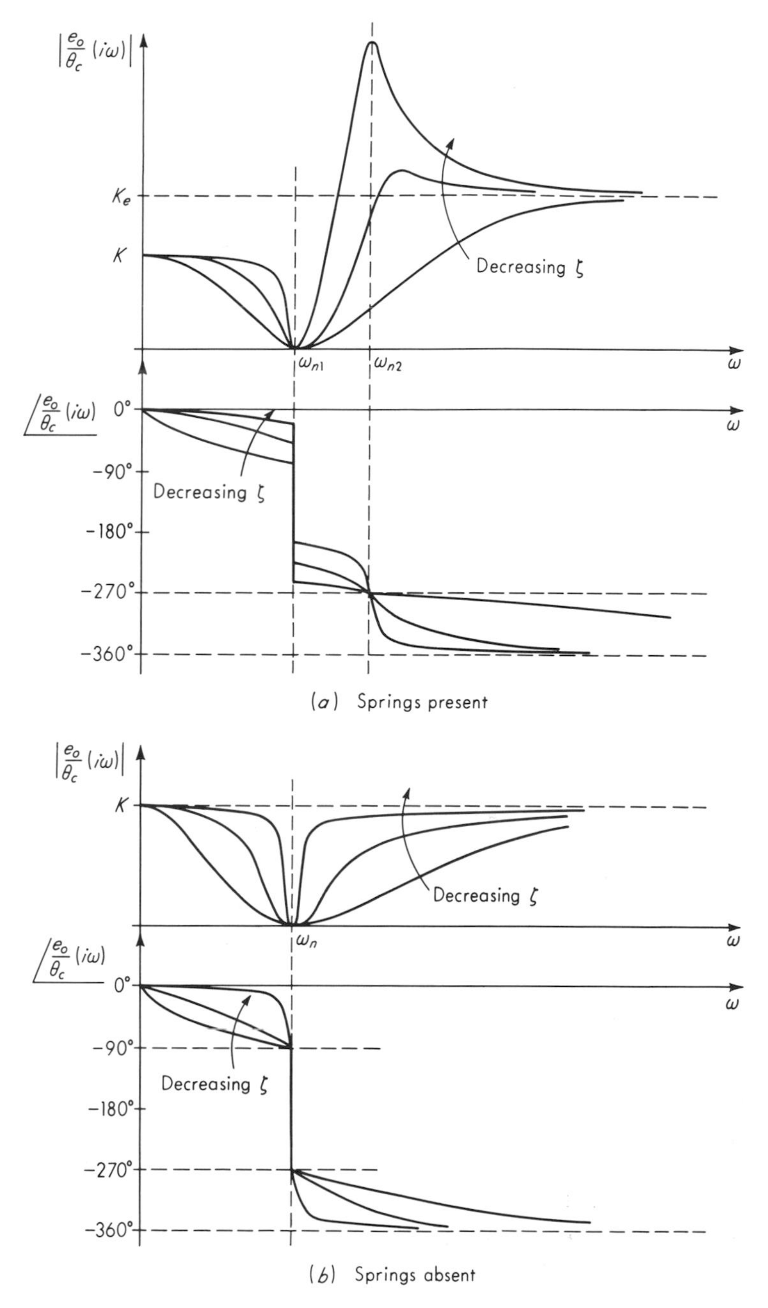

Fig. 4.61. *Frequency response of pendulum sensor.*

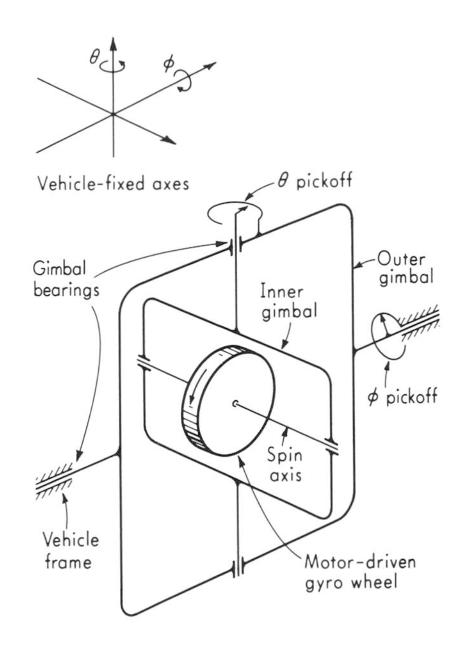

Fig. 4.62. *Free gyroscope (two-axis position gyro).*

4.12

Gyroscopic (Absolute) Angular-displacement and Velocity Sensors While gyroscopic instruments have been used in limited numbers and rather restricted applications (gyrocompasses in ships and aircraft, turn-and-bank indicators in aircraft, etc.) since around World War I, developments during and after World War II have brought them to an extreme degree of refinement, and their use in large numbers is now common in military systems.[1] Increased commercial and industrial applications have also resulted from the availability of equipment generated by military development programs. A recent estimate[2] of the value of total gyro production is placed at 121 million dollars per year.

Perhaps the simplest gyro configuration is the free gyro shown in Fig. 4.62. These instruments are used to measure the absolute angular displacement of the vehicle to which the instrument frame is attached. A single free gyro can measure rotation about two perpendicular axes, such as the angles θ and ϕ. This can be accomplished because the axis of the spinning gyro wheel remains fixed in space (if the gimbal bearings are frictionless) and thus provides a reference for the relative-motion transducers. If the angles to be measured do not exceed about 10°, the readings of the relative-displacement transducers give directly the absolute rota-

[1] Sidney Lees (ed.), "Air, Space, and Instruments," p. 32, McGraw-Hill Book Company, New York, 1963.

[2] *Control Engineering*, p. 77, May, 1963.

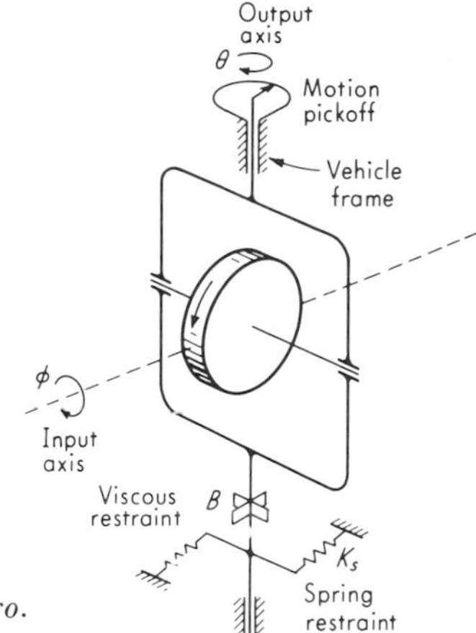

Fig. 4.63. *Single-axis restrained gyro.*

tions with good accuracy. For larger rotations of both axes, however, there is an interaction effect between the two angular motions, and the transducer readings do *not* accurately represent the absolute motions of the vehicle. The free gyro is also limited to relatively short-time applications (less than about 5 min) since gimbal-bearing friction causes gradual drift (loss of initial reference) of the gyro spin axis. A constant friction torque T_f causes a drift (precession) of angular velocity ω_d given by

$$\omega_d = \frac{T_f}{H_s} \tag{4.149}$$

where H_s is the angular momentum of the spinning wheel. It is clear that a high angular momentum is desirable in reducing drift. A typical drift rate is about 0.5 degree/min for each axis.

Rather than using free gyros to measure two angles in one gyro (thus requiring two gyros to define completely the required three axes of motion), most recent high-performance systems utilize the so-called single-axis or constrained gyros. Here a single gyro measures a single angle (or angular rate); therefore three gyros are required to define the three axes. This approach avoids the coupling or interaction problems of free gyros, and the constrained (rate-integrating) gyros can be constructed with exceedingly small drift. We shall here consider two common types of constrained gyros, the rate gyro and the rate-integrating gyro. The rate gyro measures absolute angular velocity and is widely used to generate stabilizing signals in vehicle-control systems. The rate-integrating gyro measures absolute angular displacement and thus is used as a fixed reference in navigation and attitude-control systems. The configuration of a rate gyro is shown in Fig. 4.63; the rate-integrating gyro is functionally identical except that it has no spring restraint.

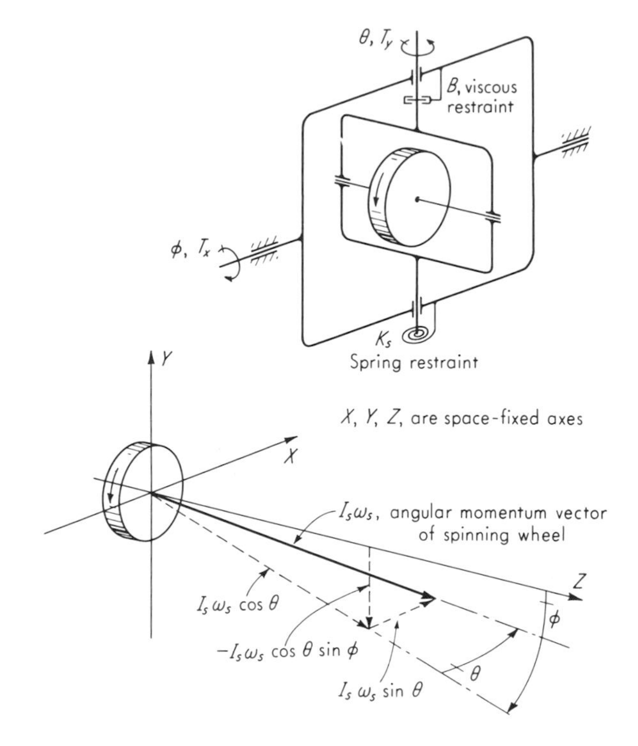

Fig. 4.64. *Gyro analysis.*

While a general analysis of gyroscopes is exceedingly complex, useful results for many purposes may be obtained relatively simply by considering small angles only. This assumption is satisfied in many practical systems. Figure 4.64 shows a gyro whose gimbals (and thus the angular-momentum vector of the spinning wheel) have been displaced through small angles θ and ϕ. We shall apply Newton's law

$$\sum \text{torques} = \frac{d}{dt}(\text{angular momentum}) \qquad (4.150)$$

to the x and y axis components of angular momentum. This angular momentum is made up of two parts, one part due to the spinning wheel and another part (due to the motion of the wheel, case, gimbals, etc.) which would exist even if the wheel were not spinning. The latter part depends on (for the x axis) the angular velocity $d\phi/dt$ and the moment of inertia I_x of everything that rotates when the outer gimbal turns in its bearing. For the y axis it depends on $d\theta/dt$ and the moment of inertia I_y of everything that rotates when the inner gimbal turns in its bearing. The external applied torques T_x and T_y are included to provide for the possibility of bearing friction and also for intentionally applied torques from small electromagnetic "torquers" which are used in some systems to cause desired precessions for control or correction purposes. The inertias I_x and I_y (which are about *space-fixed* axes) actually change when θ and ϕ change, but this effect is negligible for small angles. Also, the exact equations would contain terms in the *products* of inertia as well as the moments of inertia, but these are again negligible due to the small angles and also to the inherent symmetry of gyro structures. With the above qualifications we may write for the x axis

$$T_x = \frac{d}{dt}\left(H_s \sin\theta + I_x \frac{d\phi}{dt}\right) \tag{4.151}$$

and for the y axis

$$T_y - B\frac{d\theta}{dt} - K_s\theta = \frac{d}{dt}\left(-H_s \cos\theta \sin\phi + I_y \frac{d\theta}{dt}\right) \tag{4.152}$$

We now assume H_s is a constant (the gyro wheel is driven by a constant-speed motor) and $\cos\theta = 1$, $\sin\theta = \theta$, $\sin\phi = \phi$ to get

$$T_x = H_s\frac{d\theta}{dt} + I_x\frac{d^2\phi}{dt^2} \tag{4.153}$$

$$T_y - B\frac{d\theta}{dt} - K_s\theta = -H_s\frac{d\phi}{dt} + I_y\frac{d^2\theta}{dt^2} \tag{4.154}$$

These are two simultaneous linear differential equations with constant coefficients relating the two inputs T_x and T_y to the two outputs θ and ϕ. Writing these equations in operator form, we can treat them as algebraic equations to solve for ϕ and θ as desired. For ϕ we get

$$\phi = \frac{(I_yD^2 + BD + K_s)T_x - (H_sD)T_y}{D^2[I_xI_yD^2 + BI_xD + (H_s^2 + I_xK_s)]} \tag{4.155}$$

Since ϕ depends on both T_x and T_y, transfer functions may be obtained by considering each input separately and then using superposition. Letting $T_y = 0$, we get

$$\frac{\phi}{T_x}(D) = \frac{I_yD^2 + BD + K_s}{D^2[I_xI_yD^2 + BI_xD + (H_s^2 + I_xK_s)]} \triangleq G_1(D) \tag{4.156}$$

and letting $T_x = 0$ gives

$$\frac{\phi}{T_y}(D) = -\frac{H_s}{D[I_xI_yD^2 + BI_xD + (H_s^2 + I_xK_s)]} \triangleq G_2(D) \tag{4.157}$$

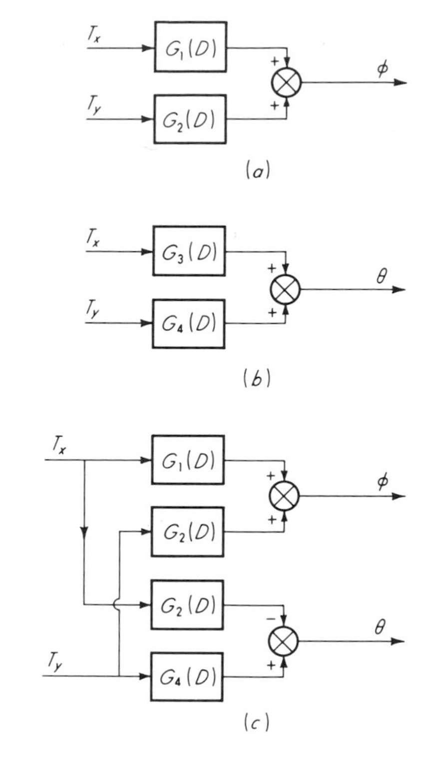

Fig. 4.65. *Gyro block diagrams.*

This leads to the block diagram of Fig. 4.65*a*. Similar analysis for θ gives

$$\frac{\theta}{T_x}(D) = \frac{H_s}{D[I_xI_yD^2 + BI_xD + (H_s^2 + I_xK_s)]} \triangleq G_3(D) = -G_2(D) \tag{4.158}$$

and
$$\frac{\theta}{T_y}(D) = \frac{I_x}{I_xI_yD^2 + BI_xD + (H_s^2 + I_xK_s)} \triangleq G_4(D) \tag{4.159}$$

leading to the block diagram of Fig. 4.65*b*. An overall block diagram may then be constructed as in Fig. 4.65*c*.

The above results are of quite general applicability to gyro systems of various configurations as long as the small-angle requirement is met. For single-axis rate and rate-integrating gyros considerable simplification of the above results is possible. In these applications (see Fig. 4.63) the input is the *motion* ϕ. A torque T_x also exists, accompanying this motion, but it is usually of no interest since it generally is so small as not to affect the motion ϕ, which is the rotation of a (usually large) vehicle. The angle θ is an indication of the angle ϕ (rate-integrating gyro) or angular velocity $\dot{\phi}$ (rate gyro); thus we should like to have transfer functions relating θ to ϕ. The torque T_y (neglecting bearing friction) is zero for this application unless a torquer is used for some special purpose. The desired θ-ϕ relation may then easily be obtained by solving Eqs. (4.156) and (4.158) for T_x and setting them equal. The result is

$$\frac{\theta}{\phi}(D) = \frac{H_s D}{I_y D^2 + BD + K_s} \tag{4.160}$$

For a rate gyro we then have the second-order response

$$\frac{\theta}{D\phi}(D) = \frac{\theta}{\dot{\phi}}(D) = \frac{K}{D^2/\omega_n{}^2 + 2\zeta D/\omega_n + 1} \tag{4.161}$$

where

$$K \triangleq \frac{H_s}{K_s} \quad \text{rad/(rad/sec)} \tag{4.162}$$

$$\omega_n \triangleq \sqrt{\frac{K_s}{I_y}} \quad \text{rad/sec} \tag{4.163}$$

$$\zeta \triangleq \frac{B}{2\sqrt{I_y K_s}} \tag{4.164}$$

A high sensitivity is achieved by large angular momentum H_s and soft spring K_s, although low K_s gives a low ω_n. Natural frequencies of commercially available rate gyros are of the order of 10 to 100 cps. Damping ratio is usually set at 0.3 to 0.7. Large angular momentum is obtained in small size by using high-speed (often 24,000 rpm) motors to spin the gyro wheel. Full-scale ranges of about ± 10 to $\pm 1{,}000$ degree/sec are readily available. Resolution of a ± 10-degree/sec-range instrument is of the order of 0.005 degree/sec. In some high-performance rate gyros the mechanical spring is replaced by an "electrical-spring" arrangement similar to that used in the servo accelerometer of Fig. 4.55.

To measure all three components (roll, pitch, and yaw) of angular velocity in a vehicle, an arrangement of three rate gyros, such as in Fig. 4.66, may be employed. It should be pointed out that in a rate gyro only the output angle θ must be kept small. This requires use of a very sensitive motion pickoff, but such are available. The input angle ϕ may be indefinitely large since no matter how large it gets the spin angular-momentum vector is always perpendicular (except for a small error due to nonzero θ) to the input angular-velocity vector. A roll-rate gyro in an aircraft thus will function correctly even if the aircraft rolls completely over. Of course the angular-velocity components measured are those about the *vehicle* axes rather than space-fixed axes, but these are usually the velocities desired for stabilization signals in control systems.

To obtain a rate-integrating gyro one merely removes the spring restraint from the configuration of Fig. 4.63. Equation (4.160) then

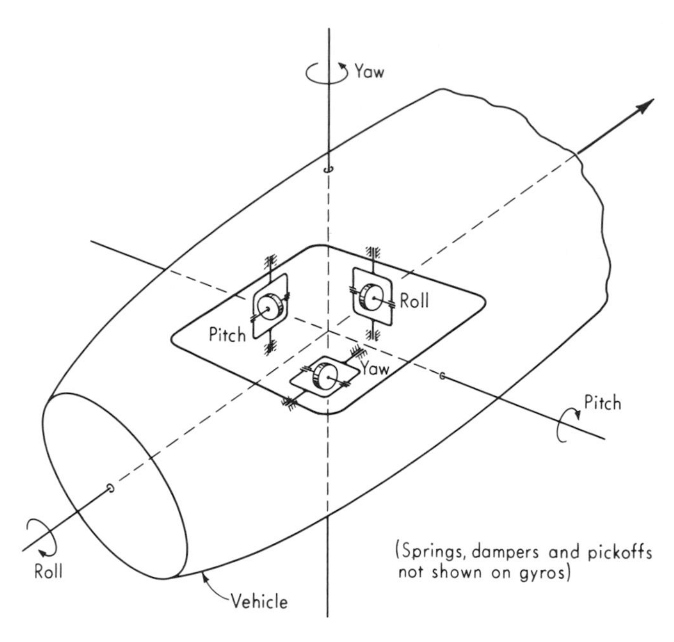

Fig. 4.66. *Three-axis rate-gyro package.*

becomes

$$\frac{\theta}{\phi}(D) = \frac{K}{\tau D + 1} \tag{4.165}$$

where

$$K \triangleq \frac{H_s}{B} \quad \text{rad/rad} \tag{4.166}$$

$$\tau \triangleq \frac{I_y}{B} \quad \text{sec} \tag{4.167}$$

We see that the output angle θ is a direct indication of the input angle ϕ according to a standard first-order response form. High sensitivity again requires high H_s, a typical value being 5×10^5 g-cm²/sec. Low damping B also increases sensitivity but at the expense of speed of response, as shown by Eq. (4.167). For a K of 15 a τ of 0.006 sec is typical of high-performance units.[1] Increase of K to 440 raises τ to 0.17 sec. The rate-integrating gyro is the basis of highly accurate inertial navigation systems where it is used as a reference to maintain so-called stable platforms in a fixed attitude within a vehicle while the vehicle moves arbitrarily. This is done by using the motion signals from the gyros to drive servomechanisms which maintain the platform in a fixed angular orientation. The

[1] General Precision Inc., Kearfott Div., Little Falls, N.J.

system accelerometers are mounted on this platform, and their double-integrated output is an accurate measure of vehicle motion along the three orthogonal axes since the platform always moves parallel to its initial orientation. The rate-integrating gyro has been subjected to extremely intensive and extensive engineering development to bring its performance characteristics to remarkable levels while maintaining small size and weight and great resistance to rugged environments. This has been accomplished by painstaking attention to minute mechanical, thermal, and electrical details which would be completely negligible in a less sophisticated device. While the performance of the best current instruments is generally classified information, drift rates less than 0.01 degree/hr are published in the open literature.

In this short section we can only indicate a few major concepts while many significant details are neglected. A few samples[1,2] of the voluminous literature are mentioned here while more will be found in the Bibliography of this chapter.

Problems

4.1 For a steel gage block of 1-in. length, what temperature change is required to cause a length change of 1 μin.?

4.2 Derive Eq. (4.1).

4.3 The output of a potentiometer is to be read by a recorder of 10,000 ohms input resistance. Nonlinearity must be held to 1 percent. A family of potentiometers having a thermal rating of 5 watts and resistances ranging from 100 to 10,000 ohms in 100-ohm steps is available. Choose from this family the potentiometer that has the greatest possible sensitivity and also meets the other requirements. What is this sensitivity if the potentiometers are single-turn (360°) units?

4.4 If a potentiometer changes resistance because of temperature changes, what effect does this have on motion measurements?

4.5 A 10-in.-stroke wire-wound translational potentiometer is excited with 100 volts. The output is read on an oscilloscope with a "sensitivity" of 0.5 mv/cm. It would appear that measurements to the nearest 0.0001 in. are easily possible. Explain why this is not so.

4.6 What resolution is possible with a 60-turn wire-wound potentiometer using appropriate gearing?

4.7 Explain methods whereby one can experimentally determine the moment of inertia and the starting and running friction torque of a rotary potentiometer.

4.8 In Fig. P4.1 a potentiometer whose moving part weighs 0.01 lb_f measures the displacement of a spring-mass system subjected to a step input. The measured natural frequency is 30 cps. If the spring constant and mass M of the system are unknown, can the true natural frequency be deduced from the above data? Suppose an additional 0.01-lb_f weight is attached to the poten-

[1] B. Lichtenstein, "Technical Information for the Engineer, Gyros," General Precision Inc., Kearfott Div., Little Falls, N.J.

[2] Savet, *op. cit.*

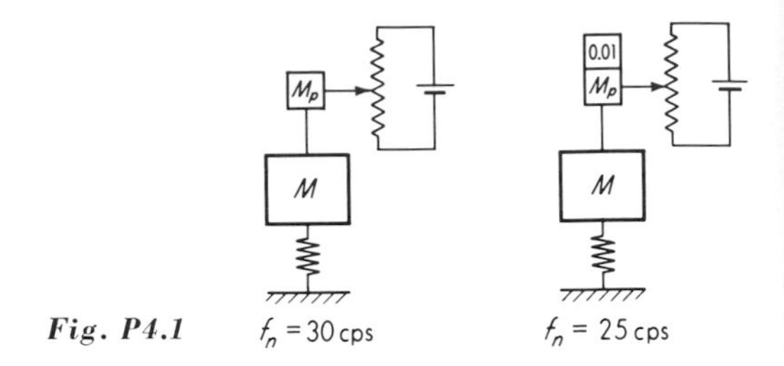

Fig. P4.1

tiometer and the test repeated, giving a 25-cps frequency. Calculate the true natural frequency of the system, that is, the frequency before the potentiometer was attached.

4.9 Find C_{gi} for the system of Fig. 4.8 if a viscous damping effect is included.

4.10 Explain why increasing the cross-sectional area of the end loops in foil-type strain gages reduces transverse sensitivity.

4.11 If, in the discussion following Eq. (4-28), dynamic strains only in the range 0 to 10,000 cps need be measured, explain how and to what extent the noise voltage may be reduced.

4.12 In a Wheatstone bridge, leg 1 is an active strain gage of Advance alloy and 120 ohms resistance, leg 4 is a similar dummy gage for temperature compensation, and legs 2 and 3 are fixed 120-ohm resistors. The maximum gage current is to be 0.030 amp.

a. What is the maximum permissible d-c bridge excitation voltage? (Use this value in the remaining parts of this problem.)

b. If the active gage is on a steel member, what is the bridge output voltage per 1,000 psi of stress?

c. If temperature compensation were *not* used, what bridge output would be caused by the active gage increasing temperature by 100F° if the gage is bonded to steel? What stress value would be represented by this voltage? Thermal-expansion coefficients of steel and Advance alloy are 6.5×10^{-6} and 14.9×10^{-6} in./(in.-F°), respectively. The temperature coefficient of resistance of Advance is 6×10^{-6} ohm/(ohm-F°).

d. Compute the value of a shunt calibrating resistor that would give the same bridge output as 10,000-psi stress in a steel member.

4.13 From Eq. (4.38), find an expression for the frequency at which zero phase shift occurs.

4.14 Perform an analysis similar to that leading to Eq. (4.38), assuming output loaded with R_m, for the following:

a. The circuit of Fig. 4.15*a*

b. The circuit of Fig. 4.15*b*

c. The circuit of Fig. 4.15*c*

d. The circuit of Fig. 4.15*d*

4.15 In Fig. 4.17*c*, let x_i be a periodic motion with a significant frequency content up to 500 cps, and let the excitation frequency be 10,000 cps. The output voltage e_o is connected to an oscillograph galvanometer, which is a second-order system with $\zeta = 0.65$ and a natural frequency of 1,000 cps. Will this combination result in a satisfactory measurement system? Justify your answer with numerical results.

4.16 Air exhibits a dielectric breakdown at fields of about 50,000 volts/in. What limitation does this impose on the ultimate sensitivity of a capacitance transducer such as in Fig. 4.25?

4.17 In Eq. (4.50), suppose a flat amplitude ratio within 5 percent down to 20 cps is required. What is the minimum allowable τ? If $A = 0.5$ in.^{2} and x_0 is 0.005 in., what value of R is needed?

4.18 A piezoelectric transducer has a capacitance of 1,000 pf and K_q of 10^{-5} coul/in. The connecting cable has a capacitance of 300 pf while the oscilloscope used for readout has an input impedance of 1 megohm paralleled with 50 pf.

a. What is the sensitivity (volts/in.) of the transducer alone?

b. What is the high-frequency sensitivity (volts/in.) of the entire measuring system?

c. What is the lowest frequency that can be measured with 5 percent amplitude error by the entire system?

d. What value of C must be connected in parallel to extend the range of 5 percent error down to 10 cps?

e. If the C value of part d is used, what will the system high-frequency sensitivity be?

4.19 A piezoelectric transducer has an input

$$x_i = At \qquad 0 \le t < T$$
$$x_i = 0 \qquad T < t < \infty$$

Solve the differential equation to find e_o. For $t = T^-$, find the error [(ideal value of e_o) − (actual value of e_o)]. Approximate this error by using the truncated series

$$e^{-T/\tau} \approx 1 - \frac{T}{\tau} + \frac{1}{2}\left(\frac{T}{\tau}\right)^2$$

Express this approximate error as a percentage of the ideal value of e_o. What must T/τ be if the error is to be 5 percent? For this value of T/τ, evaluate the error caused by truncating the series. (Use the theorem on the remainder of an alternating series.)

4.20 Analyze the nozzle-flapper displacement pickup of Fig. P4.2, using the simple incompressible relations. Explain the advantages of this configuration.

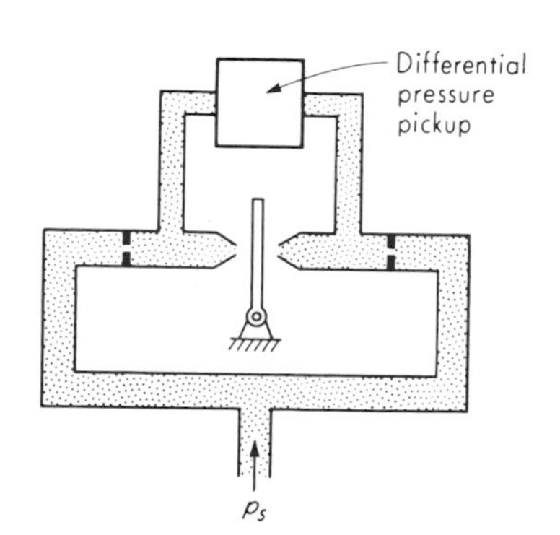

Fig. P4.2

4.21 Express in binary form the decimal numbers 27, 6,382, 9.125, 9.126.

4.22 Express in decimal form the binary numbers 10111001., 1001001111., 101011.-1100.

4.23 Express in Datex binary-coded decimal the numbers 137, 9,764, and 42.

4.24 Prove Eq. (4.112).

4.25 Derive Eq. (4.113).

4.26 In the variable-capacitance velocity pickup shown in Fig. P4.3, prove that the current i is directly proportional to the angular velocity $d\theta/dt$. Since voltage signals are more readily manipulated, how might the current signal be transduced to a proportional voltage? Does your method of doing this affect the basic operation? What must be required if the basic operation is to be only slightly affected?

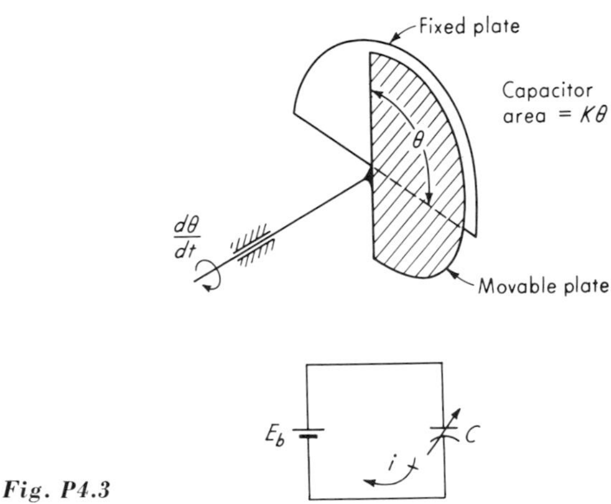

Fig. P4.3

4.27 Construct logarithmic frequency-response curves for a seismic-displacement pickup with $\zeta = 0.3$ and $\omega_n/2\pi = 10$ cps. For what range of frequencies is the amplitude ratio flat within 1 db?

4.28 Make a comprehensive list and explain the action of modifying and/or interfering inputs for the systems of the following:

a. Fig. 4.3 *g.* Fig. 4.36
b. Fig. 4.8 *h.* Fig. 4.44
c. Fig. 4.13 *i.* Fig. 4.45
d. Fig. 4.25 *j.* Fig. 4.48
e. Fig. 4.31 *k.* Fig. 4.55
f. Fig. 4.34 *l.* Fig. 4.63

4.29 Derive $(\theta_o/\theta_i)(D)$ for Fig. 4.49*b*.

4.30 Construct logarithmic frequency-response curves for a piezoelectric accelerometer with $K_q/C\omega_n^2 = 0.001$ volt/(in./sec^2), $\tau = 0.10$ sec, $\omega_n/2\pi = 10{,}000$ cps, and $\zeta = 0$. Will this accelerometer be satisfactory for shock measurements of half-sine pulses with a duration of 0.05 sec? If not, suggest needed changes.

4.31 Explain, giving a sketch, how the principle of the system of Fig. 4.55 can be adapted to the measurement of angular acceleration. Your device must *not* be sensitive to translational acceleration.

4.32 In the system of Fig. 4.55, if K_s and B are made zero, $(\theta/T_n)(D) = 1/JD^2$. Obtain $(e_o/\ddot{x}_i)(D)$ for this situation. What is the defect in this system? To remedy this, electrical "damping" may be introduced by adding a circuit with a transfer function as shown in Fig. P4.4. Obtain $(e_o/\ddot{x}_i)(D)$ for this arrangement. Why must $\tau_1 > \tau_2$?

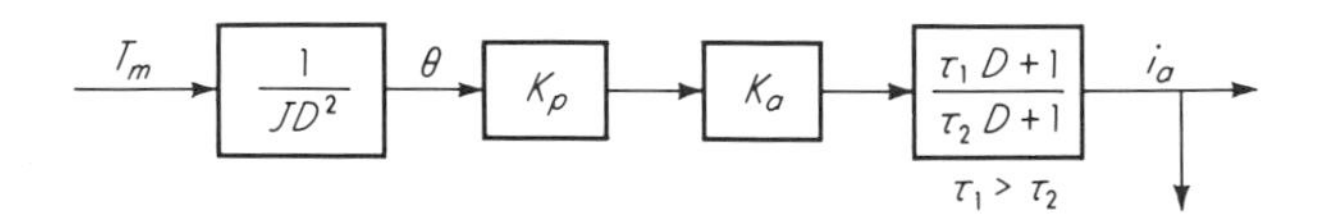

Fig. P4.4

4.33 An accelerometer used in an inertial navigation system has a so-called bias error such that there is a small output signal even when the input acceleration is exactly zero. If this bias signal is equivalent to $10^{-5}g$ of acceleration, what *position* (displacement) error does it cause over a time interval of 2 days? Assume travel along a straight line.

4.34 When a seismic-displacement pickup is used in its proper frequency range, what is an adequate mechanical model (masses, springs, dashpots) for its loading effect on the measured system?

4.35 In the Jerkmeter of Fig. 4.59, replace ω_{n1} and ζ_1 by their values in terms of J, B, and K_s, thus rewriting Eq. (4.139). Suppose a sensitivity $Mr/K_iK_{ce} =$ 0.05 volt/(ft/sec³) is required. Assume temporarily that $B = K_s = 0$ and neglect the fact that this system would be unstable. Assume also that J is due mainly to M; thus take $J = Mr^2$.

a. Find the numerical value of K_pK_{ae}/r needed to give a flat amplitude ratio for $(e_o/D^3x_i)(i\omega)$ within 5 percent over the range 0 to 5 cps.

b. If $r = 0.1$ ft and $K_p = 57.3$ volts/rad, find K_{ae}.

c. Suppose now that $M = 0.01$ lb$_m$. Find K_iK_{ce}.

d. If B and K_s are not zero, find the value of BK_s needed to put the system just on the margin of instability. (Use the Routh criterion.)

e. Let the design value of BK_s be 10 times the value of part *d*. If $K_s =$ 0.275 ft-lb$_f$/rad, what is B?

f. With the above values of B and K_s, recheck the amplitude ratio at 5 cps. Does it meet the 5 percent requirement?

g. To correct the situation found in *f*, make $K_{ae} = 2{,}490$. Recheck the amplitude ratio at 5 cps. Recheck the stability.

h. To regain the stability lost in part *g*, reduce M to 0.001 lb$_m$. Recheck the amplitude ratio at 5 cps.

i. Recheck the overall system static sensitivity. How much external amplification is now needed to return to the required 0.05 volt/(ft/sec³)?

4.36 Derive Eqs. (4.143) to (4.147). Give a physical explanation of the "notch-filter" effect. Explain the apparent discontinuity in phase angle.

4.37 Find the steady-state response of the system of Fig. 4.60 to a constant horizontal acceleration.

4.38 A commercial version of the system of Fig. 4.60 has a pendulum made as shown in Fig. P4.5. Where is the center of buoyancy of this pendulum? Where is the center of mass? Will the buoyant force tend to cause an output? Why? Derive a relation showing the requirements for completely unloading the pivot bearing. (This "floating" reduces bearing friction and thus system threshold.)

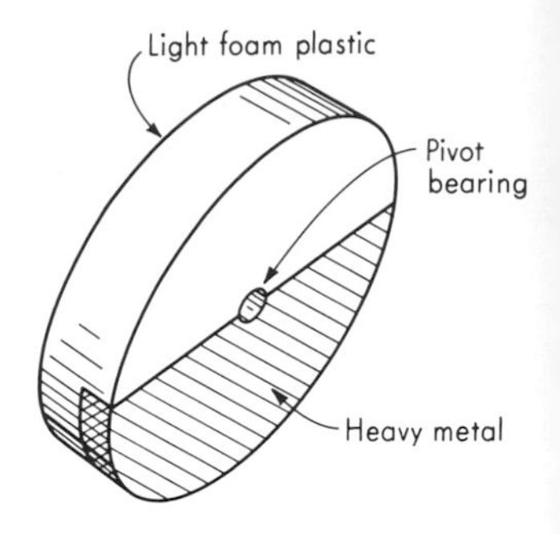

Fig. P4.5 Completely immersed in liquid

4.39 Reduce Eqs. (4.156) to (4.159) to standard form, defining appropriate K's, ζ's, and ω_n's.

4.40 Equation (4.157) can be written as

$$[I_x I_y D^2 + B I_x D + (H_s^2 + I_x K_s)]\phi = -\frac{H_s}{D} T_y$$

For a gyro with no spring nor damping, suppose T_y is a unit impulse. Solve Eqs. (4.157) and (4.159) for ϕ and θ. The combined sinusoidal motion of ϕ and θ causes the spin axis to rotate in space. This motion is called nutation. Describe it qualitatively and define its frequency. What is the effect on the above results if damping is present?

4.41 A rate gyro is mounted on a long thin missile, which is quite flexible in bending, as in Fig. P4.6. Bending vibrations cause the slope at the gyro location to go

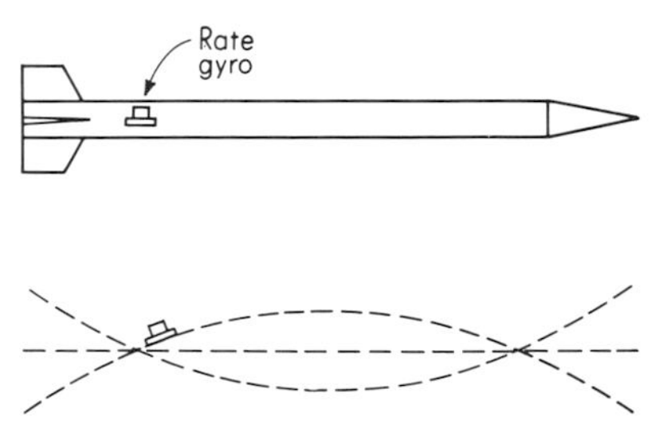

Fig. P4.6 Mode shape of bending vibrations

through sinusoidal oscillations of 0.1° amplitude at 50 cps. What maximum angular velocity will the gyro feel due to vibration? If the gross (rigid-body) rotation of the missile (which is what the gyro is *intended* to measure) is 10 rad/sec, what percent of the total gyro signal is due to the vibration? Where could one relocate the gyro to minimize this problem? If the fastest rigid-body motions expected are 1 cps, what other solution is possible?

4.42 In a rate gyro, the steady-state output/input ratio $\theta/\dot{\phi}$ is not strictly linear because the angular-momentum vector does not remain perpendicular to the angular-velocity vector $\dot{\phi}$ when θ rotates away from zero.

a. What is the maximum allowable θ if this nonlinearity is not to exceed 1 percent?

b. If a rate gyro requires $\omega_n = 100$ rad/sec and if $I_y = 0.0013$ in.-lb$_f$-sec^2, what must the spring constant K_s be?

c. If the maximum input rate $\dot{\phi}$ is 10 rad/sec, what must H_s be if $\theta_{\max}$ is the value found in part *a*? Use K_s from part *b*.

d. If the spin motor runs at 24,000 rpm and if the wheel is a solid cylinder of a length equal to its diameter and is made of a material with a specific weight of 0.3 lb$_f$/in.^{3}, what are the required dimensions of the wheel? Use all necessary numbers from the previous parts of the problem.

Bibliography

1. J. W. Dally and W. F. Riley: "Experimental Stress Analysis," McGraw-Hill Book Company, New York, 1965.
2. H. G. Buchbinder: Precision Potentiometers, *Electromech. Design,* p. 21, January, 1964.
3. K. H. Hardman: Conductive Plastic Precision Pots, *Electromech. Design,* p. 44, October, 1963.
4. J. F. Blackburn (ed.): Potentiometers, "Components Handbook," chap. 8, McGraw-Hill Book Company, New York, 1948.
5. High Temperature Potentiometers, *Electromech. Design,* p. 41, January, 1960.
6. T. L. Foldvari and K. S. Lion: Capacitive Transducers, *Instr. Control Systems,* p. 77, November, 1964.
7. Electro-optical Moire Fringe Transducer, *Electromech. Design,* p. 26, December, 1964.
8. W. H. Kliever: Measure Position Digitally, *Control Eng.,* p. 107, November, 1956.
9. J. O. Morin: 6 Transducers for Precision Position Measurement, *Control Eng.,* p. 107, May, 1960.
10. J. H. Brown: Measure Motion to 0.0001 In. without Friction or Wear, *Control Eng.,* p. 50, April, 1955.
11. G. E. Bowie: Measurement of Displacements in Contact-Stress Experiments, *ASD Tech. Rept.* 61-450, Wright-Patterson Air Force Base, Ohio, October, 1961.
12. T. G. Baxter: Measurement of Pier Tilt with a Quartz Torsion Fiber Pendulum, *Jet Propulsion Lab. Tech. Rept.* 32-44, Jet Propulsion Laboratory, Pasadena, Calif., 1960.
13. W. H. Faulkner and J. G. Wood: Thickness Measuring Devices for Sheet and Web Materials, *Automation,* p. 67, July, 1962.
14. K. V. Olsen: On the Standardization of Surface Roughness Measurements, *B & K Tech. Rev.,* B & K Instruments, Cleveland, Ohio, no. 3, 1961.
15. Precise Amplitude Measurements, *Mech. Eng.,* p. 65, February, 1963.
16. Length Calibration, National Bureau of Standards, *ISA J.,* p. 75, April, 1963.
17. W. Kinder: Comparator Measures up to 40″ with Accuracy of 0.4 Microinch, *Instr. Control Systems,* p. 123, April, 1963.
18. E. G. Loewen: Positioning System Spaces Lines to within $\frac{1}{10}$ Microinch, *Control Eng.,* p. 95, May, 1963.
19. W. R. Ketterer and R. H. Schuman: Photocell Technique for Linear Measurements, *Electro-Technol.,* p. 120, November, 1963.
20. B. Sternlicht: An Indirect Method of Film-thickness Measurement in Fluid-Film Bearings, *ISA Trans.,* vol. 2, no. 1, p. 28, January, 1963.

21. J. B. Bryan and G. I. Boyadjieff: Measuring Surface Finish, A State-of-the-Art Report, *Mech. Eng.*, p. 42, December, 1963.
22. F. Farago: Measuring the Critical Profile of Barrel Roller Bearings with Microinch Sensitivity, *Gen. Motors Eng. J.*, p. 17, January–February–March, 1964.
23. R. Zito: Nuclear-Resonance Sensing of Mechanical Motion, *Electro-Technol.*, p. 43, June, 1964.
24. K. G. Overbury: Temperature in Length Measurement, Sandia Corp., Albuquerque, N.Mex., February, 1962.
25. D. H. Parkes: The Application of Microwave Techniques to Noncontact Precision Measurement, *ASME Paper* 63-WA-346, 1963.
26. F. H. London: Laser Interferometer, *Instr. Control Systems*, p. 87, November, 1964.
27. J. G. Collier and G. F. Hewitt: Film-thickness Measurements, *ASME Paper* 64-WA/HT-41, 1964.
28. E. V. Sundt: Touchless Tachometry, *Electromech. Design*, p. 36, May, 1964.
29. R. Zito: Velocity Sensing for Spacecraft Docking, *Space/Aeron.*, p. 90, December, 1963.
30. J. Frey: The A-C Tachometer, *Electro-Technol.*, p. 88, August, 1963.
31. R. L. Pike: Measurement of Low Angular Rates, *Instr. Control Systems*, p. 83, December, 1962.
32. A. L. Fisher: Ball and Disk Read Angular Velocity Directly, *Control Eng.*, p. 125, November, 1962.
33. L. E. Bollinger and K. E. Kissell: Measurement of Detonation-wave Velocities, *ISA J.*, p. 170, May, 1957.
34. Shaoue Ezekiel: Towards a Low-level Accelerometer, *NASA*, *CR*-56941, 1964.
35. P. K. Chapman: A Cryogenic Test-Mass Suspension for a Sensitive Accelerometer, *NASA*, *N*64-27883, 1964.
36. Pressure Sensitivity of Accelerometers and Cables, Wilcoxon Research, Bethesda, Md.
37. Mercury Drop Measures Space-vehicle Acceleration, *Machine Design*, p. 28, Mar. 29, 1962.
38. H. R. Judge: Performance of Donner Linear Accelerometer Model 4310, *Space Technol. Lab. Tech. Note* 60-0000-09117, Space Technology Laboratories, Los Angeles, Calif., 1960.
39. A. Castle: Accelerometer Scribes Vector-Force Signatures, *Control Eng.*, p. 105, March, 1965.
40. R. R. Bouche: High Frequency Response and Transient Motion Performance Characteristics of Piezoelectric Accelerometers, Endevco Corp., Pasadena, Calif., 1961.
41. C. K. Stedman: Some Characteristics of Gas Damped Accelerometers, Statham Instruments Inc., Los Angeles, Calif., 1958.
42. H. B. Sabin: 17 Ways to Measure Acceleration, *Control Eng.*, p. 106, February, 1961.
43. R. P. Bowen: Calibrating Vibration Pickup Calibrators, *ISA J.*, p. 58, March, 1951.
44. V. B. Corey: Measuring Angular Acceleration with Linear Accelerometers, *Control Eng.*, p. 79, March, 1962.
45. K. E. Pope: A New Double-integrating Accelerometer, *Control Eng.*, p. 97, November, 1958.
46. K. N. Sergeyev: Investigation of Acceleration Pickup Having Filtering Properties, *NASA*, *N*64-23543, 1964.
47. D. K. Phillips: Balanced Beam Improves Angular Accelerometer, *Control Eng.*, p. 91, July, 1964.

48. S. Rubin: Design of Accelerometers for Transient Measurement, *J. Appl. Mech.*, p. 509, December, 1958.
49. A. Degenholtz: Optical-Wedge Technique for Measuring Angular Vibration, *Machine Design*, p. 167, March 12, 1964.
50. Design and Construction of a Lunar Seismograph, *NASA*, *N*63-18290, 1963.
51. G. B. Foster: Non-contacting Self-calibrating Vibration Transducer, *Instr. Control Systems*, p. 83, December, 1963.
52. B. D. Van Deusen: Analysis of Vehicle Vibration, *ISA Trans.*, vol. 3, no. 2, p. 138, April, 1964.
53. D. F. Wilkes and C. E. Kreitler: The Long Period Horizontal Air Bearing Seismometer, Sandia Corp., Albuquerque, N.Mex., SCTM74-62(13), 1962.
54. J. M. Slater: Exotic Gyros, *Control Eng.*, p. 92, November, 1962.
55. Traverse Meter Uses Gyroscope Sensor, *Machine Design*, p. 163, Apr. 13, 1961.
56. R. H. Cherwin: Ball Bearings for Precision Gyros, *Control Eng.*, p. 79, August, 1959.
57. A. W. Lane et al.: Achieving Extremely Accurate Non-floated Gyros, *Aero/Space Eng.*, p. 43, January, 1959.
58. H. Stern: Which Rate Gyro to Use, *Control Eng.*, p. 79, February, 1958.
59. C. S. Draper et al.: The Floating Integrating Gyro, *Aeron. Eng. Rev.*, p. 46, June, 1956.
60. Inertial Gyro Test, *Electromech. Design*, p. 18, November, 1960.
61. R. E. Barnaby et al.: Control of Thermal Drift in Floated Gyroscopes, *Sperry Eng. Rev.*, Sperry Gyroscope Co., Great Neck, N.Y., p. 36, September, 1961.
62. W. G. Wing: Fluid Rotor Gyros, *Control Eng.*, p. 105, March, 1963.
63. G. C. Newton: Vibratory Rate Gyros, *Control Eng.*, p. 95, June, 1963.
64. E. H. Ernst: Basic Theory-Particle Gyroscope, *Electro-Technol.*, p. 12, August, 1963.
65. H. L. Kreitzburg: Compensating Gyro Drifts, *Control Eng.*, p. 113, November, 1963.
66. H. W. Knoebel: The Electric Vacuum Gyro, *Control Eng.*, p. 70, February, 1964.
67. Gimbal-less Gyro, *Mech. Eng.*, p. 59, May, 1964.
68. S. Redner and F. Zandman: Experimental Stress Analysis, *Ind. Res.*, p. 67, May, 1965.
69. L. H. Ravitch: Some Applications of Stress Analysis Techniques in Improving Casting Designs, *Gen. Motors Eng. J.*, p. 22, October–November–December, 1958.
70. S. S. Manson: Thermal Stresses, Measurements by Photoelasticity, *Machine Design*, p. 143, Nov. 26, 1959.
71. F. Zandman: Stress Analysis with a Photoelastic Coating, *Metal Progr.*, p. 111, November, 1960.
72. F. B. Stern: Strain Sensitive Ceramic Base Brittle Coatings, *Machine Design*, p. 147, May 29, 1958.
73. G. Gerard and H. Tramposch: Photothermoelastic Investigation of Transient Thermal Stresses in a Multiweb Wing Structure, *J. Aerospace Sci.*, p. 783, December, 1959.
74. W. R. Campbell: Performance Tests of Wire Strain Gages: I. Calibration Factors in Tension, *NACA*, *Tech. Note* 954, 1944.
75. P. K. Stein: Pulsing Strain-gage Circuits, *Instr. Control Systems*, p. 128, February, 1965.
76. P. K. Stein: Strain Gages, "Measurement Engineering," Stein Engineering Services, Inc., Phoenix, Ariz.

77. S. S. Manson: Thermal Stresses in Design-Strain Gage Applications, *Machine Design*, p. 683, Nov. 12, 1959.
78. A. Kaufman: Performance of Electrical-Resistance Strain Gages at Cryogenic Temperatures, *NASA, Tech. Note, D*-1663, 1963.
79. R. H. Kemp et al.: Application of a High-temperature Static Strain Gage to the Measurement of Thermal Stresses in a Turbine Stator Vane, *NACA, Tech. Note* 4215, 1958.
80. S. S. Manson: Thermal Stresses in Design-Strain-Gage Measurements, *Machine Design*, p. 109, Oct. 29, 1959.
81. New Strain Gages for the Space Age, *ISA J.*, p. 50, February, 1959.
82. J. Gunn and E. Billinghurst: Magnetic Fields Affect Strain Gages, *Control Eng.*, p. 109, August, 1957.
83. Semiconductor Strain Gage Handbook, Baldwin-Lima-Hamilton Corp., Waltham, Mass.
84. R. J. Whitehead: Protective Coating for Strain Gages, *ISA J.*, p. 71, March, 1964.
85. R. Shiver and W. Putman: Measuring Dynamic Strain on High-speed Turbine Wheels, *Instr. Control Systems*, p. 118, September, 1962.
86. A. J. Bush: Soldered-cap Strain Gages, *Machine Design*, p. 163, Nov. 8, 1962.
87. C. E. Mathewson: The "Dimensionless" Strain Gage, *Instr. Control Systems*, p. 1870, October, 1961.
88. D. Post: The Moiré Grid-Analyzer Method for Strain Analysis, *Experimental Mechanics*, pp. 368–377, November, 1965.

5
Force, torque, and shaft power measurement

5.1

Standards and Calibration Force is defined by the equation $F = MA$; thus a standard for force depends on standards for mass and acceleration. Mass is considered a fundamental quantity, and its standard is a cylinder of platinum-iridium, called the International Kilogram, kept in a vault at Sèvres, France. Other masses (such as national standards) may be compared with this standard by means of an equal-arm balance, with a precision of a few parts in 10^9 for masses of about 1 kg. Tolerances on various classes of standard masses available from the National Bureau of Standards vary with the magnitude of the mass and may be found in its publications.[1]

Acceleration is not a fundamental quantity but rather is derived from length and time, two fundamental quantities whose standards were discussed in Chap. 4. The acceleration of gravity, g, is a convenient standard which can be determined with an accuracy of about 1 part in 10^6 by measuring the period and effective length of a pendulum or by determining the change with time of the speed of a freely falling body. The actual value of g varies with location and also slightly with time (in a periodic predictable fashion) at a given location. It may also change (slightly) unpredictably because of local geological activity. The so-called standard value of g refers to the value at sea level and 45° latitude and is numerically 980.665 cm/sec². The value at any latitude ϕ degrees may be computed from

$$g = 978.049(1 + 0.0052884 \sin^2 \phi - 0.0000059 \sin^2 2\phi) \qquad \text{cm/sec}^2 \tag{5.1}$$

while the correction for altitude h in meters above sea level is

$$\text{Correction} = -(0.00030855 + 0.00000022 \cos 2\phi)h + 0.000072 \left(\frac{h}{1{,}000}\right)^2 \qquad \text{cm/sec}^2 \tag{5.2}$$

When the numerical value of g has been determined at a particular locality, the gravitational force (weight) on accurately known standard masses may be computed to establish a standard of force. This is the basis of the so-called "dead-weight" calibration of force-measuring systems. The current National Bureau of Standards capability for such calibrations is an inaccuracy of about 1 part in 5,000 for the range of 10 to 1 million lb_f. Above this range, direct dead-weight calibration is not presently available. Rather, proving rings[2] or load cells of a capacity of 1 million lb_f or less are calibrated against dead weights and then the unknown force is applied to a multiple array of these in parallel. The range 1 to 10 million lb_f is covered by such arrangements with somewhat reduced accuracy. At the low-force end of the scale, the accuracy of standard masses ranges from about 1 percent for a mass of 10^{-5} lb_m to 0.0001 percent for the 0.1 to 10-lb_m range to 0.001 percent for a 100-lb mass. The accuracy of *force* calibrations using these masses must be somewhat

[1] T. W. Lashof and L. B. Macurdy, Precision Laboratory Standards of Mass and Laboratory Weights, *Natl. Bur. Std. (U.S.), Circ.* 547, sec. 1, 1954.

[2] Proving Rings for Calibrating Testing Machines, *Natl. Bur. Std. (U.S.), Circ.* C454, 1946.

less than the quoted figures because of error sources in the experimental procedure.

The measurement of torque is intimately related to force measurement; thus torque standards as such are not necessary, force and length being sufficient to define torque. The power transmitted by a rotating shaft is the product of torque and angular velocity. Angular-velocity measurement was treated in Chap. 4.

5.2

Basic Methods of Force Measurement An unknown force may be measured by the following means:

1. Balancing it against the known gravitational force on a standard mass, either directly or through a system of levers
2. Measuring the acceleration of a body of known mass to which the unknown force is applied
3. Balancing it against a magnetic force developed by interaction of a current-carrying coil and a magnet
4. Transducing the force to a fluid pressure and then measuring the pressure
5. Applying the force to some elastic member and measuring the resulting deflection

In Fig. 5.1, method 1 is illustrated by the analytical balance, the pendulum scale, and the platform scale. The analytical balance, while simple in principle, requires careful design and operation to realize its maximum performance. The beam is designed so that the center of mass is only slightly (a few thousandths of an inch) below the knife-edge pivot and thus barely in stable equilibrium. This makes the beam deflection (which in sensitive instruments is read with an optical micrometer) a very sensitive indicator of unbalance. For the low end of a particular instrument's range the beam deflection is often used as the output reading rather than attempting to null by adding masses or adjusting the arm length of a poise weight. This approach is faster than nulling but requires that the deflection-angle unbalance relation be accurately known and stable. This relation tends to vary with the load on the balance, because of deformation of knife edges, etc., but careful design can keep this to a minimum. For highly accurate measurements the buoyant force due to the immersion of the standard mass in air must be taken into account. Also, the most sensitive balances must be installed in temperature-controlled chambers and manipulated by remote control to reduce the effects of the operator's body heat and convection currents. Typically,

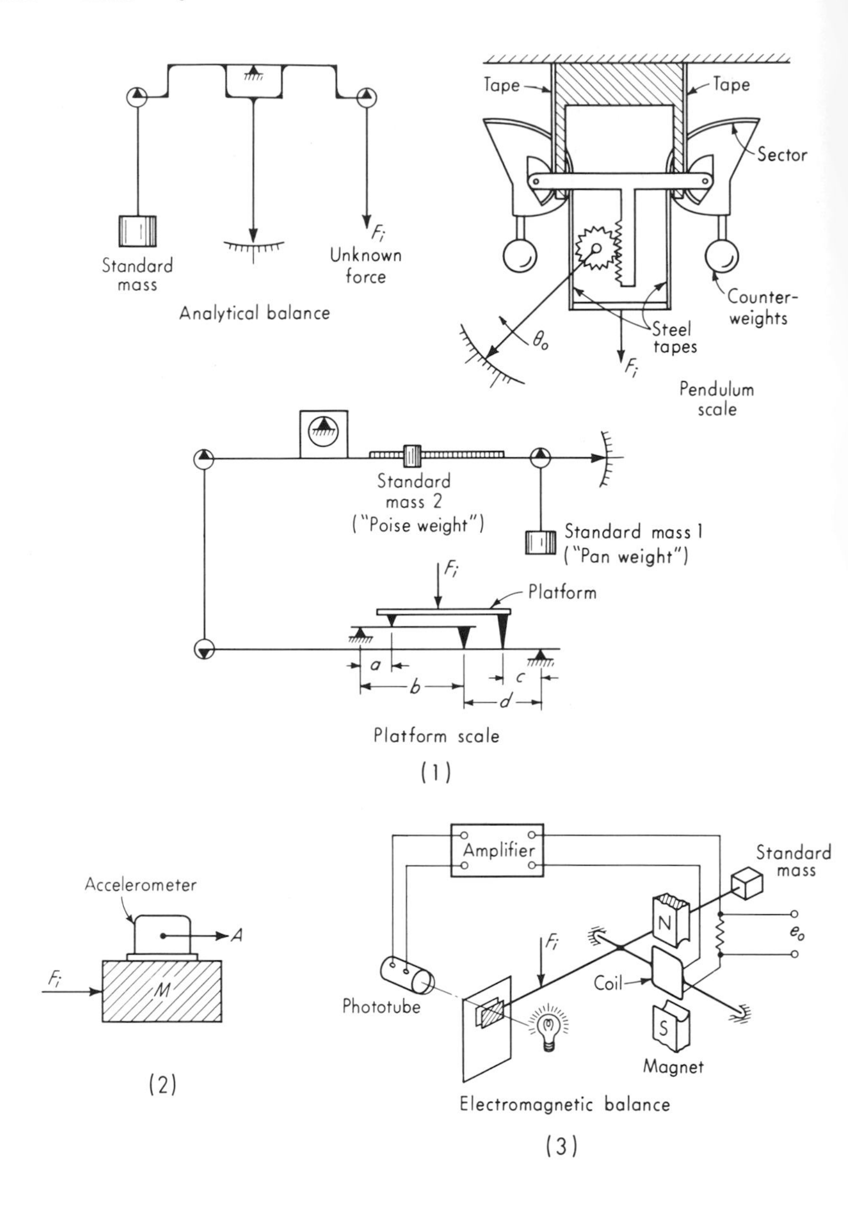

Standard mass
F_i
Unknown force
Analytical balance
Tape
Tape
Sector
Counter-weights
Steel tapes
θ_o
F_i
Pendulum scale
Standard mass 2
("Poise weight")
Standard mass 1
("Pan weight")
F_i
Platform
a
b
c
d
Platform scale
(1)
Accelerometer
A
F_i
M
(2)
Amplifier
Standard mass
e_o
N
F_i
Coil
Phototube
S
Magnet
Electromagnetic balance
(3)

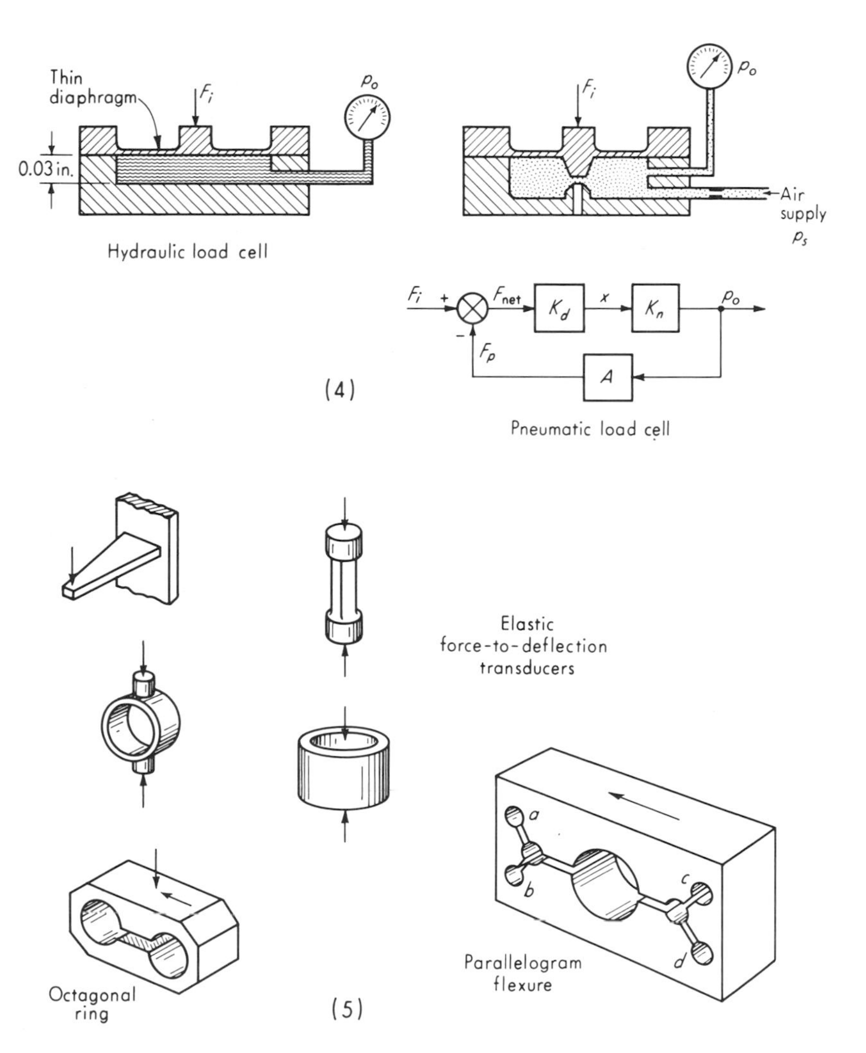

Fig. 5.1. *Basic force-measurement methods.*

a temperature difference of 1/20C° between the two arms of a balance can cause an arm-length ratio change of 1 ppm, significant in some applications. Commercially available analytical balances may be classified as follows.[1]

Description	*Range, gram*	*Resolution, gram*
Macro analytical	200–1,000	10^{-4}
Semimicro analytical	50–100	10^{-5}
Micro analytical	10–20	10^{-6}
Micro balance	less than 1	10^{-6}
Ultramicro balance	less than 0.01	10^{-7}

The pendulum scale is a deflection-type instrument in which the unknown force is converted to a torque which is then balanced by the torque of a fixed standard mass arranged as a pendulum. The practical version of this principle utilizes specially shaped sectors and steel tapes to linearize the inherently nonlinear torque/angle relation of a pendulum. The unknown force F_i may be applied directly as in Fig. 5.1 or through a system of levers, such as that shown for the platform scale, to extend the range. An electrical signal proportional to force is easily obtained from any angular-displacement transducer attached to measure the angle θ.

The platform scale utilizes a system of levers to allow measurement of large forces in terms of much smaller standard weights. The beam is brought to null by a proper combination of pan weights and adjustment of the poise-weight lever arm along its calibrated scale. The scale can be made self-balancing by adding an electrical displacement pickup for null detection and an amplifier-motor system to position the poise weight to achieve null. Another interesting feature is that, if $a/b = c/d$, the reading of the scale is independent of the location of F_i on the platform. Since this is quite convenient, most commercial scales provide this feature by use of the suspension system shown or others that allow similar results.

While analytical balances are used almost exclusively for "weighing" (really determining the *mass* of) objects or chemical samples, platform and pendulum scales are also employed for force measurements such as those involved in shaft power determinations with dynamometers. All three instruments are intended mainly for static force measurements.

Method 2, the use of an accelerometer for force measurement, is of somewhat limited application since the force determined is the *resultant* force on the mass. Often *several* unknown forces are acting, and they cannot be separately measured by this method.

[1] F. Baur, The Analytical Balance, *Ind. Res.*, p. 64, July–August, 1964.

The electromagnetic balance[1,2] (method 3) utilizes a photoelectric null detector, amplifier and torquing coil in a servosystem to balance the difference between the unknown force F_i and the gravity force on a standard mass. This instrument is mainly a competitor of the mechanical analytical balance and is used for the same types of applications. Within its weight range (full scale is presently limited to 1 g or less), its advantages relative to mechanical balances are ease of use, less sensitivity to environment, faster response, smaller size, and ease of remote operation. Also, the electrical output signal is convenient for continuous recording and/or automatic-control applications.

Method 4 is illustrated in Fig. 5.1 by hydraulic and pneumatic load cells. Hydraulic cells[3] are completely filled with oil and usually have a preload pressure of the order of 30 psi. Application of load increases the oil pressure which is read on an accurate gage. Electrical pressure transducers can be used to obtain an electrical signal. The cells are very stiff, deflecting only a few thousandths of an inch under full load. Capacities to 100,000 lb_f are available as standard while special units up to 10 million lb_f are obtainable. Accuracy is of the order of 0.1 percent of full scale; resolution is about 0.02 percent.

The pneumatic load cell shown uses a nozzle-flapper transducer as a high-gain amplifier in a servo loop. Application of force F_i causes a diaphragm deflection x which in turn causes an increase in pressure p_o since the nozzle is more nearly shut off. This increase in pressure acting on the diaphragm area A produces an effective force F_p which tends to return the diaphragm to its former position. For any constant F_i the system will come to equilibrium at a specific nozzle opening and corresponding pressure p_o. The static behavior is given by

$$(F_i - p_oA)K_dK_n = p_o \qquad (5.3)$$

where

$$K_d \triangleq \text{diaphragm compliance, in./lb}_f \qquad (5.4)$$

$$K_n \triangleq \text{nozzle-flapper gain, psi/in.} \qquad (5.5)$$

Solving for p_o, we get

$$p_o = \frac{F_i}{1/K_dK_n + A} \qquad (5.6)$$

Now K_n is not strictly constant but varies somewhat with x, leading to a nonlinearity between F_i and p_o. However, in practice, the product K_dK_n

[1] L. Cahn, Electromagnetic Weighing, *Instr. Control Systems*, p. 107, September, 1962.

[2] Cahn Instrument Co., Paramount, Calif.

[3] A. H. Emery Co., New Canaan, Conn.

is very large so that $1/K_d K_n$ is made negligible compared with A, giving

$$p_o = \frac{F_i}{A} \tag{5.7}$$

which is linear since A is constant. As in any feedback system, dynamic instability limits the amount of gain that can actually be used. A typical supply pressure p_s is 60 psi, and since the maximum value of p_o cannot exceed p_s, this limits F_i to somewhat less than $60A$. Commercial[1] load cells operating on this general principle (with refinements) have an accuracy of about 0.5 percent of full scale and a deflection under full load of less than 0.001 in. and come in full-scale ranges of 5 to 5,000 lb_f. The air consumption is of the order of 0.1 ft^3/min of free air.

While all the previously described force-measuring devices are intended mainly for static or slowly varying loads, the elastic deflection transducers of method 5 are widely used for both static and dynamic loads of frequency content up to many thousand cycles per second. While all are essentially spring-mass systems with (intentional or unintentional) damping, they differ mainly in the geometric form of "spring" employed and in the displacement transducer used to obtain an electrical signal. The displacement sensed may be a gross motion, or strain gages may be judiciously located to sense force in terms of strain. Bonded strain gages have been found particularly useful in force measurements with elastic elements. In addition to serving as force-to-deflection transducers, some elastic elements also perform the function of resolving vector forces or moments into rectangular components. An example, the parallelogram flexure[2] of Fig. 5.1, is extremely rigid (insensitive) to all applied forces and moments except in the direction shown by the arrow. A displacement transducer arranged to measure motion in the sensitive direction will thus measure only that component of an applied vector force which lies along the sensitive axis. The action of this flexure may perhaps be most easily visualized by considering it as a four-bar linkage with pin joints at a, b, c, and d.

Because of the importance of elastic force transducers in modern dynamic measurements, we shall devote a considerable portion of this chapter to their consideration. Although they may differ widely in detail construction, their dynamic-response form is generally the same, and so we shall treat an idealized model representative of all such transducers in the next section.

[1] A. H. Emery Co., New Canaan, Conn.
[2] Flex-Cell, Fluidyne Engineering Corp., Minneapolis, Minn.

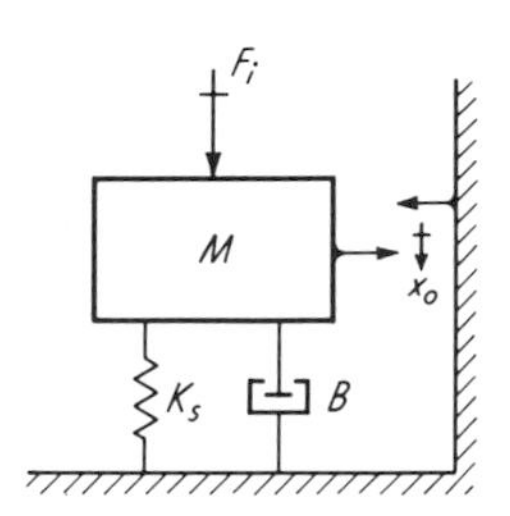

Fig. 5.2. *Elastic force transducer.*

5.3

Characteristics of Elastic Force Transducers Figure 5.2 shows an idealized model of an elastic force transducer. The relationship between input force and output displacement is easily established as a simple second-order form:

$$F_i - K_s x_o - B\dot{x}_o = M\ddot{x}_o \tag{5.8}$$

$$\frac{x_o}{F_i}(D) = \frac{K}{D^2/\omega_n^2 + 2\zeta D/\omega_n + 1} \tag{5.9}$$

where

$$\omega_n \triangleq \sqrt{\frac{K_s}{M}} \tag{5.10}$$

$$\zeta \triangleq \frac{B}{2\sqrt{K_s M}} \tag{5.11}$$

$$K \triangleq \frac{1}{K_s} \tag{5.12}$$

For transducers that do not measure a gross displacement but rather use strain gages bonded to the "spring" K_s, the output strain ϵ may be substituted for x_o if K_s is reinterpreted as force per unit strain rather than force per unit deflection. In many transducers a distinct and separate "spring" and "mass" cannot be distinguished because the elasticity and inertia are distributed rather than lumped. In these cases, for design purposes the natural frequency must be calculated from the appropriate formulas[1,2,3] for the geometric shapes involved rather than by employing Eq. (5.10). Once the transducer is constructed, its lowest natural frequency can generally be found experimentally, as can ζ, which is usually

[1] C. M. Harris and C. E. Crede (eds.), "Shock and Vibration Handbook," vol. 1, chap. 7, McGraw-Hill Book Company, New York, 1961.

[2] J. P. Den Hartog, "Mechanical Vibrations," 4th ed., pp. 431–433, McGraw-Hill Book Company, New York, 1956.

[3] R. K. Mitchell, Some Considerations in the Design of Elastic Force Transducers, M.Sc. Thesis, The Ohio State University, Mechanical Engineering Department, 1965.

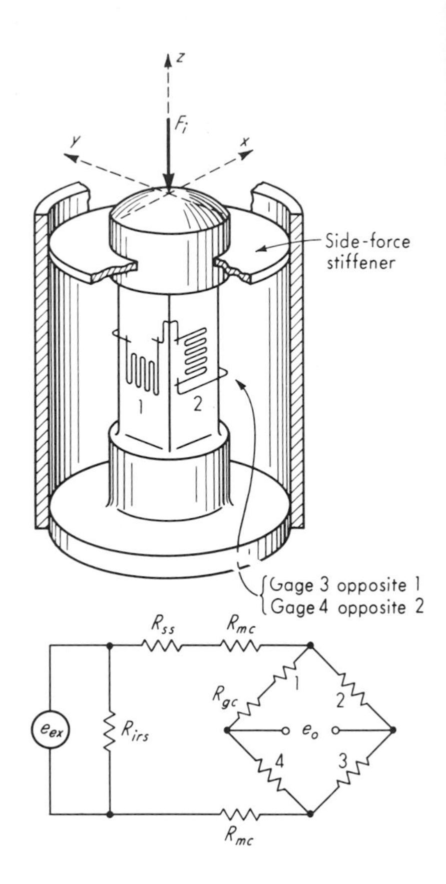

Fig. 5.3 *Strain-gage load cell.*

small and difficult to calculate. The sensitivity K is generally available theoretically from strength-of-materials or elasticity formulas, whether it relates to a gross deflection or a local unit strain. Once the transducer is constructed, it should be given an overall calibration relating electrical output to force input since none of the theoretical formulas is sufficiently accurate for this purpose.

Since the dynamic response of second-order instruments has been fully discussed previously, we shall concentrate mainly on details peculiar to specific force transducers.

Bonded-strain-gage transducers. A typical construction for a strain-gage load cell for measuring compressive forces is shown in Fig. 5.3. (Cells to measure both tension and compression require merely the addition of suitable mechanical fittings at the ends.) The load-sensing member is short enough to prevent column buckling under the rated load. Foil-type metal gages are bonded on all four sides; gages 1 and 3 sense the direct stress due to F_i and gages 2 and 4 the transverse stress due to Poisson's ratio μ. This arrangement gives a sensitivity $2(1 + \mu)$ times that achieved with a single active gage in the bridge. It also provides primary temperature compensation since all four gages are (at least for steady temperatures) at the same temperature. Furthermore, the arrangement is insensitive to bending stresses due to F_i being applied off center or at an angle. This can be seen by replacing an off-center force by an equivalent on-center force and a couple. The couple can be resolved into x and y components which cause bending stresses in the gages. If the gages are carefully placed so as to be symmetrical, the bending stresses in gages 1 and 3 will be of opposite sign, and by the rules of bridge circuits the net output e_o due to bending will be zero. Similar arguments hold for gages 2 and 4 and for bending stresses due to F_i being at an angle. The side-force stiffener plate also reduces the effects of angular forces since it is very stiff in the radial (x,y) direction but very soft in the z direction.

The deflection under full load of such load cells is of the order of 0.001 to 0.015 in., indicating their high stiffness. The natural frequency is often not quoted since it is frequently determined almost entirely by the mass of force-carrying elements external to the transducer. This is especially true in the many applications where the load cell is used for weighing purposes. The high stiffness also implies a low sensitivity. To increase sensitivity (in low-force cells where it is needed) without sacrificing column stability and surface area for mounting gages, a hollow (square on the outside, round on the inside) load-carrying member may be employed.

To achieve the high accuracy (0.3 to 0.1 percent of full scale) required in many applications, additional temperature compensation is needed. This is accomplished by means of the temperature-sensitive resistors R_{gc} and R_{mc} shown in Fig. 5.3. These resistors are permanently attached internal to the load cell so as to assume the same temperature as the gages. The purpose of R_{gc} is to compensate for the slightly different temperature coefficients of resistance of the four gages. The purpose of R_{mc} is to compensate for the temperature dependence of the modulus of elasticity of the load-sensing member. That is, although one wishes to measure force, the gages sense strain; thus any change in the modulus of elasticity will give a different strain (and thus a different e_o) even though the force is the same. Since all metals change modulus somewhat with temperature, this effect causes a sensitivity drift. The resistance R_{mc} compensates for this by changing the excitation voltage actually applied to the bridge by just the right amount to counteract the modulus effect.

Two additional (non-temperature-sensitive) resistors are often found in commercial load cells. They are R_{ss}, which is adjusted to standardize the sensitivity for a nominal e_{ex} to a desired value, and R_{irs}, which is used to adjust the input resistance to a desired value.

A family[1] of load cells covering the full-scale range from 10 to 250,000 lb_f has a full-scale output of about 35 mv, nonlinearity as good as 0.05 percent of full scale, hysteresis and nonrepeatability as good as 0.02 percent of full-scale, best temperature-induced zero drift of 0.0015 percent of full scale/F°, and best temperature-induced sensitivity drift of 0.0008 percent/F° over the range 15 to 115°F. These cells all use 350-ohm gages and 12 volts bridge excitation. The 10-lb_f cell is a cylinder about 3.5 in. in diameter and length (weight 2.5 lb_f) while the 250,000-lb_f cell is a cylinder 12 in. in diameter by 24 in. long (weight 400 lb_f).

When adequate sensitivity cannot be achieved by use of tension/compression members, configurations employing bending stresses may be helpful. These generally provide more strain per unit applied force but at the expense of reduced stiffness and thus natural frequency. Of the many possibilities, the cantilever beam and proving ring are shown in Fig. 5.4. The cantilever-beam gage arrangement provides four times the sensitivity of a single gage, temperature compensation, and insensitivity to x and y components of force if identical gages and perfect symmetry are assumed. The proving-ring transducer also is inherently temperature-compensated. For F_i as a tension load, gages 1 and 3 are in compression and 2 and 4 are in tension; thus, with the bridge arrangement shown, these effects are all additive, giving a large output.

When maximum output is desired for any strain-gage transducer one should consider the possible use of low-modulus materials (such as aluminum) to increase strain per unit force, several gages (or one high-resistance gage) per bridge leg (if space allows), and the intentional introduction of stress concentrations at the gage locations. However, such techniques also present associated problems. Low modulus reduces stiffness and natural frequency, and some low-modulus materials have excessive hysteresis and low fatigue life. Stress concentrations also lower fatigue life, and their effect may be difficult to calculate for design purposes.

Unbonded-strain-gage transducers. The unbonded strain gage described in Chap. 4 can be used for measurement of small forces by applying them directly to the strain wires and for large forces by using the strain gage to measure displacement of an elastic member. A family[2] of such instruments has full-scale ranges of ± 0.15 oz to $\pm 1{,}000$ lb_f,

[1] Baldwin-Lima-Hamilton Corp., Waltham, Mass.
[2] Statham Instruments Inc., Los Angeles, Calif.

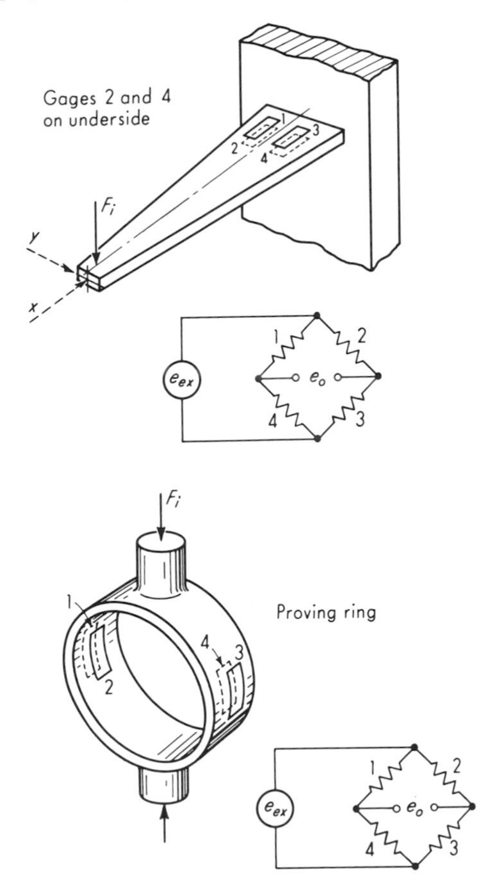

Fig. 5.4. *Beam and ring transducers.*

deflection of 0.015 to 0.0015 in. under full load, full-scale output of ± 20 to ± 53 mv, and nonlinearity and hysteresis less than ± 1 percent of full scale, with temperature-compensated units having a thermal sensitivity shift of 0.01 percent/F° from -65 to $+250$°F and a thermal zero shift of 0.01 percent of full scale/F° from -65 to $+250$°F.

Differential-transformer transducers. A family[1] of load cells which use a differential transformer to measure deflection of a machined-from-solid dual diaphragm spring of alloy steel (see Fig. 5.5) has the following characteristics: full-scale ranges from ± 1 to $\pm 5{,}000$ lb_f, full-load deflection of 0.005 in., full-scale output of 0.2 volt for 10-volt/3,000-cps excitation, nonlinearity 0.2 percent of full scale, repeatability 0.1 percent of full scale, and thermal zero or sensitivity shift of 0.02 percent of full scale/F°.

[1] Daytronic Corp., Dayton, Ohio.

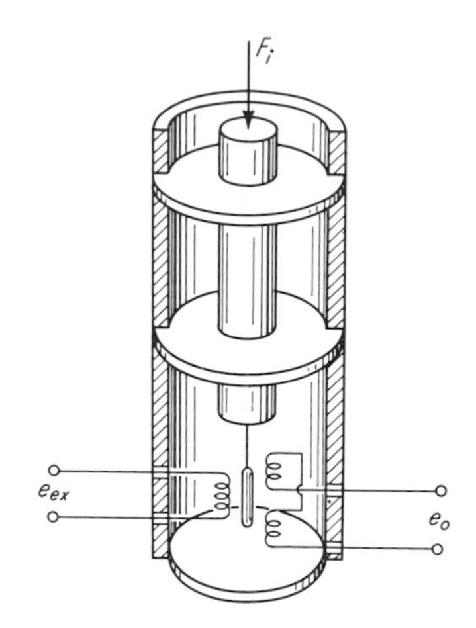

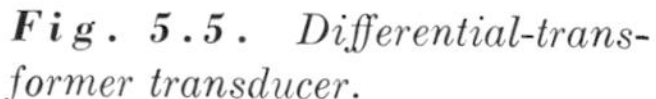
Fig. 5.5. *Differential-transformer transducer.*

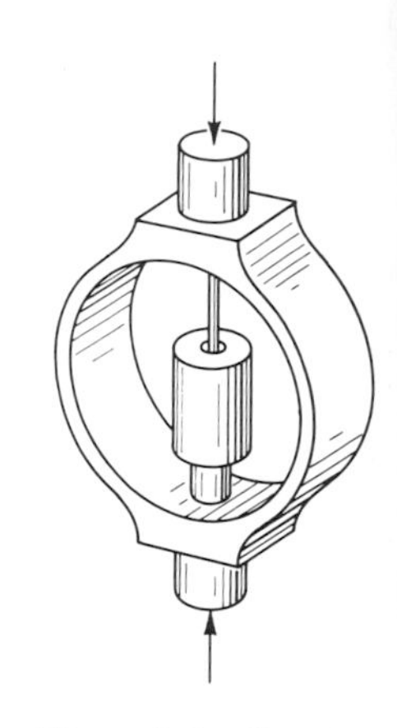
Fig. 5.6. *Proving ring transducer.*

The same manufacturer also makes a proving-ring transducer with differential-transformer measurement of the deflection (see Fig. 5.6). The characteristics of this unit are full-scale range ± 10 to $\pm 10{,}000$ lb_f, full-load deflection of 0.025 in., full-scale output of 2 volts for 10-volt/3,000-cps excitation, nonlinearity 0.25 percent of full scale, repeatability 0.1 percent of full scale, and thermal zero or sensitivity shift of 0.03 percent of full scale/F°.

Differential-transformer force transducers are also available[1] in the range ± 1 to ± 100 g with full-load deflection of 0.01 in. and 0.35-volt full-scale output.

Piezoelectric transducers. These force transducers have the same form of transfer function as piezoelectric accelerometers. They are intended for dynamic force measurement only, although some types (quartz pickup with electrometer charge amplifier) have sufficiently large τ to allow short-term measurement of static forces and static calibration. Commercially available pickups are designed primarily for compressive loading and exhibit high stiffness of the order of 5 to 35×10^6 lb_f/in.

A quartz-pickup/amplifier combination[2] has five switch-selectable full-scale ranges of 0.125, 1.25, 12.5, 125, and 500 lb_f with corresponding sensitivities of 400, 140, 19, 2, and 0.05 mv/lb_f when a 1-ft cable length is used. The natural frequency of the pickup itself is 60,000 cps, and non-

[1] Sanborn Co., Waltham, Mass.
[2] Kistler Instrument Corp., North Tonawanda, N.Y.

linearity is 1 percent. This pickup/amplifier combination will hold a static reading for several minutes, thus allowing static calibration. The good temperature characteristics of quartz give an operating range of -400 to $+500°F$.

A family[1] of force gages using synthetic-ceramic piezoelectric elements covers the full-scale range of 100 to 5,000 lb_f with corresponding sensitivities of 300 to 20 mv/lb_f and stiffness of 5 to 35×10^6 lb_f/in. The natural frequency of the pickups themselves is of the order of 25,000 cps, while an added mass of M lb_m gives frequencies ranging from $7{,}000/\sqrt{M}$ to $18{,}300/\sqrt{M}$ cps. Nonlinearity is ± 1 percent and sensitivity changes ± 10 percent over the temperature range -30 to $+230°F$. The low-frequency response is 5 percent down at 2 cps if the pickup output is connected to 1,000-megohms input resistance.

Piezoelectric force pickups tend to be sensitive to side loading, and most manufacturers recommend special precautions to minimize this. Numerical sensitivities to side loads are not often quoted on instrument specification sheets. One manufacturer[2] offering specially designed pickups resistant to side loading quotes transverse sensitivity as less than 7 percent of axial sensitivity.

Variable-reluctance/FM-oscillator digital systems. While the electrical signal from any force transducer can be converted to digital form by suitable equipment, some types of pickups are specifically designed with this in mind. We here discuss briefly one such type[3] used in rocket-engine testing where digital data provide advantages of accuracy and ease of performing computations automatically. The elastic member is a proving ring with a two-arm variable-reluctance bridge displacement transducer. The signal from this transducer is used to change the frequency of a frequency-modulated (FM) oscillator. The frequency change (from some base value) of this oscillator is directly proportional to displacement (and thus to force). A digital reading of force is accomplished by applying the output signal of the oscillator to an electronic counter over a known time interval. The total number of pulses accumulated is thus a digital measure of force. In a typical unit a change of force from minus full scale to plus full scale causes the frequency to change from 10,000 to 12,500 cps. For a 1-sec counting period a full-scale force will thus cause a counter reading 1,250 above the base value of 11,250.

Computing advantages of such a system arise from the desire to know the total impulse of the rocket engine and to be able to add and sub-

[1] Endevco Corp., Pasadena, Calif.

[2] Wilcoxon Research, Bethesda, Md.

[3] Daystrom-Wiancko Co., Pasadena, Calif.

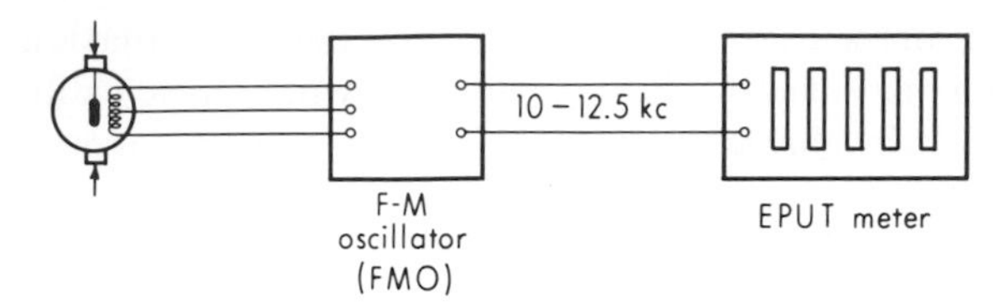

Force measurement

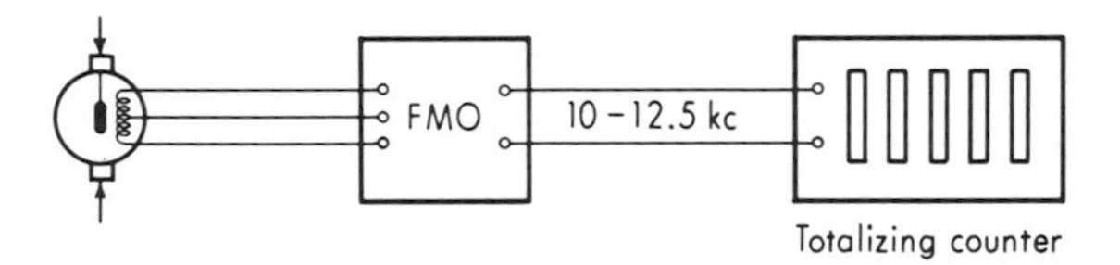

Total impulse measurement

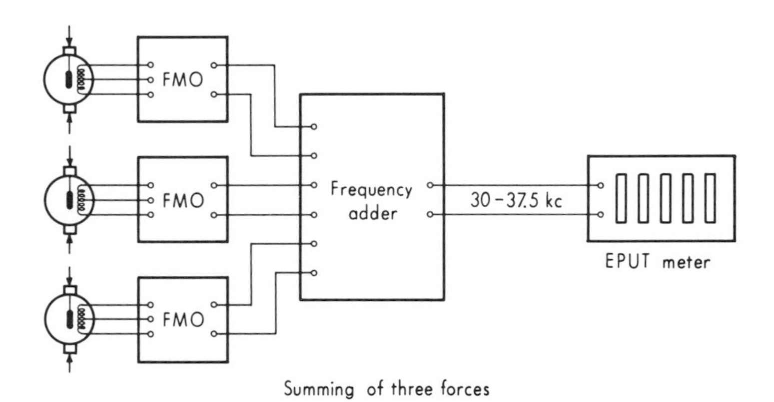

Summing of three forces

Fig. 5.7. *Digital force measurement, integration, and summing.*

tract various forces in multicomponent test stands. The total impulse is the integral of force with respect to time; this is simply the total number of counts over the desired integrating time. When several forces must be added or subtracted, each has its own force pickup and 10,000 to 12,500-cps oscillator. The oscillator output signals are combined in an electronic adder unit which produces a single output whose frequency is the sum (or difference) of the input frequencies. The output of the adder unit is a signal of a frequency of 30,000 to 37,500 cps; counting this over a timed interval gives the algebraic sum of the measured forces. Integration of this sum signal can also be easily accomplished by the same method used for a single signal. Figure 5.7 shows block diagrams of these systems.

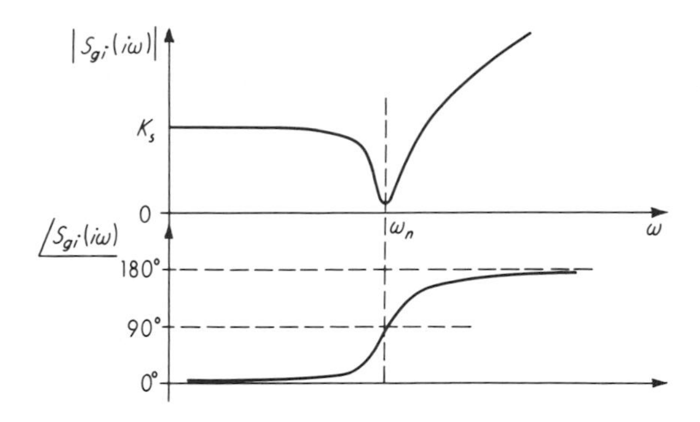

Fig. 5.8. *Stiffness of force transducers.*

Loading effects. Since force is an effort variable, the pertinent loading parameter is either stiffness or impedance, and the associated flow variable is either displacement or velocity. Stiffness is perhaps more convenient for elastic force sensors since impedance would be infinite for the static case. The generalized input stiffness S_{gi} of the system of Fig. 5.2 is given by

$$S_{gi}(D) = \frac{F_i}{x}(D) = K_s\left(\frac{D^2}{\omega_n{}^2} + \frac{2\zeta D}{\omega_n} + 1\right) \tag{5.13}$$

Recall that for small error due to loading, S_{gi} must be sufficiently large compared with S_{go}, the generalized output stiffness of the system being measured. The frequency characteristic $S_{gi}(i\omega)$ is shown in Fig. 5.8 for a particular (small) value of ζ. Note that, near ω_n, $S_{gi}(i\omega)$ becomes very small. However, force pickups are generally used only for $\omega \ll \omega_n$; thus $S_{gi}(i\omega)$ is, in most cases, adequately approximated as simply K_s.

5.4
Resolution of Vector Forces and Moments into Rectangular Components

In a number of important practical applications the force or moment to be measured is not only unknown in magnitude but also of unknown and/or variable direction. Outstanding examples of such situations are "balances" for measuring forces on wind-tunnel models,[1,2] dynamometers (force gages) for measuring cutting forces in

[1] P. K. Stein, "Measurement Engineering," vol. 1, p. 431, Stein Engineering Services, Inc., Phoenix, Ariz., 1964.

[2] C. C. Perry and H. R. Lissner, "The Strain Gage Primer," p. 212, McGraw-Hill Book Company, New York, 1955.

machine tools,[1] and thrust stands[2] for determining forces of rocket engines. Elastic force transducers of either the bonded-strain-gage or gross-deflection variety are employed in these applications. Ingenious use of various types of flexures for isolating and measuring the various force components characterizes the design of these devices. Depending on the degree to which the force or moment direction is unknown, force-resolving systems of varying degrees of complexity may be devised. The most general situation (measurement of three mutually perpendicular force components and three mutually perpendicular moment components) is regularly accomplished with high accuracy.

Figure 5.9 shows a six-component thrust stand[2] used in testing rocket engines. The load cells 1, 2, and 3 are mounted at the corners of an equilateral triangle, and load cells 4, 5, and 6 are in the sides of a concentric smaller equilateral triangle. The engine to be tested is rigidly fastened at the common center of both triangles and produces a force of unknown magnitude and direction (which can be expressed in terms of components F_x, F_y, and F_z) and a moment of unknown magnitude and direction (which can be expressed in terms of components M_x, M_y, and M_z). The rocket forces are transmitted from the mounting plate to the rigid foundation through the six load cells and their associated flexures. The action of the suspension system is most clearly seen if one considers each flexure as a pin joint. Pin joints are not actually used because of their lost motion and friction. A static analysis of the force system gives the following results which allow calculation of the rocket forces and moments from the measured load-cell forces and stand dimensions:

$$F_x = F_1 + F_2 + F_3 \tag{5.14}$$

$$F_y = \frac{(F_5 - F_4) - (F_4 - F_6)}{2} \tag{5.15}$$

$$F_z = \frac{\sqrt{3}\,(F_5 - F_6)}{2} \tag{5.16}$$

$$M_x = \frac{-d_1(F_4 + F_5 + F_6)}{2\sqrt{3}} \tag{5.17}$$

$$M_y = d_2\frac{(F_1 - F_2) - (F_3 - F_1)}{2\sqrt{3}} \tag{5.18}$$

$$M_z = d_2\frac{(F_3 - F_2)}{2} \tag{5.19}$$

[1] E. G. Loewen, E. R. Marshall, and M. C. Shaw, Electric Strain Gage Tool Dynamometers, *Proc. Soc. Exptl. Stress Anal.*, vol. 8, no. 2, 1951.

[2] The Design of High-accuracy Thrust Stands and Calibrators, WEC-BA-7D, Daystrom-Wiancko Co., Pasadena, Calif., 1961.

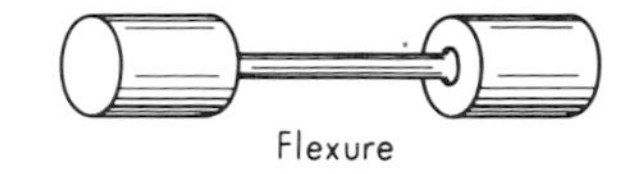

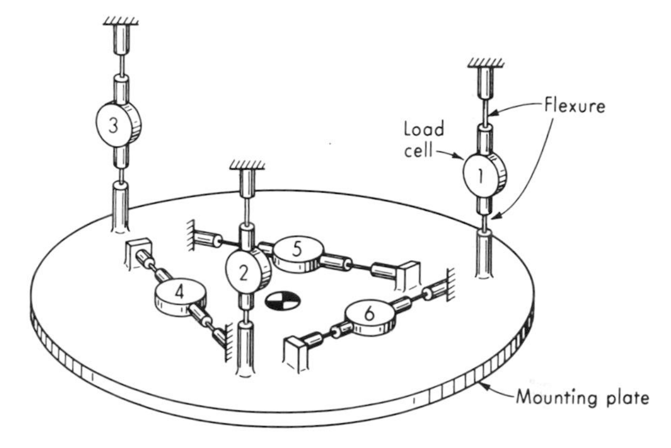

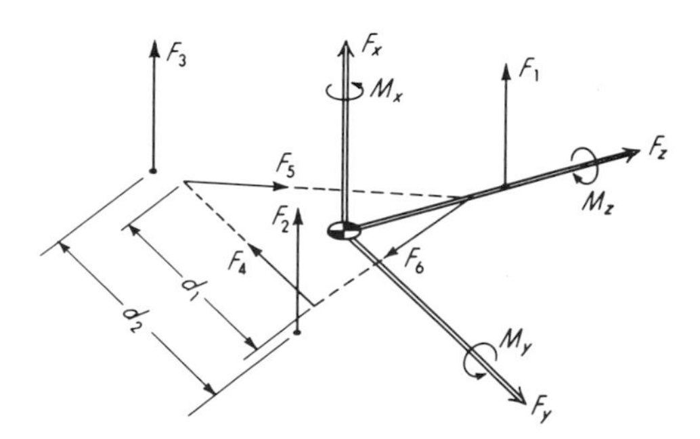

Fig. 5.9. *Six-component load frame.*

The indicated additions and subtractions of forces are (in the thrust stand described above) performed automatically by digital counters and adders since the load cells used are the variable-reluctance/FM-oscillator type described in the previous section.

A combination of bonded strain gages, Wheatstone-bridge circuits, and flexible elements of various geometries has proved a versatile tool in the development of multicomponent-force pickups of small size and high natural frequency. Figure 5.10 shows a beam with three separate bridge circuits of gages arranged to measure the three rectangular components of an applied force. All bridge circuits are temperature-compensated and respond only to the intended component of force; however, the point of

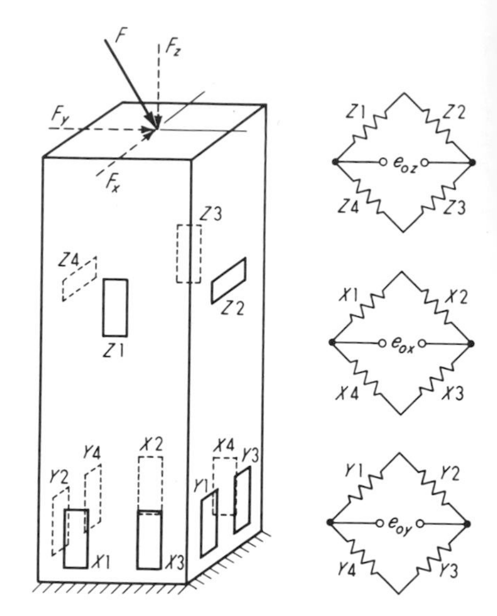

Fig. 5.10. Resolution of vector forces.

application of the force must be at the center of the beam cross section. An eccentricity in the y direction, for example, would give F_z a moment arm, causing bending stresses in the y direction that would be indistinguishable from those due to F_y. If side loads F_x and F_y are present, the end of the beam will deflect, causing just such eccentricities; thus beam stiffness must be adequate to keep deflection sufficiently low. This stiffness tends, of course, to reduce sensitivity.

While many strain-gage pickup configurations are possible and have been used, the octagonal ring[1] of Fig. 5.1 has particularly interesting properties. The reference gives information on its design and use.

5.5

Torque Measurement on Rotating Shafts Measurement of the torque carried by a rotating shaft is of considerable interest for its own sake and also as a necessary part of shaft power measurements. Torque transmission through a rotating shaft generally involves both a source of power and a sink (power absorber or dissipator), as in Fig. 5.11. Torque measurement may be accomplished by mounting either the source or the sink in bearings ("cradling") and measuring the reaction force F and arm length L, or the torque in the shaft itself is measured in terms of the angular twist or strain of the shaft (or a torque sensor coupled into the shaft).

The cradling concept is the basis of most shaft power dynamometers.

[1] N. H. Cook and E. Rabinowicz, "Physical Measurement and Analysis," p. 162, Addison-Wesley, Publishing Company, Inc., Reading, Mass., 1963.

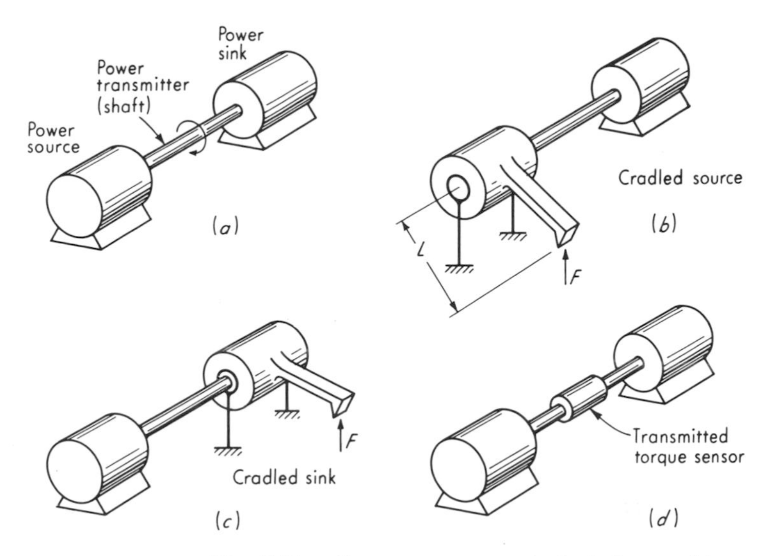

Fig. 5.11. *Torque measurement of rotating machines.*

These are used mainly for measurements of steady power and torque, using pendulum or platform scales to measure F. A free-body analysis of the cradled member reveals error sources due to friction in the cradle bearings, static unbalance of the cradled member, windage torque (if the shaft is rotating), and forces due to bending and/or stretching of power lines (electric, hydraulic, etc.) attached to the cradled member. To reduce frictional effects and also to make possible dynamic torque measurements, the cradle-bearing arrangement may be replaced by a flexure pivot with strain gages to sense torque,[1] as in Fig. 5.12. The crossing point of the

[1] Lebow Associates, Oak Park, Mich.

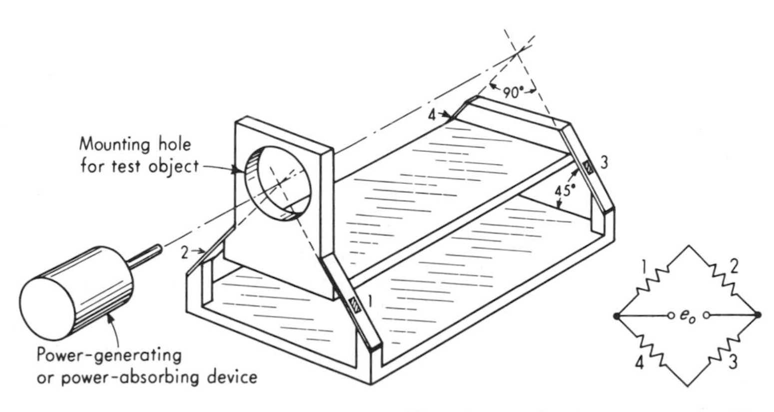

Fig. 5.12. *Strain-gage torque table.*

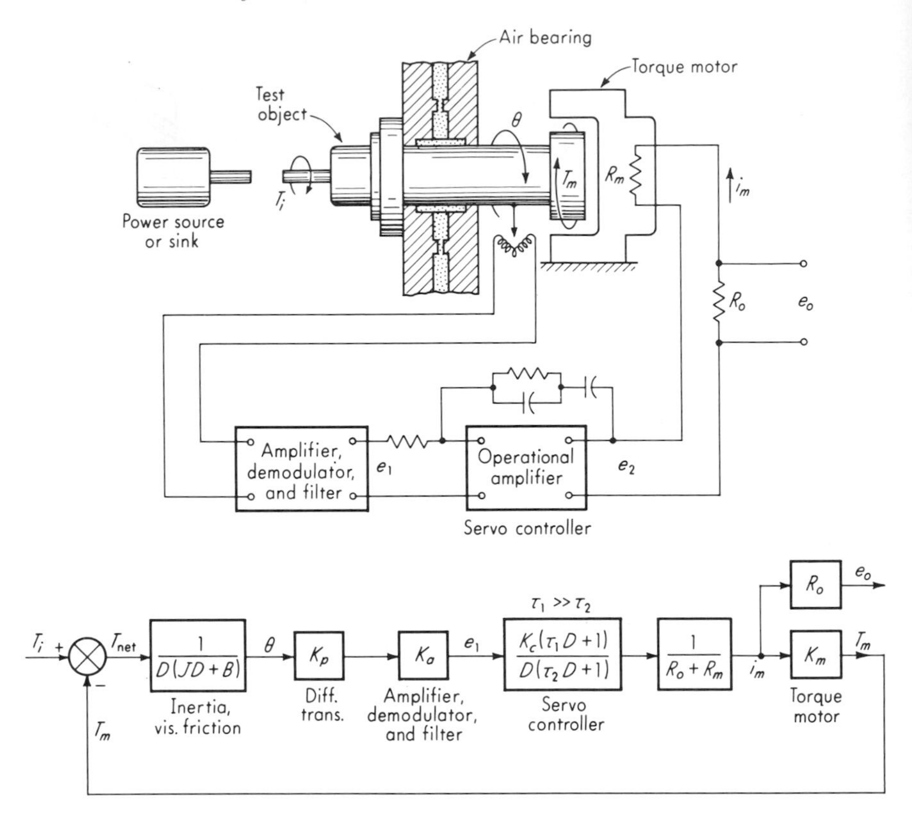

Fig 5.13. *Feedback torque sensor.*

flexure plates defines the effective axis of rotation of the flexure pivot. Angular deflection under full load is typically less than 0.5°. This type of cross-spring flexure pivot is relatively very stiff in all directions other than the rotational one desired, just as in an ordinary bearing. The strain-gage bridge arrangement also is such as to reduce the effect of all forces other than those related to the torque being measured. Speed-torque curves for motors may be obtained quickly and automatically with such a torque sensor by letting the motor under test accelerate an inertia from zero speed up to maximum while measuring speed with a d-c tachometer.[1] The torque and speed signals are applied to an *X-Y* recorder to give automatically the desired curves.

Another variation (see Fig. 5.13) on the cradle principle is found

[1] B. Hall, Motor Tests Using *X-Y* Recorders, *Electro-Technol.*, p. 116, May, 1964.

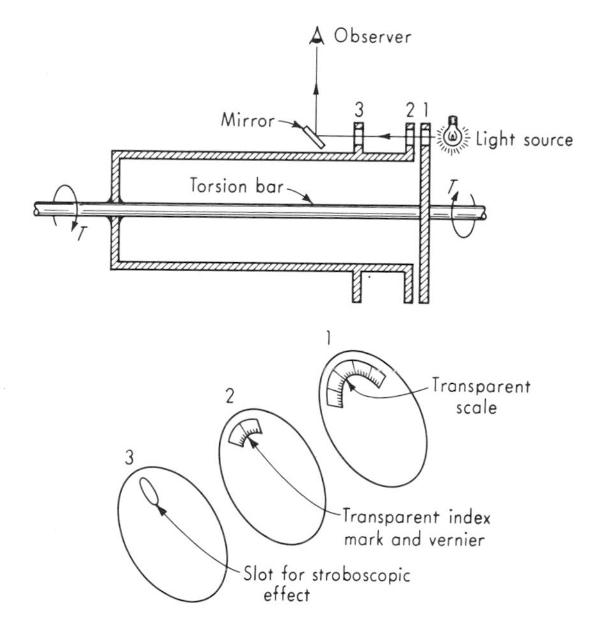

Fig. 5.14. *Torsion-bar dynamometer.*

in a null-balance torquemeter using feedback principles to measure small torques in the range 0 to 10 oz-in. In this device[1] the test object is mounted on a hydrostatic air-bearing table to reduce bearing friction to exceedingly small values. Any torque on the test object tends to cause rotation of the air-bearing table but this rotation is immediately sensed by a differential-transformer displacement pickup. The output from this pickup is converted to direct current and amplified to provide the coil current of a torque motor which applies opposing torque to keep displacement at zero. The amount of current required to maintain zero displacement is a measure of torque and is read on a meter. The servo loop uses integral control[2] to give zero displacement for any constant torque. Approximate derivative control is also used to give stability. The threshold of this air-bearing system is less than 0.0005 oz-in. while the torque/current nonlinearity is 0.001 oz-in. The overall system behaves approximately as a second-order system with a natural frequency of about 10 cps and damping ratio of 0.7 when no test object is present on the table.

[1] McFadden Electronics, South Gate, Calif.

[2] E. O. Doebelin, "Dynamic Analysis and Feedback Control," p. 223, McGraw-Hill Book Company, New York, 1962.

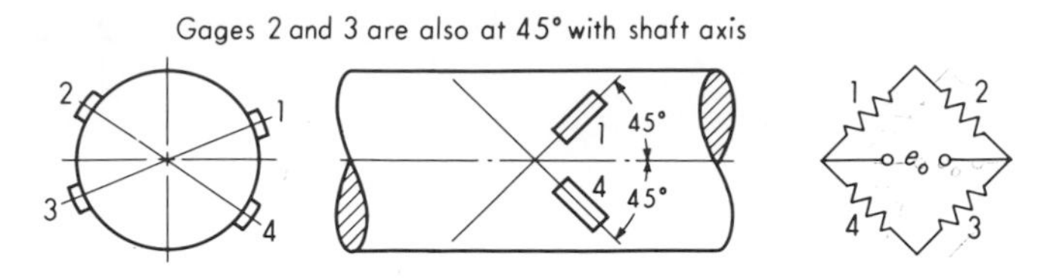

Fig. 5.15. *Strain-gage torque measurement.*

The use of elastic deflection of the transmitting member for torque measurement may be accomplished by measuring either a gross motion or a unit strain. In either case, a main difficulty is the necessity of being able to read the deflection while the shaft is rotating. Figure 5.14 illustrates a torsion-bar torquemeter using optical methods of deflection measurement. The relative angular displacement of the two sections of the torsion bar can be read from the calibrated scales because of the stroboscopic effect of intermittent viewing and the persistence of vision. The desire for electrical output signals and for the ability to measure rapidly varying torque has led to the development of strain-gage torquemeters. The problem of getting bridge power onto the shaft and output signals off the shaft has a number of possible solutions. If rotation is slow and the total angle turned through is small, one may simply let the connecting wires wrap around the shaft. For continuous rotation, slip rings and brushes may be used, or a subminiature telemetry system may be attached to the rotating shaft and the signals sent to a stationary receiver by radio.

The strain-gage bridge configuration generally used to measure torque is shown in Fig. 5.15. This arrangement (assuming accurate gage placement and matched gage characteristics) is temperature-compensated and insensitive to bending or axial stresses. The gages must be precisely at 45° with the shaft axis, and gages 1 and 3 must be diametrically opposite, as must gages 2 and 4. Accurate gage placement is facilitated by the availability of special rosettes in which two gages are precisely oriented on one sheet of backing material. In some cases the shaft already present in the machine to be tested may be fitted with strain gages. In other cases a different shaft or a commercial torquemeter must be used to get the desired sensitivity or other properties. Placement of the gages on a square, rather than round, cross section of the shaft (see Fig. 5.16) has some advantages. The gages are more easily and accurately located and more firmly bonded on a flat surface. Also, the corners of a square section in torsion are stress-free and thus provide a good location for solder joints between lead wires and gages. These joints are often a

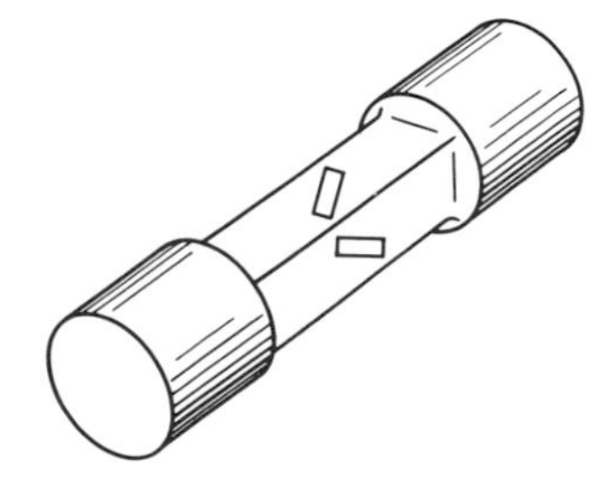

Fig. 5.16. *Square shaft-torque element.*

source of fatigue failure if located in a high-stress region. Also, for equivalent strain/torque sensitivity, a square shaft is much stiffer in bending than a round one, thus reducing effects of bending forces and raising shaft natural frequencies.

The torque of many machines, such as reciprocating engines, is not smooth even when the machine is running under "steady-state" conditions. If one wishes to measure the average torque so as to calculate power, the higher frequency response of strain-gage torque pickups may be somewhat of a liability since the output voltage will follow the cyclic pulsations and some sort of averaging process must be performed to obtain average torque. If exceptional accuracy is not needed, the low-pass filtering effect of a d-c meter used to read e_o may be sufficient for this purpose. In the cradled arrangements of Fig. 5.11 (used in many commercial dynamometers for engine testing, etc.) the inertia of the cradled member and the low frequency response of the platform or pendulum scales used to measure F perform the same averaging function.

Commercial strain-gage torque sensors are available with built-in slip rings and speed sensors. A family[1] of such devices covers the range 50 to 100,000 in.-lb_f with full-scale output of about 40 mv. The smaller units may be used at speeds up to 24,000 rpm, the largest to 4,000 rpm. Torsional stiffness of the 50-in.-lb_f unit is about 4,000 in.-lb_f/rad while the 100,000-in.-lb_f unit has 9.5×10^6 in.-lb_f/rad. Nonlinearity is 0.1 percent of full scale while temperature effect on zero is 0.002 percent of full scale/F° and temperature effect on sensitivity is 0.002 percent/F° over the range 30 to 150°F.

The dynamic response of elastic deflection torque transducers is essentially slightly damped second-order, with the natural frequency usually determined by the stiffness of the transducer and the inertia of the parts connected at either end. Damping of the transducers themselves is usually not attempted, and any damping present is due to bearing friction, windage, etc., of the complete test setup.

[1] Lebow Associates, Oak Park, Mich.

5.6

Shaft Power Measurement (Dynamometers) The accurate measurement of the shaft power input or output of power-generating, -transmitting, and -absorbing machinery is of considerable interest. While the basic measurements, torque and speed, have already been discussed, their practical application to power measurement will be considered briefly here. The term dynamometer is generally used to describe such power-measuring systems although it is also used as a name for elastic force sensors.

The type of dynamometer employed depends somewhat on the nature of the machine to be tested. If the machine is a power generator, the dynamometer must be capable of absorbing its power. If the machine is a power absorber, the dynamometer must be capable of driving it. If the machine is a power transmitter or transformer, the dynamometer must provide both the power source and the load.

Perhaps the most versatile and accurate dynamometer is the d-c electric type. Here a d-c machine is mounted in low-friction trunnion bearings (see Fig. 5.11*b*) and provided with field and armature control circuits.[1,2] This machine can be coupled to either power-absorbing or power-generating devices since it may be connected as either a motor or a generator. When it is used as a generator, the generated power is dissipated in resistance grids or recovered for use. Modern control circuits allow accurate control of dynamometer speed under varying load torque (100 percent change in load torque causes about 0.5 percent steady-state speed change). Control of dynamometer torque is also possible, although it is less used and less accurate. A speed change of about 50 percent of top speed may be expected to cause a torque change of 5 percent of maximum torque when torque control is used. The d-c dynamometer can be adjusted to provide any torque from zero to the maximum design value for speeds from zero to the so-called base speed of the machine. This is the speed at which the maximum torque develops the maximum design horsepower. At speeds above base speed, torque must be progressively reduced so as to maintain horsepower less than the design maximum. The controllability of the d-c dynamometer lends itself particularly to modern automatic load and speed programming applications.

Figure 5.17 illustrates such a situation. Tape recordings of engine torque and speed measured under actual driving conditions for an auto-

[1] P. S. Potts and P. T. Schuerman, How to Choose Electric Dynamometers, *Machine Design*, p. 102, June 27, 1957.

[2] R. F. Knudsen, A Discussion of Present Day Dynamometers, *Gen. Motors Eng. J.*, p. 18, October–November–December, 1957.

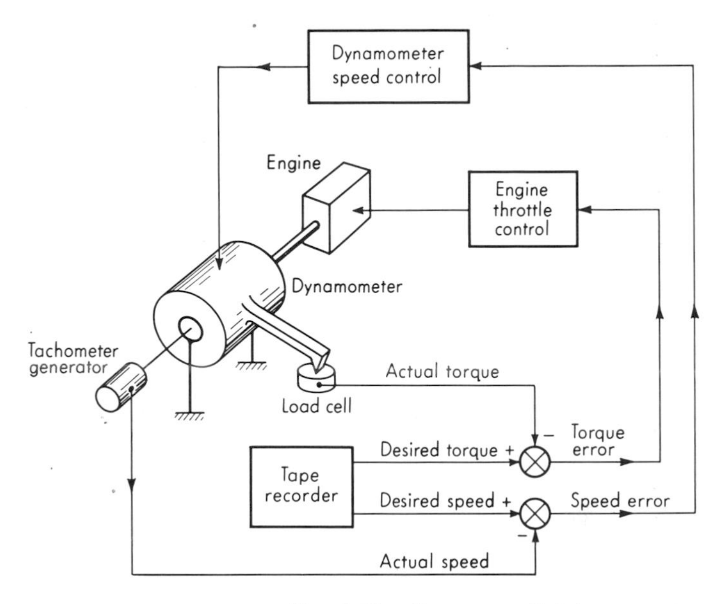

Fig. 5.17. Servo-controlled dynamometer.

mobile are used to reproduce these conditions in the laboratory engine test. Two feedback systems are used to control engine speed and torque. A tachometer generator speed signal from the dynamometer is compared with the desired speed signal from the tape recorder; if the two are different, the dynamometer control is automatically adjusted to change speed until agreement is reached. Actual engine torque is obtained from a load cell on the dynamometer and compared with the desired torque from the tape recorder. If these do not agree, the error signal actuates the engine throttle control in the proper direction. Both these systems operate simultaneously and continuously to force engine speed and torque to follow the tape-recorder commands.

Dynamometers capable only of absorbing power include the eddy-current brake (inductor dynamometer) and various mechanical brakes employing dry friction (Prony brake) and fluid friction (air and water brakes). The eddy-current brake is easily controllable by varying a d-c input, but it cannot produce any torque at zero speed and only small torque at low speeds. However, it is capable of higher power and speed than a d-c dynamometer. The power absorbed is carried away by cooling water circulated through the air gap between rotor and stator. The Prony brake is a simple mechanical brake in which friction torque is manually adjusted by varying the normal force with a handwheel. Torque is available at zero speed but operation may be jerky because

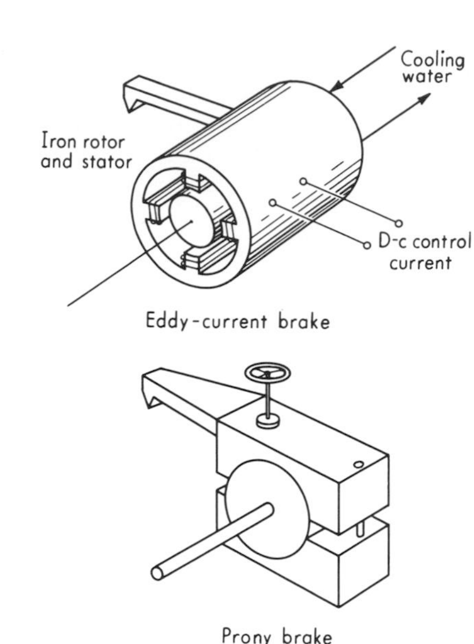

Fig. 5.18. Absorption dynamometers.

of the basic nature of dry friction. Water and air brakes utilize the churning action of paddle wheels or vanes rotating inside a fluid-filled casing to absorb power. A flow of air or water through the device is maintained for cooling purposes. No torque is available at zero speed and only small torques at low speeds. High speed and high power can be handled well, however, some water brakes being rated at 10,000 hp at 30,000 rpm.

In all the above power-measurement applications torque and speed are separately measured and then power is manually calculated. This calculation can be performed automatically in a number of ways since the basic operation (multiplication) can be accomplished physically in various ways. An interesting scheme using the properties of bridge circuits is shown in Fig. 5.19. Speed is measured with a d-c tachometer generator, and this voltage is applied as the excitation of a strain-gage load cell used to measure torque. Since bridge output is directly proportional to excitation voltage and also directly proportional to torque, the voltage e_o is actually an instantaneous power signal.

Another ingenious solution[1] of this problem is shown in Fig. 5.20. A torsion bar carrying the torque to be measured has a permanent-magnet a-c generator (alternator) coupled at either end. Each alternator puts out an a-c voltage of amplitude proportional to shaft speed and

[1] Waugh Div., Van Nuys, Calif.

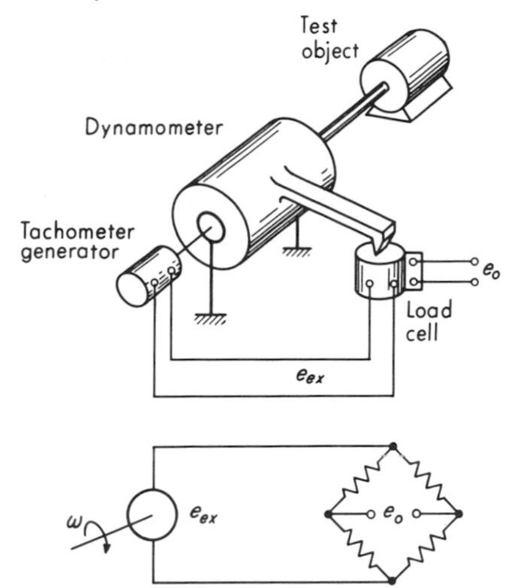

Fig. 5.19. *Instantaneous power measurement.*

frequency equal to shaft speed. When the torsion bar is unloaded, the two alternator rotor-stator positions are mechanically adjusted so that the a-c output voltages are exactly out of phase. If the two alternators are now connected electrically in series, the net output will be zero at any shaft speed so long as no torque is present. When the shaft carries torque, it twists, causing a phase shift between the two alternators and a net

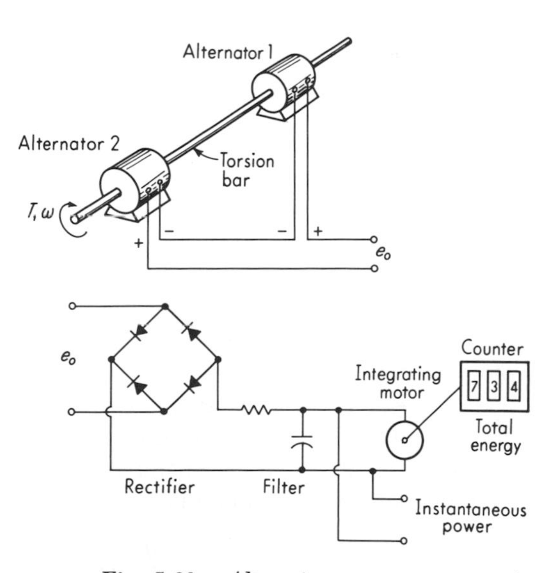

Fig. 5.20. *Alternator power measurement.*

output voltage whose amplitude is proportional to the product of torque and speed. This may be seen from the following analysis:

$$\text{Alternator 1 output} = K_\omega \omega \sin \omega t \tag{5.20}$$
$$\text{Alternator 2 output} = K_\omega \omega \sin (\omega t + \phi) \tag{5.21}$$

where

$$K_\omega \triangleq \text{peak volts/(rad/sec)} \tag{5.22}$$
$$\phi = K_t T \tag{5.23}$$
$$K_t \triangleq \text{rad/in.-lb}_f \tag{5.24}$$
$$T \triangleq \text{torque} \tag{5.25}$$

The net output of the series-connected alternators is

$$\text{Net output} = e_o = K_\omega \omega[\sin \omega t - \sin (\omega t + K_t T)] \tag{5.26}$$
$$e_o = K_\omega \omega[\sin \omega t - (\sin \omega t \cos K_t T + \cos \omega t \sin K_t T)] \tag{5.27}$$

The twist angle $\phi = K_t T$ is very small, and so $\cos K_t T \approx 1.0$ and $\sin K_t T \approx K_t T$. Thus

$$e_o = -K_\omega \omega K_t T \cos \omega t \tag{5.28}$$

and e_o is a sine wave of amplitude proportional to ωT and thus to power. In the instrument described above this a-c voltage is rectified and filtered to produce a proportional direct current of 20 volts full scale. Also, if total energy over a time period is desired, an integrator is available to integrate the d-c voltage. This is in the form of a precise d-c motor whose speed is directly proportional to the voltage applied to it from the horsepower meter. If speed is proportional to horsepower, total revolutions (read by an ordinary mechanical counter) give total energy.

5.7

Gyroscopic Force and Torque Measurement The use of gyroscopic principles in the measurement of force and torque is not widespread and thus was not considered earlier in this chapter. A short discussion is included here because of general interest and potential application. The method is particularly useful if one wishes to obtain the time integral of a force or torque since the gyro supplies a mechanical rotation proportional to it. This concept has found practical application in a gyroscopic integrating mass flowmeter which will be discussed in more detail in the chapter on flow measurement.

From Chap. 4, the transfer function relating an applied torque T_x to an output axis rotation θ is

$$\frac{\theta}{T_x}(D) = \frac{H_s/I_x}{D[I_y D^2 + BD + (H_s^2/I_x + K_s)]}$$

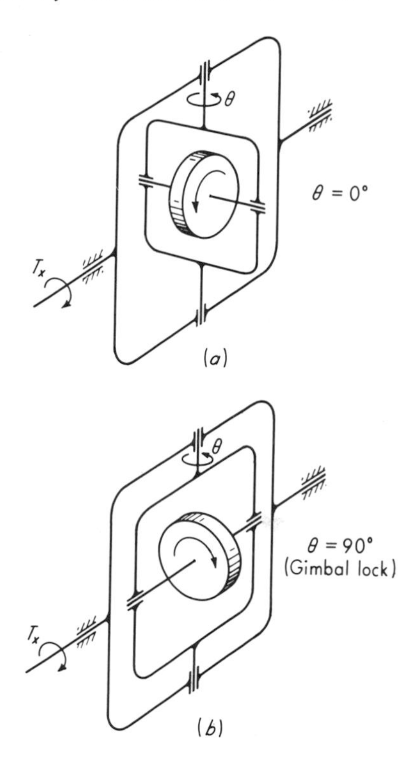

Fig. 5.21. *Gyroscopic torque measurement.*

For a free gyro, as in Fig. 5.21*a*, B and K_s are effectively zero, giving

$$\frac{\dot{\theta}}{T_x}(D) = \frac{K}{D^2/\omega_n^2 + 1} \qquad (5.29)$$

where

$$K \triangleq \frac{1}{H_s} \qquad \text{(rad/sec)/in.-lb}_f \qquad (5.30)$$

$$\omega_n \triangleq \sqrt{\frac{H_s^2}{I_x I_y}} \qquad (5.31)$$

While the second-order term of Eq. (5.29) is undamped, small damping (friction) in bearings, etc., is always present to prevent sustained oscillation. Also, if T_x is applied gradually, oscillations will not be started. Thus a constant torque T_x will produce a precessional angular velocity $\dot{\theta}$ in direct proportion according to $\dot{\theta} = KT_x$. This will be true only so long as θ is small, however, and since a constant torque produces a constant *velocity*, θ must eventually become large. This will ultimately (when θ reaches 90°) lead to what is called "gimbal lock." The gyroscopic precession actually depends on the component of the torque vector that is perpendicular to the spin angular-momentum vector. For small angles, this component is directly proportional to torque. For large angles, it is

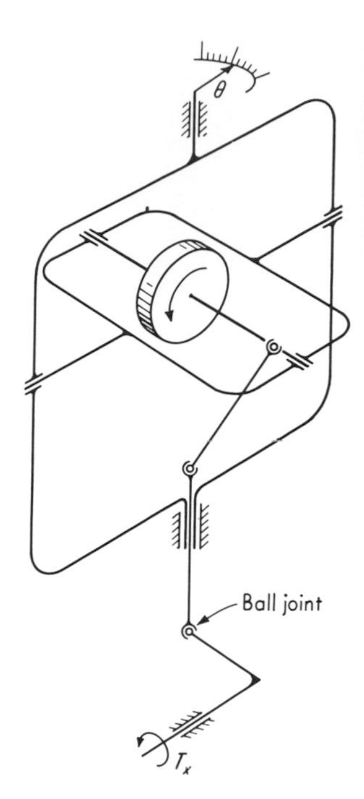

Fig. 5.22. Solution of gimbal-lock problem.

proportional to the product of torque and cos θ; this becomes smaller as θ increases and disappears completely at $\theta = 90°$. Thus the precession θ produced by a constant torque T_x gets smaller and smaller as θ approaches 90°. At 90° a torque T_x produces no precession at all; rather, the inner and outer gimbal both rotate together about the x axis.

To prevent gimbal lock and thus achieve a useful torque-sensing instrument, a simple mechanical solution is available. The requirement that torque vector and spin angular-momentum vector always be perpendicular is met by the configuration of Fig. 5.22. The equation $\dot{\theta} = KT_x$ now holds for *any* angle θ, and one can measure torque by measuring $\dot{\theta}$ or measure the time integral of torque by means of a simple revolution counter attached to the θ shaft.

Problems

5.1 Compute the value of g at your local latitude and altitude. What is the percent deviation from the standard value 980.665?

5.2 What is the percentage change in weight (compared with sea level) of a mass located above the equator at an altitude of 500 miles?

5.3 An object with a volume of 10 in.^{3} is weighed on an equal-arm balance. The standard mass required for balance is 1 lb_m and has a volume of 3 in.^{3}. What is the value of the correction necessary for air buoyancy?

5.4 A brass balance beam has a length of exactly 1 m at 60°F, and the pivot is perfectly centered to give an equal-arm balance. If one end of the beam comes to 80°F and there is a uniform temperature gradient to the other end at 60°F, what inequality in arm length results?

5.5 What general form of dynamic response would you expect from the systems of Fig. 5.1-1? Why?

5.6 Prove that the reading is independent of location of F_i if $a/b = c/d$ in the platform scale of Fig. 5.1-1.

5.7 If, in Fig. 5.1-2, $M = 1\ \text{lb}_m$ and $A = 20g$, what is the net force on M? If a friction force which is unknown but less than 1 lb_f may be present, what error may be expected in F_i?

5.8 Carry out a simplified, linear dynamic analysis of the system of Fig. 5.1-3.

5.9 Carry out a linear dynamic analysis and an accuracy/stability tradeoff study of the pneumatic load cell of Fig. 5.1-4. Use $(x/F_{\text{net}})(D) = K_d/(D^2/\omega_n^2 + 2\zeta D/\omega_n + 1)$ and $(p_o/x)(D) = K_n/(\tau D + 1)$.

5.10 In Fig. 5.3 if F_i is eccentric and also angularly misaligned a torque is produced. Does this affect the bridge output? Explain.

5.11 A load cell deflects 0.005 in. under its full-scale load of 1,000 lb_f. It is used to measure force on a machine-tool slide which weighs 500 lb_f. Estimate the highest frequency of force that may be accurately measured.

5.12 Derive Eqs. (5.14) to (5.19).

5.13 From the block diagram of Fig. 5.13, obtain the transfer function $(e_o/T_i)(D)$, assuming τ_2 is small enough to neglect. Investigate the effect of system parameters on dynamic accuracy and stability.

5.14 In Fig. 5.13 prove that $\theta = 0$ for any constant value of T_i.

5.15 Prove that the arrangement of Fig. 5.15 is insensitive to axial or bending stresses.

5.16 Prove that for equivalent strain/torque sensitivity a square shaft is stiffer in bending than a round one.

5.17 A torque sensor with torsional stiffness of 1,000 in.-lb_f/rad is coupled between an electric motor and a hydraulic pump. The moments of inertia of the motor and pump are each 0.01398 in.-lb_f-sec^2. If the motor has a small oscillatory torque component at 60 cps, will this measuring system be satisfactory? Explain what torsional stiffness is needed if the response at 60 cps is to be no more than 105 percent of the static response. The amount of damping is unknown.

5.18 Suppose the tachometer generator in the system of Fig. 5.19 puts out 6 volts/1,000 rpm and the load cell produces 0.05 mv/(lb_f-volt excitation). What will be the power calibration factor for e_o in horsepower per millivolt if the arm length is 1 ft?

5.19 In the system of Fig. P5.1:

a. For $F = 0$ and heat off, $R_1 = R_2 = R_3 = R_4$ and $e_o = 0$.

b. The gage factor of the gages is +2.0, and the temperature coefficient of resistance of the gages is positive.

c. The modulus of elasticity of the beam decreases with increased temperature.

d. The thermal-expansion coefficient of the gage is greater than that of the beam.

e. Assume the gage temperature is the same as that of the beam *immediately* beneath it.

At time $t = 0$, an upward force F is applied and maintained constant thereafter. After oscillations have died out, at a later time t_1 the radiant-heat source is

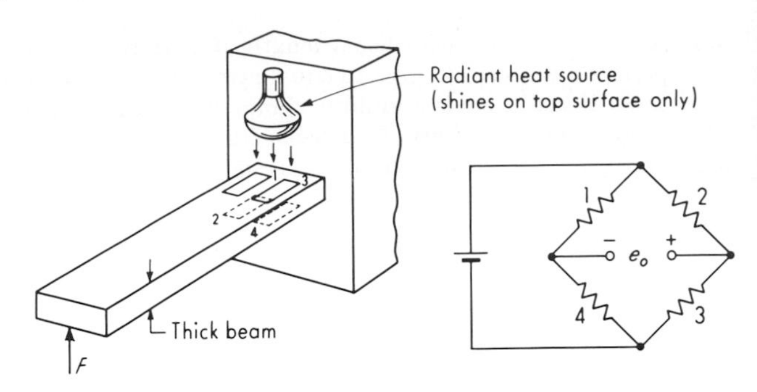

Fig. P5.1

turned on and left on thereafter. Sketch the general form of e_o versus t, justifying clearly by detailed reasoning the shape you give the curve.

5.20 It is necessary to design a strain-gage thrust transducer for small experimental rocket engines which are roughly in the shape of a cylinder 6 in. in diameter by 12 in. long. The following information is given:

a. Weight of motor and mounting bracket, 20 lb_f.
b. Maximum steady thrust, 50 lb_f.
c. Oscillating component of thrust, ± 10 lb_f maximum.
d. Oscillating components of thrust up to 100 cps must be measured with a flat amplitude ratio within ± 5 percent.
e. A recorder with a sensitivity of 0.1 volt/in., frequency response flat to 120 cps, and input resistance of 10,000 ohms is available.
f. Thrust changes of 0.5 lb_f must be clearly detected.
g. Gages with a resistance of 120 ohms and a gage factor of 2.1 are available. They are 0.5×1.0 in. in size.
h. An amplifier (to be placed between transducer and recorder) is available with a gain up to 1,000.

Design the transducer so as to require a minimum of amplifier gain. If damping is employed, calculate the required damping coefficient B but do not design the damper. Use the cantilever-beam arrangement of Fig. P5.2.

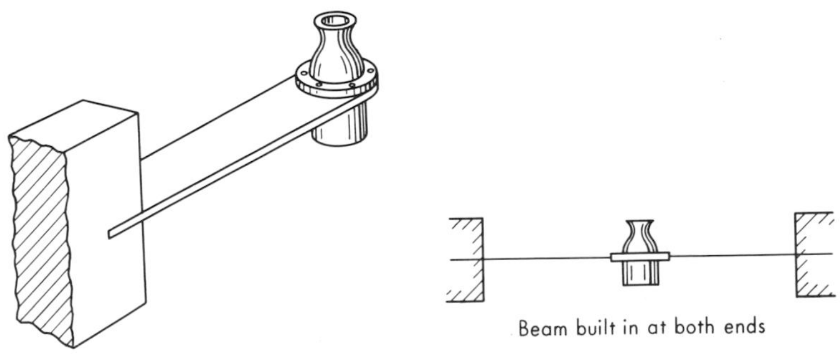

Fig. P5.2 *Fig. P5.3*

5.21 Repeat Prob. 5.20 using the configuration of Fig. P5.3.

5.22 Repeat Prob. 5.20 using the configuration of Fig. P5.4.

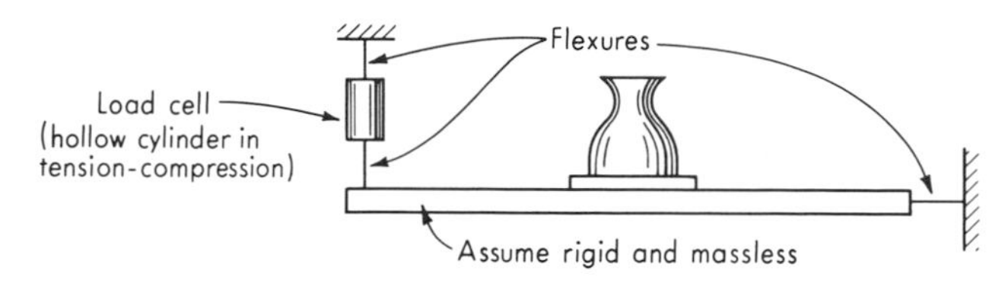

Fig. P5.4

Bibliography

1. A. Krsek and M. Tiefermann: Optical Torquemeter for High Rotational Speeds, *NASA Tech. Note*, D-1437, October, 1962.
2. H. E. Lockery: Applying the Strain-gage Torque Transducer, *ISA J.*, p. 65, March, 1962.
3. Torque-gauge Without Sliprings, *Electromech. Design*, p. 6, November, 1959.
4. J. Guthrie: Lever-shaft Torque Measurement, *Instr. Control Systems*, p. 116, August, 1964.
5. D. Ettleman and M. Hoberman: Torquemeters, *Machine Design*, p. 134, Feb. 28, 1963.
6. O. Dahle: Heavy Industry Gets a New Load Cell, *ISA J.*, p. 32, August, 1959.
7. F. M. Ryan: Automatic Weighing for Solids, *Control Eng.*, p. 103, September, 1962.
8. D. W. Kennedy: Weighing Scales Couple to Computer, *Control Eng.*, p. 83, July, 1962.
9. F. A. Ludewig: Digital Force Transducer, *Control Eng.*, p. 107, June, 1961.
10. K. Harris: Servo-balanced Supply Tank Measures Nozzle Thrust, *Control Eng.*, p. 115, February, 1960.
11. Planetary Gearing in Torquemeter Does Away with Sliding Contacts, *Machine Design*, p. 135, Sept. 1, 1960.
12. E. T. Gay: Precision Weighing with Platform Scales, *Tool Engr.*, June, 1959.
13. S. Edwards: Dynamic Measurement of Vehicle Front Wheel Loads Using a Special Purpose Transducer, *Gen. Motors Eng. J.*, p. 15, October–November–December, 1964.
14. S. Hejzlar: Backweighing Error in Scales, *Instr. Control System*, p. 95, February, 1965.
15. Apparatus Measures Very Small Thrusts, *NASA Brief* 64-10284, 1964.
16. J. A. Bierlein: Methods of Measuring Thrust, *ARS J.*, p. 128, May–June, 1953.
17. L. E. Stone: Criteria for Design and Use of an Internal Strain Gage Floating-frame Balance, *ISA Trans.*, p. 152, April, 1965.
18. R. L. Small: Belt Scales, *ISA J.*, p. 65, May, 1965.
19. V. C. Plane: Total Impulse Measuring System for Solid-propellant Rocket Engine, Rocketdyne Div., Canoga Park, Calif., *Rept.* R-5638, 1964.
20. R. W. Postma: Pulse Thrust Measuring Transducer (with Accelerometer Dynamic Compensation), Rocketdyne Div., Canoga Park, Calif., *Rept.* R-6044, 1965.

6 Pressure and sound measurement

6.1

Standards and Calibration Pressure is not a fundamental quantity but rather is derived from force and area, which in turn are derived from mass, length, and time, the latter three being fundamental quantities whose standards have been discussed earlier. Pressure "standards" in the form of very accurate instruments are available, however, for calibration of less accurate instruments. However, these "standards" depend ultimately on the fundamental standards for their accuracy. The basic standards[1,2,3] for pressures ranging from medium vacuum (about 10^{-1} mm Hg) up to several hundred thousand pounds per square inch are in the form of precision mercury columns (manometers) and dead-weight piston gages. For pressures in the range 10^{-1} to 10^{-3} mm Hg the McLeod vacuum gage is considered the standard. For pressures below 10^{-3} mm Hg a pressure-dividing technique using flow through a succession of accurate orifices to relate the low downstream pressure to a higher upstream pressure (which is accurately measured with a McLeod gage) is presently in use.[4]

This technique can be further improved by substituting a Schulz hot-cathode or radioactive ionization vacuum gage for the McLeod gage. Each of these must be calibrated against a McLeod gage at one point (about 9×10^{-2} mm Hg) but their known linearity is then used to extend their accurate range to much lower pressures.[5] This procedure is fairly well accepted to about 10^{-7} mm Hg at present, but the great interest in the high vacuums (10^{-12} mm Hg and less) of space environments will undoubtedly lead to continuing improvements.

The inaccuracies of the above-mentioned pressure standards range from about ± 4 percent at 10^{-7} mm Hg to ± 1 percent at 10^{-3} mm Hg to ± 0.1 percent at 10^{-1} mm Hg to a peak of ± 0.001 percent at 1 atm and down again to ± 0.1 percent at 200,000 psi. Since the above-mentioned pressure standards are also pressure-measuring instruments (of the highest quality and used under carefully controlled conditions), their operating principles and characteristics will not be discussed here since they are adequately covered later.

6.2

Basic Methods of Pressure Measurement Since pressure can usually be easily transduced to force by allowing it to act on a known area, the basic methods of measuring force and pressure are essentially the same, except for the high-vacuum region where a variety of special methods not directly related to force measurement are necessary. These special

[1] D. P. Johnson and D. H. Newhall, The Piston Gage as a Precise Pressure-measuring Instrument, *Instr. Control Systems*, p. 120, April, 1962.

[2] Errors in Mercury Barometers and Manometers, *Instr. Control Systems*, p. 121, March, 1962.

[3] 2″ Range Hg Manometer, *Instr. Control Systems*, p. 152, September, 1962.

[4] J. R. Roehrig and J. C. Simons, Calibrating Vacuum Gages to 10^{-9} Torr, *Instr. Control Systems*, p. 107, April, 1963.

[5] J. C. Semons, On Uncertainties in Calibration of Vacuum Gages and the Problem of Traceability, "Transactions of 10th National Vacuum Symposium," p. 246, The Macmillan Company, New York, 1963.

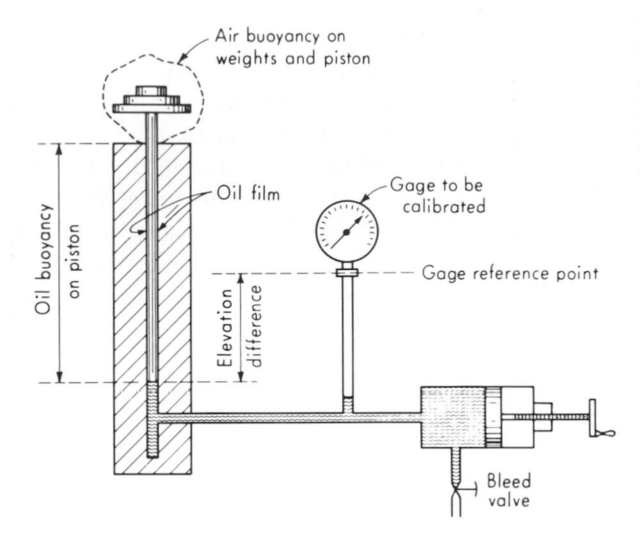

Fig. 6.1. Dead-weight-gage calibrator.

methods will be described in the section on vacuum measurement. Other than the special vacuum techniques, most pressure measurement is based on comparison with known dead weights acting on known areas or on the deflection of elastic elements subjected to the unknown pressure. The dead-weight methods are exemplified by manometers and piston gages while the elastic deflection devices take many different forms.

6.3

Dead-weight Gages and Manometers Figure 6.1 shows the basic elements of a dead-weight or piston gage. Such devices are used mainly as standards for the calibration of less accurate gages or transducers. The gage to be calibrated is connected to a chamber filled with fluid whose pressure can be adjusted by means of some type of pump and bleed valve. The chamber also connects with a vertical piston-cylinder to which various standard weights may be applied. The pressure is slowly built up until the piston and weights are seen to "float," at which point the fluid "gage" pressure (pressure above atmosphere) must equal the dead weight supported by the piston, divided by the piston area.

For highly accurate results, a number of refinements and corrections are necessary. The frictional force between the cylinder and piston must be reduced to a minimum and/or corrected for. This is generally accomplished by rotating either the piston or the cylinder. If there is no axial relative motion, this rotation should reduce the axial effects of dry friction to zero. There must, however, be a small clearance between the piston and the cylinder and thus an axial flow of fluid from the high-pressure end to the low. This flow produces a viscous shear force tending to support part of the dead weight. This effect can be estimated from theoretical calculations;[1] however, it varies somewhat with pressure since the piston and cylinder deform under pressure, thereby changing the clearance. The clearance between the piston and cylinder also raises the question of which area is to be used in computing pressure. The effective area is generally taken as the average of the piston and cylinder areas. Further corrections are needed for temperature effects on areas of piston and cylinder, air and pressure-medium buoyancy effects, local gravity conditions, and height differences between the lower end of the piston and the reference point for the gage being calibrated. Special designs and techniques allow use of dead-weight gages for pressures up to several hundred thousand pounds per square inch.

Since the piston assembly itself has weight, conventional dead-weight gages are not capable of measuring pressures lower than the piston weight/area ratio ("tare" pressure). This difficulty is overcome by the tilting-piston gage[2] in which the cylinder and piston can be tilted from vertical through an accurately measured angle, thus giving a continuously adjustable pressure from 0 psig up to the tare pressure. The described gage uses nitrogen or other inert gas as the pressure medium and covers the range 0 to 600 psig, having two interchangeable piston-cylinders and 14 weights. The accuracy is 0.01 percent of reading in the range 0.3 to 15 psig and 0.015 percent of reading in the range 2 to 600 psig. The tilting feature is used for the ranges 0 to 0.3 and 0 to 2.0 psig; higher pressures are obtained in increments by the addition of discrete weights.

Dead-weight gages may be used for absolute- rather than gage-pressure measurement by placing them inside an evacuated enclosure at (ideally) 0-psia pressure. Since the degree of vacuum (absolute pressure) inside the enclosure must be known, this really requires an additional independent measurement of absolute pressure.

The manometer, in its various forms, is closely related to the piston gage since both are based on the comparison of the unknown pressure force with the gravity force on a known mass. The manometer differs, however, in that it is self-balancing, is a deflection rather than a null instrument, and has continuous rather than stepwise output. The accuracies of dead-weight gages and manometers of similar ranges are quite comparable; however, manometers become unwieldy at high pres-

[1] R. J. Sweeney, "Measurement Techniques in Mechanical Engineering," p. 104, John Wiley & Sons, Inc., New York, 1953.

[2] Ruska Instrument Corp., Houston, Texas.

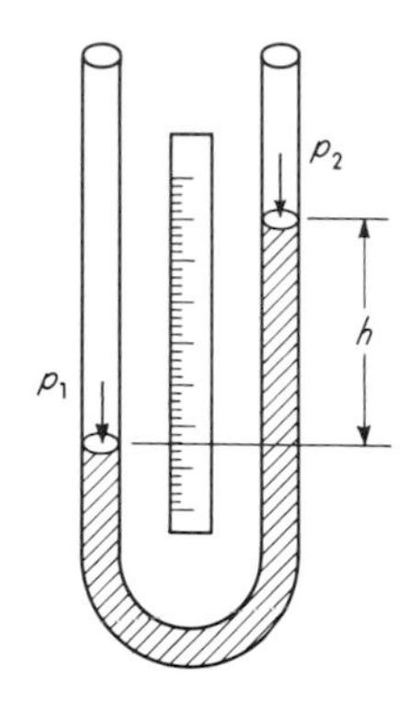

Fig. 6.2. *U-tube manometer.*

sures because of the long liquid columns involved. The U-tube manometer of Fig. 6.2 is usually considered the basic form and has the following relation between input and output for static conditions:

$$h = \frac{p_1 - p_2}{\rho g} \tag{6.1}$$

where $g \triangleq$ local gravity
$\rho \triangleq$ mass density of manometer fluid

If p_2 is atmospheric pressure, then h is a direct measure of p_1 as a gage pressure. Note that the cross-sectional area of the tubing (even if not uniform) has no effect. At a given location (given value of g) the sensitivity depends only on the density of the manometer fluid. Water and mercury are the most commonly used fluids. To realize the high accuracy possible with manometers, a number of corrections must often be applied. When visual reading of the height h is employed, the engraved-scale's temperature expansion must be considered. The variation of ρ with temperature for the manometer fluid used must be corrected and the local value of g determined. Additional sources of error are found in the nonverticality of the tubes and the difficulty in reading h due to the meniscus formed by capillarity. Considerable care must be exercised in order to keep inaccuracies as small as 0.005 in. Hg for the overall measurement.[1]

A number of practically useful variations on the basic manometer principle are shown in Fig. 6.3. The *cistern* or *well-type manometer* is widely used because of its convenience in requiring reading of only a single leg. The well area is made very large compared with the tube; thus the zero level moves very little when pressure is applied. Even this small error is compensated by suitably distorting the length scale. However,

[1] A. J. Eberlein, Laboratory Pressure Measurement Requirements for Evaluating the Air Data Computer, *Aeron. Eng. Rev.*, p. 53, April, 1958.

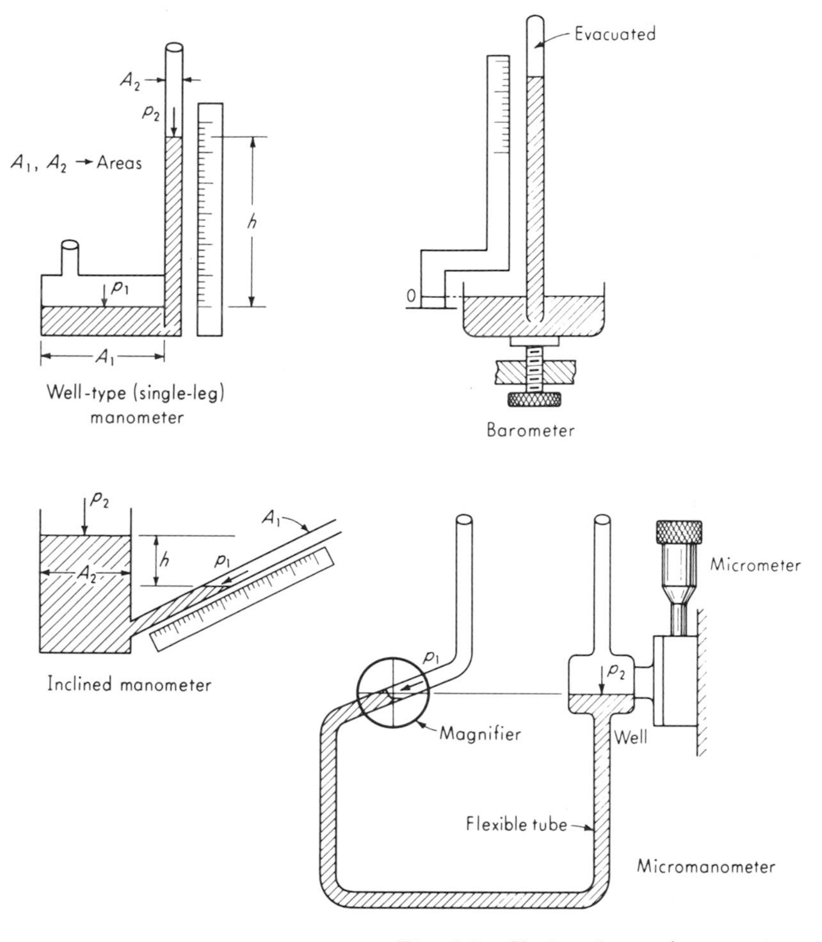

Fig. 6.3. *Various forms of manometers.*

such an arrangement, unlike a U tube, is sensitive to nonuniformity of the tube cross-sectional area and is thus considered somewhat less accurate.

Since manometers inherently measure the pressure *difference* between the two ends of the liquid column, if one end is at zero absolute pressure then h is an indication of absolute pressure. This is the principle of the *barometer* of Fig. 6.3. Although a "single-leg" instrument, high accuracy is achieved by setting the zero level of the well at the zero level of the scale before each reading is taken. The pressure in the evacuated portion of the barometer is not really absolute zero but rather the vapor pressure of the filling fluid, mercury, at ambient temperature. This is about 10^{-4} psia at 70°F and is usually negligible as a correction.

To increase sensitivity, the manometer may be tilted with respect to gravity, thus giving a greater motion of liquid along the tube for a given vertical height change. The *inclined manometer* (draft gage) of Fig. 6.3 exemplifies this principle. Since this is a single-leg device, the calibrated scale is corrected for the slight changes in well level so that rezeroing of the scale for each reading is not required.

The accurate measurement of extremely small pressure differences is accomplished with the *micromanometer*, a variation on the inclined-manometer principle. In Fig. 6.3 the instrument is initially adjusted so that when $p_1 = p_2$ the meniscus in the inclined tube is located at a reference point given by a fixed hairline viewed through a magnifier. The reading of the micrometer used to adjust well height is now noted. Application of the unknown pressure difference causes the meniscus to move off the hairline, but it can be restored to its initial position by raising or lowering the well with the micrometer. The difference in initial and final micrometer readings gives the height change h and thus the pressure. Instruments using water as the working fluid and having a range of either 10 or 20 in. of water can be read to about 0.001 in. of water.[1] In another instrument[2] in which the inclined tube (rather than the well) is moved and which uses butyl alcohol as the working fluid, the range is 2 in. of alcohol, and readability is 0.0002 in. This corresponds to a resolution of 6×10^{-6} psi.

While manometers are generally read visually by a human operator, it is possible to construct servosystems that will "track" the motion of the liquid column and provide a mechanical and/or electrical signal proportional to the pressure. Such arrangements allow the use of manometers for measuring varying pressures, are much faster than visual reading, reduce human errors, provide signals usable in control or computing systems, and may provide automatic temperature corrections. The sensing of the liquid level in the tube may be accomplished with a light source and photocell or by a differential transformer. The differential transformer has the advantage of allowing use of (nonmagnetic) stainless-steel tubes (in place of glass) for high-pressure work.

Such a "servomanometer"[3] is shown schematically in Fig. 6.4. The core of the differential transformer is fastened to a small float which rests on the surface of the manometer liquid. The coils are concentric with the tube and are positioned vertically by a perforated steel tape, $\frac{1}{2}$ in. wide and 0.005 in. thick, running over sprockets. Whenever the core is not at the null position of the coils, a voltage is applied to the amplifier

[1] Meriam Instrument Co., Cleveland, Ohio.

[2] Flow Corp., Arlington, Mass.

[3] Exactel Instrument Co., Mountain View, Calif.

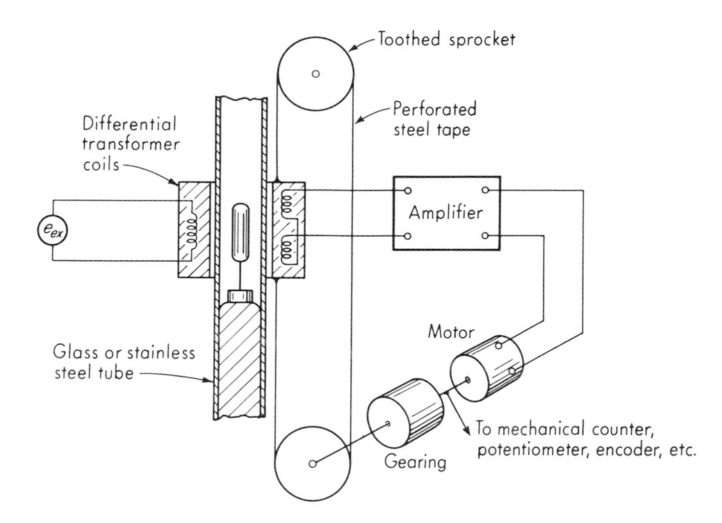

Fig. 6.4. *Servomanometer.*

(and thus to the motor), causing the coils to drive toward null. Thus the coils tend to track the core and thereby the liquid-column position. Since coil motion is kinematically related to motor rotation, this rotation is proportional to pressure and may be read out by a variety of means, depending on the application. Both U-tube and single-leg instruments are available. A U tube requires two servo followers, one for each leg. The subtraction necessary to calculate pressure is performed mechanically by applying the two motor-shaft rotations to a gear differential. Such servomanometers have a resolution of 0.0005 in. and an overall accuracy of a few thousandths of an inch. Following speed is of the order of 100 in./min or more, and automatic temperature correction is provided.

Manometer dynamics. While manometers are used mainly for static measurements, their dynamic response is sometimes of interest. The general problem of the oscillations of liquid columns is an interesting (and rather difficult) question in fluid mechanics and has received considerable attention in the literature.[1,2] Here we take a considerably simplified view of the problem which, however, gives results of practical interest. In the U-tube configuration of Fig. 6.5*a* the unknown pressures p_1 and p_2 are exerted by a gas whose inertia and viscosity may be considered

[1] J. F. Ury, Viscous Damping in Oscillating Liquid Columns, Its Magnitude and Limits, *Intern. J. Mech. Sci.*, vol. 4, p. 349, 1962.

[2] P. D. Richardson, Comments on Viscous Damping in Oscillating Liquid Columns, *Intern. J. Mech. Sci.*, vol. 5, p. 415, 1963.

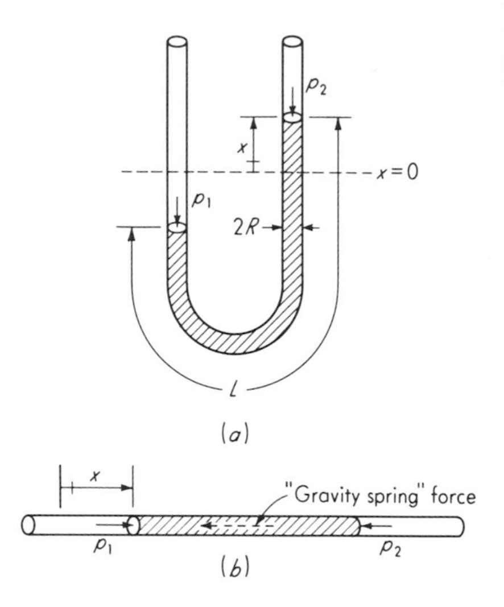

Fig. 6.5. Manometer model.

negligible compared with the manometer liquid. If the pressures vary with time, the reading of the manometer varies with time; we are interested in the fidelity with which the manometer reading follows the pressure variation. The motion of the manometer liquid in the tube is caused by the action of various forces. If we consider the manometer liquid in its entirety as a free body and search for forces acting on it, the following forces come to mind:

1. The gravity force (weight) distributed uniformly over the whole body of fluid
2. A drag force due to motion of the fluid within the tube and related to the wall shearing stress
3. The forces on the two ends of the free body due to the pressures p_1 and p_2
4. Distributed normal pressure of the tube on the fluid
5. Surface-tension effects at the two ends of the body of fluid

A detailed analysis of all these effects would lead to rather complex and unwieldy mathematics. Fortunately, useful results may be obtained by a simplified analysis. The initial step is the assumption that the system shown in Fig. 6.5*b* is dynamically equivalent to that of Fig. 6.5*a*. The "gravity spring" force of Fig. 6.5*b* is explained as follows: In Fig. 6.5*a*, whenever $x \neq 0$, there is an unbalanced gravity force acting on the liquid column, tending to restore the level to $x = 0$. The magnitude of this force is $-2\pi R^2 x\gamma$, where $\gamma \triangleq$ manometer-fluid specific weight, in $\text{lb}_f/\text{in.}^3$. We see that this force is proportional to the displacement x and always opposes it; thus it has all the characteristics of a spring force. When the liquid column is "straightened out" in Fig. 6.5*b*, we must include this "gravity spring" force in our equivalent system if we are to preserve the analogy of the two configurations. In comparing Fig. 6.5*b* with Fig. 6.5*a*, we note that any effects of flow curvature in the 180° bend are lost, but these will probably be small if the diameter of the bend is large compared with the inside diameter of the manometer tube and if the total length L is large compared with the bend length. We shall also neglect any surface-tension effects at the ends of the column. This is usually a good assumption if the column is long relative to its diameter.

In addition to the gravity-spring force and the pressure forces due to p_1 and p_2, the liquid column is also subjected to a drag force at the interface between the liquid and the wall of the tube. This drag force is equal to the wall shearing stress times the surface area of the liquid column. The motion of the liquid in the tube may be thought of as an unsteady pipe flow. We shall assume that at any instant of time the wall shearing stress can be computed from the instantaneous velocity of the liquid, *using the formulas commonly used for steady pipe flows.*

The flow of liquid in the tube may occur in the laminar, transition, or turbulent regimes. Let us first assume laminar flow prevails. The pressure drop Δp due to pipe friction for both laminar and turbulent flow is given by

$$\Delta p = f\frac{\gamma L V_{av}^2}{2gd} = f\left(\frac{L}{d}\right)\frac{\rho V_{av}^2}{2g_0} \tag{6.2}$$

where
$g_0 \triangleq$ mass unit conversion factor, lb_m/slug
$\rho \triangleq$ fluid density, lb_m/ft^3
$f \triangleq$ friction factor
$\gamma \triangleq$ fluid specific weight, local, lb_f/ft^3
$V_{av} \triangleq$ average velocity
$g \triangleq$ local gravity acceleration, ft/sec^2
$d \triangleq$ diameter of pipe
$L \triangleq$ pipe length

The wall shearing stress τ_0 is given by

$$\tau_0 = \Delta p\,\frac{d}{4l} \tag{6.3}$$

Thus

$$\tau_0 = f\frac{\gamma V_{av}^2}{8g} = f\frac{\rho V_{av}^2}{8g_0} \tag{6.4}$$

For laminar flow the friction factor is given by

$$f = \frac{64\mu g}{d\gamma V_{av}} = \frac{64}{dV_{av}\rho/\mu g_0} \tag{6.5}$$

$$\text{so that} \quad \tau_0 = \frac{4\mu V_{av}}{R} \tag{6.6}$$

where $R \triangleq d/2$.

This result can also be obtained directly from the laminar velocity distribution

$$V = V_c\left[1 - \left(\frac{r}{R}^2\right)\right] \tag{6.7}$$

where $V \triangleq$ velocity at radius r
$V_c \triangleq$ center-line velocity

The velocity gradient is

$$\frac{dV}{dr} = -\frac{V_c}{R^2}(2r) \tag{6.8}$$

which becomes at the wall

$$\left.\frac{dV}{dr}\right|_{r=R} = -\frac{2V_c}{R} \tag{6.9}$$

Shearing stress is given by

$$\tau = \mu\frac{dV}{dr} \tag{6.10}$$

and so the magnitude of the wall shearing stress τ_0 is

$$\tau_0 = \frac{4\mu V_{av}}{R}$$

since $V_c = 2V_{av}$ for laminar flow in circular pipes.

We are now in a position to apply Newton's law to the system of Fig. 6.5*b*. The average flow velocity V_{av} corresponds to $\dot{x}$, the first derivative of x with respect to time. Considering the entire body of liquid as a free body and taking the effective mass of the moving liquid as four-thirds of its actual mass, based on the kinetic energy of steady laminar flow, we can write for motion in the x direction

$$\pi R^2(p_1 - p_2) - 2\pi R^2\gamma x - 2\pi RL\frac{4\mu\dot{x}}{R} = \frac{4}{3}\frac{\pi R^2 L\gamma}{g}\ddot{x} \tag{6.11}$$

This reduces to

$$\frac{2\ddot{x}}{3g/L} + \frac{4\mu L}{R^2\gamma}\dot{x} + x = \frac{1}{2\gamma}p \tag{6.12}$$

where we have defined $p \triangleq p_1 - p_2$. In operator form, this becomes

$$\left(\frac{2D^2}{3g/L} + \frac{4\mu L}{R^2\gamma}D + 1\right)x = \frac{1}{2\gamma}p \tag{6.13}$$

The operational transfer function relating output x to input p is

$$\frac{x}{p}(D) = \frac{1/2\gamma}{\dfrac{2D^2}{3g/L} + \dfrac{4\mu L}{R^2\gamma}D + 1} \tag{6.14}$$

which is of the form

$$\frac{x}{p}(D) = \frac{K}{D^2/\omega_n^2 + 2\zeta D/\omega_n + 1} \tag{6.15}$$

where

$$K \triangleq \frac{1}{2\gamma} \qquad \text{in./psi} \qquad (6.16)$$

$$\omega_n \triangleq \sqrt{\frac{3g}{2L}} \qquad \text{rad/sec} \qquad (6.17)$$

$$\zeta \triangleq 2.45\mu \frac{\sqrt{gL}}{R^2\gamma} \qquad (6.18)$$

We note from the above that the manometer is a second-order instrument. The numerical values of the parameters are usually such that $\zeta < 1.0$; that is, the instrument is underdamped.

Since laminar flow was assumed in carrying out the above analysis, we should try to estimate a typical Reynolds number to see under what conditions laminar flow occurs. As a numerical example, we take a mercury manometer with

$$L = 26.5 \text{ in.}$$
$$R = 0.13 \text{ in.}$$
$$\mu = 2.18 \times 10^{-7} \text{ lb}_f\text{-sec/in.}^2$$
$$\gamma = 0.491 \text{ lb}_f\text{/in.}^3$$

Suppose that we wish to check our theoretical results by measuring ζ and ω_n experimentally for a step-function input. Computing ζ and ω_n from Eqs. (6.17) and (6.18) gives $\zeta = 0.007$ and $\omega_n = 4.7$ rad/sec. Since $\zeta = 0.007$ represents a *very lightly* damped system, we can estimate the maximum flow velocity by assuming no damping at all. A second-order system with no damping executes pure sinusoidal oscillations when subjected to a step-function input. Its motion would thus be given by

$$x = X \sin \omega_n t \qquad (6.19)$$

where X is the size of the step function. The velocity $\dot{x}$, which is the same as the average flow velocity, would be

$$\dot{x} = \omega_n X \cos \omega_n t \qquad (6.20)$$

and its maximum value would thus be $\omega_n X$. The Reynolds number for steady pipe flow is given by

$$N_R = \frac{2\gamma R V_{av}}{g\mu} \qquad (6.21)$$

and the critical value for transition from laminar to turbulent flow is 2,100. Since $V_{av} = \omega_n X$, it should be clear that there is a maximum-size step function that can be used without exceeding $N_R = 2{,}100$. This limiting value X_m is given by

$$2{,}100 = \frac{2\gamma R \omega_n X_m}{g\mu} \qquad (6.22)$$

which in this example gives

$$X_m = \frac{(2{,}100)(386)(2.18 \times 10^{-7})}{(2)(0.491)(0.13)(4.7)} = 0.30 \text{ in.} \tag{6.23}$$

Thus, to ensure laminar flow at all times during the oscillation the step input can be no larger than 0.30 in. Suppose we wish to measure ζ and ω_n by simple visual methods—ζ from the size of the first overshoot and ω_n by counting and timing cycles. This requires much larger step inputs for reasonable accuracy. Therefore we must investigate the effect of the presence of turbulent flow on our analysis.

Suppose that we decide that a step input X_m of 5 in. Hg will be sufficiently large to allow accurate measurements of ζ and ω_n. The maximum Reynolds number would then be

$$N_R = \frac{5}{0.30}(2{,}100) = 35{,}000 \tag{6.24}$$

For steady turbulent flow in smooth pipes with $3{,}000 < N_R < 100{,}000$ the Blasius equation for friction factor is

$$f = \frac{0.316}{(N_R)^{0.25}} \tag{6.25}$$

The turbulent wall shearing stress is then given by Eq. (6.4) as

$$\tau_0 = \frac{0.0378\gamma^{0.75}\mu^{0.25}V_{\text{av}}{}^{1.75}}{g^{0.75}R^{0.25}} \tag{6.26}$$

Using the numerical values of this particular example, we get

$$\tau_0 = 9.18 \times 10^{-6} V_{\text{av}}{}^{1.75} \qquad \text{lb}_f/\text{in.}^2 \tag{6.27}$$

For laminar flow the comparable expression is

$$\tau_0 = 6.71 \times 10^{-6} V_{\text{av}} \qquad \text{lb}_f/\text{in.}^2 \tag{6.28}$$

The most significant difference between Eqs. (6.27) and (6.28) is that (6.27) represents a nonlinear relation between shear stress and velocity whereas (6.28) is linear. This means that when the force due to wall shearing stress is substituted into Newton's law the result is a nonlinear differential equation because of the term $(\dot{x})^{1.75}$. This nonlinear equation cannot be solved except by use of analog computers or approximate numerical methods. Thus, the presence of turbulent flow leads to mathematical difficulties.

In working with oscillations of systems with nonlinear damping terms similar to the $(\dot{x})^{1.75}$ of this problem, engineers have developed an approximate method of analysis which is quite useful. This approach is based on the observation that, while the linear damping term $\dot{x}$ and the nonlinear term, such as $(\dot{x})^{1.75}$, are quite different mathematically, the general *form* of the oscillation in the two systems is not radically different in experimental tests. If the linear system is excited by a sinusoidal exciting force it will respond with a sinusoidal motion, whereas the nonlinear systems's motion will not be purely sinusoidal. However, observation shows that the deviation from pure sinusoidal motion is usually quite small. Using these facts as a basis, we might then reason as follows: If a system with nonlinear damping is executing steady oscillations of fixed amplitude, during each cycle the damping force will dissipate a certain amount of energy. If we know from experience that the wave form of the nonlinear oscillation is nearly sinusoidal, we can compute approximately the energy dissipation per cycle. This is done as follows: Suppose there exists a steady oscillation of amplitude Y and frequency ω. If we assume the wave form to be sinusoidal, we can write

$$y = Y \sin \omega t \tag{6.29}$$

and

$$\dot{y} = Y\omega \cos \omega t \tag{6.30}$$

Now, in general, the instantaneous power is the product of instantaneous force and instantaneous velocity. The power dissipation due to damping is thus the product of velocity and damping force. If the damping force is a known function of velocity $f(\dot{y})$, we can write

$$\text{Instantaneous power dissipation} = \dot{y}f(\dot{y}) \tag{6.31}$$

and the energy dissipated per cycle will be given by

$$\int_{\text{one cycle}} \dot{y}f(\dot{y})\, dt \tag{6.32}$$

For a linear damping the function $f(\dot{y})$ is just $B\dot{y}$, where B is a constant. The energy dissipation per cycle is thus

$$\int_0^{2\pi/\omega} (Y\omega \cos \omega t)(BY\omega \cos \omega t)\, dt = \pi B\omega Y^2 \tag{6.33}$$

For the nonlinear damping due to turbulent flow the function $f(\dot{y})$ is $C(\dot{y})^{1.75}$, where C is a constant. The energy dissipation per cycle for this nonlinear damping is

$$\int_0^{2\pi/\omega} (Y\omega \cos \omega t)C(Y\omega \cos \omega t)^{1.75}\, dt \tag{6.34}$$

This is equal to

$$C(Y\omega)^{2.75} \int_0^{2\pi/\omega} (\cos \omega t)^{2.75}\, dt \tag{6.35}$$

In evaluating the integral in (6.35), we must use physical reasoning, because when $\cos \omega t$ becomes *negative* the quantity $(\cos \omega t)^{2.75}$ is not defined in terms of real numbers. Physical reasoning, however, tells us that the physical processes occurring during the first quarter cycle $(0 \leq t \leq \pi/2\omega)$ give exactly the same energy dissipation as those occurring during the other three quarters of the cycle. Thus, we can integrate over only the first quarter and multiply by 4 to get the total energy dissipation. During the first quarter cycle $\cos \omega t$ is always positive, and so no mathematical difficulties arise. This amounts to saying that, to agree with the known physical facts, integral (6.34) should really be written as

$$\int_0^{2\pi/\omega} |Y\omega \cos \omega t| \, [C(Y\omega)^{1.75}|(\cos \omega t)|^{1.75}] \, dt \qquad (6.36)$$

with the absolute-value signs as shown.

Evaluation of integral (6.36) for one quarter cycle is most easily done by plotting and using a planimeter to get the area under the curve. By defining $\theta \triangleq \omega t$, integral (6.36) can be written as

$$CY^{2.75}\omega^{1.75} \int_0^{2\pi} (\cos \theta)^{2.75} \, d\theta \qquad (6.37)$$

Plotting and planimetering give the energy dissipation per cycle for nonlinear damping as

$$2.50C\omega^{1.75}Y^{2.75} \qquad (6.38)$$

Having obtained the above results, we now define the *equivalent linear damping* as that linear damping which would dissipate exactly the same energy per cycle as the nonlinear damping at a given frequency and amplitude. Thus we set (6.33) equal to (6.38) and get

$$\pi B_e \omega Y^2 = 2.50C\omega^{1.75}Y^{2.75} \qquad (6.39)$$

$$B_e \triangleq \text{equivalent linear damping}$$

$$B_e = 0.796C(\omega Y)^{0.75} \qquad (6.40)$$

Since C is the constant that multiplies $(\dot{y})^{1.75}$, in the manometer problem we have

$$\text{Damping force} = 2\pi RL\tau_0 = \frac{0.237R^{0.75}\gamma^{0.75}\mu^{0.25}L\dot{x}^{1.75}}{g^{0.75}} \qquad (6.41)$$

and thus

$$C = \frac{0.237\gamma^{0.75}R^{0.75}L\mu^{0.25}}{g^{0.75}} \qquad (6.42)$$

Now Eq. (6.18) can be written as

$$\zeta = \frac{2.45\mu L\sqrt{g}}{R^2\gamma\sqrt{L}} = \frac{0.0974B\sqrt{g}}{R^2\gamma\sqrt{L}} \qquad (6.43)$$

since $B = 8\pi\mu L$ for the linear system. We can now define the equivalent linear damping ratio ζ_e by substituting B_e from Eqs. (6.40) and (6.42) in (6.43):

$$\zeta_e \triangleq \frac{0.0184\sqrt{L}(\mu/\gamma g)^{0.25}}{R^{1.25}}(\omega Y)^{0.75} \qquad (6.44)$$

This shows clearly the dependence of ζ_e on the frequency and amplitude of the oscillation. For turbulent flow the value of ω_n is also somewhat different since the velocity profile tends to be more nearly square rather than parabolic. If the velocity is assumed uniform over the cross section, the effective mass becomes equal to the actual mass since all particles have the same velocity. If the turbulent damping, though larger than the laminar, is assumed to be still quite small, it is reasonable to expect that it will have little effect on the frequency. We shall therefore compute ω_n for turbulent flow by neglecting the nonlinear damping completely where it would appear in Eq. (6.11). We then get

$$\omega_n = \sqrt{\frac{2g}{L}} \qquad (6.45)$$

for turbulent flow.

Many assumptions were made in the above analysis. The formulas for steady pipe flow were used for an unsteady situation. In an oscillating flow, velocity actually goes to zero twice each cycle, no matter how great the amplitude or frequency; thus one wonders whether part of such a cycle is turbulent and part laminar. In the analysis above, turbulent equations were used for the whole cycle. The nonlinear differential equation containing $(\dot{x})^{1.75}$ actually has no closed-form analytical solution. Thus what is the meaning of a ζ_e and ω_n attached to such a process? Such questions and others may be at least partially resolved by more complex analyses or experimental studies. To provide some idea of the degree of validity of our simplified analysis some experimental results will be given. They were obtained at The Ohio State University by undergraduate students who study manometer dynamics in a simple experiment in a measurement course.

The experiment consists in part of suddenly releasing an air pressure applied to a mercury manometer and observing the resulting oscillations. The process is slow enough that ζ_e can be estimated from the size of the first overshoot and ω_n by counting and timing cycles with a stopwatch. The manometers used have the numerical values quoted earlier in this section. A step pressure input of 10 in. Hg ($x = 5$ in.) is used; thus turbulent flow may be expected. If laminar flow were assumed, the theoretical values would be $\zeta = 0.007$ and $\omega_n = 4.7$ rad/sec. For turbulent flow, ω_n becomes 5.4 rad/sec. To calculate ζ_e from Eq. (6.44) the frequency and amplitude of the oscillation must be known. For a step input the frequency is the damped natural frequency rather than ω_n; however, these are practically identical for the small damping present and so we use ω_n. We experimentally measure ζ_e from the first overshoot, and so the proper amplitude to use in the theoretical calculation might be an average of the initial amplitude and that at the first overshoot. Only the initial amplitude (5 in.) is known, however, and so this is used. Equation (6.44) then gives $\zeta_e = 0.082$. By timing and counting cycles (about eight cycles of the decaying oscillation can be easily measured) the experimental value of ω_n is 5.2 rad/sec. This lies between the values calculated for turbulent and laminar flow and thus is not unreasonable since several of the eight cycles used were of quite low amplitude due to decay of the oscillation. The first overshoot is about 4.05 in., giving an experimental value of ζ of 0.067. Therefore the theoretical estimate of 0.082 is fairly good. If we now use the experimental value of ω (5.2) and the average amplitude of the first half cycle (4.53) in Eq. (6.44), the predicted ζ_e is 0.074, which compares even more favorably. Based on even these limited results, a certain amount of confidence in the theoretical predictions is established.

6.4

Elastic Transducers While a wide variety of flexible metallic elements might conceivably be used for pressure transducers, the vast majority of practical devices utilize one or another form of Bourdon tube,[1,2] diaphragm,[1,3,4] or bellows[1,5] as their sensitive element, as shown in Fig. 6.6. The gross deflection of these elements may directly actuate a pointer/scale readout through suitable linkages or gears, or the motion may be transduced to an electrical signal by one means or another. Strain gages bonded to diaphragms are also widely used to measure local strains that are directly related to pressure.

The Bourdon tube is the basis of many mechanical pressure gages and is also widely used in electrical transducers by measuring the output displacement with potentiometers, differential transformers, etc. The basic element in all the various forms is a tube of noncircular cross section. A pressure difference between the inside and outside of the tube (higher pressure inside) causes the tube to attempt to attain a circular cross section. This results in distortions which lead to a curvilinear translation of the free end in the C type and spiral and helical types and an angular rotation in the twisted type, which motions are the output. The theo-

[1] D. M. Considine (ed.), "Process Instruments and Controls Handbook," sec. 3, McGraw-Hill Book Company, New York, 1957.

[2] R. W. Bradspies, Bourdon Tubes, *Giannini Tech. Notes,* Giannini Corp., Duarte, Calif., January–February, 1961.

[3] Pressure Capsule Design, *Giannini Tech. Notes,* November, 1960.

[4] C. K. Stedman, The Characteristics of Flat Annular Diaphragms, *Statham Instr. Notes,* Statham Instruments Inc., Los Angeles, Calif.

[5] R. Carey, Welded Diaphragm Metal Bellows, *Electromech. Design,* p. 22, August, 1963.

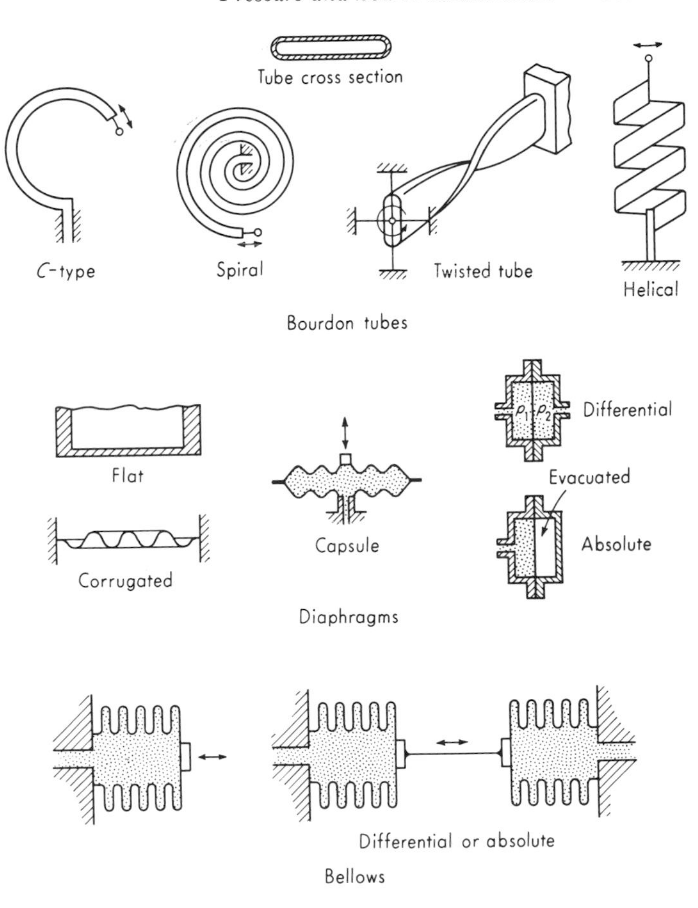

Fig. 6.6. *Elastic pressure transducers.*

retical analysis of these effects is difficult, and practical design at present still makes use of considerable empirical data. The C-type Bourdon tube has been used up to about 100,000 psi. The spiral and helical configurations are attempts to obtain more output motion for a given pressure and have been used mainly below about 1,000 psi. The twisted tube shown has a crossed-wire stabilizing device which is stiff in all radial directions but soft in rotation. This reduces spurious output motions due to shock and vibration.

Flat diaphragms are widely used in electrical transducers either by sensing the center deflection with some displacement transducer or by bonding strain gages to the diaphragm surface. The full-scale deflection at the center must be less than about one-third the diaphragm thickness if nonlinearity of less than 5 percent is desired. The pressure-deflection formula for a flat diaphragm with edges clamped is

$$p = \frac{16Et^4}{3R^4(1-\mu^2)}\left[\frac{y_c}{t} + 0.488\left(\frac{y_c}{t}\right)^3\right] \qquad (6.46)$$

where $p \triangleq$ pressure difference across diaphragm
$E \triangleq$ modulus of elasticity
$t \triangleq$ diaphragm thickness
$\mu \triangleq$ Poisson's ratio
$R \triangleq$ diaphragm radius
$y_c \triangleq$ center deflection

For small deflections, $(y_c/t)^3$ is negligible compared with y_c/t, and linear behavior may be expected since bending stresses predominate. At larger deflections, a stretching action is added to the bending, stiffening the diaphragm and contributing the $(y_c/t)^3$ term. If local strain rather than center deflection is measured, a similar nonlinear effect is noted. For strictly mechanical elements, larger linear deflections than those allowed above are often needed; they may be obtained by using corrugated diaphragms or capsules, as in Fig. 6.6. Metallic bellows serve a similar function. Bellows and diaphragms are most often used for relatively low pressures (less than a few hundred pounds per square inch) except for the diaphragm-type electrical transducers, which are available up to several thousand pounds per square inch.

Typical characteristics of electrical pressure pickups. Some numerical characteristics of the most common types of electrical pressure transducers will now be briefly reviewed. Since they are all basically spring-mass systems with intentional or unintentional damping, their dynamic behavior is of standard second-order form, just as for force transducers. An important point to note, however, is that under actual operating conditions the values of ω_n and ζ are greatly affected by the configuration of associated piping connections and the characteristics of the fluid medium. Thus numerical values of ω_n and ζ supplied by instrument manufacturers usually refer to the instrument's behavior in ambient air and may be greatly different under actual operating conditions. In some cases, instrument volume and tubing flow resistance are large enough so that the pressure felt by the elastic element follows with a first-order lag the pressure to be measured. This lag may be so large that the second-order spring-mass dynamics is completely overshadowed and the overall instrument response becomes first-order. These and similar problems will be treated in Sec. 6.6. Most instruments using various combinations of elastic elements and electrical displacement transducers are available in versions to measure gage, differential, or absolute pressure.

Pressure pickups using resistive potentiometers for motion measurement are not generally intended for measurement of extremely fast pressure changes, and their natural frequencies are not often quoted, although rise time for a step input may be. The rather large motion required by potentiometers leads to a relatively large internal volume and volume change. A family[1] of differential-pressure transducers employing a capsule diaphragm of NI-SPAN C (an alloy widely used for its constancy of elastic modulus with temperature) has full-scale ranges of 2 to 100 psi, nonlinearity ± 0.6 percent, hysteresis and friction ± 2 percent, and a temperature error ± 1 percent over the range -65 to $+200°$F. The standard potentiometer has 5,000 ohms resistance and a power rating of about 0.8 watt at room temperature. Acceleration sensitivity is of the order of 0.08 percent of full scale/g in the worst direction. The differential-pressure action is obtained by placing the capsule inside a pressure-tight case. One pressure is applied inside the capsule and the other to the space between the capsule and case. The fluid inside the capsule may be any one compatible with NI-SPAN-C; however, the space between the capsule and case contains the potentiometer and electrical leads so that only clean, nonconductive, noncorrosive fluids may be used. The 90 percent rise time for the step input of air pressure in the capsule is of the order of 40 to 70 msec. The internal case volume is 85 cm^3, the capsule volume 6 cm^3, and the full-scale volume change 4 cm^3.

A family[1] of absolute-pressure transducers using a helical Bourdon tube has full-scale ranges of 500 to 10,000 psia, total error due to nonlinearity, friction, hysteresis, resolution, and repeatability of ± 1.2 percent (± 2.2 percent for the temperature range -100 to $+200°$F), and 63 percent response time of 6 msec. Acceleration sensitivity is 0.05 percent/g; the size is a 1-in.-diameter cylinder 2.7 in. long, and weight is 4 oz.

Potentiometer-type pickups generally exhibit a finite operating life. The two pickups mentioned above have a minimum life of about 50,000 and 25,000 full-scale cycles, respectively. Other potentiometer pickups have lives to a million cycles or more.

Pressure pickups using unbonded-strain-gage transducers are available for a wide range of applications. They generally employ the center deflection of a diaphragm as the mechanical input. Both flush-diaphragm and chamber types are made. Full-scale ranges from 0.01 to several thousand pounds per square inch can be obtained. A typical family[2] of flush-diaphragm absolute-pressure transducers has ranges from 5 to 1,000 psia, full-scale output of 56 mv with 7-volts excitation, nonlinearity and hysteresis less than 0.75 percent of full scale, thermal sen-

[1] Bourns Inc., Riverside, Calif.

[2] Statham Instruments Inc., Los Angeles, Calif.

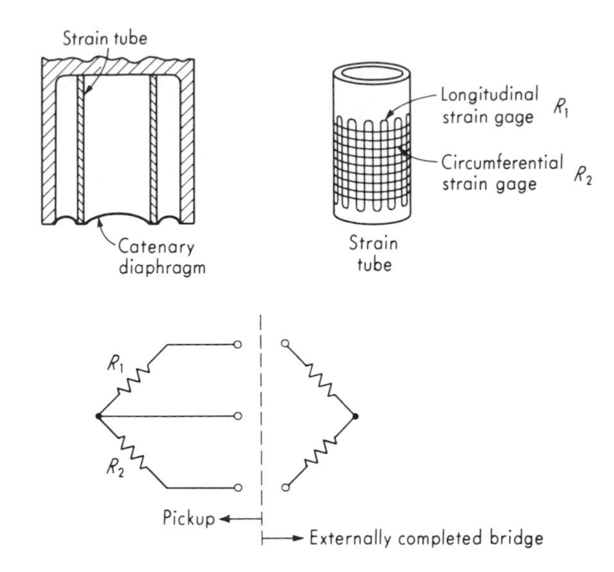

Fig. 6.7. *Strain-tube pressure pickup.*

sitivity shift of 0.01 percent/F° and thermal zero shift of 0.01 percent of full scale/F° for the range −65 to +250°F, a natural frequency of 3,500 to 25,500 cps, and acceleration sensitivity 0.25 to 0.01 percent of full scale/g.

Bonded-strain-gage pressure pickups employ a number of pressure/strain transducing schemes. The most direct method bonds the gages directly to the diaphragm. Another applies the force from the diaphragm to a proving ring. A third[1] employs a catenary diaphragm to apply compression loads to a hollow strain tube on which the strain gages are mounted (see Fig. 6.7). A typical pickup of the last-mentioned type has full scale of 1,000 psi, nonlinearity ±1 percent of full scale, natural frequency of 45,000 cps, acceleration sensitivity 0.01 percent/g, thermal zero shift 0.02 percent of full scale/F°, and full-scale output of 50 mv. A subminiature evacuated-capsule absolute-pressure pickup[2] having one active flat diaphragm with a four-arm strain-gage bridge has an overall size of 0.25-in. diameter and 0.02-in. thickness. The pressure-sensitive area is 0.028 in.^{2}, ranges are 2 to 100 psia, natural frequency 20,000 cps, nonlinearity and hysteresis ±1 percent, thermal zero shift 0.1 percent of full scale/F°, thermal sensitivity shift 0.05 percent/F°, and full-scale output of 0.4 to 4 mv. The above-mentioned pickups all use metallic strain gages of the wire or foil type. Pickups using semiconductor gages

[1] Norwood Controls, Norwood, Mass.
[2] Scientific Advances, Inc., Columbus, Ohio.

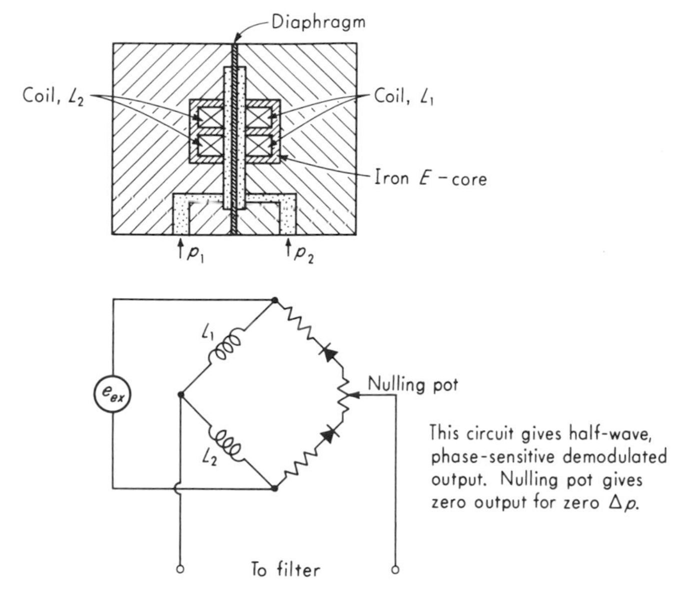

Fig. 6.8. *Variable-inductance pickup.*

are also available; they generally have considerably greater output, of the order of 0.25 volt full scale.

Variable-inductance pressure pickups are available in several forms. Often a magnetic stainless-steel diaphragm serves as the moving "iron" between two E coils arranged in a half-bridge circuit. One such pickup[1] has the interesting feature of interchangeable diaphragms, giving full-scale ranges of ± 1, 5, 25, 100, and 500 psi in a single transducer. This pickup can be used for gage or differential pressure since both sides of the diaphragm may be exposed to corrosive liquids or gases. The nonlinearity is 0.5 percent, full-scale output is 1.5 volts at 3,000 cps, thermal zero shift 0.01 percent of full scale/F°, and thermal sensitivity shift 0.02 percent/F°, both from -65 to $+250$°F. The pressure cavity has a volume of 0.004 in.^{3} while the full-scale volume change is 0.0003 in.^{3}. Natural frequency for the softest diaphragm is 5,000 cps, with acceleration sensitivity 0.2 percent/g, while the stiffest diaphragm has 40,000 cps, with 0.003 percent/g. Figure 6.8 shows the construction of such a transducer.

Piezoelectric pressure pickups have the same form of dynamic response as piezoelectric accelerometers. Except for certain quartz pickups used with electrometer amplifiers, they do not give an output for

[1] Pace Engineering Co., North Hollywood, Calif.

a static pressure. They generally have very high natural frequencies and little damping. A flush diaphragm is generally used to apply the pressure force to the piezoelectric element. A quartz pickup/amplifier combination[1] specifically designed for measurements in shock tubes has switch-selectable full-scale ranges of 10, 100, 1,000, and 5,000 psi, responds to steady pressures, and has full-scale output of 0.5 to 2.4 volts, natural frequency of 150,000 cps, nonlinearity of 1 percent, and acceleration sensitivity of 0.02 psi/g.

A capacitance-type pickup[2] using the capacitor as part of a tuned 25-Mc oscillating circuit has, together with its associated electronics, the following characteristics: full-scale ranges from 5 to 50,000 psi (absolute, gage, or differential units are available), natural frequencies from 33,000 cps for the 5-psi model to over 350,000 cps for the 50,000-psi model, operating temperature of -65 to $+250°F$, nonlinearity and hysteresis 0.75 percent of full scale, repeatability 0.15 percent, and full-scale output of 5 volts.

Analysis of diaphragm-type bonded-strain-gage transducer. To illustrate some general concepts involved in pressure-transducer design, we here consider the flat-diaphragm type since it lends itself to theoretical calculation. Such a diaphragm, clamped at the edges and subjected to a uniform pressure difference p, has at any point on the low-pressure surface a radial stress s_r and a tangential stress s_t given by the following formulas (see Fig. 6.9):

$$s_r = \frac{3pR^2\mu}{8t^2}\left[\left(\frac{1}{\mu}+1\right)-\left(\frac{3}{\mu}+1\right)\left(\frac{r}{R}\right)^2\right] \tag{6.47}$$

$$s_t = \frac{3pR^2\mu}{8t^2}\left[\left(\frac{1}{\mu}+1\right)-\left(\frac{1}{\mu}+3\right)\left(\frac{r}{R}\right)^2\right] \tag{6.48}$$

where $\mu \triangleq$ Poisson's ratio

The deflection at any point is given by

$$y = \frac{3p(1-\mu^2)(R^2-r^2)^2}{16Et^3} \tag{6.49}$$

Equations (6.47) to (6.49) all give linear relations between stress or deflection and pressure and are accurate only for sufficiently small pressures. Equation (6.46) may be used to estimate the degree of nonlinearity. The stress situation on the diaphragm surface is fortunate because both tension and compression stresses exist simultaneously.

[1] Kistler Instrument Corp., North Tonawanda, N.Y.
[2] Omega Dynamics Corp., Pasadena, Calif.

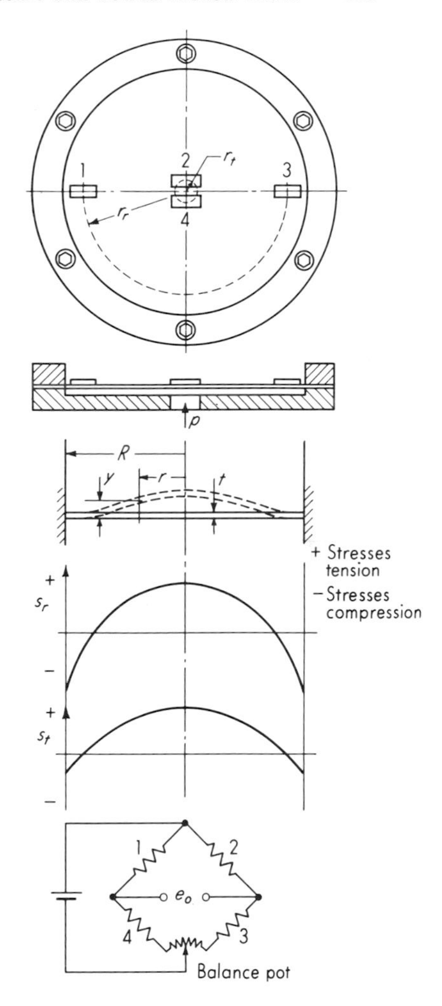

Fig. 6.9. *Diaphragm-type strain-gage pressure pickup.*

This allows use of a four-active-arm bridge in which all effects are additive (giving large output) and also gives temperature compensation. Gages 2 and 4 are placed as close to the center as possible and oriented to read tangential strain since it is maximum (positive) at this point. Gages 1 and 3 are oriented to read radial strain and placed as close to the edge as possible since radial strain has its maximum negative value at that point. The laws of bridge circuits show that the pressure effects on all four gages are additive. In computing the overall sensitivity, Eqs. (6.47) and (6.48) cannot be used directly to determine the strains "seen" by the gages since the diaphragm surface is in a state of *biaxial* stress and *both* the radial and tangential stress contribute to the radial or tangential strain at any point. The general biaxial stress-strain relation gives

$$\epsilon_r = \frac{s_r - \mu s_t}{E} \tag{6.50}$$

$$\epsilon_t = \frac{s_t - \mu s_r}{E} \tag{6.51}$$

Once the gage strains are calculated, the individual gage ΔR's are obtained from the gage factors, and e_o can then be determined from the bridge-circuit sensitivity equations.

If the transducer is to be used for dynamic measurements, its natural frequency is of interest. A diaphragm has an infinite number of natural frequencies; however, the lowest is the only one of interest here. For a clamped-edge diaphragm vibrating in a vacuum (no fluid-inertia effects) the lowest natural frequency is given by

$$\omega_n = \frac{10.21}{R^2}\sqrt{\frac{gEt^2}{12\gamma(1-\mu^2)}} \quad \text{rad/sec} \tag{6.52}$$

where $\gamma \triangleq$ local specific weight of diaphragm material, $\text{lb}_f/\text{in.}^3$
$g \triangleq$ local gravity, in./sec^2

A number of factors may make the actual operating value of ω_n different from the prediction of Eq. (6.52). The edge clamping is never perfectly rigid; any softness tends to lower ω_n. If the diaphragm is not perfectly flat, tightening the clamping bolts may cause a slight (perhaps imperceptible) "wrinkling," tending to stiffen the diaphragm and raise ω_n. If the diaphragm is used to measure liquid pressures, the inertia of the liquid tends to lower ω_n, especially if a small-diameter tube connects the pressure source to the diaphragm. When it is used with gases, the volume of gas "trapped" behind the diaphragm may act as a stiffening spring, raising ω_n.

A pressure transducer of the above type constructed for use in a transducer research project at The Ohio State University serves as a numerical example for the above discussion. This transducer used a phosphor-bronze diaphragm with

$$E = 16 \times 10^6 \text{ lb}_f/\text{in.}^2 \qquad R = 1.830 \pm 0.002 \text{ in.}$$
$$\mu = \tfrac{1}{3} \qquad r_t = 0.15 \pm 0.01 \text{ in.}$$
$$\gamma = 0.32 \text{ lb}_f/\text{in.}^3 \qquad r_r = 1.52 \pm 0.01 \text{ in.}$$
$$t = 0.0454 \pm 0.0003 \text{ in.}$$

The strain gages were SR-4 type A-7 wire gages with a gage factor of 1.97 ± 2 percent and a resistance of 119.5 ± 0.3 ohms. The bridge was excited with 7.5 volts direct current, and the transducer was designed for a full-scale range of 10 psi. The theoretically calculated sensitivity was

0.516 mv/psi. Static calibration gave 0.513 mv/psi and indicated a maximum nonlinearity of 2 percent of full scale. The theoretically calculated natural frequency was 924 cps, and the experimental value (with atmospheric air on both sides of the diaphragm) was 897.

6.5

Force-Balance Transducers Feedback or null-balance principles may be applied to pressure measurement in a manner similar to that employed for force measurement. Figure 6.10*a* shows a pneumatic/mechanical type, and electromechanical methods are used in Fig. 6.10*b*. High loop gain in these servosystems gives good linearity and accuracy. The block diagrams give the static relations only; a dynamic analysis is necessary to determine the limit set on gain by stability requirements. The operation of these instruments is left for the reader to deduce from the schematic and block diagrams. Previous discussions of feedback-type instruments may be helpful.

6.6

Dynamic Effects of Volumes and Connecting Tubing We have mentioned the possible strong effect of fluid properties and "plumbing" configurations on the dynamic behavior of pressure-measuring systems. In this section some of these problems will be investigated. It should first be pointed out that, if maximum dynamic performance is to be attained, a flush-diaphragm transducer mounted directly at the point where a pressure measurement is wanted should be used if at all possible (see Fig. 6.11). Any connecting tubing or volume chambers will degrade performance to some extent. The fact that this degradation is studied in this section indicates that in many practical circumstances a flush-diaphragm transducer is not applicable.

Liquid systems, heavily damped, slow-acting. In the system of Fig. 6.12, the spring-loaded piston represents the flexible element of the pressure pickup. For the present analysis, the only pertinent characteristic of the pressure pickup is its volume change per unit pressure change, C_{vp} in.^{3}/psi. This can be calculated or measured experimentally. For systems that are heavily damped (a criterion for judging this will be given shortly) and subjected to relatively slow pressure changes, the inertia effects of both the fluid and the moving parts of the pickup are negligible compared with viscous and spring forces. We shall show that under these conditions the measured pressure p_m follows the desired pres-

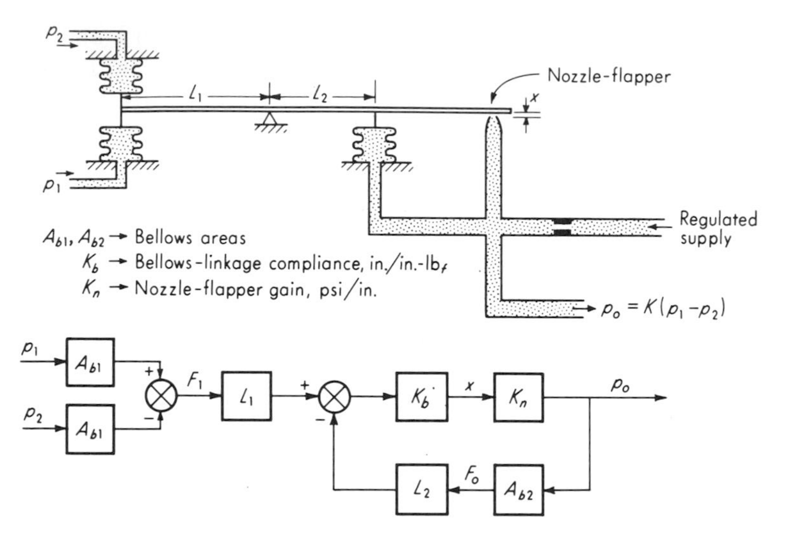

Force-balance differential-pressure transducer

(a)

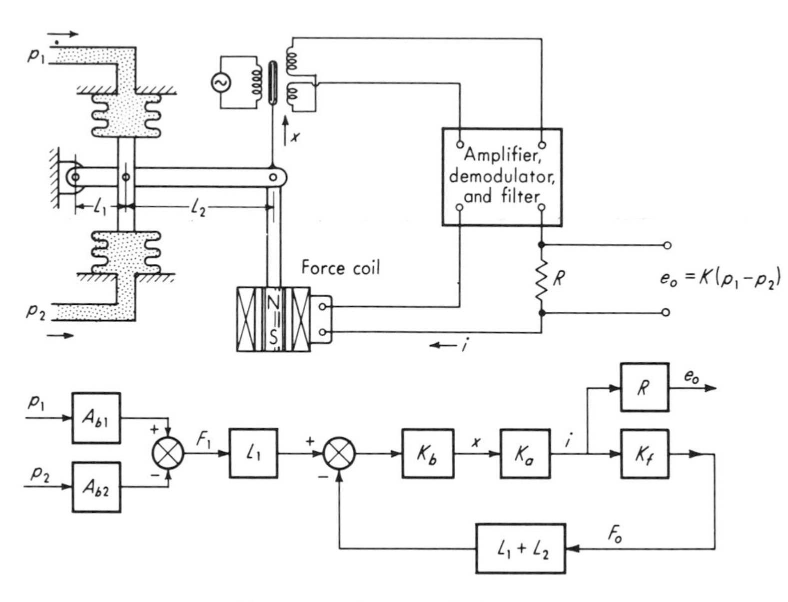

Electromagnetic pressure balance

(b)

Fig. 6.10. *Null-balance pressure sensors.*

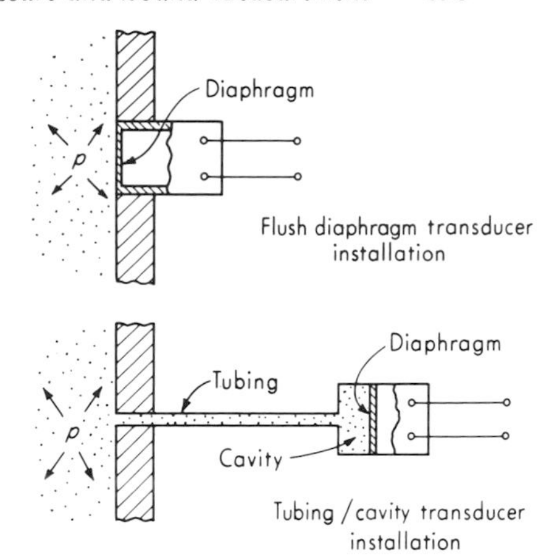

Fig. 6.11. *Transducer installation types.*

sure p_i with a first-order lag. For steady laminar flow in the tube we have

$$p_i - p_m = \frac{32\mu L V_{t,\text{av}}}{d_t^2} \tag{6.53}$$

where $\mu \triangleq$ fluid viscosity
$V_{t,\text{av}} \triangleq$ average flow velocity in tube

While this equation is exact only for steady flow, it holds quite closely for slowly varying velocities. During a time interval dt a quantity of liquid enters the chamber. This is given by

$$dV = \text{volume entering} = \frac{\pi d_t^2}{4} V_{t,\text{av}}\, dt = \frac{\pi d_t^4 (p_i - p_m)\, dt}{128\mu L} \tag{6.54}$$

Any volume added or taken away results in a pressure change dp_m given by

$$dp_m = \frac{dV}{C_{vp}} = \frac{p_i - p_m}{\tau}\, dt \tag{6.55}$$

and thus

$$\tau \frac{dp_m}{dt} + p_m = p_i \tag{6.56}$$

where

$$\tau \triangleq \frac{128\mu L C_{vp}}{\pi d_t^4} \tag{6.57}$$

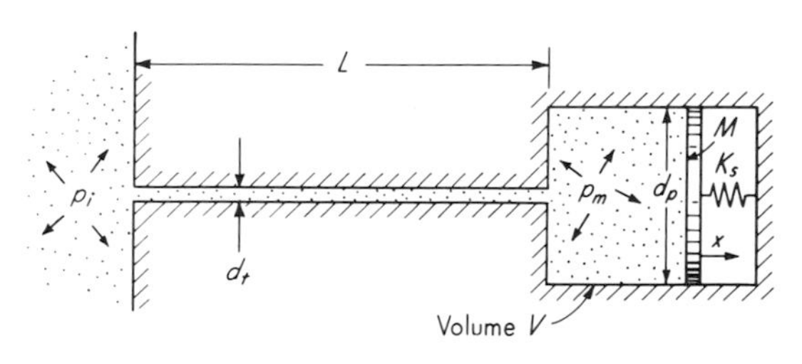

Fig. 6.12. *Transducer/tubing model.*

The response is thus seen to follow a standard first-order form. To keep τ small, the tubing length should be short and the diameter large.

The viewpoint of this analysis is that any sudden changes in p_i will cause much more gradual changes in p_m and thus the pickup spring-mass system will not be able to manifest its natural oscillatory tendencies. Under such conditions an overall first-order response from the tubing/transducer system may be expected. To obtain some numerical estimate of the conditions required for such behavior, we later carry out an analysis which includes inertia effects. This will lead to a second-order type of response, and when ζ of this model is greater than about 1.5, the simpler first-order model may be employed with fair accuracy, at least (in terms of frequency response) for frequencies less than ω_n.

The model used above predicts that, for a change in p_i, p_m starts to change immediately. This cannot be true since a pressure wave propagates through a fluid at finite speed, the velocity of sound, for small disturbances. There is thus a dead time τ_{dt} equal to the distance traversed divided by the speed of sound. For liquids and reasonable tube lengths this delay is usually small enough to ignore completely. The speed of sound, v_s, in a fluid contained in a nonrigid tube is given by

$$v_s = \sqrt{\frac{E_L}{\rho_L}} \sqrt{\frac{1}{1 + 2R_t E_L / t E_t}} \tag{6.58}$$

where $E_L \triangleq$ bulk modulus of fluid
$\rho_L \triangleq$ mass density of fluid
$R_t \triangleq$ tube inside radius
$t \triangleq$ tube wall thickness
$E_t \triangleq$ tube-material modulus of elasticity

If this dead time is significant in a practical problem, it may be included in the transfer function as

$$\frac{p_m}{p_i}(D) = \frac{e^{-\tau_{dt} D}}{\tau D + 1} \tag{6.59}$$

with fair accuracy.

Finally, it should also be pointed out that the result [Eq. (6.56)] of this analysis may also be applied to systems using gases rather than liquids if the elastic pressure-sensing element is sufficiently soft so that its volume change per unit pressure change is much larger than that due to gas compressibility.

Liquid systems, moderately damped, fast-acting. When the motions of the liquid and the pickup elastic element are rapid, their inertia is no longer negligible. An analysis[1] of this situation using energy

[1] G. White, Liquid Filled Pressure Gage Systems, *Statham Instr. Notes*, Statham Instruments Inc., Los Angeles, Calif., January–February, 1949.

methods is available. In the system of Fig. 6.12, any change in pressure p_m must be accompanied by a volume change; this in turn requires an inflow or outflow of liquid through the tube. If the tube is of small diameter compared with the equivalent piston diameter of the pickup, the tube flow will be at a much higher velocity than the piston velocity, and the kinetic energy of the liquid in the tube may be a large (sometimes major) part of the total system kinetic energy. This increase in kinetic energy is equivalent to adding mass to the piston and ignoring the fluid inertia, and the analysis below calculates just how much mass should be added to give the same effect as the fluid inertia. This added mass lowers the system natural frequency and thereby degrades dynamic response.

To find the equivalent piston/spring configuration for a given transducer, the volume change per unit pressure change must be equal for both systems. This gives

Transducer volume change = equivalent piston volume change

$$pC_{vp} = \frac{\pi^2 d_p{}^4 p}{16K_s} \qquad (6.60)$$

$$\frac{d_p{}^4}{K_s} = \frac{16C_{vp}}{\pi^2} \qquad (6.61)$$

Thus d_p and K_s for the equivalent system can have any values that satisfy Eq. (6.61). Also, the natural frequency of each system with no fluid present must be equal. This gives

$$\omega_{n,t} = \sqrt{\frac{K_s}{M}} \qquad (6.62)$$

$$\frac{K_s}{M} = \omega_{n,t}{}^2 \qquad (6.63)$$

where $\omega_{n,t} \triangleq$ transducer natural frequency. Again, any values of K_s and M that satisfy Eq. (6.63) may be used. Thus, to define the equivalent system, only C_{vp} and $\omega_{n,t}$ need be known; they can be found from experiment if theoretical formulas are unavailable.

Since we have just shown how the equivalent system is related to the real system, we can now proceed with an analysis of the equivalent system. The volume change dV is related to the piston motion dx by

$$dV = \frac{\pi d_p{}^2\,dx}{4} \qquad (6.64)$$

Thus

$$\frac{dV}{dt} = \frac{\pi d_p{}^2}{4}\frac{dx}{dt} \qquad (6.65)$$

and

$$\frac{\pi}{4} d_t{}^2\,V_{t,\mathrm{av}} = \frac{\pi d_p{}^2}{4}\frac{dx}{dt} \qquad (6.66)$$

where $\quad V_{t,\mathrm{av}} =$ average flow velocity in tube

We then get
$$V_{t,\mathrm{av}} = \left(\frac{d_p}{d_t}\right)^2 \frac{dx}{dt} \tag{6.67}$$

Next we assume laminar flow in the tube, with the parabolic velocity profile characteristic of steady flow:

$$V_t = V_{t,c}\left[1 - \left(\frac{r}{R}\right)^2\right] \tag{6.68}$$

where $\quad V_t \triangleq$ velocity at radius r
$V_{t,c} \triangleq$ center-line velocity
$R \triangleq$ tube inside radius

The kinetic energy of an annular element of fluid (density ρ) of thickness dr at radius r is

$$d(KE) = \frac{(2\pi r\,dr)L\rho V_t^2}{2} \tag{6.69}$$

Substitution of Eq. (6.68) and integration give the fluid kinetic energy as

$$KE = \frac{\pi\rho L V_{t,\mathrm{av}}^2\, d_t^2}{6} \tag{6.70}$$

For a square velocity profile the kinetic energy would be

$$KE_s = \frac{\pi\rho L V_{t,\mathrm{av}}^2\, d_t^2}{8} \tag{6.71}$$

The actual velocity profile will be somewhere between parabolic and square. Even if laminar flow exists, the velocity profile is nonparabolic except for steady flow.[1] Turbulent flow gives a rather square profile. We shall assume Eq. (6.70) to hold here; however, Eq. (6.71) can be carried through with equal ease to "bracket" the correct value. The rigid mass M_e, attached to M, which would have the same kinetic energy as the fluid, is given by

$$\frac{M_e}{2}\left(\frac{dx}{dt}\right)^2 = \frac{\pi\rho L d_p^4}{6d_t^2}\left(\frac{dx}{dt}\right)^2 \tag{6.72}$$

$$M_e = \frac{\pi\rho L d_p^4}{3d_t^2} \tag{6.73}$$

The natural frequency of the transducer/tubing system is then

$$\omega_n = \sqrt{\frac{K_s}{M + M_e}} = \sqrt{\frac{1}{M/K_s + M_e/K_s}} = \sqrt{\frac{1}{1/\omega_{n,t}^2 + 16\rho L C_{vp}/3\pi d_t^2}} \tag{6.74}$$

[1] C. K. Stedman, Alternating Flow of Fluid in Tubes, *Statham Instr. Notes,* Statham Instruments Inc., Los Angeles, Calif., January, 1956.

To keep ω_n as high as possible, L and C_{vp} must be as small as possible and d_t as large as possible. In many practical cases, $M_e \gg M$, allowing simplification of Eq. (6.74) to

$$\omega_n = \sqrt{\frac{3\pi d_t^2}{16\rho L C_{vp}}} \tag{6.75}$$

We next calculate the damping ratio of the transducer/tubing system. The transducer itself is assumed to have negligible damping; thus the only damping is due to the fluid friction in the tube. We again assume the validity of the steady laminar-flow relations to calculate the pressure drop due to fluid viscosity as $32\mu L V_{t,\mathrm{av}}/d_t^2$. The force on the piston due to this pressure drop is the damping force $B(dx/dt)$; thus

$$\frac{\pi d_p^2}{4}\frac{32\mu L V_{t,\mathrm{av}}}{d_t^2} = B\frac{dx}{dt} \tag{6.76}$$

and, since $V_{t,\mathrm{av}} = (dx/dt)(d_p/d_t)^2$,

$$B = 8\pi\mu L\left(\frac{d_p}{d_t}\right)^4 \tag{6.77}$$

Then, using the general formula for the damping ratio of a spring-mass-dashpot system, we get

$$\zeta = \frac{B}{2\sqrt{K_s(M + M_e)}} = \frac{64\mu L C_{vp}}{\pi d_t^4 \sqrt{1/\omega_{n,t}^2 + 16\rho L C_{vp}/3\pi d_t^2}} \tag{6.78}$$

If $M_e \gg M$, this simplifies to

$$\zeta = \frac{16\sqrt{3/\pi}\,\mu\sqrt{L C_{vp}/\rho}}{d_t^3} \tag{6.79}$$

The above theory has been partially checked experimentally in an M.Sc. thesis by Fowler.[1] The transducer used was the diaphragm strain-gage instrument whose numerical parameters were given at the end of Sec. 6.4. The liquid used was water, and ω_n and ζ were found from step-function tests using a bursting cellophane diaphragm to obtain a sudden 10-psi release of pressure. Tubing of 0.042- to 1.022-in. inside diameter and 4 to 32 in. in length was studied. Conditions were such that turbulent flow probably existed much of the time; thus use of Eq. (6.71) was indicated. The value of C_{vp} was calculated from theory as 0.00441 in.^{3}/psi while an experimental calibration gave 0.00500. For all cases in which oscillation occurred and thus ω_n could be accurately measured, it was found that ω_n was accurately predicted within about 5 to 10 percent. To show the severity of the performance degradation,

[1] R. L. Fowler, An Experimental Study of the Effects of Liquid Inertia and Viscosity on the Dynamic Response of Pressure Transducer-Tubing Systems, M.Sc. Thesis, The Ohio State University, Mechanical Engineering Department, 1963.

the natural frequency (which was 897 cps with no tubing and in air) became about 60 cps with a 1.022-in.-diameter tube 10 in. in length and about 3 cps with a 0.092-in. tube 32 in. long. However, the prediction of ζ was much less satisfactory, being invariably too low. The small-diameter long tubes (which encourage laminar flow) gave the best correlation with theory but even there errors of 100 percent occurred. For example, the 0.042-in. tube 10 in. long had ζ of about 1.0 while theory predicted 0.66. The poor agreement can undoubtedly be charged to turbulent flow and energy losses due to expansion and contraction (end effects) at the tubing ends.

Fowler developed corrections for these effects which resulted in much better agreement. However, the turbulent flow and end effects are nonlinear damping mechanisms; thus the meaning of ζ is confused for both theoretical predictions and experimental measurements. The use of smaller step functions should give lower velocities and thus more nearly laminar flow. This effect was checked for a 0.092-in.-diameter 10-in. long tube. A 10-psi step function gave $\zeta = 0.34$; a 5-psi step function gave 0.22. The theoretical value was 0.065, suggesting that for sufficiently small inputs the theory might be quite accurate. Detailed discussion of these problems may be found in the reference.

When ζ calculated from Eq. (6.78) or (6.79) is 1.5 or greater, the simpler first-order model of Eq. (6.56) may be adequate. To show the relationship of these two analyses, recall that in the second-order form $K/[D^2/\omega_n^2 + 2\zeta D/\omega_n + 1]$ the inertia effect resides in the D^2 term. Neglecting this gives $K/[2\zeta D/\omega_n + 1]$. If one calculates $2\zeta/\omega_n$ from Eqs. (6.74) and (6.78), he will find it numerically equal to τ of Eq. (6.57).

Gas systems, with tube volume small fraction of chamber volume. When, in the configuration of Fig. 6.12, a gas is the fluid medium, the compressibility of the gas in the volume V becomes the major spring effect when the pressure pickup is at all stiff. It is reasonable, then, to assume that the volume V is enclosed by rigid walls ($C_{vp} = 0$). The majority of practical problems involve rather low frequencies. This allows treatment as a lumped-parameter system with fluid properties considered constant along the length of the tube for small disturbances. The validity of this viewpoint is shown by noting that pressure-wave propagation in the gas follows the general law of wave motion

$$\lambda = \text{wavelength} = \frac{\text{velocity of propagation}}{\text{frequency}} = \frac{c}{f} \tag{6.80}$$

Our lumped-parameter assumption says that the gas in the tube moves as a unit, as opposed to wave motion *within* the tube. For any given

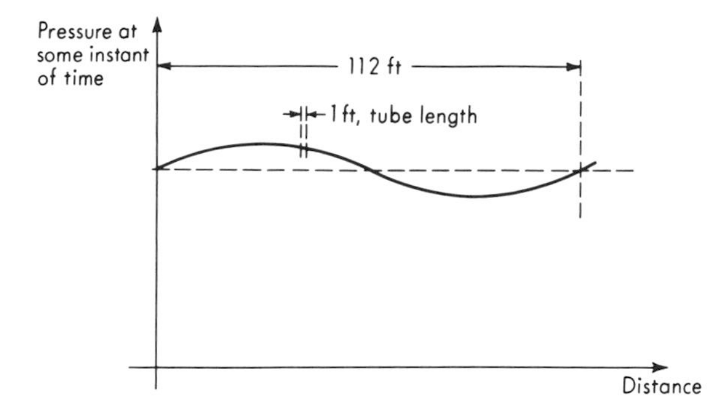

Fig. 6.13. *Justification of lumped-parameter analysis.*

tube length L, for sufficiently low frequencies, the lumped-parameter approach becomes valid. For example, the velocity of pressure waves in standard air is about 1,120 fps. If an oscillation of 10-cps frequency exists, its wavelength must be 112 ft/cycle. This means that the spacewise variation of fluid pressure due to wave motion has a wavelength of 112 ft (see Fig. 6.13). For a tube of, say, 1-ft length, the variation of pressure (and thus of density, etc.) from one end of the tube to the other, due to wave motion, is very small. That is, there is negligible *relative* motion of particles in the tube; they all move together, as a unit. While the above aspect of wave motion is neglected, the dead time due to the finite speed of propagation can be relatively easily taken into account (approximately) by multiplying the second-order transfer function (which we shall obtain shortly) by $e^{-\tau_{dt}D}$, where

$$\tau_{dt} = \frac{L}{v_s} \tag{6.81}$$

$$\text{Sound velocity} = v_s = \sqrt{\frac{\gamma p}{\rho}} \tag{6.82}$$

where $\gamma \triangleq$ ratio of specific heats
$\rho \triangleq$ mass density
$p \triangleq$ average pressure

This dead time is negligible unless very long lines (several hundred feet, as found in some industrial pneumatic control systems) are used. For ambient air, a 100-ft line has a dead time of about 0.1 sec. Thus if ω_n of the second-order response is, for example, 3 rad/sec, the effect of the dead time on the overall response will be of relatively slight importance (see Fig. 6.14).

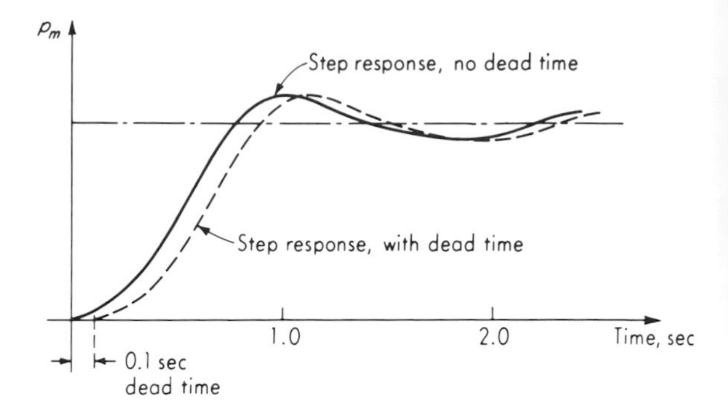

Fig. 6.14. *Effect of small dead time.*

The analysis[1] we carry out below is valid only for small pressure changes, the system becoming nonlinear for large disturbances. Steady-laminar-flow formulas are used for calculating fluid friction. However, the effective mass is taken equal to the actual mass (rather than the four-thirds actual mass given by the parabolic velocity profile). We shall follow the reference in this respect; use of the $\frac{4}{3}$ factor is easily incorporated if one wishes to bracket a more correct value. Numerically the effect is rather small in any case.

The analysis consists merely of applying Newton's law to the "slug" of gas in the tube. We assume that initially $p_i = p_m = p_0$ when p_i changes slightly in some way. From here on, the symbols p_i and p_m are taken to mean the *excess* pressures over and above p_0. The force due to the pressure p_i is $\pi p_i d_t^2/4$. The viscous force due to the wall shearing stress is $8\pi\mu L\dot{x}_t$, where x_t is the displacement of the slug of gas in the tube. If the slug of gas moves into the volume V an amount x_t, the pressure p_m will increase. We assume this compression occurs under adiabatic conditions. The adiabatic bulk modulus E_a of a gas is given by

$$E_a \triangleq -\frac{dp}{dV/V} = \gamma p \qquad (6.83)$$

The displacement x_t causes a volume change $dV = \pi d_t^2 x_t/4$. This in turn causes a pressure excess $p_m = \pi E_a d_t^2 x_t/4V$. The force due to this pressure excess is $\pi^2 E_a d_t^4 x_t/16V$. Newton's law then gives

$$\frac{\pi p_i d_t^4}{4} - 8\pi\mu L\dot{x}_t - \frac{\pi^2 E_a d_t^4 x_t}{16V} = \frac{\pi d_t^2 L\rho}{4}\ddot{x}_t \qquad (6.84)$$

[1] G. J. Delio, G. V. Schwent, and R. S. Cesaro, Transient Behavior of Lumped-constant Systems for Sensing Gas Pressures, *NACA Tech. Note* 1988, 1949.

and, since $p_m = \pi E_a d_t^2 x_t / 4V$,

$$\frac{4L\rho V}{\pi E_a d_t^2}\ddot{p}_m + \frac{128\mu L V}{\pi E_a d_t^4}\dot{p}_m + p_m = p_i \tag{6.85}$$

This is clearly the standard second-order form, and so we define

$$\omega_n \triangleq \frac{d_t}{2}\sqrt{\frac{\pi E_a}{L\rho V}} \tag{6.86}$$

$$\zeta \triangleq \frac{32\mu}{d_t^3}\sqrt{\frac{VL}{\pi E_a \rho}} \tag{6.87}$$

Since E_a and ρ both vary during pressure changes, ζ and ω_n are not really constants; that is, the system is nonlinear. For small-percentage pressure variations around the original equilibrium pressure p_0, however, ζ and ω_n vary only slightly and the behavior is nearly linear. In calculating ζ and ω_n, E_a and ρ are computed by using pressure p_0.

When L becomes very short, Eq. (6.86) predicts a very large ω_n. In practice, this will not occur since even when $L = 0$ there is some air (close to the opening in the volume) which has appreciable velocity and, therefore, kinetic energy. Theory shows that this end effect may be taken into account by using for L in Eq. (6.86) an effective length L_e given by

$$L_e = L\left(1 + \frac{8}{3\pi}\frac{d_t}{L}\right) \tag{6.88}$$

In most cases the term $(8/3\pi)(d_t/L)$ will be completely negligible compared with 1.0. However, if $L = 0$ (tube degenerates into simply a hole in the side of volume V), one can still compute an ω_n since then $L_e = (8/3\pi)d_t$. The computation of damping for this case is not straightforward and will not be discussed here.

Gas systems, with tube volume comparable to chamber volume. When the volume of the tube becomes a significant part of the total volume of a system, compressibility effects are no longer restricted to the volume chamber alone and the above formulas become inaccurate. More refined analyses[1] give the following formulas for ζ and ω_n:

$$\omega_n = \frac{\sqrt{\gamma p/\rho}}{L\sqrt{1/2 + V/V_t}} \tag{6.89}$$

$$\zeta = \frac{64\mu L^2}{\pi d_t^4 \sqrt{\gamma p \rho}}\sqrt{\frac{1}{2} + \frac{V}{V_t}} \tag{6.90}$$

where $V_t \triangleq$ tubing volume

[1] J. O. Hougen, O. R. Martin, and R. A. Walsh, Dynamics of Pneumatic Transmission Lines, *Control Eng.*, p. 114, September, 1963.

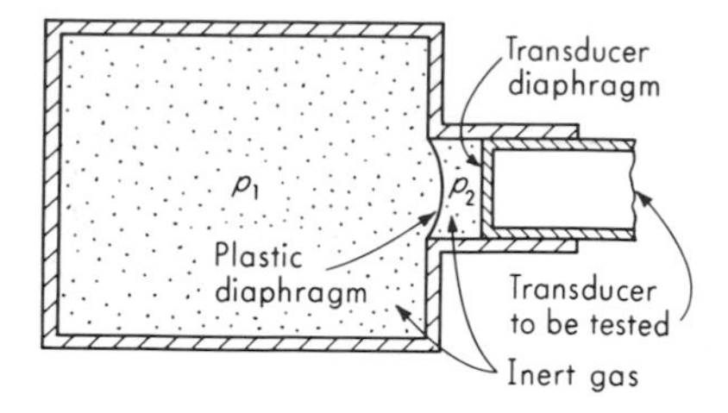

Fig. 6.15. *Step-test apparatus.*

If, in these formulas, $V/V_t \gg \frac{1}{2}$ (tubing volume negligible compared with chamber volume), the term $\frac{1}{2}$ may be neglected and the formulas become identical to (6.86) and (6.87).

Conclusion. The results of this section are to be thought of as practical working relations. The general problem treated here is quite complex and has been the subject of many intricate analyses and experimental studies, some of which will be found in the bibliography of this chapter. Most of the difficulties encountered are in the area of very high frequencies, where the lumped-parameter models used in this section are inadequate and give faulty predictions.

6.7

Dynamic Testing of Pressure-measuring Systems To determine the regions of accuracy of theoretical predictions or to find accurate numerical values of system dynamic characteristics for critical applications, recourse must be made to experimental testing. This commonly takes the form of impulse, step, or frequency-response tests, with step-function tests being perhaps the most common. A comprehensive review[1] of this subject is available. Here we can mention only a few high points.

For step-function tests of systems in which natural frequencies are not greater than about 1,000 cps, the bursting of a thin diaphragm subjected to gas pressure is often satisfactory. A general rule for step testing is that the rise time of the "step" function must be less than about one fourth of the natural period of the system tested if it is to excite the natural oscillations. Thus a 1,000-cps system requires a step with a rise time of 0.25 msec or less. Figure 6.15 shows schematically the principle of such devices. The pressures p_1 and p_2 are each individually adjustable. The volume containing p_2 is much smaller than that containing p_1; thus when the thin plastic diaphragm is ruptured by a solenoid-actuated

[1] J. L. Schweppe et al., Methods for the Dynamic Calibration of Pressure Transducers, *Natl. Bur. Std. (U.S.), Monograph* 67, 1963.

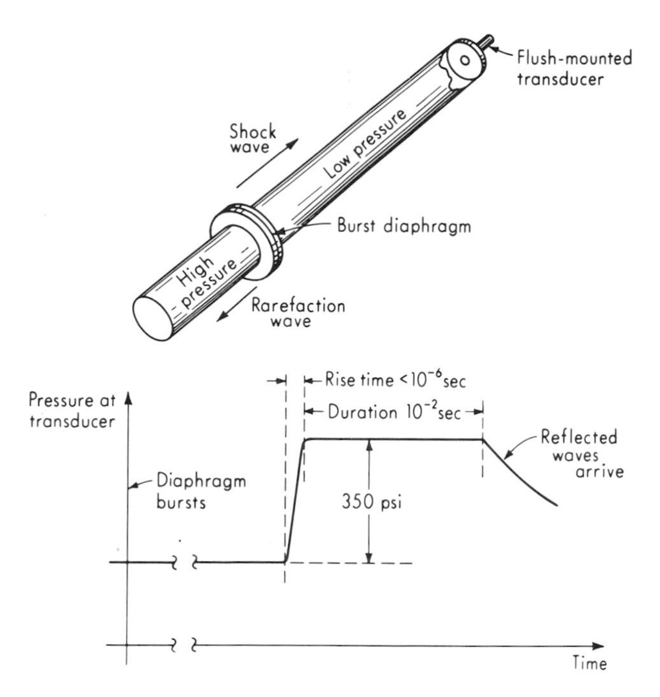

Fig. 6.16. *Shock tube.*

knife, the pressure p_2 rises to p_1 very quickly. If a decreasing step function is wanted, p_2 can be made larger than p_1. Construction and operation of such devices are quite simple, and they have been widely used in their range of applicability.

For pickups of natural frequency greater than 1,000 cps, the simple burst-diaphragm testers are not capable of exciting the natural oscillations, and the pickup output is simply an accurate record of the terminated-ramp pressure input. To achieve sufficiently short pressure-rise times the shock tube is widely used. Figure 6.16 shows a sketch of such a device. A thin diaphragm separates the high-pressure and low-pressure chambers, and the transducer to be tested is mounted flush with the end of the low-pressure chamber. When the diaphragm is caused to burst, a shock wave travels toward the low-pressure end at a speed that may greatly exceed the speed of sound (5,000 fps is not unusual). From one side of this shock front to the other there is a pressure change of the order of 2 to 1 over a distance which may be of the order of 10^{-4} in. (At the same time a rarefaction wave travels from the diaphragm toward the high-pressure end.) When the shock front reaches the end of the tube where the transducer is mounted, it is reflected as a shock wave with more than twice the pressure difference of the original shock wave. The transducer is thus exposed to a very sharp ($\sim 10^{-8}$ sec) pressure rise which is maintained constant for a short interval before various reflected waves arrive to confuse the picture. The length of this interval may be controlled to a certain extent by proper proportioning and operation of the shock tube. Some numerical characteristics of a typical shock tube[1] are: high-pressure chamber 7 ft long, low-pressure chamber 15 ft long, tubing inside dimensions 1.4 in. square (wall thickness $\frac{1}{4}$ in.), maximum high pressure 600 psi, operating fluid air, maximum pressure step 350 psi, burst diaphragm 0.001- to 0.005-in.-thick Mylar plastic, and duration of constant pressure 0.01 sec. For a pressure pickup of, say, 100,000-cps natural frequency, a pressure-rise time of less than 0.25×10^{-5} sec is required. This is readily met by the tube described above. The 0.01-sec step duration would give time for about 1,000 cycles of oscillation of a 100,000-cps pickup—more than adequate to determine the dynamic characteristics.

A simple impulse-type test method applicable to flat, flush-diaphragm transducers utilizes a small steel ball dropped onto the diaphragm. The impact excites the natural oscillations which are recorded and analyzed for natural frequency and damping ratio. Although this input is a concentrated force rather than a uniform pressure, the results have been found[2] to correlate quite well with shock-tube tests. Another impulse method[3] uses the shock wave created by discharging 25,000 volts across a spark gap. The spark gap and transducer are located about 3 in. apart in open air. A pressure impulse of 0.2 μsec rise time and 100 psi peak can be obtained.

Figure 6.17 shows one method of constructing a frequency-response tester using liquid as the pressure medium. The vibration shaker applies a sinusoidal force of adjustable frequency and amplitude to the piston/diaphragm to create sinusoidal pressure in the liquid-filled chamber. Such vibration shakers are readily available in industry and cover a wide range of force and frequency. The average pressure about which oscillations take place may be adjusted by regulating the bias pressure on the air side of the cylinder. Since it is not usually possible accurately and reproducibly to predict the pressure actually produced by such an arrangement as "frequency and/or amplitude is varied," it is customary to mount a reference transducer at a location where it will experience the same pressure as the transducer under test. The reference transducer must have a known flat frequency response beyond any frequencies to be tested. This

[1] R. Bowersox, Calibration of High-frequency-response Pressure Transducers, *ISA J.*, p. 98, November, 1958.

[2] W. C. Bentley and J. J. Walter, Transient Pressure Measuring Methods Research, *Princeton Univ. Aeron. Eng. Dept. Rept.* 595g, p. 103, 1963.

[3] Omega Dynamics Corp., Pasadena, Calif.

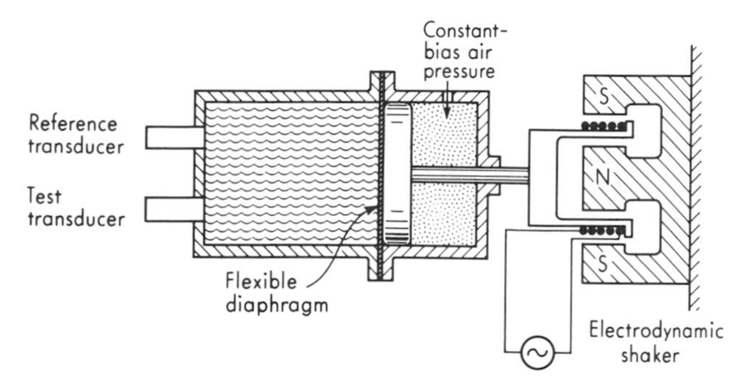

Fig. 6.17. *Sinusoidal test apparatus for liquid.*

can be determined by some independent method, such as a shock tube. In testing another transducer, one merely calculates the amplitude ratio and phase shift between the reference and test transducer to determine the test-transducer frequency response.

A different approach[1] to frequency testing, using a flow-modulating principle and gas as the fluid medium, is shown in Fig. 6.18. A chamber is supplied with compressed gas from a constant-pressure source through a small inlet passage. The gas is exhausted to the atmosphere through an outlet passage whose area is modulated approximately sinusoidally with time. This is accomplished by rotating a disk containing holes in front of the exhaust port so that outflow periodically is cut on and off. This produces a periodic (nearly sinusoidal) variation in chamber pressure which is measured by both a reference transducer and the test transducer. Varying the speed of the rotating disk changes the frequency. The amplitude of pressure oscillation of such a device drops off with frequency. For

[1] Bentley and Walter, *op. cit.*, p. 63.

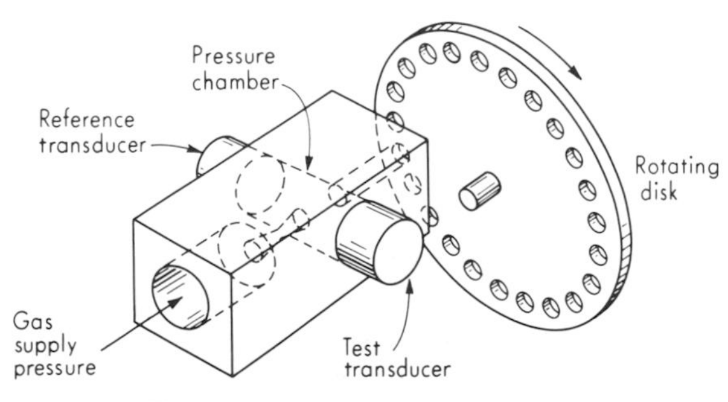

Fig. 6.18. *Sinusoidal test apparatus for gas.*

the system described, with helium gas at a supply pressure of 121 psia, the peak-to-peak pressure amplitude goes from about 15 psi at 1,000 cps to about 2 psi at 11,000 cps. In addition to this reduction in amplitude, increase in frequency also leads into the range of resonant acoustical frequencies of the chamber. When these resonances occur one cannot depend on the pressure being uniform throughout the chamber. This uniformity is a necessity when using a method based on comparison of a reference transducer with the test transducer. The acoustic resonant frequencies depend on the chamber size (smaller chambers have higher frequencies) and the speed of sound in the gas (higher sound speed gives higher frequencies). The use of helium in the above example is based on this last consideration. A system of the above type was found to be usable for dynamic calibration up to about 10,000 cps.

In summary, it should be pointed out that the dynamic testing of very-high-frequency pressure pickups involves a number of complicating factors. It has been found extremely difficult to generate high-frequency pressure sine waves that also are of large enough amplitude to give a relatively noise-free transducer output signal. Small-amplitude pressure waves (such as are applicable to sound-measuring systems) can be relatively easily produced with loudspeaker-type systems but their amplitude is far below the levels needed for pressure pickups whose full-scale range is tens, hundreds, or thousands of pounds per square inch. Step testing with a shock tube has thus been widely used since the fast rise time and large pressure steps result in a transient input with strong high-frequency content. The pickups themselves present problems since at the high natural frequencies involved many complex wave-propagation and reflection effects make the response deviate considerably from the simple second-order model. Also, these pickups generally have little damping ($\zeta \approx 0.01$ to 0.04) which makes them particularly prone to ringing at their natural frequency if any sharp transients occur.

6.8

High-pressure Measurement Pressures up to about 100,000 psi can be measured fairly easily with strain-gage pressure cells or Bourdon tubes. Bourdon tubes for such high pressures have nearly circular cross sections and thus give little output motion per turn. To get a measurable output, the helical form with many turns is generally used. Inaccuracy of the order of 1 percent of full scale may be expected with a temperature error of an additional 2 percent/100F°. Strain-gage pressure cells can be temperature-compensated to give 0.25 percent error over a large temperature range.

For fluid pressures above 100,000 psi, electrical gages based on the

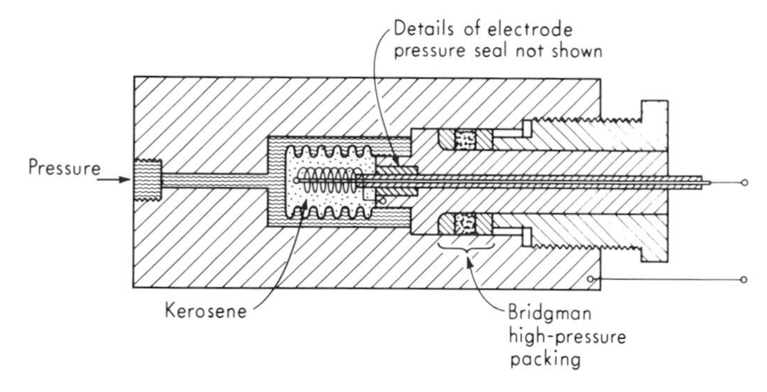

Fig. 6.19. *Very-high-pressure transducer.*

resistance change of Manganin or gold-chrome wire with hydrostatic pressure are generally utilized.[1] Figure 6.19 shows a typical gage. The sensitive wire is wound in a loose coil, one end of which is grounded to the cell body and the other end brought out through a suitable insulator. The coil is enclosed in a flexible, kerosene-filled bellows which transmits the measured pressure to the coil. The resistance change, which is linear with pressure, is sensed by conventional Wheatstone-bridge methods. Pertinent characteristics of the common wire materials are as follows:

	Pressure sensitivity, (ohm/ohm)/psi	*Temperature sensitivity, (ohm/ohm)/F°*	*Resistivity, ohm-cm*
Manganin	1.69×10^{-7}	1.7×10^{-5}	45×10^{-6}
Gold chrome	0.673×10^{-7}	0.8×10^{-6}	2.4×10^{-6}

Although its pressure sensitivity is lower, gold chrome is preferred in many cases because of its much smaller temperature error. This is particularly significant since the kerosene used in the bellows will experience a transient temperature change when sudden pressure changes occur, because of adiabatic compression or expansion. The response of the wire resistance to pressure changes is practically instantaneous; however, the accompanying temperature change will cause a transient error if temperature sensitivity is too high. Gages of the above type are commercially available with full scale up to 200,000 psi and inaccuracy of 0.1 to 0.5 percent. They have also been used successfully for much higher pressures on a special-application basis.

[1] W. H. Howe, The Present Status of High Pressure Measurement and Control, *ISA J.*, p. 77, March, 1955.

The measurement of local contact pressures between rolling elements in gears, cams, and bearings may possibly be accomplished by depositing a thin strip of Manganin or gold chrome onto the surface as a pressure transducer. Preliminary studies[1] of such a technique, using a Manganin element 0.001 in. wide and 5×10^{-6} in. thick, have been reported.

6.9

Low-pressure (Vacuum) Measurement[2]

Two commonly used units of vacuum measurement are the torr and the micron. One torr is a pressure equivalent to 1 mm Hg at standard conditions; one micron is 10^{-3} torr. Manometers and bellows gages are usable to about 0.1 torr, Bourdon gages to 10 torrs, and diaphragm gages to 10^{-3} torr. Below these ranges, other types of vacuum gages are necessary.

The McLeod gage. The McLeod gage is considered a vacuum standard since the pressure can be computed from the dimensions of the gage. It is not directly usable below about 10^{-4} torr; however, pressure-dividing techniques (see Sec. 6.1) allow its use as a calibration standard for considerably lower ranges. The multiple-compression technique[3] is also being studied to extend its range. The inaccuracy of McLeod gages is rarely less than 1 percent and may be much higher at the lowest pressures.

Of the many variations of McLeod gages, we here consider only the most basic. The principle of all McLeod gages is the compression of a sample of the low-pressure gas to a pressure sufficiently high to read with a simple manometer. Figure 6.20 shows the basic construction. By withdrawing the plunger, the mercury level is lowered to the position of Fig. 6.20*a*, admitting the gas at unknown pressure p_i. When the plunger is pushed in, the mercury level goes up, sealing off a gas sample of known volume V in the bulb and capillary tube A. Further motion of the plunger causes compression of this sample, and motion is continued until the mercury level in capillary B is at the zero mark. The unknown pressure is then calculated, using Boyle's law, as follows:

$$p_i V = p A_t h \tag{6.91}$$

$$p = p_i + h\gamma \tag{6.92}$$

$$p_i = \frac{\gamma A_t h^2}{V - A_t h} \approx \frac{\gamma A_t h^2}{V} \quad \text{if } V \gg A_t h \tag{6.93}$$

[1] F. K. Orcutt, Elastohydrodynamic Lubrication-Experimental Investigation, Mechanical Technology Inc., Latham, N.Y., *Rept.* 64TR6, 1964.

[2] S. Dushman, "Scientific Foundations of Vacuum Technique," chap. 6, John Wiley & Sons, Inc., New York, 1949.

[3] W. Kreisman, Extension of the Low Pressure Limit of McLeod Gages, *NASA*, *CR*-52877.

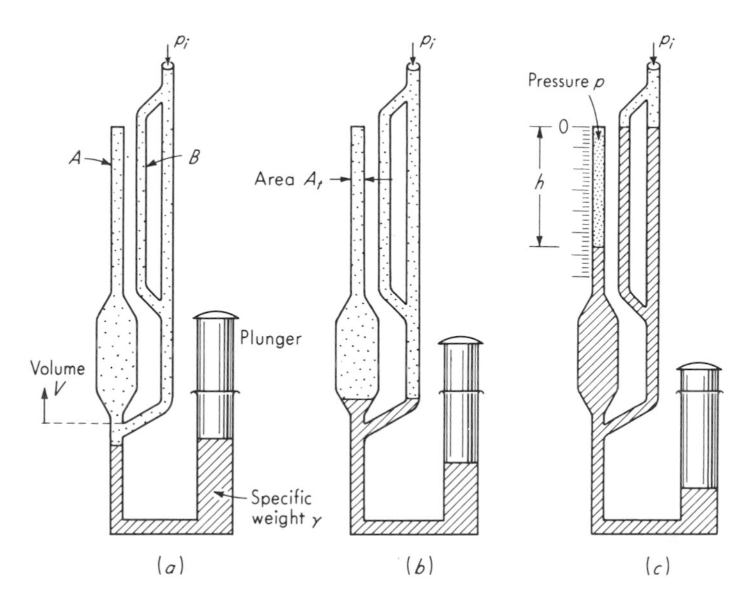

Fig. 6.20. *McLeod gage.*

In using a McLeod gage it is important to realize that if the measured gas contains any vapors that are condensed by the compression process the pressure will be in error. Except for this effect, the reading of the McLeod gage is not influenced by the composition of the gas. Only the Knudsen gage shares this desirable feature of composition insensitivity. The main drawbacks of the McLeod gage are the lack of a continuous output reading and the limitations on the lowest measurable pressures. When it is used to calibrate other gages, a liquid-air cold trap should be used between the McLeod gage and the gage to be calibrated to prevent the passage of mercury vapor.

The Knudsen gage. Although the Knudsen gage is little used at present, it is discussed briefly since it is relatively insensitive to gas composition and thus gives promise of development into a standard for pressures too low for the McLeod gage. In Fig. 6.21 the unknown pressure p_i is admitted to a chamber containing fixed plates heated to absolute temperature T_f, which temperature must be measured, and a spring-restrained movable vane whose temperature T_v must also be known. The spacing between the fixed and movable plates must be less than the mean free path of the gas whose pressure is being measured. The kinetic theory of gases shows that gas molecules rebound from the heated plates with greater momentum than from the cooler movable vane, thus giving a net

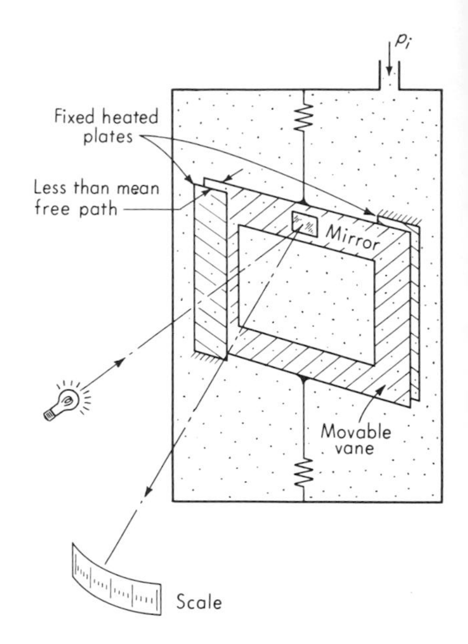

Fig. 6.21. *Knudsen gage.*

force on the movable vane which is measured by the deflection of the spring suspension. Analysis shows that the force is directly proportional to pressure for a given T_f and T_v, following a law of the form

$$p_i = KF/\sqrt{T_f/T_v - 1}$$

where F is force and K is a constant. The Knudsen gage is insensitive to gas composition except for the variation of accommodation coefficient from one gas to another. The accommodation coefficient is a measure of the extent to which a rebounding molecule has attained the temperature of the surface. This effect results, for example, in a 15 percent change in sensitivity between helium and air. Knudsen gages at present cover the range from about 10^{-8} to 10^{-2} torr.

Momentum-transfer (viscosity) gages. For pressures less than about 10^{-2} torr the kinetic theory of gases predicts that the viscosity of a gas will be directly proportional to the pressure. The viscosity may be measured, for example, in terms of the torque required to rotate, at constant speed, one concentric cylinder within another. (For pressures greater than about 1 torr, the viscosity is independent of pressure.) The variation of viscosity with pressure is different for different gases; thus gages based on this principle must be calibrated for a specific gas. While gages based on viscosity principles can measure to about 10^{-7} torr, such ranges are characteristic of laboratory-type equipment requiring great care in its use.

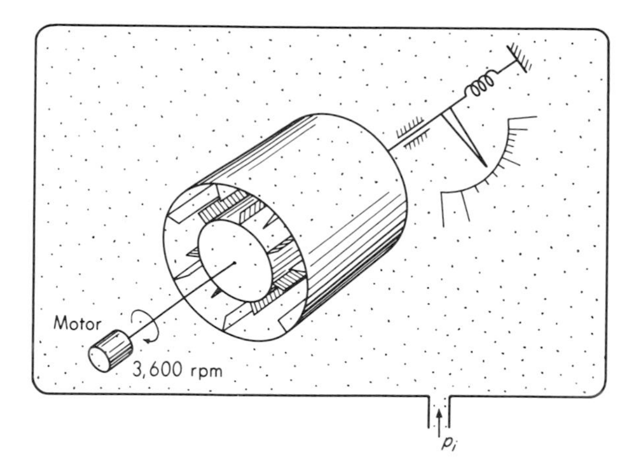

Fig. 6.22. *Momentum gage.*

A typical commercial gage,[1] shown schematically in Fig. 6.22, is calibrated for dry air and covers the range from 0 to 20 torrs. The range from 0 to 0.01 torr occupies about 10 percent of the total scale. The scale is nonlinearly calibrated because most of the range is above 10^{-2} torr and viscosity here is not proportional to pressure. To enable readings above 1 torr, where viscosity tends to become pressure-independent, bladed wheels rather than smooth concentric cylinders are used in this gage. These wheels cause a turbulent momentum exchange which is pressure-dependent above 1 torr, extending the useful range to 20 torrs. To reach the quoted lower limit (10^{-7} torr) of viscosity gages the construction of Fig. 6.22 is not used. Rather, the rate of decay of delicately constructed vibrating systems subjected to the damping effects of the gas is determined and pressure inferred from this.

Thermal-conductivity gages. Just as for viscosity, when the pressure of a gas becomes low enough that the mean free path of molecules is large compared with the pertinent dimensions of the apparatus, a linear relation between pressure and thermal conductivity is predicted by the kinetic theory of gases. For a viscosity gage the pertinent dimension is the spacing between the relatively moving surfaces. For a conductivity gage it is the spacing between the hot and cold surfaces. Again, when the pressure is increased sufficiently, conductivity becomes independent of gas pressure. The transition region between dependence and nondependence of viscosity and thermal conductivity on pressure is approximately the range 10^{-2} to 1 torr for apparatus of a size convenient to construct.

[1] General Electric Co.

The application of the thermal-conductivity principle is complicated by the simultaneous presence of another mode of heat transfer between the hot and cold surfaces, namely, radiation. Most gages utilize a heated element supplied with a constant energy input. This element assumes an equilibrium temperature when heat input and losses by conduction and radiation are just balanced. The conduction losses vary with gas composition and with gas pressure; thus for a given gas the equilibrium temperature of the heated element becomes a measure of pressure, and this temperature is what is actually measured. If the radiation losses are a major part of the total, pressure-induced conductivity changes will cause only a slight temperature change, giving poor sensitivity. Analysis shows that radiation losses may be minimized by using surfaces of low emissivity and by making the cold-surface temperature as low as practical. Since conduction and radiation losses depend on *both* the hot- and cold-surface temperatures, the cold surface may be maintained at a known constant temperature if overall accuracy warrants this measure. A further source of error is in the heat-conduction loss through any solid supports by which the heated element is mounted. The relative importance of the above-mentioned effects varies with the details of construction of the gage. The most common types of conductivity gages are the thermocouple, resistance thermometer (Pirani), and thermistor.

Figure 6.23 shows in schematic form the basic elements of a thermocouple vacuum gage. The hot surface is a thin metal strip whose temperature may be varied by changing the current passing through it. For a given heating current and gas the temperature assumed by the hot surface depends on pressure; this temperature is measured by a thermocouple welded to the hot surface. The cold surface here is the glass tube which usually is near room temperature. The accuracy of such gages is usually not high enough to warrant measurement or correction for changes in room temperature. Thermocouple gages of one type or another are available to measure in the range 10^{-4} to 1 torr.

In the resistance-thermometer (Pirani) gage the functions of heating and temperature measurement are combined in a single element. A typical construction is shown in Fig. 6.24. The resistance element is in the form of four coiled tungsten wires connected in parallel and supported inside a glass tube to which the gas is admitted. Again the cold surface is the glass tube. Two identical tubes are generally connected in a bridge circuit as shown. One of the tubes is evacuated to a very low pressure and then sealed off while the other has the gas admitted to it. The evacuated tube acts as a compensator to reduce the effect of bridge-excitation-voltage changes and temperature changes on the output reading. Current flowing through the measuring element heats it to a temperature depending on the gas pressure. The electrical resistance of the element

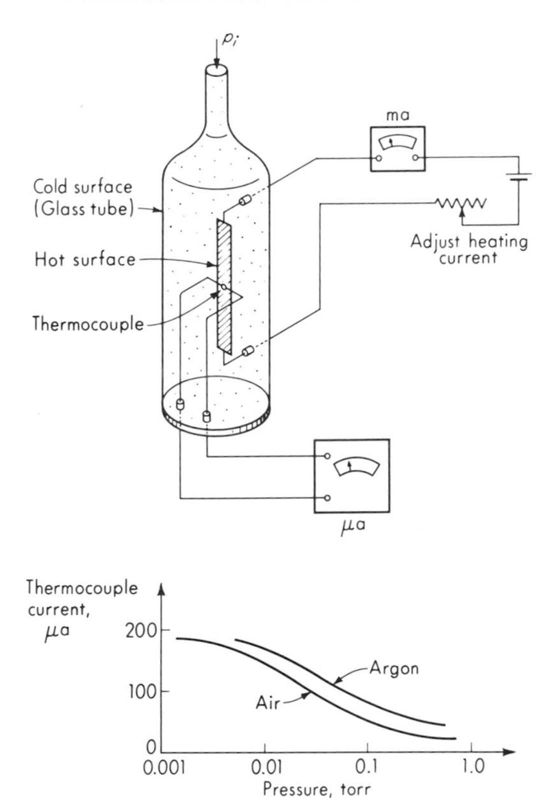

Fig. 6.23. *Thermocouple gage.*

changes with temperature, and this resistance change causes a bridge unbalance. The bridge is generally used as a deflection rather than a null device. To balance the bridge initially, the pressure in the measuring element is made very small and the balance pot set for zero output. Any changes in pressure will cause a bridge unbalance. Of course the gage must be calibrated against some standard. Calibration is nonlinear and varies from one gas to another. Pirani gages cover the range from about 10^{-5} to 1 torr.

Thermistor vacuum gages operate on the same principle as the Pirani gage except that the resistance elements are temperature-sensitive semiconductor materials called thermistors rather than metals such as tungsten or platinum. Thermistor gages are used in the range 10^{-4} to 1 torr.

Ionization gages. An electron passing through a potential difference acquires a kinetic energy proportional to the potential difference. When

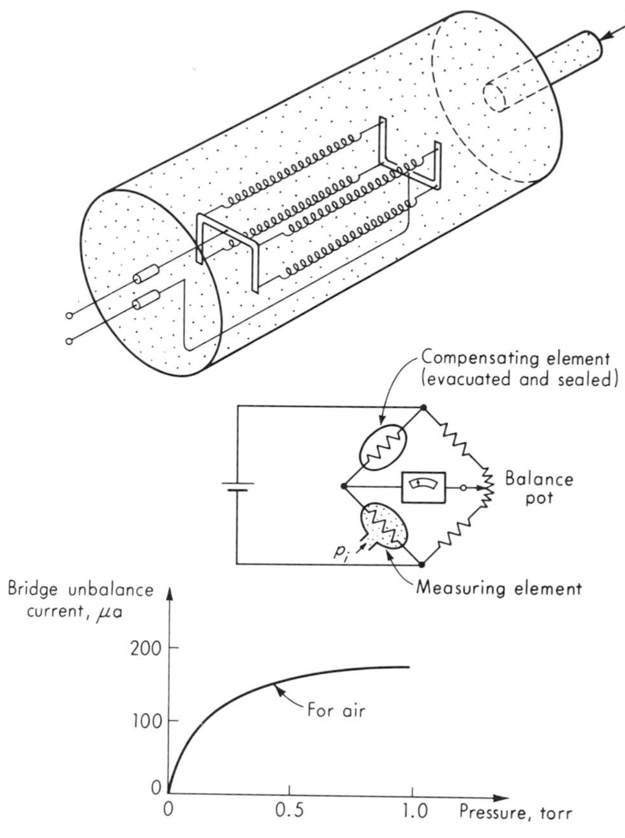

Fig. 6.24. *Pirani gage.*

this energy is large enough, if the electron strikes a gas molecule there is a definite probability that it will drive an electron out of the molecule, leaving it as a positively charged ion. In an ionization gage a stream of electrons is emitted from a cathode. Some of these strike gas molecules and knock out secondary electrons, leaving the molecules as positive ions. For normal operation of the gage the secondary electrons are a negligible part of the total electron current; thus, for all practical purposes, electron current i_e is the same whether measured at the emitting point (cathode) or the collecting point (anode). The number of positive ions formed is directly proportional to i_e and directly proportional to the gas pressure. If i_e is held fixed (as in most gages) the rate of production of positive ions (ion current) is, for a given gas, a direct measure of the number of gas molecules per unit volume and thus of the pressure. The positive ions are attracted to a negatively charged electrode which collects them and

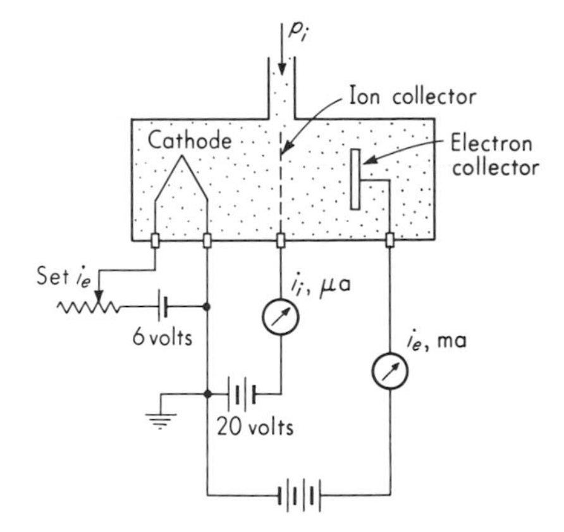

Fig. 6.25. *Ionization gage.*

carries the ion current. The "sensitivity" S of an ionization gage is defined by

$$S \triangleq \frac{i_i}{pi_e} \tag{6.94}$$

where $i_i \triangleq$ ion current, gage output
$i_e \triangleq$ electron current
$p \triangleq$ gas pressure, gage input

According to our usual definition of sensitivity as output/input, the "sensitivity" would be Si_e rather than S. But the definition of Eq. (6.94) makes "sensitivity" independent of i_e and dependent only on gage construction. This allows comparison of the "sensitivity" of different gages without reference to the particular i_e being used. A main advantage of ionization gages in general is their linearity; that is, the sensitivity S is constant for a given gas over a wide range of pressures.

Figure 6.25 shows the basic elements of a hot-cathode ionization gage. The emission of electrons is due to the heating of the cathode. Some disadvantages of hot-cathode gages are filament burnout if exposed to air while hot, decomposition of some gases by the hot filament, and contamination of the measured gas by gases forced out of the hot filament. Hot-cathode gages cover the range from 10^{-10} to 1 torr.

The Philips cold-cathode gage[1] overcomes the problems associated with a high-temperature filament by the use of a cold cathode and a high accelerating potential (2,000 volts). A superimposed magnetic field causes the electrons ejected from the cathode to travel in long helical paths to the anode. The long path results in more collisions with gas molecules

[1] J. M. Lafferty and T. A. Vanderslice, Vacuum Measurement by Ionization, *Instr. Control Systems*, p. 90, March, 1963.

and thus a greater ionization. Philips gages are used in the range 10^{-5} to 10^{-2} torr.

For the lowest pressures, hot-cathode and cold-cathode gages of the magnetron type[1] are available. They are useful down to about 10^{-13} torr. Mass spectrometers[1] are employed for even lower pressures; they allow identification of the partial pressures of components in gas mixtures.

6.10

Sound Measurement The measurement of air-borne and water-borne sound is of increasing interest to engineers. Air-borne-sound measurements are important in the development of less noisy machinery and equipment, in diagnosis of vibration problems, and in the design and test of sound-recording and-reproducing equipment. In large booster rockets the sound pressures produced by the exhaust are large enough to cause fatigue failure of metal panels because of vibration ("acoustic fatigue"). Water-borne sound has been applied in underwater direction and range-finding equipment (sonar). Since most sound transducers (microphones and hydrophones) are basically pressure-measuring devices, it is appropriate to consider them briefly in this chapter.

The basic definitions of sound are in terms of the magnitude of the fluctuating component of pressure in a fluid medium. The *sound pressure level* is defined by

$$SPL \triangleq \text{sound pressure level} \triangleq 20 \log_{10} \frac{p}{0.0002} \quad \text{decibels (db)} \tag{6.95}$$

where

$$p \triangleq \text{root-mean-square (rms) sound pressure, microbar } (\mu b) \tag{6.96}$$

$$1\ \mu b = 1\ \text{dyne/cm}^2 = 1.45 \times 10^{-5}\ \text{psi} \tag{6.97}$$

The rms value of the fluctuating component of pressure is employed because most sounds are random signals rather than pure sine waves. The value 0.0002 μb is an accepted standard reference value of pressure against which other pressures are compared by Eq. (6.95). Note that when $p = 0.0002\ \mu b$ the sound pressure level is 0 db. This value was selected somewhat arbitrarily but represents the average threshold of hearing for human beings, if a 1,000-cps tone is used. That is, the 0-db level was selected as the lowest pressure fluctuation normally discernible by human beings. Since 0 db is about 3×10^{-9} psi, the remarkable sensitivity of the human ear should be apparent. The decibel (logarithmic) scale is used as a convenience because of the great ranges of sound pressure level

[1] J. M. Lafferty and T. A. Vanderslice, Vacuum Measurement by Ionization, *Instr. Control Systems*, p. 90, March, 1963.

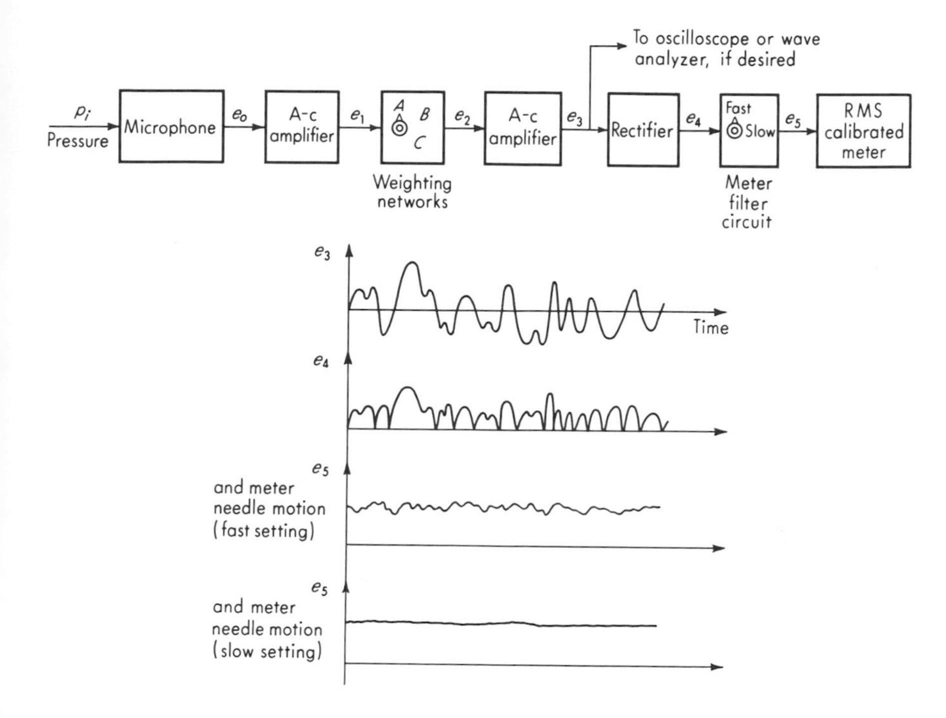

Fig. 6.26. *Sound-level meter.*

of interest in ordinary work. For example, an office with tabulating machines may have an *SPL* of 74 db (1 μb). The average human threshold of pain is 144 db. Sound pressures close to large rocket engines are the order of 170 db (1 psi). One atmosphere (14.7 psi) is 194 db. The range from the lowest to the highest pressures of interest is thus of the order of 10^{-9} to 1, a tremendous range.

The sound-level meter. The most commonly used instrument for routine sound measurements is the sound-level meter. This is actually a measurement *system* made up of a number of interconnected components. Figure 6.26 shows a typical arrangement. The sound pressure p_i is transduced to a voltage by means of the microphone. Microphones generally employ a thin diaphragm to convert pressure to motion. The motion is then converted to voltage by some suitable transducer, usually a capacitance, piezoelectric, or moving-coil type. Microphones usually have a "slow leak" (capillary tube) connecting the two sides of the diaphragm to equalize the average pressure (atmospheric pressure) and prevent bursting of the diaphragm. This is necessary because the (slow) hour-to-hour and day-to-day changes in atmospheric pressure are much greater than the sound-pressure fluctuations to which the microphone must respond. The presence of this leak dictates that microphones will not respond to constant or slowly varying pressures. This is usually no problem since many measurements involve a human response to the sound, and this is known to extend down to only about 10 to 20 cps. Thus the microphone frequency response need go only to this range, not to zero frequency.

The output voltage of the microphone is generally quite small and also at a high impedance level; thus an amplifier of high input impedance and gain is used at the output of the microphone. This can be a relatively simple a-c amplifier since response to static or slowly varying voltages is not required. Capacitor microphones often use for the first stage of the amplifier a cathode follower built right into the microphone housing. This close coupling reduces stray capacitance effects by eliminating cables at the high-impedance end.

Following the first amplifier are the weighting networks. They are electrical filters whose frequency response is tailored to approximate the frequency response of the average human ear. The ear does not interpret equal sound pressure levels at, say, 500 and 1,000 cps as being equally loud. Thus if we are trying to measure the "loudness" of a sound as heard by a person, a flat or uniform frequency response in the measuring instrument is not wanted. Measurements on human hearing have established the shape of the ear's frequency response fairly well; these shapes can be approximated by suitable filter networks. An additional complication is the nonlinearity of the ear's response in that the shape of its frequency-response curve varies with the *amplitude* of the sound. Thus the weighting network for low-level sound should be different from that for high. Most sound-level meters have three different settings (A,B,C) for the weighting networks. The A setting is used for sounds of 55-db level or below, the B for 55 to 85 db, and C for above 85 db. The C setting corresponds to a flat overall system frequency response and thus is used when an actual measurement of sound pressure level (rather than a human response to the sound) is desired. Readings taken on the A or B settings are called *sound level,* not sound pressure level.

The output of the weighting network is further amplified and an output jack provided to lead this signal to an oscilloscope (if observation of the wave form is desired) or to a wave analyzer (if the frequency content of the sound is to be determined). If only the overall sound magnitude is desired, the rms value of e_3 must be found. While true rms voltmeters are available, their expense is usually not justifiable in an ordinary sound-level meter. Rather, the *average* value of e_3 is determined by rectifying and filtering and then the meter scale is *calibrated* to read rms values. This procedure is exact for pure sine waves since there is a precise relation between the average value and the rms value of a sine wave. For nonsinusoidal waves this is not true, but the error is generally small enough to be acceptable for relatively unsophisticated work. The filtering is accomplished both by a simple low-pass RC filter and the low-pass meter dynamics. Some meters have a slow and fast response switch which changes the filtering. The "slow" position gives a steady, easy-to-read needle position but masks any short-term variations in the signal. If these short-term variations are of interest, they may be visually observed on the meter by switching it to fast response. While the meter is actually reading the rms value of e_3 (and thus of p_i), it is calibrated in decibels since Eq. (6.95) establishes a definite relation between sound pressure in microbars and decibels.

Microphones. While the design of microphones is a specialized and complex field with a large technical literature, we can here point out some of the main considerations. Frequency response is still of major interest; however, the effects on frequency response of sound wavelength and direction of propagation are aspects of dynamic behavior not regularly encountered in other measurements. The *pressure response* of a microphone refers to the frequency response relating the actual (uniform) sound pressure existing at the microphone diaphragm to the output voltage of the microphone. The pressure response of a given microphone may be estimated theoretically or measured experimentally by one of a number of accepted methods.[1]

What is usually desired is the *free-field response* of the microphone. That is, what is the relation between the microphone output voltage and the sound pressure that existed at the microphone location *before* the microphone was introduced into the sound field? The reason that the microphone distorts the pressure field is that its acoustical impedance is radically different from that of the medium (air) in which it is immersed. In fact, for most purposes, the microphone (including its diaphragm) may be considered as a rigid body. Sound waves impinging on this body give rise to complex reflections that depend on the frequency, the direction of propagation of the sound wave, and the microphone size and shape. When the wavelength of the sound wave is very large compared with the microphone dimensions (low frequencies), the effect of reflections is negligible for any angle of incidence between the diaphragm and the wave-propagation direction, and the free-field response is the same as the pressure response. At very high frequencies, where the wavelength is much smaller than the microphone dimensions, the microphone acts as an

[1] P. V. Bruel and G. Rasmussen, Free Field Response of Condenser Microphones, *B & K Tech. Rev.*, B & K Instruments Inc., Cleveland, Ohio, no. 1, January, 1959; no. 2, April, 1959.

infinite wall and the pressure at the microphone surface [for waves propagating perpendicular to the diaphragm (0° angle of incidence)] is twice what it would be if the microphone were not there. For waves propagating parallel to the diaphragm (90° incidence angle) the average pressure over the diaphragm surface is zero, giving no output voltage. Between the very low and very high frequencies the effect of reflections is quite complicated and depends on sound wavelength (frequency), microphone size and shape, and angle of incidence.

For simple geometrical shapes such as spheres and cylinders, theoretical results are available.[1] Experiments on actual microphones give results such as those shown in Fig. 6.27. Note that for sufficiently low frequencies (below a few thousand cycles per second) there is little change in pressure due to the presence of the microphone; also the angle of incidence has little effect. This flat frequency range can be extended by reducing the size of the microphone; however, smaller size tends to reduce sensitivity. The size effect is directly related to the relative size of the microphone and the wavelength of the sound. The wavelength λ of sound waves in air is roughly $13{,}000/f$ in., where f is frequency in cycles per second. When λ becomes comparable to the microphone-diaphragm diameter, significant reflection effects can be expected. For example, a 1-in.-diameter microphone would not be expected to have good response much above 13,000 cps. (These limitations can be relaxed to some extent by clever use of acoustical-mechanical techniques.[2])

The lower part of Fig. 6.27 shows a curve labeled "random incidence." This refers to the response to a diffuse sound field where the sound is equally likely to come to the microphone from any direction, the waves from all directions are equally strong, and the phase of the waves is random at the microphone position. Such a field may be approximated by constructing a room with highly irregular walls and placing reflecting objects of various sizes and shapes in it. A source of sound placed in such a room gives rise to a diffuse sound field at any point in the room. Microphones calibrated under such conditions are of interest because many sound measurements take place in enclosures which, while not giving perfect random incidence, certainly do not give pure plane waves.

Microphone calibrations may give the pressure response and also the free-field response for selected incidence angles, usually 0 and 90°. Figure 6.28 shows typical curves.

[1] L. Beranek, "Acoustic Measurements," chap. 3, John Wiley & Sons, Inc., New York, 1949.

[2] Gunnar Rasmussen, Miniature Pressure Microphones, *B & K Tech. Rev.*, B & K Instruments Inc., Cleveland, Ohio, no. 1, 1963.

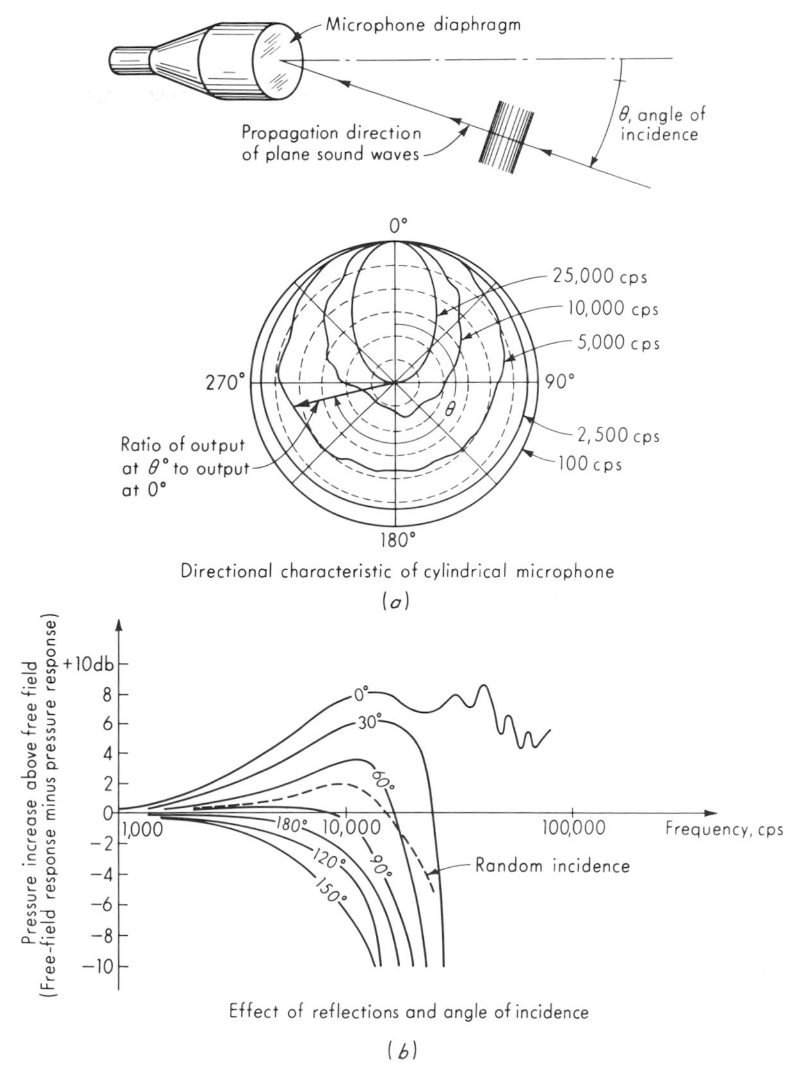

Fig. 6.27. *Microphone response characteristics.*

Pressure response of a capacitor microphone. Of the several types of microphones in common use the capacitor type is generally considered to be capable of the highest performance. Figure 6.29 shows in simplified fashion the construction of a typical capacitor microphone. The pressure response is found by assuming a uniform pressure p_i to exist all around the microphone at any instant of time. This is actually the case for

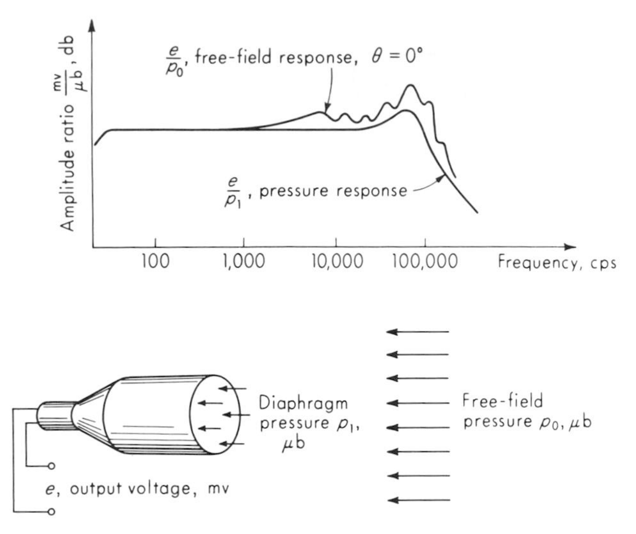

Fig. 6.28. *Free-field and pressure response.*

sufficiently low sound frequencies but reflection and diffraction effects distort this uniform field at higher frequencies, as pointed out earlier.

The diaphragm is generally a very thin metal membrane which is stretched by a suitable clamping arrangement. Diaphragm thickness ranges from about 0.0001 to 0.002 in. The diaphragm is deflected by the sound pressure and acts as the moving plate of a capacitance displacement transducer. The other plate of the capacitor is stationary and may contain properly designed damping holes. Motion of the diaphragm causes air flow through these holes with resulting fluid friction and energy dissipation. This damping effect is utilized to control the resonant peak of the diaphragm response. A diaphragm actually has many natural frequencies; however only the lowest is of interest here. For frequencies near or below the lowest natural frequency the diaphragm behaves essentially like a simple spring-mass-dashpot second-order system and may be analyzed as such.

A capillary air leak is provided to give equalization of steady (atmospheric) pressure on both sides of the diaphragm to prevent diaphragm bursting. For varying (sound) pressures the capillary-volume system results in the varying component of pressure acting *only* on the outside of the diaphragm and thus causing the desired diaphragm deflection.

The variable capacitor is connected into a simple series circuit with a high resistance R and "polarized" with a d-c voltage E_b of about 200 volts. This polarizing voltage acts as circuit excitation and also determines the neutral (zero-pressure) diaphragm position because of the elec-

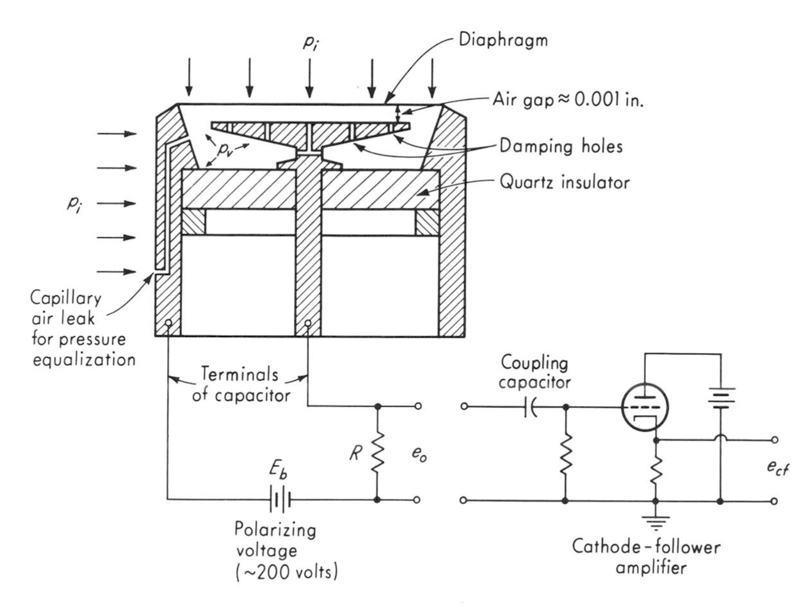

Fig. 6.29. *Capacitor microphone.*

trostatic attraction force between the capacitor plates. For a constant diaphragm deflection no current flows through R and no output voltage e_o exists; thus there is no response to static pressure differences across the diaphragm. For dynamic pressure differences a current *will* flow through R and an output voltage exists. The voltage e_o is usually applied to the input of a capacitance-coupled cathode-follower amplifier. A cathode follower always has a gain less than 1. Thus the purpose of the amplifier is not to increase the voltage level. Rather, the cathode follower has a high input impedance to prevent loading of the microphone, which has a high output impedance. Since the output impedance of the cathode follower is low, its output signal e_{cf} may be coupled into long cables and low-impedance loads without loss of signal magnitude.

The first step in the analysis involves determination of the effective force tending to deflect the diaphragm in terms of the pressure p_i. The relation between p_i and the pressure p_v in the microphone internal volume V may be obtained from Eq. (6.85). We shall neglect the inertia term since in microphones the viscous effect predominates and the filtering effect of the capillary is significant only at low frequencies. Thus we get

$$\tau_l \dot{p}_v + p_v = p_i \tag{6.98}$$

where

$$\tau_l \triangleq \text{leak time constant} \triangleq \frac{128\,\mu L V}{\pi E_a d_t^4} \tag{6.99}$$

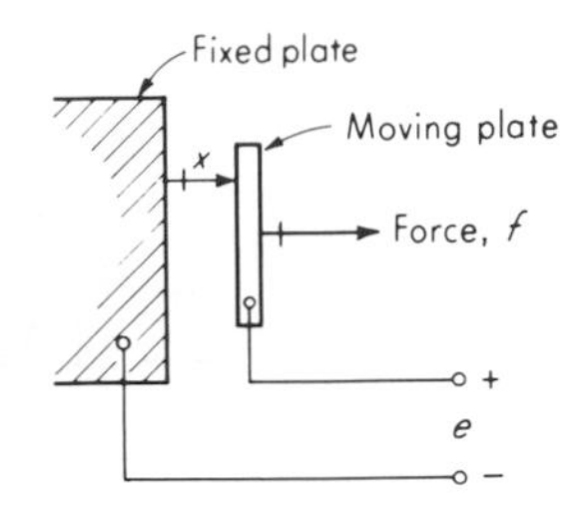

Fig. 6.30. *Moving-plate capacitor.*

Now the deflection of the diaphragm is due to the *difference* between p_i and p_v. Operationally,

$$p_v = \frac{p_i}{\tau_l D + 1}$$

and thus
$$p_i - p_v = p_i\left(1 - \frac{1}{\tau_l D + 1}\right) = \frac{\tau_l D}{\tau_l D + 1}\, p_i \qquad (6.100)$$

The total force f_d on the diaphragm is $A_d(p_i - p_v)$, where A_d is the diaphragm area, giving

$$\frac{f_d}{p_i}(D) = \frac{A_d \tau_l D}{\tau_l D + 1} \qquad (6.101)$$

The frequency response of this shows clearly that $f_d \to 0$ as frequency $\to 0$; thus slow pressure changes do not result in forces tending to burst the diaphragm. However, the time constant τ_l must be small enough so that $(f_d/p_i)(i\omega) \approx A_d$ for all frequencies above about 10 cps, the lowest-frequency sound pressures usually of interest.

The next step requires study of the electromechanical-energy-conversion process in a moving-plate capacitor. While the moving "plate" (diaphragm) of the microphone is not flat, we shall analyze the situation for a flat plate for reasons of simplicity. One can always find a flat-plate capacitor that is equivalent to the diaphragm capacitor in the sense that the capacitance variation with plate separation is the same (at least for small motions) for both. Considering Fig. 6.30, we recall that

$$\text{Energy stored by a capacitor} = \frac{q^2}{2C} = \frac{Ce^2}{2} \qquad (6.102)$$

where $q \triangleq$ charge
$e \triangleq$ voltage
$C \triangleq$ capacitance

We wish to show that the two plates attract each other with a force f. The capacitance of a parallel-plate capacitor whose area A is large compared with the plate separation x is given very closely by

$$C = \frac{\epsilon A}{x} \qquad (6.103)$$

where $\epsilon \triangleq$ permittivity of material between plates
$= 8.86 \times 10^{-12}$ farad/m for vacuum or dry air

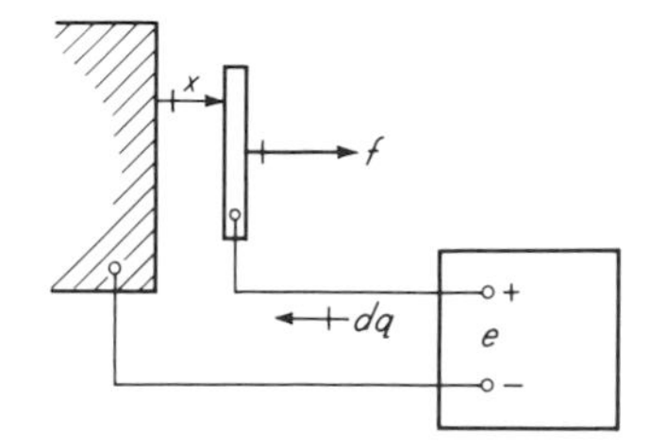

Fig. 6.31. *Capacitor with external circuit.*

We now suppose the capacitor is charged and then open-circuited so that q must remain constant. If the plates are now separated an additional amount dx, we may write

$$\text{Original energy } (x = x_0) = \frac{q^2}{2C_0} = \frac{q^2 x_0}{2\epsilon A} \tag{6.104}$$

$$\text{Final energy } (x = x_0 + dx) = \frac{q^2}{2C_f} = \frac{q^2(x_0 + dx)}{2\epsilon A} \tag{6.105}$$

The energy change is thus $q^2 dx/2\epsilon A$. Since energy is conserved in this system, it must have required a force f on the plate to cause the motion dx, since then mechanical work $f\,dx$ would have been done and converted into electrical energy $(q^2\,dx)/2\epsilon A$. The force f may thus be calculated from

$$\frac{q^2\,dx}{2\epsilon A} = f\,dx$$

$$f = \frac{q^2}{2\epsilon A} = \frac{\epsilon A e^2}{2x^2} \tag{6.106}$$

For air, with e in volts and A and x in any consistent units, this becomes

$$f = 0.99 \times 10^{-12}\,\frac{Ae^2}{x^2} \qquad \text{lb}_f \tag{6.107}$$

As an example, if $e = 200$ volts, $A = 1$ in.^{2}, and $x = 0.001$ in., the force is 0.04 lb_f.

If the capacitor is connected to an external circuit as in Fig. 6.31, we can show Eq. (6.106) still holds as follows: The work done in moving a charge dq through a potential difference e is $e\,dq$. Then, by conservation of energy,

$$f\,dx + e\,dq = d(\text{stored energy}) = d\left(\frac{Ce^2}{2}\right) \tag{6.108}$$

Then,

$$f = -e\,\frac{dq}{dx} + \frac{d}{dx}\left(\frac{Ce^2}{2}\right) = -e\,\frac{d}{dx}\,(Ce) + \frac{d}{dx}\left(\frac{Ce^2}{2}\right) \tag{6.109}$$

$$f = -e\left(C\,\frac{de}{dx} + e\,\frac{dC}{dx}\right) + Ce\,\frac{de}{dx} + \frac{e^2}{2}\frac{dC}{dx} \tag{6.110}$$

$$f = -\frac{e^2}{2}\frac{dC}{dx} = -\frac{e^2}{2}\left(-\frac{\epsilon A}{x^2}\right) = \frac{\epsilon A e^2}{2x^2} \tag{6.111}$$

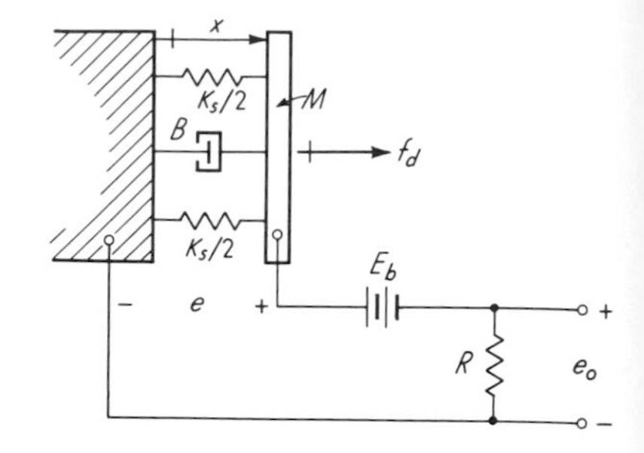

Fig. 6.32. *Microphone model.*

We next model the microphone as in Fig. 6.32. The mass M and spring K_s must be such as to give the same natural frequency as the lowest natural frequency of the diaphragm. The dashpot B must be such as to give the same resonant peak as in the microphone's measured pressure response. The capacitor plate area and air gap (with no external forces acting) must be such as to give the same capacitance as is measured for the microphone under similar conditions. The spring constant K_s and capacitor dimensions must also be such that the force f_d causes a capacitance variation equal (at least for small motions) to that caused in the actual microphone by a pressure difference $p_i - p_v = f_d/A_d$. If all the above conditions are met, the simplified model of Fig. 6.32 will respond essentially the same as the microphone itself. While the equivalent system described is defined in terms of experimental measurements on an existing microphone, microphone designers have available theoretical formulas for estimating these parameters *before* a new microphone is built.

Assuming that the equivalent system is a reasonable model, we can proceed with the analysis. With no force f_d applied and with the capacitor uncharged, the mass will assume an equilibrium position x_{fl}, where x_{fl} is the free length of the springs. If the polarizing voltage E_b is now applied, the moving plate will experience an attractive force and will move to a new position x_0 such that the spring force and electrostatic force just balance (see Fig. 6.33). Now, when pressure force f_d is applied, motion will take place around x_0 as an operating point. To find x_0, we can write

$$K_s(x_{fl} - x_0) = \frac{\epsilon A E_b^2}{2x_0^2} \tag{6.112}$$

This equation in x_0 has two positive solutions x_0 and x_0' for a practical case. The solution (equilibrium position) x_0' is unstable in the sense that any slight motion away from x_0' results in *further* motion away from this point. The desired (stable) equilibrium position is x_0, where small disturbances

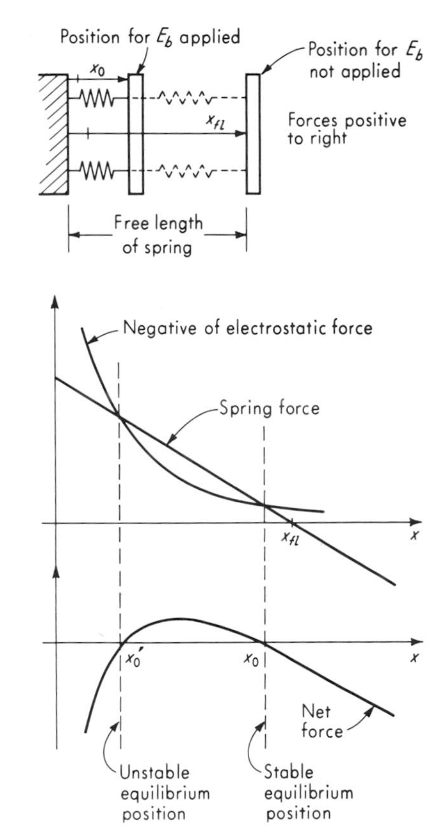

Fig. 6.33. *Determination of equilibrium points.*

from equilibrium give rise to forces tending to restore equilibrium. The microphone must thus be designed to operate at x_0 rather than x_0'.

We now apply Newton's law to the mass M to get

$$-B\frac{dx}{dt} + K_s(x_{fl} - x) - \frac{\epsilon A e^2}{2x^2} + f_d = M\frac{d^2x}{dt^2} \qquad (6.113)$$

The electrostatic-force term makes this differential equation nonlinear. For small changes in e and x from the equilibrium operating point this nonlinear term may be linearized approximately with good accuracy. This may be done by employing only the linear terms of a Taylor-series expansion of the nonlinear function. That is, if in general

$$z = z(x,y)$$

then

$$z \approx z(x_0,y_0) + \left.\frac{\partial z}{\partial x}\right|_{\substack{x=x_0\\y=y_0}}(x - x_0) + \left.\frac{\partial z}{\partial y}\right|_{\substack{x=x_0\\y=y_0}}(y - y_0) \qquad (6.114)$$

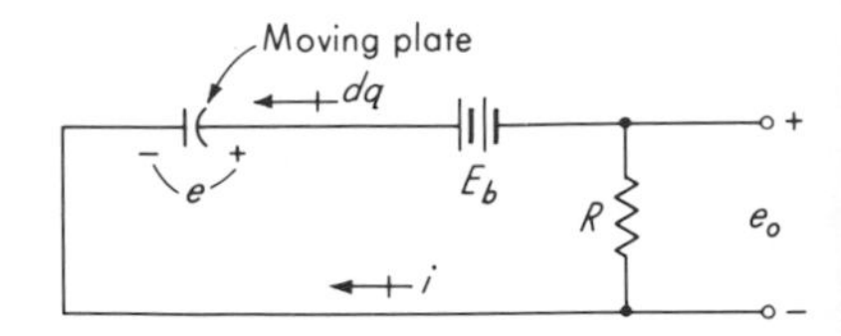

Fig. 6.34. *Circuit analysis.*

In this specific case, the nonlinear function is e^2/x^2; thus

$$\frac{e^2}{x^2} \approx \frac{E_b^2}{x_0^2} + E_b^2\left(-\frac{2}{x_0^3}\right)(x - x_0) + \frac{1}{x_0^2}2E_b(e - E_b) \tag{6.115}$$

We now define $x_1 \triangleq x - x_0$
$e_o \triangleq e - E_b$

to get

$$\frac{e^2}{x_2} \approx \frac{E_b^2}{x_0^2} - \frac{2E_b^2}{x_0^3}x_1 + \frac{2E_b}{x_0^2}e_o \tag{6.116}$$

Also

$$K_s(x_{fl} - x) = K_s(x_{fl} - x_1 - x_0) = -K_s x_1 + \frac{\epsilon A E_b^2}{2x_0^2} \tag{6.117}$$

Equation (6.113) may now be written as

$$-B\frac{dx_1}{dt} - K_s x_1 + \frac{\epsilon A E_b^2}{2x_0^2} - \frac{\epsilon A}{2}\left(\frac{E_b^2}{x_0^2} + \frac{2E_b}{x_0^2}e_o - \frac{2E_b^2}{x_0^3}x_1\right) + f_d = M\frac{d^2x_1}{dt^2} \tag{6.118}$$

Bringing in Eq. (6.101), we may write

$$\left[MD^2 + BD + \left(K_s - \frac{\epsilon A E_b^2}{x_0^3}\right)\right]x_1 + \frac{\epsilon A E_b}{x_0^2}e_o = \frac{A_d\tau_l D}{\tau_l D + 1}p_i \tag{6.119}$$

This equation contains two unknowns, x_1 and e_o; thus an additional equation must be found before a solution can be effected. This can be found from an analysis of the circuit of Fig. 6.34 as follows:

$$e_o = e - E_b = iR = -\frac{dq}{dt}R \tag{6.120}$$

$$q = Ce = \frac{\epsilon A e}{x} \tag{6.121}$$

Equation (6.121) may be linearized as

$$q \approx \frac{\epsilon A E_b}{x_0} - \frac{\epsilon A E_b}{x_0^2}x_1 + \frac{\epsilon A}{x_0}e_o \tag{6.122}$$

Then, approximately,

$$\frac{dq}{dt} = -\frac{\epsilon A E_b}{x_0^2}\frac{dx_1}{dt} + \frac{\epsilon A}{x_0}\frac{de_o}{dt} \tag{6.123}$$

and

$$\frac{e_o}{R} = -\frac{dq}{dt} = \frac{\epsilon A E_b}{x_0^2}\frac{dx_1}{dt} - \frac{\epsilon A}{x_0}\frac{de_o}{dt} \tag{6.124}$$

thus finally giving

$$-\frac{\epsilon A E_b R}{x_0^2}Dx_1 + \left(1 + \frac{\epsilon A R}{x_0}D\right)e_o = 0 \tag{6.125}$$

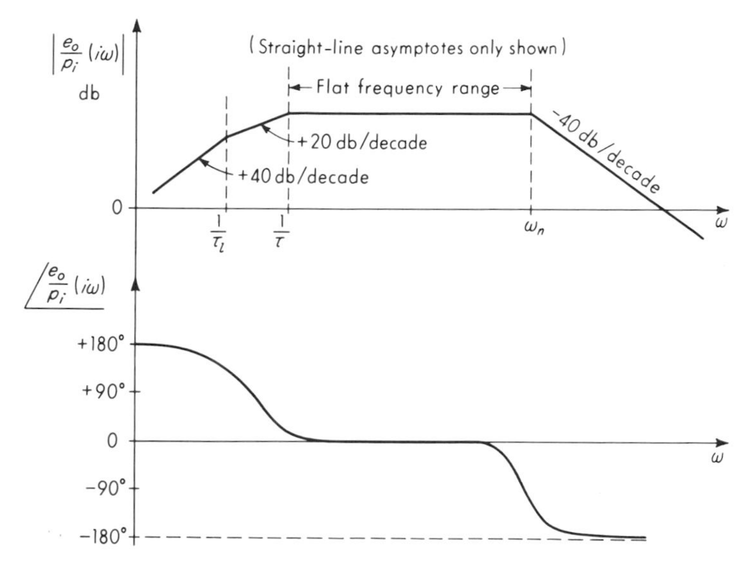

Fig. 6.35. *Microphone frequency response.*

Since we are primarily interested in e_o rather than x_1, Eq. (6.125) may be combined with (6.119) to eliminate x_1 and get

$$\left[\frac{M\tau_e}{K_e}D^3 + \left(\frac{M}{K_e} + \frac{B\tau_e}{K_e}\right)D^2 + \left(\frac{B}{K_e} + \tau_e + \frac{\tau_e{}^2E_b{}^2}{x_0{}^2RK_e}\right)D + 1\right]e_o = \frac{A_dE_b\tau_e}{K_ex_0}\frac{\tau_lD^2}{\tau_lD+1}p_i \tag{6.126}$$

where

$$K_e \triangleq K_s - \frac{\epsilon AE_b{}^2}{x_0{}^3} \tag{6.127}$$

$$\tau_e \triangleq \frac{\epsilon AR}{x_0} \tag{6.128}$$

The cubic left-hand side is not readily factored until numerical values are known. In general, one gets two complex roots and one real root. This leads to a transfer function of the form

$$\frac{e_o}{p_i}(D) = \frac{KD^2}{(\tau_lD+1)(\tau D+1)(D^2/\omega_n{}^2 + 2\zeta D/\omega_n + 1)} \tag{6.129}$$

The frequency response of the microphone is then as shown in Fig. 6.35. The sensitivity in the flat range is typically of the order of 1 to 5 mv/μb while the low-frequency cutoff is about 1 to 10 cps, though lower values are possible. The upper limit of frequency can be extended well beyond the range of human hearing; 100,000 cps is not unattainable.

Problems

6.1 For the system of Fig. 6.1:

a. By what factor must the actual weight of steel weights be multiplied to correct for air buoyancy?

b. What correction must be applied to the platform weight to account for oil buoyancy if the piston is immersed 5 in. and has a diameter of 0.2 in.? Take oil specific weight as 50 lb_f/ft^3.

c. If, in part *b*, air, rather than oil, is the pressure medium, what would the correction be when the gage pressure is 100 psig and temperature is 70°F? Make an estimate, assuming constant air temperature and pressure varying linearly from the high-pressure end of the piston to atmospheric at the top end.

6.2 A well-type mercury manometer used to measure water flow rate is shown in Fig. P6.1 for zero flow rate ($p_1 = p_2$). Derive a relation between $p_1 - p_2$ and h for this configuration.

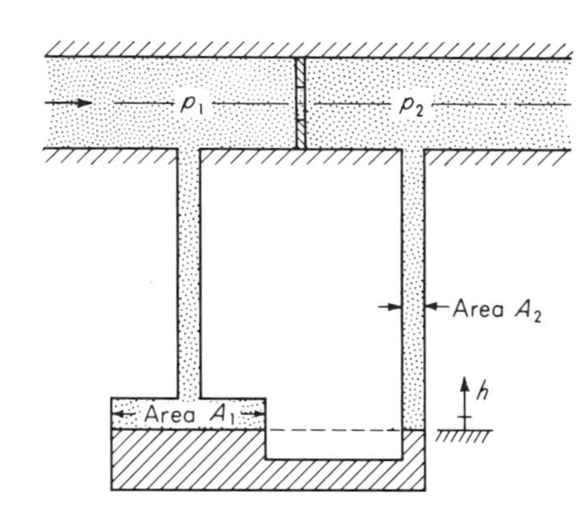

Fig. P6.1

6.3 For the inclined manometer of Fig. 6.3, derive a relation between $p_1 - p_2$ and displacement reading along the calibrated scale.

6.4 Estimate the largest step change that will give linear behavior in a water manometer with $L = 26.5$ in. and $R = 0.13$ in. What are ζ and ω_n for this manometer? If a step change five times the value found above is applied, estimate ζ_e and ω_n for this situation.

6.5 In Eq. (6.44), get an expression for $d\zeta_e/\zeta_e$ by taking the log of both sides and then differentiating. If L, μ, γ, g, R, ω, and Y are each in error by 1 percent, what is the percent error in ζ_e?

6.6 From Eq. (6.46), plot p versus y_c/t for $0 \leq y_c/t \leq 1$ and $E = 25 \times 10^6$ psi, $t = 0.04$ in., $R = 4.0$ in., and Poisson's ratio $= 0.26$.

6.7 Design a pressure pickup and bridge circuit such as that in Fig. 6.9 to meet the following requirements:

Maximum pressure = 100 psig
Natural frequency in vacuum = 10 cps minimum
Maximum nonlinearity by Eq. (6.46) = 3 percent
Full-scale output = 10 mv minimum
Diaphragm material is stainless steel.
Strain gages with 350 ohms resistance, gage factor of 2, and size 0.3 by 0.3 in. are to be used.

6.8 A pressure pickup as in Fig. 6.9 has the following characteristics:

$R = 3.0$ in. $E = 28 \times 10^6$ psi Gage resistance = 120 ohms
$r_r = 2.5$ in. $\mu = 0.26$ Gage factor = 2.0
$r_t = 0.5$ in. $\gamma = 0.3$ lb$_f$/in.^{3} Battery voltage = 5.0 volts
$t = 0.05$ in.

a. Calculate the sensitivity in mv/psi.
b. What is the natural frequency in vacuum?
c. Based on Eq. (6.46), what is the maximum allowable pressure for 2 percent nonlinearity? What is the voltage output at this point?

6.9 Explain in words the operation of the system of Fig. 6.10*a*. Derive an equation relating p_o to $p_1 - p_2$ and show how linearity is achieved even if K_n varies.

6.10 Explain in words the operation of the system of Fig. 6.10*b*. Derive an equation relating e_o to $p_1 - p_2$.

6.11 Perform a linearized dynamic analysis of the system of Fig. 6.10*a* and discuss stability/accuracy tradeoff.

6.12 Perform a linearized dynamic analysis of the system of Fig. 6.10*b* and discuss stability/accuracy tradeoff.

6.13 From Eq. (6.57), compute τ for a system with $C_{vp} = 0.4$ cm^3/psi, $d_t = 0.10$ in., $L = 10$ ft, and $\mu = 0.001$ lb$_f$-sec/ft^2. Using Eq. (6.58), find the dead time associated with this system if the tube is of steel with a wall thickness of 0.02 in., $E_L = 100{,}000$ psi, and the fluid specific weight is 0.03 lb$_f$/in.^{3}. Is this dead time significant relative to τ?

6.14 A pressure transducer has a natural frequency in vacuum of 5,000 cps and $C_{vp} = 0.0003$ in.^{3}/psi. It is used with a liquid of specific weight 0.04 lb$_f$/in.^{3} and viscosity 0.0005 lb$_f$-sec/ft^2. The tubing inside diameter is 0.2 in., and its length is 5 ft. Find ω_n and ζ of the combined transducer/tubing system.

6.15 The pressure pickup of Prob. 6.14 has an internal volume of 0.004 in.^{3}. If it is used with the same tubing as in Prob. 6.14 but if the fluid medium is changed to air at 100 psia and 100°F, what are the values of ζ and ω_n?

6.16 If the transducer of Prob. 6.15 is used with tubing of 1-in. length and 0.1-in. inside diameter, what will ω_n and ζ be?

6.17 Compute the resistance change of 100-ohm coils of Manganin and gold chrome for 50,000-psi pressure and 100°F temperature changes.

6.18 Design a capillary leak for a microphone such that frequencies of 10 cps and above will be measured with an amplitude-ratio error of no more than 10 percent. Assume standard atmospheric air and a leak length L of 1 in. Find the required leak diameter d_t. The microphone has an internal volume of 0.5 in.^{3}. What will be the amplitude ratio for atmospheric pressure drifts of 2-cph frequency?

Bibliography

1. J. B. Damrel: Quartz Bourdon Gage, *Instr. Control Systems,* p. 87, February, 1963.
2. W. R. Myers: The Electromanometer, *Instr. Control Systems,* p. 116, April, 1962.
3. R. R. Koolman: Reference Pressure Cells, *Instr. Control Systems,* p. 123, February, 1962.
4. Dead-weight Testers, *Instr. Control Systems,* p. 126, April, 1962.
5. R. C. Schumacher: Automatic Pressure Calibration Systems, *Instr. Control Systems,* p. 83, February, 1964.
6. H. Norville: A Dead Weight Pressure Balance with Extended Range to 5000 psig, *NASA, N*-64-33684, 1964.

7. S. Siegel: Pressure Calibration Circuits, *Instr. Control Systems*, p. 116, February, 1965.
8. R. C. Cerni: Measuring 100 Pressures in 15 Seconds, *Control Eng.*, p. 89, July, 1964.
9. R. I. Kreisler: Rocket Propellant Manometer System, *ISA J.*, p. 55, October, 1962.
10. W. W. Willmarth: Wall Pressure Fluctuations in a Turbulent Boundary Layer, *NACA, Tech. Note* 4139, 1958.
11. P. K. Stein: Measuring Fluctuating Pressure, *Instr. Control Systems*, p. 156, September, 1964.
12. Y. T. Li: Pressure Transducers for Missile Testing and Control, *ISA J.*, p. 81, November, 1958.
13. L. R. Voss et al.: Recent Developments in Balanced-diaphragm Pressure Transducers, *ISA J.*, p. 348, September, 1955.
14. A. F. Welch et al.: Auxiliary Equipment for the Capacitor-type Transducer, *ISA J.*, p. 548, December, 1955.
15. E. J. Rogers: Semiconductor Pressure Transducer Features Mechanical Compensation, *Instr. Control Systems*, p. 128, April, 1963.
16. D. B. Clark: Rare-earth Pressure Transducers, *Instr. Control Systems*, p. 93, February, 1963.
17. J. C. Sanchez: Semiconductor Strain-gage Pressure Sensors, *Instr. Control Systems*, p. 117, November, 1963.
18. Y. T. Li: Two-cylinder Transducer Has Straight Line Response, *Control Eng.*, p. 151, April, 1962.
19. Y. Kobashi et al: Improvements of a Pressure Pickup for the Measurement of Turbulence Characteristics, *J. Aerospace Sci.*, p. 149, February, 1960.
20. T. Wrathall: Measuring Impact Pressures on Re-entering Missile Nose Cones. *ISA J.*, p. 54, October, 1959.
21. G. E. Reis: Theoretical Examination of Variable Reluctance Diaphragm Gage, Sandia Corp., Albuquerque, N.Mex., *SCR*-162, 1960.
22. Capacitive Pressure Sensors, *Instr. Control Systems*, p. 119, May, 1962.
23. P. Smelser: Pressure Measurements in Cryogenic Systems, National Bureau of Standards, Boulder, Colo.
24. H. Chelner: High Frequency Semiconductor Probe Pressure Transducer, *AIAA Paper* 64-508, 1964.
25. J. H. Thomson: Torsion Bar Pressure Transducer, *Electromech. Design*, p. 46, June, 1964.
26. D. S. Johnson: Design and Application of Piezoceramic Transducers to Transient Pressure Measurements, *NASA*, *N*-63-18139, 1962.
27. R. E. Engdahl: Pressure Measuring Systems for Closed Cycle Liquid Metal Facilities, *NASA*, *CR*-54140.
28. J. A. Haner: Pressure/Displacement Transducer, *Instr. Control Systems*, p. 107, November, 1964.
29. D. D. Keough et al: Piezoresistive Pressure Transducer, *ASME Paper* 64-WA/PT-5, 1964.
30. Pressure Transducer $\frac{3}{8}$-inch in Size Can be Faired into Surface, *NASA*, *Brief* 64-10021, 1964.
31. Welded Pressure Transducer Made as Small as $\frac{1}{8}$-inch Diameter, *NASA*, *Brief* 63-10429, 1963.
32. Improved Variable-reluctance Transducer Measures Transient Pressure, *NASA*, *Brief* 63-10321, 1963.
33. H. B. Jones et al.: Transient Pressure Measurements in Liquid Propellant Rocket Thrust Chambers, *ISA Trans.*, p. 117, April, 1965.

34. R. L. Ledford and W. E. Smotherman: Miniature Transducers for Pressure and Heat Transfer Rate Measurements in Hypervelocity Wind Tunnels, *ISA Trans.*, p. 133, April, 1965.
35. P. S. Lederer and R. O. Smith: An Experimental Technique for the Determination of the Fidelity of the Dynamic Response of Pressure Transducers, *Natl. Bur. Std.* (*U.S.*), *Rept.* 7862, 1963.
36. R. O. Smith: A Liquid-medium Step-function Pressure Calibrator, *ASME Paper* 63-WA-263, 1963.
37. J. L. Schweppe: Calibration of Pressure Transducers with Aperiodic Input-function Generators, *ISA Trans.*, p. 72, January, 1964.
38. W. C. Bentley and J. J. Walter: Dynamic Response Testing of Transient Pressure Transducers for Liquid Propellent Rocket Combustion Chambers, *NASA*, *CR*-51995, 1963.
39. W. E. Amend: Dynamic Performance of Pressure Transducers in Shock and Detonation Tubes, *NASA*, *N*-65-13313.
40. D. Baganoff: Pressure Gauge with One-tenth Microsecond Risetime for Shock Reflection Studies, *Rev. Sci. Instr.*, p. 288, March, 1964.
41. E. L. Davis: The Measurement of Unsteady Pressures in Wind Tunnels, *AGARD Rept.* 169, March, 1958.
42. R. Oldenburger and R. E. Goodson: Hydraulic Line Dynamics, *NASA*, *CR*-52148, 1963.
43. T. R. Stalzer and G. J. Fiedler: Criteria for Validity of Lumped-parameter Representation of Ducting Air-flow Characteristics, *ASME Trans.*, p. 833, May, 1957.
44. F. Nagao and M. Ikegami: Errors of an Indicator Due to a Connecting Passage, *Bull. JSME*, vol. 8, no. 29, 1965.
45. A. L. Ducoffe and F. M. White: The Problem of Pneumatic Pressure Lag, *ASME Trans.*, p. 234, June, 1964.
46. A. S. Iberall: Attenuation of Oscillatory Pressures in Instrument Lines, *Natl. Bur. Std.* (*U.S.*), *Res. Paper* RP2115, July, *ASME Trans.*, 1950.
47. F. Nagao et al.: Influence of the Connecting Passage of a Low Pressure Indicator on Recording, *Bull. JSME*, vol. 6, no. 21, 1963.
48. C. B. Schuder and G. C. Blunck: The Driving Point Impedance of Fluid Process Lines, *ISA Trans.*, p. 39, January, 1963.
49. R. P. Benedict: The Response of a Pressure-sensing System, *ASME Paper* 59-A-289, 1959.
50. R. J. Martin and D. S. Moseley: Analysis of the Effect of Pulsations on the Response of Mercurial-type Differential-pressure Recorders, *ASME Trans.*, p. 1343, October, 1958.
51. J. E. Broadwell and A. G. Hammitt: Transient Response of Fluid Systems, *J. Aerospace Sci.*, July, 1962.
52. A. F. D'Souza and R. Oldenburger: Dynamic Response of Fluid Lines, *ASME Paper* 63-WA-73, 1963.
53. R. Oldenburger and R. E. Goodson: Simplification of Hydraulic Line Dynamics by Use of Infinite Products, *ASME Paper* 62-WA-55, 1962.
54. J. D. Regetz: Experimental Determination of the Dynamic Response of a Long Hydraulic Line, *NASA*, *Tech. Note* D-576, 1960.
55. W. Lewis et al.: Study of the Effect of a Closed-end Side Branch on Sinusoidally Perturbed Flow of Liquid in a Line, *NASA*, *Tech. Note* D-1876, 1963.
56. I. Taback: The Response of Pressure Measuring Systems to Oscillating Pressures, *NACA*, *Tech. Note* 1819, 1949.
57. F. T. Brown: The Transient Response of Fluid Lines, *ASME Paper* 61-WA-143, 1961.

58. C. B. Schuder and R. C. Binder: The Response of Pneumatic Transmission Lines to Step Inputs, *ASME Trans.*, p. 578, December, 1959.
59. High Pressure Measurement, *Mech. Eng.*, p. 76, February, 1963.
60. D. H. Newhall and L. H. Abbott: High Pressure Measurement, *Instr. Control Systems*, p. 232, February, 1961.
61. W. H. Howe: High-pressure Measurement and Control, *Control Eng.*, p. 53, April, 1955.
62. A. J. Yerman: The Tunnel Diode as an FM Hydrostatic Pressure Sensor, *ASME Paper* 63-WA-264, 1963.
63. R. J. Melling: Ionization Vacuum Gage Measures Absolute Pressures up to 1 mm Hg, *Instr. Control Systems*, p. 119, September, 1964.
64. Vacuum Instrumentation, *Instr. Control Systems*, p. 110, September, 1964.
65. J. M. Lafferty and T. A. Vanderslice, Vacuum Measurement by Ionization, *Instr. Control Systems*, p. 90, March, 1963.
66. A. P. Flanick and J. Ainsworth: A Thermistor Pressure Gage, *NASA*, *Tech. Note* D-504, 1960.
67. J. M. Benson: Calibrating Thermal Conductivity Gauges, *Instr. Control Systems*, p. 115, September, 1964.
68. J. P. Walsh: Molecular Vacuum Gages, *Instr. Control Systems*, p. 106, August, 1963.
69. W. Kreisman: Extension of the Low Pressure Limit of McLeod Gages, *NASA*, *CR*-52877, 1963.
70. M. P. Hnilicka: Extreme High Vacuum, *Ind. Res.*, p. 36, September, 1964.
71. J. M. Benson: Thermal Conductivity Vacuum Gages, *Instr. Control Systems*, p. 98, March, 1963.
72. D. Alpert: Theoretical and Experimental Studies of the Underlying Processes and Techniques of Low Pressure Measurement, *NASA*, *N*-64-17582, 1964.
73. P. J. Bryant et al.: Extreme Vacuum Technology, *NASA*, *CR*-84, 1964.
74. Vacuum Instrumentation, *Instr. Control Systems*, p. 113, October, 1964.
75. Precision Gage Measures Ultrahigh Vacuum Levels, *NASA*, *Brief* 63-10597, 1963.
76. Absolute Pressure Gage Feasibility Study, *NASA*, *CR*-58075, 1963.
77. J. Gavis: Vacuum Gage Systems, *NASA*, *N*-64-28208, 1964.
78. R. W. Roberts: Ultrahigh Vacuum Technology, General Electric Co., Schenectady, N.Y., *Rept.* 64-RL-3644C, 1964.
79. R. W. Roberts: An Outline of Vacuum Technology, General Electric Co., Schenectady, N.Y., *Rept.* 64-RL-3394C, 1964.
80. L. T. Melfi and P. R. Yeager: A Method for Calibration of Gas-composition Sensitive Pressure Gages in Condensible Vapors, *NASA*, *Tech. Note* D-2567, 1965.
81. F. Feakes et al.: Gauge Calibration Study in Extreme High Vacuum, *NASA*, *CR*-167, 1965.

7 Flow measurement

7.1

Local Flow Velocity, Magnitude, and Direction In many experimental studies of fluid flow phenomena it is necessary to determine the magnitude and/or direction of the flow-velocity vector at a "point" in the fluid and how this varies from point to point. That is, a description of the flow field is desired. While the conditions at a mathematical *point* are not susceptible to direct measurement, the *average* conditions over a small area or volume can be determined with suitable instruments.

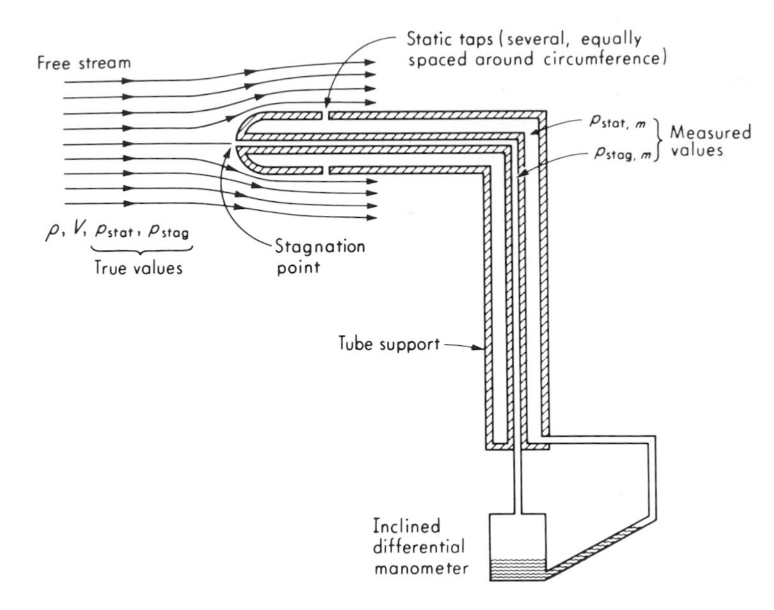

Fig. 7.1. *Pitot-static tube.*

Velocity magnitude from pitot-static tube. In some situations the direction of the velocity vector is known with sufficient accuracy without taking any measurements. If the direction is not known, it may be found in several ways discussed later. Let us assume the direction is known, so that a pitot-static tube may be properly aligned with this direction, as in Fig. 7.1. Assuming steady one-dimensional flow of an incompressible frictionless fluid, we can derive the well-known result

$$V = \sqrt{\frac{2(p_{stag} - p_{stat})}{\rho}} \tag{7.1}$$

where $V \triangleq$ flow velocity
$\rho \triangleq$ fluid mass density
$p_{stag} \triangleq$ stagnation or total pressure, free stream
$p_{stat} \triangleq$ static pressure, free stream

In an actual pitot-static tube, deviations from the ideal theoretical result of Eq. (7.1) arise from a number of sources. If ρ is accurately known, the errors can be traced to inaccurate measurement of p_{stag} and p_{stat}.

The static pressure is usually the more difficult to measure accurately. The difference between true (p_{stat}) and measured ($p_{stat,m}$) values of static pressure may be due to the following:

1. Misalignment of the tube axis and velocity vector. This exposes the static taps to some component of velocity.
2. Nonzero tube diameter. Streamlines next to the tube must be longer than those in undisturbed flow, indicating an increase in velocity. This is accompanied by a decrease in static pressure, making the static taps read low. A similar (and possibly more severe) effect occurs if a tube is inserted in a duct whose cross-sectional area is not much larger than that of the tube.
3. Influence of stagnation point on the tube-support leading edge. This higher pressure causes the static pressure upstream of the leading edge also to be high. If the static taps are too close to the support, they will read high because of this effect. Note that this error and that of item 2 above tend to cancel. By proper design, effective cancellation may be achieved[1] (see also Prandtl pitot tube. Fig. 2.17*b*). Figure 7.2 shows the nature of both these errors as revealed by experimental tests.[2]

An important application of the pitot-static tube is found in aircraft and missiles. Here the stagnation- and static-pressure readings of a tube fastened to a vehicle are used to determine the airspeed and Mach number while the static reading alone is used to measure altitude. If altitude is to be measured with an error of 100 ft, the static pressure must be accurate to 0.5 percent.[1] To achieve this accuracy, methods for compensating errors of the type mentioned in items 1, 2, and 3 above have been developed and reported.[3] An interesting and useful result of these studies is a simple method for reducing error due to angular misalignment. It was found that by locating the static-pressure taps as in Fig. 7.3 the measurement was essentially insensitive to angle of attack for the range $-2° < \alpha < 12°$ and Mach numbers in the range 0.4 to 1.2. While this method, as shown, is effective only for misalignment in the particular plane shown, it can be extended to arbitrary directions of misalignment by designing the probe with a single vane to rotate about its longitudinal axis and automatically locate the taps 37.5° from the cross-flow stagnation point.[4] It is also possible to design multiple-vaned probes[4] mounted on a gimbal system with complete rotational freedom. These probes

[1] V. S. Ritchie, Several Methods for Aerodynamic Reduction of Static-pressure Sensing Errors for Aircraft at Subsonic, Near-sonic, and Low Supersonic Speeds, *NASA, Tech. Rept.* R-18, 1959.

[2] R. G. Folsom, Review of the Pitot Tube, *Trans. ASME,* p. 1450, October, 1956.

[3] Ritchie, *op. cit.*

[4] F. J. Capone, Wind-tunnel Tests of Seven Static-pressure Probes at Transonic Speeds, *NASA, Tech. Note* D-947, 1961.

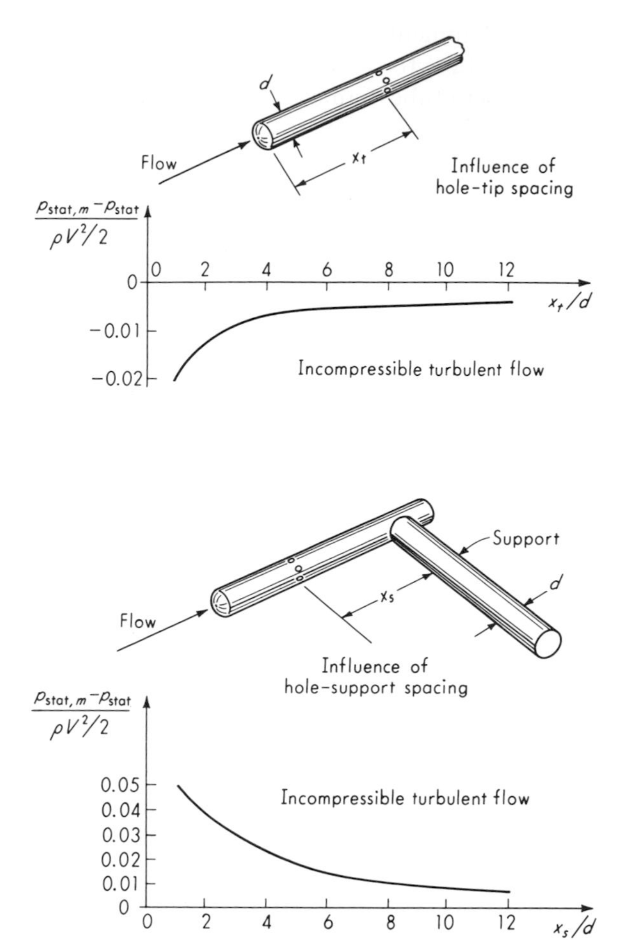

Fig. 7.2. *Static-pressure errors.*

act in a fashion similar to a weather vane (except that they have angular freedom about two axes) and thus align themselves with the velocity vector. A conventional ring of evenly spaced static taps then gives accurate readings. By measuring the rotations of the gimbals, such probes also provide information on the *direction* of the velocity.

While errors in the stagnation pressure are likely to be smaller than those in the static pressure, several possible sources of error are present, namely:

1. Misalignment. This situation prevents formation of a true stagnation point at the measuring hole since the velocity is not zero. Tubes

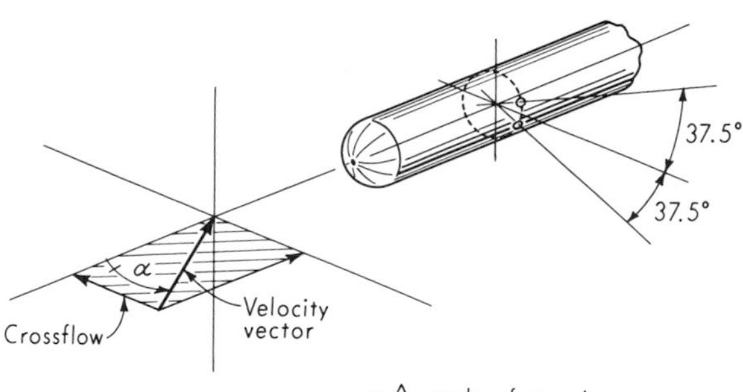

Fig. 7.3. *Probe insensitive to misalignment*

of special design have been developed which exhibit considerable tolerance to misalignment.[1,2] Figure 7.4 shows an example of such a tube which has an error less than 1 percent of the velocity pressure $\rho V^2/2$ for misalignments up to $\pm 38°$ for velocities from low subsonic to Mach 2. Conventional tubes not specifically designed for misalignment insensitivity may show 1 percent errors at only 5 or 10°.

[1] W. Gracey, D. E. Coletti, and W. R. Russell, Wind-tunnel Investigation of a Number of Total-pressure Tubes at High Angles of Attack, Supersonic Speeds, *NACA, Tech. Note* 2261, January, 1951.

[2] W. Gracey, W. Letko, and W. R. Russell, Wind-tunnel Investigation of a Number of Total-pressure Tubes at High Angles of Attack, Subsonic Speeds, *NACA, Tech. Note* 2331, April, 1951.

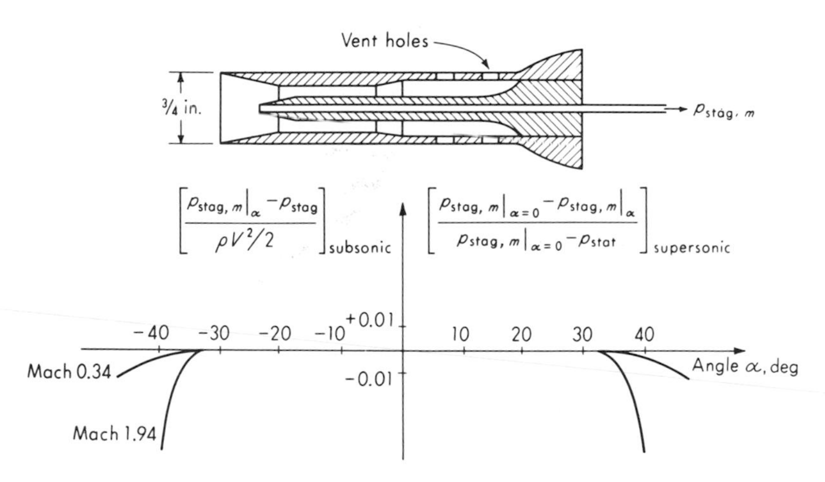

Fig. 7.4. *Special stagnation probe.*

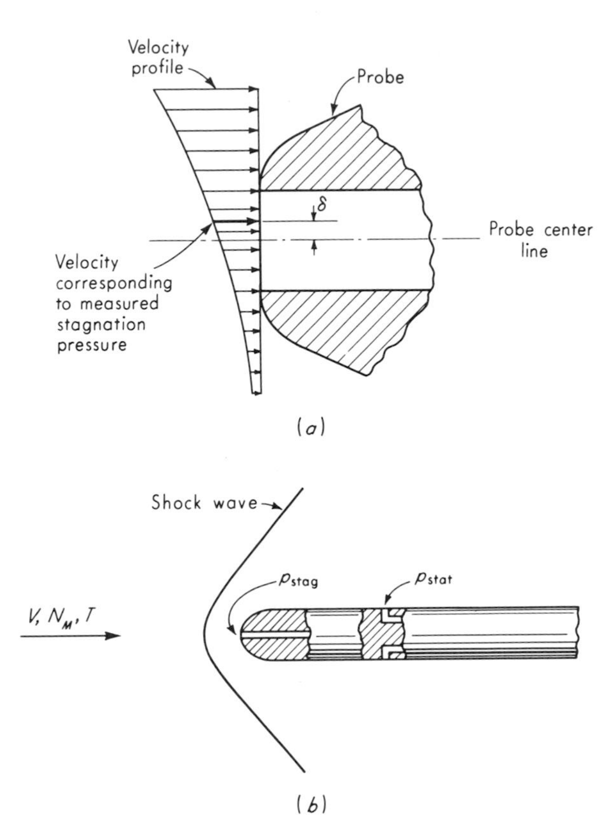

Fig. 7.5. *Nonuniform-velocity profile and supersonic probe.*

2. Two- and three-dimensional velocity fields. When the velocity is not uniform, a probe of finite size intercepts streamlines of different velocities and the stagnation pressure measured corresponds to some sort of average velocity (see Fig. 7.5*a*). For the two-dimensional situation of Fig. 7.5*a*, if one knew the displacement δ, he could assign the measured stagnation pressure (and thus velocity) to a specific point in the flow. Some limited data[1] on this problem are available.
3. Effect of viscosity. Equation (7.1) assumes the fluid to be frictionless. At sufficiently low Reynolds number the viscosity of the fluid exerts a noticeable additional force at the stagnation hole, causing the stagnation pressure to be higher than predicted by Eq. (7.1). This effect can be taken into account by introducing a correction

[1] Folsom, *op. cit.*, p. 1451.

factor C as follows:

$$p_{stag,m} = p_{stat} + \frac{C\rho V^2}{2} \qquad (7.2)$$

For negligible viscosity effects, $C = 1.0$ and Eq. (7.2) is the same as (7.1). For a given probe, the factor C is a function of Reynolds number only and may be found theoretically for simple probe shapes such as spheres and cylinders. A typical result[1] for a cylindrical probe is

$$C = 1 + \frac{4}{N_R} \qquad 10 < N_R < 100 \qquad (7.3)$$

where Reynolds number $\triangleq N_R \triangleq \frac{V\rho r}{\mu}$

$r \triangleq$ probe radius

$\mu \triangleq$ fluid viscosity

Equation (7.3) shows that the effect is about 4 percent of the velocity pressure $\rho V^2/2$ at $N_R = 100$. At $N_R = 10$, however, the effect is 40 percent. Theory and experimental tests[1] show that viscosity corrections are rarely needed for $N_R > 500$, no matter what the shape of the probe.

When a pitot-static tube is used in a compressible fluid, Eq. (7.1) no longer applies, although it may be sufficiently accurate if the Mach number is low enough. For subsonic flow (Mach number $N_M < 1$) the velocity is given by[2]

$$V = \sqrt{\frac{2k}{k-1}\frac{p_{stat}}{\rho_{stat}}\left[\left(\frac{p_{stag}}{p_{stat}}\right)^{(k-1)/k} - 1\right]} \qquad (7.4)$$

where $$k \triangleq \frac{\text{specific heat at constant pressure}}{\text{specific heat at constant volume}} = \frac{C_p}{C_v} \qquad (7.5)$$

Measurement of free-stream density ρ_{stat} requires knowledge of static temperature, which may itself be a difficult measurement. Equation (7.4) may be rewritten as

$$p_{stag} = p_{stat}\left[1 + \frac{k-1}{2}\left(\frac{V}{c}\right)^2\right]^{k/(k-1)} \qquad (7.6)$$

where $$c \triangleq \text{acoustic velocity} = \sqrt{\frac{kp_{stat}}{\rho_{stat}}} = \sqrt{kgRT} \qquad (7.7)$$

and $g \triangleq$ gravitational acceleration

$T \triangleq$ free-stream static temperature

$R \triangleq$ gas constant

[1] Folsom, *op. cit.*, p. 1453.

[2] R. C. Binder, "Advanced Fluid Dynamics and Fluid Machinery," p. 51, Prentice-Hall, Inc., Englewood Cliffs, N.J., 1951.

The right side of Eq. (7.6) may be expanded in a power series to give

$$p_{stag} = p_{stat} + \left(\rho_{stat}\frac{V^2}{2}\right)\left(1 + \frac{N_M^2}{4} + \frac{2-k}{24}N_M^4 + \cdots\right) \tag{7.8}$$

where

$$N_M \triangleq \frac{V}{c} \tag{7.9}$$

Since the Mach number of an incompressible fluid is zero, Eq. (7.8) shows that p_{stag} is higher for compressible than for incompressible flow. Also, if N_M is sufficiently small, Eq. (7.8) is closely approximated by Eq. (7.1).

For supersonic flow ($N_M > 1$) a compression shock wave forms ahead of the pitot tube. Between this shock wave and the tube end the velocity is subsonic. This subsonic velocity is then reduced to zero at the tube stagnation point (see Fig. 7.5*b*). Analysis[1] gives the following formula for computing the free-stream Mach number and thereby the velocity:

$$\frac{p_{stag}}{p_{stat}} = N_M^2\left(\frac{k+1}{2}\right)^{k/(k-1)}\left[\frac{2kN_M - k + 1}{N_M^2(k+1)}\right]^{1-1/(k-1)} \tag{7.10}$$

The measurement of stagnation and static pressures may be combined in a single probe, as in Fig. 7.1, or two separate probes, one for stagnation and one for static, may be employed. Figure 7.6 shows several examples[2] of commonly used forms. The wedge static-pressure probe of Fig. 7.6*a* also can be used to measure velocity direction in a single plane. When the two static taps read equal pressures the wedge is aligned with the flow. This probe is usable for both subsonic and supersonic flow. At Mach 0.9 the sensitivity to misalignment is about 1.5 in. of water per angular degree. The total (stagnation) probes of Fig. 7.6*b* and *c* are also intended for both sub- and supersonic flow. The simple tube is insensitive to misalignment up to about $\pm 20°$ while the venturi shielded tube is good to $\pm 50°$. The boundary-layer probe is usable up to Mach 1.0 and is insensitive to misalignment up to $\pm 5°$. Boundary-layer thickness can be measured with such a probe with an error of the order of 0.002 in. The probe and associated pressure-measuring equipment have a long time lag because of the small flow passage (0.001 in.) at the probe tips.

Velocity direction from yaw tube, pivoted vane, and servoed sphere. In addition to laboratory studies of flow processes in fluid machinery, ducting, etc., flow-velocity direction information is also of

[1] Binder, *op. cit.*, p. 52.

[2] Aero Research Instrument Co., Chicago, Ill.

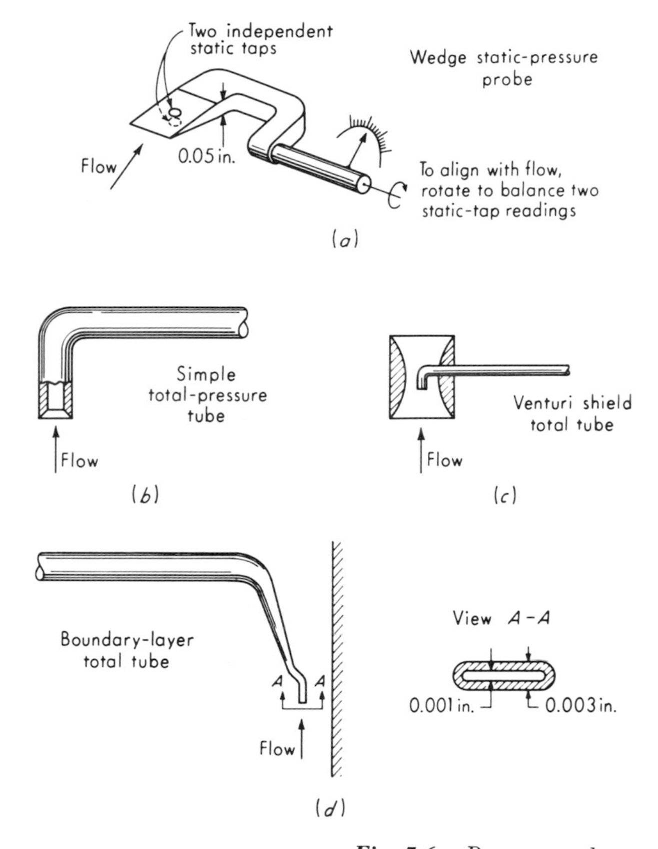

Fig. 7.6. *Pressure probes.*

interest in flight vehicles[1] where angle-of-attack measurements are utilized in attitude measurement and control, stability augmentation, and gust alleviation systems.

So-called yaw tubes[2] of one form or another are conventionally employed to determine the direction of local flow velocity. Perhaps the simplest form, useful for finding the angular inclination in one plane only, is shown in Fig. 7.7*a*. Taps 1 and 3 are connected to a differential-pressure instrument that reads zero when the tube is aligned with the flow. A central tap 2 is often included to read the stagnation pressure after alignment is attained (valid only if the angle of attack is zero). The

[1] H. H. Koelle, "Handbook of Astronautical Engineering," p. 13-33, McGraw-Hill Book Company, New York, 1961.

[2] Aero Research Instrument Co., Chicago, Ill.

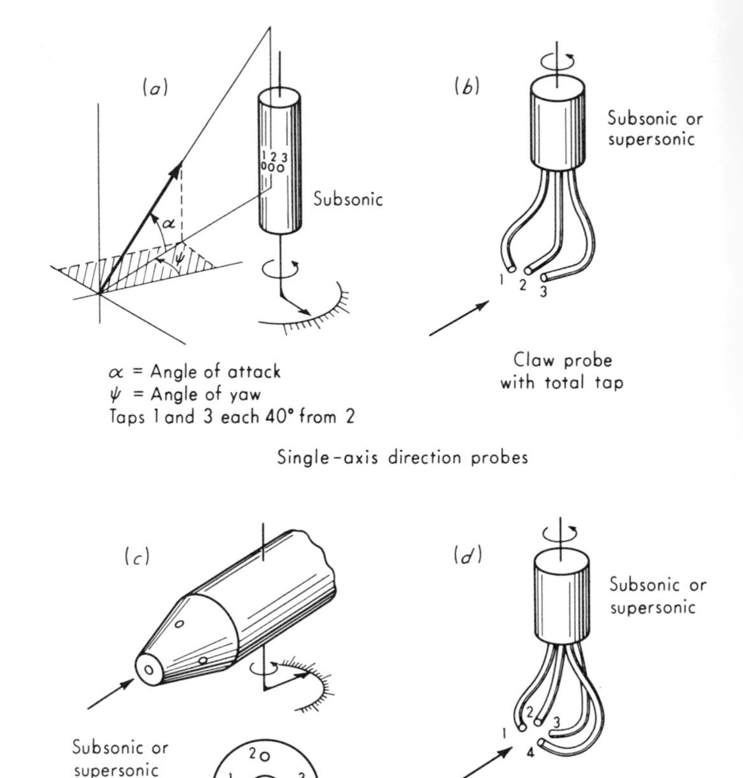

Fig. 7.7. *Flow-direction probes.*

claw tube of Fig. 7.7*b* operates on similar principles but may be used in regions where the flow direction changes greatly since its sensing holes may be located very close together. The two-axis probes of Fig. 7.7*c* and *d* could conceivably be designed to allow rotation about each axis; however, the complexity and size of such a design are generally prohibitive. Thus probe operation consists of rotation about the probe axis to balance taps 1 and 3. Then pressures 2 and 4 are each measured, and calibration charts give the angle of attack. Tap 5 does not read stagnation pressure directly; this can be obtained from calibration charts. Any of these probes may be made automatically self-aligning by using the pressure difference $p_1 - p_3$ as the error signal in a servosystem which

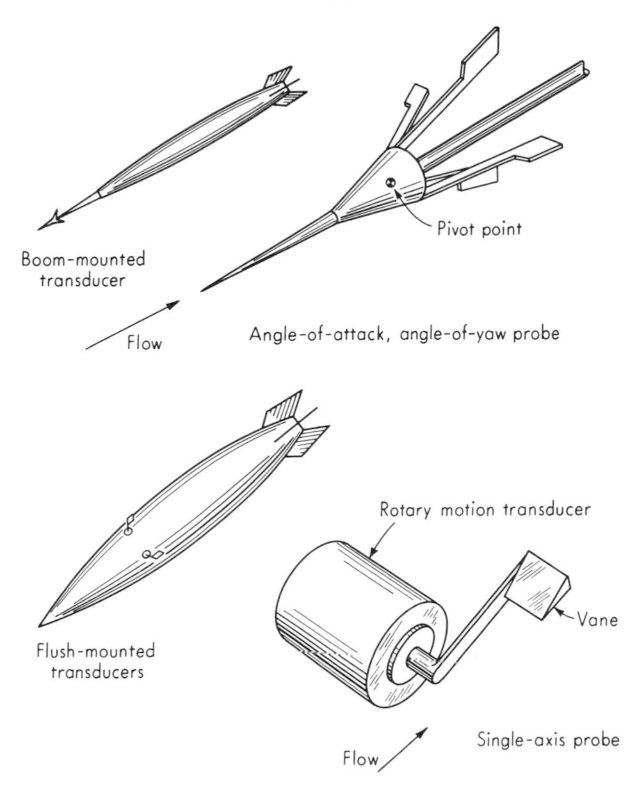

Fig. 7.8. *Vane-type probes.*

rotates the probe until a null is achieved. The details of a system of this type are shown in Fig. 7.9.

Determination of angles of attack and yaw aboard flight vehicles is often accomplished with vane-type probes as in Fig. 7.8. These devices are essentially one- or two-axis weather vanes with suitable damping to reduce oscillation and with motion pickups to provide electrical angle signals. A dynamic analysis of these devices is available.[1] Limitations of this type of device for certain high-speed, high-altitude applications have led to the development of the servoed-sphere type of sensor shown in Fig. 7.9. The one shown was developed[2] for the X-15 rocket research aircraft. A servo-driven sphere is continuously and automatically aligned with the velocity vector by means of two independent servo-systems using the differential-pressure signals $p_1 - p_2$ and $p_3 - p_4$ as

[1] G. J. Friedman, Frequency Response Analysis of the Vane-type Angle of Attack Transducer, *Aero/Space Eng.*, p. 69, March, 1959.

[2] Northrop Corp., Nortronics Div., *Rept.* NORT60-46.

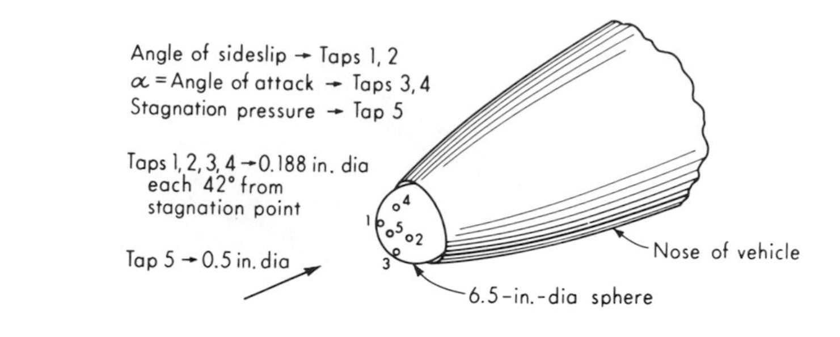

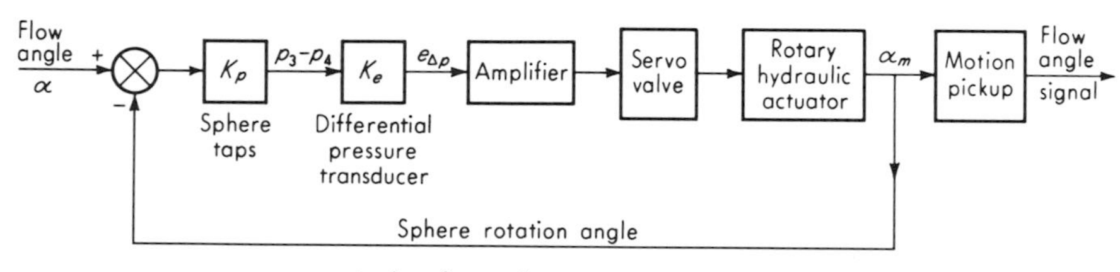

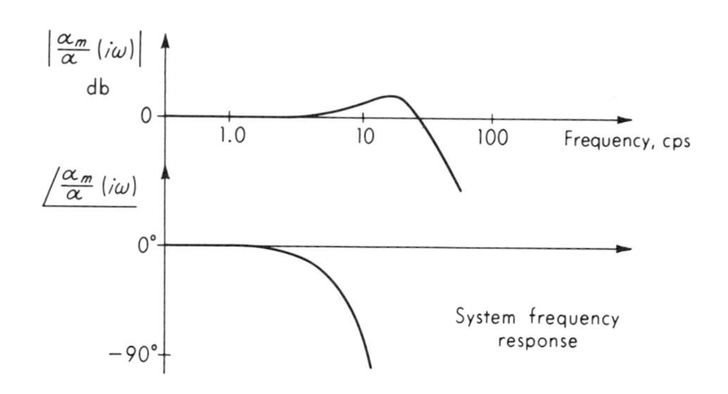

Fig. 7.9. *Servoed-sphere probe.*

error signals. A fifth tap measures the stagnation pressure. Block diagrams and frequency response of a single axis are given in Fig. 7.9. The angle-of-attack axis is designed for the range -10 to $+40°$ while the sideslip axis covers $\pm 20°$. Static inaccuracy of angle measurement is 0.25°.

Dynamic wind-vector indicator.[1] Figure 7.10 shows a transducer that measures the magnitude and direction of flow velocity in terms of the drag force exerted on a hollow sphere. The drag force F_d on a body is

[1] Flow Corp., Cambridge, Mass.

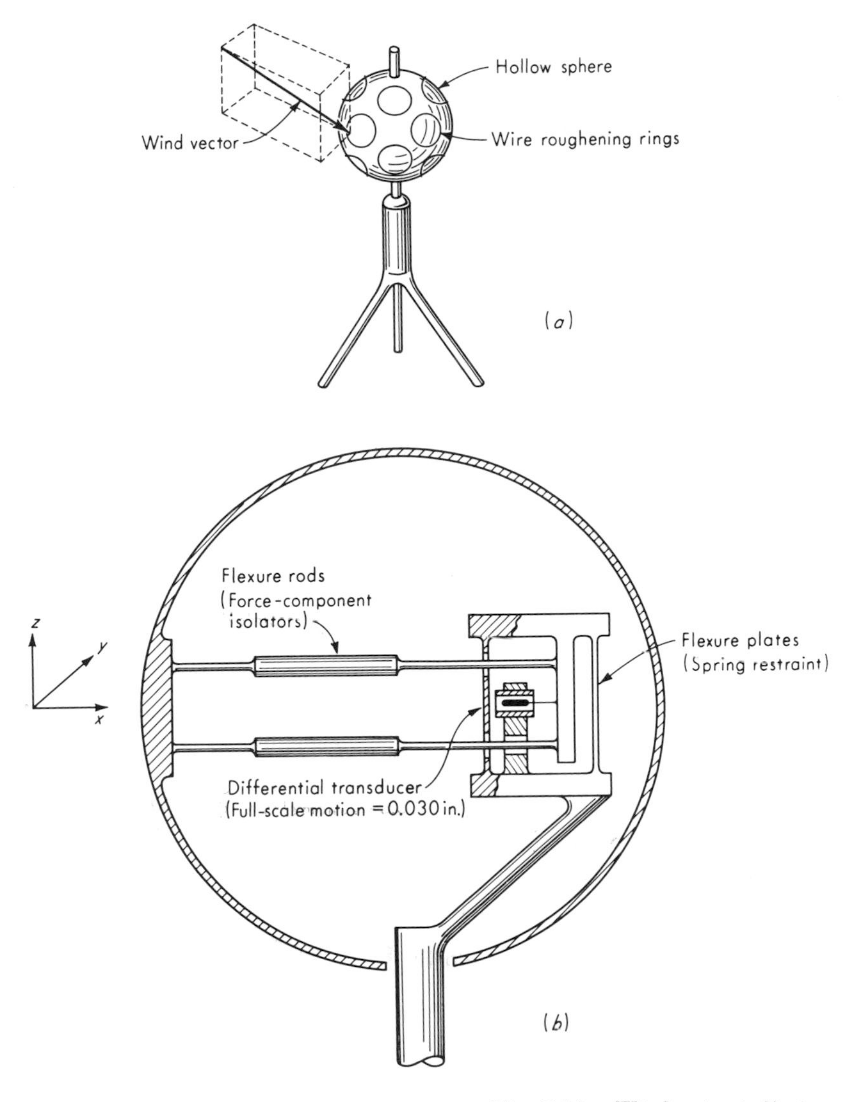

Fig. 7.10. *Wind-vector indicator.*

given by

$$F_d = C_d \frac{A \rho V^2}{2} \qquad (7.11)$$

where $C_d \triangleq$ drag coefficient of body
$= 0.567$ for these transducers
$A \triangleq$ projected area of body

Clearly, if C_d, A, and ρ are known, V may be measured by measuring F_d. If all directional components of V are to be equally effective in producing drag force, a body with spherical symmetry must be employed. If this is done, measurement of the x, y, and z components of F_d completely defines the magnitude and direction of V. Since the drag coefficient of a smooth sphere is somewhat dependent on the Reynolds number (and thus V), wire roughening rings are attached to the sphere surface to ensure turbulence. The drag coefficient of the roughened sphere is constant over the entire design range of V for a given transducer. The separation of the total drag force into three rectangular components is accomplished by mounting the sphere on a force-resolving flexure assembly. Figure 7.10*b* shows the x-axis mechanism of this structure. (The y and z axes are identical except for their orientation.) The force components are applied to flexure-plate springs to produce proportional motions which are then measured by differential transformers. (A later model uses strain gages.) A transducer designed for general meteorological work and missile ground support at launching sites uses a 1-ft-diameter sphere, covers the velocity range 0 to 100 mph, has flat frequency response 0 to 6 cps, and has a static inaccuracy of 2 percent of full scale. Auxiliary computing and display equipment is available to show total vector velocity magnitude and angular orientation in two planes visually on oscilloscopes. Transducer models designed for use in water are also available for oceanographic studies such as those of ocean currents.

Hot-wire and hot-film anemometers. Hot-wire anemometers are commonly made in two basic forms, the constant-current type and the constant-temperature type. Both utilize the same physical principle but in different ways. In the constant-current type, a fine resistance wire carrying a fixed current is exposed to the flow velocity. The wire attains an equilibrium temperature when the i^2R heat generated in it is just balanced by the convective heat loss from its surface. The circuit is designed so that the i^2R heat is essentially constant; thus the wire temperature must adjust itself to change the convective loss until equilibrium is reached. Since the convection film coefficient is a function of flow velocity, the equilibrium wire temperature is a measure of velocity. The wire temperature can be measured in terms of its electrical resistance. In the constant-temperature form, the current through the wire is adjusted to keep the wire temperature (as measured by its resistance) constant. The current required to do this then becomes a measure of flow velocity.

For equilibrium conditions one can write an energy balance for a hot wire as

$$I^2R_w = K_c hA(T_w - T_f) \qquad (7.12)$$

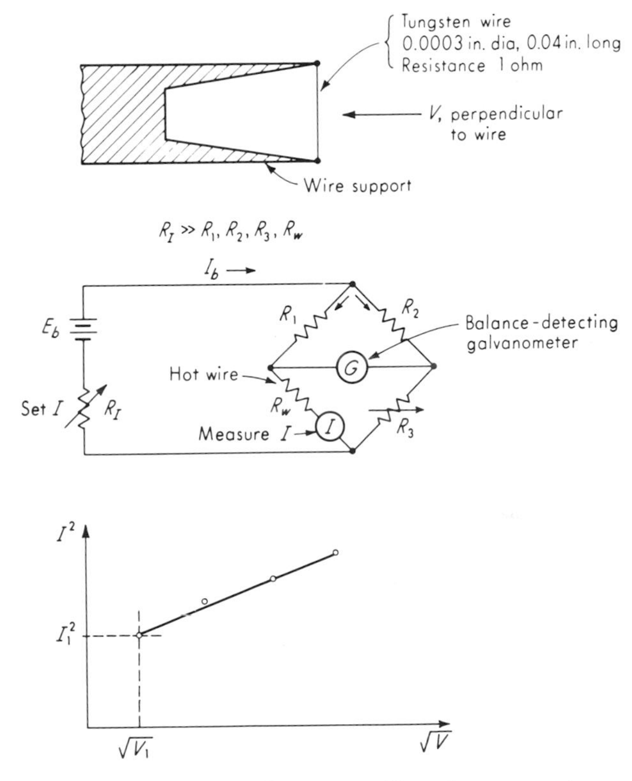

Fig. 7.11. *Hot-wire anemometer.*

where
$I \triangleq$ wire current
$R_w \triangleq$ wire resistance
$K_c \triangleq$ conversion factor, thermal to electrical power
$T_w \triangleq$ wire temperature
$T_f \triangleq$ temperature of flowing fluid
$h \triangleq$ film coefficient of heat transfer
$A \triangleq$ heat-transfer area

Now h is mainly a function of flow velocity for a given fluid density. For a range of velocities, this function has the general form

$$h = C_0 + C_1\sqrt{V} \qquad (7.13)$$

For the measurement of average (steady) velocities, the constant-temperature mode of operation is often used. Figure 7.11 shows a possible circuit arrangement. For accurate work a given hot-wire probe must be calibrated in the fluid in which it is to be used. That is, it is exposed to *known* velocities (measured accurately by some other means) and its output recorded over a range of velocities. In the circuit of Fig. 7.11, the current through R_w stays essentially constant even when R_w changes because R_I is of the order of 2,000 ohms while R_1, R_2, R_3, and R_w are much less, of the order of 1 to 20 ohms. In calibration, V is set at some known value V_1. Then R_I is adjusted to set hot-wire current I at a value low enough to prevent wire burnout but high enough to give adequate sensitivity to velocity. The resistance R_w will come to a definite temperature and resistance. The resistor R_3 is then adjusted to balance the bridge. This adjustment is essentially a measurement of wire temperature, which is held fixed at all velocities. The first point on the calibration curve is thus plotted as I_1^2, $\sqrt{V_1}$. Now V is changed to a new value, causing wire temperature and R_w also to change and thus unbalancing the bridge. Then R_w, and thereby wire temperature, are restored to their original values by adjusting I (by means of R_I) until bridge balance is restored (R_3 is *not* changed). The new current I and the corresponding V may then be plotted on the calibration curve and this procedure repeated for as many velocities as desired.

Once calibrated, the probe can be used to measure unknown velocities by adjusting R_I until bridge balance is achieved, reading I, and getting the corresponding V from the calibration curve. This assumes that the measured fluid is at the same temperature and pressure as for the calibration. Correction methods for varying temperature and pressure are fairly simple but will not be gone into here. For the above constant-temperature mode of operation, Eqs. (7.12) and (7.13) can be combined to give

$$I^2 = \frac{K_c A(T_w - T_f)(C_0 + C_1\sqrt{V})}{R_w} \triangleq C_2 + C_3\sqrt{V} \qquad (7.14)$$

indicating that the calibration curve of Fig. 7.11 should be essentially a straight line. This is borne out by experimental tests.

While the above described measurement of steady velocities is of some practical interest, perhaps the main application of hot-wire instruments is the measurement of rapidly fluctuating velocities, such as the turbulent components superimposed on the average velocity. Both constant-current and constant-temperature techniques are used; we first consider the constant-current operation. Figure 7.12 shows the basic arrangement. The current can again be assumed constant at a value I even if R_w changes, since $R_I \gg R_w$. Let us suppose the velocity is constant at a value V_0. This will cause R_w to assume a constant value, say R_{w0}, and a voltage IR_{w0} will appear across R_w. Now, we let the velocity V fluctuate about the value V_0 so that $V = V_0 + v$, where v is the fluctuating component. This will result in R_w varying so that $R_w = R_{w0} + r_w$, where r_w is the varying component. Now, during a time interval dt,

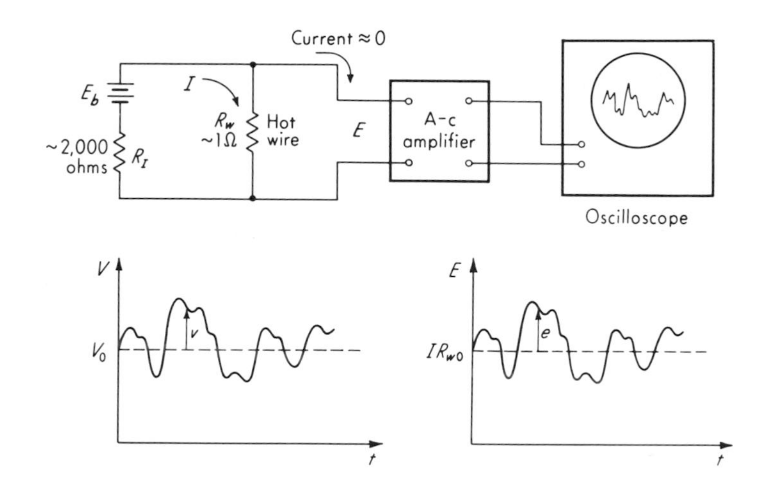

Fig. 7.12. *Velocity-fluctuation measurement.*

we may write for the wire

$$\text{(Electrical energy generated)} - \text{(energy lost by convection)} = \text{energy stored in wire} \qquad (7.15)$$

The energy lost by convection is given by

$$K_c A(T_w - T_f)(C_0 + C_1\sqrt{V})\,dt \qquad (7.16)$$

while the wire temperature T_w may be related to its resistance by

$$T_w = K_{tr}(R_{w0} + r_w) \qquad (7.17)$$

where K_{tr} is the reciprocal of a temperature coefficient of resistance. The term $C_0 + C_1\sqrt{V}$ may be approximately linearized for small changes in V with good accuracy as follows:

$$f(V) = C_0 + C_1\sqrt{V} \approx (C_0 + C_1\sqrt{V_0}) + \left.\frac{\partial f}{\partial V}\right|_{V=V_0}(V - V_0) \qquad (7.18)$$

$$C_0 + C_1\sqrt{V} \approx (C_0 + C_1\sqrt{V_0}) + K_v v \qquad (7.19)$$

Equation (7.15) then becomes

$$I^2(R_{w0} + r_w)\,dt - K_c A(T_w - T_f)(C_0 + C_1\sqrt{V_0} + K_v v)\,dt = MC\,dT_w \qquad (7.20)$$

where $M \triangleq$ mass of wire
$C \triangleq$ specific heat of wire

Now,

$$I^2R_{w0} + I^2r_w - K_cA[K_{tr}(R_{w0} + r_w) - T_f](C_0 + C_1\sqrt{V_0} + K_v v) = MCK_{tr}\frac{dr_w}{dt} \tag{7.21}$$

and since $$I^2R_{w0} - K_cA(K_{tr}R_{w0} - T_f)(C_0 + C_1\sqrt{V_0}) = 0 \tag{7.22}$$

because this represents the initial equilibrium state, we get

$$I^2r_w - K_cAK_{tr}r_w(C_0 + C_1\sqrt{V_0}) - K_cAK_{tr}r_wK_v v - K_cA(K_{tr}R_{w0} - T_f)K_v v = MCK_{tr}\frac{dr_w}{dt} \tag{7.23}$$

The term $K_cAK_{tr}K_vr_wv$ may be neglected relative to the other terms since it contains the product r_wv of two small quantities. Now the voltage across R_w is $IR_w = I(R_{w0} + r_w)$. The fluctuating component of this is Ir_w, which we shall call e. Equation (7.23) then leads to

$$\frac{e}{v}(D) = \frac{K}{\tau D + 1} \tag{7.24}$$

where $$K \triangleq \frac{-K_vK_cAI(K_{tr}R_{w0} - T_f)}{K_cK_{tr}A(C_0 + C_1\sqrt{V_0}) - I^2} \quad \text{volts/fps} \tag{7.25}$$

$$\tau \triangleq \frac{MCK_{tr}}{K_cK_{tr}A(C_0 + C_1\sqrt{V_0}) - I^2} \quad \text{sec} \tag{7.26}$$

We see that the voltage follows the flow velocity with a first-order lag. The time constant τ cannot be reduced much below 0.001 sec in actual practice, which would limit the flat frequency response to less than 160 cps. This is quite inadequate for turbulence studies since frequencies of 50,000 cps and more are of interest. This limitation is overcome by use of electrical dynamic compensation. Circuits whose frequency response just makes up the deficiency in the hot wire itself are employed as in Fig. 7.13.[1] The overall system then has a flat response to almost 100,000 cps. The main difficulty in applying this technique is that the correct compensation depends on τ whose value is not known and varies with flow conditions. The next paragraph explains a method[1] of solving this difficulty.

The basic idea of the scheme is to force a square-wave current through the hot wire while it is exposed to the flow to be studied (see Fig. 7.14). We shall show that the output-voltage response to this *current* signal has exactly the same time constant as the response to the flow-*velocity* signal. Thus if the compensation can be adjusted to be correct for the current signal it will also be correct for the velocity input. The "correctness" of the adjustment may be judged by the degree to

[1] Flow Corp., Cambridge, Mass.

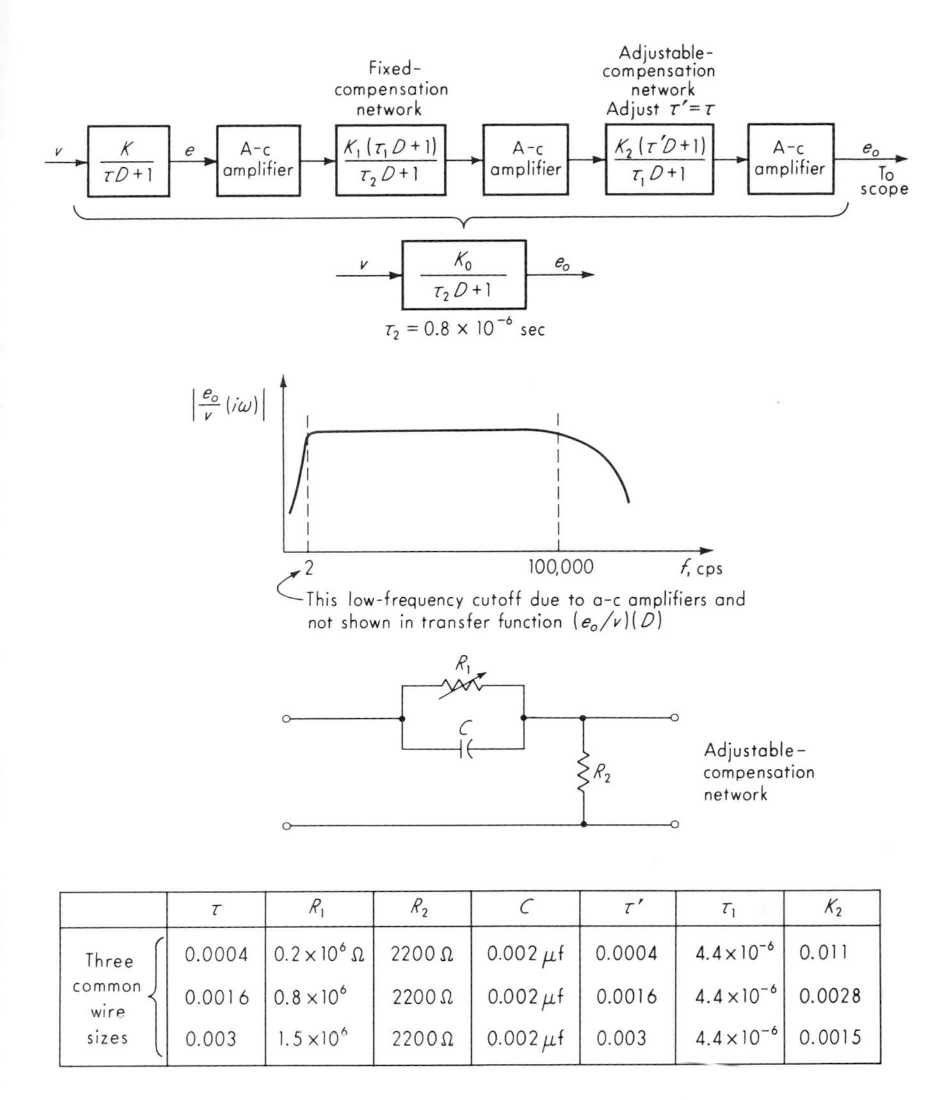

	τ	R_1	R_2	C	τ'	τ_1	K_2
Three common wire sizes	0.0004	$0.2 \times 10^6\,\Omega$	$2200\,\Omega$	$0.002\,\mu$f	0.0004	4.4×10^{-6}	0.011
	0.0016	0.8×10^6	$2200\,\Omega$	$0.002\,\mu$f	0.0016	4.4×10^{-6}	0.0028
	0.003	1.5×10^6	$2200\,\Omega$	$0.002\,\mu$f	0.003	4.4×10^{-6}	0.0015

Fig. 7.13. *Dynamic compensation.*

which the output voltage corresponds to a square wave. In the circuit of Fig. 7.14 a good approximation to linear behavior may be expected for small input signals (current or velocity); thus the superposition principle will apply, and the effects of current and velocity inputs may be considered separately. If the square-wave current is turned off, R_I adjusted to give the desired hot-wire current I_0, and R_b adjusted to balance the bridge, then $R_a/R_{w0} = R_r/R_b$, and the voltage $E_{B1,B2} = 0$. Now we

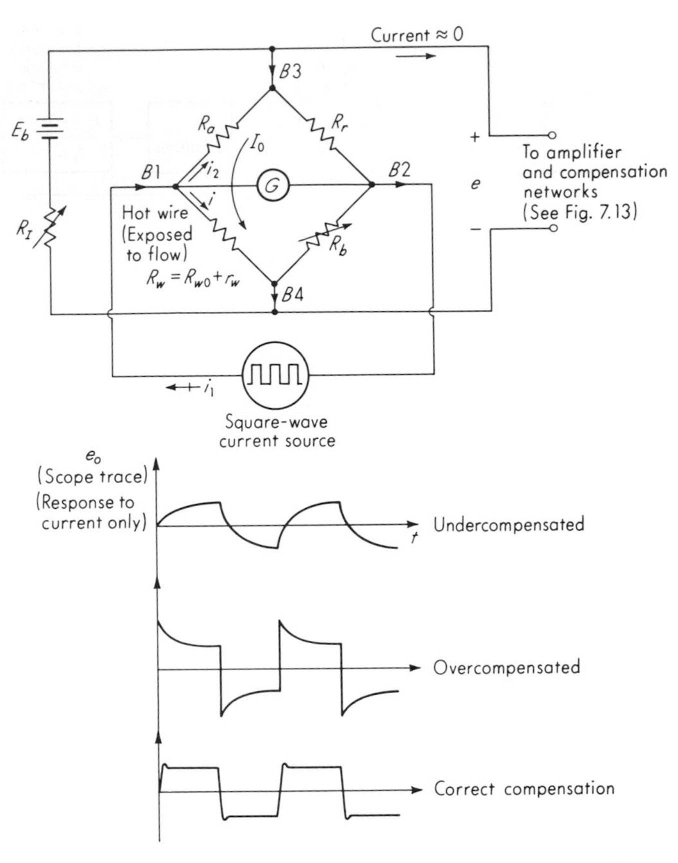

Fig. 7.14. Compensation adjustment scheme.

let the square-wave current i_1 be turned on, causing a current i, which we calculate to be $i_1(R_a + R_r)/(R_a + R_r + R_w + R_b)$, to flow through R_w and a current $i_2 = i_1(R_w + R_b)/(R_a + R_r + R_w + R_b)$ to flow through R_a and R_r. Equation (7.15) may be applied to this situation to give

$$(I_0 + i)^2(R_{w0} + r_w) - K_cA[K_{tr}(R_{w0} + r_w) - T_f](C_0 + C_1\sqrt{V_0}) = MCK_{tr}\frac{dr_w}{dt} \qquad (7.27)$$

Now, since $I_0^2R_{w0} = K_cA(K_{tr}R_{w0} - T_f)(C_0 + C_1\sqrt{V_0})$, and neglecting $2I_0ir_w$ and $i^2(R_{w0} + r_w)$ since they are products of small quantities, Eq. (7.27) reduces to

$$\frac{r_w}{i}(D) = \frac{K_i}{\tau D + 1} \qquad (7.28)$$

where
$$\tau \triangleq \frac{MCK_{tr}}{K_c K_{tr} A (C_0 + C_1 \sqrt{V_0}) - I_0^2} \qquad \text{sec} \tag{7.29}$$

$$K_i \triangleq \frac{2I_0 R_{w0}}{K_c K_{tr} A (C_0 + C_1 \sqrt{V_0}) - I_0^2} \qquad \text{ohms/amp} \tag{7.30}$$

Let us now calculate e, the varying component of the voltage appearing across $B3$ and $B4$, which will be the input to the amplifiers and compensating networks.

$$e = -R_a i_2 + (R_{w0} + r_w)i + I_0 r_w \approx -R_a i_2 + R_{w0} i + I_0 r_w \qquad ir_w \approx 0 \tag{7.31}$$

$$e = -R_a \frac{R_w + R_b}{R_a + R_r} i + R_{w0} i + I_0 r_w$$

$$= \frac{-R_a R_w - R_a R_b + R_{w0} R_a + R_{w0} R_r}{R_a + R_r} i + I_0 r_w \tag{7.32}$$

Now $R_{w0} R_a \approx R_w R_a$ and $R_{w0} R_r = R_a R_b$ (balanced-bridge relation); thus

$$e \approx I_0 r_w \tag{7.33}$$

Thus, finally,
$$\frac{e}{i}(D) = \frac{K_e}{\tau D + 1} \tag{7.34}$$

where
$$K_e \triangleq I_0 K_i \tag{7.35}$$

We see now that the response of the voltage e to impressed current signals has the identical time constant τ as the response to flow-velocity signals. Thus the compensating networks may be adjusted to optimize the response to current inputs and ensure optimum response for flow-velocity inputs. Since this adjustment is made while the probe is exposed to flow, the output will contain a superposition of current response and velocity response, resulting in a sometimes confusing picture, rather than the simple wave forms of Fig. 7.14. Usually, however, the compensation adjustment can be made satisfactorily.

The operation of the constant-temperature type of instrument for steady velocities was explained earlier in relation to Fig. 7.11. This mode of operation can be extended to measure both average and fluctuation components of velocity by making the bridge-balancing operation automatic, rather than manual, through the agency of a feedback arrangement. A simplified functional schematic of such a system is shown in Fig. 7.15. With zero flow velocity and the bridge excitation shut off ($i_w = 0$), the hot wire assumes the fluid temperature. The variable resistor R_3 is then manually adjusted so that $R_3 > R_w$, thereby unbalancing the bridge. When the excitation current is turned on, the unbalanced bridge produces an unbalance voltage e_e which is applied to the input of a high-gain current amplifier supplying the bridge excitation current. The current now flowing through R_w increases its temperature and thus its

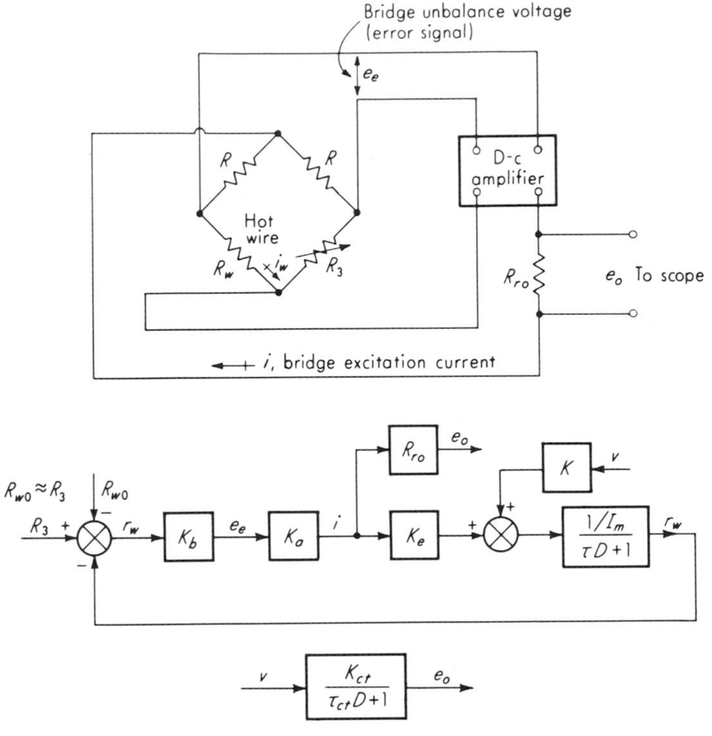

Fig. 7.15. *Constant-temperature anemometer.*

resistance. As R_w increases, it approaches R_3 and the bridge-unbalance voltage e_e decreases. If the current amplifier had an infinite gain, the bridge current required to heat R_w to match R_3 precisely could be produced with an infinitesimally small error voltage e_e and thus perfect bridge balance would be attained automatically. The current i (which is a measure of flow velocity V) produces an output voltage in passing through the readout resistor R_{ro}. Since an actual amplifier has finite gain, the bridge-unbalance voltage cannot go to zero, but it can be extremely small. This also means that R_w will be very close, but not exactly equal, to R_3.

The above described self-adjusting equilibrium will come about at *any* steady-flow velocity, not just zero, the equilibrium current in each case being a measure of velocity. If the velocity is rapidly fluctuating, perfect continuous bridge balance requires an infinite-gain amplifier. Lacking this, practical systems exhibit some lag between velocity and current. System dynamic response for small velocity fluctuations can be obtained by superimposing previously obtained results for response to velocity (wire-current constant) and response to current (flow-velocity constant), since in the constant-temperature mode both current and velocity are changing simultaneously. The effect of velocity on r_w (the varying component of R_w) is obtained from Eq. (7.24) as

$$\frac{r_w}{v}(D) = \frac{K/I_m}{\tau D + 1} \tag{7.36}$$

where I_m is the constant current about which fluctuations take place. From Eq. (7.34), the effect of current on r_w is given by

$$\frac{r_w}{i}(D) = \frac{K_e/I_m}{\tau D + 1} \tag{7.37}$$

The total effect of i and v on r_w is then, by superposition,

$$(\tau D + 1)r_w = \frac{Kv + K_e i}{I_m} \tag{7.38}$$

Now a change in r_w causes a bridge-unbalance voltage change according to

$$e_e = \frac{I_m R}{R + R_{w0}} r_w \triangleq K_b r_w \tag{7.39}$$

We assume that the amplifier has no lag and produces output current proportional to input voltage as given by

$$i = K_a e_e \tag{7.40}$$

The block diagram of Fig. 7.15 embodies the above relations which may now be manipulated to give the relation between input v and output e_o:

$$\left(\frac{K_e e_o}{R_{ro}} + Kv\right)\frac{(1/I_m)(-K_b K_a R_{w0})}{\tau D + 1} = e_o \tag{7.41}$$

This gives finally

$$\frac{e_o}{v}(D) = \frac{K_{ct}}{\tau_{ct} D + 1} \tag{7.42}$$

where

$$\tau_{ct} \triangleq \frac{\tau}{1 + K_e K_b K_a / I_m} \tag{7.43}$$

$$K_{ct} \triangleq \frac{-K K_b K_a R_0 / I_m}{1 + K_e K_b K_a / I_m} \tag{7.44}$$

Note that τ_{ct}, the time constant of the constant-temperature anemometer system, is always less than τ (the time constant of the wire itself) and in actual practice is *much* less since a very high value of amplifier gain K_a is used. As in all feedback systems, too high a loop gain will cause instability; however, careful design allows sufficiently high gain to make τ_{ct} of the order of $\frac{1}{100}$ of τ or less. A typical instrument has flat (within 3 db) frequency response to 17,000 cps when the average flow velocity is 30 fps, 30,000 cps for 100 fps, and 50,000 cps at 300 fps.

Both constant-current and constant-temperature anemometers are in use; it may be helpful to list their comparative features. In the constant-current type, the current must be set high enough to heat the wire considerably above the fluid temperature for a given average velocity. If the flow should suddenly drop to a much lower velocity or come to rest, the hot wire will burn out since the convection loss cannot match the heat generation before the wire temperature reaches the melting point. The constant-temperature type does not have this drawback because the feedback system *automatically* sets wire current to maintain the desired (safe) wire temperature for every velocity. A further advantage of the constant-temperature method lies in the nature of the dynamic compensation. In the constant-current method the compensating network must be reset (using the square-wave current) whenever the average velocity changes appreciably. Furthermore, if velocity fluctuations about the average are large (more than, say, 5 percent of the average velocity) the dynamic compensation will not be complete since the value of τ varies with V and thus the compensating network time constant τ' should be continuously and instantaneously varied, which it is not. The feedback arrangement of the constant-temperature system provides the proper compensation for velocity fluctuations of *any* size, so long as the amplifier gain is large enough to maintain nearly perfect instantaneous bridge balance. The main disadvantages of the constant-temperature scheme are the higher noise level in the electronics (which prevents measurement of very small velocity fluctuations) and the difficulty in designing sufficiently high-gain d-c amplifiers without causing instability and drift problems. Note that since the constant-temperature system uses a d-c amplifier it is usable down to zero frequency (steady velocity), whereas the constant-current (which uses a-c amplifiers) is good only to about 1 cps.

In the above analyses the fluid density ρ was assumed constant. If ρ varies, the anemometer actually measures the product ρV, that is, the local mass flow rate. Thus in the calibration curve of Fig. 7.11, a more general relationship could have been shown; that is, I^2 is really proportional to $\sqrt{\rho V}$. Thus in compressible flows where ρ varies significantly, one would have to know ρ in order to reduce anemometer readings to velocity values.

For both constant-current and constant-temperature instruments, the output voltage (assuming constant ρ) may be taken directly proportional to velocity fluctuations as long as they are a small percentage of the average velocity. Large velocity changes are inaccurately measured by the constant-current method but are properly handled by the constant-temperature scheme. However, the instrument output voltage is *not* now proportional to velocity but rather follows the basic relationship of Fig. 7.11; that is, I^2 varies linearly with $\sqrt{V}$. Since the output voltage e_o of Fig. 7.15 is proportional to I, we see that e_o varies linearly with $(V)^{\frac{1}{4}}$. It is possible to construct an electrical computing circuit (function generator)

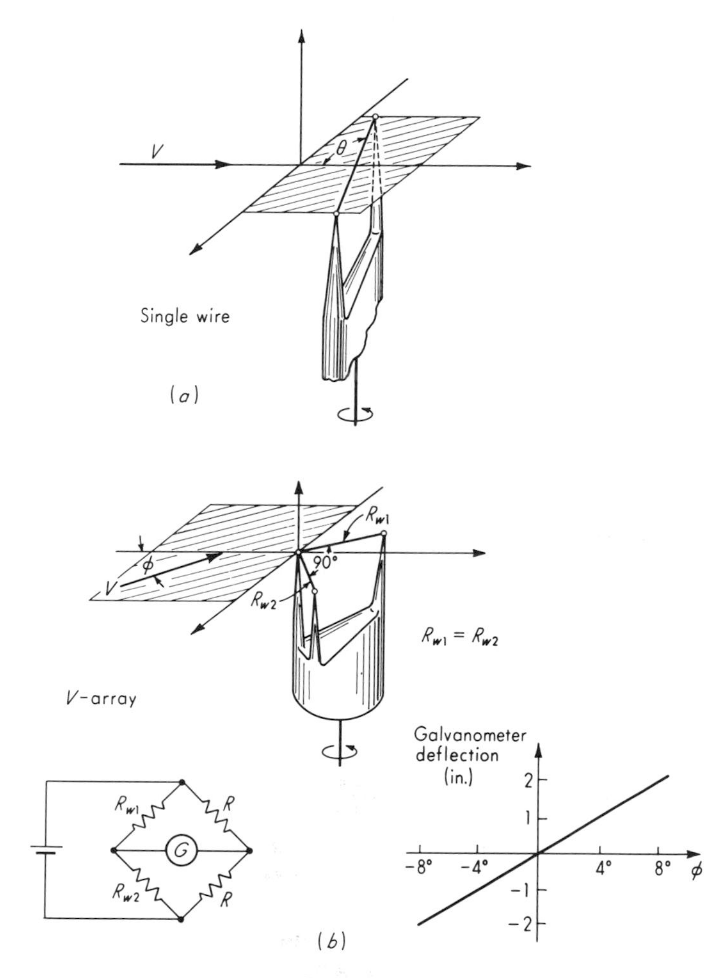

Fig. 7.16. *Flow-direction measurement.*

that will produce an output voltage proportional to the fourth power of its input voltage. If e_o is applied to the input of such a circuit, the output voltage will be proportional to velocity for both large and small changes, thus giving an easily interpretable record.

The hot-wire anemometer may be used to measure the direction of the average flow velocity in several ways.[1] It has been found that a single wire, as in Fig. 7.16*a*, responds essentially to the component of

[1] H. H. Lowell, Design and Applications of Hot-wire Anemometers for Steady-state Measurements at Transonic and Supersonic Speeds, *NACA, Tech. Note* 2117, 1950.

velocity perpendicular to it, if the angle between wire and velocity vector is between 90 and about 25°. For this range, then, the V in our derivations may be replaced by $V \sin \theta$. (For $\theta < 25°$ the heat loss is greater than predicted by $V \sin \theta$; for $\theta = 0$ it is about 55 percent of that for $\theta = 90°$.) With the arrangement of Fig. 7.11 and a rotatable probe as in Fig. 7.16*a*, the flow-direction angle (in a single plane) could be found by determining the probe-rotation angle which gives a maximum value of I. This method is quite inaccurate, however, since $\sin \theta$ changes very slowly with θ when θ is near 90°. A better procedure, if the flow angle is roughly known, is as follows: The wire is set at about 50° from the flow direction and I is measured. The probe is then rotated in the opposite direction until an angle is found at which the same I as before is measured. The bisector of the angle between the two locations then determines the flow direction. This method is more accurate since the rate of change of I with θ is a maximum near 50°. Even greater convenience and accuracy is achieved by use of a so-called V array of hot wires, as in Fig. 7.16*b*. The two hot wires R_{w1} and R_{w2} are assumed identical and form a V with an included angle which is typically 90°. They are connected into a bridge as shown. When the probe is rotated, a bridge null occurs when the flow-velocity vector is aligned with the bisector of the V. In a specific case,[1] the sensitivity of this arrangement was sufficient to determine velocity direction within about 0.5°.

Practical problems in the application of hot-wire anemometers are found in the limited strength of the fine wires and the calibration changes caused by dirt accumulations. Unless the flow is very clean, significant calibration changes can occur in a relatively few minutes of operation. Larger dirt particles striking wires may actually break them. At high speeds, wires may vibrate because of aerodynamic loads and flutter effects. While some measurements have been made in liquids, the majority of applications have been in gases.

A variation of the hot-wire anemometer intended to overcome some of its problems is the hot-film transducer. Here the resistance element is a thin film of platinum deposited on a glass base. The film takes the place of the hot wire; the required circuitry is basically similar to the constant-temperature hot-wire approach. The film transducers have great mechanical strength and may also be used at very high temperatures by constructing them with internal cooling-water passages. Various configurations of sensors are possible; Fig. 7.17 shows two possibilities.[2,3]

In addition to measurement of velocity magnitude and direction, hot-

[1] *Ibid.*

[2] DISA Electronik A/S, Herlev, Denmark.

[3] Thermo-Systems Inc., Minneapolis, Minn.

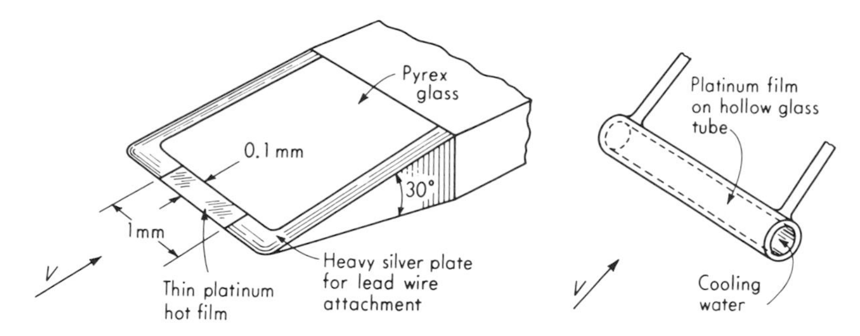

Fig. 7.17. *Hot-film anemometer.*

wire and hot-film instruments may also be adapted to measurements of fluid temperatures, turbulent shear stresses, and concentrations of individual gases in mixtures.[1] Furthermore, if several quantities (such as, say, velocity and temperature) vary simultaneously, it may be possible to extract information about each separate quantity from a hot wire with suitable auxiliary equipment.[1]

Hot-film shock-tube velocity sensors. In shock-tube experiments the propagation velocity of the shock wave down the tube must often be measured. Of the various means available for making this measurement, hot-film temperature sensors are in wide use. The passage of the shock wave past a particular section of the tube is accompanied by a step change in gas temperature. By locating thin resistance films flush with the inside of the tube, the instant of wave-front passage may be detected as a temperature (and therefore resistance) change. If two such film sensors are mounted a known distance apart, the average wave velocity may be computed from the time interval between the two sensor responses. The films are operated at constant current by the simple circuit of Fig. 7.18, and a differentiating circuit is used to sharpen the pulses for greater timing accuracy. With systems of this type,[2] shock-wave velocities have been measured with an accuracy of 1 percent for shock Mach numbers as high as 7.5 to 10.

7.2

Gross Volume Flow Rate

The total flow rate through a duct or pipe must often be measured and/or controlled. Many instruments

[1] S. Corrsin, Extended Applications of the Hot-wire Anemometer, *NACA, Tech Note* 1864, 1949.

[2] Shock Tubes, Handbook of Supersonic Aerodynamics, sec. 18, *NAVORD Rept.* 1488, vol. 6, pp. 543, 558.

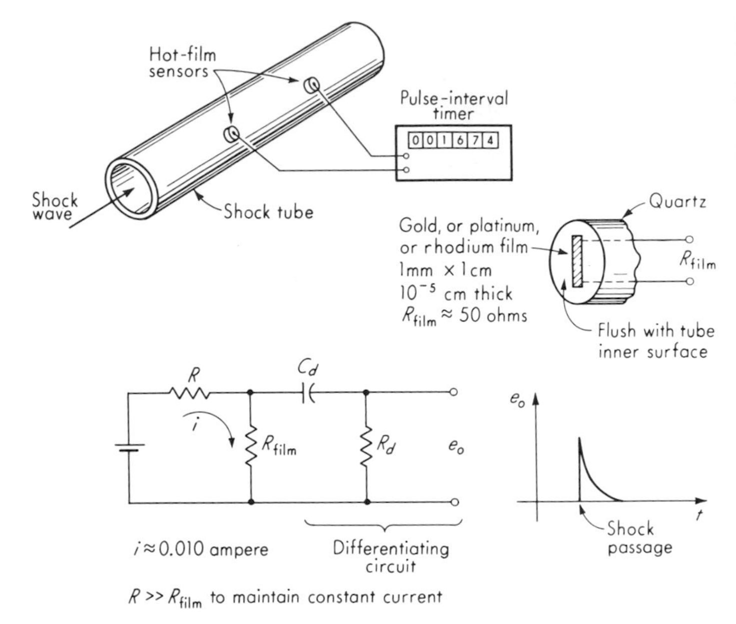

Fig. 7.18. *Thin-film velocity sensor.*

(flowmeters) have been developed for this purpose. They may be classified in various ways; a useful overall classification divides devices into those which measure volume flow rate (ft^3/time) and those which measure mass flow rate (lb_m/time). The total flow (ft^3 or lb_m) occurring during a given time interval may also be of interest. This measurement requires integration with respect to the time of the instantaneous flow rate. The integrating function may also be performed in various ways, sometimes being an integral part of the flowmetering concept and other times being performed by a general-purpose integrator more or less remote from the flowmeter.

Calibration and standards. Flow-rate calibration depends on standards of volume (length) and time or mass and time. Primary calibration is, in general, based on the establishment of steady flow through the flowmeter to be calibrated and subsequent measurement of the volume or mass of flowing fluid that passes through in an accurately timed interval. If steady flow exists, the volume or mass flow rate may be inferred from such a procedure. Any stable and precise flowmeter calibrated by such primary methods then itself becomes a secondary flow-rate standard against which other (less accurate) flowmeters may be conveniently calibrated. As in any other calibration, significant deviations of the conditions of use from

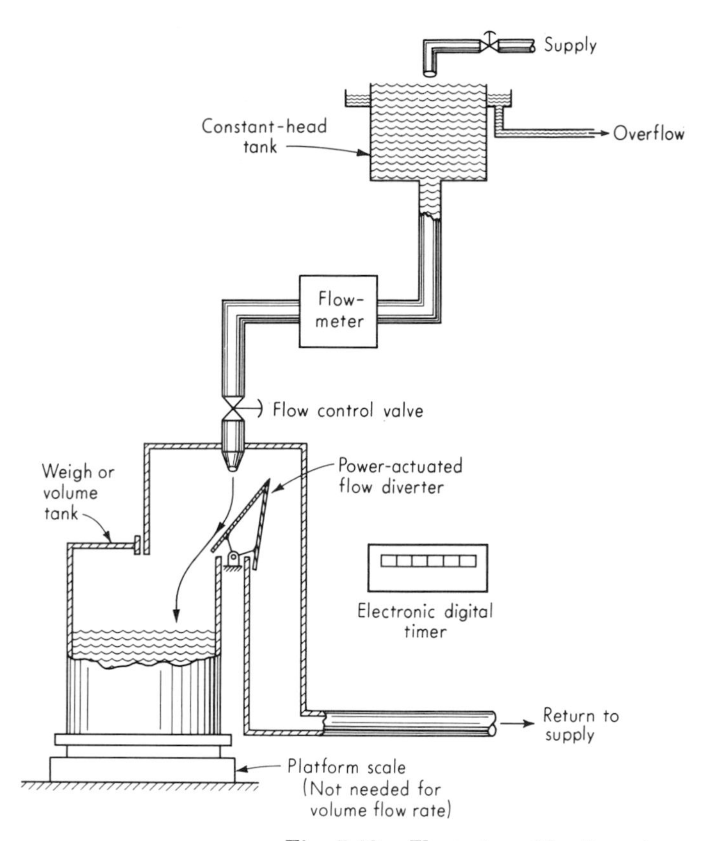

Fig. 7.19. *Flowmeter calibration setup.*

those at calibration will invalidate the calibration. Possible sources of error in flowmeters include variations in fluid properties (density, viscosity, and temperature), orientation of the meter, pressure level, and particularly flow disturbances (such as elbows, tees, valves, etc.) upstream (and to a lesser extent downstream) of the meter.

A typical calibration setup for precise primary calibration of flowmeters using liquids is shown in Fig. 7.19. A constant-head tank maintains a fixed inlet pressure to the flowmeter under test, irrespective of flow rate. (A constant-head tank may be impractical, because of the excessive elevation required, for very high inlet pressures, or because of toxicity, volatility, or flammability of certain liquids. Then a closed system using a pump to supply pressure is used. Manual or automatic control of the flow rate to maintain constancy during a given run is then necessary.) The flow rate through the meter is adjusted to the various desired values with a flow-control valve. If necessary, this valve is manipulated during the run to maintain constant flow. Constancy of flow is generally observed from the reading of the meter under calibration. Until a constant flow rate is established, the liquid is diverted from the weigh or volume tank which must be emptied (volume tank) or weighed (weigh tank) before flow into it is started. There are available volume tanks, called prover tanks, that are accurate to 0.05 to 0.1 percent of nominal volume. Platform scales accurate to 0.05 to 0.1 percent are used for mass flow measurements. When the flow diverter is suddenly moved to the tank position, a switch starts the electronic timer as the diverter passes the mid-position. Flow is continued until the tank is filled, at which time the motion of the diverter through the mid-position to the return position stops the timer. The weight or volume of liquid accumulated during the timed interval is then determined to calculate the volume or mass flow rate. With extreme attention to details, calibrations of the above type result in overall flow-rate errors of the order of a few tenths of 1 percent.[1]

The calibration of flowmeters to be used with gases can often be carried out with liquids as long as the pertinent similarity relations (Reynolds number) are maintained and theoretical density and expansion corrections are applied. If this procedure is felt to be of insufficient accuracy, a direct calibration with the actual gas to be used can be carried out by means of the *gasometer* system of Fig. 7.20. Here the gas flowing through the flowmeter during a timed interval is trapped in the gasometer bell and its volume thereby measured. Temperature and pressure measurements allow calculation of mass and conversion of volume to any desired standard conditions. By filling the bell with gas, raising it to the top, and adding appropriate weights, such a system may also be used as a gas *supply* to drive gas through a flowmeter as the bell gradually drops at a measured rate.

When the above primary calibration methods cannot be justified, comparison with a secondary standard flowmeter connected in series with the meter to be calibrated may be sufficiently accurate. Turbine flowmeters and their associated digital counting equipment have been found particularly suitable for such secondary standards. With attention to detail, such standards can closely approach the accuracy of the primary methods themselves. The Navy Primary Standards Laboratory at Pensacola, Florida, has such a system[2] with an inaccuracy of the order of 0.2 percent. For gas flow, a flow nozzle discharging air to the atmosphere can be very accurately calibrated[3] for mass flow rate by means of pitot-

[1] M. R. Shafer and F. W. Ruegg, Liquid-flowmeter Calibration Techniques, *Trans. ASME*, p. 1369, October, 1958.

[2] R. P. Bowen, Designing Portability into a Flow Standard, *ISA J.*, p. 40, May, 1961.

[3] R. J. Sweeney, "Measurement Techniques in Mechanical Engineering," p. 220, John Wiley & Sons, Inc., New York, 1953.

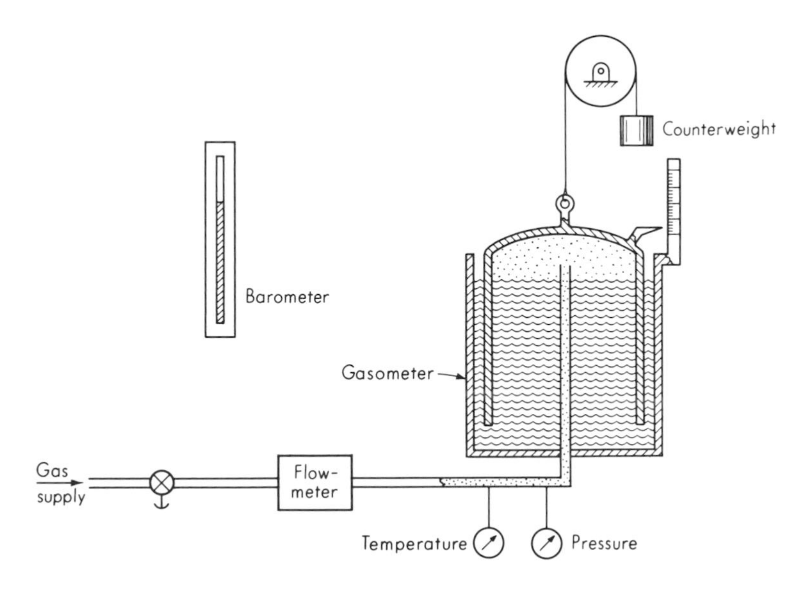

Fig. 7.20. *Gas-flow calibration.*

tube traverses at the discharge. This nozzle can then be connected in series with any flowmeter to be calibrated and used as an accurate standard. This can be done with very small error up to Mach number about 0.9.

Constant-area, variable-pressure-drop meters ("obstruction" meters). Perhaps the most widely used flowmetering principle involves placing a fixed-area flow restriction of some type in the pipe or duct carrying the fluid. This flow restriction causes a pressure drop which varies with flow rate; thus measurement of the pressure drop by means of a suitable differential pressure pickup allows flow-rate measurement. In this section we shall briefly discuss the most common practical devices that utilize this principle: the orifice, the flow nozzle, the venturi tube, the Dall flow tube, and the laminar-flow element.

The *sharp-edge orifice* is undoubtedly the most widely used flowmetering element, mainly because of its simplicity, low cost, and the great volume of research data available for predicting its behavior. A typical flowmetering setup is shown in Fig. 7.21. If one-dimensional flow of an incompressible frictionless fluid without work, heat transfer, or elevation change is assumed, theory gives the volume flow rate Q_t (ft^3/sec) as

$$Q_t = \frac{A_{2f}}{\sqrt{1 - (A_{2f}/A_{1f})^2}} \sqrt{\frac{2(p_1 - p_2)}{\rho}} \tag{7.45}$$

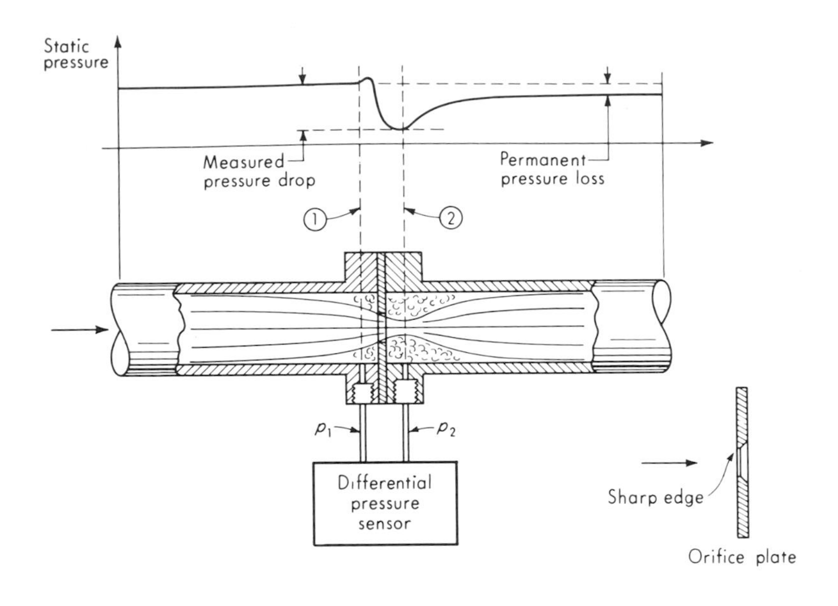

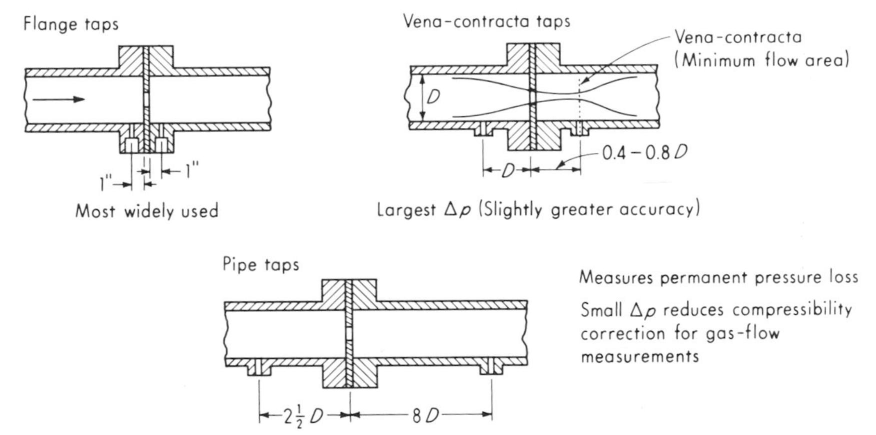

Fig. 7.21. *Orifice flowmetering.*

where $A_{1f}, A_{2f} \triangleq$ cross-section flow areas where p_1 and p_2 are measured, ft^2

$\rho \triangleq$ fluid mass density, slug/ft^3

$p_1, p_2 \triangleq$ static pressures, lb_f/ft^2

We see that measurement of Q requires knowledge of A_{1f}, A_{2f}, and ρ and measurement of the pressure differential $(p_1 - p_2)$. Actually, the real situation deviates from the assumptions of the theoretical model suffi-

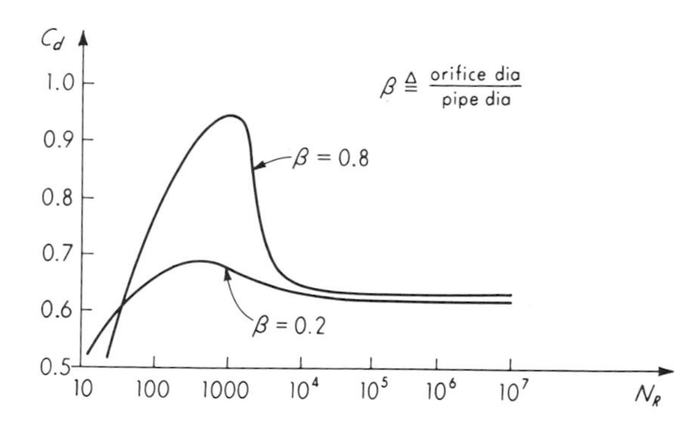

Fig. 7.22. *Variation of discharge coefficient.*

ciently to require experimental correction factors if acceptable flowmetering accuracy is to be attained. For example, A_{1f} and A_{2f} are areas of the actual flow cross section, which are *not,* in general, the same as those corresponding to the pipe and orifice diameters, which are the ones susceptible to practical measurement. Furthermore, A_{1f} and A_{2f} may change with flow rate because of flow geometry changes. Also, there are present frictional losses that affect the measured pressure drop and also lead to a permanent pressure loss. To take these factors into account, an experimental calibration to determine the actual flow rate Q_a by methods such as that of Fig. 7.19 is necessary. A discharge coefficient C_d may then be defined by

$$C_d \triangleq \frac{Q_a}{Q_t} \tag{7.46}$$

and thus

$$Q_a = \frac{C_d A_2}{\sqrt{1 - (A_2/A_1)^2}} \sqrt{\frac{2(p_1 - p_2)}{\rho}} \tag{7.47}$$

where $A_1 \triangleq$ pipe cross-section area
$A_2 \triangleq$ orifice cross-section area

The discharge coefficient of a given installation varies mainly with the Reynolds number N_R at the orifice. Thus the calibration can be performed with a single fluid, such as water, and the results used for any other fluid as long as the Reynolds numbers are the same. Variation of C_d with N_R follows typically the trend[1] of Fig. 7.22. While the above discussion would seem to indicate that each installation must be individually calibrated, this is fortunately not the case. If one is willing to

[1] "Handbook of Measurement and Control," p. 96, The Instruments Publishing Co., Pittsburgh, Pa., 1954.

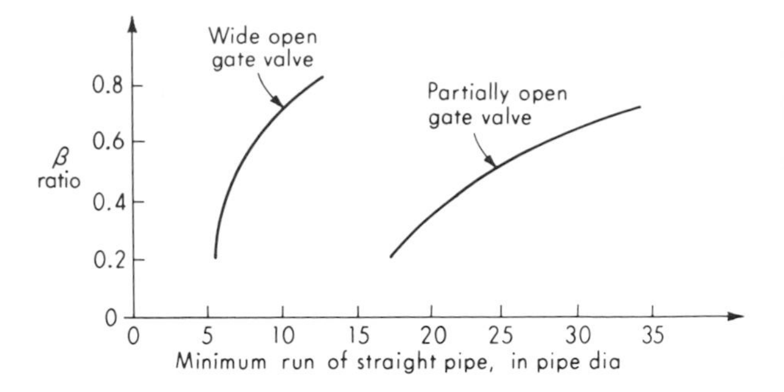

Fig. 7.23. *Effect of upstream disturbances.*

construct the orifice according to certain standard dimensions and also locate the pressure taps at specific points, then quite accurate (about 0.4 to 0.8 percent error) values of C_d may be obtained from tables or charts[1] which have been compiled based on many past experiments. Such data are available for pipe diameters 2 in. and greater, β (see Fig. 7.22) ratios of 0.2 to 0.7, and Reynolds number above 10,000. Installations exceeding these limits should be individually calibrated if high accuracy is required. It should also be noted that the standard calibration data assume no significant flow disturbances such as elbows, bends, tees, valves, etc., for a certain minimum distance upstream of the orifice. The presence of such disturbances close to the orifice can invalidate the standard data, causing errors of as much as 15 percent. Information on the minimum distances is available[3,4]; Fig. 7.23 shows a typical example. If the minimum distances are not feasible, straightening vanes[2,3,4] may be introduced ahead of the flowmeter to smooth out the flow.

Since flow rate is proportional to $\sqrt{\Delta p}$, a 10:1 change in Δp corresponds to only about a 3:1 change in flow rate. Since a given Δp-measuring instrument becomes quite inaccurate below about 10 percent of its full-scale reading, this nonlinearity typical of all obstruction meters (other than the laminar-flow element) restricts the accurate range of flow measurement to about 3 to 1. That is, a meter of this type cannot be used accurately below about 30 percent of its maximum flow rating. The square-root nonlinearity also causes difficulties in pulsating flow measure-

[1] Fluid Meters, Their Theory and Application, American Society of Mechanical Engineers, New York, 1937.

[2] "Handbook of Measurement and Control," p. 96.

[3] Fluid Meters, Their Theory and Application.

[4] P. S. Starrett and P. F. Halfpenny, The Effect of Non-standard Approach Sections on Orifices and Venturi Meters, Lockheed California Co., LR17905, 1964.

ment,[1,2] where the average flow rate (the rate to be measured) has a fluctuating component superimposed on it. Let us consider, as a simple example, a flow Q where

$$Q = Q_{av} + Q_p \sin \omega t \qquad Q_p < Q_{av} \tag{7.48}$$

and a flowmeter such that $\Delta p = KQ^2$. The Δp presented as input to the pressure-measuring system is then

$$\Delta p = K(Q_{av}^2 + 2Q_{av}Q_p \sin \omega t + Q_p^2 \sin^2 \omega t) \tag{7.49}$$

If the Δp instrument has a low-pass filtering characteristic, it will tend to read the average value of Δp. This is seen to be

$$\Delta p_{av} = K\left(Q_{av}^2 + \frac{Q_p^2}{2}\right) \tag{7.50}$$

Thus if one takes a measured Δp_{av} and computes the corresponding Q_{av} from it, using $Q_{av} = \sqrt{\Delta p_{av}/K}$, he will get a flow rate *higher* than actually existed. A further difficulty caused by the nonlinearity occurs when flow rate must be integrated to get total flow during a given time interval. The Δp signal must then be square-rooted before integration or this compensation included in the integrating device.

The orifice has the largest permanent pressure loss of any of the obstruction meters (other than the laminar-flow element); this is one of its disadvantages since it represents a power loss that must be replaced by whatever pumping machinery is causing the flow. The permanent pressure loss is given approximately by $\Delta p(1 - \beta^2)$, where Δp is the differential pressure used for flow measurement. Thus for the usual range of β(0.2 to 0.7) the permanent pressure loss ranges from $0.96\Delta p$ to $0.51\Delta p$. The actual power loss may, in fact, be quite small since the Δp recommended[3] for conventional flowmetering of liquids is only 20 to 400 in. of water (0.72 to 14.4 psi).

Orifice discharge coefficients are quite sensitive to the condition of the upstream edge of the hole. The standard orifice design requires that this edge be very sharp and also that the orifice plate be sufficiently thin relative to its diameter. Wear (rounding) of this sharp edge by long use, particularly if the fluid contains abrasive particles, can cause significant changes in the discharge coefficient. Flows that contain suspended solids

[1] A. K. Oppenheim and E. G. Chilton, Pulsating Flow Measurement—A Literature Survey, *Trans. ASME*, p. 231, February, 1955.

[2] T. Isobe and H. Hattori, A New Flowmeter for Pulsating Gas Flow, *ISA J.*, p. 38, December, 1959.

[3] L. K. Spink, "Principles and Practice of Flow Meter Engineering," The Foxboro Co., Foxboro, Mass., 1959.

also cause difficulty since the solids tend to collect behind the "dam" formed by the orifice plate and cause irregular flow. This problem can often be solved by use of an "eccentric" orifice in which the hole is at the bottom of the pipe rather than on the center line. This allows the solids to be continuously swept through. Liquids containing traces of vapor or gas may be metered if the orifice is installed in a vertical run of pipe with the flow upward. Gases containing traces of liquid may be similarly handled except that the flow should be downward.

When compressible fluids are metered, Eq. (7.47) is no longer correct. By assuming an isentropic process between states 1 and 2, the following relation may be derived[1] for compressible fluids:

$$W = C_d A_2 \sqrt{\frac{2gkp_1}{(k-1)v_1}} \sqrt{\frac{(p_2/p_1)^{2/k} - (p_2/p_1)^{(k+1)/k}}{1 - \beta^4(p_2/p_1)^{2/k}}} \tag{7.51}$$

where $W \triangleq$ weight flow rate, lb_f/sec
$k \triangleq$ ratio of specific heats (1.4 for air)
$g \triangleq$ local gravity
$v_1 \triangleq$ specific volume at state 1, ft^3/lb_f

The discharge coefficient C_d is the same for liquids or gases as long as the Reynolds number is the same. In many practical gas-flow installations the meter pressure drop is so small that the pressure ratio p_2/p_1 is 0.99 or greater. Under these conditions the simpler incompressible relation of Eq. (7.47) may be used with an error less than 0.6 percent if, for example, $\beta = 0.5$ and $k = 1.4$. Modified to give weight rather than volume flow rate, this is

$$W = C_d A_2 \sqrt{1 - \left(\frac{A_2}{A_1}\right)^2} \sqrt{\frac{2g(p_1 - p_2)}{v_1}} \tag{7.52}$$

For $p_2/p_1 < 0.99$ the error becomes greater, being 6 percent at $p_2/p_1 = 0.9$, 12 percent at 0.8, 19 percent at 0.7, and 26 percent at 0.6 (if $\beta = 0.5$ and $k = 1.4$). For such situations the more complex compressible formula must obviously be used. In flow nozzles and venturis the flow process is close enough to isentropic to allow theoretical calculation of compressibility effects from Eq. (7.51). In orifices, however, deviation from isentropic conditions is significant (greater turbulence), and an experimental compressibility factor Y is used in the equation:

$$W = Y\left[C_d A_2 \sqrt{1 - \left(\frac{A_2}{A_1}\right)^2}\right] \sqrt{\frac{2g(p_1 - p_2)}{v_1}} \tag{7.53}$$

[1] D. P. Eckman, "Industrial Instrumentation," p. 270, John Wiley & Sons, Inc., New York, 1950.

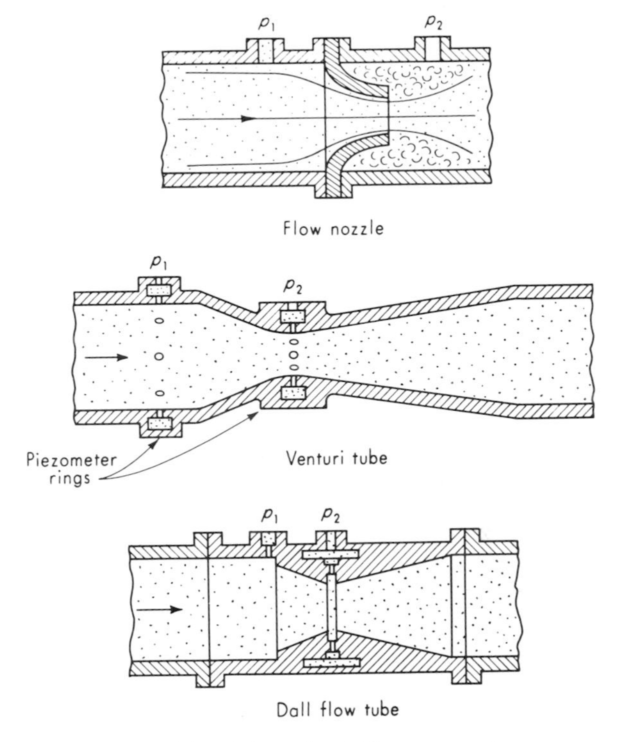

Fig. 7.24. *Variable-pressure-drop meters.*

For flange taps or vena-contracta taps[1]

$$Y = 1 - (0.41 + 0.35\beta^4)\,\frac{p_1 - p_2}{p_1}\,\frac{1}{k} \tag{7.54}$$

while for pipe taps[1]

$$Y = 1 - [0.333 + 1.145(\beta^2 + 0.7\beta^2 + 12\beta^{13})]\,\frac{p_1 - p_2}{p_1}\,\frac{1}{k} \tag{7.55}$$

These empirical formulas are accurate to ± 0.5 percent if $0.8 < p_2/p_1 < 1.0$ and the flowing fluid is a gas or vapor other than steam. For steam the accuracy is ± 1.0 percent. In sizing orifices for gas measurement, a useful rule of thumb is that the maximum Δp (in inches of water) should not exceed the upstream gage pressure in pounds per square inch.

The *flow nozzle, venturi tube*, and *Dall flow tube* (Fig. 7.24) all operate on exactly the same principle as the orifice, the significant differences lying in the numerical values of certain characteristics. Discharge coefficients

[1] Fluid Meters, Their Theory and Application.

of flow nozzles and venturis are larger than those for orifices and also exhibit an opposite trend with Reynolds numbers, varying from about 0.94 at $N_R = 10{,}000$ to 0.99 at $N_R = 10^6$. The Dall-flow-tube coefficient is more like that of an orifice, for $\beta = 0.7$, for example, going from about 0.68 at $N_R = 100{,}000$ to 0.66 at $N_R = 10^6$. Individual calibrations are generally needed on all these devices since their complicated shapes (compared with an orifice) make accurate reproduction difficult.

When comparing the permanent pressure losses of the various devices one should require that each device be producing the same measured Δp, since this would keep the accuracy constant. On this basis, the permanent pressure loss of a flow nozzle is practically identical with that of an orifice. This is because, to get the same Δp, the flow nozzle must have a smaller β ratio, and losses increase with decreasing β ratio. The venturi tube also requires a smaller β for a given Δp, but because of its streamlined form, its losses are low and nearly independent of β. The permanent pressure loss is of the order of 10 to 15 percent of the measured Δp over the range $0.2 < \beta < 0.8$; thus a venturi gives a definite improvement in power losses over an orifice and is often indicated for measuring very large flow rates, where power losses, though a small *percentage*, become economically significant in absolute value. The initial higher cost of a venturi over an orifice may thus be offset by reduced operating costs. The Dall flow tube has the unexpected (though desirable) features of a high measured Δp (similar to an orifice) and a low permanent pressure loss (similar to, and sometimes better than, a venturi). These apparently inconsistent virtues have been checked experimentally but are not fully explained theoretically.[1] The permanent pressure loss of a Dall tube is of the order of 50 percent or less of that of a venturi tube with the same Δp. Other factors to consider in choosing among the orifice, flow nozzle, venturi, or Dall tube include freedom from pressure-tap clogging due to suspended solids (venturi is best), loss of accuracy due to wear (venturi, flow nozzle, and Dall tube are better than an orifice), accuracy (venturi, when calibrated, is best), cost, and ease of changing the flow element to a different size.

Laminar-flow elements differ from the metering devices discussed above in that they are specifically designed to operate in the laminar-flow regime. Pipe flows generally are considered laminar if Reynolds number N_R is less than 2,000; however, in laminar-flow elements, considerably lower values are often designed for to ensure laminar conditions. The simplest form of laminar-flow element is merely a length of small-diameter (capillary) tubing.[2] For $N_R < 2{,}000$, the Hagen-Poiseuille viscous-

[1] I. O. Miner, The Dall Flow Tube, *Instr. Engr.*, p. 45, April, 1957.

[2] L. M. Polentz, Capillary Flowmetering, *Instr. Control Systems*, p. 648, April, 1961.

flow relation gives for incompressible fluids

$$Q = \frac{\pi D^4}{128\,\mu L}\Delta p \qquad (7.56)$$

where
$Q \triangleq$ volume flow rate, ft^3/sec
$D \triangleq$ tube inside diameter, ft
$\mu \triangleq$ fluid viscosity, $lb_f\text{-sec}/ft^2$
$L \triangleq$ tube length between pressure taps, ft
$\Delta p \triangleq$ pressure drop, lb_f/ft^2

One usually designs for $N_R \lessapprox 1{,}000$ in such a device. Extremely small flows can be measured in this way; a 3-ft length of 0.004-in.-diameter tubing measuring Δp with a 2-in. water inclined manometer gives a threshold sensitivity of about 0.000175 $in.^3/hr$ when hydrogen is flowing.[1]

A single capillary tube is capable of handling only small flow rates at laminar Reynolds numbers. To increase the capacity of laminar-flow elements, many capillaries in parallel (or their equivalent) may be employed. One commercially available variation[2] uses a large tube (about 1 in. in diameter) packed with small spheres. The passages between the spheres give the same effect as many capillary tubes. This particular instrument is designed to give a Reynolds number of 20 or less. A 5-in. length of tube gives a Δp of 20 in. of water for a 2 cm^3/min flow rate of air at 14.7 psia and 70°F. Another approach[3] uses a "honeycomb" element (see Fig. 7.25) with triangular members a few thousandths of an inch on a side and a few inches long. These devices have been used mainly to measure flow of low-pressure air, and standard models are available in ranges from 50 to 2,000 ft^3/min at pressure drops of 4 to 8 in. of water. All the above laminar elements have the advantages accruing from a linear (rather than square-root) relation between flow rate and pressure drop. These are principally a large accurate range of as much as 100 to 1 (compared with 3 or 4 to 1 for square-root devices), accurate measurement of average flow rates in pulsating flow, and ease of integrating Δp signals to compute total flow. The laminar elements also can measure reversed flows with no difficulty. They are also usually less sensitive to upstream and downstream flow disturbances than other devices that have been discussed. Their disadvantages include clogging due to dirty fluids, high cost, large size, and high pressure loss (*all* the measured Δp is lost).

[1] Polentz, *loc. cit.*

[2] A. R. Hughes, New Laminar Flowmeter, *Instr. Control Systems*, p. 98, April, 1962.

[3] Meriam Instrument Co., Cleveland, Ohio, *Tech. Note* 2A.

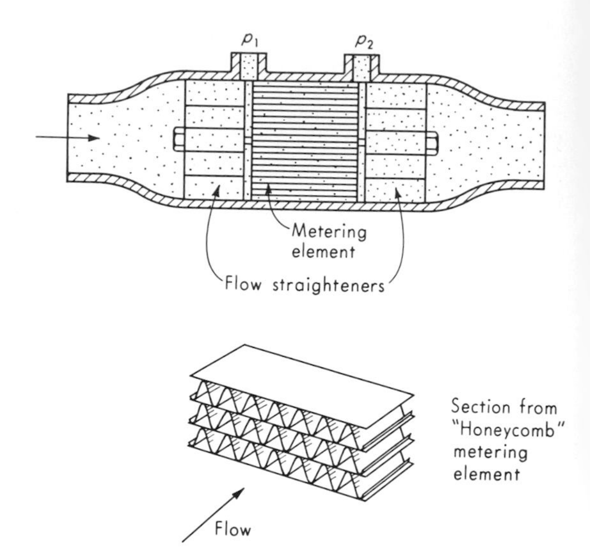

Fig. 7.25. *Laminar-flow element.*

Constant-pressure-drop, variable-area meters (rotameters). A rotameter consists of a vertical tube with tapered bore in which a "float" assumes a vertical position corresponding to each flow rate through the tube (see Fig. 7.26). For a given flow rate the float remains stationary since the vertical forces of differential pressure, gravity, viscosity, and buoyancy are balanced. This balance is self-maintaining since the meter flow area (annular area between the float and tube) varies continuously with vertical displacement; thus the device may be thought of as an

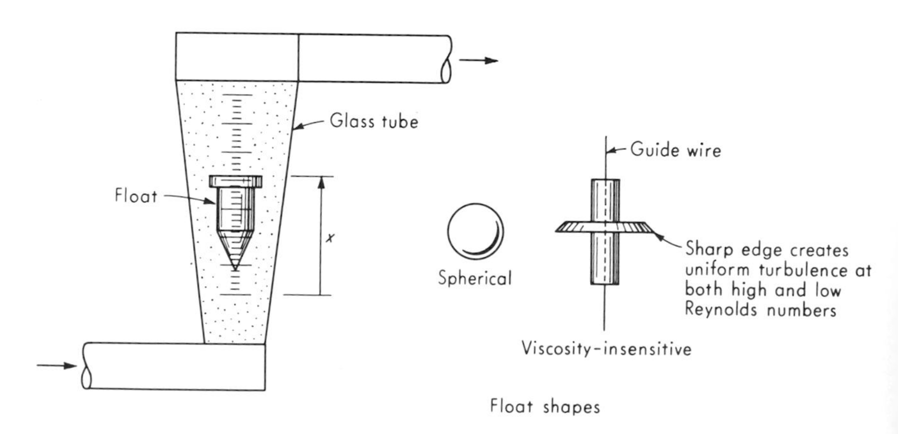

Fig. 7.26. *Rotameter.*

orifice of adjustable area. The downward force (gravity minus buoyancy) is constant and so the upward force (mainly the pressure drop times the float cross-section area) must also be constant. Since the float area is constant, the pressure drop must be constant. For a *fixed* flow area, Δp varies with the square of flow rate, and so to keep Δp *constant* for differing flow rates the area must vary. The tapered tube provides this variable area. The float position is the output of the meter and can be made essentially linear with flow rate by making the tube area vary linearly with the vertical distance. Rotameters thus have an accurate range of about 10 to 1, considerably better than square-root-type elements. Assuming incompressible flow and the above described simplified model, one can derive the result

$$Q = \frac{C_d(A_t - A_f)}{\sqrt{1 - [(A_t - A_f)/A_t]^2}} \sqrt{2gV_f \frac{w_f - w_{ff}}{A_f w_{ff}}} \qquad (7.57)$$

where
$Q \triangleq$ volume flow rate, ft^3/sec
$C_d \triangleq$ discharge coefficient
$A_t \triangleq$ area of tube, ft^2
$A_f \triangleq$ area of float, ft^2
$g \triangleq$ local gravity, ft/sec^2
$V_f \triangleq$ volume of float, ft^3
$w_f \triangleq$ specific weight of float, lb_f/ft^3
$w_{ff} \triangleq$ specific weight of flowing fluid, lb_f/ft^3

If the variation of C_d with float position is slight and if $[(A_t - A_f)/A_t]^2$ is always much less than 1, Eq. (7.57) has the form

$$Q = K(A_t - A_f) \qquad (7.58)$$

and if the tube is shaped so that A_t varies linearly with float position x, then $Q = K_1 + K_2 x$, a linear relation. The floats of rotameters may be made of various materials to obtain the desired density difference [$w_f - w_{ff}$ in Eq. (7.57)] for metering a particular liquid or gas. Some float shapes, such as spheres, require no guiding in the tube; others are kept central by guide wires or by internal ribs in the tube. Certain shapes of floats have been found to reduce viscosity effects. The tubes are often made of high-strength glass to allow direct observation of the float position. Where greater strength is required, metal tubes can be used and the float position detected magnetically through the metal wall. If a pneumatic or electrical signal related to the flow rate is desired, the float motion can be measured with a suitable displacement transducer.

Turbine meters. If a turbine wheel is placed in a pipe containing a flowing fluid, its rotary speed depends on the flow rate of the fluid. By

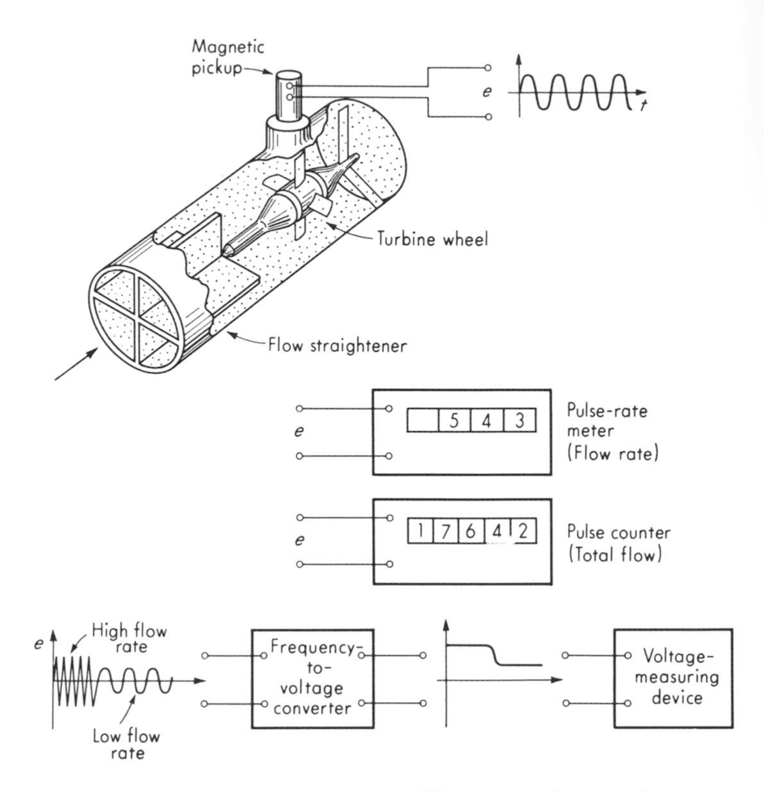

Fig. 7.27. *Turbine flowmeter*

reducing bearing friction and other losses to a minimum, one can design a turbine whose speed varies linearly with flow rate; thus a speed measurement allows a flow-rate measurement. The speed can be measured simply and with great accuracy by counting the rate at which turbine blades pass a given point, using a magnetic proximity pickup to produce voltage pulses. By feeding these pulses to an electronic pulse-rate meter one can measure flow rate, and by accumulating the total number of pulses during a timed interval the total flow is obtained. These measurements can be made very accurately because of their digital nature. If an analog voltage signal is desired, the pulses can be fed to a frequency-to-voltage converter with, however, some loss in accuracy. Figure 7.27 shows a flowmetering system of this type.

Dimensional analysis[1] of the turbine flowmeter shows that (if bearing

[1] H. M. Hochreiter, Dimensionless Correlation of Coefficients of Turbine-type Flowmeters, *Trans. ASME,* p. 1363, October, 1958.

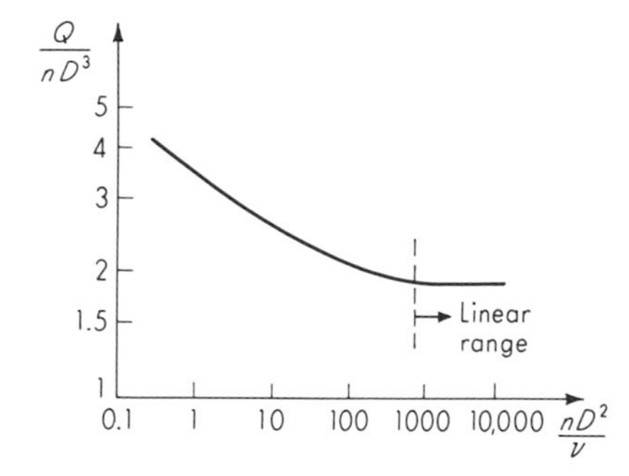

Fig. 7.28. *Turbine-flowmeter characteristic.*

friction and shaft power output are neglected) the following relation should hold:

$$\frac{Q}{nD^3} = \text{some function of } \frac{nD^2}{\nu} \qquad (7.59)$$

where $Q \triangleq$ volume flow rate, in.^{3}/sec
$n \triangleq$ rotor angular velocity, rps
$D \triangleq$ meter bore diameter, in.
$\nu \triangleq$ kinematic viscosity, in.^{2}/sec

Actually, the effect of viscosity is limited mainly to low flow rates, high flow rates being in the turbulent regime where viscosity effects are secondary. For negligible viscosity effects, a simplified analysis[1] based on strictly kinematic relationships gives the following result:

$$\frac{Q}{nD^3} = \frac{\pi L}{4D}\left[1 - \alpha^2 - \frac{2m(D_b - D_h)t}{\pi D^2}\sqrt{1 + \left(\frac{\pi D_b}{L}\right)^2}\right] \qquad (7.60)$$

where $L \triangleq$ rotor lead, in.
$\alpha \triangleq D_h/D$
$m \triangleq$ number of blades
$D_b \triangleq$ rotor-blade-tip diameter, in.
$D_h \triangleq$ rotor-hub diameter, in.
$t \triangleq$ rotor-blade thickness, in.

Equation (7.60) gives $Q = Kn$, where K is a constant for any given meter and is independent of fluid properties. This thus represents the ideal situation. Deviations from this ideal may be found from experimental calibrations,[2] such as are shown for a meter with $D = 1$ in. in Fig. 7.28. We see for sufficiently high values of nD^2/ν that Q/nD^3 becomes essentially constant as predicted by Eq. (7.60). In the particular case shown, $Q/nD^3 = 1.92$ for at least at 10-to-1 range of nD^2/ν; thus this would be a

[1] H. M. Hochreiter, Dimensionless Correlation of Coefficients of Turbine-type Flowmeters, *Trans. ASME,* p. 1363, October, 1958.
[2] *Ibid.*

useful linear operating range for this meter. The meter could be used at lower flow rates by applying corrections obtained from Fig. 7.28. However, this is usually not done since turbine meters are available in a wide range of sizes, each being linear over a different flow range. If the total flow range to be accommodated is about 10 to 1 or less, one can usually select a turbine meter that is linear in the desired range. Linearity is particularly desirable if one is totalizing pulses to get a total flow over a timed interval during which flow rate fluctuates.

Commercial turbine meters are available with full-scale flow rates ranging from about 0.1 to 30,000 gpm for liquids and 0.1 to 15,000 ft^3/min for air. Nonlinearity within the design range (usually about 10 to 1) can be as good as 0.05 percent in the larger sizes. The output voltage of the magnetic pickups is of the order of 10 mv rms at the low end of the flow range and 100 mv at the high. Pressure drop across the meter varies with the square of flow rate and is about 3 to 10 psi at full flow. Turbine meters can follow flow transients quite accurately since their fluid/mechanical time constant is of the order of 2 to 10 msec. If a frequency-to-voltage converter is used to get an analog voltage output, however, its response may be somewhat slower than this since the operating frequencies of turbine meters are of the order of 100 to 2,000 cps. That is, the frequency-to-voltage converter requires low-pass filtering which rejects frequencies somewhat below the turbine operating frequency and is thus limited in transient response also. For example, if the turbine is putting out 500 cps, the low-pass filter will have to cut off at *least* at 500 cps and probably considerably lower. If a first-order filter is designed to attenuate 20 db at 500 cps, it will have a time constant of 0.0032 sec, thus adding this much to the lag of the overall system.

Positive-displacement meters. These meters are actually positive-displacement fluid motors in which friction and inertia have been reduced to a minimum. The flow of a fluid through volume chambers of definite size causes rotation of an output shaft. This type of meter usually is used where the total flow and not the instantaneous flow rate is of interest. A simple mechanical counter records the total number of rotations, which is proportional to the total flow. While any form of positive-displacement motor could conceivably be redesigned as a flowmeter, the nutating-disk type is the most widely used. The accuracy of such devices is of the order of 1.5 percent, with the pressure drop being about 5 psi at maximum flow.

Metering pumps. A variable-displacement positive-displacement pump, if properly designed, can serve both to *cause* a flow rate and also simultaneously to *measure* it. The principle again is merely that a positive-displacement machine, except for leakage and compressibility, delivers a definite flow rate of fluid at a given speed. In most pumps of this kind the operating speed is fixed and the flow rate is varied by changing pump displacement, usually with some form of mechanical linkage. Since these pumps are often used in automatic control systems, many are designed to accept pneumatic or electrical input signals which adjust the pump displacement in a linear fashion. The flow rate of such a system can be set with an accuracy of the order of 1 percent.

Electromagnetic flowmeters. The electromagnetic flowmeter is an application of the principle of induction, shown in Fig. 7.29*a*. If a conductor of length l moves with a transverse velocity v across a magnetic field of intensity B there will be forces on the charged particles of the conductor that will move the positive charges toward one end of the conductor and the negative to the other. Thus a potential gradient is set up along the conductor, and there is a voltage difference e between its two ends. The quantitative relation among the variables is given by the well-known equation

$$e = Blv \tag{7.61}$$

where $B \triangleq$ field flux density, volt-sec/ft^2
$l \triangleq$ conductor length, ft
$v \triangleq$ conductor velocity, fps

If the ends of the conductor are connected to some external circuit that is stationary with respect to the magnetic field, the induced voltage will, in general, cause a current i to flow. This current flows through the moving conductor, which has a resistance R, causing an iR drop, so that the terminal voltage of the moving conductor becomes e-iR.

We consider now a cylindrical jet of conductive fluid with a uniform velocity profile, traversing a magnetic field as in Fig. 7.29*b*. In a liquid conductor the positive and negative ions are forced to opposite sides of the jet, giving a potential distribution as shown. The maximum voltage difference is found across the ends of a horizontal diameter and is BD_pv in magnitude. In a practical situation, the magnetic field is of limited extent, as in Fig. 7.29*c*; thus no voltage is induced in that part of the jet outside the field. Since these parts of the fluid are, however, still conductive paths, they tend partially to "short circuit" the voltages induced in the section exposed to the field; thus the voltage is reduced from the value BD_pv. If the field is sufficiently long, this effect will be slight at the center of the field length. A length of 3 diameters is usually sufficient.[1]

In a practical flowmeter, the "jet" is contained within a stationary

[1] I. C. Hutcheon, Electrical Characteristics of the Magnetic Flow Detector Head, *Instr. Eng.*, p. 1, April, 1964.

pipe. The pipe must be nonmagnetic to allow the field to penetrate the fluid and usually is nonconductive (plastic, for instance) so that it does not provide a short-circuit path between the positive and negative induced potentials at the fluid surface. This nonconductive pipe has two electrodes placed at the points of maximum potential difference. These electrodes then supply a signal voltage to external indicating or recording apparatus. Because it is impractical to make the entire piping installation nonconductive, a short length (the flowmeter itself) of nonconductive pipe must be coupled into an ordinary metal-pipe installation. Since the fluid itself is conductive, this means that there is a conductive path between the

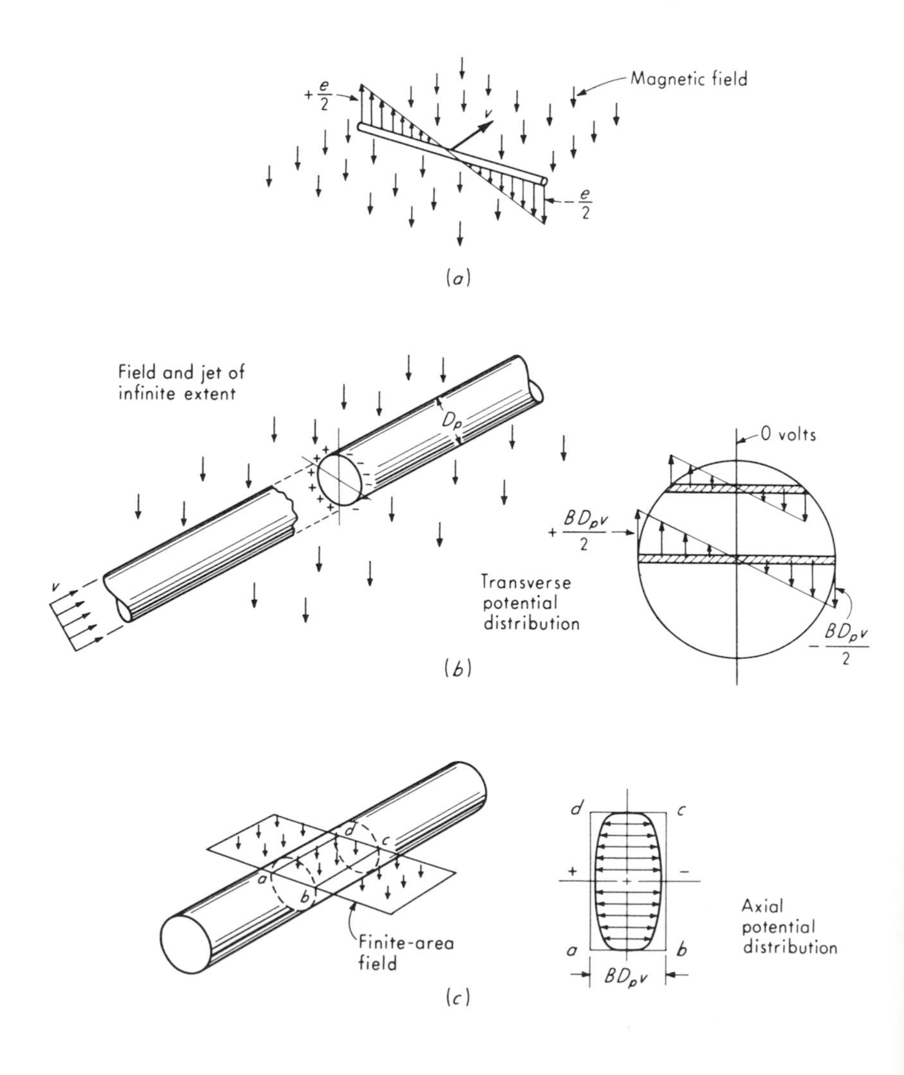

electrodes. In Fig. 7.29*d* this path is shown split into two equal parts $R_1/2$ and containing the signal source $e = BD_pv$. This resistance is not simple to calculate since it involves a continuous distribution of resistance over complex bodies. It can, however, be estimated from theory, and once a device is built it can be directly measured. The magnitude of this source resistance determines the loading effect of any external circuit connected to the electrodes.

The magnetic field used in such a flowmeter could conceivably be either constant or alternating, giving rise to a d-c or an a-c output signal, respectively. The d-c form of meter is little used for a number of reasons. First, many hydrogen-bearing or aqueous solutions exhibit a polarization effect wherein positive ions migrate to the negative electrode and disassociate, forming an insulating pocket of gaseous hydrogen. An a-c field inhibits this action. Secondly, a d-c field may distort the fluid-velocity profile by magnetohydrodynamic action. Above we assumed a uniform (square) velocity profile, whereas actual flows are never of exactly this form. It has been shown,[1] however, that, for *any* velocity profile that is symmetrical about the center line of the pipe, the voltage generated will correspond to the *average* velocity of the flow. However, *unsymmetrical*

[1] A. Kolin, An Alternating Field Induction Flowmeter of High Sensitivity, *Rev. Sci. Instr.*, vol. 16, p. 109, May, 1945.

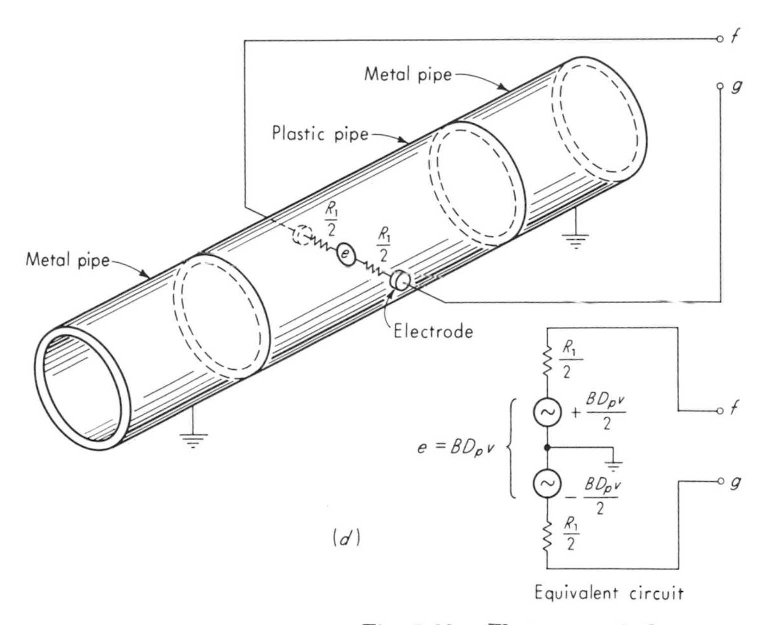

Fig. 7.29. *Electromagnetic flowmeter.*

profiles do not give correct average velocity readings and magnetohydrodynamic action can give such profiles. An a-c field (60 cps is generally adequate) has little effect on velocity profiles because fluid inertia and friction forces at 60 cps are sufficient to prevent any large fluid motions. Finally, since the flow-related voltage signals of such meters are quite small (a few millivolts), interfering voltage inputs due to thermocouple-type effects and galvanic action of dissimilar metals used in meter construction may be of a magnitude similar to the desired signal. Since these interfering inputs are generally drifts of very-low-frequency content, a 60-cps a-c system can use high-pass filtering to wash these out completely. Furthermore, the small flow signals require high amplification, which is more easily, cheaply, and reliably accomplished with a-c than with d-c amplifiers.

While a-c systems predominate, d-c types have been used in metering liquid metals, such as mercury. Here, no polarization problem exists. Also, an insulating pipe liner is not needed, since the conductivity of the liquid metal is very good relative to an ordinary metal (stainless-steel) pipe. This means that a metal pipe is not very effective as a "short circuit" for the voltage induced in a liquid-metal flow. Also, no special electrodes are necessary, the output voltage being tapped off the metal pipe itself at the points of maximum potential difference.

In a typical a-c system[1] shown in simplified form in Fig. 7.30, the field B is provided by coils operating from a 60-cps source. Since the output signal is directly proportional to B, it would seem that B must be regulated very closely to maintain constant flowmeter sensitivity. The power supplied to the field coils is not small (600 va for a typical 3-in.-diameter flowmeter), and currents of this magnitude are not easily regulated. Most systems therefore employ an instrument servo feedback system arranged to cancel the effect of any changes in field, rather than trying to keep B constant. In Fig. 7.30 a signal proportional to the field current (and thus to B) is obtained from a current transformer whose primary carries the field current. The secondary current (proportional to the field current) is applied as excitation to the motion-measuring potentiometer of an instrument servo whose input voltage is the flowmeter output voltage. The output of the servo is the angular rotation θ_p of the potentiometer, which we shall show is a direct indication of flow velocity v. When the servo is balanced, the error voltage $e_e \approx 0$ and thus

$$vD_pB = BK_BR_p\theta_p \qquad (7.62)$$

$$\theta_p = \frac{D_p}{K_BR_p} v \qquad (7.63)$$

[1] I. C. Hutcheon, Some Problems of Magnetic Flow Measurement, *Instr. Eng.*, p. 1, April, 1960.

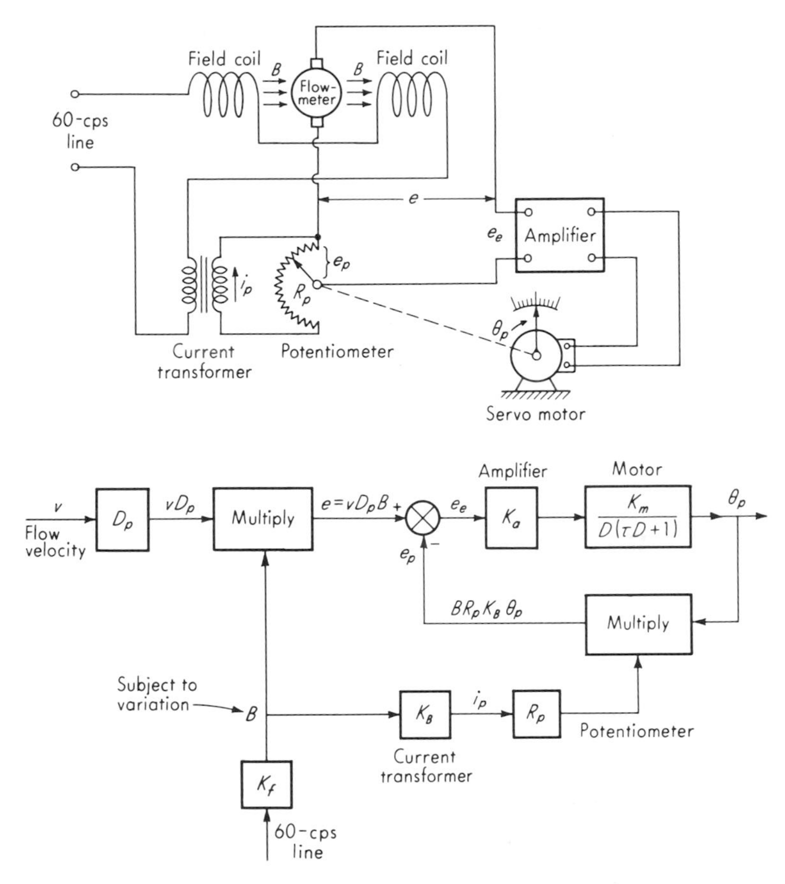

Fig. 7.30. *Magnetic flowmeter servosystem.*

We see that changes in B have no effect on the indication of such a system. A typical system produces a voltage e of about 3 mv rms if $D = 3$ in. and tap water is flowing at 100 gpm. The resistance between the electrodes is given by theory[1] as approximately $1/\sigma d$, where $\sigma \triangleq$ fluid conductivity and d = electrode diameter. For tap water $\sigma \approx 200 \times 10^{-6}$ mho/cm, and so if d is, say, $\frac{1}{4}$ in. (0.64 cm), there is a resistance of about 7,800 ohms as the internal resistance of the voltage source producing e. The signal cable and amplifier input act as a load on this source and must be of sufficiently high impedance.

A limitation of electromagnetic flowmeters is the conductivity of the metered fluid. This must be sufficiently high so that the external circuitry

[1] V. P. Head, Electromagnetic Flowmeter Primary Elements, *Trans. ASME*, p. 662, December, 1959.

is not an excessive load. For a-c flowmeters one must consider both capacitance and resistance of both the fluid and the external circuit. Capacitance causes a phase shift that leads to balancing errors in the servo. These problems can be handled fairly routinely if the fluid conductivity is greater than 20×10^{-6} mho/cm; lower values require special care but 0.1×10^{-6} mho/cm is feasible,[1] with future developments promising even lower values.

Some general features of electromagnetic flowmeters include the lack of any flow obstruction; ability to measure reverse flows; insensitivity to viscosity, density, and flow disturbances as long as the velocity profile is symmetrical; wide linear range; and rapid response to flow changes (instantaneous for a d-c system; limited by the field frequency in an a-c system).

Drag-force flowmeters. A body immersed in a flowing fluid is subjected to a drag force F_d given by

$$F_d = \frac{C_d A \rho V^2}{2} \tag{7.64}$$

where $C_d \triangleq$ drag coefficient
$A \triangleq$ cross-section area, ft^2
$\rho \triangleq$ fluid mass density, slug/ft^3
$V \triangleq$ fluid velocity, fps

For sufficiently high Reynolds number and a properly shaped body the drag coefficient is reasonably constant. Therefore, for a given density, F_d is proportional to V^2 and thus to the square of volume flow rate. The drag force can be measured by attaching the drag-producing body to a strain-gage force-measuring transducer. One type[2] uses a cantilever beam with bonded strain gages (see Fig. 7.31*a*). A hollow-tube arrangement with the gages on the outside serves to isolate the gages from the flowing fluid. If the drag body is made symmetrical, reversed flows can be measured. A main advantage of this class of flowmeters is the high dynamic response. The type just described is basically second-order with a natural frequency of 70 to 200 cps. When such a meter is used for dynamic-flow studies it is desirable to linearize the square-root characteristic. This can be done by feeding the strain-gage voltage signal to a square-root computing element whose output will then be linear with flow rate. Another[3] flow transducer of the drag type uses internal flow

[1] D. R. Lynch, A Low-conductivity Magnetic Flowmeter, *Control Eng.*, p. 122, December, 1959.

[2] Ramapo Instrument Co., Bloomingdale, N.J.

[3] Dynamic Instrument Co., Cambridge, Mass.

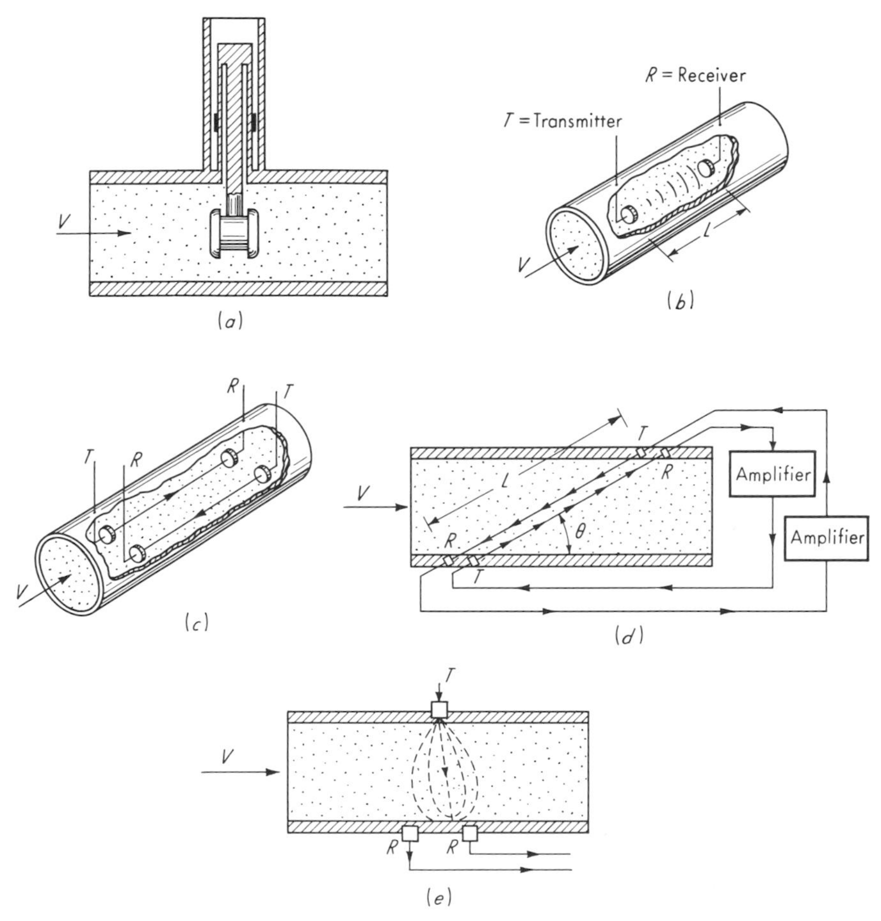

Fig. 7.31. *Drag-force and ultrasonic flowmeters.*

through an orifice plate with many holes. The force on the orifice plate is measured by an unbonded-strain-gage transducer. The natural frequency of this flowmeter is 1,500 cps.

Ultrasonic flowmeters. Small-magnitude pressure disturbances are propagated through a fluid at a definite velocity (the speed of sound) *relative to the fluid.* If the fluid also has a velocity, the *absolute* velocity of pressure-disturbance propagation is the algebraic sum of the two. Since flow rate is related to fluid velocity, this effect may be used in several ways as the operating principle of an "ultrasonic" flowmeter. The term ultrasonic refers to the fact that, in practice, the pressure disturbances usually are short bursts of sine waves whose frequency is above the range audible to human hearing, about 20,000 cps. A typical frequency might be 10 Mc (10^7 cps).

The various methods of implementing the above phenomenon all depend on the existence of transmitters and receivers of acoustic energy. A common approach is to utilize piezoelectric crystal transducers for both these functions. In a transmitter, electrical energy in the form of a short burst of high-frequency voltage is applied to a crystal, causing it to vibrate. If the crystal is in contact with the fluid, the vibration will be communicated to the fluid and propagated through it. The receiver crystal is exposed to these pressure fluctuations and responds by vibrating. The vibratory motion produces an electrical signal in proportion, according to the usual action of piezoelectric displacement transducers.

Figure 7.31*b* shows the most direct application of these principles. With zero flow velocity the transit time t_0 of pulses from the transmitter to the receiver is given by

$$t_0 = \frac{L}{c} \tag{7.65}$$

where $L \triangleq$ distance between transmitter and receiver
$c \triangleq$ acoustic velocity in fluid

For example, in water, $c \approx 5{,}000$ fps, and so, if $L = 1$ ft, $t_0 = 0.0002$ sec. If the fluid is moving at a velocity V, the transit time t becomes

$$t = \frac{L}{c + V} = L\left(\frac{1}{c} - \frac{V}{c^2} + \frac{V^2}{c^3} - \cdots\right) \approx \frac{L}{c}\frac{1 - V}{c} \tag{7.66}$$

and if we define $\Delta t \triangleq t_0 - t$

$$\Delta t \approx \frac{LV}{c^2} \tag{7.67}$$

Thus, if c and L are known, measurement of Δt allows calculation of V. While L may be taken as constant, c varies, for example, with temperature and since c appears as c^2 the error caused may be significant. Also, Δt is quite small since V is a small fraction of c. For example, if $V = 10$ fps, $L = 1$ ft, and $c = 5{,}000$ fps, $\Delta t = 0.4$ μsec, a very short increment of time to measure accurately. Since the measurement of t_0 is not directly provided for in this arrangement, the modification of Fig. 7.31*c* may be preferable. If t_1 is the transit time with the flow and t_2 is the transit time against the flow, we get

$$\Delta t \triangleq t_2 - t_1 = \frac{2VL}{c^2 - V^2} \approx \frac{2VL}{c^2} \tag{7.68}$$

This Δt is twice as large as before and also is a time increment that physically exists and may be directly measured. However, the dependence on c^2 is still a drawback.

In Fig. 7.31*d* two self-excited oscillating systems are created by using the received pulses to trigger the transmitted pulses in a feedback arrangement. The pulse repetition frequency in the forward propagating loop is $1/t_1$ while that in the backward loop is $1/t_2$. The frequency difference $\Delta f \triangleq 1/t_1 - 1/t_2$ can be measured by multiplying the two signals together to get a beat frequency. Since $t_1 = L/(c + V \cos \theta)$ and $t_2 = L/(c - V \cos \theta)$ we get

$$\Delta f = \frac{2V \cos \theta}{L} \tag{7.69}$$

which is independent of c and thus not subject to errors due to changes in c. The above analysis assumes a square velocity profile which, of course, does not occur in practice. For actual profiles, V can be replaced by V_{av} as long as the profiles are symmetrical about the pipe center line. A commercial system[1] of this type has an accuracy ± 2 percent of full scale, a linear range 20 to 1, and a capacity to 25 fps.

Another approach,[2] shown in Fig. 7.31*e*, senses the deflection of an ultrasonic beam propagated transversely to the flow. With no flow, the two receivers get equal signals; with a flow present, the beam deflects, giving one receiver a stronger signal and the other a weaker signal. Electronic circuitry forms the ratio of these two signals and this ratio is a linear function of flow rate. This arrangement is sensitive to changes in sound velocity but these are mainly dependent on temperature and are compensated for by a thermistor placed on the outside wall of the pipe.

Ultrasonic flowmeters are presently used mainly for liquids. The practical versions (Fig. 7.31*d* and *e*) have no flow obstruction and are relatively insensitive to viscosity, temperature, and density variations. Their complexity and relatively high cost somewhat limit industrial application at present.

7.3

Gross Mass Flow Rate In many applications of flow measurement mass flow rate is actually more significant than volume flow rate. As an example, the range capability of an aircraft or liquid-fuel rocket is determined by the *mass* of fuel, not the volume. Flowmeters used in fueling such vehicles should thus indicate mass, not volume. In chemical process industries also, mass flow rate is often the significant quantity.

Two general approaches are used to measure mass flow rate. One involves the use of some type of volume flowmeter, some means of density measurement, and some type of simple computer to compute mass flow

[1] Fischer and Porter Co., Hatboro, Pa.

[2] H. E. Dalke and W. Welkowitz, A New Ultrasonic Flowmeter for Industry, *ISA J.*, p. 60, October, 1960.

rate. The other, more basic approach is to find flowmetering concepts that are inherently sensitive to mass flow rate. Both methods are currently finding successful application in various detail forms.

Volume flowmeter plus density measurement. Some basic methods of fluid-density measurement are shown in Fig. 7.32. In Fig. 7.32*a* a portion of the flowing liquid is bypassed through a still well. The buoyant force on the float is directly related to density and may be measured in a number of ways, such as the strain-gage beam shown. Buoyant force is used in a different way in Fig. 7.32*b*. For each liquid density the system of three floats assumes a unique angular orientation θ. By proper choice of float densities, volumes, and angular locations, a very closely linear relation between density and θ is achieved.[1] The angle θ is measured by an angular position transducer to obtain an electrical signal. In Fig. 7.32*c* a definite volume of flowing liquid contained within the U tube is continuously weighed by a spring and pneumatic displacement transducer. Flexible couplings isolate external forces from the U tube. A pneumatic force-balance feedback system can also be used to measure the weight.[2] This minimizes deflection and thus reduces errors due to variable spring effects of flexible couplings and flexure pivots.

Figure 7.32*d* shows a method of measuring gas density using a small centrifugal blower (run at constant speed) to pump continuously a sample of the flow. The pressure drop across such a blower is proportional to density and may be measured with a suitable differential pressure pickup. Ultrasonic volume flowmeters often use an ultrasonic density-measuring technique when mass flow rate is wanted. In Fig. 7.32*e* the crystal transducer serves as an acoustic-impedance detector. Acoustic impedance depends on the product of density and speed of sound. Since a signal proportional to the speed of sound is available from the volume flowmeter, division of this signal into the acoustic-impedance signal gives a density signal. The attenuation of radiation from a radioisotope source depends on the density of the material through which the radiation passes (see Fig. 7.32*f*). Over a limited (but generally adequate) density range the output current of the radiation detector is nearly linear with density, for a given flowing fluid.[3] For gas flow, indirect measurement of density by means of computation from pressure and temperature signals (Fig. 7.32*g*) is also common. Figure 2.17*c* shows an ingenious method of accomplishing this.

In computing mass flow rate from volume flowmeter and densitometer (density-measuring device) signals, the necessary form of com-

[1] Potter Aeronautical Corp., Union, N.J.

[2] Halliburton Co., Duncan, Okla.

[3] Industrial Nucleonics Corp., Columbus, Ohio.

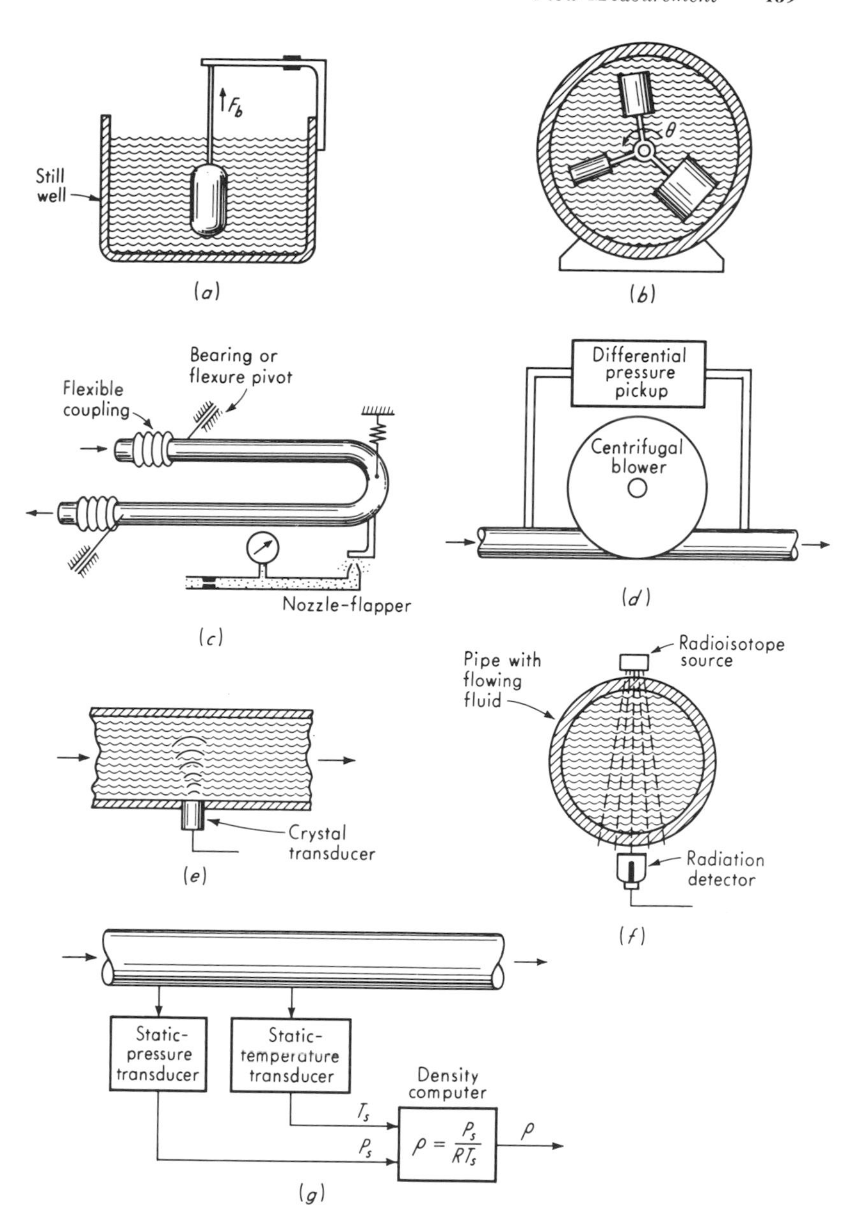

Fig. 7.32. *Fluid-density measurement.*

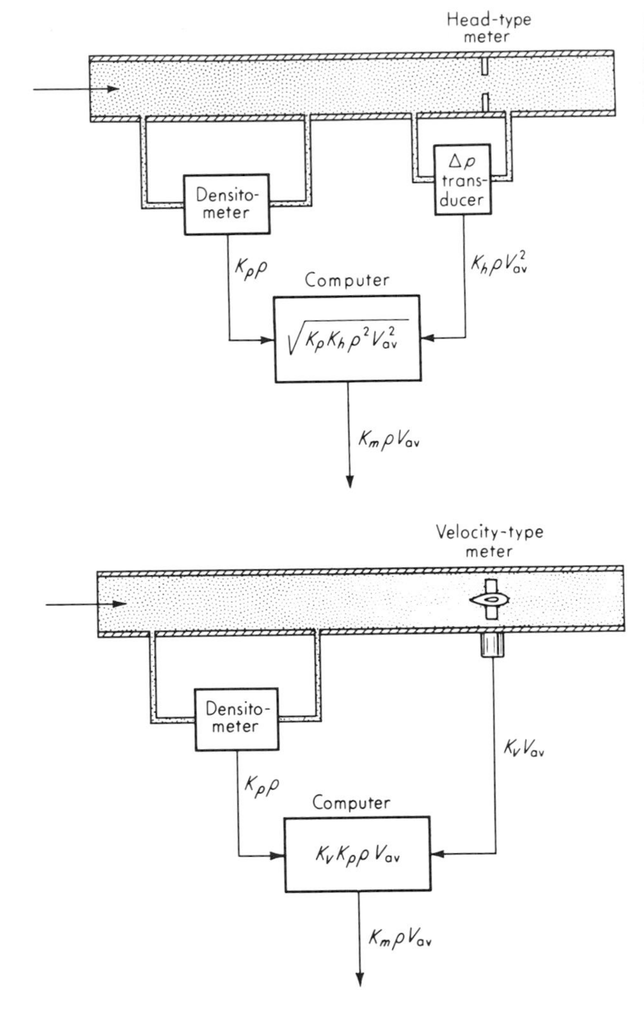

Fig. 7.33. Computed mass flow measurement.

puter varies somewhat, depending on the type of flowmeter. For so-called "head" meters (those producing a differential pressure or electrical signal proportional to ρV_{av}^2, such as an orifice) the computer multiplies the ρ signal by the ρV_{av}^2 to form $\rho^2 V^2$ and then takes the square root of this to get ρV, which is proportional to the mass flow rate. For "velocity" flowmeters, such as the turbine and electromagnetic types, the available signal is proportional to V_{av}; thus the computer must simply multiply this by the ρ signal, a square-root operation being unnecessary. Figure 7.33 shows these concepts.

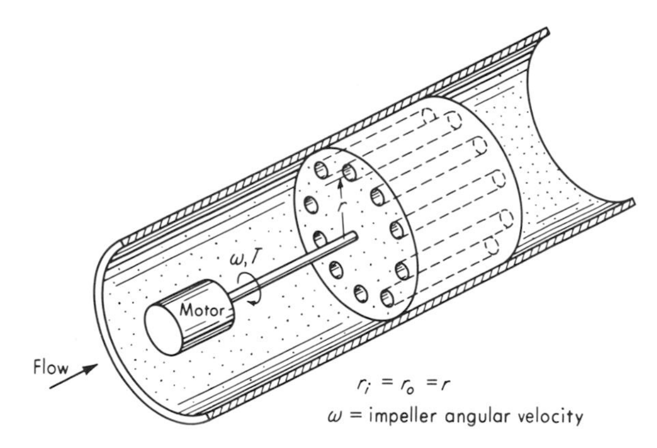

Fig. 7.34. *Angular-momentum element.*

Direct mass flowmeters. While the above indirect methods of mass-flow-rate measurement are often satisfactory and are in wide use, it is possible to find flowmetering concepts that are more directly sensitive to mass flow rate. These may have advantages with respect to accuracy, simplicity, cost, weight, space, etc., in certain applications. We shall discuss briefly some of the more common principles in terms of the practical hardware through which they have been realized.

Perhaps the most widely used principle depends on the moment-of-momentum law of turbomachines. Fluid mechanics shows that, for one-dimensional, incompressible, lossless flow through a turbine or impeller wheel, the torque T exerted by an impeller wheel on the fluid (minus sign) or on a turbine wheel by the fluid (plus sign) is given by

$$T = G(V_{ti}r_i - V_{to}r_o) \tag{7.70}$$

where
$G \triangleq$ mass flow rate through wheel, slug/sec
$V_{ti} \triangleq$ tangential velocity at inlet, fps
$V_{to} \triangleq$ tangential velocity at outlet, fps
$r_i \triangleq$ radius at inlet, ft
$r_o \triangleq$ radius at outlet, ft
$T \triangleq$ torque, ft-lb$_f$

Consider now the system of Fig. 7.34. The flow to be measured is directed through an impeller wheel which is motor-driven at constant speed. If the incoming flow has no rotational component ($V_{ti} = 0$) and if the axial length of the impeller is enough to make $V_{to} = r\omega$, the driving torque necessary on the impeller is

$$T = r^2\omega G \tag{7.71}$$

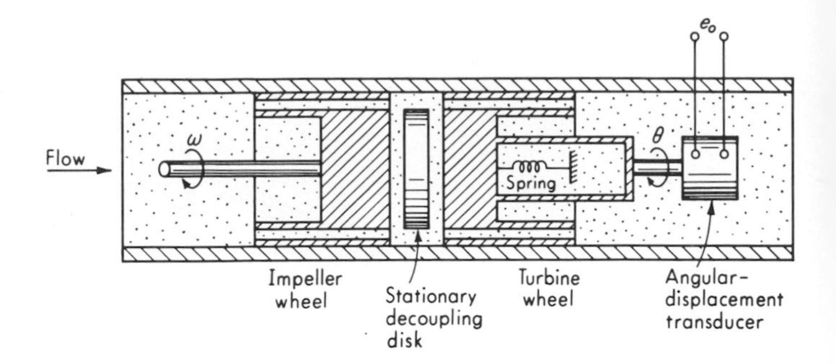

Fig. 7.35. *Angular-momentum mass flowmeter.*

Since r and ω are constant, the torque T (which could be measured in several ways) is a direct and linear measure of mass flow rate G. However, for $G = 0$, torque will *not* be zero, because of frictional effects; furthermore, viscosity changes would also cause this zero-flow torque to vary. A variation on this approach is to drive the impeller at constant *torque* (with some sort of slip clutch). Then, impeller *speed* is a measure of mass flow rate according to

$$\omega = \frac{T/r^2}{G} \tag{7.72}$$

The speed ω is now nonlinear with G but may be easier to measure than torque. If a magnetic proximity pickup is used for speed measurement, the time duration t between pulses is inversely related to ω; thus measurements of t would be linear with G.

A further variation, used in several commercial instruments, is shown in Fig. 7.35. A constant-speed motor-driven impeller again imparts angular momentum to the fluid; however no torque or speed measurements are made on this wheel. Close by, downstream, a second ("turbine") wheel is held from turning by a spring restraint. For the impeller, $V_{to} = r\omega$; furthermore, this becomes V_{ti} for the turbine. Since the turbine cannot rotate, if it is long enough axially, the angular momentum is removed and V_{to} for the turbine is zero. Then the torque on the turbine is

$$T = r^2\omega G \tag{7.73}$$

If the spring restraint is linear, the deflection θ is a direct and linear measure of G and can be transduced to an electrical signal in a number of ways. The decoupling disk reduces the viscous coupling between the impeller and turbine so that, at zero flow rate, a minimum of viscous torque is exerted on the turbine wheel.

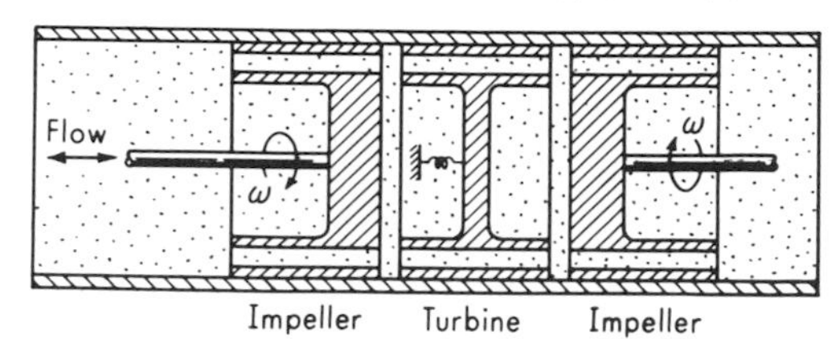

Fig. 7.36. *Bidirectional flowmeter.*

The arrangement of Fig. 7.35 handles flow in one direction only. To make the meter bidirectional, two counterrotating impellers are used, as in Fig. 7.36. The reverse flow causes a reversed turbine-deflection signal. Also, at zero flow, the viscous coupling effects tend to cancel. A drawback is the increased pressure drop due to more obstructions in the flow.

When total mass flow over a time interval, rather than instantaneous rate, is desired, the output signal must be integrated. An ingenious method[1] for doing this, using the basic arrangement of Fig. 7.35, is shown schematically in Fig. 7.37. The torque signal is brought through the

[1] General Electric Co., West Lynn, Mass.

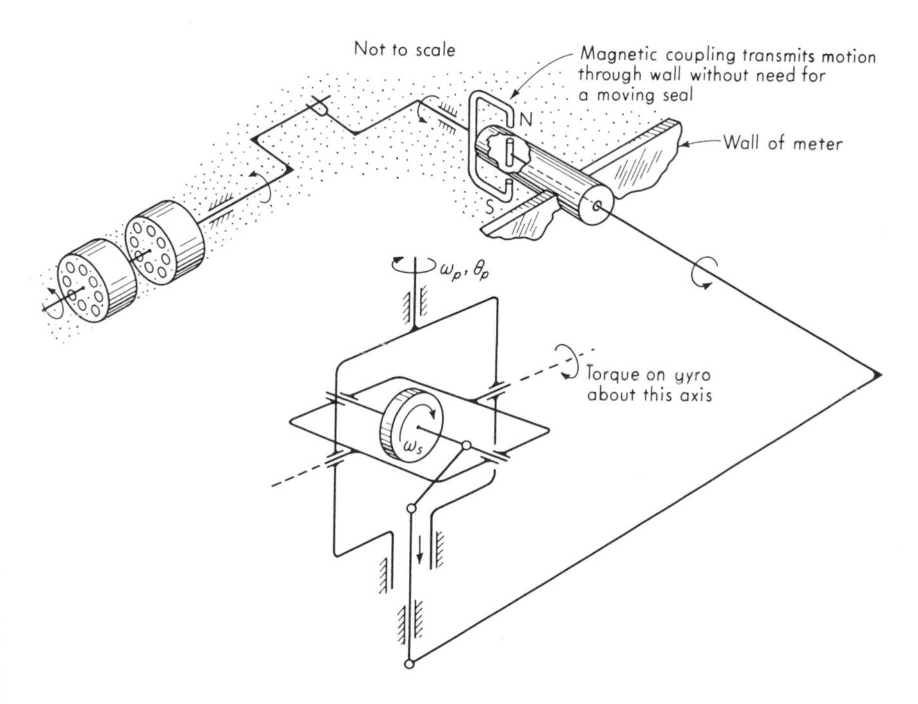

Fig. 7.37. *Gyro integrating flowmeter.*

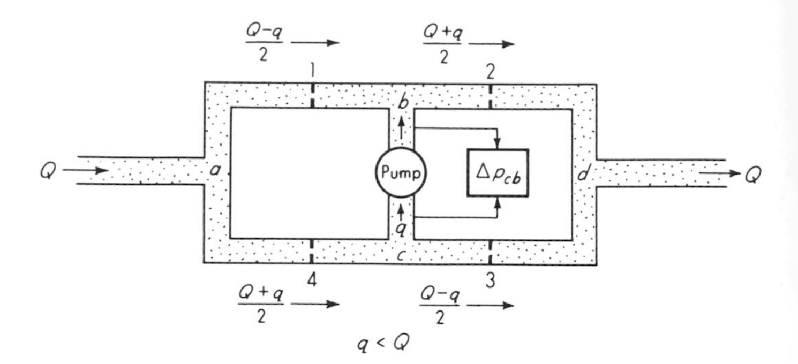

Fig. 7.38. Bridge-circuit flowmeter.

wall of the flowmeter by a magnetic coupling and then applied through suitable linkages to a gyro as shown. The precession velocity ω_p is proportional to the torque and inversely proportional to the gyro spin angular velocity ω_s. If the flowmeter impeller motor and the gyro spin motor are driven from the same a-c source, any frequency changes will affect both speeds equally. Since torque is directly proportional to impeller speed whereas ω_p is inversely proportional to ω_s, the effect on ω_p of an a-c-source frequency change is made zero. If ω_p is directly proportional to G, then θ_p, the total angle turned through in a given time, will be the integral of G or the total mass flow. This total angle is measured with a simple revolution counter. Meters of this type for both gases and liquids have an accurate operating range from 10 to 133 percent of nominal flow rating. Accuracy in this range is ± 1 percent of reading while repeatability is 0.25 percent of reading.

A number of other meters use an angular-momentum principle in different ways. They include the vibrating gyroscope meter,[1] the Coriolis meter,[1] the rotating gyroscope meter,[1] the twin-turbine meter,[1] and the S-tube meter.[2]

An interesting mass flowmeter[3] for liquids, based on quite a different principle, is shown schematically in Fig. 7.38. For a given fluid the pressure drop across an orifice is proportional to ρQ^2. In Fig. 7.38, four identical orifices are connected into a "bridge circuit." A positive-displacement pump of fixed displacement runs at constant speed and volume flow rate q. The pressure rise across this pump is Δp_{cb}, which pressure difference is the output signal of the flowmeter. The flow rates through

[1] C. M. Halsell, Mass Flowmeters, *ISA J.*, p. 49, June, 1960.

[2] J. Haffner, A. Stone, and W. K. Genthe, Novel Mass Flowmeter, *Control Eng.*, p. 69, October, 1962.

[3] Flo-Tron Inc., Paterson, N.J.

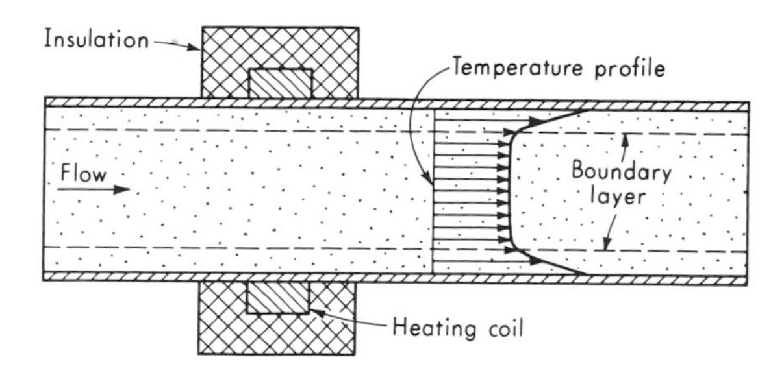

Fig. 7.39. *Boundary-layer principle.*

the individual orifices must be as shown, because of symmetry. The output signal Δp_{cb} is

$$\Delta p_{cb} = K\rho\left(\frac{Q+q}{2}\right)^2 - K\rho\left(\frac{Q-q}{2}\right)^2 = Kq\rho Q = K_1 G \qquad (7.74)$$

Thus Δp_{cb} is linear with mass flow rate G. The "constant" K_1 includes the orifice discharge coefficient; thus all orifice flow rates must be maintained at high enough Reynolds numbers so that the discharge coefficients of all orifices are equal and do not vary when Q varies. Various other arrangements of orifices, pumps, and pressure pickups have been devised; they give the same overall result but have relative advantages and disadvantages in other respects.

Our final example of mass flowmeters uses heat-transfer principles. In Fig. 7.39 an electric heating coil is transferring heat to a fluid flowing inside a pipe. If the pipe wall is a good thermal conductor and quite thin and if heat losses are minimized by insulation, the temperature drop across the boundary layer for turbulent flow is given by

$$\Delta T = \frac{K_1 P_h}{h} \qquad (7.75)$$

where $K_1 \triangleq$ constant (conversion factor and heat-transfer area)
$P_h \triangleq$ heater power
$h \triangleq$ film conductance of boundary layer

For turbulent flow the film conductance is given by

$$k = 0.023\left[\frac{k^{0.6}c^{0.4}}{D^{0.2}\mu^{0.4}}\right]G^{0.8} \qquad (7.76)$$

where $k \triangleq$ fluid thermal conductivity
$c \triangleq$ fluid specific heat
$D \triangleq$ pipe diameter
$\mu \triangleq$ fluid absolute viscosity
$G \triangleq$ mass flow rate

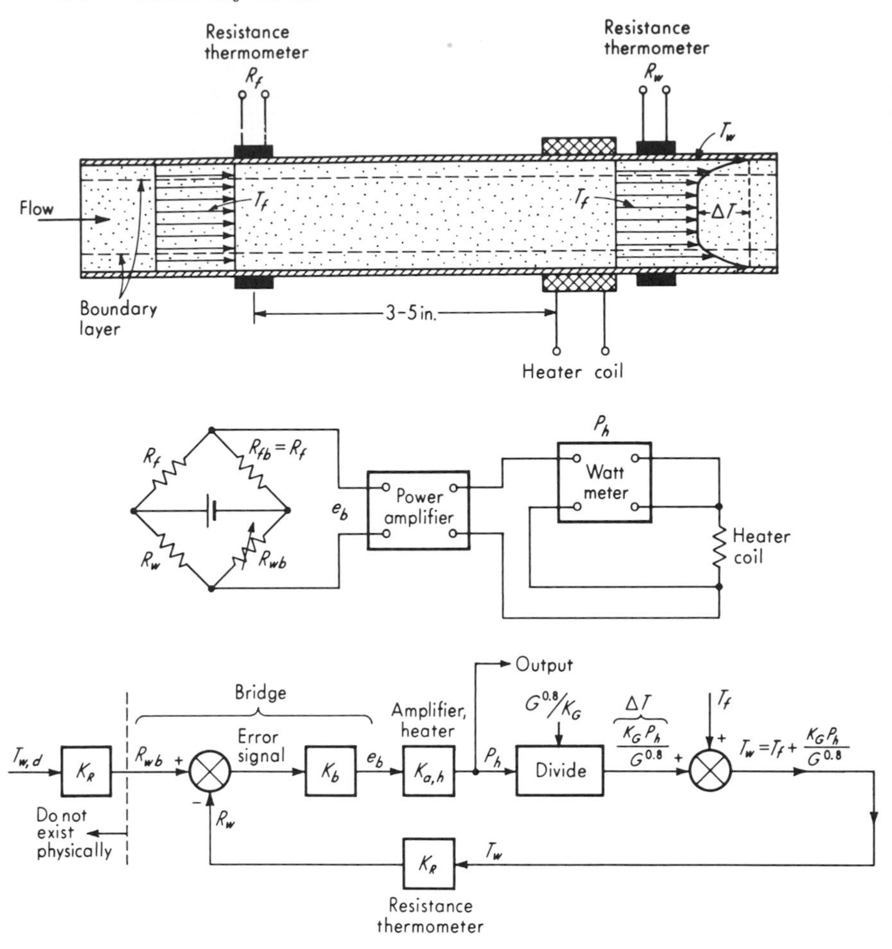

Fig. 7.40. *Boundary-layer flowmeter.*

In Eq. (7.76), if the bracketed quantity is constant or can be compensated for, we see that h is given by $h = K_2G^{0.8}$. Then from Eq. (7.75) we get

$$P_h = \frac{\Delta T K_2}{K_1} G^{0.8} \tag{7.77}$$

If the heater power P_h is adjusted to keep ΔT always constant, P_h becomes a direct (and almost linear) measure of mass flow rate G. This is the principle of the boundary-layer (electrocaloric) flowmeter.[1]

It would seem that to measure ΔT one would require a temperature

[1] J. H. Laub, Measuring Mass Flow with the Boundary-layer Flowmeter, *Control Eng.*, p. 112, March, 1957.

probe in the core of the flowing fluid. Fortunately this complication is unnecessary; the pipe-wall temperature 3 to 5 in. upstream of the heater is very close to the fluid core temperature downstream of the heater. The adjustment of heater power to maintain a constant ΔT (about 2°F is used) is accomplished continuously and automatically by a feedback system shown simplified in Fig. 7.40. A wattmeter reading P_h can be calibrated in mass-flow-rate units for a given fluid. Its reading is the output signal of the flowmeter. A typical value of P_h is about 40 watts if 12,000 lb/hr of water is flowing in a 1.5-in.-diameter pipe.

The feedback-system operation may be explained as follows: If the incoming fluid has a fixed temperature T_f, the resistance R_f will be constant. The bridge resistor R_{fb} will also be constant and made equal to R_f. Suppose that R_{wb} is set equal to R_w; then the bridge is balanced and no heater power is being supplied. If the temperature coefficient of R_w is known, the change in R_w corresponding to the desired ΔT can be calculated. Now R_{wb} is changed by this amount, unbalancing the bridge and thus turning on the heater power. This raises R_w toward the new value of R_{wb}, tending to rebalance the bridge. Perfect balance ($R_w = R_{wb}$) cannot be achieved, since then the heater power would have to be zero. If the gain of the feedback system is high, however, nearly perfect bridge balance is possible without making the heater power zero. With the bridge in this condition, an increase, say, in mass flow rate G results in a momentary decrease in T_w (and thus R_w), unbalancing the bridge in the direction to increase heater power, increase T_w, and thus maintain the desired ΔT. A decrease in G results in the opposite action, again tending to maintain ΔT fixed. If ΔT is maintained fixed for all flow rates, Eq. (7.77) shows that the wattmeter reading (system output signal) is a direct indication of mass flow rate.

Problems

7.1 Water flows in a 1-in.-diameter pipe at 10 fps. If a pitot-static tube of 0.5-in.-diameter is inserted, what velocity will be indicated? Assume one-dimensional frictionless flow. Find the pitot-static-tube diameter needed to reduce the above error to 1 percent.

7.2 For the system of Fig. 7.10, what diameter sphere is needed to obtain a 1-lb force from a 50-mph wind when atmospheric pressure is 14.7 psi and temperature is 70°F? If the 50-mph wind is assumed to be the full-scale value and if 0.05-in. full scale deflection of the differential transformer is desired, what must be the flexure-plate spring constant? If the total moving mass is assumed to be due to the spherical shell alone and if the shell is made of $\frac{1}{16}$-in.-thick aluminum, estimate the usable frequency range of this instrument.

7.3 If, in the system of Fig. 7.15, the amplifier lag is not neglected, so that $(i/e_e)(D) = K_a/[(\tau_1 D + 1)(\tau_2 D + 1)]$, find the value of K_a that will put the feedback system just on the margin of instability (use the Routh criterion). If $\tau = 0.001$ sec and $\tau_1 = \tau_2 = 0.000001$, what is the maximum allowable value of $K_b K_a K_e / I_m$

for marginal stability? If a value of $K_bK_aK_e/I_m$ about one-fifth of that giving marginal stability can be used and if Eq. (7.43) is assumed applicable under these conditions, what percentage improvement of τ_{ct} as compared with τ may be achieved?

7.4 The frequency response of a hot-wire anemometer system also determines its ability to resolve *spacewise* variations in velocity. That is, at a given instant of time, if the velocity pattern in a flow were "frozen," the velocity component in a certain direction would be different at different stations along the line of travel of the gross flow. When this velocity structure is swept past a hot wire by the average velocity, it requires adequate frequency response to resolve the spacewise velocity variations. Consider a simple spacewise variation wherein velocity deviation from average is given by $v = v_0 \sin(2\pi x/\lambda)$, where λ is the wavelength in inches and x is displacement in inches along the flow direction. If the average flow velocity is V_0 in./sec, find an expression for the smallest wavelength λ_{min} that can be resolved by a system whose flat frequency response extends to f_0 cps. Plot λ_{min} versus V_0 for a system with f_0 of 100,000 cps.

7.5 Analyze the error in flow-rate measurement caused by thermal expansion of an orifice plate.

7.6 A capillary-tube laminar-flow element is needed to measure water flow of 0.01 in.^{3}/min at 70°F. A flowmeter pressure drop of 3 in. of water is desired. If the element is designed for a Reynolds number of 500, what length and diameter of tubing are needed?

7.7 Using the simplified model discussed in the text, derive Eq. (7.57).

7.8 Using the assumptions discussed following Eq. (7.57), one can write for the weight flow rate

$$W = K_1(A_t - A_f)\sqrt{w_{ff}(w_f - w_{ff})} \qquad \text{lb}_f/\text{sec}$$

where K_1 includes all the other constants in Eq. (7.57). To make the weight-flow indication relatively insensitive to changes in fluid density w_{ff}, the float density w_f should be twice the density of the flowing fluid. Show the truth of this statement. Hint: Set $\partial W/\partial w_{ff} = 0$.

7.9 Classify, according to the categories of Sec. 2.5, the compensation method used in the system of Fig. 7.30.

7.10 Outline the procedure you would use to design a drag-force flowmeter of the type shown in Fig. 7.31*a*. The given specifications include static sensitivity, dynamic response, flow-velocity range to be covered, allowable size, and fluid-density range to be covered.

7.11 Perform a dynamic analysis on the system of Fig. 7.32*a* to obtain the transfer function relating fluid density as an input to strain-gage bridge voltage as an output. What is the effect of changes in liquid level in the still well on the output signal? What is the effect of thermal (volume) expansion of the float? If the entire assembly is aboard a vehicle that is accelerating vertically, explain the effect on the output.

7.12 Perform a static analysis on the system of Fig. 7.32*b*, obtaining a relation between fluid density as an input and rotation angle θ as an output. List modifying and/or interfering inputs for this instrument.

7.13 Perform a static analysis on the system of Fig. 7.32*c* to get a relation between fluid density as an input and nozzle-flapper pressure as output. List modifying and/or interfering inputs for this instrument.

7.14 Modify the system of Fig. 7.32*c* using a feedback principle (null-balance system) to keep vertical deflection nearly zero for all densities. A bellows may be used to provide the rebalancing force.

7.15 For the system of Fig. 7.35, suppose that the full-scale flow rate is 10 lb_m/sec, an impeller speed of 100 rpm is used, and $r = 1.0$ in. If full-scale transducer rotation is to be 20°, what torsional spring constant is required? If this spring constant must be increased to improve dynamic response, what design changes are possible to achieve this?

7.16 Intuitively, what would you expect the dynamic response of the system of Fig. 7.35 to be? What design parameters would have a major influence on this response and in what way? Devise an experimental technique for subjecting this instrument to an approximate step input.

7.17 List modifying and/or interfering inputs for the system of Fig. 7.40.

Bibliography

1. Standards for Discharge Measurement with Standardized Nozzle and Orifices, *NACA, TM*-952, 1940.
2. V. P. Head: Electromagnetic Flowmeter Primary Elements, *ASME Trans.*, p. 660, December, 1959.
3. D. R. Lynch: A Low-conductivity Magnetic-Flowmeter, *Control Eng.*, p. 122, December, 1959.
4. V. Cushing et al.: Development of an Electromagnetic Induction Flowmeter for Cryogenic Fluids, *NASA, N*-64-20708, 1964.
5. J. A. Shercliff: "The Theory of Electromagnetic Flow Measurement," Cambridge University Press, New York, 1962.
6. Seawater Voltage Tells Submarine Speed, *Machine Design*, p. 28, Feb. 18, 1965.
7. S. Blechman: Techniques for Measuring Low Flows, *Instr. Control Systems*, p. 82, October, 1963.
8. E. W. Miller: Turbine Gas-flow Sensor, *Instr. Control Systems*, p. 105, January, 1962.
9. H. J. Evans: Turbine Flowmeter for Gases, *Instr. Control Systems*, p. 103, March, 1964.
10. R. D. Wood: Steam Measurement by Orifice Meter, *Instr. Control Systems*, p. 135, April, 1963.
11. R. B. Crawford: A Broad Look at Cryogenic Flow Measurement, *ISA J.*, p. 65, June, 1963.
12. D. Shichman and B. S. Johnson: Tap Location for Segmental Orifices, *Instr. Control Systems*, p. 102, April, 1962.
13. P. J. Klass: Laser Flowmeter, *Aviation Week*, Jan. 11, 1965.
14. R. W. Henke: Positive Displacement Meters, *Control Eng.*, p. 56, May, 1955.
15. T. Isobe and H. Hattori: A New Flowmeter for Pulsating Gas Flow, *ISA J.*, p. 38, December, 1959.
16. A. K. Oppenheim and E. G. Chilton: Pulsating-flow Measurement—A Literature Survey, *ASME Trans.*, p. 231, February, 1955.
17. J. R. Musham and B. G. Lewis: Direct Reading Flow Rate Meter for Low Flow Rates, *Control Eng.*, p. 115, December, 1961.
18. L. Gess: Common Troubles with Head Flowmeters, *ISA J.*, p. 58, February, 1958.
19. R. Shapcott: How to Select Flowmeters, *ISA J.*, p. 272, July, 1957.
20. H. E. Wingo: Thermistors Measure Low Liquid Velocities, *Control Eng.*, p. 131, October, 1959.
21. W. D. Hamilton: Flow Elements from Tubing Elbows, *ISA J.*, p. 61, July, 1963.
22. E. G. Keshock: Comparison of Absolute- and Reference-system Methods of Measuring Containment Vessel Leakage Rates, *NASA, Tech. Note* D-1588, 1964.

23. Ball Bearing Used in Design of Rugged Flowmeter, *NASA, Brief* 64-10170, 1964.
24. Meter Accurately Measures Flow of Low-conductivity Fluids, *NASA, Brief* 63-10280, 1963.
25. E. L. Upp: Flowmeters for High-pressure Gas, *Instr. Control Systems*, p. 151, March, 1965.
26. Turbine Flow Sensors, *Instr. Control Systems*, p. 123, March, 1965.
27. A. Haalman: Pulsation Errors in Turbine Flowmeters, *Control Eng.*, p. 89, May, 1965.
28. R. Siev: Mass Flow Measurement, *Instr. Control Systems*, p. 966, June, 1960.
29. Heat Transfer Flowmeter Has No Pressure Drop, *Space/Aeron.*, p. 259, January, 1964.
30. Mass Flow by Temperature Measurement, *Instr. Control Systems*, p. 95, March, 1964.
31. C. M. Holsell: Mass Flowmeters, *ISA J.*, p. 49, June, 1960.
32. L. N. Mortenson: Mass Flowmeter Calibration, *Instr. Control Systems*, p. 133, March, 1964.
33. J. Haffner et al.: Novel Mass Flowmeter, *Control Eng.*, p. 69, October, 1962.
34. G. T. Gebhardt: What's Available for Measuring Mass Flow, *Control Eng.*, p. 90, February, 1957.
35. G. F. Battista: The Use of Momentum Effects in Liquid Flow Measurement, U.S. Naval Air Turbine Test Station, Trenton, N.J., *NATTS-ATL-TN*-26, 1963.
36. G. Bloom: Low Flow Mass Flow Meter, *Instr. Control Systems*, p. 117, March, 1965.
37. E. C. Evans and G. W. Ray: Gas Mass Flow Rate Measurement to 0.1%, *Instr. Control Systems*, p. 105, March, 1965.
38. J. W. Freshour: Mass Flow Measurement of Cryogens, *Instr. Control Systems*, p. 97, March, 1965.
39. J. C. Pemberton: Flow Measurement in Rotating Machinery, *Instr. Control Systems*, p. 105, March, 1964.
40. T. J. Larson and L. D. Webb: Calibrations and Comparisons of Pressure-type Airspeed-Altitude Systems of the X-15 Airplane from Subsonic to High Supersonic Speeds, *NASA, Tech. Note* D-1724, 1963.
41. W. Gracey et al.: Wind-tunnel Investigation of a Number of Total-pressure Tubes at High Angles of Attack, *NACA, Tech. Note* 2261, 1951.
42. F. J. Capone: Wind-tunnel Tests of Seven Static-pressure Probes at Transonic Speeds, *NASA, Tech. Note* D-947, 1961.
43. R. S. Ritchie: Several Methods for Aerodynamic Reduction of Static-pressure Sensing Errors for Aircraft at Subsonic, Near-sonic, and Low Supersonic Speeds, *NASA, Tech. Rept.* R-18, 1959.
44. A. O. Pearson and H. A. Brown: Calibration of a Combined Pitot-static Tube and Vane-type Flow Angularity Indicator at Transonic Speeds and at Large Angles of Attack or Yaw, *NACA, RM*-L52F24, 1952.
45. J. M. Savino and A. J. Hilovsky: On the Use of Single Total- and Static-pressure Probes to Measure the Average Mass Velocity in Thin Rectangular Channels, *NASA, Tech. Note* D-2212, 1964.
46. W. Gracey: Measurement of Static Pressure on Aircraft, *NASA, Rept.* 1364, 1958.
47. W. H. Reed: Dynamic Response of Rising and Falling Balloon Wind Sensors with Application to Estimates of Wind Loads on Launch Vehicles, *NASA, Tech. Note* D-1821, 1963.
48. W. F. Van Tassell and C. E. Covert: Relaxation Effects on the Interpretation of Impact-probe Measurements, *J. Aerospace Sci.*, p. 147, February, 1960.

49. F. S. Sherman: New Experiments on Impact-pressure Interpretation in Supersonic and Subsonic Rarefied Air Streams, *NACA, Tech. Note* 2995, 1953.
50. New Anemometer Has Fast Response, Measures Dynamic Pressure Directly, *NASA, Brief* 63-10530, 1963.
51. L. V. Baldwin and V. A. Sandborn: Hot-wire Calorimetry: Theory and Application to Ion Rocket Research, *NASA, Tech. Rept.* R-98, 1961.
52. H. P. Grant: Hot-Wire in Liquid Flow Measurement, *Flow Corp.*, Cambridge, Mass., *Bull.* 89.
53. L. Kovasznay: Calibration and Measurement in Turbulence Research by Hot-wire Method, *NACA, TM*-1130, 1947.
54. H. L. Dryden and A. M. Kuethe: The Measurement of Fluctuations of Air Speed by the Hot-wire Anemometer, *NACA, Rept.* 320, 1929.
55. W. G. Spangenberg: Heat-loss Characteristics of Hot-wire Anemometers at Various Densities in Transonic and Supersonic Flow, *NACA, Tech. Note* 3381, 1955.
56. L. Kovasznay: Development of Turbulence-measuring Equipment, *NACA, Rept.* 1209, 1954.
57. V. A. Sandborn and J. C. Laurence: Heat Loss from Yawed Hot Wires at Subsonic Mach Numbers, *NACA, Tech. Note* 3563, 1955.
58. C. E. Shepard: A Self-excited, Alternating Current, Constant-temperature Hot-wire Anemometer, *NACA, Tech. Note* 3406, 1955.
59. W. G. Rose: Some Corrections to the Linearized Response of a Constant-temperature Hot-wire Anemometer Operated in a Low-speed Flow, *ASME Paper* 62-WA-11, 1962.
60. G. P. Katys: "Continuous Measurement of Unsteady Flow," The Macmillan Company, New York, 1964.

8
Temperature and heat-flux measurement

8.1

Standards and Calibration The International Measuring System sets up independent standards for only four fundamental quantities: length, time, mass, and temperature. Standards for all other quantities are basically derived from these. We have previously discussed the standards of length, time, and mass; let us now consider the temperature standard.[1] It should first be noted that temperature is fundamentally

[1] A. G. McNish, Fundamentals of Measurement, *Electro-Technol.* (*New York*), p. 114, May, 1963.

different in nature from length, time, and mass in that it is an intensive quantity whereas the others are extensive. That is, if two bodies of like length are "combined" the total length is twice the original; the same is true for two time intervals or two masses. However, the combination of two bodies of the same temperature results in exactly the same temperature. Thus the idea of a standard unit of mass, length, or time that can be indefinitely divided or multiplied to generate any arbitrary magnitude of these quantities cannot be carried over to the concept of temperature. Also, even though statistical mechanics relates temperature to the mean kinetic energies of molecules, these kinetic energies (which are dependent only on mass, length, and time standards for their description) are not at present measurable. Thus an *independent* temperature standard is required.

The fundamental meaning of temperature, just as for all basic concepts of physics, is not easily given. For most purposes the zeroth law of thermodynamics gives a useful concept. This is that, for two bodies to be said to have the same temperature, they must be in thermal equilibrium; that is, when thermal communication is possible between them, no change in the thermodynamic coordinates of either occurs. The zeroth law says that two bodies each in thermal equilibrium with a third body are in thermal equilibrium with each other. Then, by definition, the bodies are all at the same temperature. Thus if one can set up a reproducible means of establishing a range of temperatures, unknown temperatures of other bodies may be compared with the standard by subjecting any type of "thermometer" successively to the standard and the unknown temperatures and allowing equilibrium to occur in each case. That is, the thermometer is calibrated against the standard and thereafter may be used to read unknown temperatures.

In choosing the means of defining the standard temperature scale, one could conceivably employ any of the many physical properties of materials that vary reproducibly with temperature. For instance, the length of a metal rod varies with temperature. To define a temperature scale numerically one must choose a reference temperature and also state a rule for defining the difference between the reference and other temperatures. (Mass, length, or time measurements do *not* require universal agreement on a reference point at which each quantity is assumed to have a particular numerical value. Every centimeter, for example, in a meter is the same as every other centimeter.)

Suppose we take a copper rod 1 m long, place it in an ice-water bath which we have taken as our reference temperature source, and measure its length. Let us choose to call the ice-bath temperature 0°. We are now free to define any rule we wish to fix the numerical value to be assigned to all lower and higher temperatures. Suppose we decide that each additional 0.01 mm of expansion will correspond to $+1.0°$ on our temperature scale and each 0.01 mm of contraction to $-1.0°$. If the expansion phenomenon were reproducible, such a temperature scale would, in principle, be perfectly acceptable as long as everyone adhered to it. Would it be correct to say that each degree of temperature on this scale was "equal" to every other degree? That depends on what one means by "equal." If "equal" means that each degree causes the same amount of expansion of the copper rod, then all degrees are equal. If, instead, one considers the expansion of, say, iron rods, then equal amounts of expansion would not, in general, be caused by a 1° (copper scale) change from -6 to $-5°$ as by a 1° change from 100 to 101°. Or, consider conduction heat transfer in, for example, silver. If a temperature difference from 100 to 200° causes a given heat-transfer rate, will the same rate be caused by a temperature difference from -50 to $+150$? The answer is, in general, no.

The point of the above discussion is that, while our arbitrarily defined temperature scale is, in principle, as good as any other such scale based on some material property, its graduations have no particular significance with regard to physical laws *other* than the one used in the definition. One measures temperature for some reason, such as computing thermal expansion, heat-transfer rate, electrical conductivity, gas pressure, etc. The forms of the equations used to make such calculations depend on the nature of the standard used to define temperature. A temperature scale that gives a simple form to thermal-expansion equations may give complex forms to all other physical relations involving temperature. Since this difficulty is common to *all* standards based on the properties of a particular substance, a way of defining a temperature scale independent of *any* substance is desirable.

The thermodynamic temperature scale[1] proposed by Lord Kelvin in 1848 provides the theoretical base for a temperature scale independent of any material property and is based on the Carnot cycle. Here a perfectly reversible heat engine transfers heat from a reservoir of infinite capacity at temperature T_2 to another such reservoir at T_1. If the heat taken from reservoir 2 is Q_2 and that supplied to reservoir 1 is Q_1, for a Carnot cycle, $Q_2/Q_1 = T_2/T_1$; this may be taken as a *definition* of temperature ratio. If, also, a number is selected to describe the temperature of a chosen fixed point, the temperature scale is completely defined. At present, the fixed point is taken as the triple point (the state at which solid, liquid, and vapor phases are in equilibrium) of water because this is the most reproducible state known. The number assigned to this point is

[1] F. W. Sears, "Thermodynamics, Kinetic Theory and Statistical Mechanics," p. 116, Addison-Wesley Publishing Company, Inc., Reading, Mass., 1950.

273.16°K (°K = degrees Kelvin) since this makes the temperature interval from the ice point (273.15°K) to the steam point equal to 100K°. This would thus coincide with the previously established centigrade (now called Celsius) scale as a matter of convenience.[1]

While the Kelvin absolute thermodynamic scale is ideal in the sense that it is independent of any material properties, it is not physically realizable since it depends on an ideal Carnot cycle. Fortunately it can be shown[2] that a temperature scale defined by a constant-volume or constant-pressure gas thermometer using an ideal gas is *identical* with the thermodynamic scale. A constant-volume gas thermometer keeps a fixed mass of gas at constant volume and measures the pressure changes caused by temperature changes. The perfect-gas law then gives the fact that temperature ratios are identical to pressure ratios. The constant-pressure thermometer keeps mass and pressure constant and measures volume changes caused by temperature changes. Again the perfect-gas law says that temperature ratios are identical to volume ratios. These ratios are identical with those of the thermodynamic scale; thus if the same fixed point (the triple point of water) is selected for the reference point, the two scales are numerically identical. However, there is now the problem that the ideal gas is a mathematical model, not a real substance, and therefore the gas thermometers described above cannot actually be built and operated.

To obtain a physically realizable temperature scale, *real* gases must be used in the gas thermometers; the readings must be corrected, as well as possible, for deviation from ideal-gas behavior; and then the resulting values accepted as a definition of the temperature scale. The corrections for non-ideal-gas behavior are obtained for a constant-volume gas thermometer as follows: The thermometer is filled with a certain mass of gas and mercury is added until the desired volume is achieved (see Fig. 8.1). Suppose that this is done with the system at the ice-point temperature. The gas pressure is measured; let us call it p_{i1}. The system is then raised to the steam-point temperature, causing volume expansion. By adding more mercury, however, the volume can be returned to the original value. The pressure will now be higher; we shall call it p_{s1}. For an ideal gas, the ratio of the steam-point and ice-point temperatures would also be given by the pressure ratio p_{s1}/p_{i1}. If one repeats this experiment but uses a different mass of gas, thus giving different ice-point and steam-point pressures, p_{i2} and p_{s2}, he finds that $p_{s1}/p_{i1} \neq p_{s2}/p_{i2}$. This is a manifestation of the nonideal behavior of the gas; an ideal gas would have $p_{s1}/p_{i1} = p_{s2}/p_{i2}$.

[1] *Ibid.*, p. 8.
[2] *Ibid.*, p. 116.

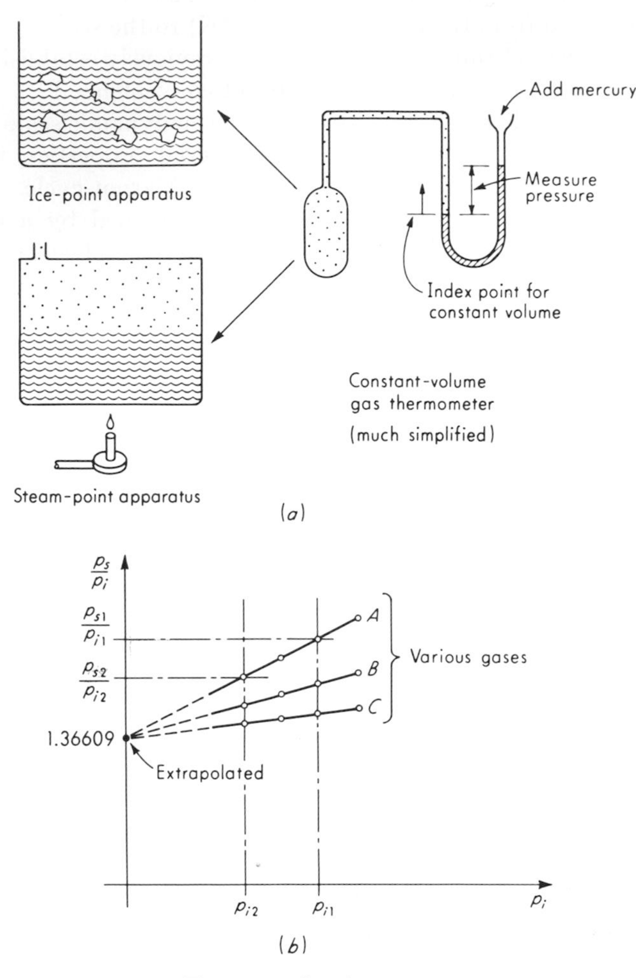

Fig. 8.1. *Gas-thermometer temperature scale.*

Real gases approach ideal-gas behavior if their pressure is reduced toward zero; thus one repeats the above experiment with successively smaller masses of gas, generating the curve A of Fig. 8.1b. Since one cannot use zero mass of gas, the zero-pressure point on this curve must be obtained by extrapolation. This zero-pressure point is taken as the true value of the pressure ratio corresponding to the steam-point/ice-point temperature ratio. If this experiment is repeated with *different* gases (B, C in Fig. 8.1b), all the curves intersect at the same point, showing that the procedure is independent of the type of gas used. Actual results give the numerical value $p_s/p_i = 1.36609 \pm 0.00004$. If one takes $T_s/T_i = p_s/p_i$, choice of a numerical value for any chosen reference point (such as calling $T_i = 273.15°K$) completely fixes the entire temperature scale. Such a scale, unfortunately, is not practical for day-to-day temperature measurements since the procedures involved are extremely tedious and time-consuming. Also, gas thermometers actually have a lower precision (repeatability) than some other temperature-measuring devices, such as resistance thermometers. This situation led to the acceptance in 1927 of the International Practical Temperature Scale (IPTS) which, with some minor revisions, is the temperature standard today.

The International Practical Temperature Scale is set up to conform as closely as practical with the thermodynamic scale. At the triple point of water the two scales are in exact agreement, by definition. Five other primary fixed points are used. They are the boiling points of liquid oxygen (−182.970°C), water (100°C), and sulfur (444.600°C) and the melting points of silver (960.8°C) and gold (1063.0°C), all at standard atmospheric pressure. The thermodynamic temperatures of these states were all determined as accurately as possible by gas thermometry. Interpolation between these fixed points is accomplished by use of a platinum resistance thermometer for the range −182.970 to 630.5°C and by a platinum/platinum–10 percent rhodium thermocouple for 630.5 to 1063.0°C. (The melting point of antimony, 630.5°C, is a secondary fixed point.) The equation[1] used to calculate Celsius temperature from the measured resistance of a platinum thermometer is, for the range −182.970° to 0°C,

$$t = \frac{R_t - R_0}{R_{100} - R_0} 100 + \delta \left(\frac{t}{100} - 1 \right) \frac{t}{100} + \beta \left(\frac{t}{100} - 1 \right) \left(\frac{t}{100} \right)^3 \tag{8.1}$$

where $t \triangleq$ Celsius (centigrade) temperature
$R_t \triangleq$ resistance at t
$R_0 \triangleq$ resistance at ice point
$R_{100} \triangleq$ resistance at steam point

For the range 0 to 630.5°C, the equation is

$$t = \frac{R_t - R_0}{R_{100} - R_0} 100 + \delta \left(\frac{t}{100} - 1 \right) \frac{t}{100} \tag{8.2}$$

The constant δ is a characteristic of a particular thermometer and is found by solving for δ in Eq. (8.2) when t is the sulfur point, 444.600°C.

[1] R. P. Benedict, Temperature and Its Measurement, *Electro-Technol. (New York)*, p. 71, July, 1963.

The constant β in Eq. (8.1) is similarly obtained when t in Eq. (8.1) is the oxygen point, $-182.970°C$. From 630.5 to 1063.0°C the thermocouple interpolation formula is

$$E = a + bt + ct^2 \qquad (8.3)$$

where E is the net emf of the thermocouple with one junction at the ice point and the other at t. The constants a, b, and c are found by solving the set of three simultaneous equations formed when Eq. (8.3) is applied at the antimony, silver, and gold points. When these constants have been found for the particular thermocouple, temperature t can be calculated from measured voltage E by Eq. (8.3).

For temperatures below the oxygen point the International Practical Temperature Scale is not defined. However, the National Bureau of Standards has standards extending down to about 2°K, although they are not all internationally accepted. They include the acoustical interferometer, the helium-vapor-pressure thermometer, and the platinum resistance thermometer.

Above the gold point the International Practical Temperature Scale *is* defined and uses a narrow-band radiation pyrometer ("optical" pyrometer) and the Planck equation to establish temperatures. The formula used is

$$\frac{J_t}{J_{Au}} = \frac{e^{1.438/\lambda(t_{Au}+273.15)} - 1}{e^{1.438/\lambda(t+273.15)} - 1} \qquad (8.4)$$

where $t_{Au} \triangleq$ gold-point temperature, 1063.0°C
$\lambda \triangleq$ effective wavelength of pyrometer

The quantity J_t/J_{Au} is measurable with the pyrometer and is the ratio of the spectral radiance of a blackbody at temperature t to one at temperature t_{Au}. Since λ can be determined for a given pyrometer, Eq. (8.4) allows calculation of t when J_t/J_{Au} has been measured. In principle, this method can be applied to arbitrarily high temperatures but in practice few reliable results above 4000°C are known.

The highest meaningful temperatures, existing in the interior of stars and for short times in atomic explosions, are inferred from kinetic theory to be in the range 10^7 to 10^9°K. Definition of temperature, much less measurement, is difficult at these extremes, although spectroscopic methods have given useful results. At the other extreme, temperatures of 10^{-6}°K have been produced by using the concept of nuclear cooling. Magnetic susceptibility of certain materials has been used to measure temperatures in the extremely low ranges.

The question of the accuracy of temperature standards may be considered from two viewpoints. First, how closely can the International

Practical Temperature Scale be reproduced and second, how closely does it agree with the thermodynamic absolute scale? The highest reproducibility of the International Practical Temperature Scale occurs at the triple point of water, which can be realized with a precision of a few ten-thousandths of a degree, giving an accuracy of about 1 ppm. For either lower or higher temperatures the accuracy falls off. At 100°K it is about 1 part in 2×10^4, at 10°K about 1 part in 10^3, and at 1°K about 1 part in 300. Above the triple point, accuracy is about 1 part in 10^5 at 800°K, drops sharply to about 1 part in 5,000 at 1500°K, and then gradually decreases to 1 part in 500 at 4000°K and 1 part in 15 at 10,000°K. The above statements refer to the International Practical Temperature Scale within its defined range and to the best available standards outside this range.[1,2] The question of agreement between the various empirical scales (such as the International Practical Temperature Scale) and the absolute thermodynamic scale involves the fact that, in general, the thermodynamic scale is considerably less reproducible than the empirical scales. For example, the steam-point temperature is reproducible to 0.0005° with a platinum resistance thermometer but to only 0.02° with a gas thermometer. The disagreement between the International Practical Temperature Scales and the absolute thermodynamic scale has been estimated in centigrade degrees as[3]

$$\frac{t}{100}\left(\frac{t}{100} - 1\right)[0.04106 - 7.363(10^{-5})t] \qquad 0° < t < 444.6°\text{C} \qquad (8.5)$$

where t is the Celsius temperature. The error is seen to be zero at $t = 0$ and 100°C and has a maximum value of about 0.14C° near $t = 400$°C.

Calibration of a given temperature-measuring device is generally accomplished by subjecting it to some established fixed-point environment, such as the melting and boiling points of standard substances, or by comparing its readings with those of some more accurate (secondary standard) temperature sensor which has itself been calibrated. The latter is generally accomplished by placing the two devices in intimate thermal contact in a constant-temperature-controlled bath. By varying the temperature of the bath over the desired range (allowing equilibrium at each point), the necessary corrections are determined. Accurate resistance thermometers, thermocouples, or mercury-in-glass expansion thermometers are generally useful as secondary standards. Fixed-point standards using the melting points of various metals and the triple point of water are commercially available.

[1] McNish, *op. cit.*, p. 113.

[2] "High Temperature Technology," p. 34, McGraw-Hill Book Company, 1960.

[3] "Temperature, Its Measurement and Control in Science and Industry," vol. 2, p. 93, Reinhold Publishing Corporation, New York, 1955.

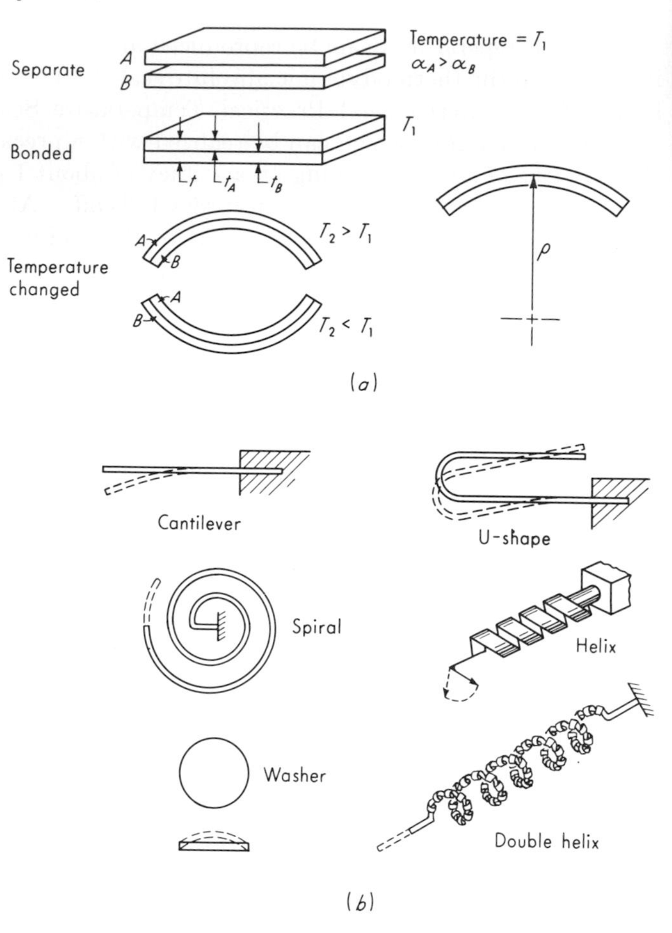

Fig. 8.2. Bimetallic sensors.

8.2

Thermal-expansion Methods A number of practically important temperature-sensing devices utilize the phenomenon of thermal expansion in one way or another. The expansion of solids is employed mainly in bimetallic elements by utilizing the differential expansion of bonded strips of two metals. Liquid expansion at essentially constant pressure is used in the common liquid-in-glass thermometers. Restrained expansion of liquids, gases, or vapors results in a pressure rise which is the basis of pressure thermometers.

Bimetallic thermometers. If two strips of metals A and B with different thermal-expansion coefficients α_A and α_B but at the same temperature (Fig. 8.2a) are firmly bonded together, a temperature change causes a differential expansion and the strip, if unrestrained, will deflect into a uniform circular arc. Analysis[1] gives the relation

$$\rho = \frac{t[3(1 + m)^2 + (1 + mn)(m^2 + 1/mn)]}{6(\alpha_A - \alpha_B)(T_2 - T_1)(1 + m)^2} \tag{8.6}$$

where $\rho \triangleq$ radius of curvature
$t \triangleq$ total strip thickness, 0.0005 in. $< t <$ 0.125 in. in practice
$n \triangleq$ elastic modulus ratio, E_B/E_A
$m \triangleq$ thickness ratio, t_B/t_A
$T_2 - T_1 \triangleq$ temperature rise

In most practical cases, $t_B/t_A \approx 1$ and $n + 1/n \approx 2$, giving

$$\rho \approx \frac{2t}{3(\alpha_A - \alpha_B)(T_2 - T_1)} \tag{8.7}$$

Combination of this equation with appropriate strength-of-materials relations allows calculation of the deflections of various types of elements in practical use. The force developed by completely or partially restrained elements can also be calculated in this way. Accurate results require the use of experimentally determined factors[2] which are available from bimetal manufacturers.

Since there are no practically usable metals with negative thermal expansion, the B element is generally made of Invar, a nickel steel with a nearly zero $[1.7 \times 10^{-6}$ in./(in.-C°)] expansion coefficient. While brass was originally employed, a variety of alloys are now used for the high-expansion strip, depending on the mechanical and electrical characteristics required. Details of materials and bonding processes are in some cases considered trade secrets. A wide range of configurations has been developed to meet application requirements (Fig. 8.2*b*).

Bimetallic devices are used for temperature measurement and also very widely as combined sensing and control elements in temperature-control systems, mainly of the on-off type. They are also used as overload cutout switches in electrical apparatus by allowing the current to flow through the bimetal, heating and expanding it and causing a switch to open when excessive current flows. Further applications are found as temperature-compensating devices[3] for various instruments that have temperature as a modifying or interfering input. The mechanical motion proportional to temperature is used to generate an opposing

[1] S. G. Eskin and J. R. Fritze, Thermostatic Bimetals, *Trans. ASME*, p. 433, July, 1940.

[2] General Plate Division, Attleboro, Mass., *Bull.* PR750.

[3] R. Gitlin, How Temperature Effects Instrument Accuracy, *Control Eng.*, April, May, June, 1955.

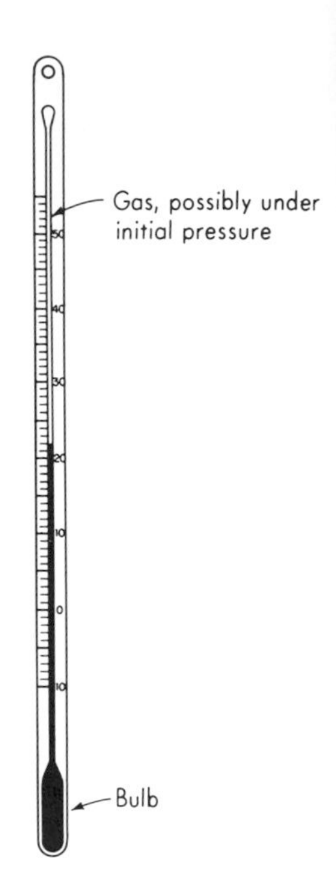

Fig. 8.3. *Liquid-in-glass thermometer.*

compensating effect. The accuracy of bimetallic elements varies greatly, depending on the requirements of the application. Since the majority of control applications are not extremely critical, requirements can be satisfied with a rather low-cost device. For more critical applications, performance can be much improved. The working temperature range is about from −100 to 1000°F. Inaccuracy of the order of 0.5 to 1 percent of scale range may be expected in bimetal thermometers of high quality.

Liquid-in-glass thermometers.[1,2] The well-known liquid-in-glass thermometer is adaptable to a wide range of applications by varying the materials of construction and/or configuration. Mercury is the most common liquid used at intermediate and high temperatures; its freezing point of −38°F limits its lower range. The upper limit is in the region of 1000°F and requires use of special glasses and an inert-gas fill in the capillary space above the mercury (see Fig. 8.3). Compression of the gas helps to prevent separation of the mercury thread and raises the liquid boiling point. For low temperatures, alcohol is usable to −80°F,

[1] J. F. Swindells, Calibration of Liquid-in-glass Thermometers, *Natl. Bur. Std. (U.S.), Circ.* 600, 1959.

[2] M. F. Behar, "Handbook of Measurement and Control," p. 25, The Instruments Publishing Co., Pittsburgh, Pa., 1954.

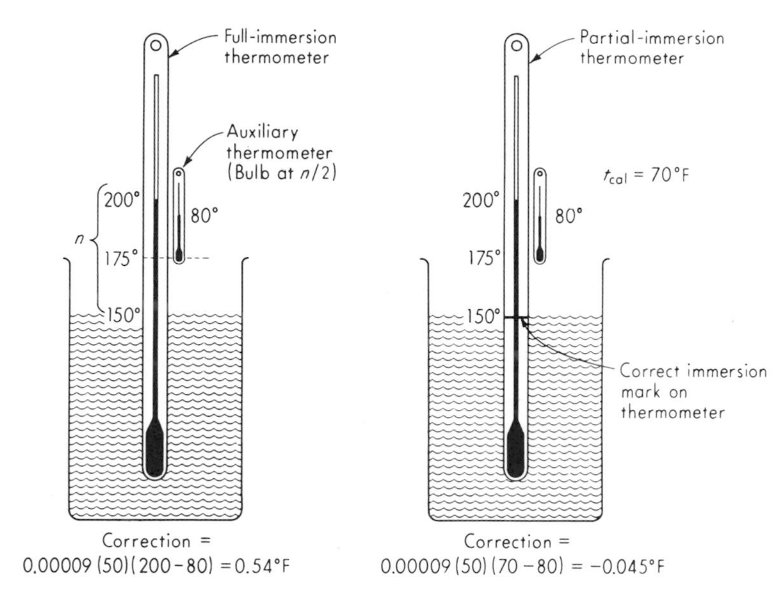

Fig. 8.4. *Full- and partial-immersion thermometers.*

toluol to −130°, pentane to −330°, and a mixture of propane and propylene giving the lower limit of −360°.

Thermometers are commonly made in two types: total immersion and partial immersion. Total-immersion thermometers are calibrated to read correctly when the liquid column is completely immersed in the measured fluid. Since this may obscure the reading, a small portion of the column may be allowed to protrude with little error. Partial-immersion thermometers are calibrated to read correctly when immersed a definite amount and with the exposed portion at a definite temperature. They are inherently less accurate than full-immersion types. If the exposed portion is at a temperature different from that at calibration a correction must be applied. Corrections for full- and partial-immersion thermometers used at conditions other than those intended are most accurately determined by the use of a special "faden" thermometer[1] designed to measure the average temperature of the emergent stem. If such a thermometer is not available, the correction may be estimated by suspending a small auxiliary thermometer close to the stem of the thermometer to be corrected, as in Fig. 8.4. This auxiliary thermometer estimates the mean temperature of the emergent stem. When a partial-immersion thermometer is used at correct immersion but with a surround-

[1] Swindells, *op. cit.*

ing air temperature different from its original calibration condition, the correction may be calculated from

$$\text{Correction} = 0.00009n(t_{cal} - t_{act}) \qquad \text{F}^\circ \qquad (8.8)$$

where $n \triangleq$ number of scale degrees equivalent to emergent stem length, F°

$t_{cal} \triangleq$ air temperature at calibration, °F

$t_{act} \triangleq$ actual air temperature at use, °F (from auxiliary thermometer)

When a total-immersion thermometer is used at partial immersion the same formula may be used except that $t_{cal} - t_{act}$ is replaced by (main-thermometer reading) − (auxiliary-thermometer reading). For Celsius thermometers the constant 0.00009 becomes 0.00016.

The accuracy obtainable with liquid-in-glass thermometers depends on instrument quality, temperature range, and type of immersion. For full-immersion thermometers the best instruments, when calibrated, are capable of errors as small as 0.4F° (range −328 to 32°F), 0.05F° (range −69 to 32°F), 0.04F° (range 32 to 212°F), 0.4F° (range 212 to 600°F), and 0.8F° (range 600 to 950°F). Errors in partial-immersion types may be several times larger even after corrections have been applied for air-temperature variations. All the above figures refer to the ultimate performance attainable with the best instruments and great care in application. Errors in routine day-to-day measurements may be much larger.

For measuring small temperature changes from some chosen value with high accuracy and sensitivity the Beckman thermometer (Fig. 8.5) is available. Its high sensitivity requires a very large bulb. An impractically long capillary would also be required in order to cover a sizable temperature range with a single instrument. This difficulty is overcome by providing means for transferring mercury from the main bulb to an auxiliary bulb, so that at any chosen temperature the mercury thread is within the range of the capillary scale. The usual range is only 5 or 6C°, and the scale may be graduated in 0.01C° intervals. Measurement of the difference between the original set temperature and the measured temperature may be made with an error as little as 0.002 to 0.005C° if sufficient care is taken.

Pressure thermometers.[1,2] Pressure thermometers consist of a sensitive bulb, an interconnecting capillary tube, and a pressure-measuring

[1] D. M. Considine (ed.), "Process Instruments and Controls Handbook," p. 68, McGraw-Hill Book Company, New York, 1957.

[2] Behar, *op. cit.*, p. 29.

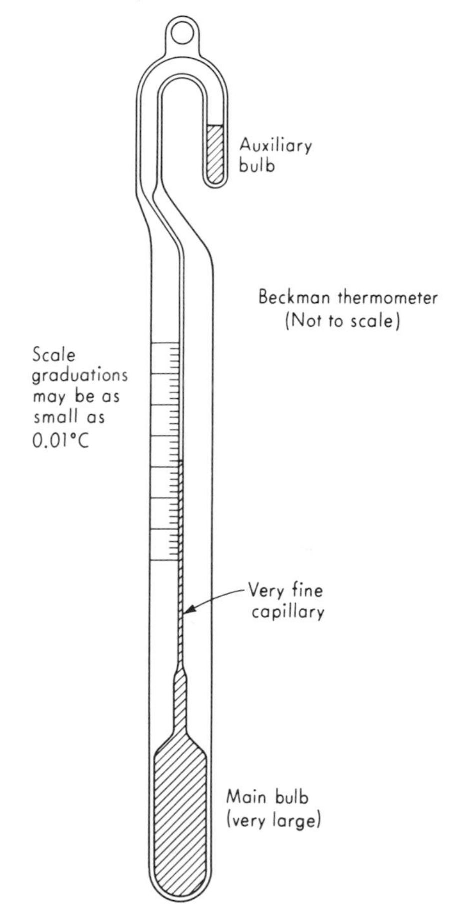

Fig. 8.5. *Beckman thermometer.*

device such as a Bourdon tube, bellows, or diaphragm (Fig. 8.6). When the system is completely filled with a liquid (mercury and xylene are common) under an initial pressure, the compressibility of the liquid is often small enough relative to the pressure gage $\Delta V/\Delta p$ that the measurement is essentially one of volume change. For gas or vapor systems the reverse is true, and the basic effect is one of pressure change at constant volume.

Capillary tubes as long as 200 ft may be used for remote measurement. Temperature variations along the capillary and at the pressure-sensing device generally require compensation, except in the vapor-pressure type, where pressure depends only on the temperature at the liquid's free surface, located at the bulb. A common compensation scheme using an auxiliary pressure sensor and capillary is shown in Fig. 8.7. The motion of the compensating system is due to the interfering effects only

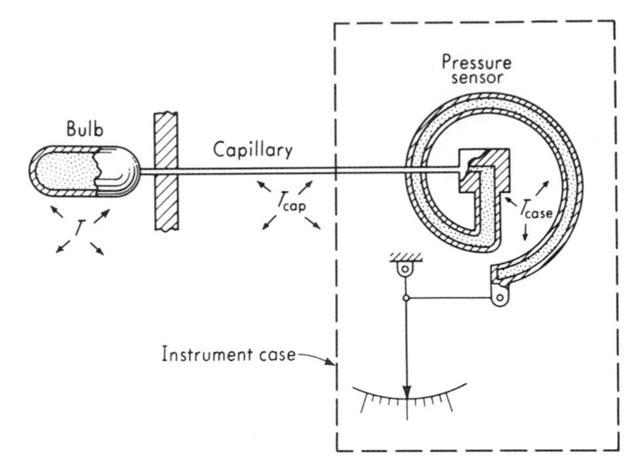

Fig. 8.6. *Pressure thermometer.*

and is subtracted from the total motion of the main system, resulting in an output dependent only on bulb temperature. The "trimming" capillary (which may be lengthened or shortened) allows the volume to be changed to attain accurate case compensation by experimental test. Bimetal elements are also used to obtain case and partial capillary compensation.

Liquid-filled systems cover the range −150 to 750°F with xylene or a similar liquid and −38 to 1100°F with mercury. Response is essentially linear over ranges up to about 300°F with xylene and 1000°F with mercury. Elevation differences between the bulb and pressure sensor different from those at calibration may cause slight errors. Gas-filled

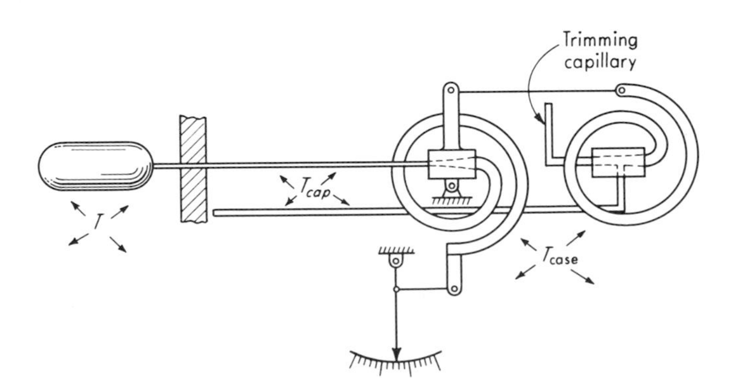

Fig. 8.7. *Compensation methods.*

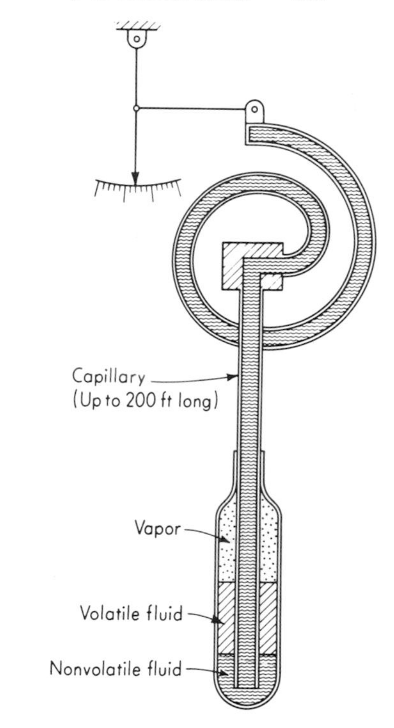

Fig. 8.8. *Vapor-pressure thermometer.*

systems operate over the range −400 to 1200°F with linear ranges as great as 1000°F; errors due to capillary temperature variations are usually small enough not to justify compensation. Case compensation is accomplished with bimetal elements. Vapor-pressure systems are usable in the range −40 to 600°F. The calibration is strongly nonlinear; special linearizing linkages are needed if linear output is required. Characteristics of the system vary, depending on whether the bulb is hotter than, colder than, or equal in temperature to the rest of the system, since this determines where liquid and vapor will exist. The most versatile arrangement is shown in Fig. 8.8, where the volatile-liquid surface is *always* in the bulb. Capillary and case corrections are not needed in such a device since the vapor pressure of a liquid depends only on the temperature of its free surface. Commonly used volatile liquids include ethane (vapor pressure changes from 20 to 600 psig for a temperature change from −100 to +80°F), ethyl chloride (0 to 600 psig for 40 to 350°F), and chlorobenzene (0 to 60 psig for 275 to 400°F). The accuracy of pressure thermometers under the best conditions is of the order ±0.5 percent of the scale range. Adverse environmental conditions may increase this error considerably.

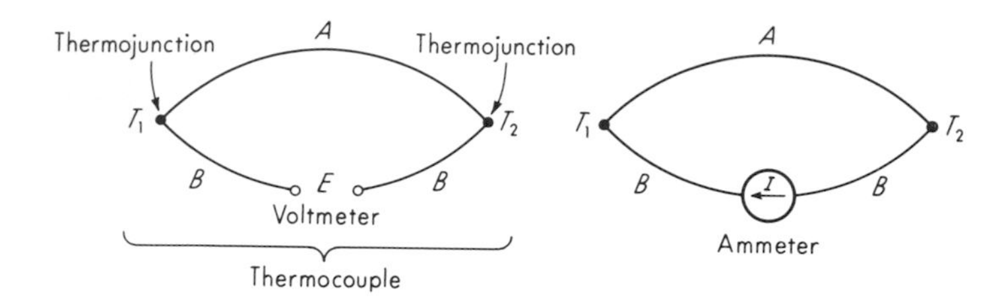

Fig. 8.9. Basic thermocouple.

8.3

Thermoelectric Sensors (Thermocouples) If two wires of different materials A and B are connected in a circuit as shown in Fig. 8.9, with one junction at temperature T_1 and the other at T_2, an infinite-resistance voltmeter detects an electromotive force E, or if an ammeter is connected a current I is measured. The magnitude of the voltage E depends on the materials and the temperatures T_1 and T_2. The current I is simply E divided by the total resistance of the circuit, including the ammeter resistance. If current is allowed to flow, electrical power is developed; this comes from a heat flow from the surroundings to the wires. A direct conversion of heat energy to electrical energy is thus obtained. The effect is reversible, so that forcing a current from an external source through a thermoelectric circuit will cause heat flows to and from the circuit. While here we are concerned with this thermoelectric effect only as a means of sensing temperatures, modern developments in materials have made the principle of practical application in electric power generation, heating, and cooling, though at present only on a small scale.

The overall relation between voltage E and temperatures T_1 and T_2, which is the basis of thermoelectric temperature measurement, is called the Seebeck effect. The temperatures T_1 and T_2 refer to the junctions themselves, whereas when using a thermocouple one is trying to measure the temperature of some body in contact with the thermojunction. These two temperatures are not exactly the same if current is allowed to flow through the thermojunction, since then heat is generated or absorbed at the junction, which must thus be hotter or colder than the surrounding medium whose temperature is being measured. This heating and cooling are related to the Peltier effect.[1] If the thermocouple voltage is measured with a potentiometer, no current flows and Peltier heating and cooling are not present. When a millivoltmeter is used, current flows, and heat

[1] P. H. Dike, "Thermoelectric Thermometry," Leeds & Northrup Co., Philadelphia, Pa., 1954.

is absorbed at the hot junction (requiring it to become cooler than the surrounding medium) while heat is liberated at the cold junction, making it hotter than its surrounding medium. These heating and cooling effects are proportional to the current and fortunately are completely negligible[1] when the current is that produced by the thermocouple itself in a practical millivoltmeter circuit.

Another reversible heat-flow effect, the Thomson effect,[1] influences the temperature of the conductors between the junctions rather than the junctions themselves. When current flows through a conductor having a temperature gradient (and thus a heat flow) along its length, heat is liberated at any point where the current flow is in the same direction as the heat flow, while heat is absorbed at any point where these are opposite. Since this effect also depends on current flow, it is not present if a potentiometer is used. Even if a millivoltmeter is used, the effect of the heat flows on conductor temperature is completely negligible. Finally it should be noted that in any current-carrying conductor I^2R heat is generated, raising the circuit temperature above its local surroundings. Again, potentiometric voltage measurements are not susceptible to this error. Errors in millivoltmeter circuits are usually negligible also but can be estimated if heat-transfer conditions are known.

The above physical effects can be analyzed[2] on a macroscopic scale by classical thermodynamics, with fewer assumptions by irreversible thermodynamics, and qualitatively on a microscopic basis by solid-state physics. Thermodynamic approaches are based on the two experimentally observed reversible energy-conversion processes, the Peltier and Thomson effects, and neither require nor give any explanation of the basic atomic mechanisms. The total emf produced is made up of a part due to the Peltier effect, which is localized at each junction, and a (usually much smaller) part caused by the Thomson effect, which is distributed along each conductor between the junctions. The Peltier emf's are assumed proportional to the junction temperature while the Thomson emf's are proportional to the difference between the squares of the junction temperatures. For the total voltage, the equation takes the form

$$E = C_1(T_1 - T_2) + C_2(T_1^2 - T_2^2) \qquad (8.9)$$

Copper/constantan thermocouples, for example, give

$$E = 37.5(T_1 - T_2) - 0.045(T_1^2 - T_2^2) \qquad (8.10)$$

where $E \triangleq$ total voltage, μv

$T_1, T_2 \triangleq$ absolute junction temperatures, °K

[1] *Ibid.*

[2] R. R. Heikes and R. W. Ure, "Thermoelectricity," Interscience Publishers, Inc., New York, 1961.

Unfortunately the assumptions made in the analyses leading to Eq. (8.9) are not exactly satisfied in practice; thus equations such as (8.10) can *not* usually be used to predict accurately temperatures from measured voltages. Rather, a given thermocouple material must be calibrated over the complete range of temperatures in which it is to be used. In this calibration only the overall voltage is of interest, and the separate contributions of Peltier and Thomson effects are not determined. Temperature measurement by thermoelectric means is thus based entirely on empirical calibrations and the application of so-called thermoelectric "laws" which experience has shown to hold. These laws, quoted below, are adequate for analysis of most practical thermocouple circuits. In those cases where the circuit configuration does not lend itself to direct application of these laws, alternative approaches[1] are available.

The laws of thermocouple behavior may be stated as follows:

1. The thermal emf of a thermocouple with junctions at T_1 and T_2 is totally unaffected by temperature elsewhere in the circuit if the two metals used are each homogeneous (Fig. 8.10*a*).
2. If a third homogeneous metal C is inserted into either A or B (see Fig. 8.10*b*), as long as the two new thermojunctions are at like temperatures, the net emf of the circuit is unchanged irrespective of the temperature of C away from the junctions.
3. If metal C is inserted between A and B at one of the junctions, the temperature of C at any point away from the AC and AB junctions is immaterial. So long as the junctions AC and AB are both at the temperature T_1, the net emf is the same as if C were not there (Fig. 8.10*c*).
4. If the thermal emf of metals A and C is E_{AC} and that of metals B and C is E_{CB}, the thermal emf of metals A and B is $E_{AC} + E_{CB}$ (Fig. 8.10*d*).
5. If a thermocouple produces emf E_1 when its junctions are at T_1 and T_2, and E_2 when at T_2 and T_3, it will produce $E_1 + E_2$ when the junctions are at T_1 and T_3 (Fig. 8.10*e*).

These laws are of great importance in the practical application of thermocouples. The first one states that the lead wires connecting the two junctions may be safely exposed to an unknown and/or varying temperature environment without affecting the voltage produced. Laws 2 and 3 make it possible to insert a voltage-measuring device into the circuit actually to measure the emf rather than just talking about its

[1] P. Stein, "Measurement Engineering," vol. I, chap. 18, Stein Engineering Services, Inc., Phoenix, Ariz., 1964.

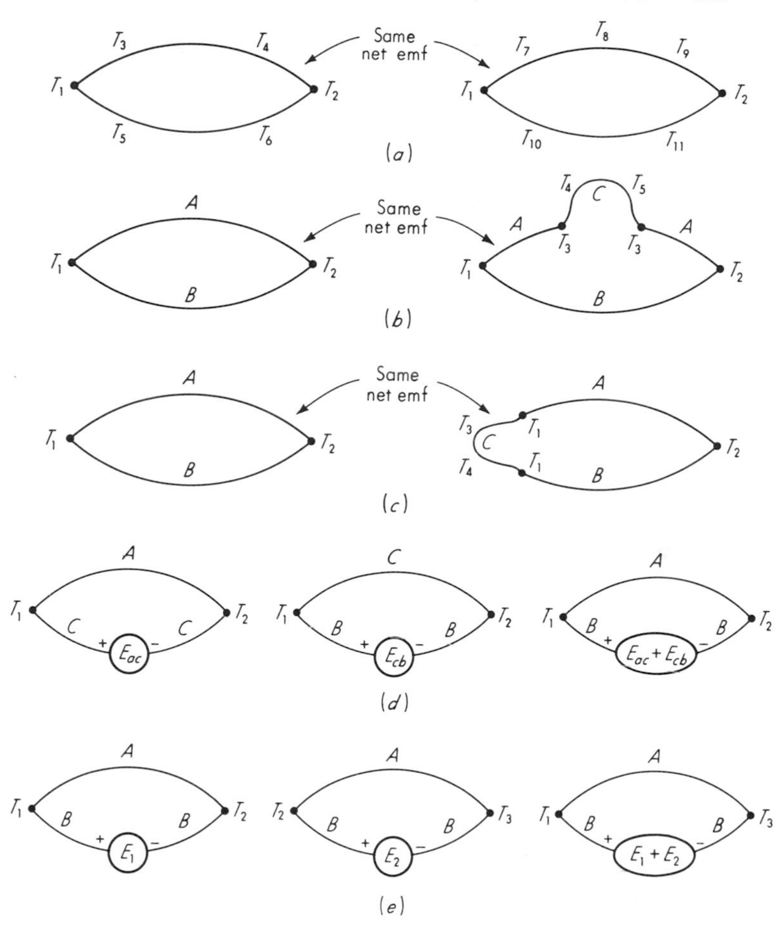

Fig. 8.10. *Thermocouple laws.*

existence. That is, the metal C represents the internal circuit (usually all copper in precise instruments) between the binding posts of a millivoltmeter or potentiometer. The instrument can be connected in two ways, as shown in Fig. 8.10*b* and *c*. Law 3 also shows that thermocouple junctions may be soldered or brazed (thereby introducing a third metal) without affecting the readings. The fourth law shows that all possible pairs of metals need not be calibrated since the individual metals can each be paired with *one* standard (platinum is used) and calibrated. Any other combinations can then be *calculated;* calibration is not necessary.

In considering the fifth law we should note that, in using a thermocouple to measure an unknown temperature, the temperature of one of the thermojunctions (called the reference junction) must be known by some independent means. A voltage measurement then allows one to get the temperature of the other (measuring) junction from calibration tables. These calibration tables were obtained by maintaining the reference junction at a fixed known value (usually 32°F, the ice point), varying the measured junction over the desired range of temperatures (known by some independent means), and recording the resulting voltages. Thus most calibration tables are based on the reference junction being at the ice point. When a thermocouple is used, the reference junction may or may not be at the ice point. If it is, the calibration table may be used directly to find the measuring-junction temperature. If it is not, the fifth law allows use of the standard table as follows: Suppose the reference junction is at 70°F and the voltage reading is 1.23 mv. In Fig. 8.10*e* we take $T_1 = 32°F$, $T_2 = 70°F$, and T_3 is unknown. We can look up E_1 directly in the table; suppose it is 0.71 mv. Now E_2 is the measured value 1.23 mv; thus $E_1 + E_3 = 1.94$ mv. The unknown temperature can be found by looking up the temperature value corresponding to 1.94 mv in the standard table; it is 100°F.

Common thermocouples. Thermojunctions formed by welding, soldering, or merely pressing the two materials together give identical voltages. If current is allowed to flow, the currents may be different since the contact resistance differs for the various joining methods. Welding (either gas or electric) is most widely used although both silver solder and soft solder (low temperatures only) are used in copper/constantan couples.

While many materials exhibit the thermoelectric effect to some degree, only a small number of pairs are in wide use. They are platinum/rhodium, Chromel/Alumel, copper/constantan, and iron/constantan. Each of these pairs exhibits a combination of properties that suit it to a particular class of applications. Since the thermoelectric effect is somewhat nonlinear, the sensitivity varies with temperature. The maximum sensitivity of any of the above pairs is about $60\mu V/C°$ for copper/constantan at 350°C. Platinum/platinum-rhodium is the least sensitive: about $6\mu V/C°$ between 0 and 100°C.

The accuracy of the common thermocouples may be stated in two different ways. If one uses standard thermocouple wire (which is *not* individually calibrated by the manufacturer) and makes up a thermocouple to be used without calibration, he is relying on the wire manufacturer's quality control to limit deviations from the published calibration tables. These tables give the *average* characteristics, not those of a particular batch of wire. Platinum/platinum-rhodium is the most accurate; error is of the order of ± 0.25 percent of reading. Copper/constantan gives ± 0.5 percent or ± 1.5F° (whichever is larger) between -75

and 200°F and ±0.75 percent between 200 and 700°F. Chromel/Alumel gives ±5F° (32 to 660°F) and ±0.75 percent (660 to 2300°F). Iron/constantan has $\pm 66 \mu V$ below 500°F and ±1.0 percent from 500 to 1500°F. If higher accuracies are needed, the individual thermocouple may be calibrated. An indication of the achievable accuracy is available from National Bureau of Standards listings[1] of the results the Bureau will guarantee. At the actual calibration points the error ranges from 0.05 to 0.5C°. Interpolated points are less accurate: 0.1 to 1.0C°, except platinum/platinum-rhodium at 1450°F, 2.0 to 3.0C°. The realization of this potential accuracy in applying such calibrated thermocouples to practical temperature measurement is, of course, dependent on the application conditions and is rarely possible.

Platinum/platinum-rhodium thermocouples are used mainly in the range 0 to 1500°C. The main features of this combination are its chemical inertness and stability at high temperatures in oxidizing atmospheres. Reducing atmospheres cause rapid deterioration at high temperatures as the thermocouple metals are contaminated by absorbing small quantities of other metals from nearby objects (such as protecting tubes). This difficulty, causing loss of calibration, is unfortunately common to most thermocouple materials above 1000°C.

Chromel ($Ni_{90}Cr_{10}$)/Alumel ($Ni_{94}Mn_3Al_2Si_1$) couples are useful over the range −200 to +1300°C. Their main application, however, is from about 700 to 1200°C in nonreducing atmospheres. The temperature/voltage characteristic is quite linear for this combination (see Fig. 8.11).

Copper/constantan ($Cu_{57}Ni_{43}$) is used at temperatures as low as −200°C; its upper limit is about 350°C because of the oxidation of copper above this range. Iron/constantan is the most widely used thermocouple for industrial applications and covers the range −150 to +1000°C. It is usable in oxidizing atmospheres to about 760°C and reducing atmospheres to 1000°C.

Thermocouple manufacturers have a wealth of experience concerning the application of thermocouples to diverse temperature-measuring problems and should be consulted if special types of problems are foreseen in a particular case.

Reference-junction considerations. For the most precise work, reference junctions should be kept in a triple-point-of-water apparatus[2] whose temperature is 0.01 ± 0.0005C°. Such accuracy is rarely needed, and an ice bath is much more commonly used. A carefully made ice bath is reproducible to about 0.001C° but a poorly made one may have an error of

[1] Thermocouple Calibration, *Instr. Control Systems*, p. 1663, September, 1961.
[2] Transonics Inc., Lexington, Mass.

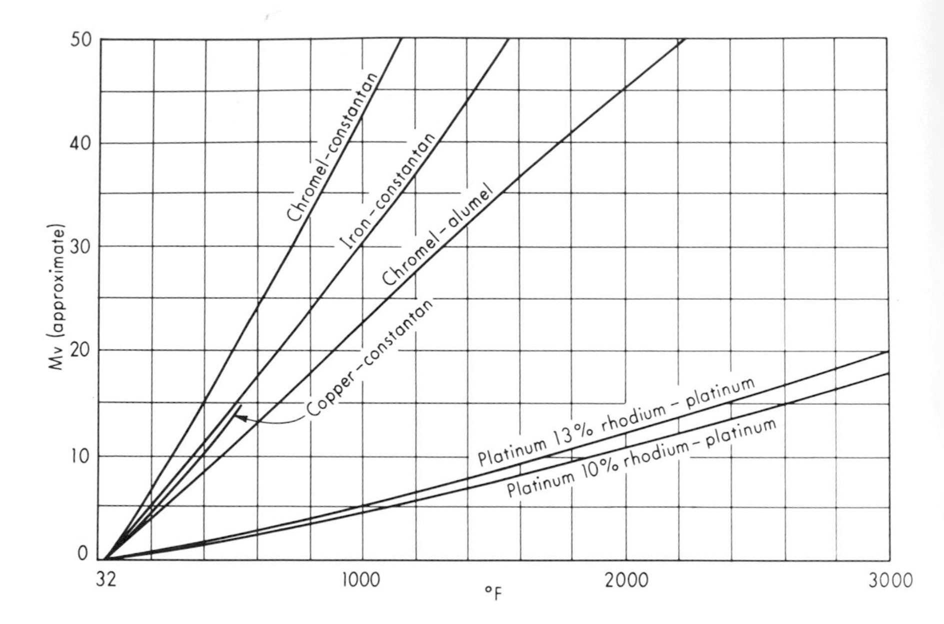

Fig. 8.11. *Thermocouple temperature/voltage curves.*

1C°.[1] Figure 8.12 shows one method of constructing an ice-bath reference junction. The main sources of error are insufficient immersion length and an excessive amount of water in the bottom of the flask. Automatic ice baths that use the Peltier cooling effect as the refrigerator, rather than relying on externally supplied ice (which must be continually replenished), are available with an accuracy of 0.05C°.[1] These systems use the expansion of freezing water in a sealed bellows as the temperature-sensing element that signals the Peltier refrigerator when to turn on or off by displacing a microswitch.

Since low-power heating is more easily obtained than low-power cooling, some reference junctions are designed to operate at a fixed temperature higher than any expected ambient. A feedback system operates an electrical heating element to maintain a constant temperature in an enclosure containing the reference junctions. Since the reference junction is not at 32°F the thermocouple-circuit net voltage must be corrected by adding the reference-junction voltage before the measuring-junction temperature can be found. This correction is, however, a constant.

In some situations the reference junction is allowed to assume

[1] C. L. Feldman, Automatic Ice-point Thermocouple Reference Junction, *Instr. Control Systems*, p. 101, January, 1965.

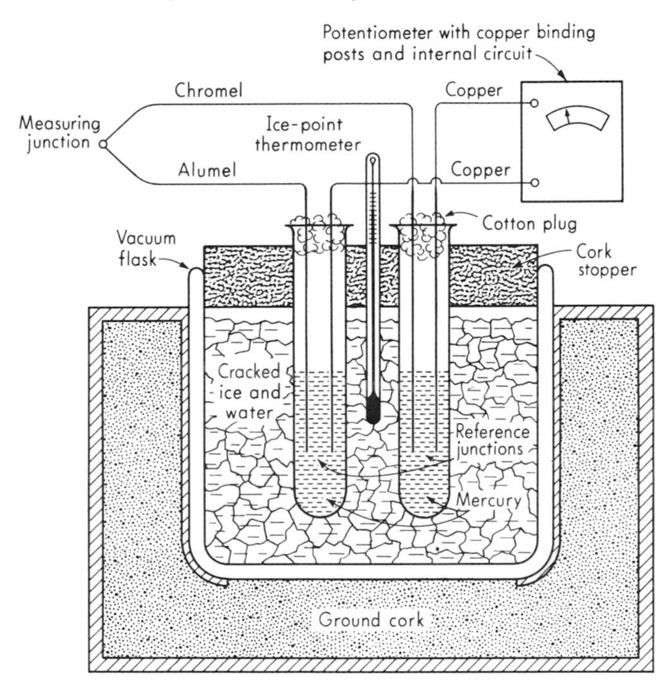

Fig. 8.12. *Ice-bath reference junction.*

ambient temperature. Knowledge of the ambient temperature then allows correction of the net measured voltage. This correction is made automatic in some potentiometers that are intended to measure voltages from a specific type of thermocouple. Ambient temperature is sensed by a bimetal element or a temperature-sensitive resistor. These adjust a compensating voltage in the circuit so that the voltage indicated by the instrument is the same as it would be if the reference junction were at 32°F, thus allowing a direct reading of the measuring-junction temperature.

Another approach uses two temperature-controlled ovens and an auxiliary thermocouple to provide an equivalent reference junction at 32°F.[1] Figure 8.13*b* shows the arrangement for an iron/constantan thermocouple. In Fig. 8.13*a* if the reference junctions are kept at 130°F, when the measuring junction is at 32°F the output voltage is 2.82 mv. To make the system act as if the reference junctions were at 32°F an opposing voltage of 2.82 mv must be put in series. This is done by connecting a Chromel/constantan thermocouple as shown in Fig. 8.13*b*. This produces an opposing voltage which will be 2.82 mv if the second oven is set at

[1] Pace Engineering Co., North Hollywood, Calif.

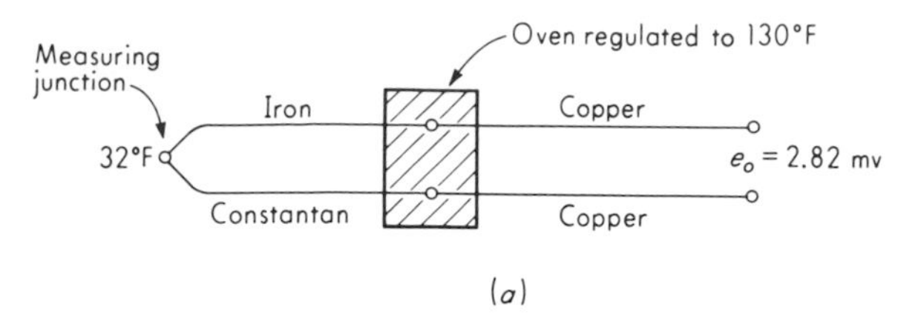

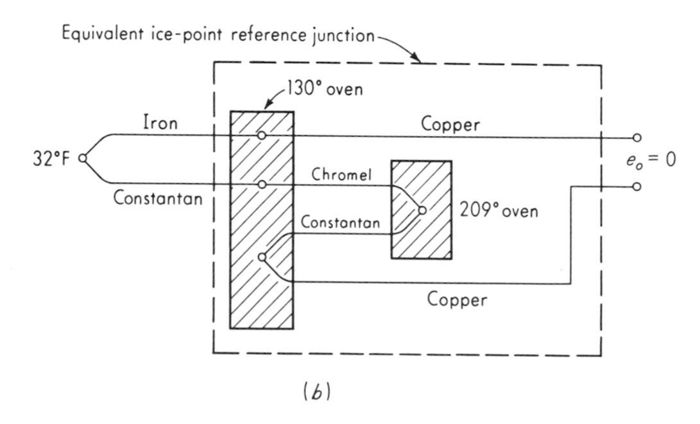

Fig. 8.13. *Oven reference junctions.*

209°F. Systems of this type maintain a reference temperature within ±0.2F° for long-term unattended operation.

Special materials, configurations, and techniques. Increasing interest in high-temperature processes in jet and rocket engines and nuclear reactors has led to requirements for reliable temperature sensors in the range 2000 to 4500°F. New thermocouples developed for these applications include rhodium-iridium/rhodium,[1] tungsten/rhenium,[1] and boron/graphite.[2] Rhodium-iridium is usable to about 4000°F under proper conditions and has a sensitivity of the order of $6\mu V/\mathrm{C}°$. Various alloys of tungsten and rhenium may be used up to 5000°F under favorable conditions and have about the same sensitivity at the highest temperatures as rhodium-iridium. Boron/graphite has a high sensitivity (about $40\mu V/\mathrm{C}°$) and is usable for short times up to 4500°F.

An alternative solution to high-temperature problems may be found in various cooling schemes. Two such[3] in actual use are shown in Fig. 8.14. In Fig. 8.14*a* the hot-gas flow whose temperature is to be measured

[1] P. D. Freeze, Review of Recent Developments of High-temperature Thermocouples, *ASME Paper* 63-WA-212, 1963.

[2] Astro Industries Inc., Santa Barbara, Calif., *Bull.* BGT-1, 1963.

[3] *NASA*, *SP*-5015, p. 128, 1964.

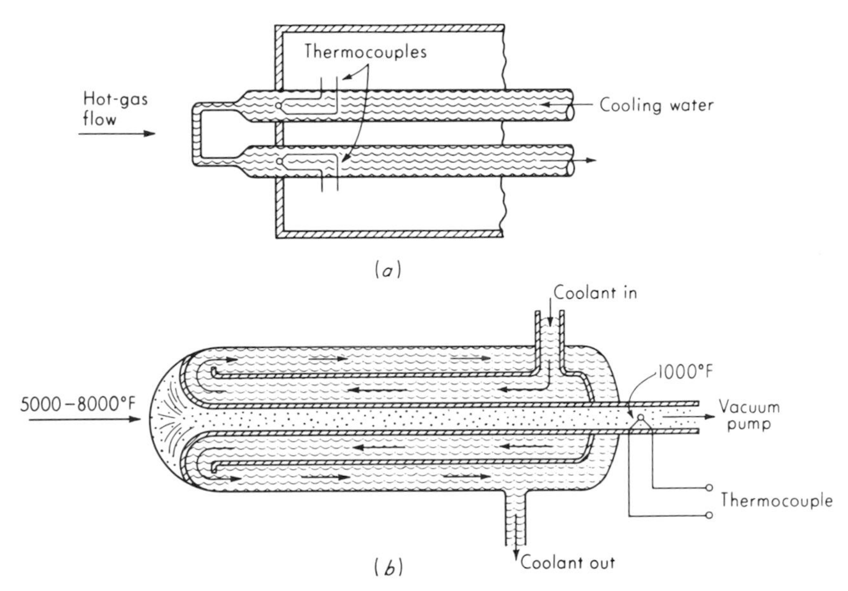

Fig. 8.14. *Cooled thermocouples.*

impinges on a small tube carrying cooling water, causing a temperature rise of about 100F°. If heat-transfer coefficients are known, measurements of water flow rate, temperature, and temperature rise allow calculation of the hot-gas temperature. Figure 8.14*b* shows another approach wherein the hot gas is aspirated through a heat exchanger, cooling it to about 1000°F. Knowledge of heat-transfer characteristics and flow rates again allows calculation of the hot-gas temperature. Such methods have been used in the range 5000 to 8000°F.

Figure 8.15 shows in simplified fashion the principle of a pulse-cooling technique[1] which allows use of Chromel/Alumel thermocouples (melting point 2550°F) to measure temperatures up to 7000°F. The measuring junction is kept at a low temperature by a cooling air flow. When this flow is shut off by a solenoid valve the thermocouple starts to heat up, following the first-order equation

$$\tau \frac{dT_{tc}}{dt} + T_{tc} = T_{gas} \tag{8.11}$$

where $\tau \triangleq$ thermocouple time constant (assumed known)
$T_{tc} \triangleq$ thermocouple temperature
$T_{gas} \triangleq$ hot-gas temperature

[1] A. F. Wormser and R. A. Pfuntner, Pulse Technique Extends Range of Chromel-Alumel to 7000°F, *Instr. Control Systems,* p. 101, May, 1964.

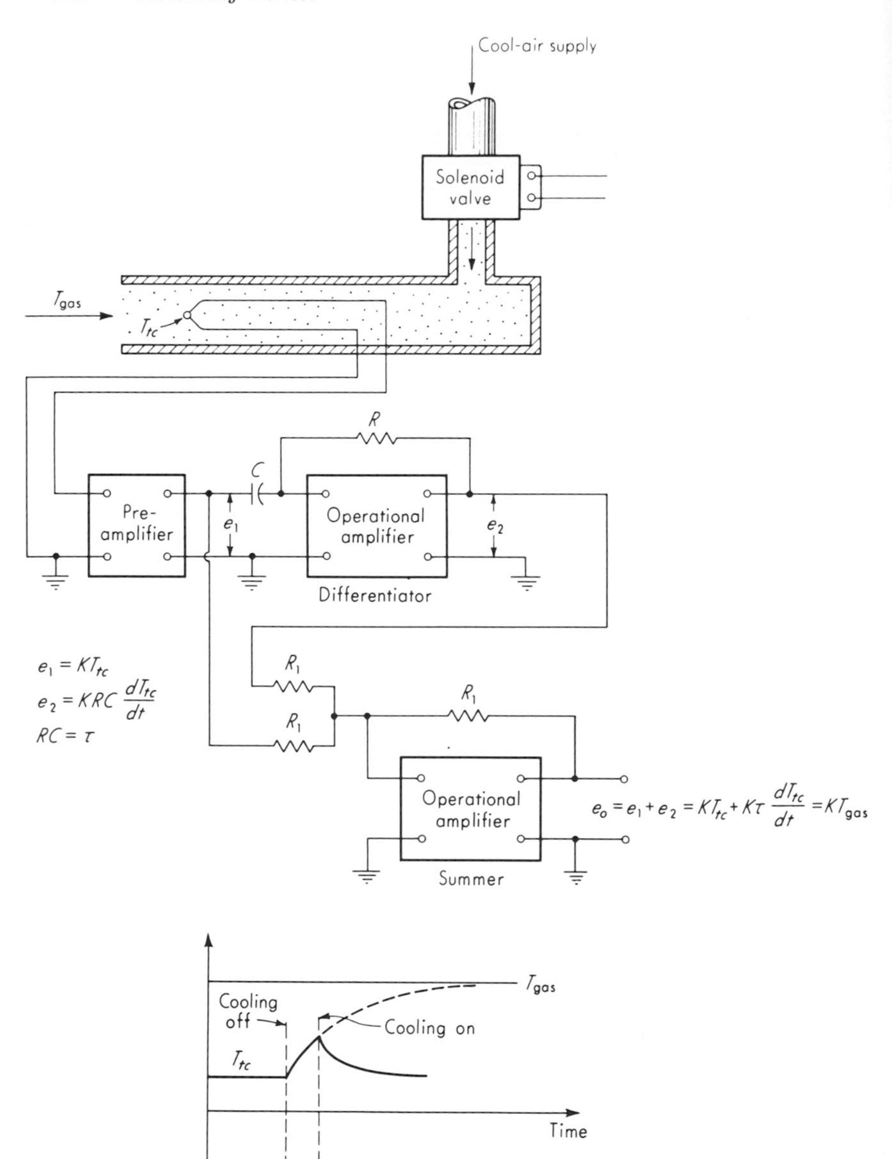

Fig. 8.15. *Pulsed-thermocouple technique.*

This equation shows that T_{gas} can be *computed* any time after the cooling is shut off if dT_{tc}/dt is known. A voltage proportional to dT_{tc}/dt can be obtained by use of the differentiating circuit shown and, when summed with a voltage proportional to T_{tc}, provides a signal proportional to T_{gas}. This signal is theoretically available immediately after the cooling is shut off; however, in practice, the cooling is left off a finite interval during which the value of T_{gas} is recorded. The cooling is turned on again before the thermocouple is overheated. In the actual system of the quoted reference, additional computing elements also compute the numerical value of τ; thus this need not be known beforehand.

The measurement of rapidly changing internal and surface temperatures of solid bodies may be accomplished with arrangements such as those in Fig. 8.16. The main requirements in such applications are that the thermojunction be of minimum size and be precisely located and that any materials placed into the wall have thermal properties identical with those of the wall so that temperature distributions are not distorted. In Fig. 8.16*a*[1] the thermojunction is formed by drawing an abrasive tool, such as a file or emery cloth, across the end of the sensing tip. This action flows metal from one thermocouple element to the other since the 0.0002-in. mica insulation is easily bridged over, thus forming numerous microscopic hot-weld thermojunctions. Subsequent erosion or abrasive action forms new thermal junctions continuously as the tip wears away. Such thermocouples have time constants as small as 10^{-5} sec and are available in materials usable to 5000°F and 10,000-psi pressure. In Fig. 8.16*b*[2,3] two thermojunctions are formed by plating a thin rhodium film over the end of a coaxial pair of thermocouple metals. Since the rhodium/metal *A* and rhodium/metal *B* junctions are at the same temperature, the third metal (rhodium) has no effect. The plating is performed by vacuum evaporation and results in a rhodium layer 10^{-4} to 10^{-5} in. in thickness. Theoretical calculations indicate the time constant of such a probe is of the order of 0.3 μsec.

Thermocouples in common use are made from wires ranging from about 0.020 to 0.1 in. in diameter, the larger diameters being required for long life in severe environments. Since speed of response, conduction and radiation errors, and precision of junction location all are improved by the use of smaller wire, very-fine-wire thermocouples are used in special applications requiring these attributes and where lack of ruggedness is not a

[1] Nanmac Corp., Indian Head, Md.

[2] D. Bendersky, A Special Thermocouple for Measuring Transient Temperatures, *Mech. Eng.*, p. 117, February, 1953.

[3] MO-RE', Inc., Bonners Springs, Kans.

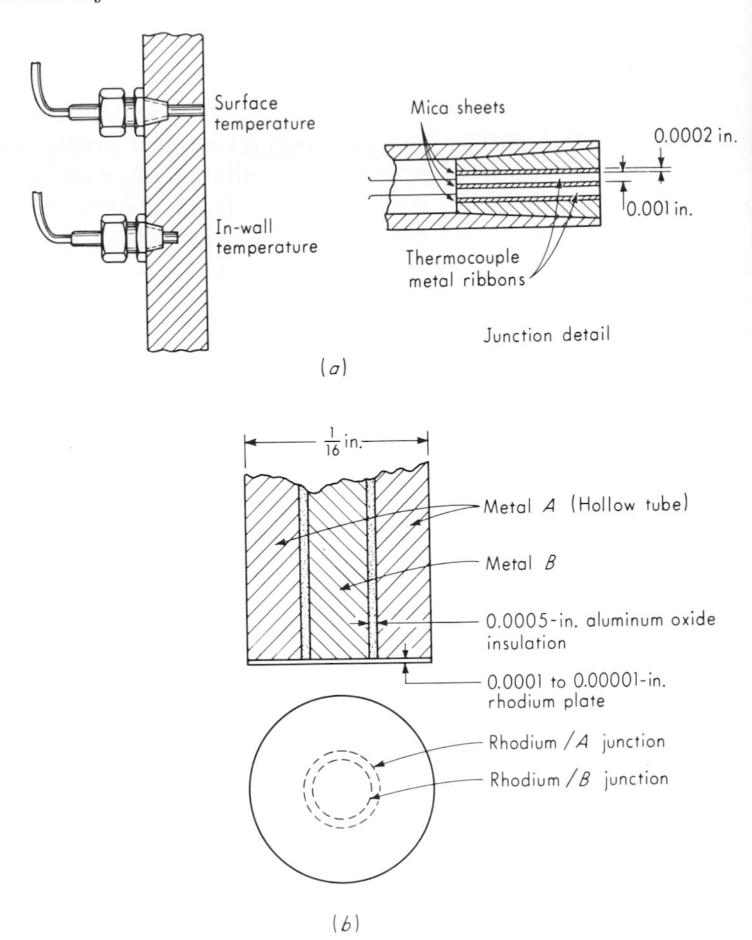

Fig. 8.16. *High-speed thermocouples.*

serious drawback. Such couples are available ready-made[1,2] in most common materials and wire sizes from 0.0005 to 0.015 in. in diameter. The time constant of an iron/constantan couple of 0.0005-in.-diameter wire for a step change from 200 to 100°F in still water is about 0.001 sec.

Several thermocouples may be connected together in series or parallel to achieve useful functions (Fig. 8.17). The series connection with all measuring junctions at one temperature and all reference junctions at another is used mainly as a means of increasing sensitivity.[3] Such an

[1] Omega Engineering Inc., Springdale, Conn.

[2] Baldwin-Lima-Hamilton, Waltham, Mass., *Tech. Data* 4336-1.

[3] R. A. Schnurr, Thermopiles Aid in Measuring Heat Rejection, *Gen. Motors Eng. J.*, p. 8, April–May–June, 1963.

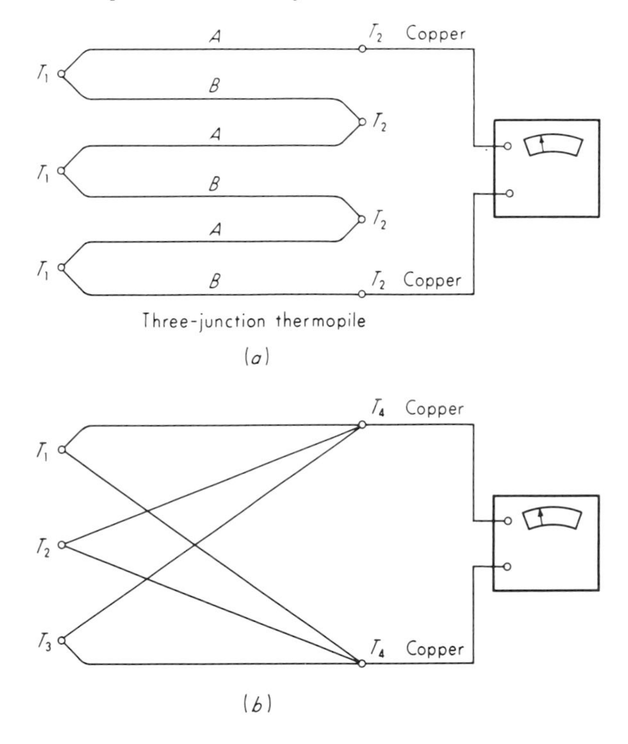

Fig. 8.17. *Multiple-junction thermocouples.*

arrangement is called a thermopile and for n thermocouples gives an output n times as great as a single couple. A commercially available[1] Chromel/constantan thermopile has 25 couples and produces about 1 mv/F°. Common potentiometers can resolve 1 μv, thus making such an arrangement sensitive to 0.001F°. The parallel combination generates the same voltage as a single couple if all measuring and reference junctions are at the same temperatures. If the measuring junctions are at different temperatures and the thermocouples are all the same resistance, the voltage measured is the average of the individual voltages. The temperature corresponding to this voltage is the average temperature only if the thermocouples are linear over the temperature range being measured.

8.4

Electrical-resistance Sensors The electrical resistance of various materials changes in a reproducible manner with temperature, thus forming the basis of a temperature-sensing method. Materials in actual use

[1] Science Products Corp., Dover, N.J.

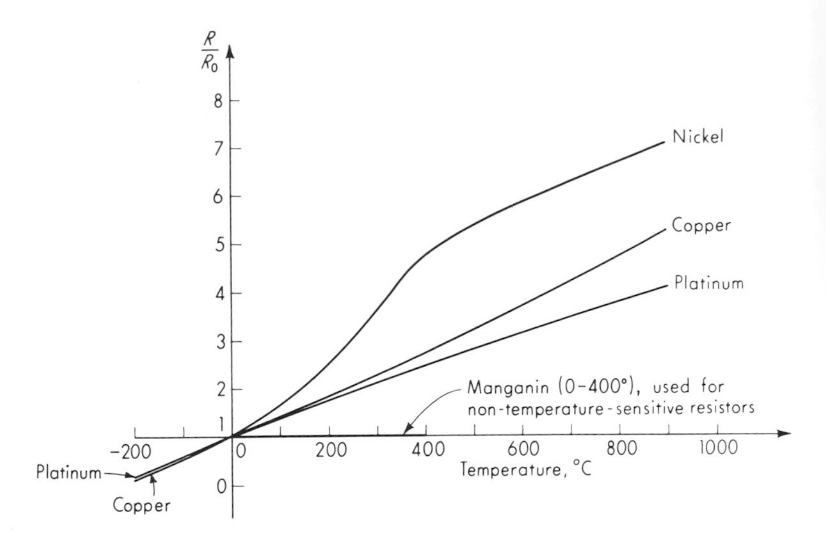

Fig. 8.18. *Resistance/temperature curves.*

fall into two main classes: conductors (metals) and semiconductors. Conducting materials historically came first and have traditionally been called resistance thermometers. Semiconductor types appeared more recently and have been given the generic name thermistor. Any of the various established techniques of resistance measurement may be employed to measure the resistance of these devices, with the Wheatstone bridge being the most common.

Conductive sensors (resistance thermometers). The variation of resistance R with temperature T for most metallic materials can be represented by an equation of the form

$$R = R_0(1 + a_1T + a_2T^2 + \cdots + a_nT^n) \qquad (8.12)$$

where R_0 is the resistance at temperature $T = 0$ (see Fig. 8.18). The number of terms necessary depends on the material, the accuracy required, and the temperature range to be covered. Platinum, nickel, and copper are the most commonly used and generally require, respectively, two, three, and three of the a constants for a highly accurate representation. Tungsten and nickel/iron alloys are also in use. Only constant a_1 may often be used since quite respectable linearity may be achieved over limited ranges. Platinum,[1] for instance, is linear within ± 0.4 percent over the ranges -300 to -100°F and -100 to $+300$°F, ± 0.3 percent from

[1] Platinum Resistance Thermometers, Transducer Handbook 1, Trans-Sonics Inc., Burlington, Mass.

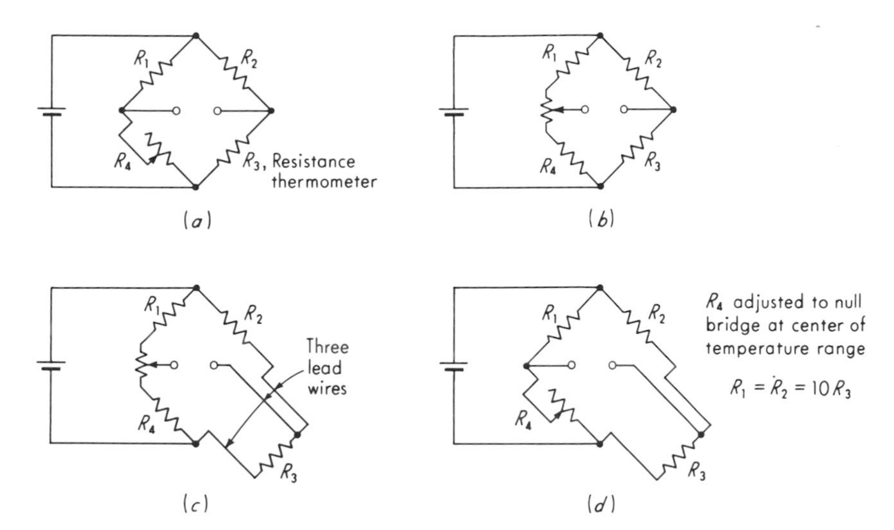

Fig. 8.19. *Resistance-thermometer bridge circuits.*

0 to 300°F, ±0.25 percent from −300 to −200°F, ±0.2 percent from 0 to 200°F, and ±1.2 percent from 500 to 1500°F.

Sensing elements are made in a number of different forms.[1] For measurement of fluid temperatures the winding of resistance wire may be encased in a stainless-steel bulb to protect it from corrosive liquids or gases. Open-type pickups expose the resistance winding directly to the fluid (which must be noncorrosive) and give faster response. Various flat grid windings are available for measuring surface temperatures of solids. These may be taped, welded, or cemented onto the surface. Thin deposited films of platinum[2] are also used in place of wire windings. Surface-temperature transducers affixed to bodies may exhibit spurious output due to interfering strain inputs.[3] These strains may be due to loading of the structure or differential expansion.

Bridge circuits used with resistance temperature sensors may employ either the deflection mode of operation or the null (manually or automatically balanced) mode. If the null method is used, resistor R_4 in Fig. 8.19*a* is varied until balance is achieved. When the highest accuracy is required, the arrangement of Fig. 8.19*b* is preferred since the (variable and unknown) contact resistance in the adjustable resistor has no influence on the resistance of the bridge legs. If long lead wires subjected to temperature varia-

[1] Platinum Resistance Thermometers, Tranducer Handbook 1, Trans-Sonics Inc., Burlington, Mass.

[2] Microdot Inc., South Pasadena, Calif.

[3] A. B. Kaufman, Bonded-wire Temperature Sensors, *Instr. Control Systems*, p. 103, May, 1963.

tions are unavoidable, errors due to their resistance changes may be canceled by use of the Siemens three-lead circuit of Fig. 8.19*c*. Three lead wires of identical length and material exhibit identical resistance variations, and since one of these leads is in each of legs 2 and 3, their resistance changes cancel. Resistance change in the third wire has no effect on bridge balance since it is in the null detector circuit for null-mode operation. For deflection operation its effect is negligible if the indicating instrument draws little current.

While the resistance/temperature variation of the sensing element may be quite linear, the output voltage signal of a bridge used in the deflection mode is not necessarily linear for large percentage changes of resistance. Unlike strain gages, the resistance change of resistance thermometers for full-scale deflection may be quite large. Typically, a 500-ohm platinum element may exhibit 100 ohms change over its design range. For a bridge with four equal arms this would cause severe nonlinearity; however, by making the fixed arms R_1, R_2 of considerably higher resistance (say 10:1) than R_3 and R_4 and by balancing the bridge at the middle of the temperature range rather than at one end, good linearity may be achieved (Fig. 8.19*d*). Typically,[1] a platinum element covering a range from 0 to 100°C using the 10:1 resistance ratio mentioned above gives a nonlinearity of only 0.5C°. For nickel elements, whose resistance/temperature variation is quite nonlinear, this nonlinearity and the bridge nonlinearity can be made nearly to cancel by proper design[2] since the two effects are of opposite directions.

Resistance-thermometer bridges may be excited with either a-c or d-c voltages. The direct or rms alternating current through the thermometer is usually in the range 2 to 20 ma. This current causes an I^2R heating which raises the temperature of the thermometer above its surroundings, causing the so-called self-heating error. The magnitude of this error depends also on heat-transfer conditions and is usually quite small. A 450-ohm platinum element of open construction carrying 25-ma current has a self-heating error of 0.2F° when immersed in liquid oxygen. Actually, by using an unsymmetrical pulse type of excitation voltage whose rms (heating) value is small compared with its peak value, quite large instantaneous currents (and thus large peak output voltages) may be obtained without significant self-heating. Such pulse excitation voltages (Fig. 8.20) can be obtained by commutating a d-c source; this also allows time sharing of the bridge among several resistance sensors. As much as

[1] Temperature Recording from Platinum Resistance Sensors, Brush Instruments, Cleveland, Ohio.

[2] D. R. Mack, Linearizing the Output of Resistance Temperature Gages, *Exptl. Mechanics*, p. 122, April, 1961.

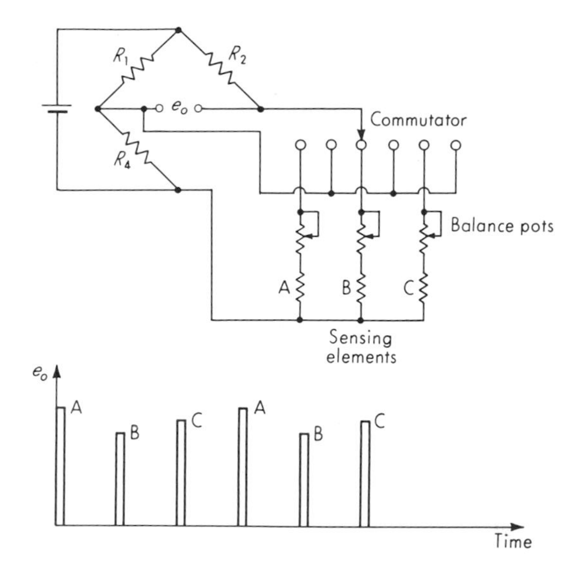

Fig. 8.20. *Pulse-excitation technique.*

5 volts full-scale bridge output signal can be obtained from resistance sensors used in this way.[1]

Resistance-thermometer elements range in resistance from about 10 ohms to as high as 25,000 ohms. Higher-resistance elements are less affected by lead-wire and contact resistance variations, and since they generally produce large voltage signals, spurious thermoelectric emf's due to joining of dissimilar metals are also usually negligible. Platinum is used from −450 to 1850°F, copper from −320 to 500°F, nickel from −320 to 800°F, and tungsten from −450 to 2000°F. Average temperatures may be measured using resistance thermometers as in Fig. 8.21*a*, while differential temperature is sensed by the arrangement of Fig. 8.21*b*.[2] Differential-temperature measurements to an accuracy of 0.05C° have been accomplished in a nuclear-reactor-coolant heat-rise application.[3]

Semiconductor sensors (thermistors). The earlier types of semiconductor resistance temperature sensors were made of manganese, nickel, and cobalt oxides which were milled, mixed in proper proportions with binders, pressed into the desired shape, and sintered. These were given

[1] Platinum Resistance Thermometers, Transducer Handbook 1.

[2] Temperature Recording from Platinum Resistance Sensors.

[3] B. G. Kitchen, Precise Measurement of Process Temperature Differences, *ISA J.*, p. 39, February, 1959.

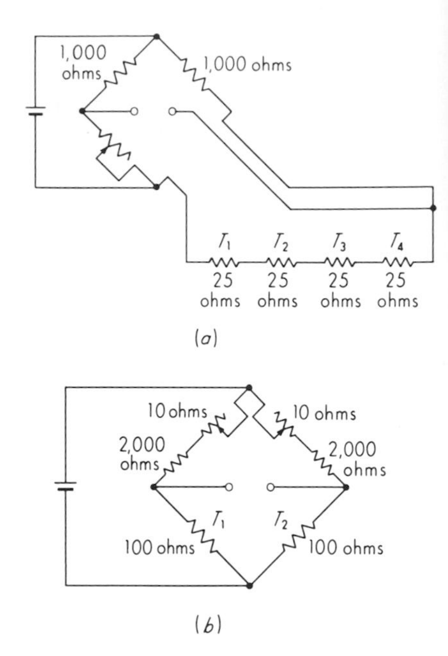

Fig. 8.21. *Average- and differential-temperature sensing.*

the name thermistor and are in wide use today. Compared with conductor-type sensors (which have a small positive temperature coefficient), thermistors have a very large negative coefficient. While some conductors (copper, platinum, tungsten) are quite linear, thermistors are very nonlinear. Their resistance/temperature relation is generally of the form

$$R = R_0 e^{\beta(1/T - 1/T_0)} \tag{8.13}$$

where

$R \triangleq$ resistance at temperature T, ohms
$R_0 \triangleq$ resistance at temperature T_0, ohms
$\beta \triangleq$ constant, characteristic of material, °K
$e \triangleq$ base of natural log
$T, T_0 \triangleq$ absolute temperatures, °K

The reference temperature T_0 is generally taken as 298°K (25°C) while the constant β is of the order of 4000. By computing $(dR/dT)/R$ we find the temperature coefficient of resistance to be given by $-\beta/T^2$ ohms/(ohm-C°). If β is taken as 4000, the temperature coefficient at room temperature (25°C) is -0.045, compared with $+0.0036$ for platinum. While the exact resistance/temperature relation varies somewhat with the particular material used and the configuration of the resistance element, Fig. 8.22 shows the general type of curve to be expected.

Thermistors are commercially available in the form of beads, probes,

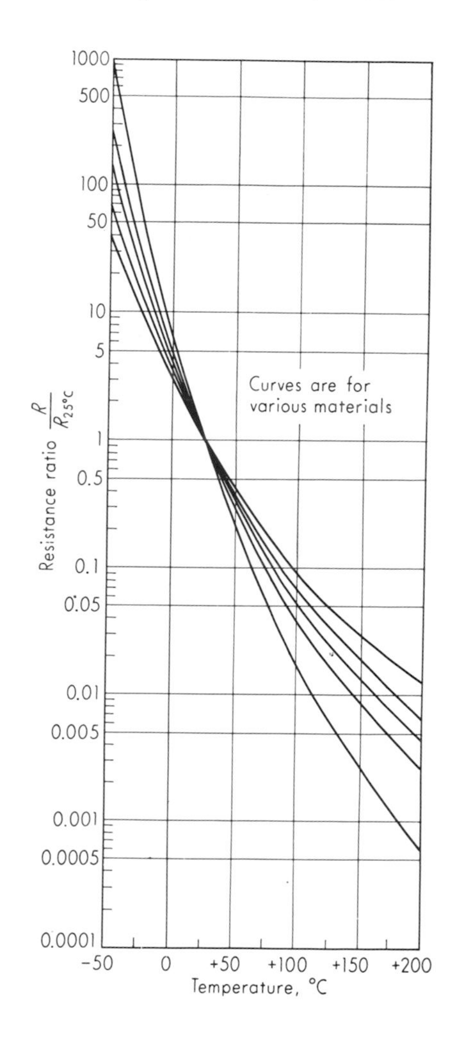

Fig. 8.22. *Thermistor resistance/temperature curves.*

disks, and rods as shown in Fig. 8.23. Beads, much used for temperature measurement, may be bare but are more often glass-coated. They may be as small as a few thousandths of an inch in diameter, giving fast response. Resistance at 25°C can vary over a wide range, from 500 ohms to several megohms. The usable temperature range is from about −420 to 1200°F; however, a single thermistor is not ordinarily used over such a large range. Glass probes have a diameter of about 0.1 in. and a length varying from $\frac{1}{4}$ to 2 in., are also widely used in temperature measurement, and have resistance properties similar to beads. Disks and rods are used more as temperature-compensating devices, time delay elements, and voltage and power controls in electronic circuits.

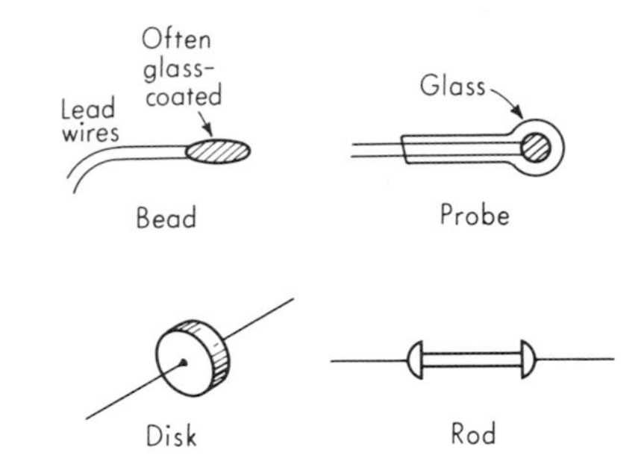

Fig. 8.23. Thermistor forms.

Other semiconductor temperature sensors include carbon resistors and silicon[1] and germanium[2] crystal elements. Carbon resistors are merely the commercial carbon-composition elements commonly used as resistance elements in radios and other electronic circuitry. The 0.1- to 1-watt rated resistors with room-temperature resistance of 2 to 150 ohms are widely used for the measurement of cryogenic temperatures in the range 1 to 20°K. From about 20°K downward these elements exhibit a large increase in resistance with decrease in temperature given by the relation[3]

$$\log_{10} R + \frac{K}{\log_{10} R} = A + \frac{B}{T} \qquad (8.14)$$

where R is the resistance at Kelvin temperature T, and A, B, and K are constants determined by calibration of the individual resistor. Reproducibility of the order of 0.2 percent is obtained in the range 1 to 20°K.

Silicon, with varying amounts of boron impurities, can be designed to have either a positive or negative temperature coefficient over a particular temperature range. The resistance/temperature relation is quite nonlinear. A typical element shows a resistance change (from the nominal value at 25°C) of -80 percent at -150°C to $+180$ percent at $+200$°C. The temperature coefficient near room temperature is of the order of $+0.7$ percent/C°. Germanium, doped with arsenic, gallium, or antimony, is used for cryogenic temperatures, where it exhibits a large decrease in resistance with increasing temperature. The relation is quite nonlinear but very reproducible, giving precise measurements within 0.001 to

[1] J. R. Pies, A New Semiconductor for Temperature Measuring, *ISA J.*, p. 50, August, 1959.

[2] J. S. Blakemore, Germanium for Low-temp Resistance Thermometry, *Instr. Control Systems*, p. 94, May, 1962.

[3] L. G. Rubin, Temperature, *Electron. Progr.*, Raytheon Corp., p. 1, autumn, 1963.

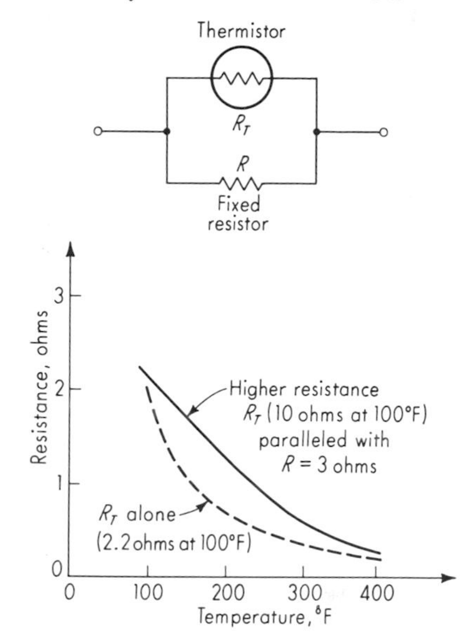

Fig. 8.24. *Thermistor linearization.*

0.0001K° near 4°K when adequate care is taken in technique. Commercially available elements cover a range from about 0.5 to 100°K, a typical unit changing resistance from 7,000 ohms at 2°K to 6 ohms at 60°K.

Circuitry[1,2] for applying the various types of semiconductor resistance sensors to temperature measurement, control, and compensation problems is essentially the same as for conductive sensors, although the greater nonlinearity makes wide temperature ranges less convenient. One technique[3,4] for reducing this nonlinearity is to shunt the thermistor with an ordinary resistor as shown in Fig. 8.24. The stability (variation of resistance/temperature characteristic with time) of early semiconductor elements was inferior to that of conductive elements. While it is unlikely that they will ever approach the excellent stability of platinum, modern semiconductor elements have quite acceptable stability for many applications. This is achieved by proper aging of the elements prior to sale by the manufacturer and continuing improvements in design and production techniques.

[1] D. S. Saulson, The Thermistor Bridge, *Electro-Technol. (New York)*, p. 73, September, 1961.

[2] O. Schwelb and G. C. Temes, Thermistor-Resistor Temperature-sensing Networks, *Electro-Technol. (New York)*, p. 71, November, 1961.

[3] F. Bennett, Designing Thermistor Temperature-correcting Networks Graphically, *Control Eng.*, November, 1955.

[4] R. W. Smith, An Evaluation of Thermistors, *Gen. Motors Eng. J.*, p. 14, October–November–December, 1960.

8.5

Radiation Methods All the temperature-measuring methods discussed up to this point require that the "thermometer" be brought into physical contact with the body whose temperature is to be measured. Also, except for the pulsed thermocouple of Fig. 8.15, the temperature sensor generally is intended to assume the same temperature as the body being measured. This means that the thermometer must be capable of withstanding this temperature, which in the case of very hot bodies presents real problems, since the thermometer may actually melt at the high temperature required. Also, for bodies that are moving, a noncontacting means of temperature sensing is most convenient. Furthermore, if one wishes to determine the temperature variations over the surface of an object, a noncontacting device can readily be "scanned" over the surface.

To solve problems of the type mentioned above, a variety of instruments based in one way or another on the sensing of radiation have been devised. These might, in general, be called radiometers; however, common usage employs terms such as radiation pyrometer, radiation thermometer, optical pyrometer, etc., to describe a particular type of instrument. Since this terminology is not standardized, one must inquire into the basic operating principle of a given instrument to be sure what its characteristics are, rather than relying on the name given the instrument.

Other important applications of infrared radiation include missile guidance, satellite attitude sensing, and infrared spectroscopy. In missile guidance (the Sidewinder missile is an outstanding example) the missile is designed to "home" on the infrared radiation emitted by the target, often the hot jet exhaust of the target aircraft's engine. A scanning system in the missile locates the target and produces error signals that steer the missile into the target. For satellite attitude sensing[1] the infrared sensors are able to distinguish the radiation from the earth, the moon, or a planet from the background of space and thus generate accurate orientation signals for control purposes. Infrared spectroscopy[2] involves the use of infrared principles for the analysis of gases, liquids, and solids to identify and determine the concentration of molecules or molecular groups.

Radiation fundamentals. Radiation-temperature sensors operate with electromagnetic radiation whose wavelengths lie in the visible and infrared portions of the spectrum. The visible spectrum is quite narrow: 0.3 to 0.72 μ ($1\ \mu = 10^{-6}$ m). The infrared spectrum is generally defined as the range from 0.72 to about 1,000 μ. Bordering the visible spectrum

[1] Barnes Engineering Co., Stamford, Conn., *Bull.* 14-003 and 0-014, 1962.

[2] D. M. Considine (ed.), "Process Instruments and Controls Handbook," pp. 6–67, McGraw-Hill Book Company, New York, 1957.

on the low-wavelength side are the ultraviolet rays, while microwaves border the infrared spectrum on the high side. Radiation-temperature-sensing devices utilize mainly some part of the range 0.3 to 40 μ.

Physical bodies (solids, liquids, gases) may emit electromagnetic radiation or subatomic particles for a number of reasons. As far as temperature sensing is concerned, we need be concerned only with that part of the radiation caused solely by temperature. Every body above absolute zero in temperature emits radiation dependent on its temperature. The ideal thermal radiator is called a blackbody. Such a body would absorb completely any radiation falling on it and also, for a given temperature, emits the maximum amount of thermal radiation possible. The law governing this ideal type of radiation is Planck's law, which states that

$$W_\lambda = \frac{C_1}{\lambda^5(e^{C_2/\lambda T} - 1)} \tag{8.15}$$

where $W_\lambda \triangleq$ hemispherical spectral radiant intensity, watts/(cm²-μ)
$C_1 \triangleq$ 37,413, (watts-μ^4)/cm²
$C_2 \triangleq$ 14,388, μ-°K
$\lambda \triangleq$ wavelength of radiation, μ
$T \triangleq$ absolute temperature of blackbody, °K

The quantity W_λ is the amount of radiation emitted from a flat surface into a hemisphere, per unit wavelength, at the wavelength λ. Equation (8.15) thus gives the distribution of radiant intensity with wavelength; that is, a blackbody at a certain temperature emits *some* radiation per unit wavelength at every wavelength from zero to infinity, but not the same amount at each wavelength. Figure 8.25 shows the curves obtained from Eq. (8.15) by fixing T at various values and plotting W_λ versus λ. The curves exhibit peaks at particular wavelengths, and the peaks occur at longer wavelengths as the temperature decreases. The area under each curve is the total emitted power and increases rapidly with temperature. Equations giving the peak wavelength λ_p and the total power W_t are

$$\lambda_p = \frac{2,891}{T} \qquad \mu \tag{8.16}$$

and

$$W_t = 5.67 \times 10^{-12}\, T^4 \qquad \text{watts/cm}^2 \tag{8.17}$$

Figure 8.26 shows the wavelength range over which 90 percent of the total power is found for various temperatures. Note that lower temperatures require measurement out to longer wavelengths.

While the concept of a blackbody is a mathematical abstraction, real physical bodies can be constructed to approximate closely blackbody behavior. Such radiation sources are needed for calibration of radiation thermometers and generally take the form of a blackened conical cavity of about 15° cone angle. The temperature is adjustable, automatically con-

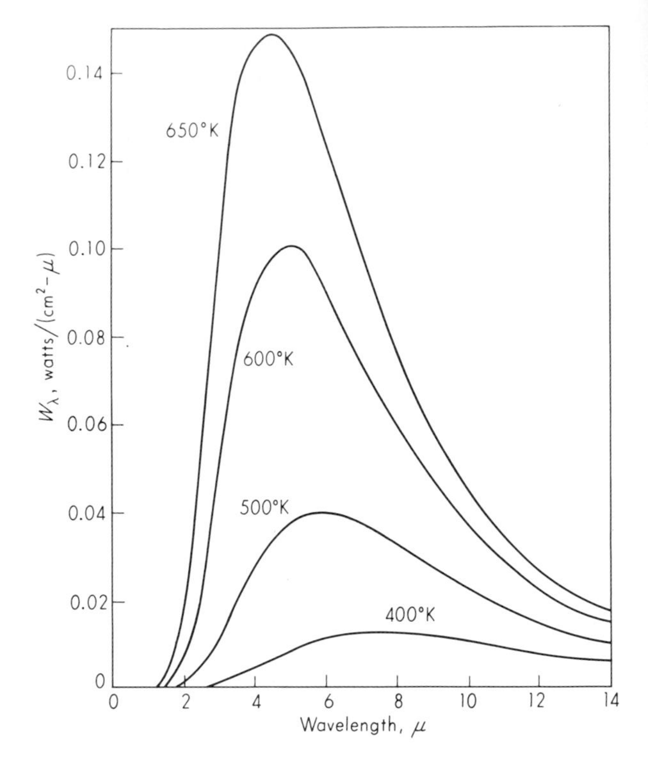

Fig. 8.25. *Blackbody radiation.*

trolled for constancy, and measured by some accurate sensor such as a platinum resistance thermometer. A typical unit[1] covers the range 500 to 1000°K with 1K° accuracy and emittance 0.99 ± 0.01 (blackbody has 1.00). While it is possible to construct a nearly perfect blackbody, the bodies whose temperatures are to be measured with some radiation-type

[1] Infrared Industries Inc., Riverside, Calif.

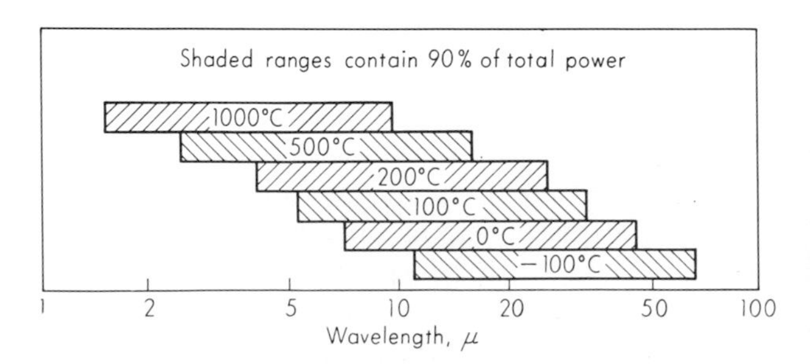

Fig. 8.26. *Power/wavelength distribution.*

instrument often deviate considerably from such ideal conditions. The deviation from blackbody radiation is expressed in terms of the emittance of the measured body. Several types of emittance have been defined to suit particular applications. The most fundamental form of emittance is the hemispherical spectral emittance $\epsilon_{\lambda,T}$. Let us call the *actual* hemispherical spectral radiant intensity of a real body at temperature T $W_{\lambda a}$ and assume it can be measured. Then $\epsilon_{\lambda,T}$ is defined as

$$\epsilon_{\lambda,T} \triangleq \frac{W_{\lambda a}}{W_\lambda} \qquad (8.18)$$

where W_λ is the blackbody intensity at temperature T. Emittance is thus dimensionless and always less than 1.0 for real bodies. In the most general case it varies with both λ and T. With the definition of Eq. (8.18), the radiation from a real body may be written as

$$W_{\lambda a} = \frac{C_1 \epsilon_{\lambda,T}}{\lambda^5(e^{C_2/\lambda T} - 1)} \qquad (8.19)$$

Similarly, the total power W_{ta} of an actual body is given by

$$W_{ta} = C_1 \int_0^\infty \frac{\epsilon_{\lambda,T}}{\lambda^5(e^{C_2/\lambda T} - 1)}\, d\lambda \qquad (8.20)$$

and if we assume that W_{ta} can be measured experimentally we may define the hemispherical total emittance $\epsilon_{t,T}$ by

$$\epsilon_{t,T} \triangleq \frac{W_{ta}}{W_t} \qquad (8.21)$$

where W_t is the blackbody total power at temperature T. Thus if $\epsilon_{t,T}$ is known, the total power of a real body is given by

$$W_{ta} = 5.67 \times 10^{-12}\, \epsilon_{t,T} T^4 \qquad \text{watts/cm}^2 \qquad (8.22)$$

If a body has $\epsilon_{\lambda,T}$ equal to a constant for all λ and at a given T, it is called a *graybody*. In this case we see that $\epsilon_{\lambda,T} \equiv \epsilon_{t,T}$. Also the curves of $W_{\lambda a}$ versus λ have exactly the same shape as for W_λ. Since many radiation thermometers operate in a restricted band of wavelengths, the hemispherical band emittance $\epsilon_{b,T}$ has been defined by

$$\epsilon_{b,T} \triangleq \frac{\int_{\lambda_a}^{\lambda_b} [\epsilon_{\lambda,T}/\lambda^5(e^{C_2/\lambda T} - 1)]\, d\lambda}{\int_{\lambda_a}^{\lambda_b} [1/\lambda^5(e^{C_2/\lambda T} - 1)]\, d\lambda} \qquad (8.23)$$

This is seen to be just the ratio of the total powers, actual and blackbody, within the wavelength interval λ_a to λ_b for bodies at temperature T. If the actual power can be measured directly, $\epsilon_{b,T}$ can be found without knowing $\epsilon_{\lambda,T}$. For a graybody, $\epsilon_{b,T} \equiv \epsilon_{\lambda,T}$.

If a radiation thermometer has been calibrated against a blackbody source, knowledge of the appropriate emittance value allows correction

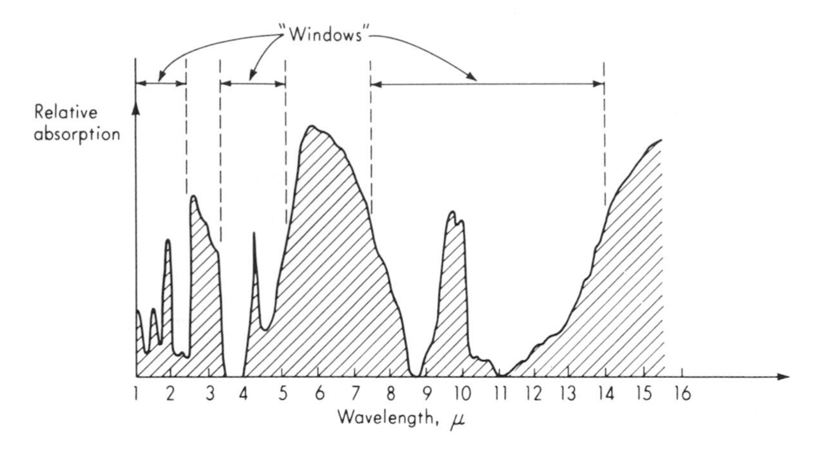

Fig. 8.27. *Atmospheric absorption.*

of its readings for nonblackbody measurements. Unfortunately, emittances are not simple material properties such as densities but rather depend on size, shape, surface roughness, angle of viewing, etc. This leads to uncertainties in the numerical values of emittances, which are one of the main problems in radiation-temperature measurement.

Another source of error is the losses of energy in transmitting the radiation from the measured object to the radiation detector. Generally the optical path consists of some gas (often atmospheric air) and various windows, lenses, or mirrors used to focus the radiation or protect sensitive elements from the environment. In atmospheric air the attenuation of radiation is due mainly to the resonance-absorption bands of water vapor, carbon dioxide, and ozone and the scattering effect of dust particles and water droplets. The combined absorption effect of H_2O, CO_2, and O_3 is roughly as shown in Fig. 8.27. Since the absorption varies with wavelength, a radiation thermometer can be designed to respond only within one of the "windows" shown, thus making it insensitive to these effects. Since the absorption varies with the thickness of the gas traversed by the radiation, the effect is not an instrument constant and cannot thus be calibrated out. The lenses used in infrared instruments must often be made of special materials, since glasses normally used for the visible spectrum are almost opaque to radiation of wavelength longer than about 2 μ. Figure 8.28 shows the variation of transmission factor of various materials with wavelength. While infrared radiation follows the same optical laws used for lens and mirror design as visible light, some materials useful for infrared wavelengths (arsenic trisulfide, for example) are opaque to visible-light wavelengths.

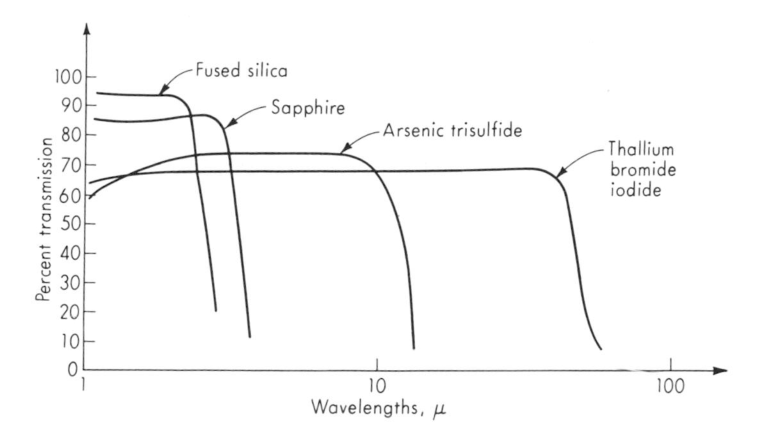

Fig. 8.28. *Optical-material spectral transmission.*

Radiation detectors. In all radiation thermometers (other than the disappearing-filament optical pyrometer) the radiation from the measured body is focused on some sort of radiation detector which produces an electrical signal. Detectors may be classified as thermal detectors or photon detectors. Thermal detectors are blackened elements designed to absorb a maximum of the incoming radiation at all wavelengths. The absorbed radiation causes the temperature of the detector to rise until an equilibrium is reached with heat losses to the surroundings. Thermal detectors actually measure this temperature, using a resistance thermometer, thermistor, or thermocouple (thermopile) principle.

Resistance-thermometer and thermistor elements are made in the form of thin films or flakes and are called bolometers. Performance criteria for both thermal and photon detectors include the time constant (most detectors behave roughly as first-order systems), the responsivity (volts of signal per watt of incident radiation), and the noise-equivalent power (N.E.P.). The noise-equivalent power gives an indication of the smallest amount of radiation that can be detected, which is limited by the inherent electrical noise level of the detector. That is, with no radiation whatever coming in, the detector still puts out a small random voltage due to various electrical noise sources within the detector itself. The amount (watts) of incoming radiation required to produce a signal just equal in strength to the noise (signal/noise ratio of 1) is called the noise-equivalent power. A low value of noise-equivalent power is thus desirable. An evaporated-nickel-film bolometer of about 35 mm^2 area has a resistance of about 100 ohms, a time constant of 0.004 sec, responsivity of 0.4 volt/watt, and noise-equivalent power of 3×10^{-9} watt. A thermistor bolometer of 0.5 mm^2 area might have 3 megohms resistance,

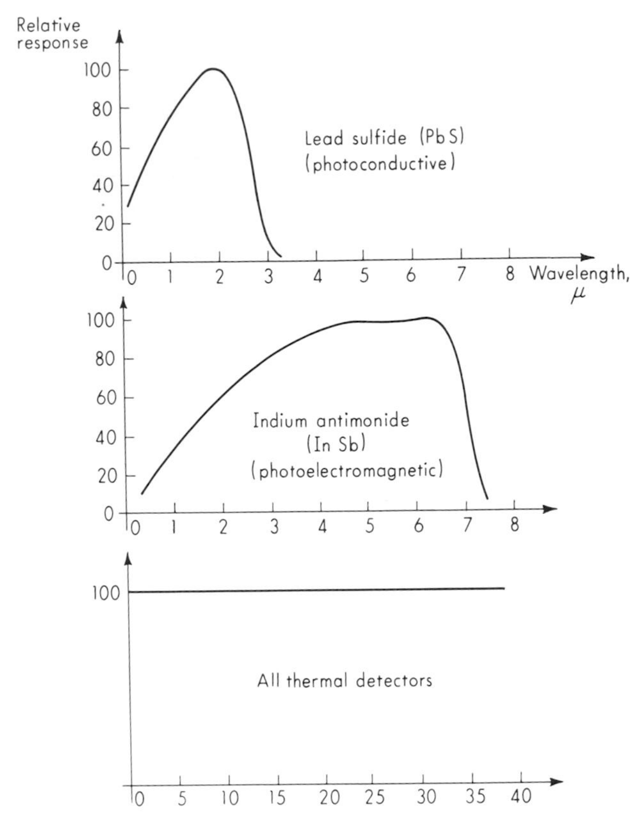

Fig. 8.29. *Radiation-detector spectral sensitivity.*

a time constant of 0.004 to 0.030 sec, responsivity of 700 to 1,200 volts/watt, and noise-equivalent power of 2×10^{-10} watt. Thermopiles of an area from 0.4 to 10 mm² have resistance of the order of 10 to 100 ohms, time constants in the range 0.005 to 0.3 sec, responsivity of 3 to 90 volts/watt, and noise-equivalent power of 2×10^{-11} to 7×10^{-10}.

In the various types of photon detectors the incoming radiation (photons) frees electrons in the detector structure and produces a measurable electrical effect. These events occur on an atomic or molecular time scale rather than on the gross time scale involved in the heating and cooling of thermal detectors. A much higher response speed is thus possible. However, photon detectors have a sensitivity that varies with wavelength; thus incoming radiation of different wavelengths is not equally treated. Typical spectral response of some common types is shown in Fig. 8.29. Photon detectors commonly in use operate in the photoconductive, photovoltaic, or photoelectromagnetic (PEM) modes.

Photoconductive types exhibit an electrical resistance that changes

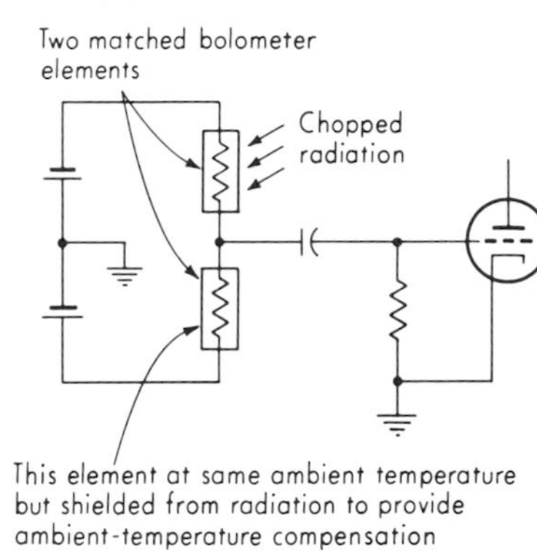

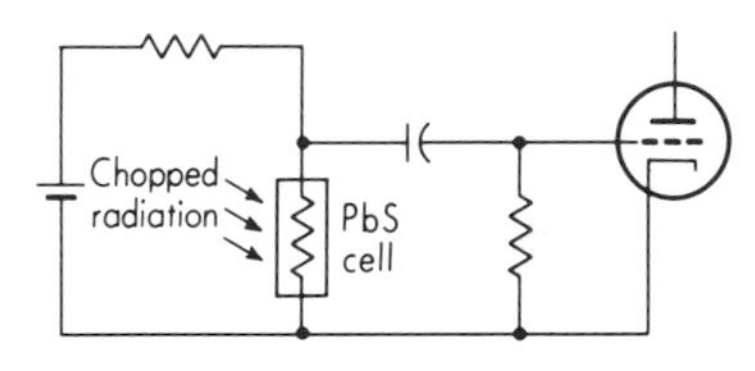

Fig. 8.30. *Basic detector circuits.*

with the incoming radiation level. Photovoltaic cells, also called barrier photocells, employ a photosensitive barrier of high resistance, deposited between two layers of conducting material. A potential difference between these two layers is built up when the cell is exposed to radiation. In photoelectromagnetic detectors the Hall effect is utilized. A semiconductor crystal is subjected to a strong magnetic field and radiation applied to one side. A potential difference is developed across the ends of the crystal. Lead sulfide photoconductive cells are by far the most used type, typical units of 1 to 35 mm^2 area having resistances of 10^5 to 2×10^6 ohms, time constants of 2 to 0.04 msec, responsivity of 5,000 to 150,000 volts/watt, and noise-equivalent power of 4×10^{-11} to 4×10^{-12} watt. An indium antimonide photoelectromagnetic cell of 100 ohms resistance may have a time constant less than 1 μsec, responsivity of 1 volt/watt, and noise-equivalent power of 10^{-9} watt.

Some type of circuit must be employed to realize a usable electrical signal (generally a voltage) from a radiation detector. Thermopile devices generally work with an uninterrupted stream of radiation and require no circuitry other than the usual reference junction, which is commonly left at ambient temperature. Bolometers and photoconductive cells often employ a chopper to interrupt the radiation at a fixed rate of the order of several hundred cycles per second. This leads to an a-c-type electrical signal and allows use of high-gain a-c amplifiers. Typical circuits for such arrangements are shown in Fig. 8.30.

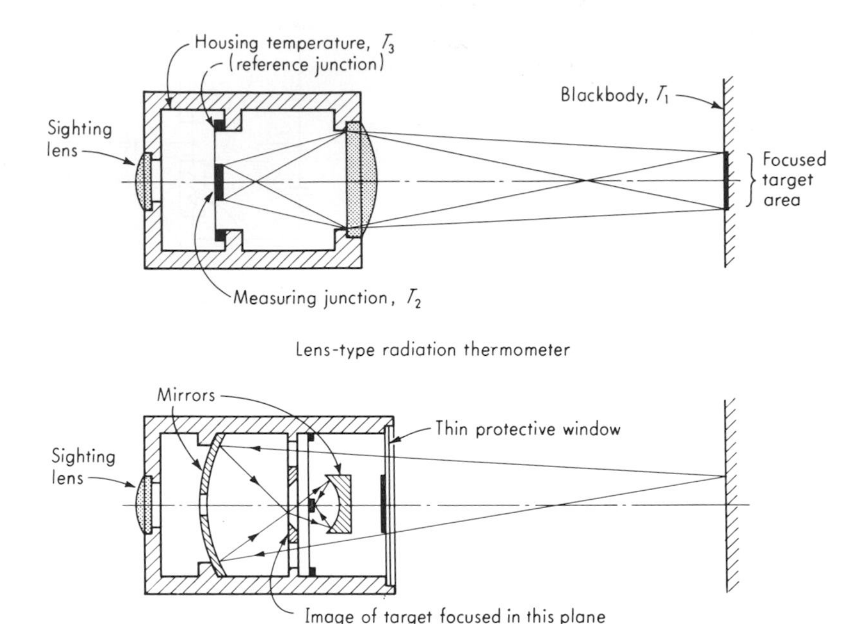

Fig. 8.31. *Lens- and mirror-type radiation thermometers.*

Unchopped (d-c) broadband radiation thermometers. We begin our study of complete radiation-sensing instruments by consideration of the most common type used in day-to-day industrial applications.[1] These instruments use a blackened thermopile as detector and focus the radiation by means of either lenses or mirrors. Figure 8.31 shows in simplified fashion the construction of this class of instruments. The reference of footnote 1 gives a very complete analysis of such devices.

Basically, for a given source temperature T_1, the incoming radiation heats the measuring junction until conduction, convection, and radiation losses just balance the heat input. The measuring-junction temperature is usually less than 40°C above its surroundings even if the source is incandescent. An oversimplified analysis gives

$$\text{Heat loss} = \text{radiant heat input}$$

$$K_1(T_2 - T_3) = K_2 T_1^4 \qquad (8.24)$$

[1] T. R. Harrison, "Radiation Pyrometry and Its Underlying Principles of Radiant Heat Transfer," John Wiley & Sons, Inc., New York, 1960.

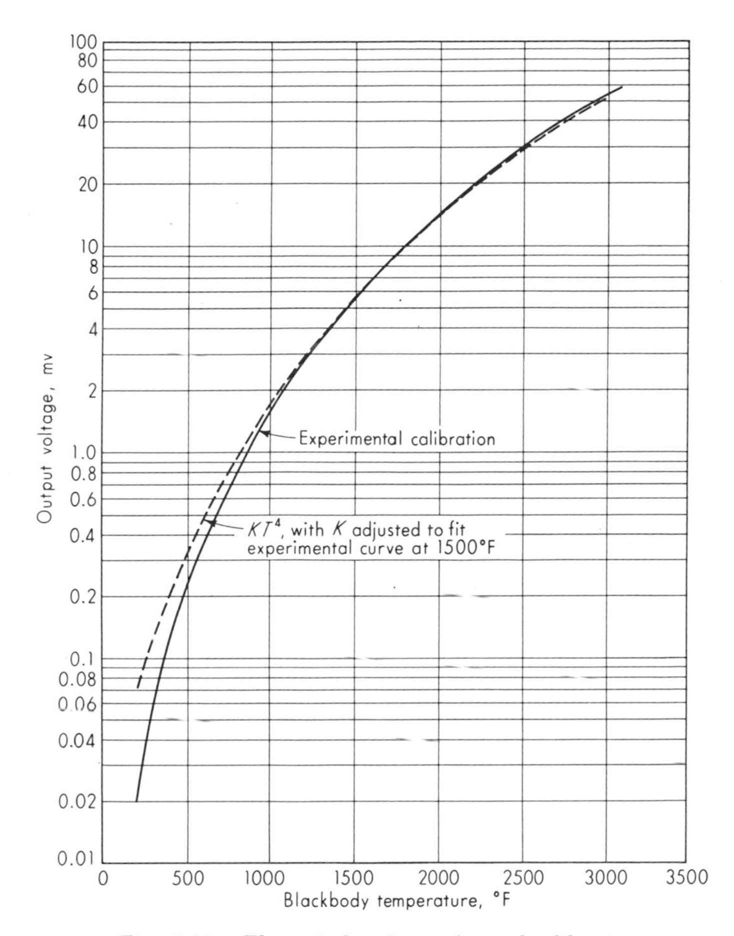

Fig. 8.32. *Theoretical and experimental calibration curves.*

If the thermocouple voltage is proportional to $T_2 - T_3$, the voltage output should be proportional to T_1^4. Figure 8.32 shows an actual calibration curve of such an instrument together with the ideal relationship. For high temperatures the agreement is quite close. The temperatures T_2 and T_3 are both influenced by the environmental temperature; thus compensation must generally be provided for this, particularly if the instrument is intended to measure low temperatures. This compensation may include a thermostatically controlled housing temperature.

The thermopiles used may have from 1 or 2 to 20 or 30 junctions. A small number of junctions has less mass and thus faster response, but lower sensitivity limits application to high temperatures. Response is

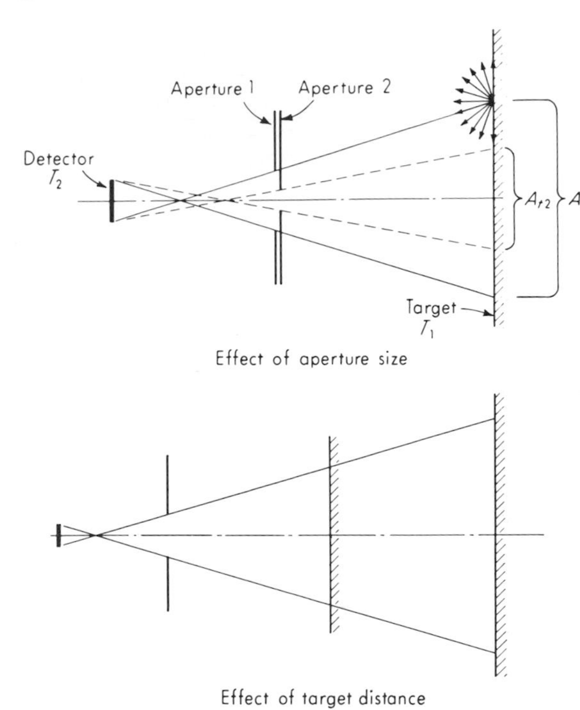

Fig. 8.33. Effect of aperture size and target distance.

roughly that of a first-order system, with time constants ranging from about 0.1 sec (high-temperature systems) to 2 sec (low-temperature systems). Instruments of this class are available to measure temperatures as low as 0°F; that is, the thermopile is actually cooler than the ambient temperature and reads a negative voltage. Theoretically there is no upper limit to the temperatures that can be measured in this way. Commercial instruments usable to 3200°F are readily available.

Conceivably, an instrument of the above type could be constructed with no focusing means, i.e., no lens or mirror. A simple diaphragm with a circular aperture (Fig. 8.33) would define the target from which radiation is received. To define smaller target areas for a given target distance, a smaller aperture could be used; however, there would be a proportionate loss in incoming radiation and thus sensitivity. The reading of such an instrument is independent of the distance between the target and the instrument, since the amount of radiation received is limited by the solid angle of the cone defined by the aperture and detector and this is always the same. However, as the target distance increases, the target area necessary to fill the cone increases. If, because of non-

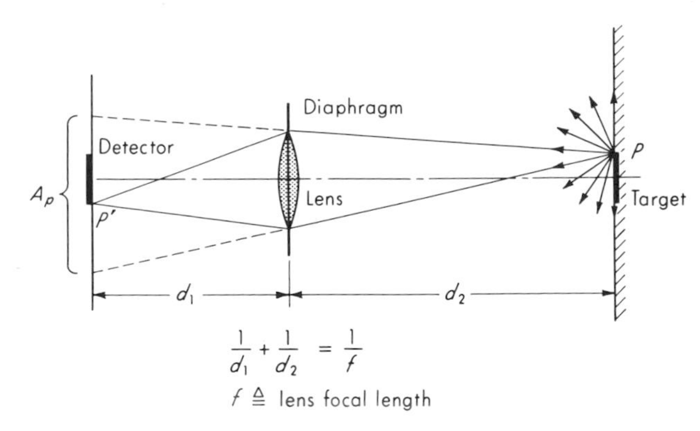

Fig. 8.34. *Advantages of focusing.*

uniformity of target temperature or small target size, one wishes to restrict the target area to small values, a very small aperture is required, giving very low sensitivity.

The basic purpose of lens or mirror systems is to overcome this restriction, thus allowing the resolution of small targets without loss of sensitivity. Thus, in Fig. 8.34, radiation emanating from the point P on the target is focused on the corresponding point P' of the detector. If a simple diaphragm, rather than a focused lens, had been used, the same amount of energy would be spread out over the area A_p, with the detector receiving only a fraction of the radiation. Commercial lens- or mirror-type instruments generally have parameters such that targets 2 ft or more distant are adequately focused with a fixed-focus lens, and the instrument calibration is independent of distance as long as the target fills the field of view. Minimum target size to fill the field of view is of the order of one-twentieth of the target distance for common instruments. For targets closer than 2 ft, focusing is necessary and may affect the calibration, depending on instrument construction and closeness of target. Target diameters of 0.1 to 0.3 in. at target distances of 4 to 12 in. are available. Since the focal length of a lens depends on the index of refraction, which in turn varies with wavelength, all wavelengths are not focused at the same point. In particular, if one focuses a lens visually (using visible light), the longer infrared wavelengths, which contribute a large portion of the total energy at lower temperatures, will be out of focus. Such an instrument may thus have to be focused by adjusting for maximum thermopile output rather than for sharpest visual definition. Another effect of lenses is selective transmission, as shown in Fig. 8.28. The use of mirrors rather than lenses is an attempt to alleviate some of these problems. However, instruments of both types are widely used with success.

Chopped (a-c) broadband radiation thermometers. A number of advantages accrue when the radiation coming from the target to the detector is periodically interrupted (chopped) at a fixed frequency; therefore many infrared systems employ this technique. When high sensitivity is needed, amplification is required, and high-gain a-c amplifiers are easier to construct than their d-c counterparts. This is usually the main reason for using choppers. Additional benefits related to ambient-temperature compensation and reference-source comparison may also be obtained. Systems employing thermal (broadband) and photon (restricted-band) detectors and choppers are in common use. We here consider those using thermal detectors.

The time constants of adequately sensitive thermopile detectors are generally too long to allow efficient use of chopping; thus the faster bolometers, usually the thermistor type, are employed. We shall consider two specific forms of this class of instruments: the blackened-chopper type and the mirror-chopper type. Figure 8.35 shows the basic elements of a blackened-chopper radiometer. A mirror focuses the target radiation on the detector; however, this beam is interrupted periodically by the chopper rotating at constant speed. Thus the detector alternately "sees" radiation from the target and the radiation from the chopper's blackened surface. For high target temperatures, sufficient accuracy may be achieved by leaving the chopper temperature at ambient. Higher accuracy, particularly at low target temperatures, is obtained by thermostatically controlling the chopper temperature. An arrangement similar to that of Fig. 8.30 is used for the detector circuit. The output voltage of the detector circuit is essentially as shown in Fig. 8.35. By amplifying this in an a-c amplifier the mean value (which is subject to drift) is discarded, and only the difference between the target and chopper radiation levels is amplified. If the chopper radiation level is considered as a known reference value, the target radiation and thus its temperature may be inferred. To provide a d-c output signal related to target temperature and suitable for recording or control purposes the a-c amplifier is followed by a phase-sensitive demodulator and filter circuit. The necessary synchronizing signal for the demodulator may be generated by placing a magnetic proximity pickup near the chopper blades. While the response time of the detector itself may be of the order of a millisecond, the chopper frequency and necessary demodulator filter time constant greatly reduce the overall system speed. High chopper speeds allow faster overall response but reduce sensitivity if the detector time constant is too large, since the detector does not have time to reach equilibrium during the time either the target or chopper is in view.

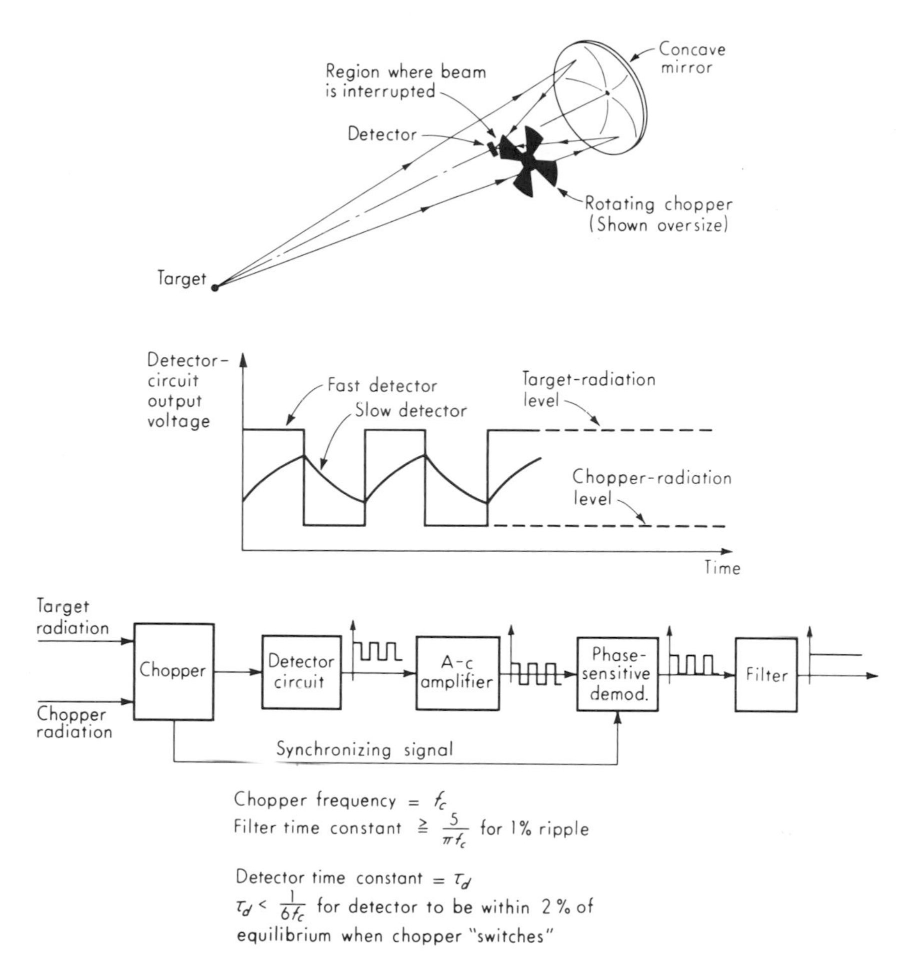

Fig. 8.35. *Blackened-chopper system.*

A typical instrument[1] uses a square thermistor detector, thus giving rise to a square field of view of size 1° by 1°. Sighting and focusing in the range 2 ft to infinity are accomplished with an attached optical telescope. The standard chopping frequency is 180 cps, leading to an overall system time constant of about 0.008 sec. Standard temperature range is from ambient to 1300°C.

In the mirror-chopper instrument[2] of Fig. 8.36 two thermistor detectors are used. Also, an accurate blackbody source whose temperature is automatically controlled and accurately measured (say by a

[1] Servo Corp. of America, New Hyde Park, N.Y.
[2] Barnes Engineering Co., Stamford, Conn.

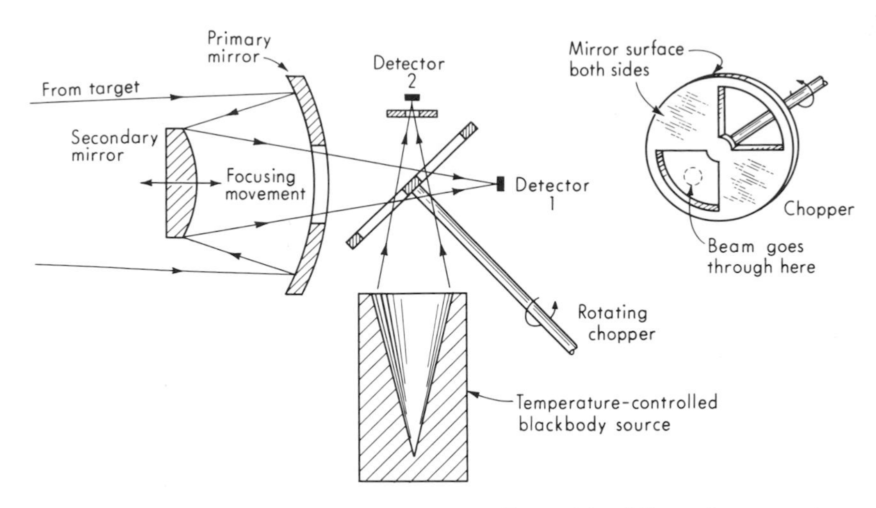

Fig. 8.36. *Mirror-chopper system.*

thermocouple) is provided within the radiometer. A chopper operating at 77 cps alternately exposes each of the two detectors to the target radiation and the blackbody source of known temperature. In Fig. 8.36 detector 1 is receiving target radiation while 2 is receiving blackbody. When the chopper disk rotates 90° so that a solid sector is in the line of sight the number 1 detector receives blackbody radiation reflected from the rear mirror surface of the chopper while number 2 receives target radiation reflected from the front surface. Circuitry similar to that of Fig. 8.35 can be employed to develop a d-c output signal related to target temperature. Such a system can be used for very accurate static measurements by using it in a null method of operation. The blackbody source temperature is adjusted until no output is obtained. Then the target and blackbody are at identical temperatures, provided the target emittance is 1.0. If the emittance is not 1.0 but is known, a correction may be applied. Such a null method makes the reading independent of detector sensitivity (which may drift) and amplifier gain. A system of this type employing 1.5- by 1.5-mm square thermistor detectors has a 0.5 by 0.5° field of view (1-in.-square target at 10 ft), an overall system time constant of 0.016 sec (bandwidth 10 cps), and will detect (signal/noise ratio = 1.0) a target temperature change of 0.4C°. By sacrificing response speed for sensitivity, heavier filtering can be switched in to give a time constant of 0.16 or 1.6 sec, with corresponding increase of resolution to 0.14 and 0.04C°. The instrument can be focused on targets from 4 ft to infinity.

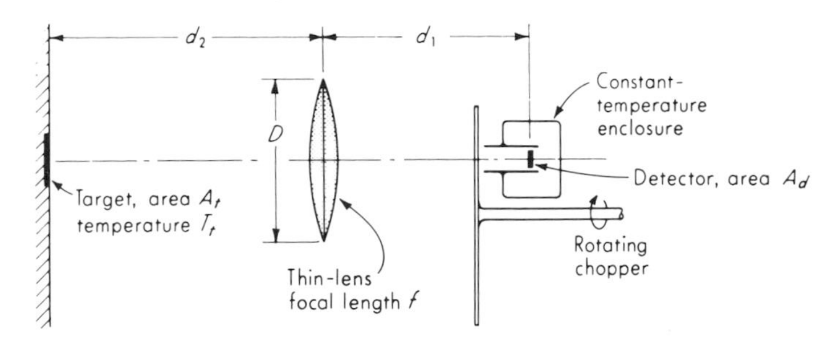

Fig. 8.37. Photon-detector system.

Chopped (a-c) selective band (photon) radiation thermometers. The use of photon detectors allows faster response speeds and may reduce sensitivity to ambient temperature. Since such detectors respond *directly* to the incident photon flux, rather than detecting a temperature change, the disturbing influence of ambient-temperature changes is restricted mainly to changes in the responsivity of the detector. When radiation is expressed in terms of photon flux rather than watts, the formulas are somewhat modified. Equation (8.15) becomes

$$N_\lambda = \frac{2\pi c}{\lambda^4(e^{C_2/\lambda T} - 1)} \tag{8.25}$$

where $N_\lambda \triangleq$ hemispherical spectral photon flux, photons/(cm^2-sec-μ)
$c \triangleq$ speed of light, 3×10^{10} cm/sec

The peak of the photon-flux curves occurs at a different wavelength from that of the radiant intensity. It is given by

$$\lambda_{p,p} = \frac{3{,}669}{T} \qquad \text{microns} \tag{8.26}$$

The total photon flux for all wavelengths is

$$N_t = 1.52 \times 10^{11} T^3 \qquad \text{photons/(cm}^2\text{-sec)} \tag{8.27}$$

Figure 8.37 shows the basic arrangement of an instrument[1] using a photon detector and a chopper. Basic optics gives

$$\frac{1}{d_1} + \frac{1}{d_2} = \frac{1}{f} \tag{8.28}$$

and

$$A_t = \frac{A_d d_2^2}{d_1^2} \tag{8.29}$$

[1] Infrared Thermometry, Infrared Industries, Santa Barbara, Calif.

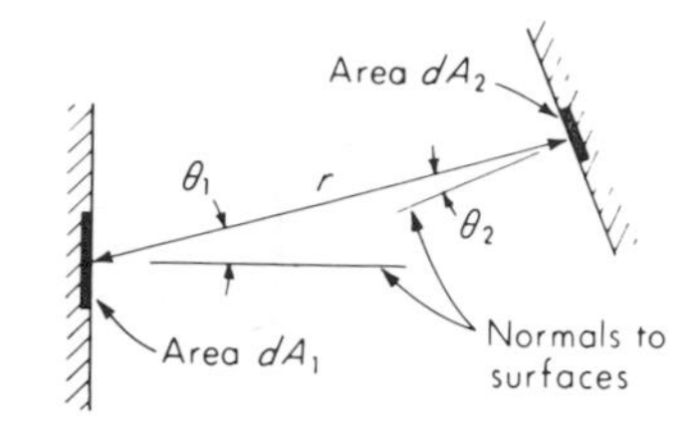

Fig. 8.38. *Basic radiation configuration.*

Combining these gives

$$A_t = \frac{A_d(d_2 - f)^2}{f^2} \tag{8.30}$$

Now if $d_2 \gg f$, we get approximately

$$A_t = \frac{A_d d_2^2}{f^2} \tag{8.31}$$

If the detector is a square with side L_d, the resolved target will be a square of side $L_d d_2/f$. For example, a detector 1 mm square used with a lens with a focal length of 75 mm requires a target of size $d_2/75$ to fill exactly the field of view.

For the general configuration of Fig. 8.38, basic radiation laws[1] give

$$\frac{\text{Radiation incident on } dA_2}{dA_2} = \frac{\cos\theta_1 \cos\theta_2}{\pi r^2} (\text{radiation emitted from } dA_1) \tag{8.32}$$

We can apply this to the configuration of Fig. 8.37 to find the radiation received over the area of the lens from the target. If d_2 is large compared with the size of target and lens (usually true), $\cos\theta_1 \approx \cos\theta_2 \approx 1$, and dA_1 and dA_2 may be replaced by A_1 and A_2 in Eq. (8.32). To simplify the analysis, let us assume that the detector has uniform spectral response from $\lambda = 0$ to $\lambda = 6.9$ and zero response beyond and that the target is a graybody with emittance ϵ. The radiation emitted by the target is then

$$A_t\epsilon \int_0^{6.9} N_\lambda \, d\lambda = A_t\epsilon E(T_t) \qquad \text{photons/sec} \tag{8.33}$$

where $$E(T_t) \triangleq \int_0^{6.9} N_\lambda \, d\lambda \qquad \text{a function of } T_t$$

The radiation received at the lens is

$$\frac{A_t\epsilon E(T_t)(\pi D^2/4)}{\pi d_2^2} = \frac{A_t\epsilon E(T_t)D^2}{4d_2^2} \tag{8.34}$$

If the lens transmits $100K_{tr}$ percent of the flux incident on it (perfect transmission has $K_{tr} = 1.0$) and if the system is focused so that the target

[1] A. I. Brown and S. M. Marco, "Introduction to Heat Transfer," p. 237, McGraw-Hill Book Company, New York, 1951.

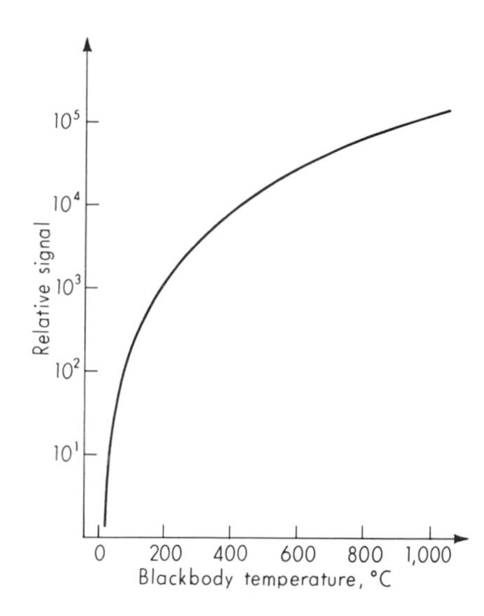

Fig. 8.39. *Indium antimonide system-response curve.*

image just fills the detector area, the photon flux density at the detector is given by

$$N_d \triangleq \frac{K_{tr} A_t \epsilon E(T_t) D^2}{4 d_2^2 A_d} \qquad \text{photons/(cm}^2\text{-sec)} \tag{8.35}$$

For the indium antimonide photoelectromagnetic detector the output voltage is proportional to N_d. Since for a focused target $A_t/A_d d_2^2 = 1/f^2$, Eq. (8.35) can be written as

$$N_d = \frac{K_{tr} D^2}{4f^2} \epsilon E(T_t) \tag{8.36}$$

The output voltage of the overall instrument is $K_{dr} K_a N_d$, where $K_{dr} \triangleq$ detector responsivity, volts/[photons/(cm²-sec)], and K_a is the amplifier gain, volts/volt. Thus

$$\text{Instrument output voltage} \triangleq e_o = \frac{K_{dr} K_a K_{tr} D^2}{4f^2} \epsilon E(T_t) \tag{8.37}$$

Note that the first factor in Eq. (8.37) is a constant of the instrument while the second $[\epsilon E(T_t)]$ is a function of target temperature and emittance only. Also, as long as the target is focused, the reading is independent of the distance from the target. The variation of $E(T_t)$ with T_t for a blackbody target can be found by experimental calibration of the overall instrument. Its general shape is shown in Fig. 8.39. Since Eq. (8.27) shows that the total photon flux varies as T^3 and since the detector output is roughly proportional to total flux, the instrument output signal varies approximately as T^3; thus, for a graybody

$$e_o \approx K\epsilon T^3 \qquad (8.38)$$

Since the instruments are calibrated against blackbody sources, for non-blackbodies a value of ϵ must be known to find T. If the value of ϵ used is in error, an error in T will result. Because of the third power law, however, errors in ϵ do not cause proportionate errors in T. Rather

$$\frac{T_{\text{actual}}}{T_{\text{assumed}}} = \left(\frac{\epsilon_{\text{assumed}}}{\epsilon_{\text{actual}}}\right)^{\frac{1}{3}} \qquad (8.39)$$

Thus if we assume $\epsilon = 0.8$ when it is really 0.6, the temperature error is only 10 percent.

An instrument[1] of the above general class which, however, uses mirror optics has a range from 100 to 2000°F (8000°F with calibrated aperture), focusing range 4 ft to infinity (18 to 60 in. optional), 0.5° field of view, output signal 10 mv full scale, and an overall time constant (in chopped mode) of 2, 20, or 200 msec. For the study of very rapid transients, the chopper may be turned off and the system operated with just the detector and a-c amplifier. Transients as brief as 10 μsec may thus be measured. Other models of this manufacturer, using lens optics and lead sulfide detectors, have target sizes as small as 0.009 in. at a target distance of 2.8 in.

Another instrument[2] of this class accepts radiation only in the wavelength band 4.8 to 5.6 μ. This band avoids the absorption bands of atmospheric water vapor and carbon dioxide, thus removing the effect of these variables on instrument response. The measurement of the surface temperature of glass is also facilitated since in this spectral range the emittance of glass is high and independent of thickness. This instrument has a range of 100 to 1000°F, target size equal to its distance divided by 57, time constant 0.2 to 0.5 sec, focusing range 17 in. to infinity, calibration accuracy 2 percent of range or 20°F (whichever is larger), resolution 0.25 percent of range or 3°F (whichever is larger), and repeatability of 0.5 percent of range or 10°F (whichever is larger).

***Monochromatic-brightness radiation thermometers* (*optical pyrometers*).** The classical form of this type of instrument is the disappearing-filament optical pyrometer. It is the most accurate of all the radiation thermometers; however, it is limited to temperatures greater than about 700°C since it requires a visual brightness match by a human operator. This instrument is used to realize the International Practical Temperature Scale above 1063°C.

[1] Infrared Industries, Santa Barbara, Calif.
[2] Ircon, Inc., Chicago, Ill.

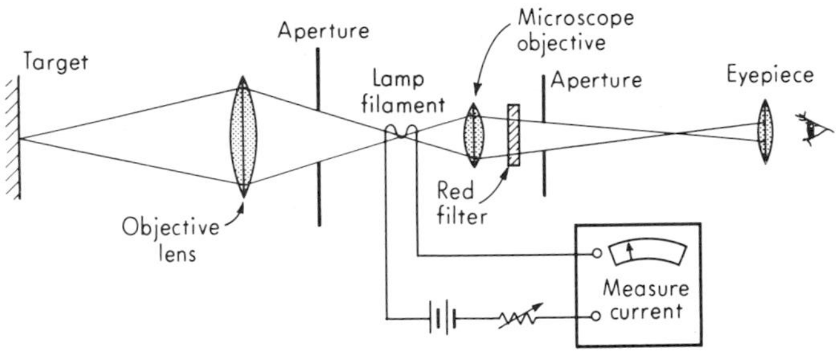

Fig. 8.40. *Disappearing-filament optical pyrometer.*

Monochromatic-brightness thermometers utilize the principle that, at a given wavelength λ, the radiant intensity ("brightness") varies with temperature as given by Eq. (8.15). In a disappearing-filament instrument (Fig. 8.40) an image of the target is superimposed on a heated tungsten filament. This tungsten lamp, which is very stable, has been previously calibrated so that when the current through the filament is known the brightness temperature of the filament is known. (Such a calibration is basically obtained by visually comparing the brightness of a blackbody source of known temperature with that of the tungsten lamp.) A red filter which passes only a narrow band of wavelengths around 0.65 μ is placed between the observer's eye and the tungsten lamp and target image. The observer controls the lamp current until the filament disappears in the superimposed target image. Then the brightness of the target and lamp are equal, and one can write

$$\frac{\epsilon_{\lambda_e} C_1}{\lambda_e{}^5(e^{C_2/\lambda_e T_t} - 1)} = \frac{C_1}{\lambda_e{}^5(e^{C_2/\lambda_e T_L} - 1)} \tag{8.40}$$

where $\epsilon_{\lambda_e} \triangleq$ emittance of target at wavelength λ_e
$\lambda_e \triangleq$ effective wavelength of filter, usually 0.65 μ
$T_t \triangleq$ target temperature
$T_L \triangleq$ lamp brightness temperature

For T less than about 4000°C, the terms $e^{C_2/\lambda_e T}$ are much greater than 1, allowing Eq. (8.40) to be simplified to

$$\frac{\epsilon_{\lambda_e}}{e^{C_2/\lambda_e T_t}} = \frac{1}{e^{C_2/\lambda_e T_L}} \tag{8.41}$$

Then

$$\epsilon_{\lambda_e} = e^{-(C_2/\lambda_e)(1/T_L - 1/T_t)} \tag{8.42}$$

and finally

$$\frac{1}{T_t} - \frac{1}{T_L} = \frac{\lambda_e \ln \epsilon_{\lambda_e}}{C_2} \tag{8.43}$$

If the target is a blackbody ($\epsilon_{\lambda_e} = 1.0$), there is no error since $\ln \epsilon_{\lambda_e} = 0$ and $T_t = T_L$. If ϵ_{λ_e} is not 1.0 but is known, Eq. (8.43) allows calculation of the needed correction. The errors caused by inexact knowledge of ϵ_{λ_e} for a particular target are not as great for an optical pyrometer as for an instrument sensitive to a wide band of wavelengths. The percent error is given by

$$\frac{dT_t}{T_t} = -\frac{\lambda_e T_t}{C_2} \frac{d\epsilon_{\lambda_e}}{\epsilon_{\lambda_e}} \tag{8.44}$$

Thus, for a target at 1000°K, a 10 percent error in ϵ_{λ_e} results in only a 0.45 percent error in T_t. The use of a monochromatic red filter aids the operator in matching the brightness of target and lamp since color effects are eliminated. Also, the target emittance need be known only at one wavelength. If ϵ_{λ_e} is exactly known, temperatures can be measured with optical pyrometers with errors of the order of 3C° at 1000°C, 6C° at 2000°C, and 40C° at 4000°C. With special optical systems, targets as small as 0.001 in. can be measured at distances of 5 or 6 in.

Because of its manual null-balance principle, the optical pyrometer is not usable for continuous-recording or automatic-control applications. To overcome this drawback, automatic brightness pryometers[1,2,3] have been developed. One model[3] of such a device uses a mirror-chopper arrangement to produce a square wave of radiation flux in which the target radiation and standard lamp radiation are alternately applied to a photomultiplier tube. The standard lamp is left at a fixed brightness; thus a null method is not used. A red filter with $\lambda_e = 0.653\ \mu$ is employed. This instrument has a range of 700 to 3000°C (though not in a single instrument), inaccuracy of 1 percent of span plus 0.3 percent of measured temperature, repeatability 0.3 percent of measured temperature, resolution 1°F, time constant 0.3 sec, target size 0.4 in. at 18 in., and a recorder output of 50 or 100 mv full scale. The above specifications refer to an instrument for commercial use; refined models for standards laboratories are expected to exceed the performance of manual pyrometers shortly and will probably replace them for the most accurate work.

Another class of brightness pyrometer which does not use a reference lamp source is also available. They are merely chopper-type radiation thermometers using photon detectors and narrow-band optical filters. One such instrument[4] has a range of 1400 to 8300°F, 0.5° field of view, focusing range 24 in. to infinity, time constant 0.1 sec (0.01 sec optional), repeatability 0.25 percent of span, inaccuracy 1 percent of span, and

[1] S. Ackerman and J. S. Lord, Automatic Brightness Pyrometer Uses a Photomultiplier "Eye," *ISA J.*, p. 48, December, 1960.

[2] J. S. Lord, Brightness Pyrometry, *Instr. Control Systems,* p. 109, February, 1965.

[3] Instrument Development Laboratories, Attleboro, Mass., *Bull.* 614.

[4] Infrared Industries, Santa Barbara, Calif.

recorder output 10 to 100 mv full scale. This instrument uses a filter centered at 0.80 μ with a bandwidth of 0.06 μ.

Two-color radiation thermometers. Since errors due to inaccurate values of emittance are a problem in all radiation-type temperature measurements, considerable attention has been given to possible schemes for alleviating this difficulty. Although no universal solution has been found, the two-color concept has met with some practical success. The basic concept requires that W_λ be determined at two different wavelengths and then the ratio of these two W_λ's be taken as a measure of temperature. For the usual conditions of practical application, the terms $e^{C_2/\lambda T}$ are much greater than 1.0, and we may write with close approximation

$$W_{\lambda 1} = \frac{\epsilon_{\lambda_1} C_1}{\lambda_1^5 e^{C_2/\lambda_1 T}} \tag{8.45}$$

$$W_{\lambda 2} = \frac{\epsilon_{\lambda_2} C_1}{\lambda_2^5 e^{C_2/\lambda_2 T}}$$

Then

$$\frac{W_{\lambda 1}}{W_{\lambda 2}} = \frac{\epsilon_{\lambda_1}}{\epsilon_{\lambda_2}} \left(\frac{\lambda_2}{\lambda_1}\right)^5 e^{(C_2/T)(1/\lambda_2 - 1/\lambda_1)} \tag{8.46}$$

For a graybody, $\epsilon_{\lambda_1} = \epsilon_{\lambda_2}$; thus

$$\frac{W_{\lambda 1}}{W_{\lambda 2}} = \left(\frac{\lambda_2}{\lambda_1}\right)^5 e^{(C_2/T)(1/\lambda_2 - 1/\lambda_1)} \tag{8.47}$$

and we see that the ratio $W_{\lambda 1}/W_{\lambda 2}$ is independent of emittance as long as it is numerically the same at λ_1 and λ_2. The wavelengths λ_1 and λ_2 are usually both in the visible range and are generally varied depending on the temperature range of the particular instrument. In one commercial instrument[1,2] the two filters are mounted on a rotating wheel so that the incoming radiation passes alternately through each on its way to a photon detector. Special electronic circuitry performs operations equivalent to taking the ratio of $W_{\lambda 1}$ and $W_{\lambda 2}$. Instruments covering the range 1400 to 4000°F are available as standard models. A recorder output of 100 mv full scale is provided.

8.6

A Digital Temperature-sensing System

While the analog outputs of the various temperature sensors described up to this point can be transformed to digital form by using any of several available analog-to-

[1] T. P. Murray and V. G. Shaw, Two-color Pyrometry in the Steel Industry, *ISA J.*, p. 36, December, 1958.

[2] Shaw Instrument Corp., Pittsburgh, Pa.

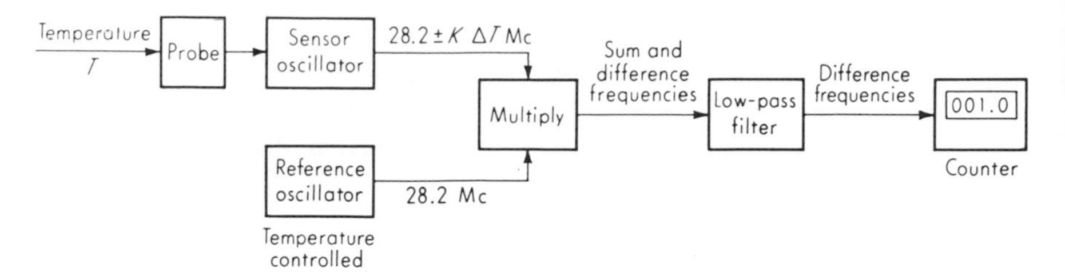

Fig. 8.41. *Digital thermometer.*

digital conversion techniques, we here describe briefly a system available as a package which accepts temperature inputs and provides digital signals and/or displays as outputs.

Electronic oscillators using piezoelectric quartz crystals as the resonant element that establishes the frequency of oscillation have been widely used for many years. For the most critical applications it has been necessary to place the crystal in a temperature-controlled oven, since the natural frequency of the crystal varies with temperature, causing drifts in oscillator frequency. This difficulty is turned to good advantage in the quartz thermometer where the crystal is placed in a probe which serves as a temperature-sensing element. Changes in probe temperature cause a frequency change in proportion. By applying the oscillator voltage to an electronic counter for a definite time interval, a direct digital reading of temperature is obtained.

A commercial version[1] of the above principle has a block diagram as in Fig. 8.41. The crystals used have a temperature coefficient of 35.4 ppm/C° which is constant within ± 0.05 percent of range over the span -40 to $+230$°C, giving a very linear response. If the sensor oscillator is designed for a frequency of 28.2 Mc when the probe is at 0°C, a 1C° change in temperature will cause a frequency change of 1,000 cps. Were this frequency change applied to a four-digit electronic counter for a sample period of 0.01 sec, the counter would read 001.0°C, and such an arrangement would have a resolution of 0.1C°. To obtain the frequency *change* as a usable signal, the sensor oscillator signal is multiplied with the signal of a 28.2-Mc reference oscillator which is temperature-controlled and thus of fixed frequency. The output of the multiplier (amplitude modulator) contains the sum and difference frequencies of the two input signals. By filtering out the high-frequency (sum) component, only the difference frequency signal remains and may be sent to the counter.

Probes used in the above system are $\frac{3}{8}$ in. in diameter and either $\frac{3}{4}$ or

[1] Hewlett-Packard Co., Dymec Div., Palo Alto, Calif.

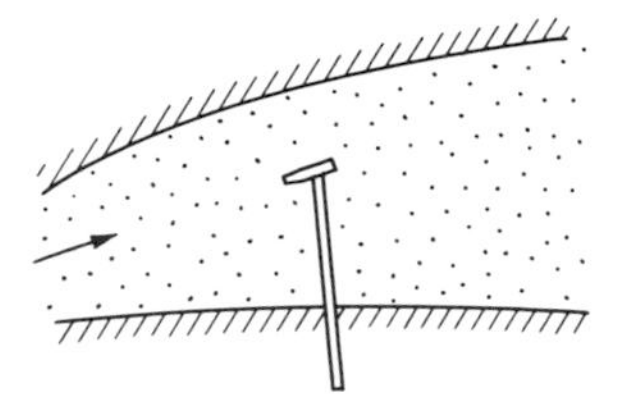

Fig. 8.42. *Probe configuration.*

9 in. long, with a time constant of 1 sec in water with a velocity of 2 fps. System resolution can be increased to 0.0001C° by increasing the sampling period to 10 sec. Differential temperature can be measured by multiplying the signals from two probes rather than one probe and the reference oscillator. Self-heating of the probe is 10 μw and gives less than 0.01C° error in water at 2 fps. A binary-coded-decimal output signal is also available.

8.7

Temperature-measuring Problems in Flowing Fluids In attempting to measure the static temperature of flowing fluids (particularly high-speed gas flows), one encounters certain types of problems irrespective of the particular sensor being used. These have to do mainly with errors caused by heat transfer between the probe and its environment and the problem of measuring static temperature of a high-velocity flow with a stationary probe.

Conduction error. Let us consider the so-called conduction error first. Figure 8.42 shows a common situation. A probe has been inserted into a duct or other flow passage and is supported at a wall. In general, the wall will be hotter or colder than the flowing fluid; thus there will be heat transfer, and this leads to a probe temperature different from that of the fluid. We shall analyze a simplified model of this arrangement to find what measures can be taken to reduce and/or correct for the error to be expected in such a case. Figure 8.43 shows the simplified model to be used in analyzing this situation. A slender rod extends a distance L from the wall. We assume the rod temperature T_r is a function of x only; it does not vary with time or over the rod cross section at a given x. A fluid of constant and uniform temperature T_f completely surrounds the rod and exchanges heat with it by convection. For a steady-state situation

$$\text{Heat in at } x = (\text{heat out at } x + dx) + (\text{heat loss at surface})$$

$$q_x = q_{(x+dx)} + q_l$$

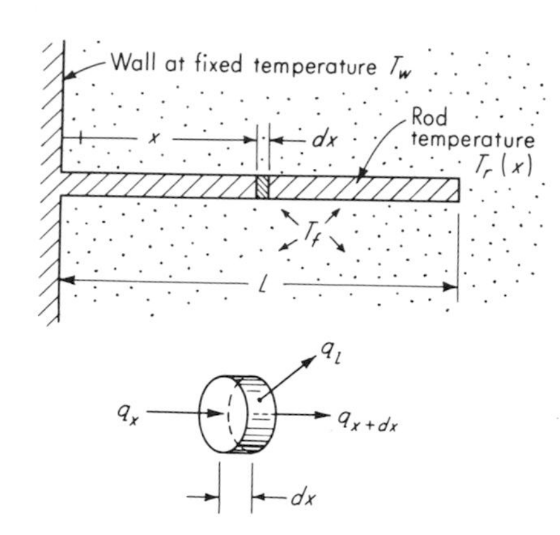

Fig. 8.43. *Conduction-error analysis.*

One-dimensional conduction heat transfer gives

$$q_x = -kA\frac{dT_r}{dx} \tag{8.48}$$

where $k \triangleq$ thermal conductivity of rod
$A \triangleq$ cross-section area

Then

$$q_{(x+dx)} = q_x + \frac{d}{dx}(q_x)\,dx = -kA\frac{dT_r}{dx} + \frac{d}{dx}\left(-kA\frac{dT_r}{dx}\right)dx \tag{8.49}$$

Now if k and A are assumed constant

$$q_{(x+dx)} = -kA\frac{dT_r}{dx} - kA\frac{d^2T_r}{dx^2}\,dx \tag{8.50}$$

We assume the heat loss by convection at the surface to be given by

$$q_l = h(C\,dx)(T_r - T_f) \tag{8.51}$$

where $h \triangleq$ film coefficient of heat transfer
$C \triangleq$ "circumference" of rod (it need not be circular)
$C\,dx =$ surface area

We then have

$$\frac{d^2T_r}{dx^2} - \frac{hC}{kA}T_r = -\frac{hC}{kA}T_f \tag{8.52}$$

If we take h and C as being constants, Eq. (8.52) is a linear differential equation with constant coefficients and is readily solved for T_r as a function of x. We need two boundary conditions to accomplish this. Clearly, $T_r = T_w$ at $x = 0$ is one such condition. The simplest assumption at $x = L$ is an insulated end; this gives $dT_r/dx = 0$ at $x = L$. Even if the end is not insulated, if L is quite large we intuitively see that the variation of T_r with x must be as in Fig. 8.44; thus $dT_r/dx \approx 0$ for $x = L$. Using

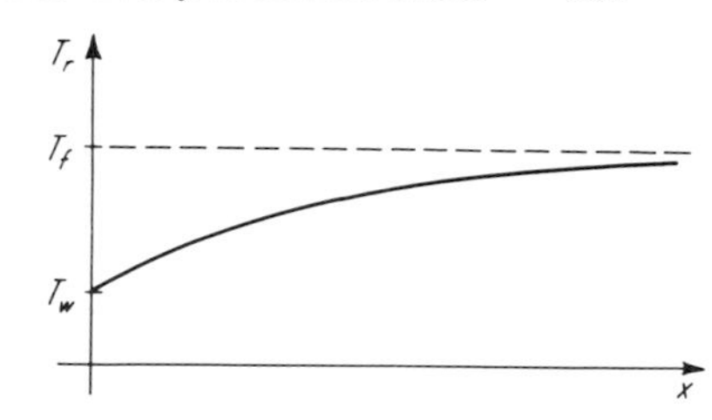

Fig. 8.44. *Temperature profile.*

these two boundary conditions with Eq. (8.52) gives

$$T_r = T_f - (T_f - T_w)\left[\left(1 - \frac{e^{mL}}{2\cosh mL}\right)e^{mx} + \frac{e^{mL}}{2\cosh mL}e^{-mx}\right] \tag{8.53}$$

where

$$m \triangleq \sqrt{\frac{hC}{kA}} \tag{8.54}$$

Since the temperature-sensing element (thermocouple bead, thermistor, etc.) is generally located at $x = L$, we evaluate Eq. (8.53) there to get

$$\text{Temperature error} = T_r - T_f = \frac{T_w - T_f}{\cosh mL} \tag{8.55}$$

Equation (8.55) may be used in two ways: to indicate how to design a probe support to minimize the error and also to allow one to calculate and correct for whatever error there might be. It is clear that the error is reduced if T_w is close to T_f. Insulating or controlling the temperature of the wall encourages this. The term cosh mL will be large if m and L are large. Thus the probe should be immersed (L is called the immersion length) as far as practical. We see that, to make m large, h should be large (high rate of convection heat transfer) and k should be small (the probe support made of insulating material). The term C/A depends on the shape of the rod. For the usual circular cross section, $C/A = 2/r$, where r is the rod radius. Thus we see that the rod should be of small cross section to reduce error.

If the boundary condition at $x = L$ is changed to a more realistic (and complicated) one in which there is convection heat transfer with a film coefficient h_e at the end, one gets at $x = L$

$$T_r - T_f = \frac{T_w - T_f}{\cosh (mL) + (h_e/mk)\sinh mL} \tag{8.56}$$

Since the error predicted by (8.56) is less than that of (8.55), the use of the simpler relation is conservative.

Radiation error. Additional error is caused by radiant-heat exchange between the temperature probe and its surroundings. This occurs

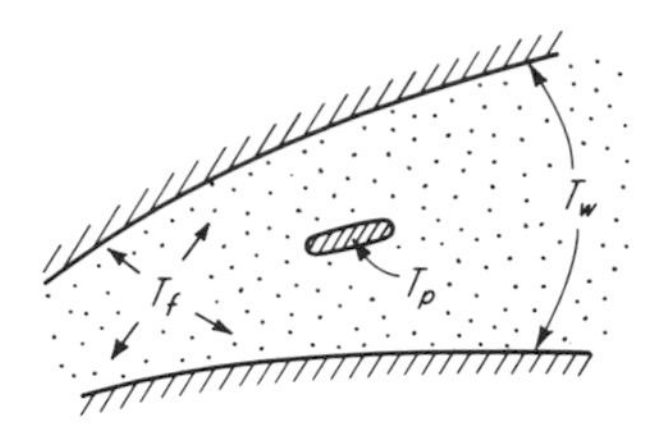

Fig. 8.45. *Radiation-error analysis.*

simultaneously with the previously studied conduction losses but we here consider it separately for simplicity. We also assume radiation exchange only between the probe and the surrounding walls, neglecting radiation of the gas itself or the absorption by the gas of radiation passing through it. Neglecting conduction losses, we may consider the probe as in Fig. 8.45. For steady-state conditions

$$\text{Heat convected to probe} = \text{net heat radiated to wall} \tag{8.57}$$

$$hA_s(T_f - T_p) = 0.174\epsilon_p A_s \left[\left(\frac{T_p}{100}\right)^4 - \left(\frac{T_w}{100}\right)^4\right] \tag{8.58}$$

where $h \triangleq$ film coefficient at probe surface, Btu/(hr-ft²-°F)
$A_s \triangleq$ probe surface area
$\epsilon_p \triangleq$ emittance of probe surface
$T_p \triangleq$ probe absolute temperature, °R
$T_w \triangleq$ wall absolute temperature, °R

Equation (8.58) assumes the radiation configuration described as "a small body completely enclosed by a larger one." The error due to radiation is given by

$$\text{Temperature error} = T_p - T_f = \frac{0.174\epsilon_p}{h}\frac{T_w^4 - T_p^4}{10{,}000} \tag{8.59}$$

By insulating the wall or controlling its temperature, error can be reduced by making the difference between T_w and T_p as small as possible. A probe surface of low emittance ϵ_p (a shiny surface) will further reduce such errors, as will a high value of heat-transfer coefficient h. To obtain a high value of h when the fluid velocity is low, the aspirated type of probe may be used. Here a high local velocity is induced at the probe by connecting a vacuum pump into the probe tubing. Equation (8.59) may also be used to calculate corrections if numerical values of the needed quantities are available.

Probes with some form or another of radiation shield are widely used to reduce radiation errors. Figure 8.46 shows a probe with a single shield. The principle of all radiation shields is to interpose between the probe and the wall a body (the shield) whose temperature is closer to the fluid temperature than is the wall. Thus the probe "sees" the shield rather than

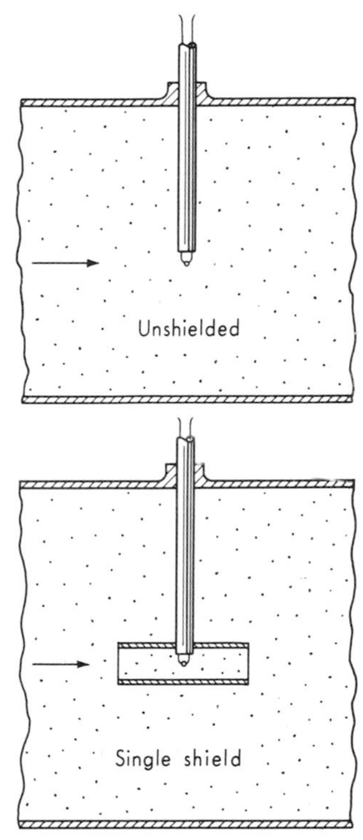

Fig. 8.46. *Radiation shielding.*

the wall, and if the shield is close to fluid temperature the probe radiation-heat loss will be small. It is easy to show that, for pure-radiation-heat transfer, interposing a single screen between a body and its surroundings will reduce the heat loss to one-half the former value, since the screen comes to a temperature $T^4_{\text{screen}} = (T^4_{\text{body}} + T^4_{\text{surroundings}})/2$. For n screens the heat loss is reduced to $1/(n + 1)$ of the unscreened value. All these results are for the simplest case where the screens and surroundings completely enclose the body. For actual probe shields, various geometrical and emittance factors complicate the situation. Also, additional heating of the shield by convection raises its temperature and reduces probe error. Experimental tests[1] with concentric circular cylinder shields (Fig. 8.47) have shown the following:

1. A significant decrease in error may be achieved by adding more shields, at least up to about four.
2. Little is gained by increasing the length/diameter ratio beyond 4:1 in attempting to reduce the unshielded angles at the open ends.

[1] W. J. King, Measurement of High Temperatures in High-velocity Gas Streams, *Trans. ASME,* p. 421, July, 1943.

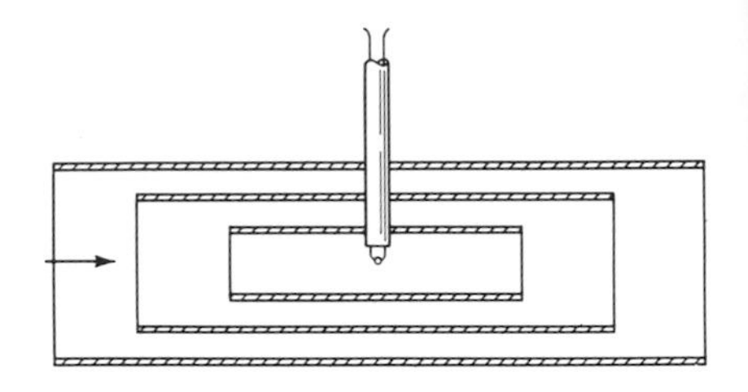

Fig. 8.47. *Multiple radiation shield.*

3. For multiple shields the spacing between shields must be sufficiently large to prevent excessive conduction heat transfer between shields and to allow high enough flow velocity for good convection from gas to shields. A double shield with only $\frac{1}{32}$-in. spacing acted almost like a single shield.

To illustrate the need for shielding, an unshielded probe exposed to 1800°F gas flow at 270 $lb_m/(ft^2\text{-min})$ may exhibit an error of 160F°. A suitable quadruple shield can reduce this to about 20F°.

Another shielding technique employs an electrically heated shield with an additional temperature sensor fastened to the shield. The heat input to the shield is adjusted until the probe sensor and shield sensor register identical temperatures. At this point, probe and shield should both be at the fluid temperature, with the heat loss from the shield to the cooler wall being replaced by the shield heater.

Velocity effects. It is often necessary to determine the static temperature of a flowing gas, since its physical properties depend on this temperature. To measure this temperature directly with a probe, however, requires that the probe be stationary with respect to the fluid; thus it must be moving at the same velocity as the fluid. Since this is usually impractical, various indirect methods of measuring static temperatures are in use. If one can measure static pressure and either density, sound velocity, or index of refraction, formulas allow calculation of static temperature. Experimental techniques based on each of these principles have been developed. However for routine measurements a different approach is generally employed. This involves placing a stationary probe in the stream and calculating the static temperature from the readings of this probe, using suitable corrections. Ideally, if a perfect gas is decelerated from free-stream velocity to zero velocity adiabatically (not necessarily isentropically) the temperature rises from the free-stream static temperature T_{stat} to the so-called stagnation or total temperature T_{stag}, where

$$\frac{T_{stag}}{T_{stat}} = 1 + \frac{\gamma - 1}{2} N_m{}^2 \qquad (8.60)$$

where $\gamma \triangleq c_p/c_v$, ratio of specific heats
$N_m \triangleq$ Mach number
$T_{stag}, T_{stat} \triangleq$ absolute temperatures

This result holds for both subsonic and supersonic flow because the shock wave that forms ahead of a probe in supersonic flow affects only the entropy and not the total enthalpy of the gas. For air, Eq. (8.60) becomes

$$\frac{T_{stag}}{T_{stat}} = 1 + 0.2N_m^2 \qquad (8.61)$$

and if $N_m < 0.22$, T_{stag} is within 1 percent of T_{stat}. Thus for sufficiently low velocities a stationary probe can be used to read static temperature directly. For higher Mach numbers, T_{stat} can be calculated from Eq. (8.60) if γ and N_m are known. Measurement of Mach numbers with a pitot-static tube is discussed in Chap. 7.

Unfortunately, real temperature probes do not attain the theoretical stagnation temperature predicted by Eq. (8.60). Even if the conduction and radiation errors discussed earlier in this section are corrected for, there remain further deviations of the actual situation from the assumed ideal. Correction for these effects is generally accomplished by experimental calibration to determine the recovery factor r of the particular probe. This is defined by

$$r \triangleq \frac{T_{stag,ind} - T_{stat}}{T_{stag} - T_{stat}} \qquad (8.62)$$

where $T_{stag,ind} \triangleq$ temperature actually indicated by probe

If r is assumed to be a known number for a given probe, combination of Eqs. (8.60) and (8.62) gives

$$T_{stat} = \frac{T_{stag,ind}}{1 + r[(\gamma - 1)/2]N_m^2} \qquad (8.63)$$

A probe that measures T_{stag} exactly would have a recovery factor of 1.0 while one that measures T_{stat} exactly would have $r = 0$.

A possible apparatus[1] for determination of r is shown in Fig. 8.48. The flow velocity in the stagnation chamber is $\frac{1}{100}$ of the nozzle flow velocity; thus measurement of tank temperature and pressure is accurately carried out under essentially zero velocity conditions. By careful design to minimize friction and heat transfer, the nozzle can be made to provide an almost perfect isentropic expansion. The validity of this assumption has been checked experimentally. For an isentropic process

[1] H. C. Hottel and A. Kalitinsky, Temperature Measurements in High-velocity Air Streams, *J. Appl. Mech.*, p. A-25, March, 1945.

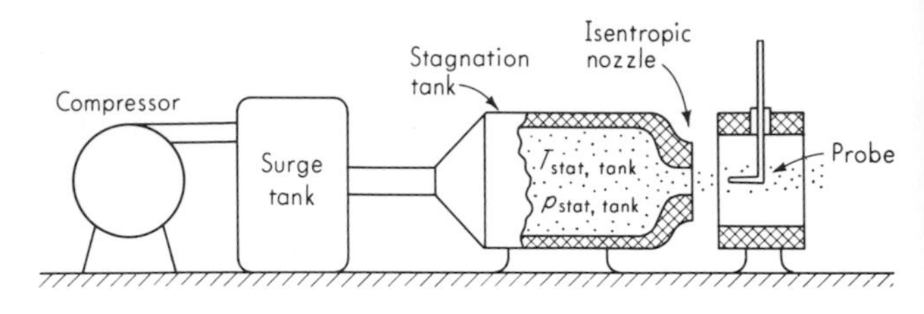

Fig. 8.48. *Recovery-factor calibration setup.*

from tank to nozzle,

$$T_{stag,\text{nozzle}} = T_{stat,\text{tank}} \tag{8.64}$$

and

$$p_{stag,\text{nozzle}} = p_{stat,\text{tank}} \tag{8.65}$$

In a free jet

$$p_{stat,\text{nozzle}} = p_{\text{atmosphere}} \tag{8.66}$$

The nozzle Mach number N_m can now be computed from the standard pitot-tube formulas since $p_{stat,\text{nozzle}}$ and $p_{stag,\text{nozzle}}$ are both known. However, the actual use of a pitot tube (with its attendant errors) is avoided since $p_{stag,\text{nozzle}}$ is obtained by measurement of $p_{stat,\text{tank}}$, and $p_{stat,\text{nozzle}}$ is obtained from a barometer reading of $p_{\text{atmosphere}}$. Once N_m is known, $T_{stat,\text{nozzle}}$ can be computed from

$$T_{stat,\text{nozzle}} = \frac{T_{stag,\text{nozzle}}}{1 + [(\gamma - 1)/2]N_m^2} = \frac{T_{stat,\text{tank}}}{1 + [(\gamma - 1)/2]N_m^2} \tag{8.67}$$

The reading of the probe itself supplies $T_{stag,ind}$. Thus one can now compute r from its definition

$$r = \frac{T_{stag,ind} - T_{stat,\text{nozzle}}}{T_{stag,\text{nozzle}} - T_{stat,\text{nozzle}}} \tag{8.68}$$

For bare thermocouple sensors the numerical value of r usually lies in the range 0.6 to 0.9, depending on the form of the junction (butt-welded, twisted, or spherical bead) and the orientation (wire parallel to flow or transverse to flow). To get a high value of r and one relatively independent of flow conditions such as velocity magnitude and direction, sensors (usually thermocouples or resistance thermometers) are built into probes that have been specifically designed to approach ideal stagnation conditions. Figure 8.49 shows two examples[1] of such probes.

[1] R. W. Ladenburg et al., "Physical Measurements in Gas Dynamics and Combustion," p. 186, Princeton University Press, Princeton, N.J., 1954.

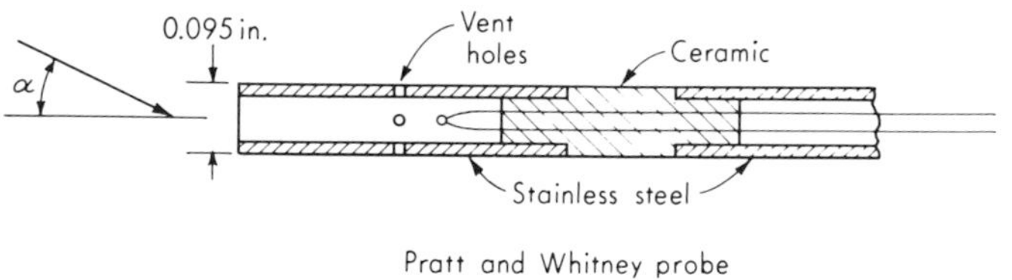

Pratt and Whitney probe

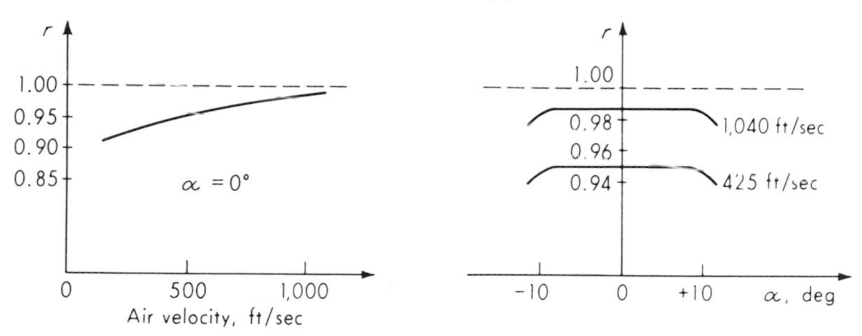

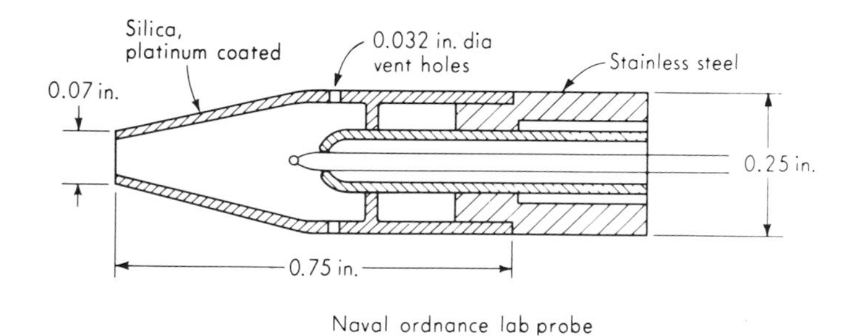

Naval ordnance lab probe

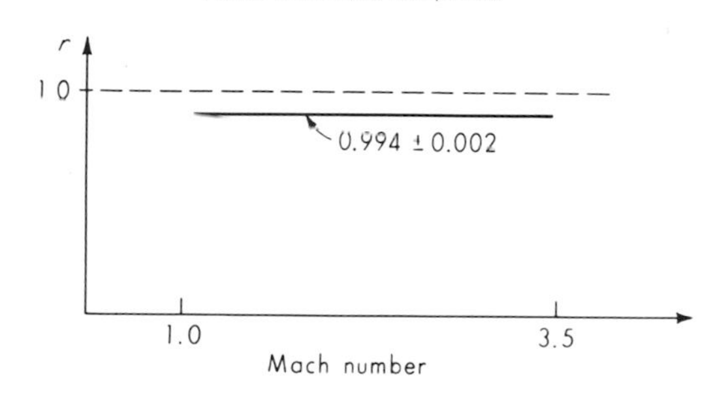

Fig. 8.49. *Stagnation-temperature probes.*

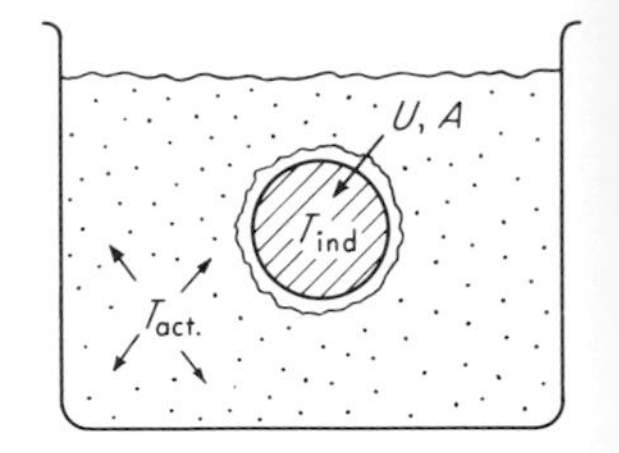

Fig. 8.50. First-order sensor model.

Desirable characteristics of probes include the following:

1. Low heat capacity in the sensing element for a fast response.
2. Conduction loss of lead wires minimized by exposing enough length of the lead to the stagnation temperature.
3. Radiation shield of low thermal conductivity and low surface emittance.
4. Vent holes provided to replenish continuously the fluid in the stagnation chamber; otherwise it would be cooled by conduction and radiation. This flow must be kept small enough, however, so that stagnation conditions are essentially preserved. The increased convection coefficient caused by the flow speeds the response and reduces the radiation error.
5. Blunt shape causes formation of a normal shock wave in supersonic flow. This increases the temperature in the boundary layer and reduces the heat loss from the probe. The shock wave also reduces the influence of misalignment.

8.8

Dynamic Response of Temperature Sensors Since the conversion from sensing-element temperature to thermal expansion, thermoelectric voltage, or electrical resistance is essentially instantaneous, the dynamic characteristics of temperature sensors are related to the heat-transfer and storage parameters that cause the sensing-element temperature to lag that of the measured medium. When a sensing element is used "bare" (not in a protective well), the model of Fig. 8.50 is often adequate. Here heat losses are neglected, resistance to heat transfer is lumped in a single element, and energy storage is lumped in a single element. Conservation of energy gives

$$UA(T_{act} - T_{ind})\,dt = MC\,dT_{ind} \tag{8.69}$$

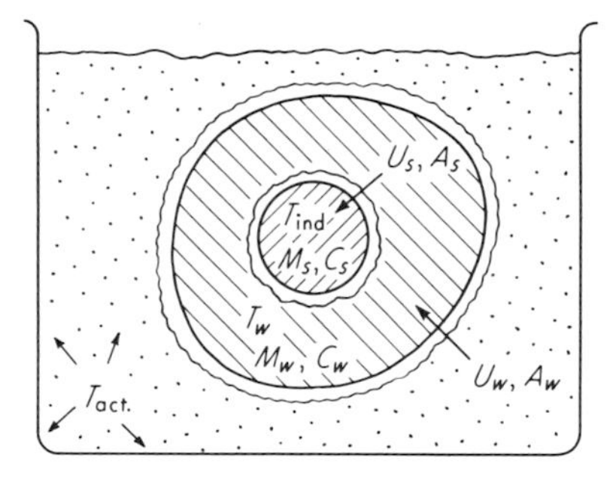

Fig. 8.51. *Second-order sensor model.*

where $U \triangleq$ overall heat-transfer coefficient, Btu/(sec-F°-in.²)
$A \triangleq$ heat-transfer area
$T_{act} \triangleq$ actual temperature of surrounding fluid
$T_{ind} \triangleq$ temperature indicated by sensor
$M \triangleq$ mass of sensing element
$C \triangleq$ specific heat of sensing element

This leads to

$$\frac{T_{ind}}{T_{act}}(D) = \frac{1}{\tau D + 1} \tag{8.70}$$

where

$$\tau \triangleq \frac{MC}{UA} \tag{8.71}$$

Clearly, speed of response may be increased by decreasing M and C and/or increasing U and A. Since U, in general, depends on the surrounding fluid and its velocity, τ is not a constant for a given sensor but rather varies with how it is used.

Since temperature sensors are often enclosed in protective wells or sheaths, a thermal model taking into account heat-transfer resistance and energy storage in the well is of practical interest. Figure 8.51 shows such a configuration. Analysis gives

$$\frac{T_{ind}}{T_{act}}(D) = \frac{1}{\tau_w\tau_s D^2 + (\tau_w + \tau_s + M_sC_s/U_wA_w)D + 1} \tag{8.72}$$

where $\tau_w \triangleq M_wC_w/U_wA_w$, time constant of well alone
$\tau_s \triangleq M_sC_s/U_sA_s$, time constant of sensor alone

We see that the addition of a well changes the form of response to second-order and increases the lag. The term M_sC_s/U_wA_w is called the coupling term between the well and the sensor. If it is small compared with $\tau_w + \tau_s$, we have approximately

$$\frac{T_{ind}}{T_{act}}(D) = \frac{1}{\tau_w\tau_s D^2 + (\tau_w + \tau_s)D + 1} = \frac{1}{\tau_w D + 1}\frac{1}{\tau_s D + 1} \tag{8.73}$$

which is just a cascade combination of the sensor and well individual dynamics.

The accuracy of the theoretical model can be increased by increasing the number of "lumps" of heat-transfer resistance and energy storage employed, the ultimate limit being an infinite number corresponding to a distributed-parameter (partial differential equation) rather than a lumped-parameter (ordinary differential equation) approach. When temperature sensors are used as measuring devices in feedback-control systems they are usually allowed to contribute no more than 30° phase lag at the frequency where the entire open-loop lag is 180°. Under such conditions they are usually adequately modeled as simple first-order systems. The best[1] time constant to use for such a model is determined by an experimental ramp-input test and is numerically the steady-state time lag observed in such a test.

When greater accuracy is needed in utilizing the results of experimental tests to determine sensor dynamics, a model[2] using three time constants and a dead time may be employed. The transfer function is then

$$\frac{T_{ind}}{T_{act}}(D) = \frac{e^{-\tau_{dt}D}}{(\tau_1 D + 1)(\tau_2 D + 1)(\tau_3 D + 1)} \tag{8.74}$$

Numerical values of τ_1, τ_2, τ_3, and τ_{dt} may be obtained from step-function response tests. For example, a thermocouple used in a heavy-duty stainless-steel well had

$$\frac{T_{ind}}{T_{act}}(D) = \frac{e^{-2.6D}}{(21.6D + 1)(2.9D + 1)(2.1D + 1)} \tag{8.75}$$

where the time constants are in seconds.

When accurate numerical values are needed, experimental tests are generally required to determine temperature probe dynamics. For simple bare thermocouples, however, extensive research and testing have provided semiempirical formulas which allow calculation of the time constant with fair accuracy. One such relation[3] useful for temperatures from 160 to 1600°F, wire diameter 0.016 to 0.051 in., mass velocity 3 to 50 $\text{lb}_m/(\text{ft}^2\text{-sec})$, and static pressure of 1 atm is

$$\tau = \frac{3{,}500\rho c d^{1.25} G^{-15.8/\sqrt{T}}}{T} \quad \text{sec} \tag{8.76}$$

[1] G. A. Coon, Responses of Temperature-sensing-element Analogs, *Trans. ASME*, p. 1857, November, 1957.

[2] J. R. Louis and W. E. Hartman, The Determination and Compensation of Temperature-sensor Transfer Functions, *ASME Paper* 64-WA/AUT-13.

[3] R. J. Moffat, How to Specify Thermocouple Response, *ISA J.*, p. 219, June, 1957.

where $\rho \triangleq$ average density of two thermocouple materials, lb_m/ft^3
$c \triangleq$ average specific heat of two thermocouple materials, Btu/(lb_m-F°)
$d \triangleq$ wire diameter, in.
$G \triangleq$ flow mass velocity, $lb_m/(ft^2\text{-sec})$
$T \triangleq$ stagnation temperature, °R

Within the above restrictions, this formula will predict τ for butt-welded junctions within about 10 percent. Another such result,[1] based on tests for a Mach-number range of 0.1 to 0.9 and a Reynolds-number range of 250 to 30,000, gives

$$\tau = \frac{4.05\rho c d^{1.50}\{1 + [(\gamma - 1)/2]N_m{}^2\}^{0.25}}{p^{0.5}N_m{}^{0.5}T^{0.18}} \tag{8.77}$$

where $\gamma \triangleq$ ratio of specific heats
$N_m \triangleq$ Mach number
$p \triangleq$ static pressure, atm

This reference also presents a comprehensive analysis of conduction and radiation errors and the effects of differences in the thermal properties of the two metals used in a thermocouple.

Dynamic compensation of temperature sensors. When environmental conditions require a rugged temperature sensor, the mass may be so high as to cause a sluggish response. It may be possible to obtain an improved overall measuring-system response by cascading an appropriate dynamic compensation device with the sensor. Such schemes have been applied in practice with considerable success and may be implemented in a number of ways.[2,3] An *RC* network such as that shown for hot-wire anemometer compensation in Chap. 7 may be used if the sensor is essentially first-order. Second order sensors may also be compensated;[3] operational-amplifier networks provide a convenient means of implementation. Since the compensation is correct only for specific sensor dynamics, changes in numerical values or form of transfer function caused by changes in operating conditions can lead to loss of compensation.

[1] M. D. Scadron and I. Warshawsky, Experimental Determination of Time Constants and Nusselt Numbers for Bare-wire Thermocouples in High-velocity Air Streams and Analytic Approximation of Conduction and Radiation Errors, *NACA, Tech. Note* 2599, 1952.

[2] C. E. Shepard and I. Warshawsky, Electrical Techniques for Compensation of Thermal Time Lag of Thermocouples and Resistance Thermometer Elements, *NACA, Tech. Note* 2703, 1952.

[3] Louis and Hartman, *op. cit.*

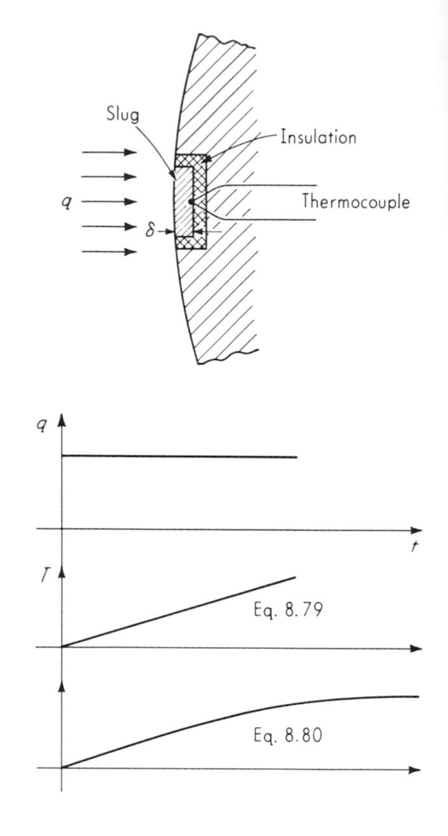

Fig. 8.52. *Slug-type heat-flux sensor.*

Increased speed of response is, in general, traded off for overall sensitivity in such compensation schemes. This can be made up by additional amplification but only up to a point; then the inherent noise level prevents further improvement. However, improvement of 100:1 or more is often possible.

8.9

Heat-flux Sensors

In recent years, requirements for measurement of local convective, radiative, or total heat-transfer rates in missile structures have led to the development of several types of heat-flux sensors. We here briefly review the operating principles and characteristics of the most common types.

Slug-type sensors. In Fig. 8.52 a slug of metal is buried in (but insulated from) the surface across which the heat-transfer rate is to be measured. Neglecting losses through the insulation and the thermocouple wires, one may write

$$\text{Heat transferred in} = \text{energy stored}$$

$$Aq\,dt = Mc\,dT \tag{8.78}$$

where $A \triangleq$ surface area of slug, in.^{2}
$q \triangleq$ local heat-transfer rate, Btu/(sec-in.^{2})
$M \triangleq$ mass of slug, lb$_m$
$c \triangleq$ specific heat of slug, Btu/(lb$_m$-F°)
$T \triangleq$ slug temperature, °F

Then

$$q = \frac{Mc}{A}\frac{dT}{dt} \tag{8.79}$$

and thus q may be determined by measuring dT/dt if Mc/A is known. Since the thermocouple reads T rather than dT/dt, a graphical, numerical, or electrical differentiation must be carried out to get q. For greater accuracy, the heat losses may be taken into account by modifying Eq. (8.79) to give

$$q = \frac{Mc}{A}\frac{dT}{dt} + K_l\,\Delta T \tag{8.80}$$

where $K_l \triangleq$ loss coefficient, Btu/(in.^{2}-sec-F°)
$\Delta T \triangleq$ temperature difference between slug and casing (usually taken as temperature rise of slug by assuming constant casing temperature)

The numerical values of Mc/A and K_l for a given sensor are determined by calibration and supplied by the manufacturer. Equation (8.79) predicts that, for a constant q, T increases linearly with time and without limit. Actually, the unavoidable heat losses eventually make dT/dt approach zero, as shown by the more correct Eq. (8.80).

The analysis of Eq. (8.78) assumes the slug is at all times at uniform temperature T throughout. This is not actually the case; thus there is a time-lag effect which has been evaluated[1] on the basis of a step input of q. A partial-differential-equation analysis leads to

$$q_m = q(1 - 2e^{-\pi^2 \alpha t/\delta^2}) \tag{8.81}$$

where $q_m \triangleq$ measured flux, using temperature at back surface of slug
$q \triangleq$ actual flux
$\alpha \triangleq$ thermal diffusivity $= k/\rho c$
$\delta \triangleq$ slug thickness
$k \triangleq$ thermal conductivity
$\rho \triangleq$ mass density

We see that a fast response requires a small value of $\delta^2 \rho c/k$.

[1] Heat Technology Laboratory, Inc., Huntsville, Ala., *Rept. HTL*-ER-4, p. 4, 1962.

Since the materials of which the sensor is made can withstand only a certain maximum temperature rise ΔT_{max}, a slug can be exposed to a given heat-transfer rate q for only a limited time t_{max}. Neglecting losses, Eq. (8.79) can be integrated to give the slug thickness δ required for a given q, ΔT_{max}, and t_{max} as

$$\delta = \frac{q t_{max}}{\rho c \, \Delta T_{max}} \tag{8.82}$$

For convective heat transfer from a gas of fixed temperature T_g, the heat flux into the slug is $h(T_g - T)$, where h is the film coefficient. Neglecting losses, we may write

$$h(T_g - T)\, dt = \rho c \delta \, dT \tag{8.83}$$

which leads to

$$\delta = \frac{h t_{max}}{\rho c \log_e \left| \dfrac{1}{1 - \Delta T_{max}/(T_g - T_i)} \right|} \tag{8.84}$$

where T_i is the initial slug temperature.

A more refined analysis[1] for the case of constant q (which takes into account that the *front* surface will overheat before the back) shows that there is an optimum value of δ in the sense that the linear (steady-state) part of the rear-surface response is the longest possible before the front surface overheats. This optimum value of δ is given by

$$\delta_{opt} = \frac{k T_{f,max}}{1.366 q} \tag{8.85}$$

where $T_{f,max}$ is the maximum allowable front-surface temperature. The time interval of linear response is found to be

$$\Delta t_{linear} = \frac{0.366 k^2 T_{f,max}^2}{\alpha q^2} \tag{8.86}$$

Figure 8.53 illustrates these concepts.

Steady-state or asymptotic sensors (Gardon gage). Figure 8.54 shows the essential features of this type of sensor which was first proposed by R. Gardon.[2] A thin constantan disk is connected at its edges to a large copper heat sink while a very thin (< 0.005-in.-diameter) copper

[1] R. H. Kirchhoff, Calorimetric Heating-rate Probe for Maximum-response-time interval, *AIAA J.*, p. 967, May, 1964.

[2] R. Gardon, An Instrument for the Direct Measurement of Intense Thermal Radiation, *Rev. Sci. Instr.*, p. 366, May, 1953.

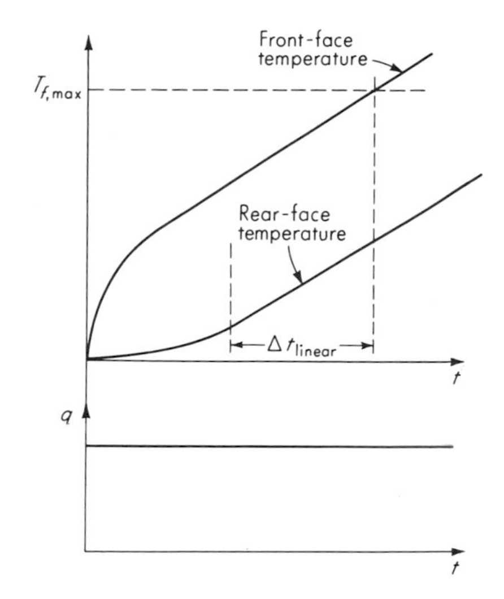

Fig. 8.53. *Slug-type-sensor response.*

wire is fastened at the center of the disk. This forms a differential thermocouple between the disk center and its edges. When the disk is exposed to a constant heat flux, an equilibrium temperature difference is rapidly established which is proportional to the heat flux. Since the thermocouple signal is now directly proportional to the heat flux, no differentiating process (such as is required in a slug-type sensor) is necessary. Furthermore, loss corrections are generally not needed nor is a thermocouple reference junction required. Instrument response[1] is

[1] Heat Technology Laboratory, *Rept. HTL*-ER-4, p. 4.

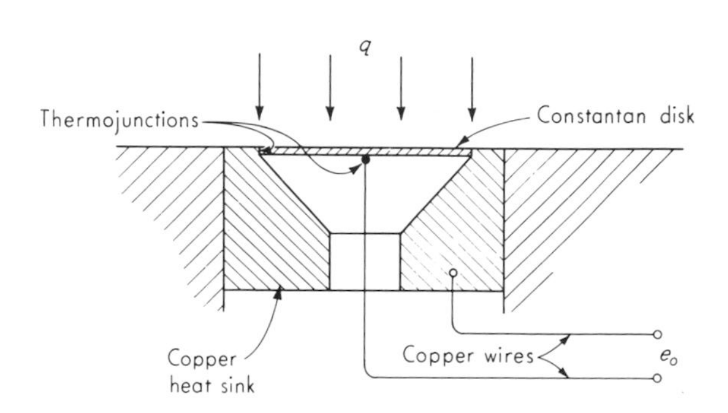

Fig. 8.54. *Gardon gage.*

approximately of the first-order type; thus

$$\frac{e_o}{q}(D) = \frac{K}{\tau D + 1} \tag{8.87}$$

where

$$K \triangleq \frac{d^2 K_e}{16\delta k} \tag{8.88}$$

$$\tau \triangleq \frac{\rho c d^2}{16k} \tag{8.89}$$

and

$d \triangleq$ diameter of disk
$\delta \triangleq$ thickness of disk
$K_e \triangleq$ thermocouple sensitivity, mv/F°
$k \triangleq$ thermal conductivity of disk
$c \triangleq$ specific heat of disk

For copper/constantan, numerical values are

$$\frac{e_o}{q}(D) = \frac{0.0308(d^2/\delta)}{(5.96d^2)D + 1} \tag{8.90}$$

where d and δ are in inches, e_o is in millivolts, q is in Btu/(sec-ft^2), and $5.96d^2$ is in seconds. Typical commercial units are available for full-scale heat fluxes of 15 to 300 Btu/(sec-ft^2), produce 10 mv full-scale output, and have time constants of 0.07 to 0.2 sec.

Application considerations. The introduction of the sensor into the wall locally alters the thermal properties of the wall and causes the measured heat flux to differ from that which would occur if the sensor were not present.[1] It is thus desirable to match, insofar as feasible, the thermal properties of sensor and wall. For a Gardon gage, there is a radial temperature gradient over the disk which, if excessive, causes a variation in local convection coefficient and thus an error. By sacrificing sensitivity (and then recovering it by external amplification if necessary) the temperature gradient and associated error may be reduced. When only the radiation component of the total flux is desired, the front of the sensor is covered with a thermally isolated sapphire window which passes the radiation flux but blocks the convective flux.

Problems

8.1 An Invar/brass cantilever bimetal (see Fig. 8.2*b*) has $t = 0.05$ in., a length of 2 in., a width of 0.5 in., and has $t_A = t_B$ and $n + 1/n \approx 2$. Estimate the end deflection for temperature changes of 30 and 60°C. If the end is held fixed, estimate the force developed for temperature changes of 30 and 60°C.

[1] D. R. Hornbaker and D. L. Rall, Thermal Perturbations Caused by Heat-flux Transducers and Their Effect on the Accuracy of Heating-rate Measurements, *ISA Trans.*, p. 123, April, 1964.

8.2 A Beckman thermometer is to have a sensitivity of 10 in./°C when using mercury in the neighborhood of room temperature. Obtain an expression relating capillary cross-section area and bulb volume to meet this requirement. Obtain expressions for the time constant if the bulb is spherical and also if it is cylindrical with a length equal to 5 diameters. Also find the ratio of these two time constants.

8.3 Sketch and explain the operation of a bimetallic compensator to replace the auxiliary pressure sensor in Fig. 8.7. Can both case and capillary compensation be obtained? Explain.

8.4 Analyze the system of Fig. 8.14*a* to obtain a steady-state relation between hot-gas temperature as an input and thermocouple voltage as an output.

8.5 Develop equations to estimate the dynamic response of the system of Fig. 8.14*a*.

8.6 Repeat Prob. 8.4 for the system of Fig. 8.14*b*.

8.7 In Fig. 8.15 if $\tau = 2.8$ sec, $T_{gas} = 5000$°F, and the thermocouple damage limit is 2000°F, how long can the cooling be left off if the steady-state cooled thermocouple temperature is 500°F?

8.8 A resistance-thermometer circuit as in Fig. 8.19*d* has $R_1 = R_2 = 10{,}000$ ohms and is to cover the temperature range 0 to 400°C. The thermometer element is platinum with a resistance of 1,000 ohms at 200°C. Plot the curve of bridge output voltage (open circuit) versus input temperature, using the data of Fig. 8.18. See Chap. 10 for bridge-circuit equations. Bridge excitation is 20 volts.

8.9 A 500-ohm resistance thermometer carries 5-ma current. Its surface area is 0.5 in.², and it is immersed in stagnant air, so that the heat-transfer coefficient is $U = 1.5$ Btu/(hr-ft²-F°). Find its self-heating error. What would be the error in water with $U = 100$ Btu/(hr-ft²-F°)?

8.10 A pulse-excited resistance thermometer (see Fig. 8.20) has an excitation voltage in the form of a rectangular pulse of 100 volts height and 0.1-sec duration. The pulse is on for 0.1 sec and off for 0.9 sec in a repetitive cycle. Compute the ratio of peak/rms voltage for this pulse. What average heating power would this voltage pulse produce in a 500-ohm resistor?

8.11 For blackbody radiation, what surface temperature is needed to radiate 1 hp/in.²?

8.12 Estimate the percentage of total power found above 10μ for blackbody radiation at $T = 400$°K.

8.13 Explain the disadvantage of a large time constant in a thermal-radiation detector using a chopper.

8.14 Derive Eq. (8.53).

8.15 Consider a subsonic air flow in a duct. Pitot-static-tube measurements give a static pressure of 100 psia and a stagnation pressure of 129.1 psia. A temperature probe with a recovery factor $r = 0.80$ extends from a 100°F wall a distance of 1 ft into the flow. The probe thermocouple reads 400°F. The probe support has a radius of 0.02 ft and a thermal conductivity of 100 Btu/(hr-ft²-F°). The end of the probe may be assumed insulated, and the surface convection coefficient is 10 Btu/(hr-ft²-F°). Radiation effects are negligible. Calculate the static temperature of the flow.

8.16 Derive Eq. (8.72).

8.17 Make and analyze a third-order model of a temperature sensor analogous to the second-order model of Fig. 8.51.

8.18 A butt-welded 0.03-in. bare-wire copper/constantan thermocouple is used to measure the temperature (near 100°F) of atmospheric air flowing at 100 fps. Estimate the time constant using both formulas available.

8.19 A copper slug-type heat-flux sensor is 0.2 in. thick. Plot its time response (q_m versus t) for a step change in q. Is this a first-order instrument? Explain. How long must one wait before q_m is 95 percent of q?

8.20 Derive Eq. (8.82).

8.21 Derive Eq. (8.84).

8.22 Compare the sensitivity and response speed of Gardon gages of like diameter and thickness but made of the following:

a. Constantan disk, copper heat sink
b. Copper disk, constantan heat sink
c. Iron disk, constantan heat sink
d. Constantan disk, iron heat sink

8.23 Sketch and explain a test setup for evaluating the step-function response of temperature sensors exposed to air flows of different velocities.

8.24 Sketch and explain a test setup for static calibration of heat-flux sensors.

8.25 Sketch and explain a test setup for step-function testing of heat-flux sensors.

Bibliography

1. H. F. Stimson: International Temperature Scale of 1948, Text Revision of 1960, *Natl. Bur. Std. (U.S.), Monograph* 37, 1961.
2. J. Nicol and C. J. Rauch: Below One Degree, *Ind. Res.*, p. 60, September, 1964.
3. J. R. Van Orsdel et al.: Development of a Vapor-pressure-operated High Temperature Sensor Device, *NASA*, *CR*-50001, 1964.
4. C. F. Alban and C. C. Perry: Maximum-work Bimetals, *Machine Design*, p. 143, Apr. 16, 1959.
5. C. F. Alban and C. C. Perry: Adjusting Performance of Thermostatic Bimetals, *Machine Design*, p. 195, May 14, 1959.
6. C. F. Alban and C. C. Perry: Optimum Design of Thermostatic Bimetal Elements, *Machine Design*, p. 119, Feb. 21, 1957.
7. J. M. Benson and R. Horne: Surface Temperature of Thin Sheets and Filaments, *Instr. Control Systems*, p. 115, October, 1962.
8. C. E. Moeller: Do Shields Improve Thermocouple Response?, *ISA J.*, p. 56, August, 1960.
9. J. L. LeMay: More Accurate Thermocouples with Percussion Welding, *ISA J.*, p. 42, March, 1959.
10. L. E. Bollinger: Thermocouple Measurements in an RF Field, *ISA J.*, p. 338, September, 1955.
11. J. C. Lachman and F. W. Kuether: Stability of Rhenium/Tungsten Thermocouples in Hydrogen Atmospheres, *ISA J.*, p. 67, March, 1960.
12. A. R. Driesner et al.: High Temperature W/W-25 Re Thermocouples, *Instr. Control Systems*, p. 105, May, 1962.
13. F. W. Kuether and J. C. Lachman: How Reliable Are the Two New High-temperature Thermocouples in Vacuum?, *ISA J.*, p. 67, April, 1960.
14. J. J. Van Drasek and B. A. Short: Conversion Formulas for Copper-Constantan Thermocouples, *Instr. Control Systems*, p. 106, February, 1965.
15. G. E. Reis et al.: A Thermocouple Unit for Measuring Transient Temperatures at Specified Locations in Metal Bodies, Sandia Corp., Albuquerque, N.Mex., *SCR*-3, 1958.
16. H. C. Jordan: Welded Thermocouple Junctions, *Instr. Control Systems*, p. 988, June, 1960.
17. Simple Circuit Continuously Monitors Thermocouple Sensor Continuity, *NASA*, *Brief* 63-10567, 1963.

18. Thermocouple Calibration, *Instr. Control Systems*, p. 1663, September, 1961.
19. M. B. Dow: Comparison of Measurements of Internal Temperatures in Ablation Material by Various Thermocouple Configurations, *NASA, Tech. Note* D-2165, 1964.
20. R. Dutton and E. C. Lee: Surface-temperature Measurement of Current-carrying Objects, *ISA J.*, p. 49, December, 1959.
21. J. Nanigian: Thermal Properties of Thermocouples, *Instr. Control Systems*, p. 87, October, 1963.
22. Unusual Thermocouples and Accessories, *Instr. Control Systems*, p. 110, June, 1963.
23. E. G. Weissenberger: Metal Sheathed Thermocouples, *Instr. Control Systems*, p. 109, May, 1963.
24. C. E. Moeller: Special Thermocouple Solves Surface Temperature Problem, *ISA J.*, p. 47, June, 1959.
25. C. M. Stover: Method of Butt Welding Small Thermocouples 0.001 to 0.01 Inch in Diameter, *Rev. Sci. Instr.*, vol. 31, no. 6, p. 605, June, 1950.
26. C. M. Stover: Method of Making Small Pointed Thermocouples, *Rev. Sci. Instr.*, vol. 32, no. 3, p. 366, March, 1961.
27. G. E. Reis: Temperature Measurements on High Speed Missiles, Sandia Corp., Albuquerque, N.Mex., *SCR*-73, 1956.
28. O. Schwelb and G. C. Temes: Thermistor-Resistor Temperature Sensing Networks, *Electro-Technol. (New York)*, p. 71, November, 1961.
29. D. S. Saulson: The Thermistor Bridge, *Electro-Technol. (New York)*, p. 73, September, 1961.
30. J. S. Blakemore: Germanium for Low-temp Resistance Thermometry, *Instr. Control Systems*, p. 94, May, 1962.
31. J. R. Pies: A New Semiconductor for Temperature Measuring, *ISA J.*, p. 50, August, 1959.
32. J. M. Janicke: Direct-reading Platinum Thermometer, *Instr. Control Systems*, p. 129, May, 1965.
33. H. N. Norton: Resistance Elements for Missile Temperatures, *Instr. Control Systems*, p. 993, June, 1960.
34. A. R. Anderson and T. M. Stickney: Ceramic Resistance Thermometers as Temperature Sensors Above 2200°R, *Instr. Control Systems*, p. 1864, October, 1961.
35. A. B. Kaufman: Cryogenic Characteristics of Alloy Wires, *Instr. Control Systems*, p. 119, March, 1964.
36. H. C. Tsien: Piston Zone Temperature Measurement, *Instr. Control Systems*, p. 105, May, 1964.
37. R. S. Benson: Measurement of Transient Exhaust Temperatures in I.C. Engines, *The Engineer*, Feb. 28, 1964.
38. T. Coor and L. Szmanz: Digital Thermometer, *Instr. Control Systems*, p. 125, May, 1965.
39. E. W. Jones: Calibration Techniques for Thermistors, *Instr. Control Systems*, p. 123, May, 1965.
40. D. B. Schneider: The Thermistor Thermometer, *Instr. Control Systems*, p. 119, May, 1965.
41. P. W. Montgomery and R. L. Lowery: Jet Temperature by IR Pyrometry, *ISA J.*, p. 61, April, 1965.
42. R. J. Thorn and G. H. Winslow: Recent Developments in Optical Pyrometry, *ASME Paper* 63-WA-224, 1963.

43. R. H. Tourin: Recent Developments in Gas Pyrometry by Spectroscopic Methods, *ASME Paper* 63-WA-252, 1963.
44. D. A. McGraw and R. G. Mathias; Application of Radiation Pyrometry to Glass-forming Processes, *Ceram. Age*, August, 1962.
45. R. K. McDonald: Infrared Radiometry, *Instr. Control Systems*, p. 1527, September, 1960.
46. R. A. Hanel: The Dielectric Bolometer, A New Type of Thermal Radiation Detector, *NASA*, *Tech. Note* D-500, 1960.
47. R. A. Hanel: A Low-resolution Unchopped Radiometer for Satellites, *NASA Tech. Note* D-485, 1961.
48. R. W. Reynolds: Infrared-radiation Reference Sources, *Electro-Technol.* (*New York*), p. 46, January, 1963.
49. G. Conn and D. Avery: "Infrared Methods," Academic Press Inc., New York 1960.
50. F. Schwarz: Infrared Detectors, *Electro-Technol.* (*New York*), p. 116, November, 1963.
51. Calibration of Thermopiles, *NASA*, *N*-64-28205, 1964.
52. Pyroelectric Detection Techniques and Materials, *NASA*, *CR*-44, 1964.
53. D. Greenshields: Spectrometric Measurements of Gas Temperatures in Arc-heated Jets and Tunnels, *NASA*, *Tech. Note* D-1960, 1963.
54. D. R. Buchele: Nonlinear-averaging Errors in Radiation Pyrometry, *NASA*, *Tech. Note* D-2406, 1964.
55. M. Weiss: High Temperature Ultraviolet Radiometer, *Instr. Control Systems*, p. 95, May, 1964.
56. E. W. Bivans: Measuring Infrared Detector Noise, *Electron. Design*, Aug. 2, 1962.
57. Reference Blackbody Is Compact, Convenient to Use, *NASA Brief* 63-10004, 1963.
58. Lunar Surface Temperature Instrument, *NASA*, *N*-64-10097, 1964.
59. A. J. Metzler and J. R. Branstetter: Fast Response, Blackbody Furnace for Temperatures to 3000°K, *Rev. Sci. Instr.*, vol. 34, no. 11, p. 1216, November, 1963.
60. H. C. Ingrao et al.: Ferroelectric Bolometer for Space Research, *NASA*, *CR*-55542, 1964.
61. E. M. Wormser: Radiation Thermometer with In-line Blackbody Reference, *Instr. Control Systems*, p. 101, December, 1964.
62. T. P. Murray and V. G. Shaw: Two-color Pyrometry in the Steel Industry, *ISA J.*, p. 36, December, 1958.
63. Calibration of Optical Pyrometers, *Instr. Control Systems*, p. 84, May, 1962.
64. B. Bernard: Flame Temperature Measurements, *Instr. Control Systems*, p. 113, May, 1965.
65. A. G. Gaydon: "The Spectroscopy of Flame," John Wiley & Sons, Inc., New York, 1957.
66. R. Looney: Method for Presenting the Response of Temperature Measuring Systems, *ASME Trans.*, p. 1851, November, 1957.
67. W. J. King: Measurement of High Temperatures in High-velocity Gas Streams, *ASME Trans.*, p. 421, July, 1943.
68. W. M. Rohsenow and J. P. Hunsaker: Determination of the Thermal Correction for a Single-shielded Thermocouple, *ASME Trans.*, p. 699, August, 1947.
69. T. M. Stickney: Recovery and Time-response Characteristics of Six Thermocouple Probes in Subsonic and Supersonic Flow, *NACA*, *Tech. Note* 3455, 1955.

70. R. C. Turner and G. D. Gordon: Thermocouple for Vacuum Tests Minimizes Error, *Space/Aeron.*, p. 256, January, 1964.
71. D. Wald: Measuring Temperature in Strong Fields, *Instr. Control Systems*, p. 100, May, 1963.
72. L. M. K. Boelter et al.: Thermocouple Conduction Error Observed in Measuring Surface Temperatures, *NACA, Tech. Note* 2427, 1951, and *Tech. Note* 1452, 1948.
73. M. Sibulkin: A Total-temperature Probe for High-temperature Boundary-layer Measurements, *J. Aerospace Sci.*, p. 458, July, 1959.
74. J. C. Faul: Thermocouple Performance in Gas Streams, *Instr. Control Systems*, p. 104, December, 1962.
75. I. Fruchtman: Temperature Measurement of Hot Gas Streams, *AIAA J.*, vol. 1, no. 8, p. 1909, 1963.
76. D. L. Goldstein and R. Scherrer: Design and Calibration of a Total-temperature Probe for Use at Supersonic Speeds, *NACA, Tech. Note* 1885, 1949.
77. R. Sandri et al.: On the Measurement of the Average Temperature of a Fluid Stream in a Tube by Means of a Special Type of Resistance Thermometer, National Research Laboratory, Ottawa, Canada, *Mech. Eng. Div. Rept. MI*-826, April, 1962.
78. R. D. Wood: A Heated Hypersonic Stagnation-temperature Probe, *J. Aerospace Sci.*, p. 556, July, 1960.
79. R. P. Benedict: Temperature Measurement in Moving Fluids, *Electro-Technol. (New York)*, p. 56, October, 1963.
80. R. V. DeLeo et al.: Measurement of Mean Temperature in a Duct, *Instr. Control Systems*, p. 1659, September, 1961.
81. C. F. Hansen et al.: Investigation of Heat Conduction in Air, *NASA, Tech. Rept.* R-27 (Nickel Film Surface Temperature Detectors), 1959.
82. R. P. Benedict: High Response Aerosol Probe for Sensing Gaseous Temperature in a Two-phase, Two-component Flow, *ASME Paper* 62-WA-317, 1962.
83. I. Warshawsky: Measurements of Rocket Exhaust-gas Temperatures, *ISA J.*, p. 91, November, 1958.
84. M. G. Holland et al.: Temperature Measurement from 2°K–400°K, *Instr. Control Systems*, p. 89, May, 1962.
85. J. Grey: Thermodynamic Methods of High-temperature Measurement, *ISA Trans.*, p. 102, April, 1965.
86. T. A. Perls and J. J. Hartog: Pyroelectric Transducers for Heat-transfer Measurements, *ISA Trans.*, p. 21, January, 1963.
87. D. L. Johnson: The Design and Application of a Steady-state Heat Flux Transducer for Aerodynamic Heat-transfer Measurements, *ISA Trans.*, p. 46, January, 1965.
88. Simple Transducer Measures Low Heat-transfer Rates, *NASA, Brief* 64-10122, 1964.
89. F. C. Stempel and D. L. Rall: Direct Heat Transfer Measurements, *ISA J.*, p. 68, April, 1964.
90. E. A. Laumann: A Steady-state Heat Meter for Determining the Heat-transfer Rate to a Cooled Surface, *NASA, N*-63-18868, 1963.
91. L. R. Hunt and R. R. Howell: Experimental Technique for Measuring Total Aerodynamic Heating Rates to Bodies of Arbitrary Shape with Results to Mach 7, *NASA, Tech. Note* D-2446, 1964.
92. R. J. Conti: Heat-transfer Measurements at a Mach Number of 2 in the Turbulent Boundary Layer on a Flat Plate Having a Stepwise Temperature Distribution, *NASA, Tech. Note* D-159, 1959.

93. R. A. Jones and J. L. Hunt: Use of Temperature-sensitive Coatings for Obtaining Quantitative Aerodynamic Heat-transfer Data, *AIAA J.*, p. 1354, July, 1964.
94. P. H. Rose and J. O. Stankevics: Heat Transfer Measurements in Partially Ionized Gases, *NASA*, *CR*-59768, 1964.
95. C. H. Liebert et al.: Application of Various Techniques for Determining Local Heat-transfer Coefficients in a Rocket Engine from Transient Experimental Data, *NASA*, *Tech. Note* D-277, 1960.
96. D. R. Beck and F. Kreith: A New Steady State Calorimeter for Measuring Heat Transfer through Cryogenic Insulation, *NASA*, *N*-64-14283, 1964.
97. L. Bogdon: High-temperature, Thin-film Resistance Thermometers for Heat Transfer Measurement, *NASA*, *CR*-26, 1964.
98. J. C. Cook and H. S. Levine: Calorimeter and Accessories for Very High Thermal Radiation Flux Measurements, *Rev. Sci. Instr.*, October, 1960.
99. L. Bogdan: Measurement of Radiative Heat Transfer with Thin-film Resistance Thermometers, *NASA*, *CR*-27, 1964.
100. R. C. Bachmann et al.: Investigation of Surface Heat-flux Measurements with Calorimeters, *ISA Trans.*, p. 143, April, 1965.

9 Miscellaneous measurements

9.1

Time, Frequency, and Phase-angle Measurement The fundamental standard of time was discussed in Sec. 4.2. The United States Frequency Standard is a cesium-beam resonator whose precision is of the order of 1 part in 10^{11}. By radio-broadcasting signals related to the frequency of the standard, the National Bureau of Standards makes these frequency and time standards available to any other laboratory equipped to receive the signals. Radio station WWV broadcasts such signals with a precision of 1 part in 10^8. Owing to errors in transmission, the signals as remotely received have a precision of about 1 part in 10^7. Recently

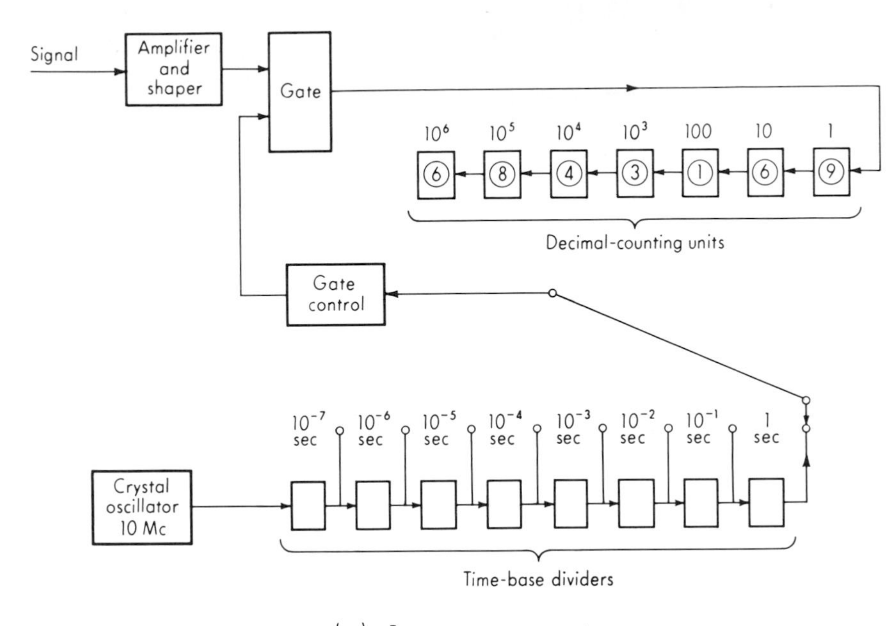

(*a*) Frequency measurement

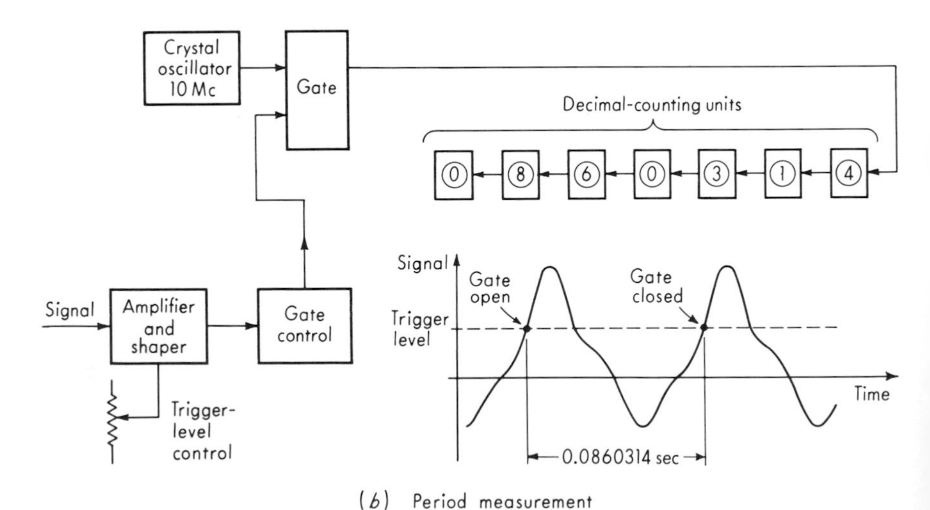

(*b*) Period measurement

Fig. 9.1. *Basic counter applications.*

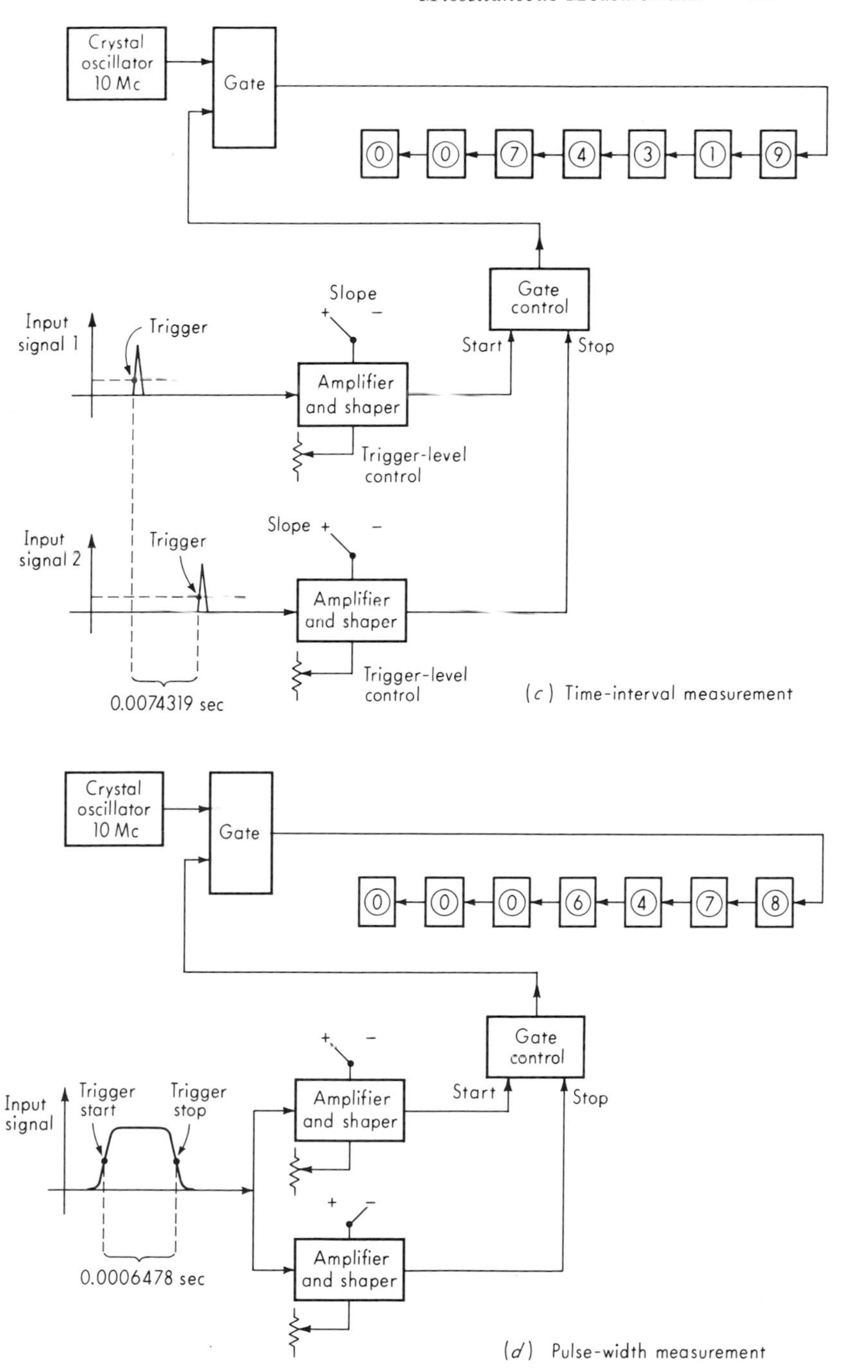

Fig. 9.1. *(Continued).*

two new stations, WWVB and WWVL, began broadcasts whose precision as received is about 1 part in 10^{10}.

Perhaps the most convenient and widely used instrument for accurate measurement of frequency and time interval is the electronic counter-timer. Figure 9.1 gives a block diagram showing the basic operation of such devices. The instrument's time and frequency standard is a piezoelectric crystal oscillator which generates a voltage whose frequency is very stable since the crystal is kept in a temperature-controlled oven. A typical frequency is 10^7 cps while the drift in frequency may be of the order of 3 parts in 10^7 per week. This gradual drift can, over a period of time, cause errors; thus highly accurate measurements require periodic recalibration of the oscillator against a suitable standard such as the radio-broadcast signals. In Fig. 9.1*a* the instrument is set up for frequency measurement of a signal whose frequency is 6,843,169 cps. This is accomplished by allowing the signal (suitably "shaped" to define each cycle more precisely) to go through a gating circuit to the decimal counting units for a precisely timed interval. This interval may be selected in 10-to-1 steps from 10^{-7} to 1 sec. Thus in the 1-sec interval used in Fig. 9.1*a* the counters accumulate 6,843,169 pulses. This mode of operation is also called EPUT (events-per-unit-time).

Sometimes it is more desirable to measure the period (rather than the frequency) of a signal. Figure 9.1*b* shows the arrangement used for this measurement. The trigger-level control is adjusted so that triggering occurs on the steepest part of the signal wave form to reduce error. There is usually provision for triggering on either a positive slope or a negative slope as desired. Since there is an inherent potential error of ± 1 count in turning the gate on and off, frequency-mode measurements are more accurate for high-frequency signals whereas period-mode measurements are more accurate for low-frequency signals. For example, a 10-cps signal measured in the frequency mode with the usual 1-sec time interval gives only 10 counts; thus an error of ± 1 count is a 10 percent error. The same signal measured in the period mode with a 10-Mc counter gives 10^6 counts and an error of only 0.0001 percent. Thus for a given counter there is some frequency below which period measurements should be used and above which frequency measurements should be employed. For a 1-sec sampling period this frequency f_0 is given by

$$f_0 = \sqrt{f_c} \qquad (9.1)$$

where $f_c \triangleq$ frequency of crystal oscillator ("clock" frequency)

Thus a 10-Mc clock-frequency counter has $f_0 = 3{,}160$ cps.

Measurements of the time interval between two events are very important in many experimental studies. The basic building blocks

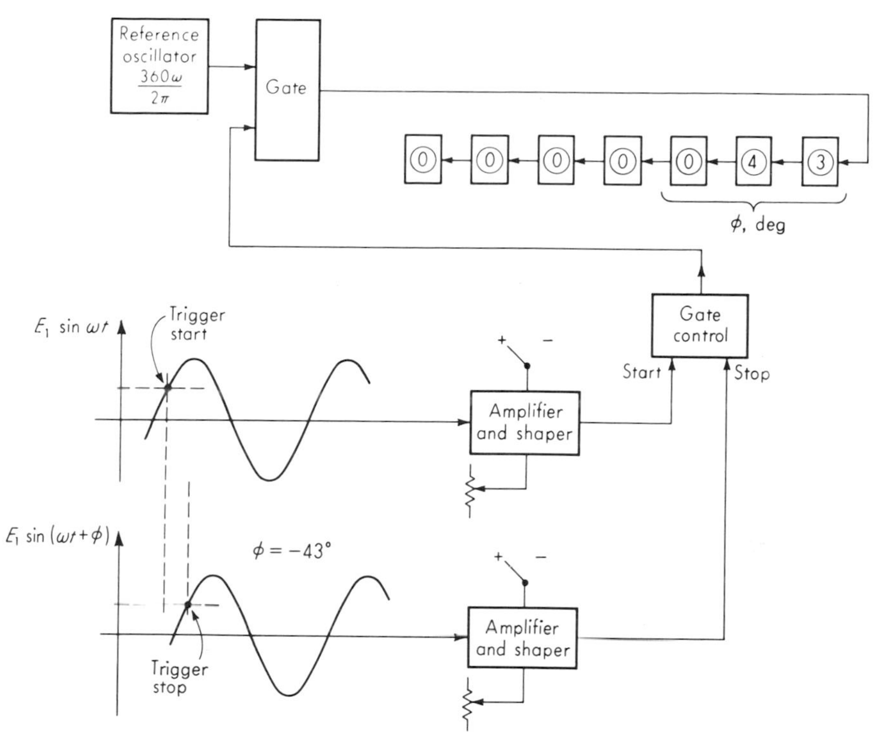

Fig. 9.2. *Phase-angle measurement.*

described above can be interconnected in a slightly different fashion, as in Fig. 9.1*c* and *d*, to accomplish this. In Fig. 9.1*c* two separate events have been transduced to electrical pulses; one event pulse is used to open the gate, and the other to close it, thereby timing the interval between them. Considerable versatility in triggering is obtained by providing trigger-level and slope controls on each input. By using the above arrangement, but only one input signal (Fig. 9.1*d*), the widths of pulses may be determined. Additional circuits are sometimes provided to send to an oscilloscope pulses that show the exact point on the incoming signals at which triggering is initiated. These are helpful in adjusting the trigger-level and slope controls and in interpreting the resulting information.

Measurement of the phase angle between two sinusoidal signals of the same frequency is often required. The experimental determination of the frequency response of some physical system is a good example of this type of measurement. A general-purpose digital counter-timer can be used for such measurements, as shown in Fig. 9.2. To use this method the amplitude of the two signals must be made equal and the triggering

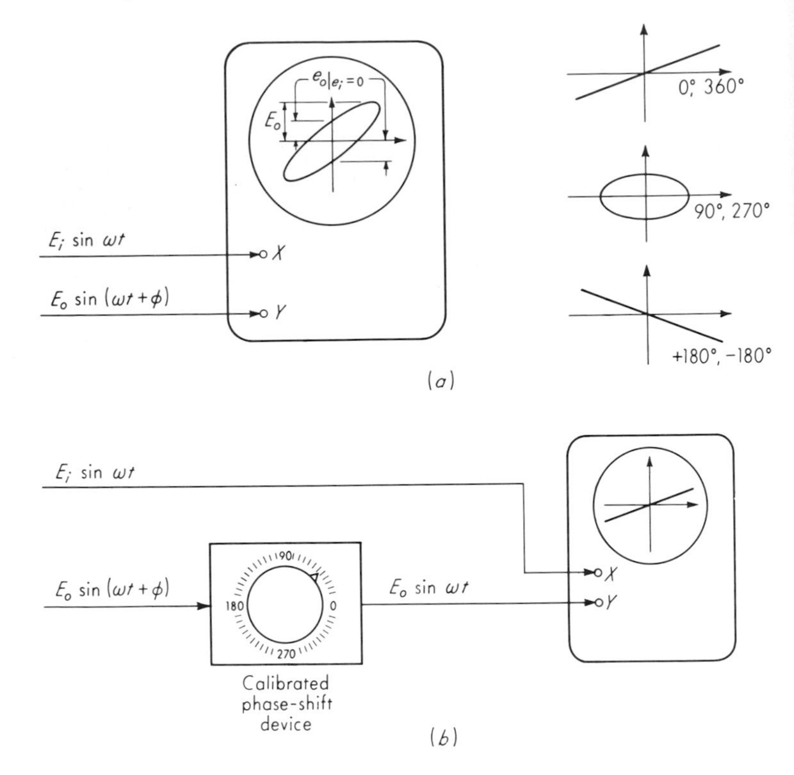

Fig. 9.3. *Phase angle from Lissajous figure.*

point of the two channels adjusted to be the same. Then the phase angle can be read directly with a resolution of 1° for the setup shown, or 0.1° if the reference frequency is set at $3{,}600 f_s$.

Another common method of phase-angle measurement involves cross-plotting the two sinusoidal signals against each other, using an X-Y plotter for very low frequencies and an oscilloscope for high frequencies. The cross plot can be shown to be an ellipse, and suitable measurements on this ellipse give the phase angle (see Fig. 9.3*a*). We have

$$e_i = E_i \sin \omega t \tag{9.2}$$

and

$$e_o = E_o \sin (\omega t + \phi) \tag{9.3}$$

If we set $t = 0$, $e_i = 0$, and $e_o = E_o \sin \phi$. Then

$$\sin \phi = \frac{e_o \Big|_{e_i=0}}{E_o} \tag{9.4}$$

Since $e_o \Big|_{e_i=0}$ has two values (+,−), the quadrant of ϕ is ambiguous;

however this can usually be resolved by visual observation of the two sine waves plotted against time (say on a dual-beam oscilloscope) or from knowledge of the system characteristics. The direction of travel of the "spot" as it plots the ellipse also resolves this difficulty but may be hard to detect at high frequencies. An alternative method employing the same basic principle but a null technique is shown in Fig. 9.3*b*. Here the calibrated phase-shift circuit is adjusted until the ellipse degenerates into a straight line (0° phase shift). The phase angle ϕ is then read directly from the phase-shifter dial.

9.2

Liquid Level Measurement and/or control of liquid level in tanks is an important function in many industrial processes and also in more exotic applications such as the operation and fueling of large liquid-fuel rocket motors. Figure 9.4 illustrates the more common methods of accomplishing this measurement.

The simple float of Fig. 9.4*a* can be coupled to some suitable motion transducer to produce an electrical signal proportional to the liquid level. Figure 9.4*b* shows a "displacer" which has negligible motion and measures the liquid level in terms of buoyant force by means of a force transducer. Since hydrostatic pressure is directly related to liquid level, the pressure-sensing schemes of Fig. 9.4*c* and *d* allow measurement of the liquid level in open and pressure vessels, respectively. In the "bubbler" or purge system of Fig. 9.4*e* the gas pressure downstream of the flow restriction is the same as the hydrostatic head above the bubble-tube end. The flow of gas is quite small; a bottle of nitrogen used as a source of pressurized gas may last six months or more.

Capacitance variation has been employed in various ways for level sensing. For essentially nonconducting liquids (conductivity less than 0.1 μmho/cm^3) the bare-probe arrangement of Fig. 9.4*f* may be satisfactory since the liquid resistance R is sufficiently high. For conductive liquids the probe must be insulated as in Fig. 9.4*g* to prevent short-circuiting of the capacitance by the liquid resistance. The measurement of the capacitance between the terminals *ab* may be accomplished in several ways. However high-frequency a-c (radio-frequency) methods offer significant advantages. Capacitance level-sensing techniques have been used with many common liquids, powdered or granular solids, liquid metals (high temperatures), liquefied gases (low temperatures), corrosive materials such as hydrofluoric acid, and in very high-pressure processes.

Figure 9.4*h* illustrates the use of radioisotopes for level measurement. Since the absorption of beta-ray or gamma-ray radiation varies with the thickness of absorbing material between the source and the detector, a

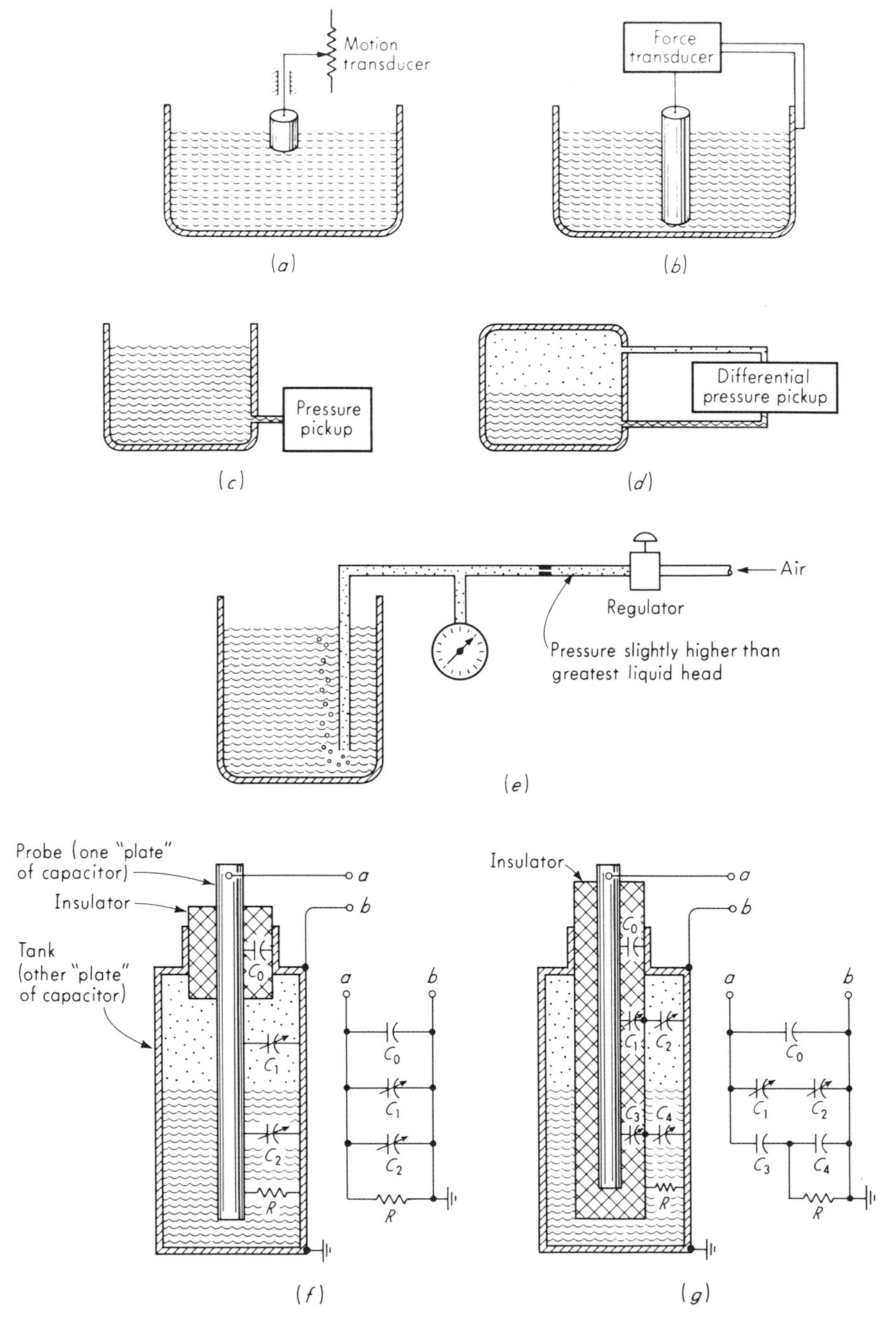

Fig. 9.4. *Liquid-level measurement.*

signal related to tank level may be developed. For analyzing such arrangements one may use the law

$$I = I_o e^{-\mu \rho x} \tag{9.5}$$

where $I \triangleq$ intensity of radiation falling on detector
$I_0 \triangleq$ intensity at detector with absorbing material not present
$e \triangleq$ base of natural logarithms
$\mu \triangleq$ mass absorption coefficient (constant for given source and absorbing material), cm^2/g
$\rho \triangleq$ mass density of absorbing material, g/cm^3
$x \triangleq$ thickness of absorbing material, cm

The gamma-ray source cesium 137 has been widely used for liquid-level measurements and has $\mu = 0.077\ cm^2/g$ for water or oil, 0.074 for aluminum, 0.072 for steel, and 0.103 for lead. For gaging a tank of water, then, if a vertical radiation path (rather than the angled one of Fig. 9.4*h*) is assumed, the variation of I with liquid height h is given by (neglecting absorption of air path)

$$I = I_0 e^{-0.077h} \tag{9.6}$$

This exponential relation of I and h is nearly linear only for sufficiently small values of h. For h as large as, say, 100 cm, the nonlinearity is quite apparent. This can be overcome by using either a radiation source or detector in the form of a strip oriented vertically rather than a "point" source or detector. Such arrangements are nonlinear for small values of $\mu \rho x_{max}$. Therefore point-to-point configurations are indicated for small ranges, whereas larger ranges require the more complex strip-to-point

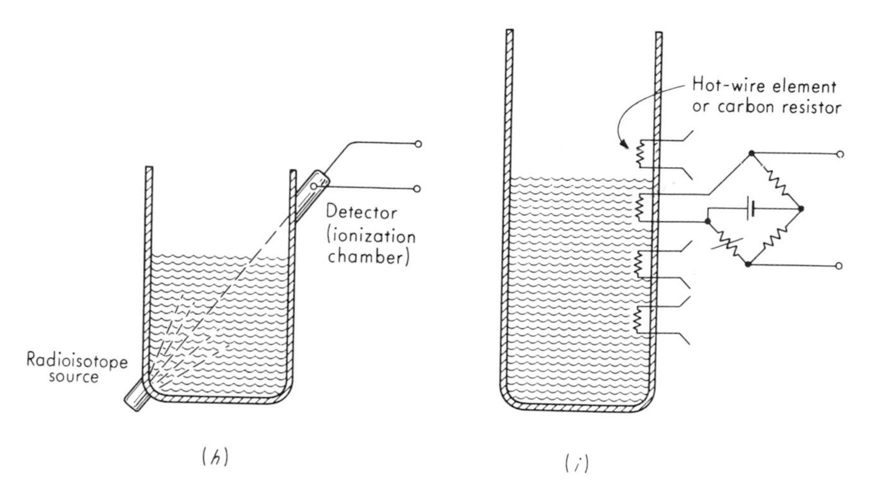

Fig. 9.4. (*Continued*)

type. For a strip source (or detector), the strength (sensitivity for a detector), can be "tailored" to vary in just the right way with position along the strip to give a linear tank-level/detector-signal relation.

Figure 9.4*i* shows the method of using hot-wire or carbon resistor elements for the measurement of liquid level in discrete increments. The basic concept is that the heat-transfer coefficient at the surface of the resistance element changes radically when the liquid surface passes it. This changes its equilibrium temperature and thus its resistance, causing a change in bridge output voltage. By locating resistance elements at known height intervals, the tank level may be measured in discrete increments. Such arrangements have been used in filling fuel tanks of large rocket engines with cryogenic liquid fuels.

9.3

Humidity Knowledge of the amount of water vapor in the air is important to the operation and/or automatic control of many industrial processes. This information may be gathered and presented in a number of ways, depending on the needs of the particular process and the measuring instrumentation used. In common use are relative humidity (ratio of water partial pressure to saturation pressure), dew-point temperature, mixing ratio or specific humidity (mass of water per unit mass of dry gas), and volume ratio (parts of water vapor per million parts of air).

The ultimate standard for calibration of humidity-measuring devices is the National Bureau of Standards standard gravimetric hygrometer. This is a strictly laboratory apparatus in which the water vapor in an air sample is absorbed by suitable chemicals and then very carefully weighed. It directly determines the mixing ratio in grams per kilogram, covers the range 0.30 to 20.0, and has an uncertainty (systematic error plus three standard deviations) of about 0.1 percent of the reading. For less critical calibrations the National Bureau of Standards uses its two-pressure humidity generator.[1] This equipment generates air/water mixtures in the dew-point range −70°C (uncertainty ±1.2C°) to +25°C (±0.1C°), relative-humidity range 10 to 98 percent at temperatures ranging from −55°C (relative-humidity uncertainty ±2.5 percent) to +40°C (relative-humidity uncertainty 0.5 percent), mixing ratio in the range 0.0013 g/kg (uncertainty ±0.0003 g/kg) to 20 g/kg (uncertainty ±0.5 percent), and volume ratio 2 ppm (±0.5 ppm) to 30,000 ppm (±0.5 percent).[2]

[1] A. Wexler and R. D. Daniels, Pressure-Humidity Apparatus, *J. Res. Natl. Bur. Std.*, vol. 48, p. 269, 1952.

[2] Humidity Calibration Service, *Instr. Control Systems*, p. 123, November, 1964.

Classically, relative humidity has been found from psychrometric charts and the temperature readings of two thermometers. One, the dry-bulb thermometer, reads the ordinary air temperature while the other, the wet-bulb, is intended to read the temperature of adiabatic saturation. The latter measurement requires that the bulb be kept wet and a suitably high (about 1,000 fpm) air velocity be maintained over the wet bulb. While these operations may be automated to a certain extent, the complexity of the calculations equivalent to the psychrometric chart hinders development of this technique into a continuous-reading instrument. Furthermore, the evaporation process at the wet bulb adds moisture to the air, thus disturbing the measured medium.

For continuous recording and/or control of relative humidity, electrical transducers of the Dunmore type are widely used. These were first developed about 1944 by F. W. Dunmore of the National Bureau of Standards and are basically a resistance element which changes resistance with relative humidity. The resistance element is constructed of a dual winding of noble-metal wires on a plastic form with a definite spacing between them. When the windings are coated with a lithium chloride solution a conducting path is formed between the windings. The electrical resistance of this path is found to vary reproducibly with the relative humidity of the surrounding air and may thus be used as a sensing element. Bridge-type resistance-measuring circuitry with a-c excitation is normally employed. The resistance/relative-humidity relation is quite nonlinear, and a single transducer generally can cover only a small range of the order of 10 percent relative humidity. Where large ranges (as great as 5 to 99 percent relative humidity) are needed, seven or eight of the transducers, each designed for a specific part of the total range, are combined in a single package. A single narrow-range sensing element may have an inaccuracy of the order of 1.5 percent relative humidity, resolution about 0.15 percent relative humidity, time constant as small as 3 sec, and size 1 in. in diameter by 2 in. long. Since these units are also sensitive to temperature, some form of temperature compensation may be required. The sensors do not add or subtract moisture or heat from their environment in significant amounts and may thus be used in sealed areas. Working temperatures in the range of -40 to $+150°F$ are possible.

Dew-point temperature can be determined by noting the temperature of a polished metal surface (mirror) when the first traces of condensation ("fogging") appear. Commercial devices in which this operation has been completely automated by means of a feedback system are available. Figure 9.5 shows the operation of a typical system.[1] The mirror (0.13-in.-diameter by 0.003-in.-thick rhodium-plated copper) is cooled by a CO_2

[1] Burton Manufacturing Co., Los Angeles, Calif.

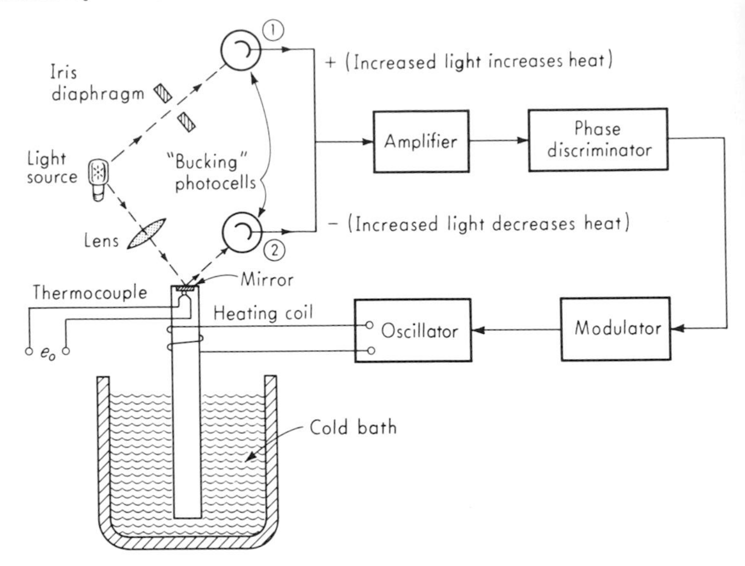

Fig. 9.5. *Automatic dew-point sensor.*

and acetone bath and heated by an induction coil. Initial adjustment of the system by a human operator consists of setting the opening of the iris diaphragm so that the difference in light received by photocells 1 and 2 is just sufficient to produce enough heating to maintain the lightest discernible fog on the mirror, which is observed with a 5-power magnifier. Since the light received by photocell 2 is reduced as the fogging of the mirror increases, the heating power is increased if the fog builds up and is decreased if the fog disappears. The system thus tends at all times to maintain automatically the degree of fog initially set into it. Dew-point temperature (mirror temperature) is continuously measured by the mirror thermocouple. An accuracy of $\pm 1\text{C}°$ over the dew-point range -58 to $+65°\text{F}$ is attained by the instrument described above.

The lithium chloride sensor mentioned above under relative humidity can be modified to give a signal related to dew-point temperature.[1] The dual wire windings are supplied with a-c power, causing a heating of the lithium chloride film. Lithium chloride shows a very sharp decrease in electrical resistance when *its* relative humidity increases above about 11 percent. Thus, when surrounded by moist air the lithium chloride momentarily absorbs moisture, its resistance drops, allowing more current to flow and more heat to be generated. This raises the temperature, driving off excess moisture, increasing the resistance, and reducing the heating.

[1] E. J. Amdur, Humidity Sensors, *Instr. Control Systems,* p. 93, June, 1963.

The sensor itself thus regulates its temperature so that the relative humidity of the lithium chloride element stays near 11 percent no matter what the moisture content of the surrounding air. It has been established that the temperature attained by the lithium chloride element, while not equal to the dew-point temperature, is directly related to it. Thus by measuring this temperature with some appropriate sensor, the dew-point temperature may be established. Probes of this type cover a dew-point range of -50 to $+160°F$ with an error of the order of 1 or 2F°.

The final humidity sensor considered here is the electrolytic type.[1] Here a continuous flow of sample gas (100 cm^3/min, regulated ± 2 percent) is passed through an analyzer tube. This tube has two platinum wires wound in a double helix on the inside of the tube. The space between the wires is coated with a strong desiccant (phosphorous pentoxide) and a d-c potential applied to the wire ends. When moisture in the sample gas is taken up by the P_2O_5, the water is electrolyzed into hydrogen and oxygen gas and a measurable electrolysis current flows. Such instruments have an inaccuracy of the order of 5 percent of full scale, ranges of 0 to 100 to 0 to 10,000 ppm, resolution better than 1 ppm, and a time constant of the order of 30 sec. They have also been adapted to the measurement of the water content of liquids and solids.

9.4

Chemical Composition In years past the chemical composition of materials was necessarily determined by taking a sample to a laboratory and performing the required chemical tests, usually with somewhat tedious procedures. Today, many important measurements of this type are made on a relatively continuous and automatic basis without the need for a human operator. The need for such measuring systems is due largely to the desire automatically to control product quality directly in terms of its chemical composition rather than inferring it from measurements of temperature, pressure, flow rate, etc. Even in manually controlled situations the desire to increase production rates while also maintaining or improving quality leads to a need for rapid analysis methods. Rapid and accurate analysis techniques are also very useful in research and development problems. These needs of industry have led to the development of a wide variety of instruments for measuring various aspects of chemical composition and related quantities.

Some examples of measurements of the above type include analysis of products of combustion, monitoring of the composition of dissolved gases in oil-well drilling mud, detection of alcohol contaminant in a heavy

[1] Consolidated Electrodynamics, Pasadena, Calif., *Bull.* 26303, June, 1963.

hydrocarbon liquid stream, detection of explosive solvents in the atmosphere within a uranium-extraction kettle, measurement of pH of industrial-waste effluent to control river pollution, determination of alloying constituents in metals, outgassing of materials under high vacuum, rocket-borne instruments for analyzing atmospheric gases at high altitudes, air-pollution studies, and analysis of anesthetic gases in blood. A wide variety of techniques[1] have been developed for handling such problems. A discussion of these methods adequate for their selection and use is beyond the scope of this text. However the references given will serve this purpose for the interested reader.

Problems

9.1 Derive Eq. (9.1).

9.2 Prove that a plot of $A_o \sin(\omega t + \phi)$ against $A_i \sin \omega t$ is an ellipse.

9.3 Analyze the system of Fig. 9.4*a* to obtain the transfer function relating liquid level h_i to float motion x_o. Neglect dry-friction effects. If the liquid level increases very slowly and the float motion is subject to a dry-friction force F_f, develop a formula to estimate the maximum steady-state error.

9.4 Assume the force transducer in Fig. 9.4*b* is of the elastic deflection type and obtain the transfer function relating liquid level h_i to force-transducer deflection x_o.

9.5 Discuss the effect of liquid density changes on the accuracy of liquid-level measurement in the systems of Figs. 9.4*a* and 9.4*b*.

9.6 For the system of Fig. 9.4*a*, discuss the effect on static and dynamic behavior of using a float that is a body of revolution but *not* a cylinder.

9.7 Repeat Prob. 9.6 for the system of Fig. 9.4*b*.

9.8 Discuss interfering and/or modifying inputs for the system of Fig. 9.4*d*. Assume the pressure pickup itself to be insensitive to such inputs.

9.9 Repeat Prob. 9.8 for the system of Fig. 9.4*e*.

Bibliography

1. S. J. Goldwater: Phase-angle Measurement in Control Systems, *Trans. Soc. Instr. Tech. (London)*, p. 100, June, 1960.
2. R. J. A. Paul and M. H. McFadden: Measurement of Phase and Amplitude at Low Frequencies, *Electron. Eng.*, vol. 31, no. 373, March, 1959.
3. F. J. Huddleston: Frequency Response by Sum or Difference, *Control Eng.*, p. 113, October, 1957.
4. Timers, *Electromech. Design*, p. 51, March, 1961.
5. Electric Timing Motors, *Electromech. Design*, p. 59, May, 1961.
6. Electronic Tuning Fork Beats Time for Accuracy, *Machine Design*, p. 30, October 27, 1960.
7. Time Interval Measurement, *Instr. Control Systems*, p. 125, September, 1962.
8. P. Young: 1 Nanosecond Time Interval Counter, *Instr. Control Systems*, p. 105, January, 1965.

[1] D. M. Considine (ed.), "Process Instruments and Controls Handbook," sec. 6, McGraw-Hill Book Company, New York, 1957.

9. A. MacMullen: Sources of Error in Phase Measurement, *Instr. Control Systems*, p. 91, January, 1965.
10. A New Approach to Precision Time Measurements, *Gen. Radio Experimenter*, General Radio Co., West Concord, Mass., February–March, 1965.
11. Correlating Time from Europe to Asia with Flying Clocks, *Hewlett-Packard J.*, Hewlett-Packard Co., Palo Alto, Calif., April, 1965.
12. Level Measurement and Control, *Instr. Control Systems*, p. 148, March, 1965.
13. N. Z. Alcock, and S. K. Ghosh: Minimizing Measurement Errors in Nuclear Gages, *Control Eng.*, p. 87, May, 1961.
14. F. W. Hannula: Use Capacitance for Accurate Level Measurement, *Control Eng.*, p. 104, November, 1957.
15. R. C. Muhlenhaupt and P. Smelser: Carbon Resistors for Cryogenic Liquid Level Measurement, *Natl. Bur. Std. (U.S.), Tech. Note* 200, 1963.
16. W. A. Olsen: An Integrated Hot Wire–Stillwell Liquid Level Sensor System for Liquid Hydrogen and Other Cryogenic Fluids, *NASA, Tech. Note* D-2074, 1963.
17. Liquid Hydrogen Level Sensors, *Instr. Control Systems*, p. 129, May, 1964.
18. R. L. Rod: Propellant Gaging and Control, *Instr. Control Systems*, p. 119, October, 1962.
19. G. H. Burger: Reliable Level Measurements for Liquid Metals, *Control Eng.*, p. 131, July, 1959.
20. F. Marton: Level Measurement and Control, *Instr. Control Systems*, p. 107, January, 1965.
21. E. Ulicki: Propellant Gaging System for Apollo Spacecraft, *Space/Aeron.*, p. 68, October, 1964.
22. D. D. Kana: A Resistive Wheatstone Bridge Liquid Wave Height Transducer, *NASA, CR*-56551, 1964.
23. L. Siegel: Nuclear and Capacitance Techniques for Level Measurement, *Instr. Control Systems*, p. 129, July, 1964.
24. N. H. Roos: Level Measurement in Pressurized Vessels, *ISA J.*, p. 55, May, 1963.
25. Wire Matrix Gages Zero-*g* Liquids, *Machine Design*, p. 10, Feb. 16, 1961.
26. C. L. Pleasance: Accurate Volume Measurement of Large Tanks, *ISA J.*, p. 56, May, 1961.
27. Moisture and Humidity, *Instr. Control Systems*, p. 121, October, 1964.
28. D. J. Fraade: Measuring Moisture in Gases, *Instr. Control Systems*, p. 100, April, 1963.
29. R. E. Fishburn: Measurement and Control of Humidity, *Automation*, p. 61, January, 1963.
30. R. M. Atkins: Wet/Dry Bulb Thermistor Hygrometer with Digital Indication, *Instr. Control Systems*, p. 111, April, 1964.

part 

Manipulation, Transmission, and Recording of Data

10 Manipulating, computing, and compensating devices

The information or data generated by a basic measuring device generally require "processing" or "conditioning" of one sort or another before they are finally presented to the observer as an indication or record. Devices for accomplishing these operations may be specific to a certain class of measuring sensors or they may be quite general-purpose. In this chapter we briefly consider those devices most often needed in building up measurement systems.

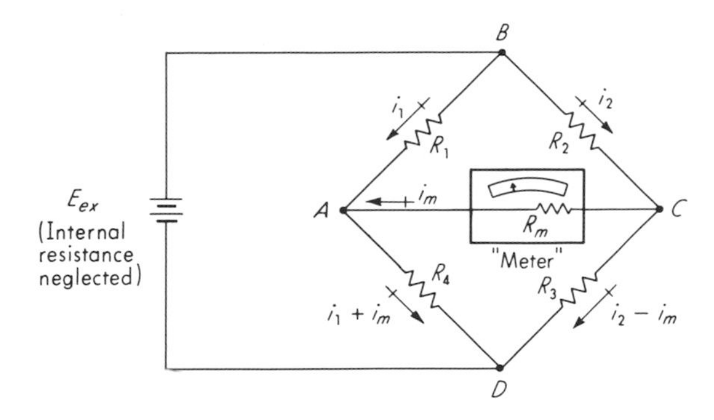

Fig. 10.1. *Basic Wheatstone bridge.*

10.1

Bridge Circuits Bridge circuits of various types are widely used for the measurement of resistance, capacitance, and inductance. Since we have seen that many transducers convert some physical variable into a resistance, capacitance, or inductance change, bridge circuits are of considerable interest. While capacitance and inductance bridges are important, the simpler resistance bridge is in the widest use, and we shall thus concentrate on it here. Adequate technical literature on all types of bridge circuits is readily available.[1]

Figure 10.1 shows a purely resistive (Wheatstone) bridge in its simplest form. The excitation voltage E_{ex} may be either d-c or a-c voltage; we here consider only direct voltage. In measurement applications, one or more of the legs of the bridge is a resistive transducer such as a strain gage, resistance thermometer, or thermistor. The basic principle of the bridge may be applied in two different ways: the null method and the deflection method. Let us assume that the resistances have been adjusted so that the bridge is balanced; that is, $e_{AC} = 0$. (It is easily shown that this requires $R_1/R_4 = R_2/R_3$.) Now we let one of the resistors, say R_1, change its resistance. This will unbalance the bridge and a voltage will appear across AC, causing a meter reading. The meter reading is an indication of the change in R_1 and can actually be used to compute this change. This method of measuring the resistance change is called the *deflection method*, since the meter deflection indicates the resistance change. In the *null method*, one of the resistors is manually adjustable. Thus if R_1 changes, causing a meter deflection, R_2 can be manually adjusted until its effect just cancels that of R_1 and the bridge

[1] E. Frank, "Electrical Measurement Analysis," chaps. 10 and 13, McGraw-Hill Book Company, New York, 1959.

is returned to its balanced condition. The adjustment of R_2 is guided by the meter reading; R_2 is adjusted so that the meter returns to its null or zero position. In this case the numerical value of the change in R_1 is directly related to the change in R_2 required to effect balance.

Both the null and deflection methods are used in practice. In the deflection method a calibrated meter is needed, and if the excitation E_{ex} changes, an error is introduced, since the meter reading is changed by changes in E_{ex}. With the null method, a calibrated variable resistor is needed, and since there is no meter deflection when the final reading is made, no error is caused by changes in E_{ex}. The deflection method gives an output voltage across terminals AC that almost instantaneously follows the variations of R_1. This output voltage can be applied to an oscilloscope (rather than the meter shown in Fig. 10.1) and thus measurements of rapid dynamic phenomena are possible. The null method, on the other hand, requires that the balancing resistor be adjusted to null the meter before a reading can be taken. This adjustment takes considerable time if done manually; even when an instrument servomechanism makes the adjustment automatically, the time required is much longer than is allowable for measuring many rapidly changing variables. Thus the choice of the null or the deflection method in a given practical case depends on the speed of response, drift, etc., required by the particular application.

In order to obtain quantitative relations governing the operation of the bridge circuit, a circuit analysis is necessary. The following information is desired:

1. What relation exists among the resistances when the bridge is balanced ($e_{AC} = 0$)? The answer to this has already been given as $R_1/R_4 = R_2/R_3$.
2. What is the sensitivity of the bridge? That is, how much does the output voltage e_{AC} change per unit change of resistance in one of the legs?
3. What is the effect of the meter internal resistance on the measurement?

We shall consider the question of bridge sensitivity first for the case where the "meter" has a very high internal resistance R_m. If this is the case, the meter current i_m will be negligible compared with the currents in the legs. This situation is closely approximated in practice in the following cases:

1. The "meter" is an oscilloscope. The internal or input resistance of a typical oscilloscope is of the order of 1 million ohms. When

the legs of the bridge are each of the order of 10,000 ohms or less, the current in the oscilloscope will be of the order of $\frac{1}{100}$ or less of the currents in the legs and may thus be neglected. In a typical strain-gage bridge, $R_1 = R_2 = R_3 = R_4 = 120$ ohms, and so the oscilloscope current would be extremely negligible.

2. The voltage e_{AC} is applied to the input of an electronic amplifier. Here again, the input resistance of amplifiers is often much higher than the resistance of the legs in the bridge, and thus the current may be treated as effectively zero. (Actually, the input to an oscilloscope is through an amplifier; thus cases 1 and 2 are not really different.)
3. The voltage e_{AC} is measured with a potentiometer. In the potentiometer method of voltage measurement the unknown voltage (such as e_{AC}) is bucked against a known and adjustable voltage of opposite polarity supplied by the potentiometer. The known voltage is adjusted until a galvanometer indicates the two bucking voltages are equal. The value is then read from the calibrated dial of the potentiometer. When the potentiometer is balanced, no current is being drawn from terminals AC. (Actually the potentiometer cannot be *perfectly* balanced since any galvanometer has a threshold sensitivity below which it cannot detect the presence of current.)

Since it appears the condition of $i_m = 0$ is quite closely approximated in many practical cases, it will be worthwhile to study this situation. We have

$$i_1 = \frac{E_{ex}}{R_1 + R_4} \tag{10.1}$$

$$i_2 = \frac{E_{ex}}{R_2 + R_3} \tag{10.2}$$

$$e_{AB} = \text{voltage rise from } A \text{ to } B = i_1 R_1 = \frac{R_1}{R_1 + R_4} E_{ex} \tag{10.3}$$

$$e_{CB} = \frac{R_2}{R_2 + R_3} E_{ex} \tag{10.4}$$

and finally

$$e_{AC} = e_{AB} + e_{BC} = e_{AB} - e_{CB} = \left(\frac{R_1}{R_1 + R_4} - \frac{R_2}{R_2 + R_3}\right) E_{ex} \tag{10.5}$$

Thus we see that the output voltage is a linear function of the bridge excitation E_{ex} but, in general, a *nonlinear* function of the resistances R_1, R_2, R_3, and R_4. If the bridge is initially balanced and then R_1, say, begins to change, the output voltage signal will *not* be directly proportional to the change in R_1. For certain practically important special cases, however, perfect linearity is possible. The best example of this is found in many strain-gage transducers in which, at the balanced condition, $R_1 = R_2 = R_3 = R_4 = R$. Also, the resistance changes are such that $+\Delta R_1 = -\Delta R_2 = +\Delta R_3 = -\Delta R_4$. We may then write

$$e_{AC} = \left[\frac{R_1 + \Delta R_1}{(R_1 + \Delta R_1) + (R_4 + \Delta R_4)} - \frac{R_2 + \Delta R_2}{(R_2 + \Delta R_2) + (R_3 + \Delta R_3)}\right] E_{ex} \tag{10.6}$$

$$e_{AC} = \frac{\Delta R_1}{R} E_{ex} \tag{10.7}$$

Clearly, Eq. (10.7) shows a strictly linear relationship of e_{AC} with ΔR_1.

Even when the above symmetry does not exist, the bridge response is very nearly linear as long as the ΔR's are small percentages of the R's. In strain gages, for example, the ΔR's rarely exceed 1 percent of the R's. Since the case of small ΔR's is of practical interest, we shall work out an expression for bridge sensitivity that is a good approximation for such a situation. From Eq. (10.5), $e_{AC} = f(R_1,R_2,R_3,R_4)$, and thus for small changes from the null condition we may write approximately

$$\Delta e_{AC} = e_{AC} = \frac{\partial e_{AC}}{\partial R_1} \Delta R_1 + \frac{\partial e_{AC}}{\partial R_2} \Delta R_2 + \frac{\partial e_{AC}}{\partial R_3} \Delta R_3 + \frac{\partial e_{AC}}{\partial R_4} \Delta R_4 \tag{10.8}$$

Now,

$$\frac{\partial e_{AC}}{\partial R_1} = E_{ex} \frac{R_4}{(R_1 + R_4)^2} \quad \text{volts/ohm} \tag{10.9}$$

$$\frac{\partial e_{AC}}{\partial R_2} = -E_{ex} \frac{R_3}{(R_2 + R_3)^2} \tag{10.10}$$

$$\frac{\partial e_{AC}}{\partial R_3} = E_{ex} \frac{R_2}{(R_2 + R_3)^2} \tag{10.11}$$

$$\frac{\partial e_{AC}}{\partial R_4} = -E_{ex} \frac{R_1}{(R_1 + R_4)^2} \tag{10.12}$$

The partial derivatives are taken as constants; thus Eq. (10.8) shows a linear relation between e_{AC} and the ΔR's.

We have explained above, in a qualitative fashion, that if the meter resistance is "high enough" the terminals AC may be thought of as an open circuit (no current i_m). It would be useful to have a more quantitative method of deciding whether the meter resistance was "high enough" and also, if it were not, how to correct for it. This will now be done.

By using Thévenin's theorem, the bridge circuit and the "meter" that loads it may be represented as in Fig. 10.2. Since we have been calling the bridge output voltage under assumed open-circuit conditions e_{AC}, this becomes the E_o of Fig. 3.22. Let us call the bridge output

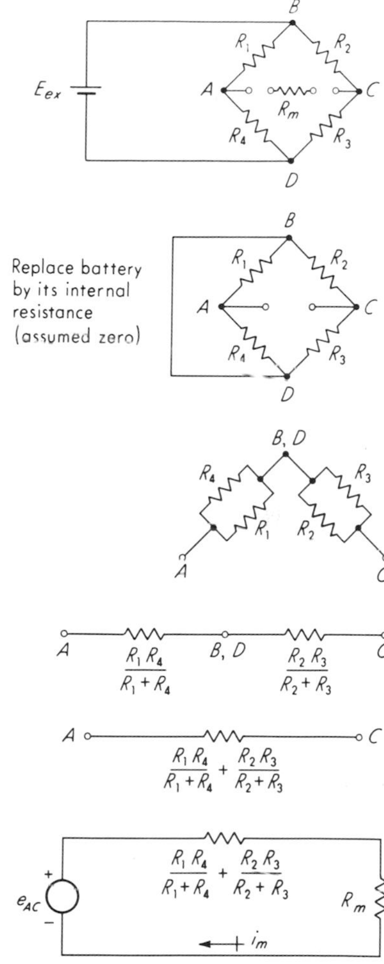

Fig. 10.2. *Thévenin analysis of bridge.*

under the actual loaded condition e_{ACL}. We can then immediately write

$$i_m = \frac{e_{AC}}{R_{\text{total}}} = E_{ex} \frac{R_1/(R_1 + R_4) - R_2/(R_2 + R_3)}{R_m + R_1R_4/(R_1 + R_4) + R_2R_3/(R_2 + R_3)} \tag{10.13}$$

Knowing i_m, we can now compute the actual voltage e_{ACL} across the meter under the condition where the meter draws current, since the voltage across the meter will be the product of the current i_m and the meter resistance R_m. Carrying this out and simplifying, we get

$$e_{ACL} = \frac{E_{ex}(R_1R_3 - R_2R_4)}{(R_1 + R_4)(R_2 + R_3) + [(R_1 + R_4)R_2R_3 + R_1R_4(R_2 + R_3)]/R_m} \tag{10.14}$$

Now
$$e_{AC} = \frac{E_{ex}(R_1R_3 - R_2R_4)}{(R_1 + R_4)(R_2 + R_3)} \tag{10.15}$$

and if we wish to display the effect of the meter resistance on the bridge output voltage we can form the ratio of e_{ACL} to e_{AC}. After some manipulation this can be shown to be

$$\frac{e_{ACL}}{e_{AC}} = \frac{1}{1 + (1/R_m)[R_2R_3/(R_2 + R_3) + R_1R_4/(R_1 + R_4)]} \tag{10.16}$$

We now have a quantitative way of assessing the effect of the meter resistance R_m on the bridge output. We see that, if $R_m = \infty$, $e_{ACL} = e_{AC}$, as one would expect. If R_m is not infinite, there will be a *reduction* in the output signal, and the magnitude of this reduction depends on the relative values of R_m and the bridge "equivalent resistance" R_e which is defined as

$$R_e \triangleq \frac{R_2R_3}{R_2 + R_3} + \frac{R_1R_4}{R_1 + R_4} \tag{10.17}$$

In terms of R_e, Eq. (10.16) becomes

$$\frac{e_{ACL}}{e_{AC}} = \frac{1}{1 + R_e/R_m} \tag{10.18}$$

Thus, if $R_m = 10R_e$,

$$\frac{e_{ACL}}{e_{AC}} = \frac{1}{1.1} = 0.91 \tag{10.19}$$

and there is a 9 percent loss in signal due to the noninfinite meter resistance. This type of loss is usually referred to as a *loading effect;* that is, the meter "loads down" the bridge and reduces its sensitivity.

The theory developed above is useful in assessing the effects of various parameters on the bridge sensitivity and could actually be used to compute the sensitivity if all quantities were known exactly. It is preferable, however, to calibrate the bridge directly by introducing a known resistance change and noting the effect on the bridge output. This known resistance change is usually introduced by means of the arrangement shown in Fig. 10.3. The resistance R_c of the calibrating resistor is accurately known. If the bridge is originally balanced with the switch open, when the switch is closed the resistance in leg 1 will change and the bridge will be unbalanced. The output voltage e_{AC} is read on the meter, and the resistance change ΔR that caused this voltage is computed from

$$\Delta R = R_1 - \frac{R_1R_c}{R_1 + R_c} \tag{10.20}$$

The bridge sensitivity is then

$$S \triangleq \frac{e_{AC}}{\Delta R} \quad \text{volts/ohm} \tag{10.21}$$

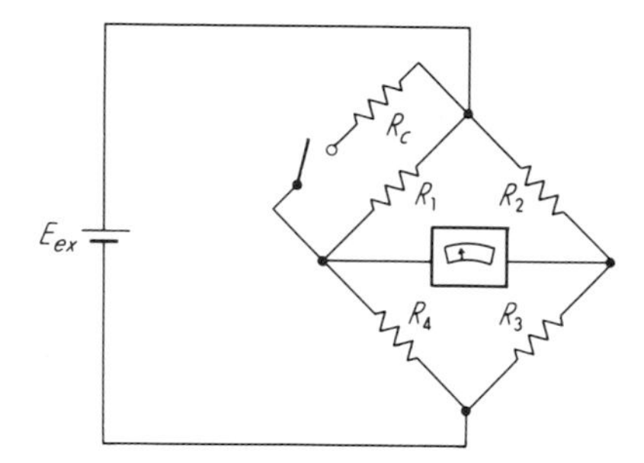

Fig. 10.3. *Shunt calibration method.*

This procedure gives an overall calibration, since the values of all the resistors and the battery voltage are taken into account.

Figure 10.1 shows a bridge circuit with the bare essentials. Often additional features are necessary or desirable for the convenience of the user. Figure 10.4 shows a versatile arrangement providing the following capabilities:

1. Variation of overall sensitivity without the need to change E_{ex}
2. Provision for adjusting the output voltage to be precisely zero when the measured physical quantity is zero even if the legs are not exactly matched
3. Shunt resistor calibration

Commercial transducers may also include additional temperature-sensitive resistors to achieve temperature compensation.

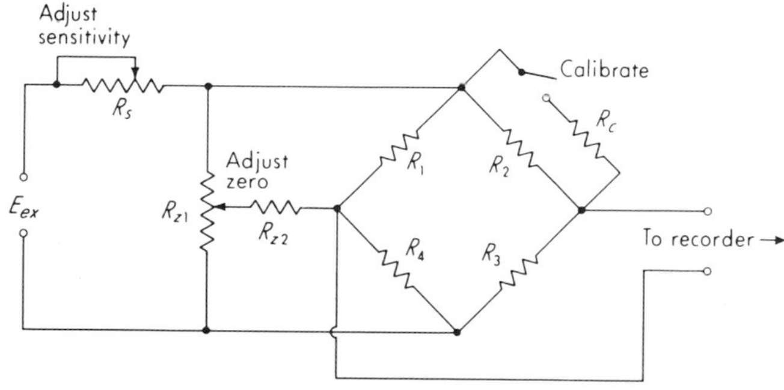

If $R_1 \approx R_2 \approx R_3 \approx R_4 < 1{,}000$ ohms (usual strain-gage transducer), then

$R_{z2} \approx 100\, R_1$

$R_{z1} \approx 25{,}000$ ohms

Fig. 10.4. *Bridge with sensitivity, balance, and calibration features.*

10.2 Amplifiers

Since the electrical signals produced by most transducers are at a low voltage and/or power level, it is often necessary to amplify them before they are suitable for transmission, further manipulation, indication, or recording. While the design of amplifiers is beyond the scope of this text, the criteria to be applied in choosing and using an amplifier for a specific measuring-system application are of interest and will be briefly reviewed here.

General performance characteristics of amplifiers. The dynamic response of an amplifier must equal or exceed that of the transducer feeding it and is usually specified as the essentially flat range of frequency response. In order not to draw much current from the transducer (thereby "loading" it and causing loss of sensitivity and/or linearity), the input impedance of the amplifier must be sufficiently high relative to the transducer output impedance. Such questions should be studied for both static and dynamic conditions in critical applications, using the general impedance methods developed in Chap. 3. Similarly, the output impedance of the amplifier must be sufficiently low relative to the input impedance of the following device (often a recorder) that loading effects at this point are not excessive. Amplifier output impedance can also be interpreted as current-supplying ability and is sometimes quoted as such. For example, a particular amplifier may be quoted as supplying a full-scale voltage of ± 10 volts to any load with resistance R_L greater than 1,000 ohms and a full-scale current of 0.010 amp to any load with resistance less than 1,000 ohms. For such an amplifier we note that the voltage gain is independent of load resistance for $R_L > 1{,}000$ but decreases progressively for $R_L < 1{,}000$, full-scale voltage output being only 1 volt, for example, if $R_L = 100$ ohms. In general, a low output impedance indicates a high current capacity.

If the input terminals of an amplifier are short-circuited so that the input voltage is exactly zero, the output voltage will *not* be precisely zero. This defect is charged to two sources, zero drift and noise. Zero drift is a slow change in the short-term mean value of the output voltage whereas noise is a fast fluctuation around the short-term mean value (see Fig. 10.5). Zero drift is generally large until the equipment is thoroughly warmed up since its origins are largely thermal. Once warm-up is over, a certain amount of random zero drift will remain. Balancing controls are generally provided to set the output voltage to zero with the input shorted; however, they must be periodically readjusted to compensate for drift. Since the drift after warm-up is quite slow, it interferes mainly with the measurement of slowly varying quantities over long time periods. Noise has its origins in a number of random processes, such as Johnson noise in

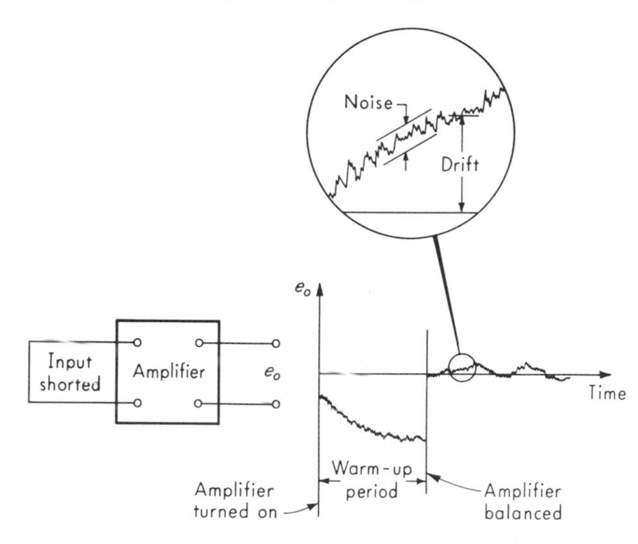

Fig. 10.5. *Amplifier noise and drift test.*

resistors; shot, partition, and gas noise in electron tubes; and transistor noise. It also arises from nonrandom sources such as power-line (60 cps) hum and chopper action in amplifiers that use choppers. Noise limits amplifier behavior mainly with respect to the threshold or smallest signal voltages that can be detected.

The overload recovery time of an amplifier is mainly of interest in multiplexed systems in which a given amplifier is time-shared by several transducers. Here the amplifier input is switched periodically from one transducer to another. This switching action may introduce large transient overloads into the amplifier, and a certain settling time is required before the amplifier output again reads correctly. A long recovery time prevents the amplifier from being switched in a rapid cycle and thus reduces the number of transducers that can be sampled by an amplifier in a given time.

In many measurement systems the amplifiers are located remotely from the transducers for one reason or another. This means that long connecting cables are necessary (several hundred feet is not unusual). A number of problems are introduced by such remote installations, the main ones being inductive pickup of noise voltages in the long cables (usually from nearby 60-cycle power lines, motors, etc.) and the inability to provide an identical ground reference point for two locations several hundred feet apart. Both these phenomena result in spurious voltages appearing at the amplifier input, and these can be very large relative to the transducer signal if adequate precautions are not taken. Shielding of the

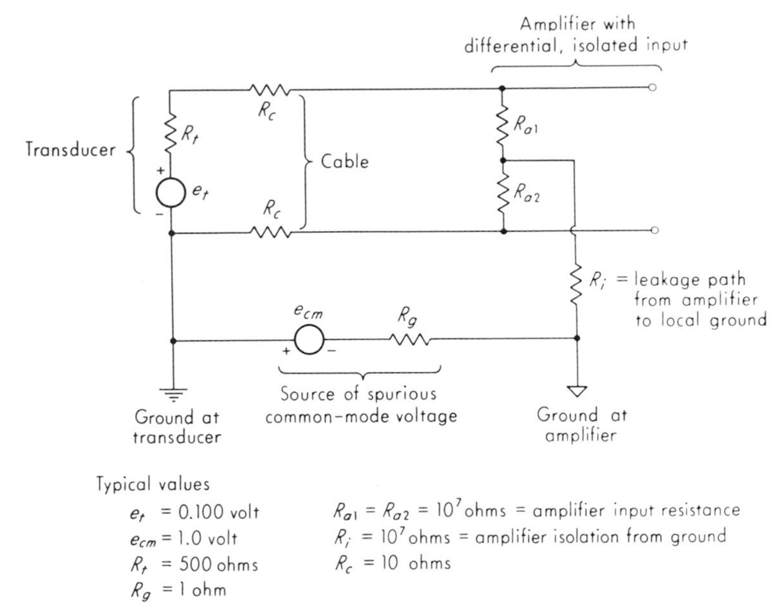

Fig. 10.6. *Differential input configuration.*

signal cables is helpful but the most critical applications also require the amplifier to have certain characteristics.

The performance specification generally used to indicate the amplifier's ability to reject the above-mentioned spurious voltages is the common-mode rejection ratio (CMRR). This is the ratio by which the undesired signals are attenuated relative to the desired signals. It depends on both the amplifier and the transducer and cable circuitry; thus any numerical value quoted for an amplifier assumes a certain input configuration. The common-mode rejection ratio also varies with the frequency of the noise voltages. Thus an amplifier may have a common-mode rejection ratio of 10^6 at direct current but only 10^4 at 60 cps. Some idea of the frequency content of the noise voltages is therefore necessary in order to evaluate an amplifier adequately. A full discussion of these questions is not possible here; however, adequate literature[1,2] is available. The amplifier features that are most significant in obtaining a desirable high value of the common-mode rejection ratio are differential

[1] A. S. Buchman, Noise Control in Low Level Data Systems, *Electromech. Design*, p. 64, September, 1962.

[2] Instrumentation Grounding and Noise Minimization Handbook, Consolidated Systems Corp., Pomona, Calif., AD612-027, *Tech. Rept.* AFRPL-TR-65-1, January, 1965.

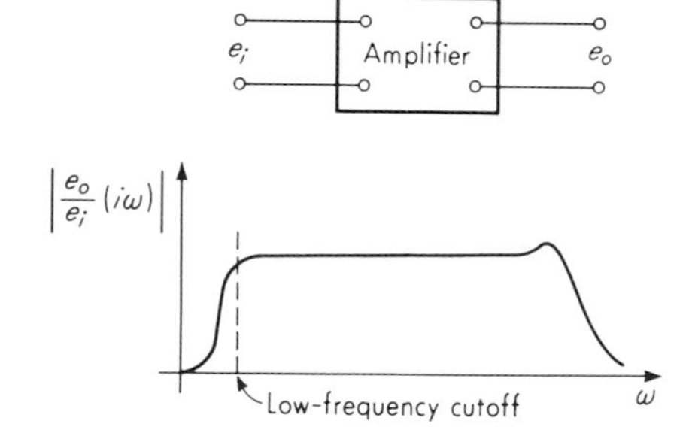

Fig. 10.7. *A-C-amplifier frequency response.*

input and "floating" (highly isolated from ground) input. These can be employed separately or, in the most critical applications, in conjunction in order to obtain a high common-mode rejection ratio. Figure 10.6 shows the circuit configuration when both methods are used. The spurious-voltage source e_{cm} can be shown[1] to act identically whether it is due to "ground loop" (lack of identical ground at two remote points) or inductive pickup in the cable; thus for simplicity Fig. 10.6 shows it only in the ground loop. Circuit calculations for the typical numerical values shown give[1]

$$\text{Desired signal at amplifier input} \approx 0.100 \text{ volt} \tag{10.22}$$

$$\text{Spurious common-mode signal at amplifier input} \approx 16.5 \times 10^{-6} \text{ volt} \tag{10.23}$$

Thus the percentage error in this case is 0.0165 percent. The above calculations assumed purely resistive circuit elements. However similar calculations can be made for more general impedances.

A-C amplifiers. We begin our discussion of specific amplifier types by considering a class we here call "a-c." These amplifiers are closely related to the amplifiers ordinarily used in radios, television sets, and sound systems; however, the specifications for instrumentation types are generally more stringent. Vacuum-tube, transistor, and hybrid (partly tube, partly transistor) designs are in common use. We choose the name a-c based on the frequency response of such instruments, shown in Fig. 10.7. Note that the amplifier is incapable of handling steady or low-frequency input signals. This is a disadvantage if slowly varying quantities must be measured but also makes such amplifiers relatively drift-free since drift is due to slowly changing effects that cannot pass through the amplifier to its output, owing to the capacitive coupling that blocks direct current.

[1] A. S. Buchman, Noise Control in Low Level Data Systems, *Electromech. Design*, p. 64, September, 1962.

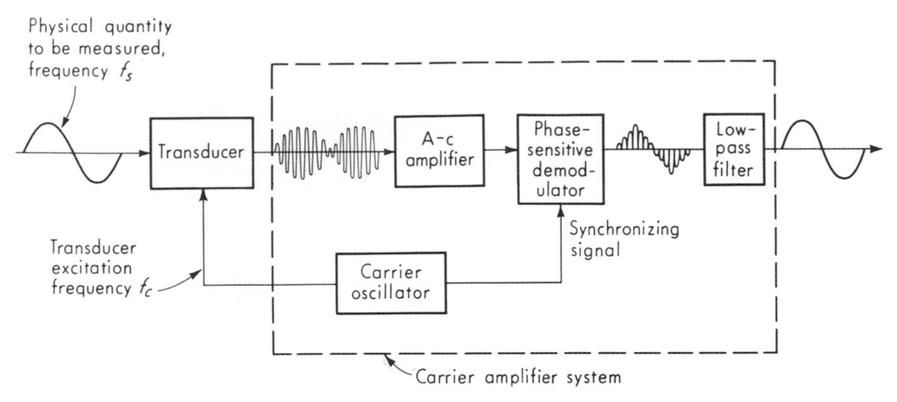

Fig. 10.8. *Carrier amplifier system.*

A typical tube-type portable battery-operated instrument[1] especially useful with strain-gage transducers has three stages of amplification. By using one, two, or all three stages a gain of 15, 150, or 2,000 may be obtained. Frequency response is flat ±5 percent from 5 to 25,000 cps while the noise level is equivalent to 10 μv at the input. Input impedance is 2 megohms; output impedance is 27,000 ohms. The high output impedance indicates this amplifier is designed to drive a load requiring little current, such as an oscilloscope.

An a-c amplifier[2] using silicon field-effect transistors has adjustable gain (10 to 1,000), frequency response flat ±1 db from 5 to 60,000 cps, input impedance 10^8 ohms shunted by 10^{-10} farad, output impedance 1,000 ohms (1-ma current available), noise level under 10 μv referred to the input, full-scale output of 1.4 volts, rise time (10 to 90 percent) of 3 μsec, and recovery time of 3 sec for a 500-volt input pulse.

Carrier amplifier systems. To extend the advantages of simplicity and lack of drift characteristic of a-c amplifiers to the measurement of steady (d-c) signals, a number of approaches have been developed. The so-called carrier amplifier systems, much used with strain-gage transducers, are one such scheme. Such systems work only with transducers (such as strain-gage bridges and differential transformers) that are excited by an a-c voltage. Figure 10.8 shows the amplitude-modulation principle used; it requires that the a-c amplifiers have a flat frequency response only over the band $f_c \pm f_s$. Because of demodulation and filtering require-

[1] Model BA-1, Ellis Associates, Pelham, N.Y.
[2] Sensonics, Inc., Kensington, Md.

ments, the carrier frequency f_c must be 6 to 10 times the highest signal frequency f_s.

A typical system[1] uses a carrier oscillator of a frequency of 5,000 cps and an amplitude of 0.1 to 5 volts rms, has a flat frequency response ± 5 percent from 0 to 1,000 cps, gain of about 3,000, input resistance of 350 ohms, output resistance of 12 ohms, maximum output voltage and current of 1.5 volts and 0.100 amp, nonlinearity ± 2 percent of full scale, noise less than 1 percent of full-scale output, and zero drift less than 0.5 percent of full scale over a 10-hr period at constant temperature. After $\frac{1}{2}$-hr warm-up, 10F° temperature changes cause less than 2 percent zero drift per hour.

D-C amplifiers. While the carrier system above will amplify "d-c" (constant) physical variables acting on a suitable transducer, it is not generally considered a d-c amplifier because it will not amplify a d-c voltage coming from an arbitrary transducer. Since this latter capability is much desired, various types of so-called d-c amplifiers have been developed.

Although the fundamental principles of both tube and transistor amplification inherently permit d-c operation, practical problems of drift and interstage coupling limited the use of such "true" d-c amplifiers in the past. Increasing demands of instrumentation systems and new developments in components and circuitry have led to the availability today of practical "true" d-c amplifiers as well as various forms of chopper and chopper-stabilized amplifiers. Chopper and chopper-stabilized types, while perhaps not accurately classified as pure d-c instruments, may practically be so considered since (on a black-box, input/output basis) they accept d-c inputs and produce amplified d-c outputs.

In a *chopper amplifier* the input signal (which may have a frequency content from zero to about one-tenth of the chopper frequency) is chopped to produce a square-wave voltage whose amplitude is proportional to the incoming signal. This modulation shifts all frequencies to a band around the chopper frequency; thus the chopped signal can be amplified with an ordinary a-c amplifier. Phase-sensitive demodulation and low-pass filtering reconstitute the input signal in amplified form at the output. Figure 10.9 shows a half-wave circuit which uses a chopper for both modulation and demodulation. More sophisticated circuits using two choppers and/or feedback techniques result in improved performance.[2]

A very-high-gain chopper amplifier[3] using a 400-cps chopper fre-

[1] Honeywell Heiland Div., Denver, Colo.

[2] P. F. Howden, A Review of Chopper Amplifiers, *Electro-Technol. (New York)*, p. 64, June, 1963.

[3] Honeywell Philadelphia Div., Philadelphia, Pa.

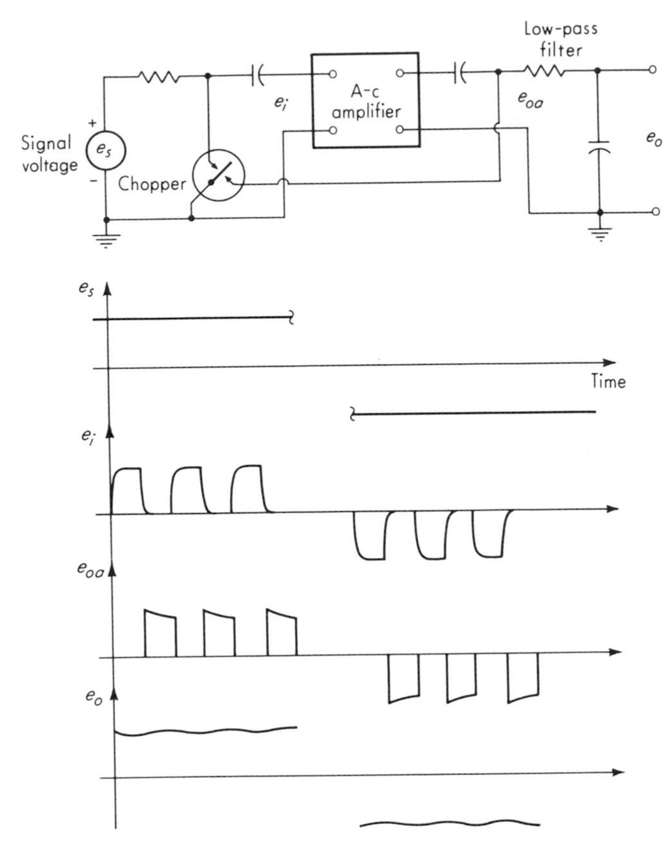

Fig. 10.9. *Chopper amplifier.*

quency has a gain adjustable from 50 to 100,000, output voltage of 5 volts into a 10,000-ohm load, maximum source resistance of 300,000 ohms, frequency response flat ± 3 db from 0 to 30 cps (for a gain of 50, reduced at higher gain), gain stability ± 0.1 percent, zero drift ± 0.5 μv (referred to the input) per month, nonlinearity 0.1 percent, noise 0.5 μv peak to peak with the input shorted, ripple ± 0.1 percent, and a common-mode rejection ratio of 10^8 at d-c and 60 cps.

By using electronic modulation techniques rather than electromechanical choppers, a sinusoidal rather than square-wave signal is obtained and much higher "chopping" frequency is possible, thus extending the frequency response of the overall system. A system of this type[1] designed to drive oscillograph galvanometers uses a 50,000-cps modulator, has a flat response of 0 to 5,000 cps and an input resistance of 35,000 ohms, pro-

[1] Alleghany Instrument Co., Cumberland, Md.

vides ±60-ma output current into 5- to 60-ohm loads, and has 10 μv rms noise referred to the input, nonlinearity 0.5 percent of full scale, and drift of 50 μv/hr referred to the input.

By careful selection of components and attention to compensation of drift-producing inputs it is today possible to produce *pure d-c amplifiers* (completely devoid of choppers or modulators) with quite respectable drift characteristics, particularly if the operating environment is relatively constant-temperature. Amplifiers of this type range from rather simple low-cost units of limited performance to those incorporating ingenious compensation schemes and meeting critical specifications. A typical unit[1] of the latter sort using silicon semiconductors has a frequency response flat within ±1 db from 0 to 20,000 cps, input impedance of 10^7 ohms, output impedance of 0.1 ohm, output voltage and current of ±10 volts and ±100 ma, nonlinearity ±0.01 percent for direct current to 2,000 cps, gain of 1,000, common-mode rejection 10^6 for direct current to 60 cps, noise referred to the input of 4 μv rms for full bandwidth (1 μv peak to peak for 10 cps bandwidth), recovery time for 10-volt overload of 0.01 sec, constant-temperature zero drift after 30-min warm-up ±0.02 percent for 200 hr, temperature-induced zero drift ±0.001 percent/C° referred to the output ±1 μv referred to the input, and an operating temperature of 0 to 50°C.

Another widely used d-c amplifier in instrumentation systems is the *chopper-stabilized d-c amplifier*. This instrument combines an essentially drift-free chopper amplifier (similar to Fig. 10.9) with a pure d-c amplifier in such a way that the drift of the overall system is reduced by a factor equal to the gain of the chopper amplifier. The arrangement also allows a frequency response that is not limited by the chopping frequency, since high-frequency signals are made to bypass the chopper amplifier and go directly through the pure d-c amplifier. Thus the wide bandwidth of the pure d-c amplifier is combined with the lack of drift of the chopper amplifier to produce a "hybrid" unit of highly desirable characteristics.

To illustrate these points briefly, we consider Fig. 10.10a, where a three-stage amplifier is shown with the drift voltage of each stage referred to its input. We can easily show that

$$\frac{e_o}{K_1K_2K_3} = e_i + e_{\text{drift},1} + \frac{e_{\text{drift},2}}{K_1} + \frac{e_{\text{drift},3}}{K_1K_2} \tag{10.24}$$

and thus compare the output due to desired input e_i with that due to drift. Note that the drift of the first stage is most important, since its effect at the output is unattenuated relative to e_i. Drift of stages 2 and 3, however, is reduced by factors K_1 and K_1K_2, respectively. Thus if a low-drift first stage of relatively high gain can be constructed, the overall

[1] Dana Laboratories, Irvine, Calif.

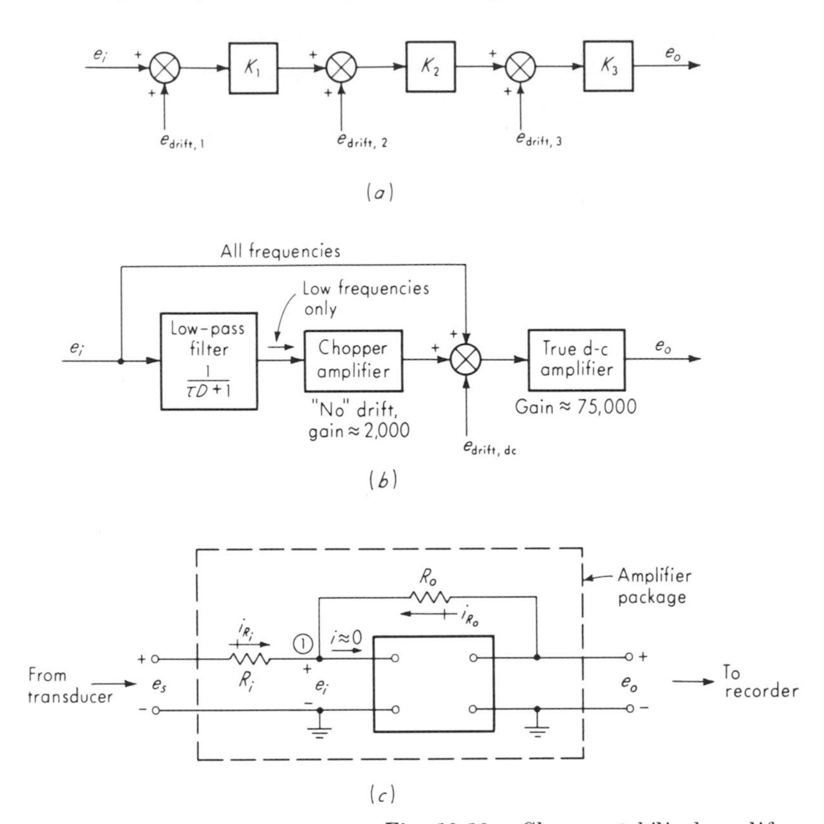

Fig. 10.10. *Chopper-stabilized amplifier.*

system drift will be small. Figure 10.10*b* shows in block-diagram form how this is realized in a typical chopper-stabilized d-c amplifier. All frequencies present in the signal e_i pass through the true d-c amplifier while only low-frequency components pass through the chopper amplifier and then through the d-c amplifier. Using the typical numerical values shown and assuming the first (chopper) stage has negligible drift, we see that the overall drift is reduced to 1/2,000 of the drift of the d-c amplifier itself. The gain for low-frequency signals is (2,000) (75,000) $= 150 \times 10^6$ while the high-frequency gain is 75,000. Since drift is a low-frequency phenomenon, the first-stage gain of 2,000 is effective in reducing drift to 1/2,000 of its former value, as noted above.

The amplifiers are not actually used in the "open-loop" configuration shown in Fig. 10.10*b* because such high gain is not needed, and by closing a feedback loop around the amplifier one can trade off some of this gain for other desirable properties such as linearity and control of input and output impedances. Thus the package sold as a chopper-stabilized instrumentation amplifier has the configuration of Fig. 10.10*c*, where the resistances R_i and R_o are chosen to give a closed-loop (e_s to e_o) gain of the order of a few thousand or less. We may analyze this circuit as follows: Assuming a high input resistance of the open-loop amplifier makes the current $i \approx 0$ and summing currents at the node 1 give

$$\frac{e_s - e_i}{R_i} = -\frac{e_o - e_i}{R_o} \tag{10.25}$$

Because of the very high open-loop gain (150×10^6 for low frequencies), the signal e_i can produce full-scale e_o (say 10 volts) without ever being greater than a few microvolts at most. Thus e_i is always negligible compared with e_s (which is of the order of millivolts) and e_o (which is of the order of volts). Equation (10.25) then becomes, to a very good degree of approximation,

$$\frac{e_o}{e_s} = -\frac{R_o}{R_i} \tag{10.26}$$

and thus the gain becomes a function of R_o and R_i only. Their numerical values can be very precisely set and maintained, giving very stable and accurate gain.

The above discussions and diagrams are simplifications intended to make clear the principles involved but, of course, neglect many details that must be considered in arriving at a practical amplifier. A typical differential-input chopper-stabilized instrument[1] has a gain of 1,000, nonlinearity ± 0.1 percent of full scale, input impedance of 10^8 ohms shunted by 700 pf, full-scale output of ± 10 volts and ± 10 ma, output resistance less than 25 ohms, drift ± 2 μv referred to the input plus ± 0.01 percent of full scale at 25°C, noise 3.5 μv rms referred to the input plus 150 μv referred to the output from 0.05 to 5,000 cps, frequency response flat within 1 db from 0 to 5,000 cps, recovery time from 500 percent overload of 300 μsec, and common-mode rejection of 10^7 at direct current and 10^6 at 60 cps.

Operational amplifiers. Operational amplifiers are actually open-loop pure d-c or chopper-stabilized amplifiers, as just described above, with very high gain. By closing appropriate feedback loops around them one can build up many kinds of useful active circuits. Such amplifiers are the basic building blocks of electronic analog computers,[2] for instance. They are also extremely useful in building special-purpose instrumentation equipment such as amplifiers, filters, integrators, etc., since a person with little electronics background can purchase these amplifiers ready-

[1] Astrodata, Anaheim, Calif.

[2] A. S. Jackson, "Analog Computation," McGraw-Hill Book Company, New York, 1960.

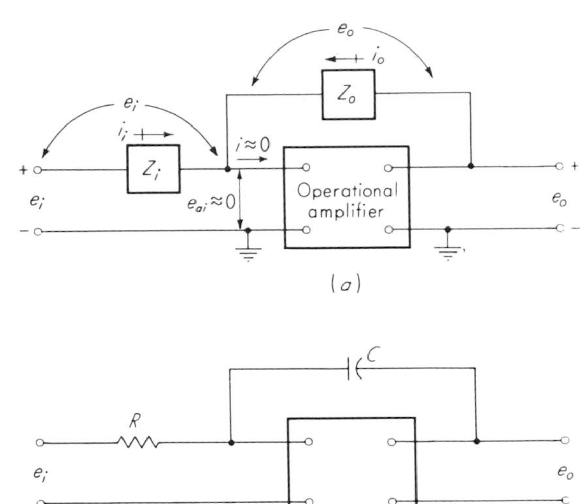

Fig. 10.11. *Operational amplifier.*

made and then add various relatively simple passive elements (resistors, capacitors, diodes, etc.) to obtain the desired functional characteristics.[1] Both linear and nonlinear operations are conveniently performed. To show the general linear operations, we consider Fig. 10.11*a*. Using assumptions as in Fig. 10.10*c*, we get

$$\frac{e_o}{e_i}(D) = -\frac{Z_o}{Z_i} \tag{10.27}$$

where

$$Z_o \triangleq \frac{e_o}{i_o}(D) = \text{operational impedance} \tag{10.28}$$

$$Z_i \triangleq \frac{e_i}{i_i}(D) = \text{operational impedance} \tag{10.29}$$

As an example, in Fig. 10.11*b*, $Z_o = 1/CD$ and $Z_i = R$, giving

$$\frac{e_o}{e_i}(D) = -\frac{1}{RCD} \tag{10.30}$$

$$e_o = -\frac{1}{RC}\int e_i\,dt \tag{10.31}$$

Thus this circuit performs the useful operation of integrating the signal e_i with respect to time. A wide variety of other operations can be per-

[1] A. S. Jackson, "Analog Computation," McGraw-Hill Book Company, New York, 1960.

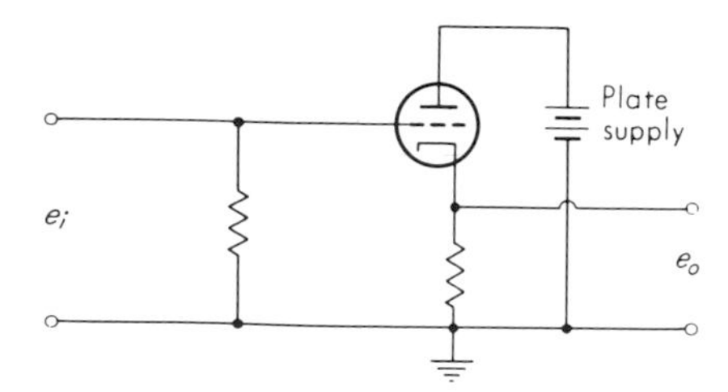

Fig. 10.12. *Cathode follower.*

formed, some of which will be discussed in the sections on filters and dynamic compensation.

Cathode followers, emitter followers. When a high-impedance transducer such as a piezoelectric crystal or capacitance displacement pickup must be coupled into a recording system of some type, the amplifier used must have a very high-input impedance if loading is to be minimized. If the amplifier is to supply appreciable current to the recorder, its output impedance must be low. The cathode-follower circuit of Fig. 10.12 provides such impedance-transformation properties. However, its gain is always less than 1 (often about 0.9); thus, if voltage amplification is needed, additional amplification must be provided. While the basic circuit of Fig. 10.12 has d-c response, many applications do not require this; thus a coupling capacitor is used at the input to reduce drift and biasing problems.

A typical unit[1] has a gain of 0.95, flat frequency response of 0.02 to 10^6 cps, input impedance of 2×10^9 ohms shunted by 12×10^{-12} farad, output impedance of 290 ohms, output voltage of 30 volts maximum, output current of 3 ma into 10,000 ohms load, and noise 75 μv rms. The transistor version of the cathode follower is called an emitter follower or source follower. Typical specifications[1] are a gain of 0.994, frequency response flat -3 db from 1 to 100,000 cps, input impedance of 10^8 ohms shunted by 20×10^{-12} farad, output impedance under 500 ohms, output voltage of 6 volts maximum, and noise of 300 μv at the output with the input shorted.

Charge amplifiers. Increasing use of piezoelectric accelerometers, pressure pickups, and load cells has led to the development of an amplifier type that offers some advantages over the usual voltage amplifier in certain applications. Such a *charge amplifier* is shown connected to a piezoelectric transducer in Fig. 10.13. The idealized form is shown in Fig. 10.13*a* where we note that an operational amplifier is used with a

[1] Columbia Research Laboratories, Woodlyn, Pa.

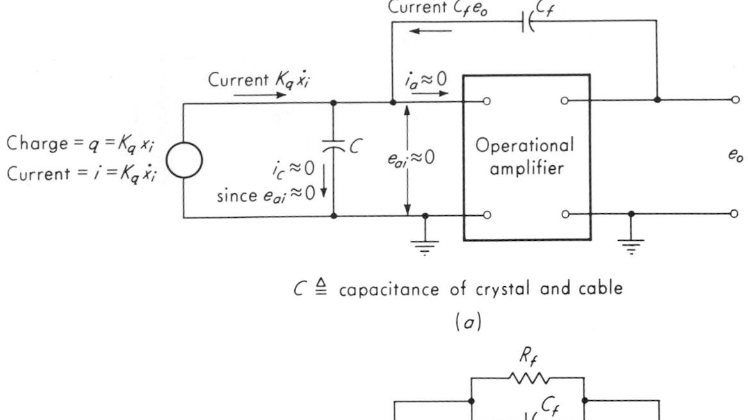

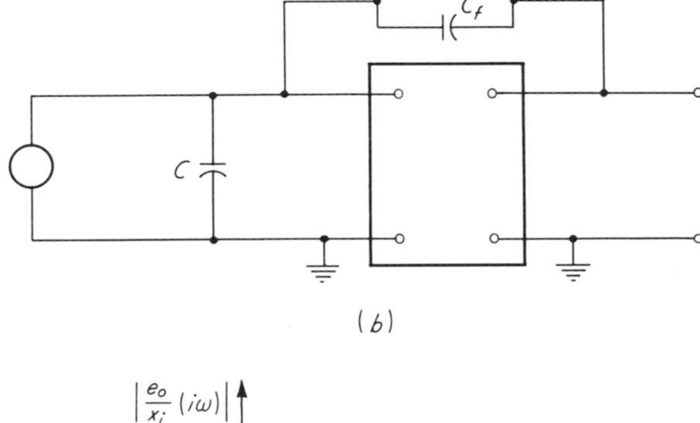

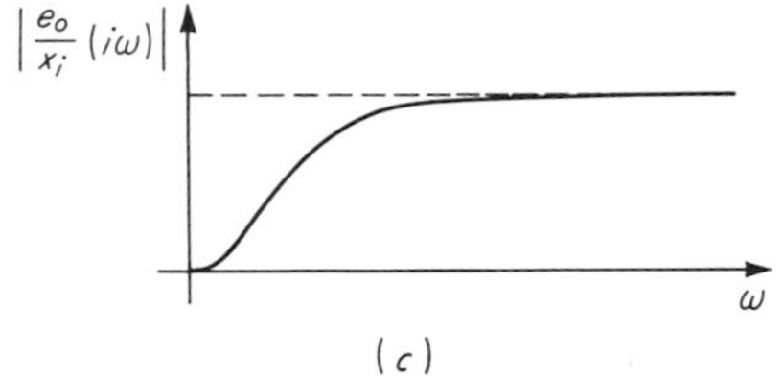

***Fig. 10.13.** Charge amplifier.*

capacitor C_f in the feedback path. Assuming, as usual, that the input voltage e_{ai} and current i_a of the operational amplifier are small enough to take as zero, we get

$$K_q D x_i = -C_f D e_o \tag{10.32}$$

$$e_o = -\frac{K_q x_i}{C_f} \tag{10.33}$$

Equation (10.33) indicates that e_o would be instantaneously and linearly related to displacement x_i without the usual loss of steady-state response associated with piezoelectric transducers and voltage amplifiers. Unfortunately this advantage is not realizable since a system constructed as in Fig. 10.13*a* would, because of noninfinite input resistance of the operational amplifier and leakage of C_f, exhibit a steady charging of C_f by the leakage current until the amplifier saturated. To overcome this problem, in the practical circuit of Fig. 10.13*b* a feedback resistance R_f is included to prevent this small leakage current from developing a significant charge on C_f. Analysis of this new circuit gives

$$\frac{e_o}{x_i}(D) = \frac{K\tau D}{\tau D + 1} \tag{10.34}$$

where

$$K \triangleq \frac{K_q}{C_f} \quad \text{volts/in.} \tag{10.35}$$

$$\tau \triangleq R_f C_f \quad \text{sec} \tag{10.36}$$

Equation (10.34) is of identical form with the transfer function of a piezoelectric transducer and a *voltage* amplifier and exhibits the same loss of static and low-frequency response. The advantages of the charge amplifier are found in Eqs. (10.35) and (10.36). We note that both the sensitivity K and time constant τ are now independent of the capacitance of the crystal itself and the connecting cable whereas with a voltage amplifier neither of these advantages is obtained. Thus long cables (often several hundred feet in practical setups) do not result in a reduced sensitivity or a variation in frequency response. These advantages and others[1] are sufficient to make the charge amplifier of practical interest in many systems. Disadvantages[2] that may arise in certain applications include a possibly poorer signal/noise ratio and a reduction in natural frequency of the transducer due to loss of stiffness caused by what amounts to a short circuit across the crystal.

When used with quartz-crystal transducers,[3] the value of C_f is from 10 to 100,000 pf and R_f is 10^{10} to 10^{14} ohms. For $C_f = 100{,}000$ pf and $R_f = 10^{14}$ ohms, $\tau = 10^6$ sec, showing that practically d-c response, allowing static calibration and measurement, is possible under these conditions. For ceramic-type transducers, C_f is from 10 to 1,000 pf and R_f from 10^8 to 10^{10} ohms, making the maximum τ about 10 sec and static measurements thus usually impractical.[4] A typical charge amplifier[4] for use with quartz-crystal transducers has adjustable gain from 0.01 to 100 mv/pcoul, output voltage to high impedance load +10, −5 volts, output current to low impedance load ±10 ma, output impedance of 100 ohms, R_f switch selectable as 10^{11} or 10^{14} ohms, frequency response from practically direct cur-

[1] D. Pennington, Charge Amplifier Applications, Endevco Corp., Pasadena, Calif., April, 1964.

[2] Wilcoxon Research, Bethesda, Md., *Wilcoxon Res. Bull.* 5.

[3] Kistler Instrument Corp., Clarence, N.Y., *Tech. Notes* 133762 and 130662.

[4] Kistler Instrument Corp., *Tech. Notes* 133762 and 130662.

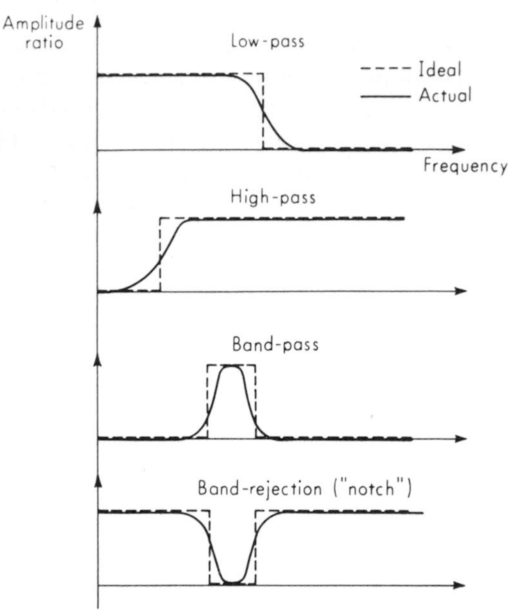

Fig. 10.14. *Basic filter characteristics.*

rent to 150,000 cps, nonlinearity 0.1 percent, and output noise of 2 mv with the input shorted.

Magnetic amplifiers. Magnetic amplifiers are not at present widely used as general-purpose instruments in research and development laboratories or data-gathering systems. For special-purpose applications in measurement and control systems their reliability, high isolation, and other advantages make them attractive, and numerous successful applications exist.[1,2]

10.3

Filters The use of frequency-selective filters to pass the desired signals and reject spurious ones has been discussed before. Figure 10.14 summarizes the most common frequency characteristics used. Filters may take many physical forms; however, the electrical form is most common and highly developed with regard to both theory and practical realization. By use of analogies, the material on electrical filters may suggest

[1] B. A. Mazzeo, A Low-level, High-accuracy D-C Magnetic Amplifier, *Elec. Mfg.*, November, 1958.

[2] Acromag Design Manual, Acromag Inc., Detroit, Mich.

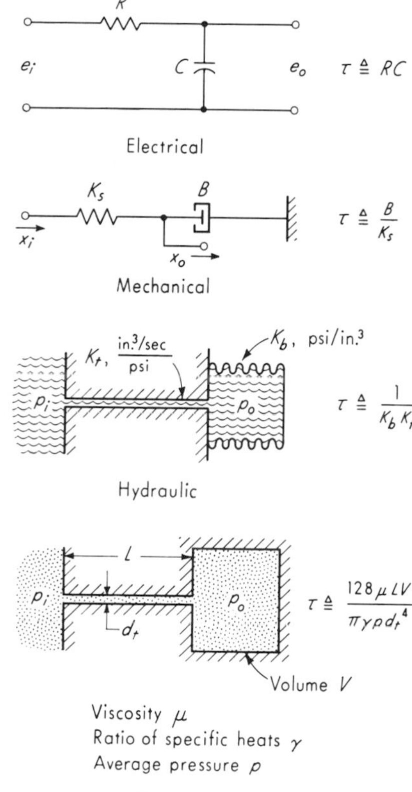

Fig. 10.15. *Low-pass filters.*

the configurations of mechanical, hydraulic, acoustical, etc., systems that will provide the desired filtering action in specific problems.

Low-pass filters. The simplest low-pass filters commonly in use are shown in several different physical forms in Fig. 10.15. They all have identical transfer functions given by

$$\frac{e_o}{e_i}(D) = \frac{x_o}{x_i}(D) = \frac{p_o}{p_i}(D) = \frac{1}{\tau D + 1} \tag{10.37}$$

Since these are all simple first-order systems the attenuation is quite gradual with frequency: 6 db/octave. This does not give a very sharp distinction between the frequencies that are passed and those that are rejected. By adding more "stages" (see Fig. 10.16*a*) the sharpness of cutoff may be increased. The use of inductance elements (Fig. 10.16*b*) may also lead to better filtering action. When inserting a filter into a

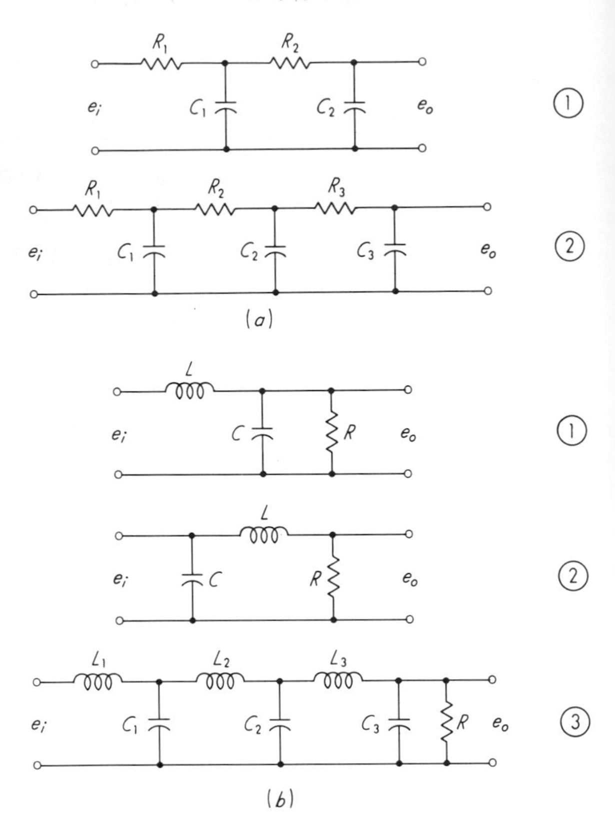

Fig. 10.16. *Sharper-cutoff low-pass filters.*

system, it is necessary to take into account possible loading effects by use of appropriate impedance analysis.

High-pass filters. Figure 10.17 shows the simplest high-pass filters, which all have the transfer function

$$\frac{e_o}{e_i}(D) = \frac{x_o}{x_i}(D) = \frac{x_o/K_{px}}{p_i}(D) = \frac{\tau D}{\tau D + 1} \tag{10.38}$$

Again, the attenuation is quite gradual, and more complex configurations are needed to obtain a more sharply defined cutoff (Fig. 10.18).

Band-pass filters. By cascading a low-pass and a high-pass filter one can obtain the band-pass characteristic (Fig. 10.19). To sharpen the rejection on either side of the passband one can simply use the sharper low- and high-pass sections mentioned above.

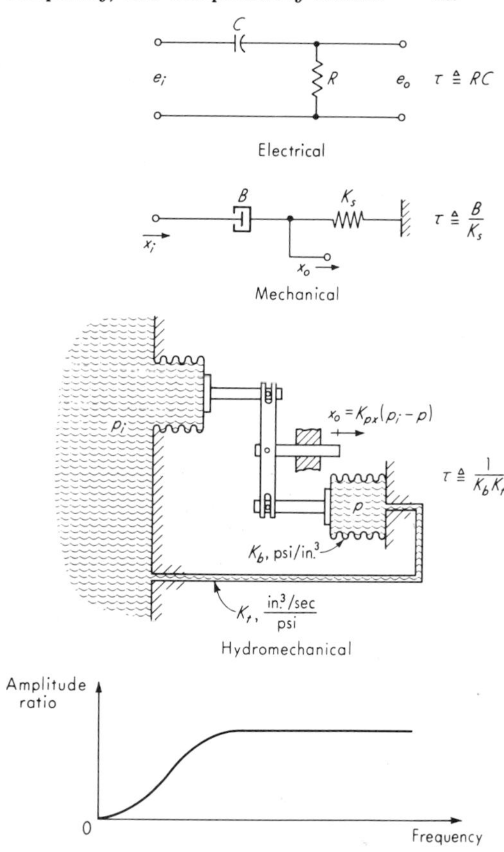

Fig. 10.17. *High-pass filters.*

Band-rejection filters. A common application of a band-rejection filter is found in the input circuits of self-balancing potentiometer and X-Y recorders. These instruments are subject to interfering 60-cps noise voltages. Since the frequency response of the overall recorder is good only to a few cycles per second, a band-rejection filter tuned to 60 cps may be employed without distorting any desired signals. Such a filter prevents noise signals from saturating the recorder's amplifiers and preventing the proper amplification of the desired signals.

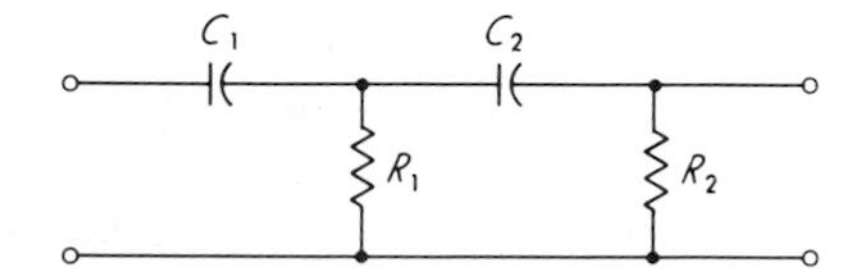

Fig. 10.18. *Sharper-cutoff high-pass filter.*

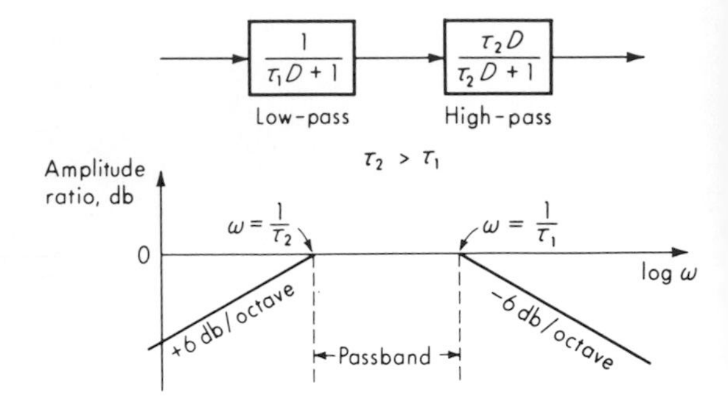

Fig. 10.19. *Band-pass filter*

Passive networks commonly used for rejection of a band of frequencies include the bridged-T and the twin-T network (Fig. 10.20). While the bridged-T does not completely reject any frequency, the twin-T can be so designed, as shown in Fig. 10.20. Equations and charts for designing these filters are available.[1]

Active filters. The filter circuits shown up to this point have all been passive networks; that is, they have no power source within them. By the use of amplifiers as buffer devices, in feedback schemes, or in operational amplifier configurations, many desirable features difficult to obtain in other ways may be achieved. While we shall not discuss here the internal construction of such active filters, we shall describe briefly the operating characteristics of some particular general-purpose laboratory-type units.[2] The frequency characteristics of these filters are adjustable over a wide range, including very low frequencies which are difficult to achieve with passive filters. Their high input impedance and low output impedance allow them to be inserted into a system without causing loading problems. One unit provides a band-pass characteristic with the low-frequency and the high-frequency cutoffs separately and continuously adjustable over the range 0.02 to 2,000 cps. The transfer function is approximately

$$\frac{e_o}{e_i}(D) = \frac{\tau_1{}^4D^4}{(\tau_1{}^2D^2 + 1.2\tau_1D + 1)^2(\tau_2{}^2D^2 + 1.2\tau_2D + 1)^2} \tag{10.39}$$

where $1/2\pi\tau_1$ is the low cutoff frequency and $1/2\pi\tau_2$ is the high. We see that the attenuation rate of this filter at the edges of the passband is

[1] J. E. Gibson and F. B. Tuteur, "Control System Components," p. 43, McGraw-Hill Book Company, New York, 1958.

[2] Krohn-Hite Corp., Cambridge, Mass.

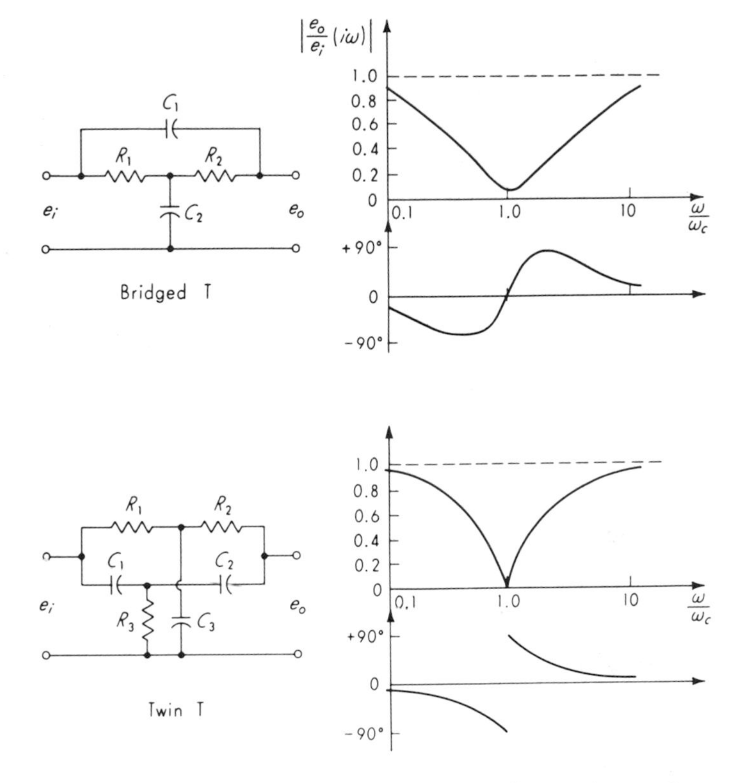

Fig. 10.20. *Band-rejection filters.*

24 db/octave. With the input shorted the noise level at the output is 0.1 to 1.0 mv. This noise may be attributed mainly to the active elements and limits the use of active filters for very low-level signals.

A band-rejection filter of the same manufacturer has a transfer function of approximately

$$\frac{e_o}{e_i}(D) = \frac{1}{(\tau_1{}^2D^2 + 1.2\tau_1 D + 1)^2} + \frac{\tau_2{}^4D^4}{(\tau_2{}^2D^2 + 1.2\tau_2 D + 1)^2} \tag{10.40}$$

where the low cutoff frequency is $1/2\pi\tau_1$ and the high $1/2\pi\tau_2$. Both are independently adjustable over the range 0.02 to 2,000 cps. The unit may also be set to give a sharp single-frequency null over the range 0.1 to 500 cps. The attenuation rate is again 24 db/octave.

A hydraulic band-pass filter for an oceanographical transducer. While filtering with electrical networks is most common, in some instances other physical forms present advantages. This section illustrates this

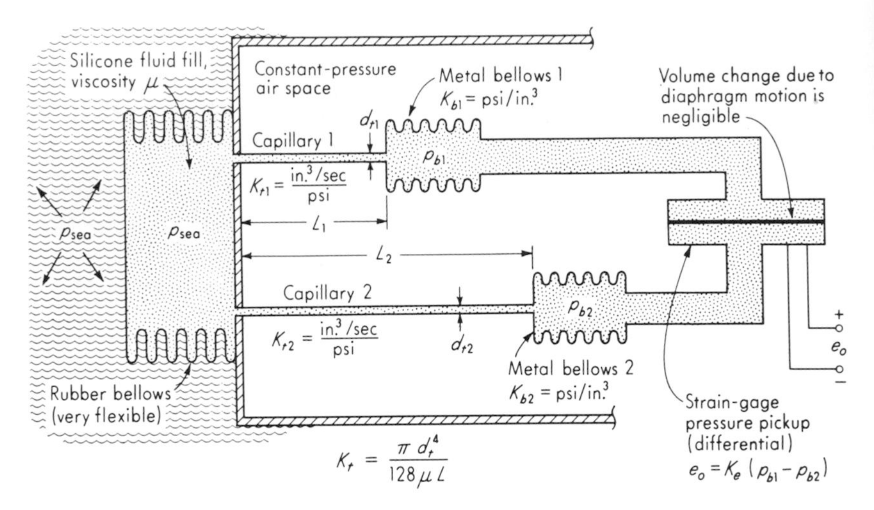

Fig. 10.21. *Hydraulic band-pass filter.*

with an example of a filter that has been successfully constructed and used.[1]

The Scripps Institution of Oceanography at La Jolla, California, uses pressure transducers in its studies of ocean-wave phenomena. A particular study required measurements of waves whose frequencies are lower than those of ordinary gravity waves (which one observes visually) and higher than those due to tides. These waves of intermediate frequency are of rather low amplitude relative to those due to tides and gravity and thus are rather difficult to measure with a pressure pickup which treats all frequencies about equally. The band-pass filter and pressure pickup of Fig. 10.21 solves this problem since it is "tuned" to the frequency range of interest, which is about 0.001 cps. Such low frequencies are very difficult to handle with electrical circuits but the hydraulic filter shown gives very good results with quite simple and reliable components.

In use, the pressure transducer is located underwater, often buried in a foot of sand for temperature insulation, with a "snorkel" tube extending up through the sand to sense water pressure. This pressure is directly related to the height of the waves passing overhead; thus a record of pressure-transducer output voltage is a record of wave activity. Analysis of the system gives

$$\frac{e_o}{p_{\text{sea}}}(D) = \frac{K_e(\tau_2 - \tau_1)D}{(\tau_1 D + 1)(\tau_2 D + 1)} \tag{10.41}$$

[1] F. E. Snodgrass, Shore-based Recorder of Low-frequency Ocean Waves, *Trans. Am. Geophys. Union*, p. 109, February, 1958.

where $K_e \triangleq$ sensitivity of differential pressure pickup, mv/psi

and

$$\tau_1 \triangleq \frac{1}{K_{t1}K_{b1}} \qquad \text{sec} \tag{10.42}$$

$$\tau_2 \triangleq \frac{1}{K_{t2}K_{b2}} \qquad \text{sec} \tag{10.43}$$

One can easily show that the frequency ω_p of peak response is given by

$$\omega_p = \sqrt{\frac{1}{\tau_1\tau_2}} \qquad \text{rad/sec} \tag{10.44}$$

the amplitude ratio M_p at this frequency is

$$M_p = \frac{K_e(\tau_2 - \tau_1)}{\tau_2 + \tau_1} \tag{10.45}$$

and the phase angle at ω_p is zero.

Filtering by statistical averaging. All the filters mentioned above are of the frequency-selective type and of course require that the desired and spurious signals occupy different portions of the frequency spectrum. When signal and noise contain the same frequencies, such filters are useless. A basically different scheme may be usefully employed under such circumstances if the following is true:

1. The noise is random.
2. The desired signal can be caused to repeat itself over and over.

If these two conditions are fulfilled, it should be clear that, if one adds up the ordinates of several samples of the total signal at like values of abscissa (time), the desired signal will reinforce itself while the random noise will gradually cancel itself. This will occur even if the frequency content of signal and noise occur in the same part of the frequency spectrum. It can be shown that the signal/noise ratio improves in proportion to the square root of the number of samples used. Thus theoretically the noise can be eliminated to any desired degree by adding up a sufficiently large number of signals. In practice, various factors prevent realization of theoretically optimum performance.

While the above procedure could be carried out manually, it may also be automated to increase speed, convenience, and accuracy. One such system[1] samples the incoming signal at 512 time points, converts the ordinates to digital values, and stores them in a digital memory. Further

[1] Noise Reduction by Digital Signal Averaging, Signal Averaging by Waveform Totalling, Signal Averaging by Waveform Comparison, Northern Scientific Inc., Madison, Wis.

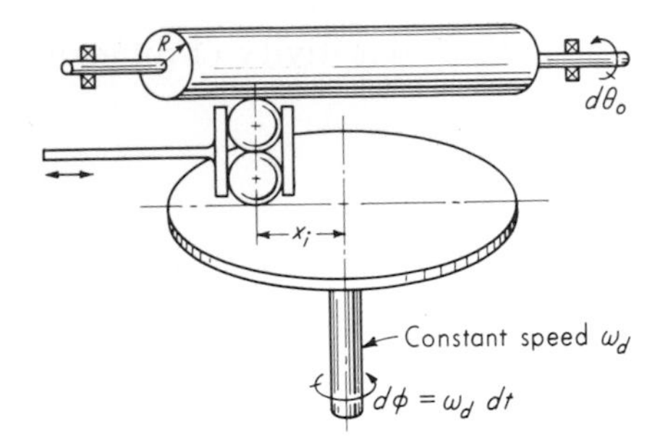

Fig. 10.22. Ball-disk integrator.

repetitions of the signal are similarly sampled with the respective ordinate values being totalized. The contents of the digital memory are presented on an oscilloscope screen, and one can actually watch the true signal "emerge" from the noise as one sample after another is put into the system.

10.4 Integration and Differentiation

Often in measurement systems it is necessary to obtain integrals and/or derivatives of signals with respect to time. Depending on the physical nature of the signal, various devices may be most appropriate. Accurate differentiation is generally harder to accomplish than integration since differentiation tends to accentuate noise (which is usually high frequency) whereas integration tends to smooth noise. Thus second and higher integrals may easily be achieved while derivatives present real difficulties.

Integration. If the signal to be integrated is already a mechanical displacement or is easily transduced to one, the *ball-and-disk integrator* of Fig. 10.22 may be used. Assuming rigid bodies and no slippage, we can show that

$$\frac{\theta_o}{x_i}(D) = \frac{\omega_d}{R}\left(\frac{1}{D}\right) \qquad (10.46)$$

and thus the rotation angle θ_o is proportional to the first time integral of the displacement x_i. A typical unit[1] has a maximum ω_d of 500 rpm, $x_i = \pm 0.75$ in., maximum input force of 2 oz, output torque of 3 in.-oz, reproducibility 0.01 percent of full scale, accuracy 0.05 percent of full scale, and expected life of 10,000 hr. This unit uses a precision-lapped tungsten carbide disk with 1-μin. surface finish and hardened-tool-steel roller and balls.

Two electromechanical means of obtaining integrals are shown in

[1] Librascope Div., Glendale, Calif.

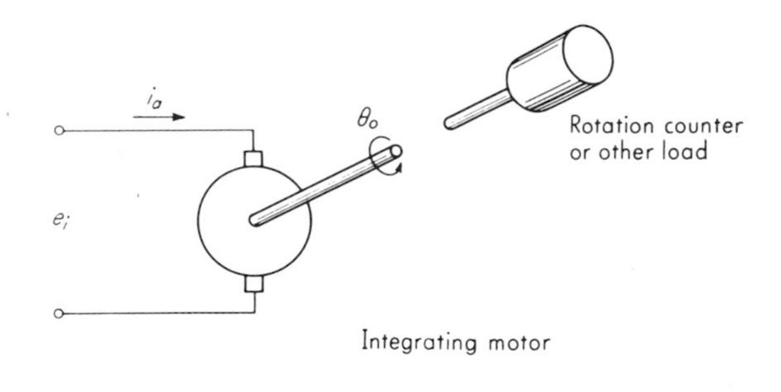

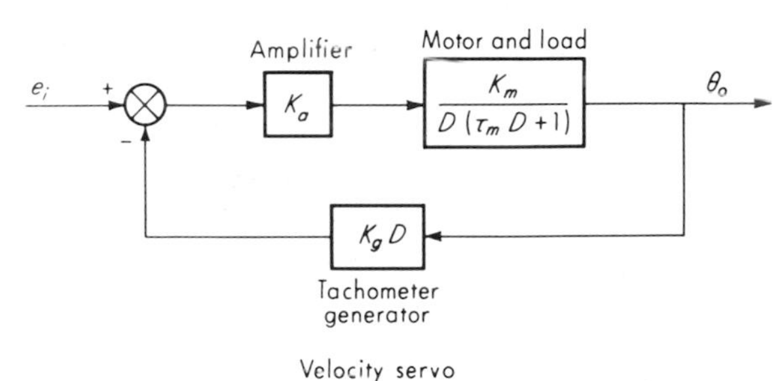

Fig. 10.23. *Electromechanical integrators.*

Fig. 10.23: the *integrating motor* and the *velocity-servo integrator.* Both these accept electrical signals as input and produce mechanical rotations in proportion to the time integral of the input voltage. The integrating motor is essentially a d-c motor with permanent magnet field in which friction, iron losses, and brush-contact voltage drop have been reduced to extremely low levels, resulting in an input-voltage/output-speed characteristic that is very linear over a wide range of input voltage. For a d-c motor with constant field,

$$\text{Armature current} = i_a = \frac{e_i - e_m}{R} \tag{10.47}$$

where

$$e_m \triangleq \text{motor back emf} = K_e\dot{\theta}_o \tag{10.48}$$

$$R \triangleq \text{armature resistance} \tag{10.49}$$

Motor torque $T_m = K_{mt}i_a$, where K_{mt} is the motor-torque constant. Thus if rotor inertia is J, we have

$$T = J\ddot{\theta}_o \tag{10.50}$$

$$K_{mt}\frac{e_i - K_e\dot{\theta}_o}{R} = J\ddot{\theta}_o \tag{10.51}$$

$$\frac{\theta_o}{\int e_i\,dt}(D) = \frac{1/K_e}{\tau D + 1} \tag{10.52}$$

$$\tau \triangleq \frac{RJ}{K_{mt}K_e} \tag{10.53}$$

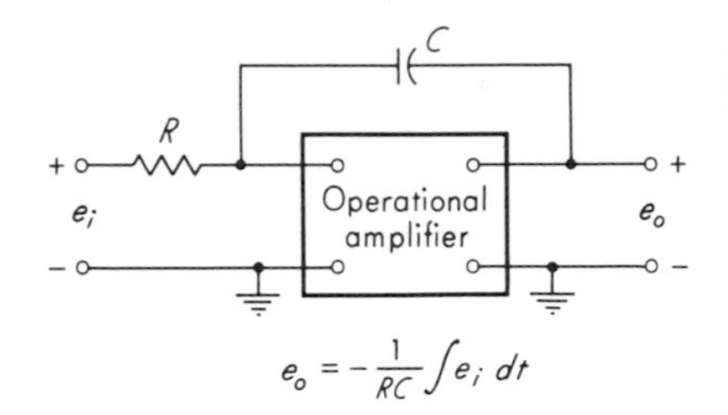

Fig. 10.24. *Electronic integrator.*

We see that the rotation angle θ_o (which can be counted by a simple mechanical counter) is a measure of the time integral of e_i with a first-order lag. A family of such instruments[1] has full-scale input voltage ranging from 1.5 to 24 volts, R of 2.8 to 700 ohms, τ of about 0.01 sec, full-scale speed of 1,885 to 1,260 rpm, and starting voltages of 4.2 to 79 mv. For a motor without any external load the nonlinearity is better than 0.3 percent of full scale from 5 to 200 percent of full-scale voltage. These motors can be used only to drive very light loads, 1.8 to 12.4 g-cm at full voltage.

For greater accuracy and the ability to drive loads requiring greater power output, the velocity-servo integrator may be employed. Analysis of the block diagram of Fig. 10.23 gives

$$\frac{\theta_o}{\int e_i\, dt}(D) = \frac{1/K_g}{\tau D + 1} \tag{10.54}$$

where

$$\tau \triangleq \frac{\tau_m}{1 + K_a K_g K_m} \tag{10.55}$$

and $1/K_g \approx K_a K_m/(1 + K_a K_m K_g)$ since K_a is very large. Such integrators achieve accuracies of 0.1 percent and better.[2]

Figure 10.24 shows the operational-amplifier type of integrator extensively used in general-purpose electronic analog computers. By use of high-quality chopper-stabilized operational amplifiers an integrator of quite low drift can be constructed in this way. Accuracies of the order of 0.1 percent for short-term operation and 1 percent over 14 hr are typical of high-quality electronic integrators of this type.[2] If higher integrals are desired, such units may be cascaded; however, drift becomes more troublesome. In addition to providing a closer approximation to true integration than the passive networks discussed in the following paragraph, the presence of the amplifier (with its own power supply) means that power can be supplied to the device following the integrator without taking any signifi-

[1] Electro Methods Ltd., Stevenage, England.
[2] W. H. Barr, Integrators, *Electromech. Design*, p. 57, October, 1961.

cant power from the device supplying the integrator. That is, operational-amplifier circuits can generally have a high input impedance and low output impedance.

All the low-pass filters of Fig. 10.15 may be used as *approximate integrators* for input signals within a restricted frequency range. This can be shown as follows:

$$\frac{e_o}{e_i}(i\omega) = \frac{1}{i\omega\tau + 1} \tag{10.56}$$

Now if $\omega\tau \gg 1$

$$\frac{e_o}{e_i}(i\omega) \approx \frac{1}{i\omega\tau} \tag{10.57}$$

and thus

$$\frac{e_o}{e_i}(D) \approx \frac{1}{\tau D} \tag{10.58}$$

$$e_o \approx \frac{1}{\tau}\int e_i\,dt \tag{10.59}$$

Thus, if the frequency spectrum of the input signal is such that $\omega\tau \gg 1$ for all significant frequencies, a good approximation to the desired integrating action is obtained. For a given τ the approximation improves as ω increases. It appears as if any ω can be accommodated by choosing τ sufficiently large. However, large τ decreases the magnitude of the output; thus this can be carried only as far as the noise level of the system permits.

If a signal is available in digital form, it may be integrated by a general-purpose digital computer by programming it for one of the approximate numerical-integration schemes such as Simpson's rule. Another type of *digital integration*, which can be carried out without use of a general-purpose computer, involves use of pulse-totalizing methods. Here the analog voltage signal is converted to a periodic voltage signal whose frequency is proportional to the input-signal amplitude (voltage-to-frequency converter). This periodic signal is then applied to an electronic counter. Thus the reading of the counter at any time is proportional to the time integral of the input signal up to that time.

Differentiation. For mechanical displacement signals the various velocity pickups, accelerometers, jerk pickup, tachometer generator, and rate gyro of Chap. 4 may be considered as differentiating devices.

All the high-pass filters of Fig. 10.17 may be used as *approximate differentiators* for input signals within a restricted frequency range, as shown by the following analysis:

$$\frac{e_o}{e_i}(i\omega) = \frac{i\omega\tau}{i\omega\tau + 1} \tag{10.60}$$

Now if $\omega\tau \ll 1$

$$\frac{e_o}{e_i}(i\omega) \approx i\omega\tau \tag{10.61}$$

$$\frac{e_o}{e_i}(D) \approx \tau D \tag{10.62}$$

$$e_o \approx \tau \frac{de_i}{dt} \tag{10.63}$$

We note here that for a given τ the approximation improves for lower values of ω. Again τ may be reduced to extend accurate differentiation to higher frequencies. However, small τ reduces sensitivity; thus noise level is limiting just as in the approximate integrators.

Use of operational amplifiers results in both approximate and "exact" differentiators of improved performance relative to the passive high-pass filters discussed above. Figure 10.25 shows some of these circuits. In Fig. 10.25*a*, analysis of this "exact" differentiator gives

$$\frac{e_o}{e_i}(D) = -RCD \tag{10.64}$$

This circuit is rarely useful because the ever-present noise (generally of high frequency relative to the desired signal) will completely swamp the desired signal at the output. All exact differentiators must suffer from this problem. It can be alleviated only by shifting to approximate differentiators which include low-pass filters to take out the effects of high-frequency noise. Figure 10.25*b* shows a common scheme which, when analyzed, gives

$$\frac{e_o}{e_i}(D) = -\frac{R_2CD}{R_1CD + 1} \tag{10.65}$$

This gives an accurate derivative for frequencies such that $R_1C\omega \ll 1$ and amplifies high-frequency noise only by an amount R_2/R_1. To attenuate noise, one must use a second-order-type low-pass filter, such as given by the circuit of Fig. 10.25*c*. Analysis of this gives

$$\frac{e_o}{e_i}(D) = -\frac{R_2C_1D}{(R_2C_2D + 1)(R_1C_1D + 1)} \tag{10.66}$$

Figure 10.25*d* shows an actual circuit[1] designed for measuring the rate of charging or discharging of batteries and using a solid-state operational amplifier. Analysis gives

$$\frac{e_o}{e_i}(D) = -\frac{10R_2C_1D}{(R_1C_1D + 1)[R_2(10C_3 + C_2)D + 1]} \tag{10.67}$$

[1] The Lightning Empiricist, Philbrick Researches Inc., Boston, Mass., October, 1963.

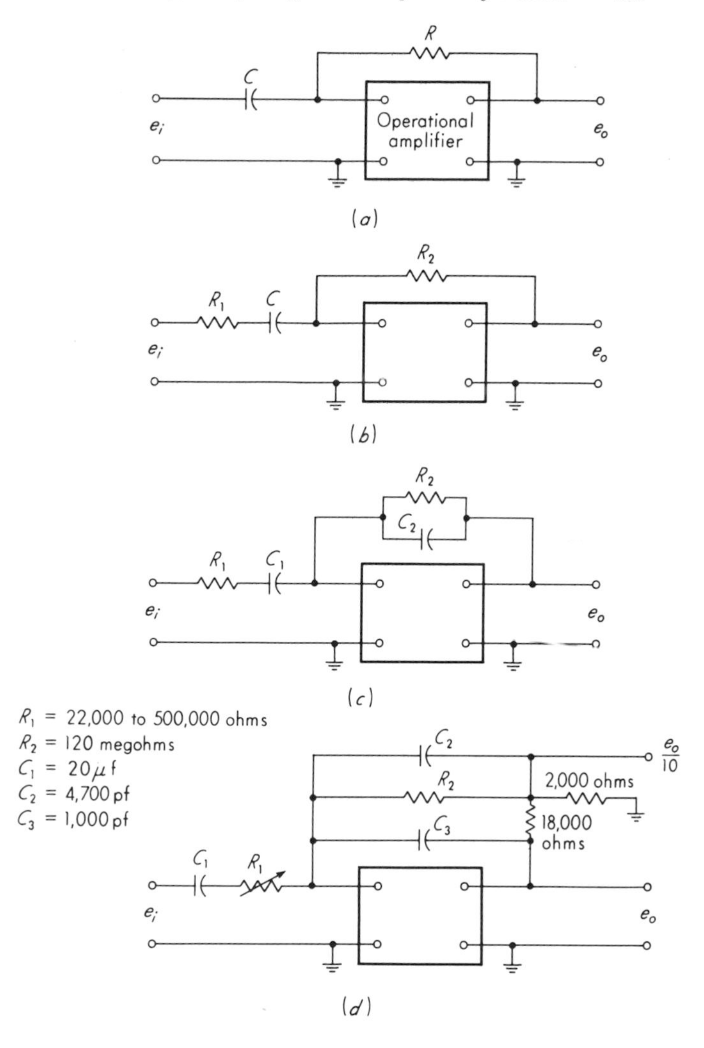

Fig. 10.25. *Electronic differentiators.*

The output is read on a meter which may be connected to e_o or $e_o/10$, depending on the size of the output. For the numerical values given and $R_1 = 22,000$

$$\frac{e_o}{e_i}(D) = -\frac{24,000D}{(0.44D + 1)(1.764D + 1)} \qquad (10.68)$$

We note that for $De_i = 10$ mv/min the output is 4 volts. If, say, $\frac{1}{2}$ mv of 60-cps noise is present at the input, the output noise is only 41 mv, which is about 1 percent of the desired output.

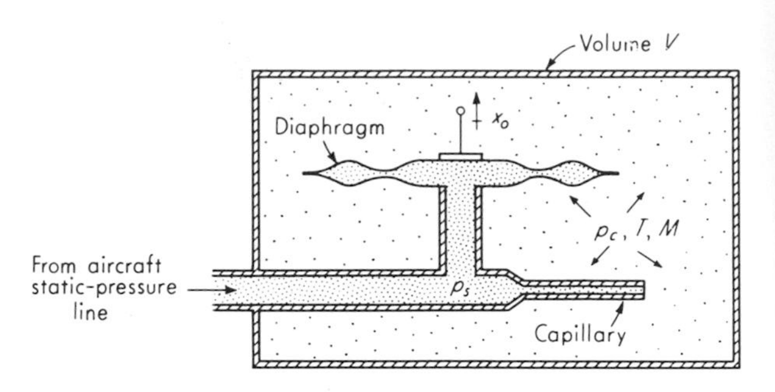

Fig. 10.26. *Rate-of-climb sensor.*

A final example of a differentiator using nonelectrical methods is the aircraft rate-of-climb indicator shown in Fig. 10.26. Since atmospheric pressure varies with altitude, a device that measures rate of change of atmospheric pressure can indicate rate of climbing or diving. While actual design requires a more critical study,[1] we here consider a simplified linear analysis to show the main features. Static pressure p_s corresponding to aircraft altitude is fed from the vehicle's static pressure probe to the input tube of the rate-of-climb indicator. Leakage through a capillary tube into the chamber of volume V occurs at a mass flow rate assumed to be $K_c(p_s - p_c)$ lb_m/sec. Air in the chamber follows the perfect-gas law $p_cV = MRT$. Motion of the output diaphragm is according to $x_o = K_d(p_s - p_c)$. Assuming K_c and T to be constant, analysis gives

$$\frac{x_o}{Dp_s}(D) = \frac{K}{\tau D + 1} \tag{10.69}$$

where

$$K \triangleq \frac{K_dV}{RTK_c} \quad \text{in./(psi/sec)} \tag{10.70}$$

$$\tau \triangleq \frac{V}{RTK_c} \quad \text{sec} \tag{10.71}$$

Thus x_o, which may be measured with any displacement transducer, is an indication of rate of change of p_s, and thereby a measure of rate of climb, if pressure is assumed to be a linear function of altitude. Since this is not exactly true, various compensating devices are needed in a practical instrument for this and other spurious effects.

10.5

Dynamic Compensation Sometimes it is not possible to obtain the desired behavior from a measuring device solely by adjusting its own

[1] D. P. Johnson, Aircraft Rate-of-climb Indicators, *NACA, Rept.* 666, 1939.

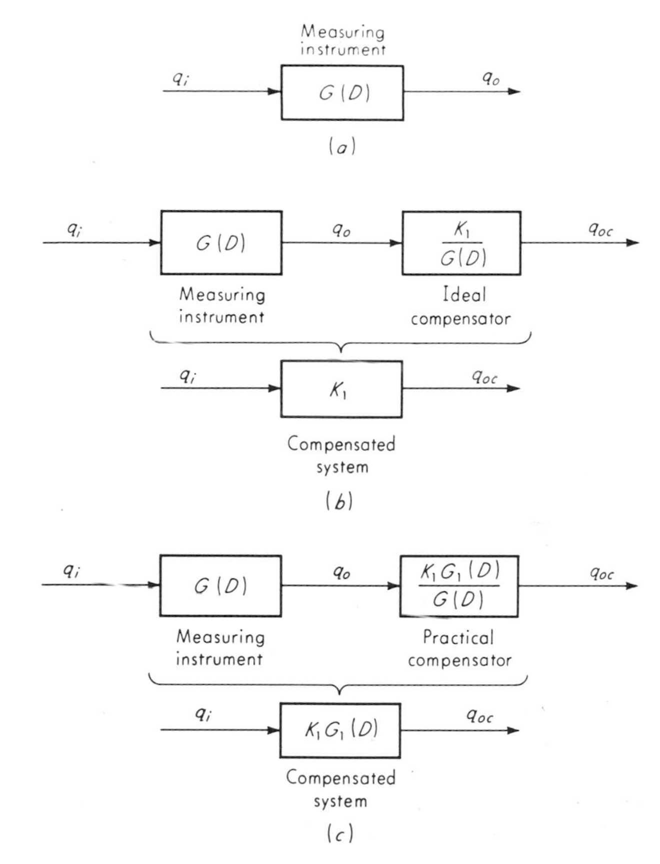

Fig. 10.27. *Generalized dynamic compensation.*

parameters. To get fast response from a thermocouple, for example, very fine wire must be used. Perhaps the vibration and temperature environment might be so severe that such a fine-wire thermocouple would be destroyed before any readings could be obtained. For this and similar situations the concept of dynamic compensation may provide a solution.

Figure 10.27 shows the general arrangement by which dynamic compensation may be employed. Ideally an instrument with transfer function $G(D)$ is cascaded with a compensator $K_1/G(D)$ and thus (if negligible loading is assumed) the overall system now has *instantaneous response* since its transfer function is just the constant K_1. This result is, of course, too good to be true, the practical difficulty lying in the construction of the compensator $K_1/G(D)$, which is generally not physically realizable, because of the need for perfect differentiating effects. While perfect compensation for instantaneous response is *not* possible, very great improve-

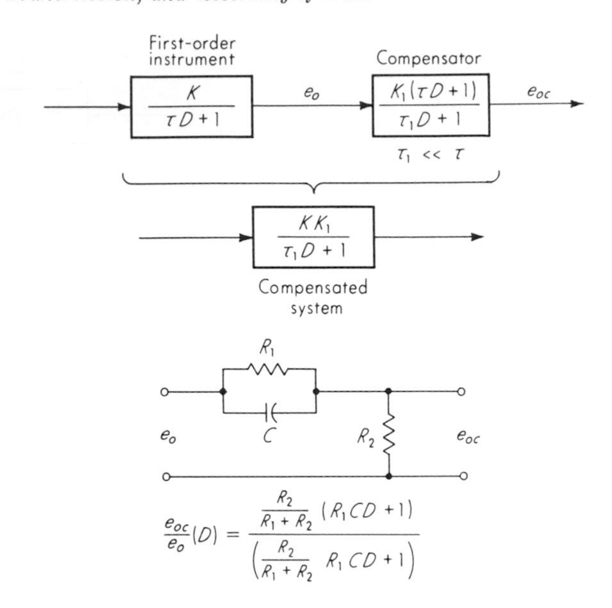

***Fig. 10.28.** Dynamic compensation for first-order system.*

ments *may* be achieved with the scheme of Fig. 10.27*c*. Here the undesirable dynamics $G(D)$ are *replaced* with more desirable ones, $G_1(D)$. This technique has, for example, been used with good success in speeding up the response of temperature-sensing elements and hot-wire anemometers. These are basically first-order instruments; thus $G(D) = K/(\tau D + 1)$. While the compensator can, in general, take any suitable physical form, because most sensors produce an electrical output, most compensators in use are electrical circuits. The compensator generally used for first-order systems takes the form shown in Fig. 10.28. Note that any increase in speed of response ($\tau_1 \ll \tau$) is paid for by a loss of sensitivity in direct proportion, since $K_1 = R_2/(R_1 + R_2) = \tau_1/\tau_2$. If this loss of sensitivity is not tolerable, additional amplification is needed. It is usually placed between the sensor and the compensator because it will then also serve to unload the two circuits from each other. While several stages of such compensation may sometimes be used, such staging cannot be carried beyond a certain point because of additional noise introduced by the amplifiers and accentuated by the compensators. However, a response speedup of the order of 100:1 is feasible and has been achieved for thermocouples and hot-wire anemometers.

The concept of dynamic compensation is theoretically applicable to any-order system. A good example of a more complex application is found in the "equalization" of vibration shaker systems. Figure 10.29*a*

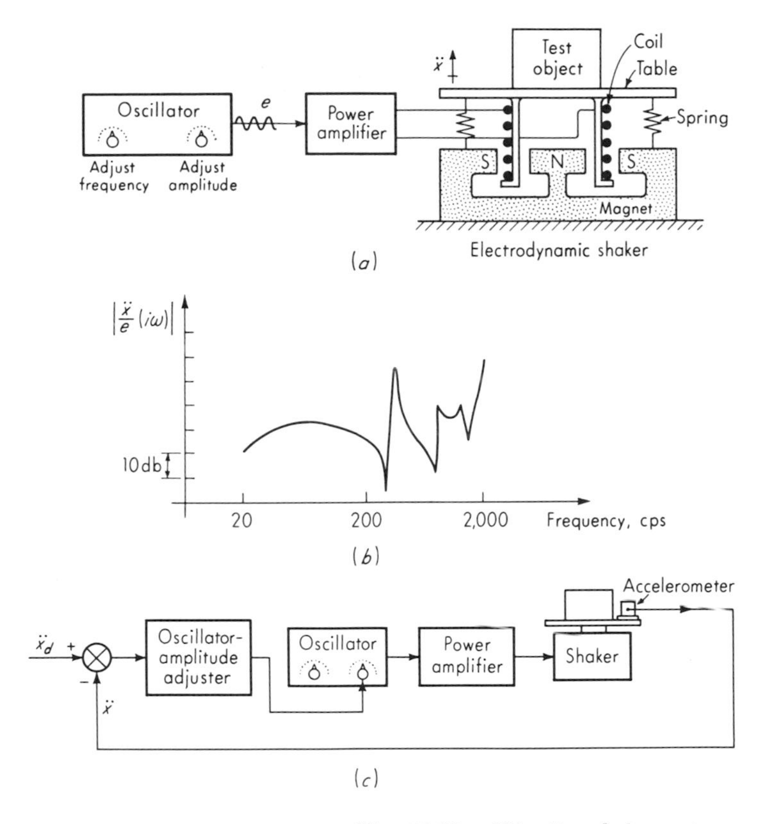

Fig. 10.29. *Vibration shaker systems.*

shows the basic arrangement of a shaker system for sinusoidal vibration testing. Many tests involve "sweeping" the frequency of the oscillator through a certain range while maintaining the acceleration amplitude constant at the test object. While it is not difficult to maintain constant the amplitude of oscillator output voltage e while sweeping through the frequency range, the acceleration $\ddot{x}$ will not be constant since the transfer function $(\ddot{x}/e)(i\omega)$ is not constant over this range. In fact, because of various resonances in the electromechanical shaker, test fixtures, and the test object itself, severely distorted frequency response is not uncommon (Fig. 10.29*b*). This difficulty can be overcome by use of a feedback scheme as in Fig. 10.29*c*. Here the actual acceleration $\ddot{x}$ is compared with the desired value $\ddot{x}_d$; if they differ, the amplitude of the oscillator is adjusted to obtain correspondence. This adjustment is performed automatically and continuously as the frequency range is swept.

When random vibration testing (see Fig. 10.30) rather than pure sinusoidal is desired, the above approach is not directly applicable since

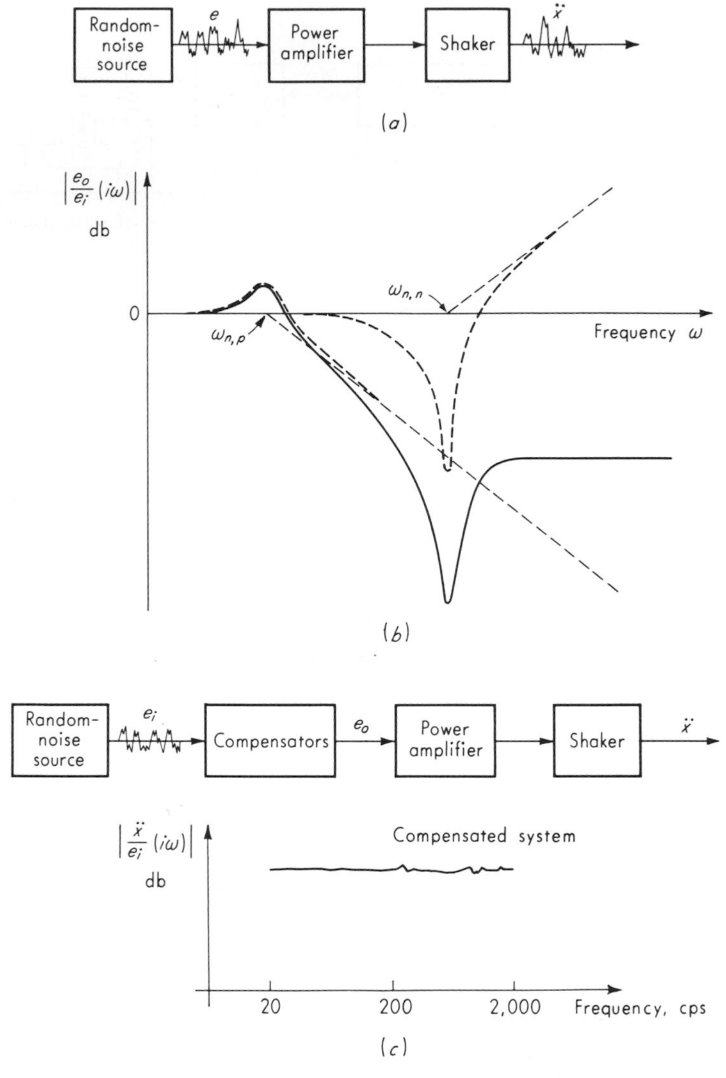

Fig. 10.30. *Dynamic compensation for vibration shaker.*

the signal $\ddot{x}$ is now a random signal and one cannot adjust the noise source in any simple fashion to force $\ddot{x}$ to have the desired frequency spectrum (the spectrum of e). One approach is to provide dynamic compensation such that the transfer function $(\ddot{x}/e)(i\omega)$ *is* flat over the desired frequency range. Then the spectrum at $\ddot{x}$ will be the same as that put in at e. The necessary dynamic compensation here is one which can put "peaks" where there are "notches" and notches where there are peaks in the curve of Fig.

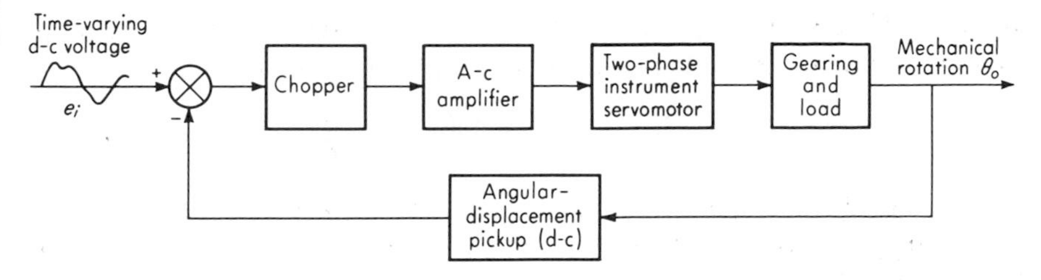

(*a*) D-c voltage-input position servo

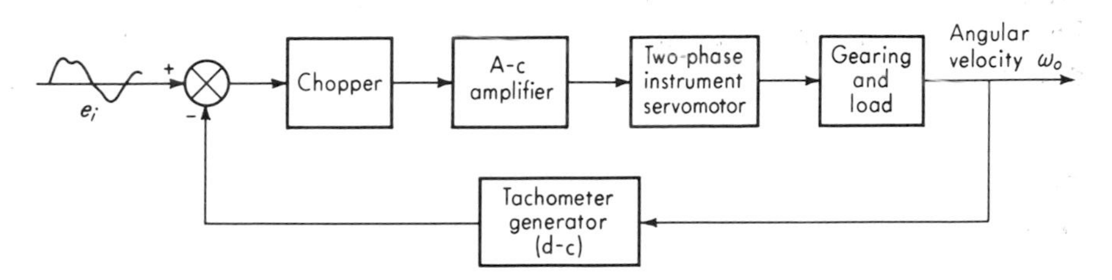

(*b*) D-c voltage-input velocity servo

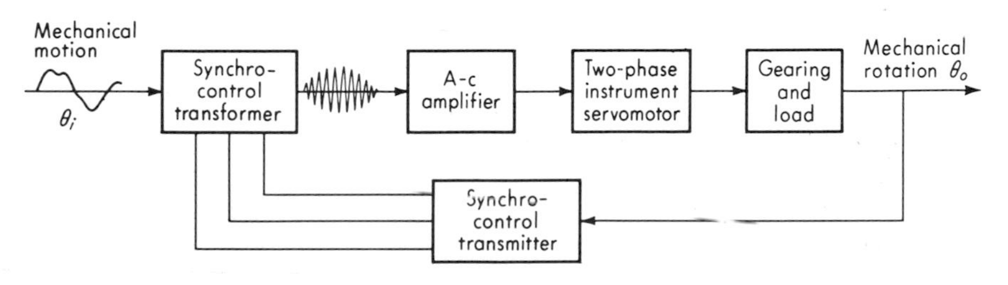

(*c*) Motion-input position servo (all a-c)

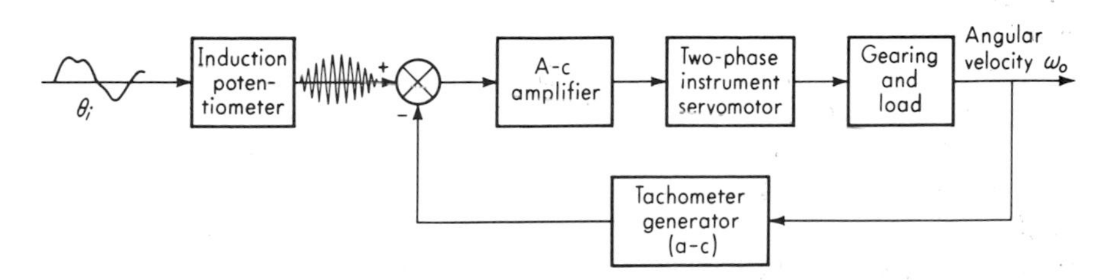

(*d*) Motion-input velocity servo (all a-c)

Fig. 10.31. *Instrument servomechanisms.*

10.29*b*. Thereby the overall curve can be made relatively flat. A compensator that will provide one peak and one notch has the form

$$\frac{e_o}{e_i}(D) = \frac{D^2/\omega_{n,n}^2 + 2\zeta_n D/\omega_{n,n} + 1}{D^2/\omega_{n,p}^2 + 2\zeta_p D/\omega_{n,p} + 1} \tag{10.72}$$

Figure 10.30*b* shows the frequency response of such a compensator in which the peak occurs at a lower frequency than the notch. The reverse is also possible if needed. Since ζ_n and ζ_p are also adjustable, the compensator can be "tailored" to cancel out exactly the undesired shaker-system dynamics. When several peaks and notches are present (as in Fig. 10.29*b*), several compensators are used; as many as 10 are not uncommon.

10.6

Instrument Servomechanisms Measurement systems often require the conversion of a low-power electrical signal or a low-power mechanical motion into an accurately proportional and (relative to the input signal) high-power mechanical motion. When frequency response beyond about 5 cps is not required, this function may often be performed by an appropriate instrument servomechanism. Figure 10.31 shows block diagrams of some of the most common types. The velocity servos are generally used to obtain the integral of the input signal, as explained in Sec. 10.4.

The d-c-input position servo, while important in its own right, also is the basis of practically all self-balancing potentiometer recorders and *X-Y* plotters. Thus we shall explain its operation in more detail. Most instrument servos utilize a-c amplifiers and two-phase a-c instrument servomotors even if the input signal is d-c. The use of a-c amplifiers is based on their freedom from drift and reasonable cost, while the use of two-phase a-c motors relates to their low friction (no brushes are needed as in a d-c motor) and controllability. Figure 10.32 briefly summarizes the operating characteristics of this type of motor. One of the phases is of fixed amplitude. The amplitude of the other phase (which must be displaced in phase by ± 90 electrical degrees from the fixed phase) controls the direction and amount of torque developed. When the controlled phase reverses polarity (goes from $+90$ to $-90°$ or vice versa) the torque reverses.

The schematic and graphs of Fig. 10.33 show how a d-c signal is converted to alternating current by a chopper and how a reversal in polarity of the d-c error signal e_{aa} results in a 180° phase shift (from $+90$ to $-90°$ or vice versa) in the motor-control phase voltage e_{dd}, thereby causing the required reversal in the direction of torque. We see that whenever

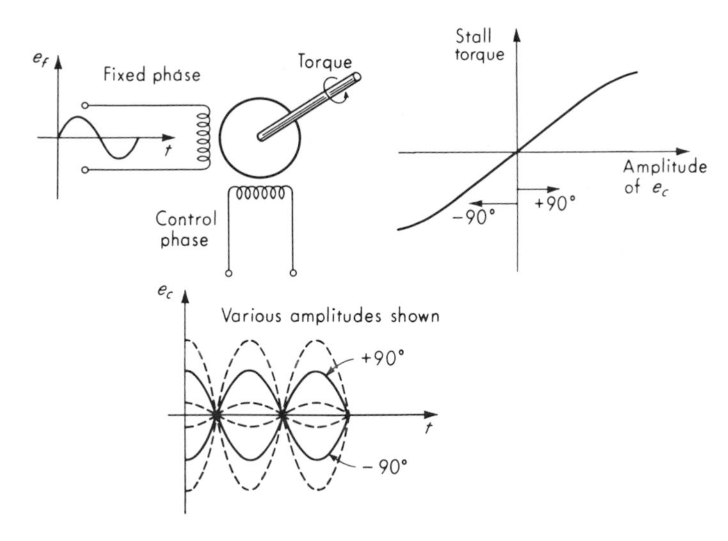

Fig. 10.32. *Two-phase servomotor.*

$e_i \neq e_p$ there will be an error voltage e_{au} which is converted to alternating current and amplified so that it tends to drive the motor in a direction to change e_p until it equals e_i. If e_i is changing, if the amplifier gain is high enough, e_p (and therefore output motion θ_o) will "track" e_i with very little error. Instrument servos regularly achieve high static accuracy, having errors as small as 0.1 to 0.2 percent of full scale in positioning θ_o as a linear function of e_i. However, their frequency response is limited by inertia of moving parts to about 5 cps or less.

10.7

Addition and Subtraction The addition or subtraction of mechanical-motion signals is generally accomplished by use of gear differentials or summing links; see Fig. 10.34*a*. Forces or pressures are summed and transduced to displacement by the schemes shown in Fig. 10.34*b*. The spring restraints that transduce force to displacement may be removed if a feedback system using a null-balance force to return deflection to zero is employed. Summing of voltage signals is accomplished by the simple series circuit or the operational-amplifier circuit shown in Fig. 10.34*c*. Subtraction rather than addition in all the above devices is obtained by simply reversing the sense of the input to be subtracted. Addition is the basic operation of digital computers; thus addition or subtraction of digital signals in binary form is easily accomplished with such equipment. Equipment for interconverting numbers in binary and decimal form is also available. Digital signals in the form of pulse

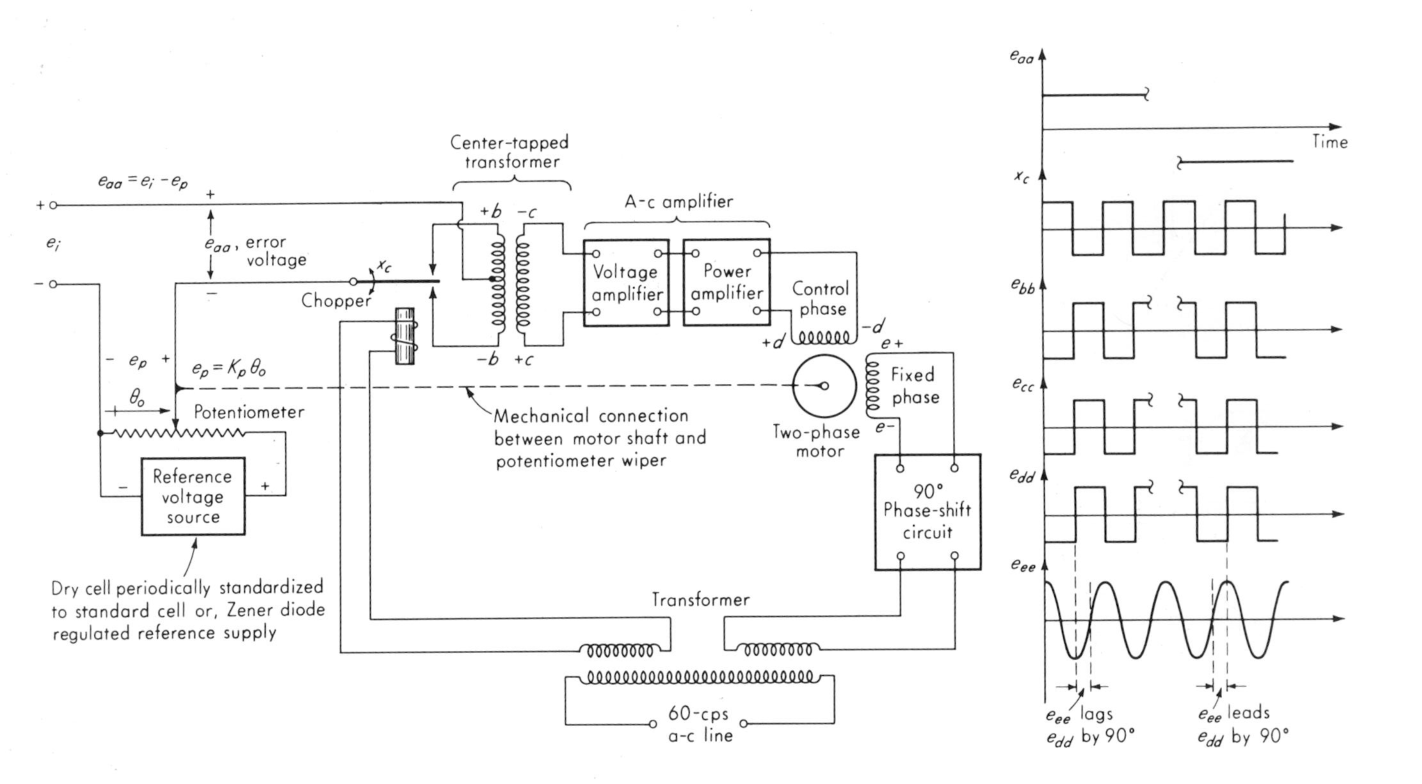

Fig. 10.33. *Position servo.*

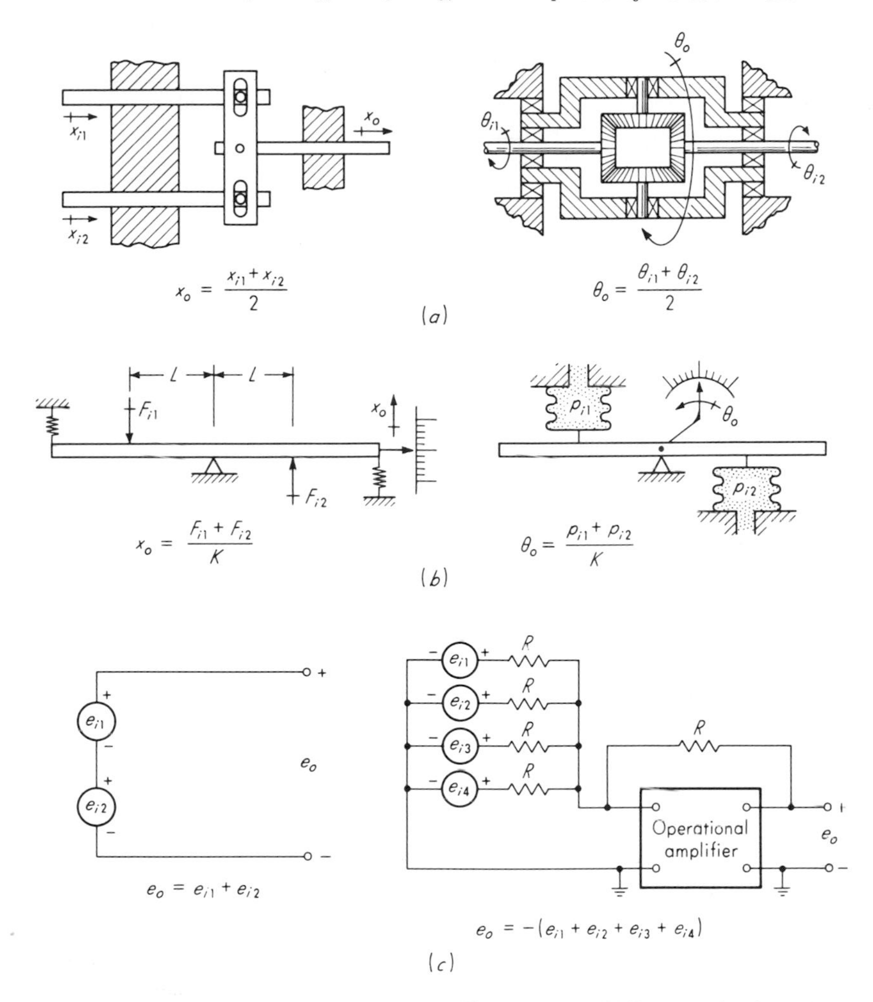

Fig. 10.34. *Addition and subtraction.*

rates may also be added and subtracted to obtain pulse rates that are the sum or difference of the input pulse rates.

10.8

Multiplication and Division When data manipulation requires multiplication or division of two variable signals a number of techniques are available, depending on the physical nature of the signals.[1] We here consider a few of the most common.

[1] S. A. Davis, 31 Ways to Multiply, *Control Eng.*, p. 36, November, 1954.

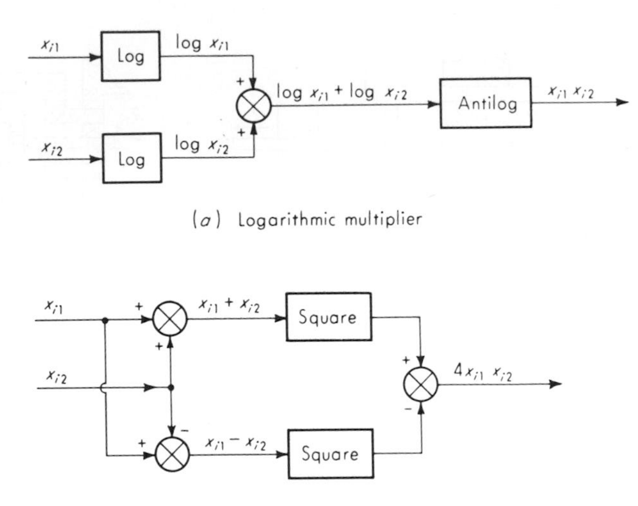

Fig. 10.35. *Multiplier methods.*

Two general methods which may be implemented either mechanically or electrically are the logarithmic method and the quarter-squares method. In the logarithmic method one first obtains logarithms of the two input signals x_{i1} and x_{i2}. For mechanical signals this can be done with special noncircular gearing[1]; electromechanical types use specially wound potentiometers, while electronic units use diode function generators. Once the two logarithms are obtained, they are added (subtracted if division is wanted) and the antilog of the sum taken (see Fig. 10.35*a*).

$$\text{Antilog}\,(\log x_{i1} + \log x_{i2}) = x_{i1}x_{i2} \qquad (10.73)$$

The devices for taking the antilog are essentially inversions of the devices for taking the log.

In the quarter-squares method one must have available devices for adding, subtracting, and squaring. Squaring can be done mechanically with linkages, cams, or noncircular gears; electromechanically with nonlinear potentiometers; and electronically with special nonlinear resistors[2] or diode function generators. If the needed operations can be performed, the action of the multiplier is given by Fig. 10.35*b*.

An all-mechanical multiplier[3] based on a similar triangles principle

[1] A. E. Lockenvitz et al., Geared to Compute, *Automation*, p. 37, August, 1955.
[2] Quadratron, Bourns Inc., Riverside, Calif.
[3] G. W. Michalec, Analog Computing Mechanisms, *Machine Design*, p. 157, Mar. 19, 1959.

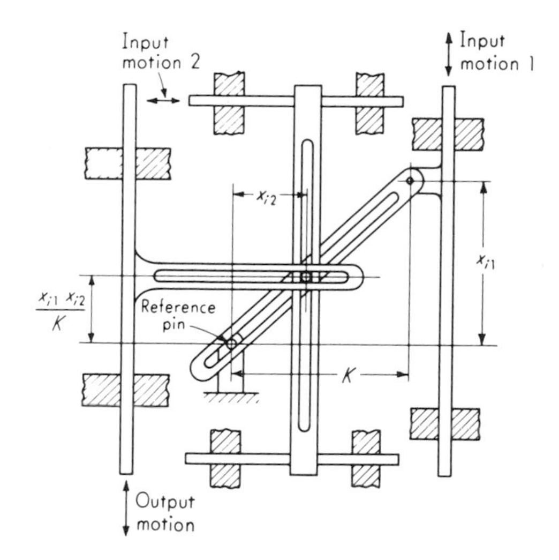

Fig. 10.36. Mechanical multiplier.

is shown in Fig. 10.36. If the motions to be multiplied are at a low power level, instrument servos may be used to drive the x_{i1} and x_{i2} input members. Similarly, if nonmotion quantities such as temperatures, pressures, etc., are to be multiplied, they may be transduced to voltages and instrument servos again used to provide the required multiplier input motions.

A Wheatstone bridge may be used for multiplication if one of the signals can be transduced to a voltage (which is used as the bridge excitation voltage) and the other can be transduced to a resistance change (see Fig. 10.37). The bridge output is then proportional to the product $q_{i1}q_{i2}$ for small percentage resistance changes.

Servo multipliers have been widely used in general-purpose analog computing installations and also find application in control and data systems in which two voltage signals must be multiplied. As shown in Fig. 10.38*a*, one voltage is applied to an instrument servo which positions a potentiometer wiper in direct proportion. The other voltage serves as the potentiometer excitation; thus the output voltage is proportional

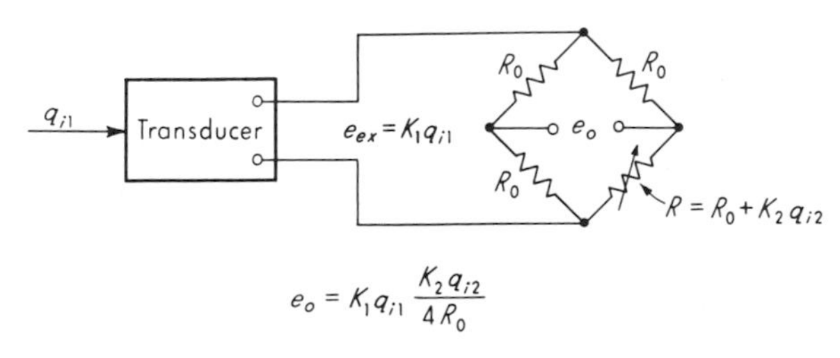

Fig. 10.37. Wheatstone-bridge multiplier.

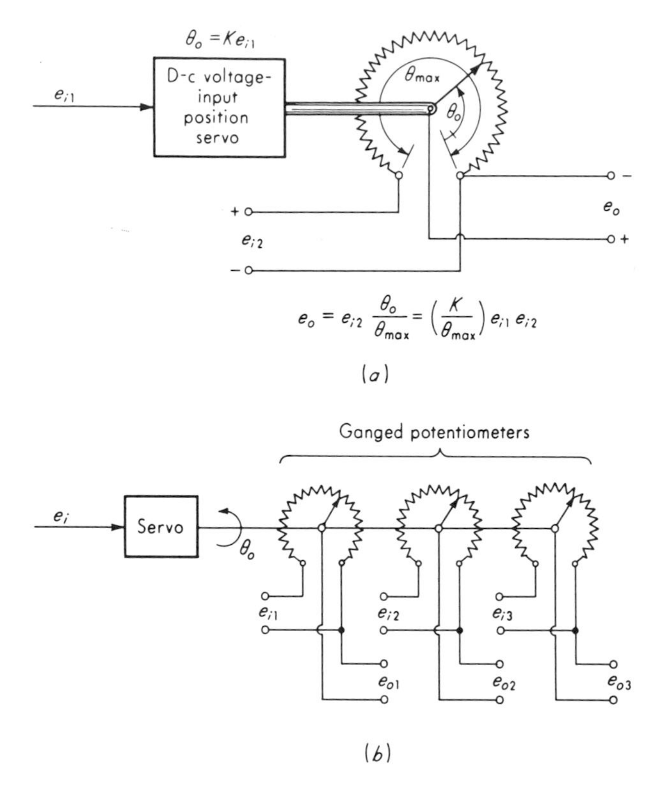

Fig. 10.38. *Servo multiplier.*

to the desired product. Such multipliers are particularly useful if several signals e_{i1}, e_{i2}, e_{i3}, etc., are each to be multiplied by one other signal e_i. Then "ganged" potentiometers all driven from the same servo shaft give the desired products (see Fig. 10.38*b*). Servo multipliers achieve high accuracy, errors being of the order of 0.1 percent for static operation. The frequency response of the servos limits their accurate dynamic operation to less than about 5 cps.

When multiplication or division of rapidly varying voltage signals is required, all-electronic techniques employing the quarter-square scheme (with diode function generators)[1] or the time-division method[1] are available. We shall not go into the details of these methods but merely quote typical performance specifications. A typical quarter-square multiplier[2] accepts full-scale inputs of ± 100 volts and produces ± 100 volts output,

[1] A. S. Jackson, "Analog Computation," McGraw-Hill Book Company, p. 474, New York, 1960.

[2] Comcor Inc., Denver, Colo.

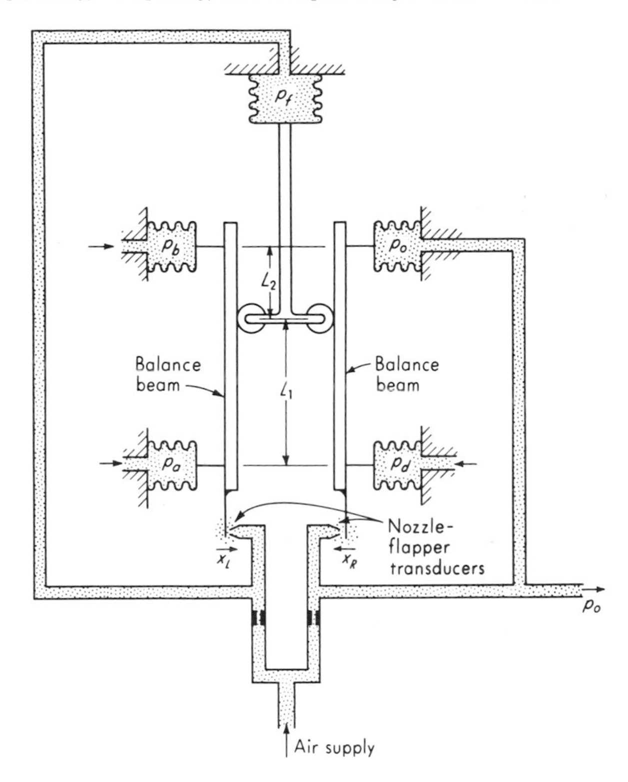

Fig. 10.39. *Sorteberg force bridge.*

has ± 100 mv static error, noise level less than 20 mv peak to peak, frequency response flat within 3 db from 0 to 20,000 cps, phase shift of 2° at 1,000 cps, zero error less than ± 20 mv with either input zero (± 2 mv with both inputs zero), and drift of 1.6 mv/F° from 70 to 95°F. A time-division multiplier[1] accepts full-scale inputs of ± 100 volts and produces ± 100 volts output, has a static inaccuracy of 0.01 percent of full scale for single-quadrant multiplication (0.05 percent in all four quadrants), noise level less than 100 mv peak, frequency response flat within 1 db from 0 to 700 cps, phase shift less than 1° at 100 cps, zero error of 40 mv with one input zero, and drift less than 100 mv for 8 hr.

Multiplication and division of pneumatic pressure signals may be accomplished by devices such as the Sorteberg force bridge[2] of Fig. 10.39. Let us assume that for some chosen initial equilibrium condition (say $p_a = p_b = p_d = p_o$, $L_1 = L_2$, and $p_f = p_{f0}$) both balance beams are

[1] Donner Scientific Co., Concord, Calif.
[2] Sorteberg Controls Co., South Norwalk, Conn.

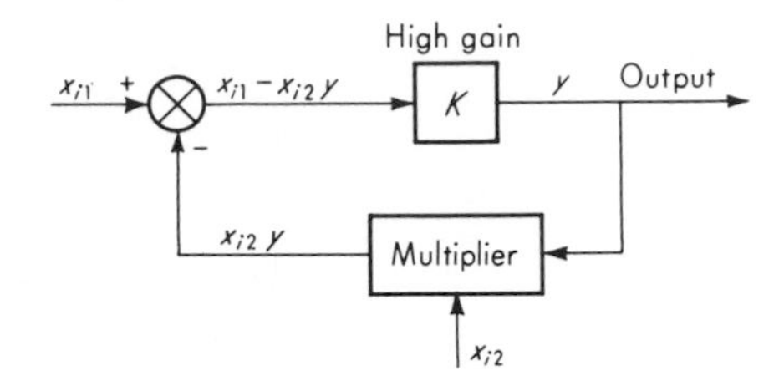

Fig. 10.40. *Feedback inversion of multiplier.*

vertical. Now, if we assume the gain of the nozzle-flapper transducers is very high, only a minute deflection x_L or x_R can cause a large change in p_f or p_o. Examination of the action of the system will show that any change in input pressure p_d that causes a beam deflection x_R will result in a change in p_o such as to return the beam to very nearly its original position. A change in p_a and/or p_b causes a change in x_L which in turn changes p_f (and thereby L_1 and L_2) so as to return the beam to balance. The change in L_1 and L_2 tends to unbalance the right-hand beam but this again tends to correct itself through a change in p_o. The overall result is thus that for any changes in the three inputs p_a, p_b, and p_d the beams always stay balanced. If this is true, we have (assuming all bellows of equal area)

$$p_bL_2 = p_aL_1 \tag{10.74}$$

$$p_oL_2 = p_dL_1 \tag{10.75}$$

and thus

$$p_o = \frac{p_bp_d}{p_a} \tag{10.76}$$

We see that the system thus multiplies p_b and p_d and divides the product by p_a. Inaccuracy of this system is of the order of 1 percent of full scale, resolution 0.1 percent, and hysteresis 0.25 percent. Dynamic response is approximately first-order with a time constant of 4 to 30 sec, depending on the length of the transmission tubing (1 to 200 ft).

To perform the operation of division, any multiplier may be connected into a high-gain feedback loop as shown in Fig. 10.40, and this procedure is regularly used with mechanical, electromechanical, and electronic multipliers. From the block diagram we have

$$K(x_{i1} - x_{i2}y) = y \tag{10.77}$$

$$x_{i1} - x_{i2}y = \frac{y}{K} \tag{10.78}$$

and if gain K is very high

$$x_{i1} - x_{i2}y \approx 0$$

giving the output y as

$$y \approx \frac{x_{i1}}{x_{i2}} \tag{10.79}$$

Dividers have essentially the same performance characteristics as

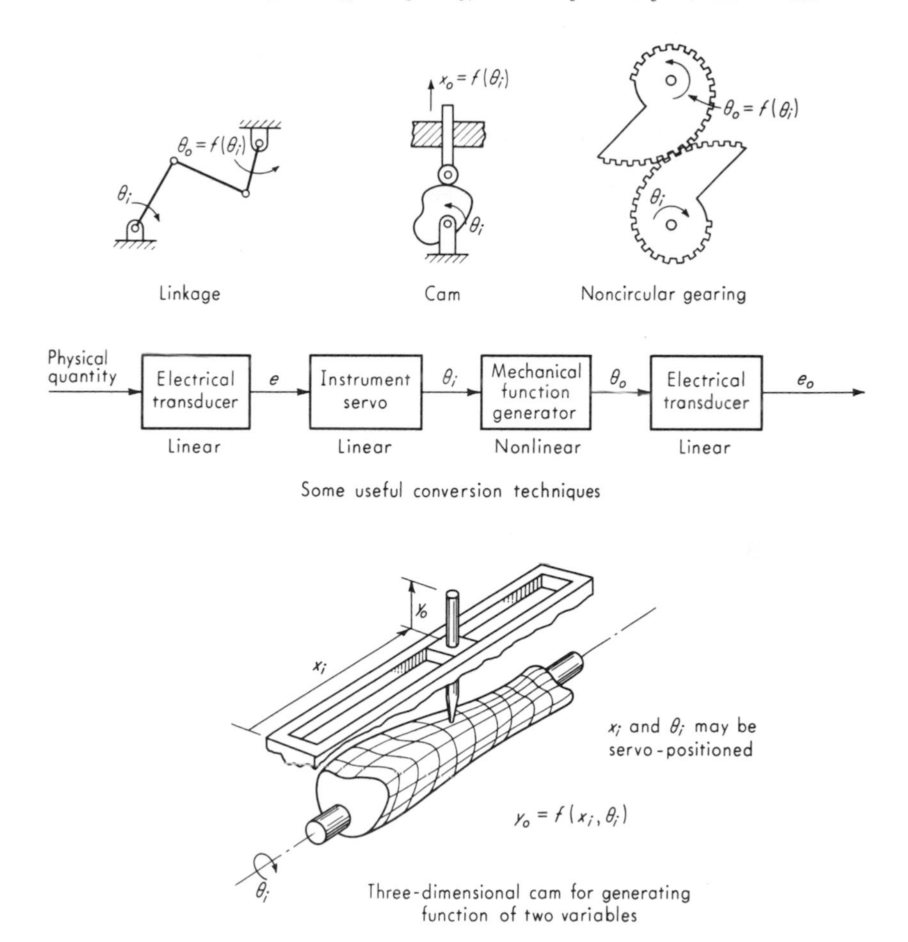

Fig. 10.41. *Mechanical function generation.*

multipliers except that, since division by zero is allowed neither mathematically nor physically, the signal representing the divisor cannot change sign (go through zero). Also, very small values of the divisor overload the device.

10.9

Function Generation When one needs to generate a specific nonlinear function of a mechanical-motion signal, the use of cams, linkages, and noncircular gears allows great freedom since almost any reasonable function can be adequately approximated by one or a combination of these methods. The use of instrument servos also allows these methods

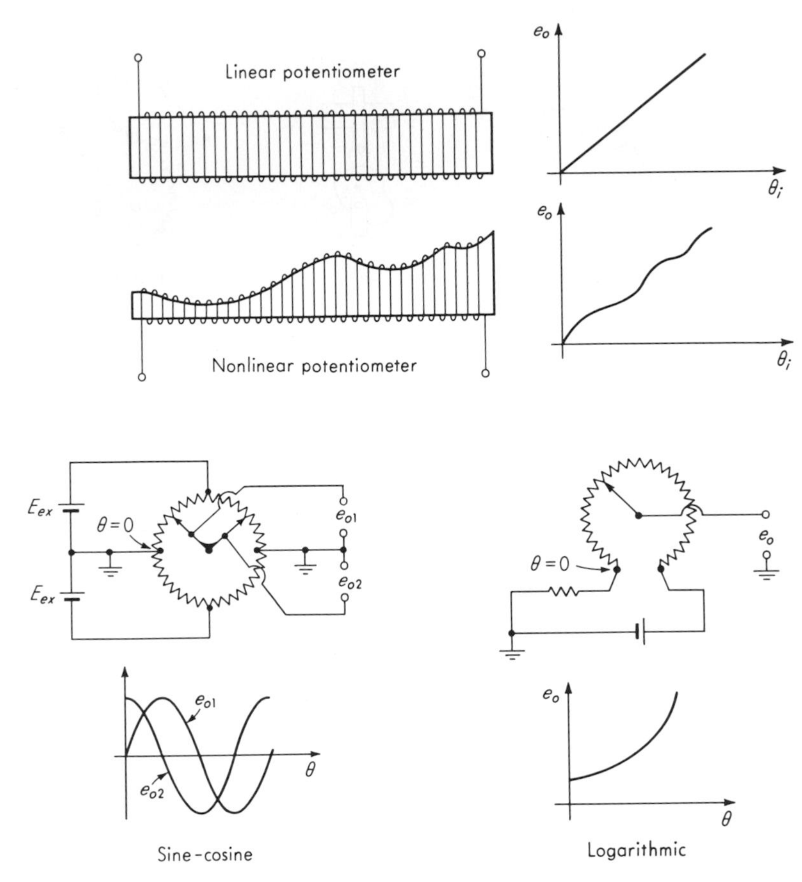

Fig. 10.42. *Nonlinear potentiometers.*

to be employed with electrical signals. If an electrical output is wanted, a motion transducer can be attached to the mechanical output member. Figure 10.41 illustrates these concepts.

Nonlinear potentiometers are also widely used in function generation. They are constructed in basically the same manner as potentiometer displacement transducers except that a specific *nonlinear* relation between θ_i and e_o is wanted rather than the linear relation desired for a motion transducer; see Fig. 10.42. A wide variety of functions is possible by distributing the resistance winding in a proper nonlinear fashion on the mandrel. Techniques have also been developed for constructing nonlinear potentiometers, using conducting plastic or deposited-film (rather than wire-wound) resistance elements. While functions of rather arbitrary form are available as special items, a small number of basic

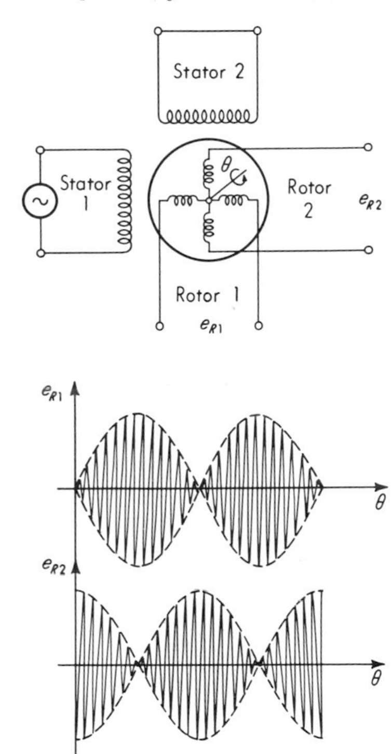

Fig. 10.43. *Resolver.*

functions are so commonly used that they are obtainable ready-made as stock items. These include sine and cosine over 360°, sine or cosine over 360°, 180° sine, 90° sine, ±75° tangent, square function, and logarithmic function. The conformity of the voltage-output/rotation-input relation to the theoretical function is of the order of 0.3 to 2 percent of full scale, depending on the type of function and the instrument quality.

When very accurate sine and/or cosine functions are needed (as in navigation and fire-control computers where resolution and composition of vectors must be performed), the use of resolvers rather than nonlinear potentiometers may be indicated. Resolvers are small a-c rotating machines similar to synchros. In general they have two stator windings and two rotor windings (see Fig. 10.43). If one of the stator windings is excited with an a-c signal of constant amplitude (60 or 400 cps is commonly used) and the other is short-circuited, rotation of the rotor through an angle θ_i from a null position gives at the two rotor windings a-c signals whose amplitudes are respectively proportional to sine θ_i and cosine θ_i. Other important computing functions[1] such as converting vehicle rotation angles to earth coordinates in navigation systems can also be performed

[1] "Resolvers," Ford Instrument Co., Long Island City, N.Y.

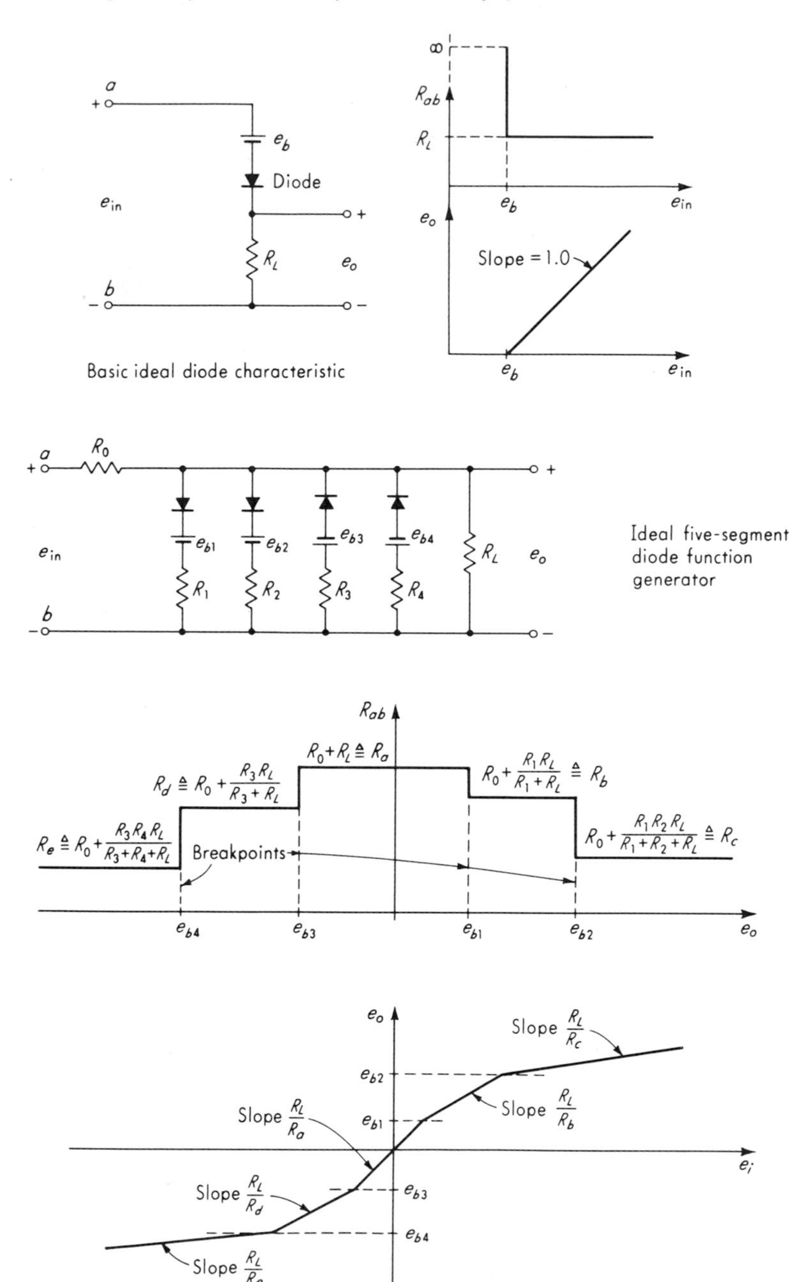

Fig. 10.44. *Diode function generator.*

by resolvers. A typical high-accuracy resolver[1] has an excitation voltage of 26 volts maximum at 400 cps, open-circuit output voltage of 0 to 26 volts, residual null voltage of 1 mv maximum, and a maximum deviation from the desired functional relation of 0.01 percent.

All the function generators discussed up to this point involve moving parts and are thus limited in speed of response. If higher speed is needed, all-electronic methods are available. The most widely used and versatile device is the diode function generator.[2] It is available in general-purpose forms which can be adjusted to fit almost any single-valued function one can draw on a piece of paper. Figure 10.44 shows the operating principle of such devices. Ideally the diode (either vacuum tube or semiconductor) is assumed to have zero resistance in the forward direction and infinite resistance in the reverse; thus it may be thought of as a switch that is open or closed, depending on the polarity of the voltage across it. While the circuit of Fig. 10.44 allows generation of monotonic functions only, commercially available units using operational amplifiers can sum two monotonic functions (one increasing, one decreasing) to obtain functions with both positive and negative slopes. Such units also allow continuous adjustment of breakpoint locations and segment slopes over wide ranges. Some units[3] also blend the straight-line-segment breakpoints with a tangent parabola of adjustable curvature. This allows generation of smoother curves with more continuous derivatives. Commercial diode function generators generally have 10 to 20 straight-line segments. A typical unit[4] having 10 segments has a full-scale output of ± 100 volts, maximum slope of 10 volts/volt, phase shift of 0.8 to 1.4° at 100 cps, frequency response down 3 db at 7,000 cps, and noise level of 20 to 100 mv.

The use of specially prepared nonlinear resistors[5] together with conventional operational amplifiers allows generation of many common functions with a relatively small amount of equipment. Being all-electronic, such methods allow high-speed operation.

10.10

Amplitude Modulation and Demodulation We have seen earlier in the text a number of examples of measurement systems in which interconversion between a-c and d-c signals was necessary and/or

[1] "Resolvers," Ford Instrument Co., Long Island City, N.Y.

[2] E. J. Galli, How Diodes Generate Functions, *Control Eng.*, p. 109, March, 1959; and p. 107, February, 1960.

[3] G. A. Philbrick Corp., Boston, Mass.

[4] Comcor Inc., Denver, Colo.

[5] Specifications and Applications of the Douglas Quadratron, Douglas Aircraft Co., Santa Monica, Calif., Feb. 27, 1963.

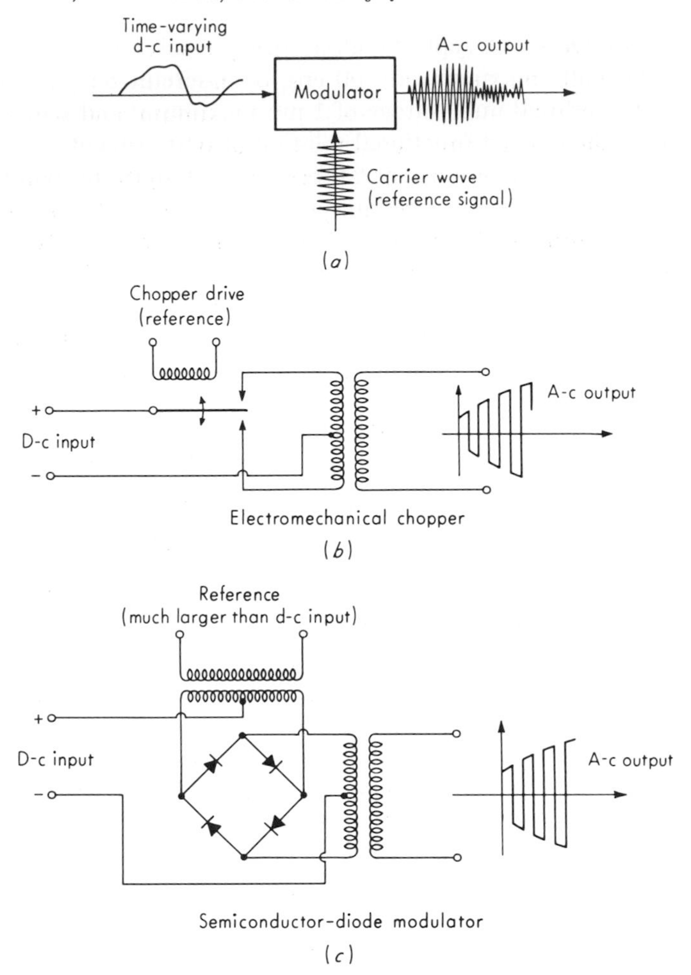

Fig. 10.45. *Amplitude modulation.*

desirable. The conversion from direct current to alternating current is a form of amplitude modulation whereas conversion from alternating to direct current is called demodulation or detection. We here give briefly a few examples of the hardware needed to accomplish these functions.

The process of modulation may be performed by a wide variety of devices[1]; however, all may be represented in block-diagram form as in Fig. 10.45*a*. In general, the frequency spectrum of the d-c input signal

[1] B. T. Barber, Servo Modulators, *Control Eng.*, August, October, November, December, 1957.

must not go beyond about 10 to 20 percent of the carrier frequency for proper operation. Electromechanical choppers (vibrators) are widely used as modulators. The carrier is a square wave, often at 60 or 400 cps and limited to less than about 1,000 cps. The output is often transformer-coupled to the following circuitry to provide electrical isolation (see Fig. 10.45*b*). When higher carrier frequencies are required, all-electronic modulators are available. Tube, transistor, and diode types are usable up to about 10,000 cps. Figure 10.45*c* shows a ring-type diode modulator. Here the a-c reference signal serves to "switch" the diodes from their conducting to nonconducting states, thus giving an action quite similar to the chopper but without moving parts.

In most measurement systems, phase-sensitive demodulation is required, if modulation was performed earlier, so as to recover the algebraic sign of the original d-c information. This requires that the reference signal used to drive the modulator must also be used in the demodulator to ensure proper "synchronization." For "square-wave"-type modulators and demodulators such as the examples of Figs. 10.45 and 10.46, if there were no time delays, attenuation, or phase shifts, the d-c signal recovered at the demodulator would be identical with that originally put into the modulator and no filtering would be required. This, of course, is not possible, and so a low-pass filter is generally required at the demodulator output to remove high-frequency components introduced by imperfect modulation and demodulation. Demodulation may be accomplished with a number of devices[1]; Fig. 10.46 shows two analogous to the modulators of Fig. 10.45.

10.11

Voltage-to-Frequency and Frequency-to-Voltage Converters

The conversion of a d-c voltage input to a periodic-wave output whose frequency is proportional to the d-c input may serve several useful functions in measurement systems. Such devices, called voltage-controlled oscillators, are widely used in FM/FM telemetry systems, since the voltage-to-frequency conversion process is a form of frequency modulation. They are also used in the integrating digital voltmeter where a d-c signal is converted into a periodic wave of proportional frequency. This wave is then applied to an electronic counter for a fixed time interval, giving a reading proportional to the average d-c voltage over the time interval. The recording of d-c voltages on magnetic tape recorders is also accomplished through the use of frequency modulation.

[1] J. E. Gibson and F. B. Tuteur, "Control System Components," p. 249, McGraw-Hill Book Company, New York, 1958.

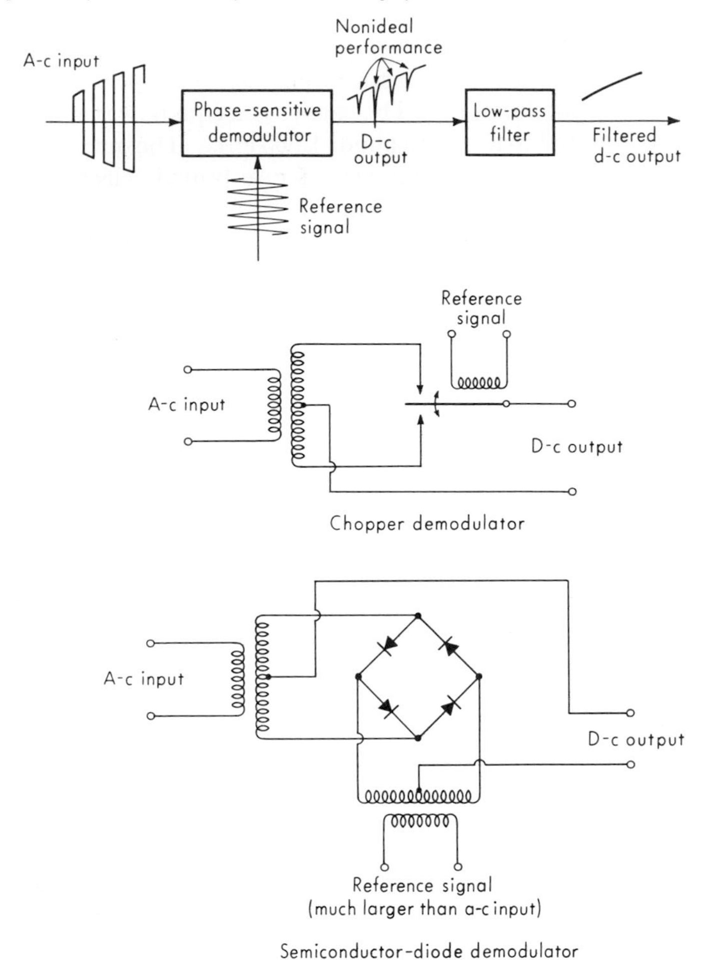

Fig. 10.46. *Phase-sensitive demodulation.*

Voltage-controlled oscillators[1] are generally of the resistance-capacitance phase-shift type or the multivibrator type. For zero input these oscillators produce a given output frequency, called the center frequency, which commonly is from 400 to 70,000 cps. Variation of the d-c input then results in frequency variation of ± 7.5 or ± 15 percent around the center frequency. Input variation is ordinarily ± 2.5 or 0 to $+5$ volts direct current; however, some units incorporate their own amplification so that ± 10-mv input causes full output-frequency deviation. Non-

[1] M. H. Nichols and L. L. Ranch, "Radio Telemetry," p. 253, John Wiley & Sons, Inc., New York, 1956.

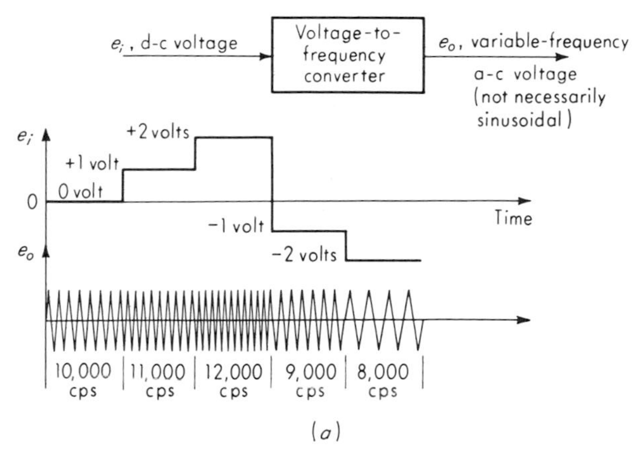

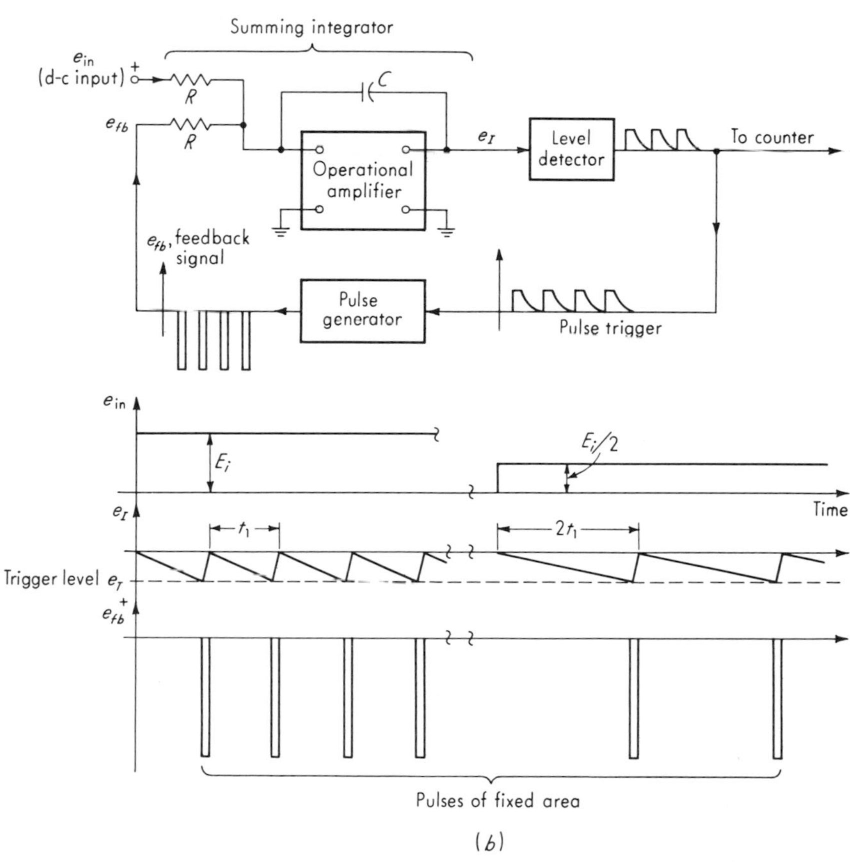

Fig. 10.47. *Voltage-to-frequency converter.*

linearity of the output-frequency/input-voltage characteristic is of the order of 0.1 to 1.0 percent.

Figure 10.47*b* shows a voltage-to-frequency converter used in an integrating digital voltmeter.[1] This device produces an output frequency proportional to the input d-c voltage, down to and including 0-volt input. That is, zero input produces zero-frequency output; thus there is no "center frequency" as described above. The operation of this instrument may be explained as follows: Using the usual operational-amplifier network analysis assumptions, we get the equation of the summing integrator as

$$e_I = -\frac{1}{RC}\int (e_{in} + e_{fb})\,dt = -\frac{1}{RC}\int e_{in}\,dt - \frac{1}{RC}\int e_{fb}\,dt \tag{10.80}$$

Now, if a constant input voltage of magnitude E_i is applied, the output e_I of the integrator will be $(-E_i/RC)t$, a negative-going ramp. This voltage is applied to a level detector that will produce a trigger pulse whenever e_I goes through a set value, say e_T. It is clear that the time required for e_I to reach e_T is inversely proportional to E_i. The trigger pulse triggers a pulse generator which produces a pulse of short duration and high amplitude whose area is accurately maintained constant. The area of this pulse, which is applied at the e_{fb} input terminal of the summing integrator, is calculated to just return e_I to zero. That is,

$$e_I \text{ due to } e_{fb} = -\frac{1}{RC}\int e_{fb}\,dt = |e_T| \tag{10.81}$$

$$e_I \text{ due to } e_{fb} = -\frac{1}{RC}\,(\text{pulse area}) = |e_T| \tag{10.82}$$

and thus

$$|\text{Pulse area}| = RC\,|e_T| \tag{10.83}$$

Since e_T is fixed and known, the pulse generator can be designed to produce pulses of the required area. When the pulse is over, the negative-going ramp due to E_i starts again and the process repeats itself over and over, producing pulses at a rate proportional to E_i. The commercial unit described above has an output frequency range of 0 to 10,000 cps and a short-term inaccuracy of ± 0.1 percent of full scale.

Frequency-to-voltage converters accept a periodic (not necessarily sinusoidal) input signal and produce a d-c output in direct proportion to input frequency. They are used, for example, to get a d-c analog signal from a-c tachometers or turbine flowmeters. When the input is nonsinusoidal the "frequency" is interpreted as the fundamental frequency or repetition rate of the wave form. Figure 10.48 shows in block-diagram form the mode of operation of a typical frequency-to-voltage converter.

[1] Hewlett-Packard Co., Palo Alto, Calif.

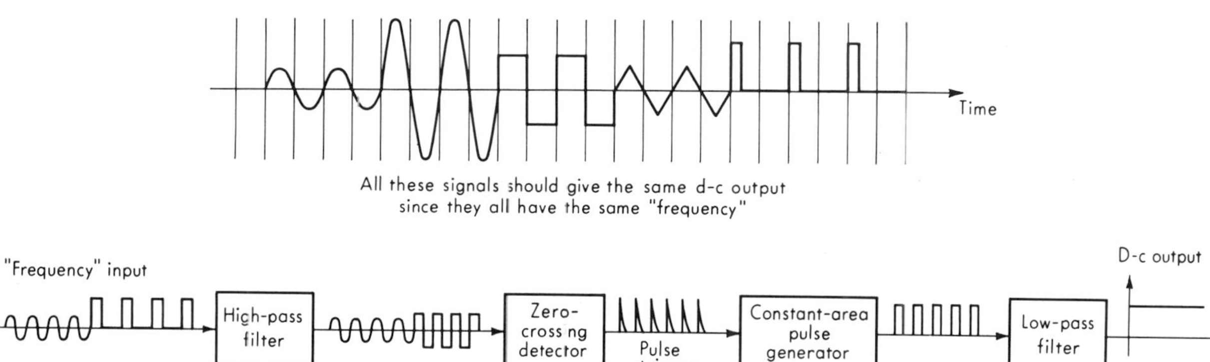

Fig. 10.48. *Frequency-to-voltage converter.*

Input wave forms are passed through a high-pass filter so that pulses that do not themselves go negative can be handled. The characteristic of a signal of any wave form that determines its repetition rate is the number of zero crossings per unit time, and so these are detected and a trigger pulse generated for each one. These pulses then trigger constant-area pulses at a rate proportional to input frequency. The average value of this pulse train is thus proportional to frequency and is obtained by low-pass filtering. This filter gives the system approximately a first-order type of response since its dynamics are the slowest of the entire system.

A commercial frequency-to-voltage converter[1] handles the frequency range 80 to 100,000 cps in nine ranges, accepts sinusoidal input signals from 5 mv rms (5 cps to 10 kc), 30 mv rms (10 to 100 kc) to 10 volts rms, pulse inputs of 25 mv 0 to peak (5 to 20,000 pps), 75 mv 0 to peak (20,000 to 100,000 pps), 5 μsec minimum pulse width, and has nonlinearity 0.025 percent of full scale, long-term drift less than 0.1 percent per week, and an effective time constant of 50 divided by the full-scale frequency range (200, 500, 1,000, 2,000, 5,000, 10 kc, 20 kc, 50 kc, 100 kc). The longest time constant is thus $50/200 = 0.25$ sec.

10.12

Analog-to-Digital and Digital-to-Analog Converters The increasing use of digital computers in measurement and control systems and the scarcity of true digital measuring devices lead to a need for analog-to-digital converters to allow analog sensors to communicate with the computer. Sometimes the digital output of the computer must be used in an analog system; thus digital-to-analog converters are also necessary.

Analog-to-digital converters may be classified broadly as shaft-angle-to-digital or voltage-to-digital types. The shaft-angle-to-digital type simply uses one of the encoders discussed at the end of Sec. 4.3 to convert the analog shaft angle into coded digital signals. If the analog signal is not already a shaft rotation, it can be converted to one by using a suitable transducer and position servo (see Fig. 10.49). Such systems can achieve very high accuracy but are limited in speed of response by the capability of the servos (less than 5 cps).

Voltage-to-digital converters take a number of different forms.[2] Figure 10.50 illustrates the operation of a time-base encoder in simplified form. Basically, the analog input signal is compared with a recurrent

[1] Vidar Corp., Mountain View, Calif.

[2] J. T. Tou, "Digital and Sampled-data Control Systems," p. 368, McGraw-Hill Book Company, New York, 1959.

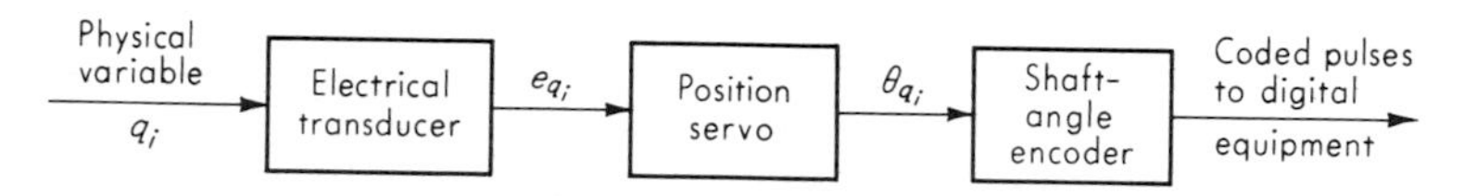

Fig. 10.49. *Shaft-angle-to-digital converter.*

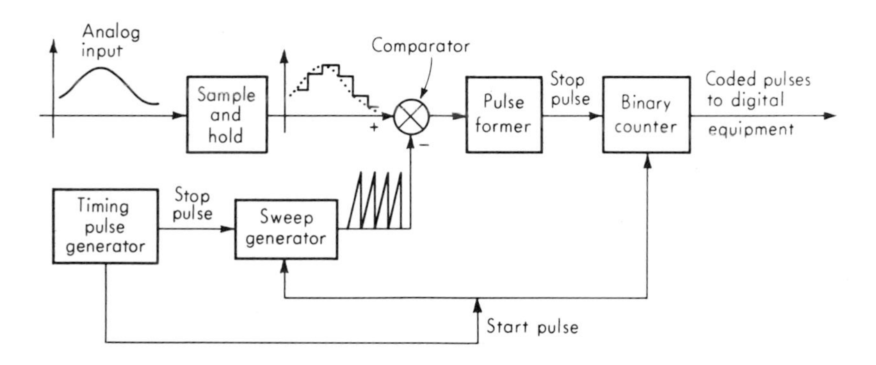

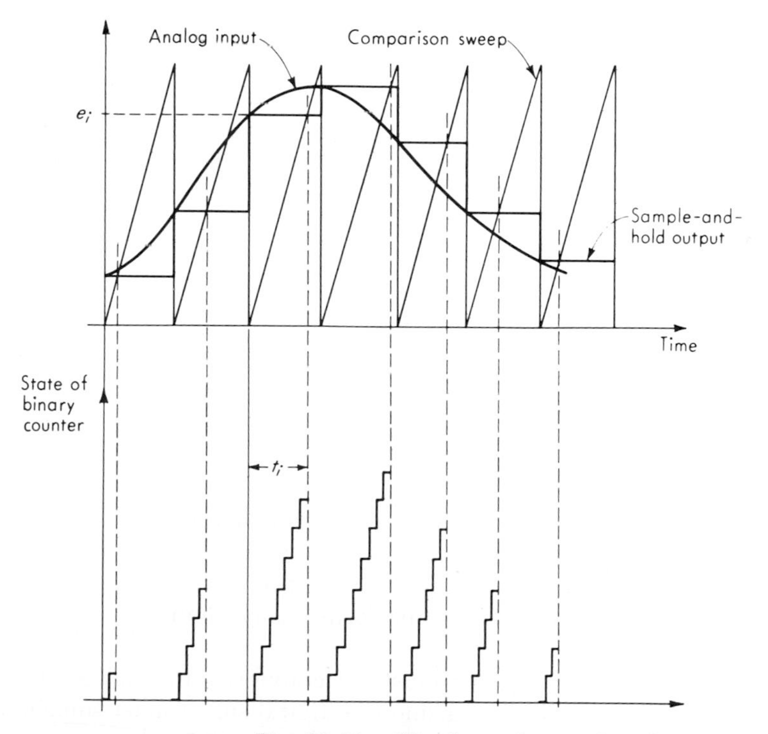

Fig. 10.50. *Time-base voltage-to-digital converter.*

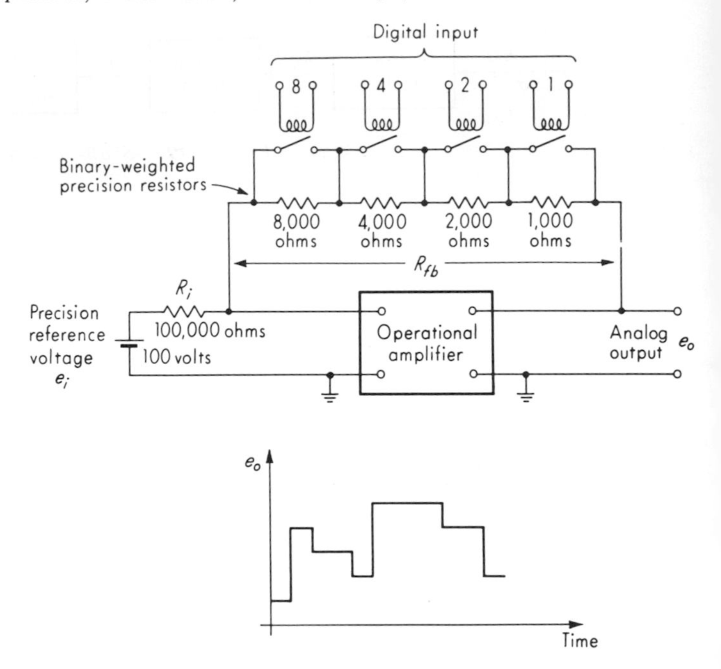

Fig. 10.51. *Digital-to-analog converter.*

sweep voltage whose peak value is larger than any analog-signal voltage expected. To make the readings of analog input occur at constant time intervals, the input is generally passed through a sample-and-hold device as shown. The output of the sample-and-hold is then compared with the sweep. A binary counter is started at the beginning of each sweep cycle and is stopped at the instant the sweep coincides with the sampled analog input. This counter generates coded pulses representing a binary number which increases in magnitude at a uniform rate. If such a counter is left on for a time interval t_i, as shown in Fig. 10.50, its "state" (binary number represented by its contents) will be proportional to t_i; since t_i is proportional to the sampled analog voltage e_i, the analog voltage will have been converted to its binary digital equivalent. The counter is reset to zero at the end of each cycle; thus the process repeats itself and generates a sequence of digital numbers representing the value of the analog signal at equally spaced intervals of time. All-electronic equipment of this type can perform many thousand such conversions per second.

When digital signals must be converted to analog, the "switch" openings or closures that define the digital signal must somehow be converted into an equivalent voltage. Figure 10.51 shows one method of accomplishing this by using an operational amplifier. The switches shown are closed when a particular digit is not present and open when it is. The feedback resistance at any time is thus the sum of the precision resistors that are not shorted out. For example, if the digital number 13 is present the switches for digits 1, 4, and 8 would be open, giving

$$e_o = -e_i \frac{R_{fb}}{R_i} = -100 \frac{13{,}000}{100{,}000} = -13 \text{ volts} \qquad (10.84)$$

The system shown can handle numbers up to 15 in steps of 1 but obviously can be expanded to get greater resolution. The analog output voltage is seen to vary in stepwise fashion but these steps can be smoothed by filtering if desired.

10.13

Signal and System Analyzers In the analysis and design of many devices and systems it is necessary to have accurate knowledge about the characteristics of the inputs to the system. Once a system has been built, its performance is often checked by studying its output. Equipment for carrying out such studies may be characterized as *signal-analysis equipment.* Closely related to this is the problem of experimentally defining the characteristics (transfer function, frequency response, etc.) of a physical system which may be too complex to analyze accurately by theory alone. Equipment for such investigations might be called *system-analysis equipment* and generally utilizes coordinated simultaneous measurements of both the system input signal and the output signal, together with suitable data processing to obtain conveniently the desired system characterization.

Perhaps the most widely used signal-analysis equipment is that which measures the frequency spectrum of a fluctuating physical quantity. The most common applications are in the field of sound and vibration where the frequency spectrum of a sound pressure, stress, acceleration, etc., may be very useful in diagnosing faults in an operating machine or system. These can be traced back to their origin by noting peaks in the frequency content at certain frequencies and then finding the machine parts that run at speeds that would produce such frequencies. While machines containing rotating and/or reciprocating parts give rise to sound and vibration signals having strong peaks at certain frequencies, a glance at an oscilloscope screen showing such a signal will generally not indicate a simple periodic wave form. Rather, the appearance will be that of a more-or-less random variation. Since this is the case, we would expect that the proper description of such a signal in the frequency domain would be in terms of the mean-square spectral density (power spectral density). While this is theoretically true, many problems of sound and vibration analysis do not require that a true mean-square spectral-density analysis be performed. Rather, somewhat simpler equipment is employed which gives sufficient information for most purposes.

Figure 10.52 shows the functional operation of a typical system. The incoming signal is passed through a narrow band-pass filter with center frequency ω_c and effective bandwidth $\Delta\omega$. This filter is tunable, so that ω_c may be varied smoothly and continuously over a given frequency range, say 20 to 20,000 cps, by manual turning of a knob or automatic drive from a constant-speed motor. This filtering operation appears to be exactly the same as that which would be performed in a true mean-square spectral-density (MSSD) analysis. However a difference exists with regard to the filter bandwidth $\Delta\omega$. In a true mean-square spectral-density analysis the output of the mean-squaring operation is divided by $\Delta\omega$. In practical mean-square spectral-density systems this division need not actually be carried out since $\Delta\omega$ is held constant for all ω_c's and thus the output recorder simply has its scale adjusted to take account of the division by a constant and known $\Delta\omega$. Circuits for obtaining a constant, small $\Delta\omega$ over a wide range of frequencies are quite complex and costly; thus many practical analyzers do not attempt to measure a true mean-square spectral density.

A filter which *can* be constructed fairly easily is the so-called constant-percentage bandwidth type. Here $\Delta\omega$ is not constant but rather varies in direct proportion to ω_c. Thus if $\Delta\omega$ is, say, 1 cps when ω_c is 100 cps it will be 10 cps when ω_c is 1,000 cps; that is, the bandwidth is 1 percent of the center frequency. Now with such a filter one can still compute a true mean-square spectral density by dividing a reading at a given ω_c by the proper $\Delta\omega$. However *this* division is not now possible by a simple fixed scaling of the output recorder, and most systems of this type do not perform it at all. The output record of such an analyzer is *not* a plot of mean-square spectral density, and even though it could be replotted manually to give it (since $\Delta\omega$ is usually known as a function of ω_c) this is not generally done since the uses to which the record is to be put do not require this level of sophistication. Thus the analyzer of Fig. 10.52 performs operations *similar* to those of a true mean-square spectral-density analysis but there is a definite difference which must be remembered. In particular, equal recorded output levels at a low and at a high frequency do *not* mean that the input signal has equal frequency content at these two points, because the true measure of frequency content is mean-square spectral density which requires division by a larger $\Delta\omega$ at the high frequency than at the low.

For applications that *do* require an accurate measurement of mean-

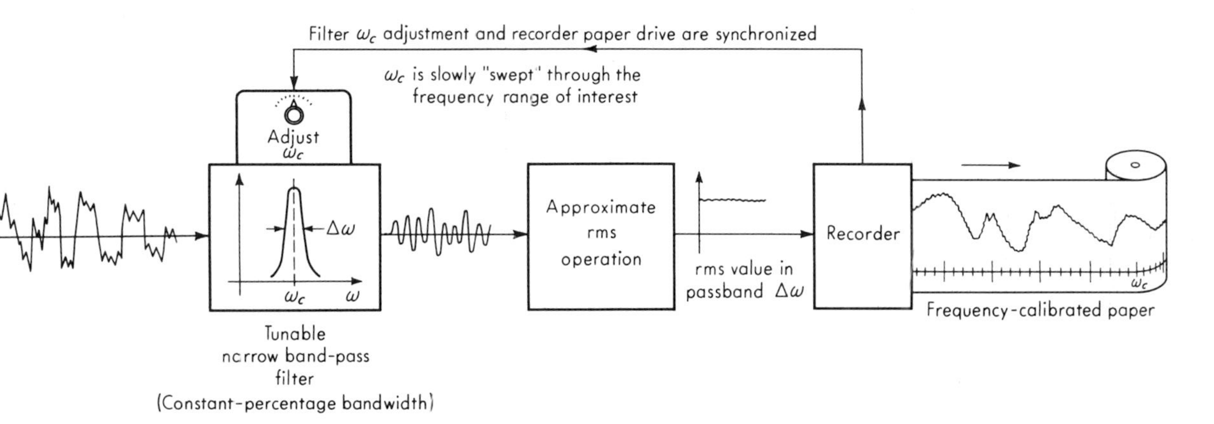

Fig. 10.52. *Frequency-spectrum analyzer.*

square spectral density, suitable analysis equipment is available, though rather costly. The block diagram of such analyzers is quite similar to that of Fig. 10.52 except that the passband $\Delta\omega$ is kept constant for all ω_c's. The value of $\Delta\omega$ can usually be selected to suit the needs of a particular problem; 2 to 100 cps bandwidths are generally available. Also, a mean-square rather than root-mean-square operation is performed, and greater pains are taken to make it accurate. The range of ω_c that can be covered is of the order of 2 to 20,000 cps. If lower or higher frequencies must be studied, a tape-recorder approach using tape speedup or slowdown is possible. Systems are also commercially available for obtaining the cross spectral density of two random signals. Their operation follows essentially the block diagram of Fig. 3.100.

While most sound and vibration analyzers describe the signal in the frequency domain, equipment working in the time domain is also available. A commercially available time-delay correlator[1] computes autocorrelation and cross-correlation functions and automatically plots the required curves. The time delay τ is continuously adjustable from 0 to 17 msec and is accomplished all-electronically without the use of a magnetic-tape loop device. In operation, the value of τ is slowly swept from zero to the maximum value, the sweep period ranging from 1 to 200 min. The averaging time [time over which the integral that defines $R(\tau)$ is computed] can be adjusted from 0.2 to 20 sec, longer averaging times requiring slower τ sweeps. The instrument puts out two d-c voltages, one proportional to τ and the other to $R(\tau)$; thus they can be applied to an X-Y plotter to plot $R(\tau)$ versus τ directly and automatically.

Commercial equipment for the frequency analysis of transient signals is also available. One such analyzer[2] is a direct analog mechanization of the Fourier-transform integrals for computing the frequency spectrum of a transient signal (see Fig. 10.53). The transient input $f(t)$ is multiplied by sine and cosine signals of adjustable frequency; the product curves are then integrated and the integrator outputs applied to d-c meters which read the real and imaginary parts of $F(i\omega)$ directly after the signals settle to their final values. Points on a plot of $F(i\omega)$ versus ω are obtained one frequency at a time by setting the desired ω into the sine and cosine generator and initiating the computing cycle. The input transient $f(t)$ must thus be reproducible since it is needed for each frequency computation. This is usually accomplished by tape recording $f(t)$ and then playing it into the analyzer over and over again, once for each frequency. A standard unit of this type can be used over the frequency range 0.01 to 1,000 cps. If $F(i\omega)$ is wanted in the $M\underline{/\phi}$ form

[1] Honeywell Denver Div., Denver, Colo.

[2] Weston-Boonshaft and Fuchs, Hatboro, Pa.

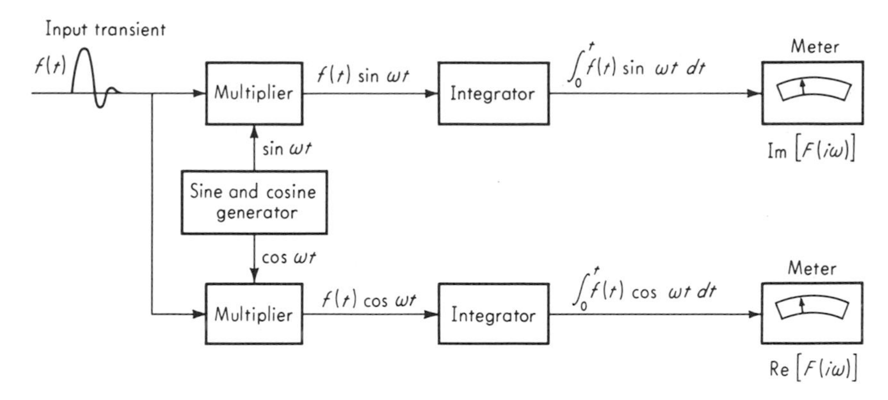

Fig. 10.53. *Fourier-transform transient analyzer.*

rather than the $a \pm ib$ form, a phase and magnitude computer is available to perform this operation.

In some studies of random signals it is desired to know the form of the amplitude-distribution function $W_1(q_i)$. Equipment based essentially on the definition of $W_1(q_i)$ associated with Fig. 3.93 is commercially available.[1,2] This instrument produces d-c voltages proportional to q_i and $W_1(q_i)$ when the signal to be analyzed is applied to its input. The value of q_i can be set manually or automatically swept slowly through a range of q_i while an X-Y plotter plots the desired curve of $W_1(q_i)$ versus q_i. Input signals in the frequency range 0 to 10,000 cps can be handled, while the range of q_i covered is ± 5 rms values.

To determine the transfer function of an unknown system experimentally (*system* analysis) one must measure and analyze both the input signal and the output signal of the system. Equipment is available to carry out these analyses for sinusoidal, transient, or random input signals. Sinusoidal (frequency-response) testing is the most common approach, and many analyzers of this type are available. Figure 10.54*a* shows the operation of a particular commercial unit.[3] The oscillator provides both the input signal for the system under test and the sine and cosine waves needed at the multipliers. Its frequency can be set over the range 0.01 to 1,000 cps on a standard unit. The frequency response $G(i\omega)$ of the unknown system is obtained one point at a time by setting the oscillator at the desired frequency ω and waiting for transients to die out. Then the two meters can be read to get $(A_o \sin \phi)/2$ and $(A_o \cos \phi)/2$, which

[1] B & K Instruments Inc., Cleveland, Ohio.

[2] H. L. Fox, Probability Density Analyzer, Bolt, Beranek and Newman Inc., Cambridge, Mass., *Rept.* 895, 1962.

[3] Weston-Boonshaft and Fuchs, Hatboro, Pa.

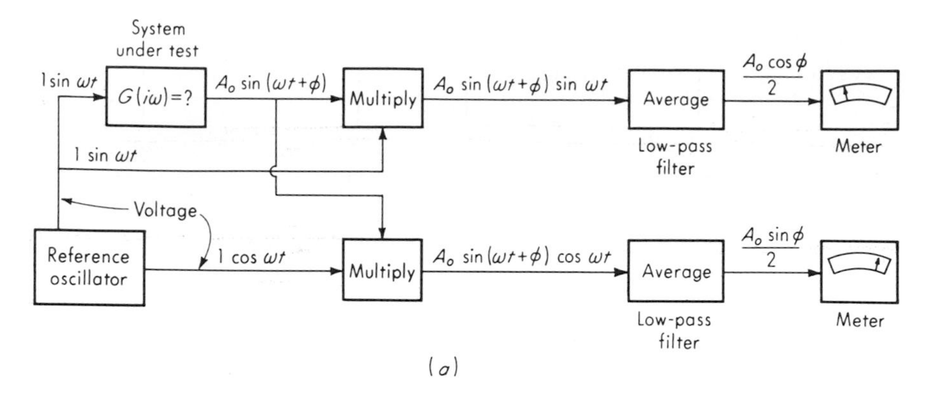

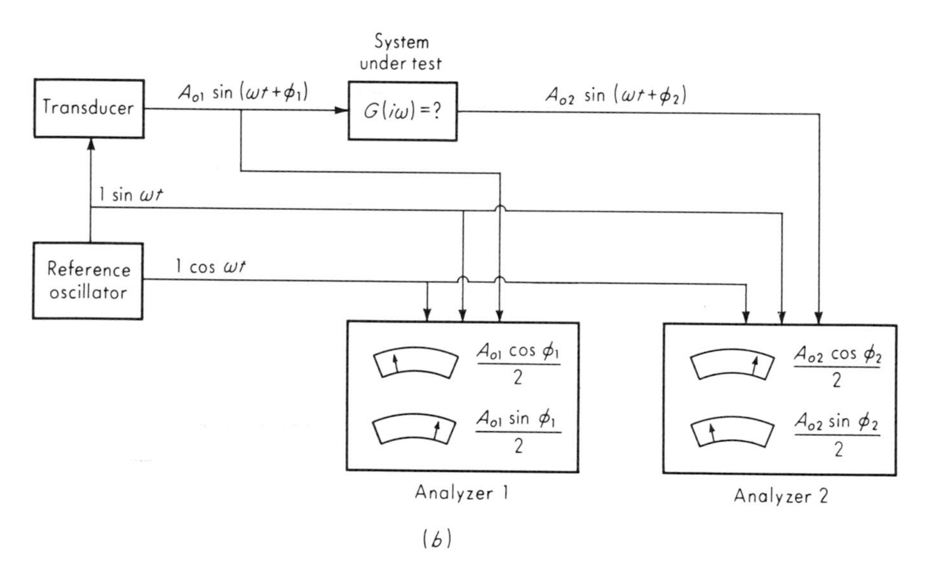

Fig. 10.54. *Frequency-response analyzer.*

allows calculation of A_o and ϕ either by hand or by an additional computer which is available. The operation of this analyzer may be understood by writing

$$[A_o \sin(\omega t + \phi)] \sin \omega t = A_o (\sin \omega t \cos \phi + \cos \omega t \sin \phi) \sin \omega t \tag{10.85}$$

Then

$$\text{Average value of } (A_o \cos \phi \sin^2 \omega t + A_o \sin \phi \sin \omega t \cos \omega t) = \frac{A_o \cos \phi}{2} + 0 \tag{10.86}$$

since the integral of $\sin \omega t \cos \omega t$ is zero over any number of complete cycles and thus its average value is zero. A similar analysis leads to $(A_o \sin \phi)/2$. If the system being tested is nonlinear, its output will not be a pure sine wave $A_o \sin(\omega t + \phi)$ but, rather, will also contain harmonics of the fundamental frequency ω. An analyzer of the type just described completely rejects such harmonics and measures only the fundamental component. This is a result of the general relation

$$\int_{-\pi}^{\pi} \cos mx \cos nx \, dx = \int_{-\pi}^{\pi} \sin mx \sin nx \, dx = \int_{-\pi}^{\pi} \cos mx \sin nx \, dx = 0 \qquad \text{if } n \neq m \tag{10.87}$$

Such behavior is particularly useful because it corresponds to determining the *describing function*[1] when the system is nonlinear. Describing function analysis is a powerful tool in nonlinear-system studies. When the input to the system under study is not an electrical voltage, the oscillator output must be transduced to the appropriate form. Since the transducer may not have perfect dynamics, *two* signals must now be analyzed by using the above procedures on each and then dividing the two complex numbers obtained to get $G(i\omega)$. This can be done sequentially, using one analyzer twice, or simultaneously if two analyzers are available (see Fig. 10.54*b*).

System analysis by transient techniques utilizes the procedure of Fig. 10.53 except that the frequency analysis must be performed on *two* transients one of which is the input $f_i(t)$ to the system and the other $f_o(t)$ which is the output. When $F_i(i\omega)$ and $F_o(i\omega)$ have been obtained over the frequency range of interest, one simply divides, point by point, the complex number $F_o(i\omega)$ by $F_i(i\omega)$ to get $G(i\omega)$, the frequency response of the unknown system. System analysis by random techniques utilizes the hardware arrangement of Fig. 3.100 and Eq. (3.39) to obtain the system frequency response from a measured spectral density and cross spectral density. Commercial systems that do this automatically and plot $G(i\omega)$ directly on an X-Y plotter are available.

A piece of equipment useful in performing a number of the analyses described in this section is the so-called tracking filter or synchronous filter. This is a narrow band-pass filter whose center frequency is adjusted electronically rather than by mechanical rotation of a knob. Such filters have two inputs: the signal to be filtered and the tuning signal. The filter is constructed so that its center frequency is equal to the frequency of the tuning signal, even if the tuning signal is changing frequency very rapidly. Commercial units[2] can "track" the tuning signal at rates

[1] E. O. Doebelin, "Dynamic Analysis and Feedback Control," p. 207, McGraw-Hill Book Company, New York, 1962.

[2] Spectral Dynamics Corp., San Diego, Calif.

in excess of 20,000 cps/sec. Such filters are useful in studying the sound and vibration of machines whose operating speed changes while the analysis is in progress. By picking off the operating speed with a magnetic pickup or a-c tachometer and using it as the tuning signal of the tracking filter, one can "slave" the center frequency to the machine operating speed. Equipment[1] is also available to tune the filter to any desired multiple of the tracking signal, continuously over a wide range. Thus if one wishes to observe, say, the third harmonic of operating speed, he can easily do so even if the speed varies during the analysis.

Problems

10.1 Derive the balanced-bridge relationship $R_1/R_4 = R_2/R_3$.

10.2 For a Wheatstone bridge, show that if $R_1 = R_2 = R_3 = R_4$ at balance, and if $\Delta R_2 = \Delta R_3 = 0$ and $\Delta R_1 = -\Delta R_4$, the output voltage is a perfectly linear function of ΔR_1, no matter how large ΔR_1 gets.

10.3 Discuss qualitatively the effect on bridge operation, for both the null method and the deflection method, of the excitation voltage source having an internal resistance.

10.4 In the system of Fig. 10.4, what considerations determine the numerical value of R_s?

10.5 In a Wheatstone bridge, $R_1 = 3{,}000$, $R_4 = 4{,}000$, $R_2 = 6{,}000$, and $R_3 = 8{,}000$ ohms at balance. Find the open-circuit output voltage if $\Delta R_1 = 30$, $\Delta R_2 = -20$, $\Delta R_3 = 40$, $\Delta R_4 = -50$, and $E_{ex} = 50$ volts. If the bridge output is connected to a meter of 20,000 ohms resistance, what will the output voltage now be?

10.6 Derive the results of Eqs. (10.22) and (10.23).

10.7 Derive Eq. (10.34).

10.8 Why does a charge amplifier essentially amount to a short circuit across the crystal?

10.9 Derive the transfer functions of the circuits of the following:

a. Figure 10.16*a*1
b. Figure 10.16*a*2
c. Figure 10.16*b*1
d. Figure 10.16*b*2
e. Figure 10.16*b*3

10.10 Derive the transfer function of the hydromechanical filter of Fig. 10.17.

10.11 In the circuit of Fig. 10.17, let e_i be supplied by a sinusoidal generator with an internal resistance of 1,000 ohms and an open-circuit voltage of 10 volts peak to peak. Also let $C = 10\ \mu$f and $R = 10$ ohms. If e_o is open-circuit, what voltage will actually appear at the e_i terminals for frequencies of 0, 100, 1,000, 10,000, and 100,000 cps?

10.12 Derive the transfer function of the filter of Fig. 10.18.

10.13 Explain how a notch filter can be used in the feedback path of a high-gain feedback system to construct a band-pass filter.

10.14 Plot logarithmic frequency-response curves of Eq. (10.39) if $1/2\pi\tau_1 = 100$ cps and $1/2\pi\tau_2 = 200$ cps.

10.15 Explain how you would plot frequency-response curves of Eq. (10.40) if numerical values were given.

[1] Spectral Dynamics Corp. San Diego, Calif.

10.16 Derive Eqs. (10.41) to (10.45). Show also that the phase angle at ω_p is zero.

10.17 Sketch the configuration of a hydraulic band-pass filter which has an amplitude-ratio attenuation of 40 db/decade on either side of the passband. This is twice the attenuation rate of the system of Fig. 10.21. Only components of the type used in Fig. 10.21 are allowed. Short "transition" regions of slope ± 20 db/decade are allowed between the flat response portion and ± 40 db/decade portions. You must derive the transfer functions to prove your "invention" works as claimed.

10.18 Derive Eq. (10.46).

10.19 Derive Eq. (10.54).

10.20 Derive Eq. (10.64).

10.21 For the system of Fig. 10.25*a*, let $e_i = e_{\text{signal}} + e_{\text{noise}}$, where $e_{\text{signal}} = 10 \sin 20t$ and $e_{\text{noise}} = 0.1 \sin 377t$. What is the signal/noise ratio before and after the differentiation?

10.22 The input to a differentiator with transfer function $(e_o/e_i)(D) = D$ is a random signal with a constant mean-square spectral density of 0.001 volt²/cps from 0 to 10,000 cps and zero elsewhere. Calculate the rms voltage at the input and at the output of the differentiator.

10.23 Derive Eq. (10.66).

10.24 Derive Eq. (10.67).

10.25 Using the system of Eq. (10.68) and the e_i of Prob. 10.21, compute the signal/noise ratio at both the input and the output.

10.26 Derive Eq. (10.69).

10.27 Discuss sensitivity/response-speed tradeoffs in the system of Fig. 10.26.

10.28 Derive the transfer function of the compensating circuit of Fig. 10.28.

10.29 Design a compensating network to speed up by a factor of 10 the response of a thermocouple with a time constant of 1 sec. Thermocouple resistance is 10 ohms and full-scale output is 5 mv. The amplifier/recorder available has maximum full-scale sensitivity of 0.1 mv and an input resistance of 100,000 ohms.

10.30 List and explain the action of all effects that tend to degrade the static accuracy of the system in Fig. 10.33.

10.31 Derive the equation of the operational-amplifier summing circuit of Fig. 10.34*c*.

10.32 Derive the operating equation of the mechanical multiplier of Fig. 10.36.

Bibliography

1. A. Miller: Bridge Circuits, *The Right Angle*, Sanborn Co., Cambridge, Mass., May and August, 1954.
2. P. R. Perino: The Effect of Transmission Line Resistance in the Shunt Calibration of Bridge Transducers, *Statham Instr. Notes* 36, Statham Instruments Inc., Los Angeles, Calif., November, 1959.
3. P. Pohl: Signal Conditioning for Semiconductor Strain Gages, *ISA J.*, p. 33, June, 1962.
4. Another Look at the Wheatstone Bridge, *Electromech. Design*, p. 36, February, 1965.
5. A. Baracz: Graph Finds Temperature Sensing Bridge Response, *Control Eng.*, p. 85, October, 1961.
6. R. B. F. Schumacher: Differential High-resistance Bridge, *ISA J.*, p. 65, April, 1965.
7. P. Perino: System Considerations for Bridge Circuit Transducers, *Statham Instr. Notes* 37, Statham Instruments Inc., Los Angeles, Calif., September, 1964.

8. G. White: Temperature Compensation of Bridge Type Transducers, *Statham Instr. Notes* 5, Statham Instruments Inc., Los Angeles, Calif., October, 1948.
9. B. B. Helfand: Summation and Averaging of Multiple Measurements by Parallel Transducer Operation, *Statham Instr. Notes* 16, Statham Instruments Inc., Los Angeles, Calif., July, 1950.
10. B. B. Helfand and J. Burns: Calibration of Resistance Bridge Transducer Circuits Under Temperature Extremes, Statham Instruments Inc., Los Angeles, Calif., *Statham Instr. Notes* 14.
11. H. E. Darling: Magnetic Amplifiers for Instrumentation, *ISA J.*, p. 58, January, 1960.
12. J. J. Rado: Input Impedance of a Chopper-modulated Amplifier, *Electro-Technol. (New York)*, p. 140, June, 1962.
13. J. DiRocco: Potentiometric Amplifiers Improve Impedance Buffering, *Control Eng.*, p. 87, July, 1962.
14. J. Minck and E. Smith: Noise Figure Measurement, *Instr. Control Systems*, p. 115, August, 1963.
15. W. R. Williams and R. C. Hawes: Vibrating Reed Electrometer, *Instr. Control Systems*, p. 112, November, 1963.
16. A. Pearlman: Selecting and Testing Solid-state Operational Amplifiers, *Instr. Control Systems*, p. 121, February, 1965.
17. R. D. Moore: Lock-in Amplifiers for Signals Buried in Noise, *Electronics*, June 8, 1962.
18. C. T. Stelzried: Loaded Parallel-T *RC* Filters, *Control Eng.*, p. 113, May, 1961.
19. G Cocquyt: Evaluating Bridged-T Networks for AC Systems, *Control Eng.*, p. 77, December, 1963.
20. A. I. Zverev: Introduction to Filters, *Electro-Technol. (New York)*, p. 61, June, 1964.
21. W. Gile: Solid-state Low-frequency Filter, *Electro-Technol. (New York)*, p. 34, September, 1964.
22. A. W. Langill: Designing Passive Compensators, *Electro-Technol. (New York)*, p. 26, January, 1965.
23. Miniature Servo Packages, *Electromech. Design*, p. 70, June, 1960.
24. J. B. Heaviside: Sources of Error in AC Servos, *Control Eng.*, p. 85, February, 1964.
25. H. J. Huttenlocker et al.: Instrument Servomechanism Systems, *Electromech. Design*, p. 37, July, 1964.
26. Miniature Servo Packages, *Electromech. Design*, p. 202, May, 1962.
27. Specifying an Instrument Servomechanism, *Electromech. Design*, p. 32, November, 1959.
28. M. Richter: A Simplified Technique in Instrument Servo Analysis, *Electromech. Design*, p. 36, February, 1962.
29. A. Svoboda: "Computing Mechanisms and Linkages," McGraw-Hill Book Company, New York, 1948.
30. T. R. Fredriksen: A Way to Design Low-loss Nonlinear Networks, *Control Eng.*, p. 117, June, 1962.
31. A. J. Baracz: How to Design a Compensating Bridge, *Control Eng.*, p. 81, March, 1965.
32. F. M. Ryan: Special Purpose Analog Computers, *Control Eng.*, p. 103, May, 1963.
33. J. T. Nichols: Zener-regulated Power Supplies, *Instr. Control Systems*, p. 2242, December, 1961.
34. J. Nagy: Zener Diode Power Supplies, *ISA J.*, p. 65, July, 1964.

11 Data transmission

When components of a measurement system are located more or less remotely from one another, it becomes necessary to transmit information between them by some sort of communication channel. There are also cases where, even though components are close together, transmission problems arise because of relative motion of one part of the system with respect to another. We shall briefly examine questions of this sort and some of the equipment commonly used to solve such problems.

The transmission of information is amenable to mathematical analysis totally disassociated from any hardware considerations, and there is a large body of technical literature on this subject. This science of communication has been extremely useful in putting the design of hardware on a rational basis, showing the tradeoffs in competitive systems, and putting theoretical limits on what can possibly be done. Its consideration, however, is beyond the scope of this text, and we shall restrict ourselves to rather qualitative, hardware-oriented discussions.

11.1
Cable Transmission of Analog Voltage Signals

Perhaps the most common situation is that in which a simple cable is used to transmit an analog voltage signal from one location to another. The accurate analysis of a cable or transmission line involves the use of a distributed-parameter (partial differential equation) approach since the properties of resistance, inductance, and capacitance are not lumped or localized.[1] Figure 11.1*a* shows the model generally used for such an analysis. An approximation suitable for low frequencies is the lumped network of Fig. 11.1*b*. The shunt conductance has been neglected here since, in practice, it is generally negligible. If the line is not too long and frequencies not too high, the crude model of Fig. 11.1*c* may be adequate. The inductance has been totally neglected, and R represents the total resistance of both conductors in the cable while C represents the total capacitance between them. These can be numerically calculated from the values per foot of length given by cable manufacturers. Typically, resistance per foot might be of the order of 0.01 ohm while capacitance would be about 30 pf/ft. Thus a 1,000-ft length of two-conductor cable has $R = 20$ ohms and $C = 30{,}000$ pf. The actual frequency response of a length of cable can be measured experimentally to get its exact characteristics. Figure 11.1*d* and *e* show some typical results.[2] The use of cables up to 7,000 ft long to transmit low-level (± 10 mv full scale) data has been accomplished.[2] However, even short cables (less than 10 ft) can cause difficulties in high-impedance transducers such as the piezoelectric type. The use of charge amplifiers rather than voltage amplifiers may be helpful in such cases.

11.2
Cable Transmission of Digital Data

When data must be transmitted very long distances (100 miles is not unusual), analog signals tend

[1] H. H. Skilling, "Electric Transmission Lines," McGraw-Hill Book Company, New York, 1951.

[2] R. L. Smith, Transmission of Low-level Voltage Over Long Telephone Cables, *NASA, Tech. Note* D-1320, p. 14, January, 1963.

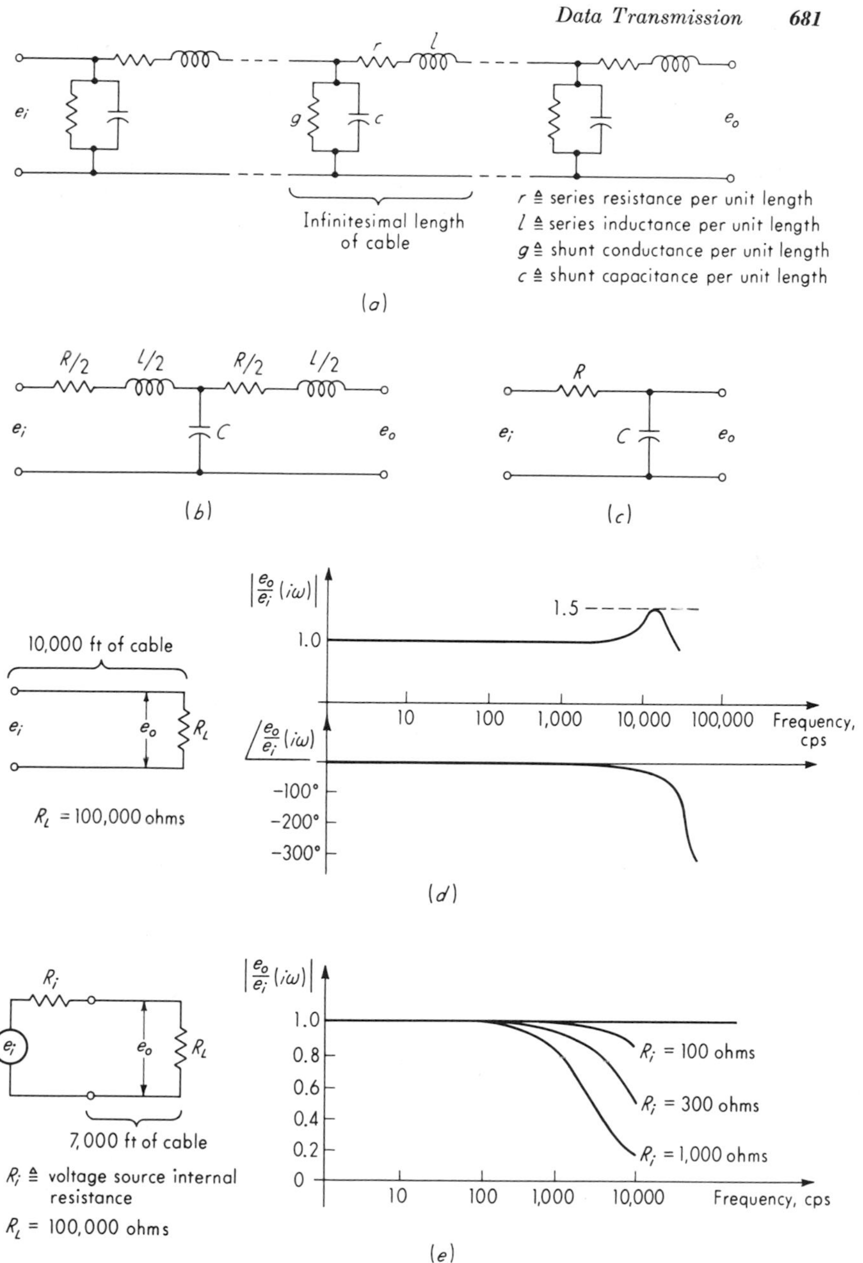

Fig. 11.1. *Cable models and response.*

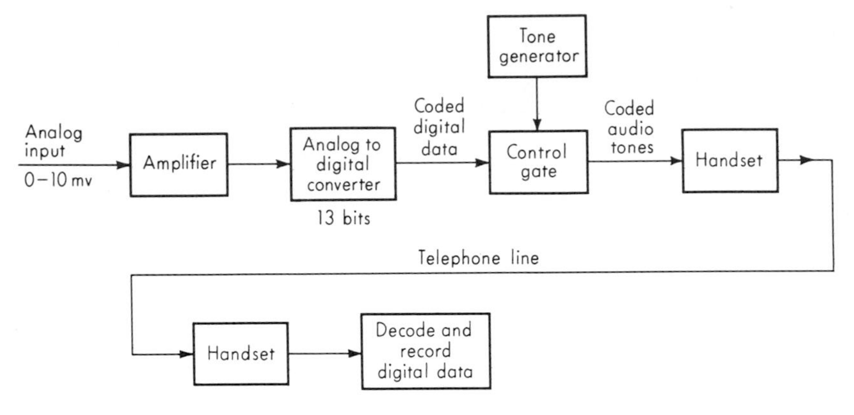

Fig. 11.2. *Telephonic digital data transmission.*

to be corrupted by the response characteristics of the transmission line and the pickup of spurious noise voltages from a number of sources. Under such conditions it may be desirable to convert the analog data to some digital form, transmit it in digital form, and then reconvert to analog form if desired. A time-honored example of digital transmission is the telegraph system, in which letters and numbers are represented by a system of coded pulses (dots and dashes).

In many cases the information can be transmitted over lines that have already been installed for other purposes. Electric power systems, for example, transmit information signals over their power transmission lines simultaneously with the transmission of 60-cycle power. It is simply necessary to keep the frequencies used sufficiently separated to allow easy filtering for elimination of unwanted signals. Telephone lines are also in wide use for data transmission. An interesting system[1] shown in block-diagram form in Fig. 11.2 uses ordinary telephone handsets together with auxiliary equipment to transmit and receive data. The telephone line is not leased; one can use *any* telephone to call any other in the usual way. Once voice contact is established, the handsets are placed into an "acoustic coupler." Here tones at 1,400 cps (representing binary 1) and 2,100 cps (representing binary 0) produced by a tone generator actuated from the analog-to-digital converter are coupled into a loudspeaker which the telephone "hears." These sounds are transmitted over the telephone line in the usual way to be received by a similar system (but working in reverse fashion) at the other end. Frequency response of

[1] M. L. Klein, Telephonic Transmission of Data, *Instr. Control Systems*, p. 99, June, 1962.

such a system is quite low but accuracy of transmission is the order of 0.1 percent.

The high accuracy of digital data transmission as compared with analog is due to the fact that the size or precise shape of a pulse in a digital system is not particularly important. Rather, the system operates on the presence or absence of some sort of pulse. Thus even rather severe degradation of pulse shape by the transmission medium will not affect the accuracy of a digital system *at all* as long as the presence or absence of a pulse can be detected.

11.3

FM/FM Radio Telemetry When interconnecting wires are not possible or desirable, data may be transmitted by radio. A number of different schemes[1,2] are in use; we here consider only one which is widely employed. The word telemetry means simply measurement at a distance and includes all forms of such systems, irrespective of the means of transmission or physical nature of the hardware.

Radio telemetry probably received its greatest impetus from the requirements of aircraft and missile flight testing during and after World War II. Considerable standardization based on the requirements of such systems has been accomplished, and our discussion here reflects this emphasis. Figure 11.3 shows the widely used FM/FM system of radio telemetry. The symbol FM/FM refers to the fact that two frequency-modulation processes are employed. In the first process, time-varying d-c voltages are converted to proportional frequencies, using voltage-to-frequency converters as described in Sec. 10.11. When used in FM/FM telemetry systems these converters are generally called subcarrier oscillators (SCO's). Instead of voltage-to-frequency converters, subcarrier oscillators may also be inductance-controlled. The inductance of a variable-inductance transducer forms part of the oscillator circuit, and changes in inductance cause proportional changes in frequency from the center frequency (see Fig. 11.4).

The standard FM/FM system of Fig. 11.3 has 18 available channels; thus 18 different physical variables may be measured and transmitted simultaneously. Figure 11.5 shows the characteristics of these 18 channels. Note that the center frequencies range from 400 to 70,000 cps. Such low frequencies cannot be practically transmitted by radio propaga-

[1] C. M. Harris and C. E. Crede (eds.), "Shock and Vibration Handbook," vol. 1, pp. 19–76, McGraw-Hill Book Company, New York, 1961.

[2] M. H. Nichols and L. L. Rauch, "Radio Telemetry," John Wiley & Sons, Inc., New York, 1956.

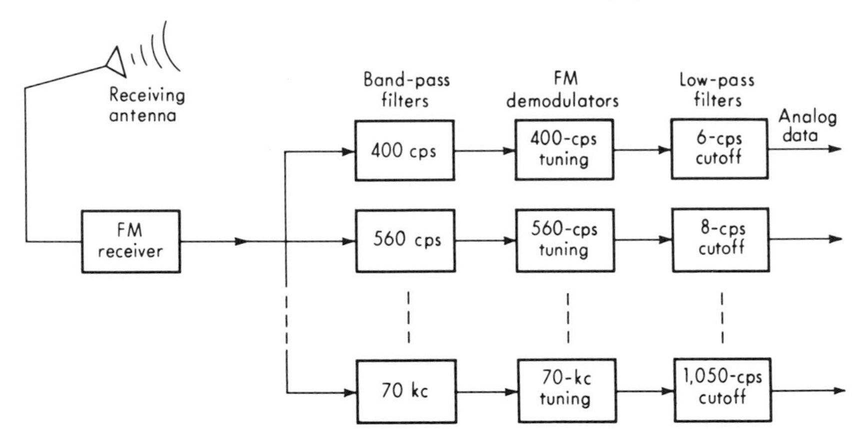

Fig. 11.3. *FM/FM radio telemetry.*

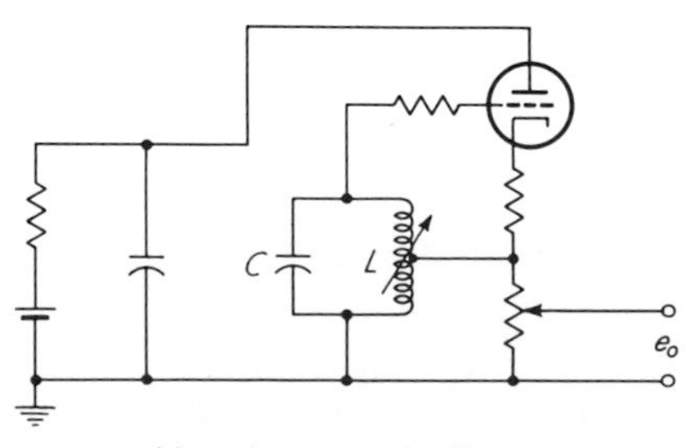

Fig 11.4. *Variable-inductance subcarrier oscillator.*

	Band	Center frequency, cps	±Full-scale frequency deviation, %	Overall frequency response, d-c to ___ cps
	1	400	7.5	6.0
	2	560	7.5	8.4
	3	730	7.5	11.0
	4	960	7.5	14
	5	1,300	7.5	20
	6	1,700	7.5	25
	7	2,300	7.5	35
	8	3,000	7.5	45
	9	3,900	7.5	59
	10	5,400	7.5	81
	11	7,350	7.5	110
	12	10,500	7.5	160
	13	14,500	7.5	220
	14	22,000	7.5	330
	15	30,000	7.5	450
	16	40,000	7.5	600
	17	52,500	7.5	790
	18	70,000	7.5	1,050
Optional bands				
Omit 15 and B	A	3,300	15.0	660
Omit 14, 16, A, C	B	4,500	15.0	900
Omit 15, 17, B, D	C	6,000	15.0	1,200
Omit 16, 18, C, E	D	7,880	15.0	1,600
Omit 17 and D	E	10,500	15.0	2,100

Fig. 11.5. *Telemetry channel characteristics.*

tion since they would require antennas of immense size because the size of an antenna must be of the order of the wavelength to be transmitted [wavelength in meters $= 3 \times 10^8/$(frequency in cps)]. Thus an additional frequency modulation to boost all frequencies into the radio-frequency range is employed. Rather than use a separate radio-frequency transmitter for each of the 18 channels (which is wasteful of the crowded RF spectrum and also requires much more equipment), the 18 channels are "mixed" (added) and sent out together over one radio-frequency channel. Two such channels, 217.550 and 219.450 Mc, are available for radio telemetry. The frequency deviation caused by any of the subcarrier-oscillator frequency variations cannot exceed ± 125 kc around either of these two frequencies. (Other radio frequencies in the range 216 to 235 Mc may also be used if available.) When the radio signals are received, the 18 channels must be reseparated by suitable band-pass filters, FM-demodulated and low-pass-filtered to reconstruct the original analog data. We see from Fig. 11.5 that by using optional band E a channel with frequency response as great as 0 to 2,100 cps is available. This is adequate for many purposes. Systems of the above type have ranges up to about 400 miles and accuracies of the order of ± 2 percent of full scale. For data of relatively low frequency content, any one of the channels may be time-shared by several transducers if a commutator is used to sample each transducer periodically. If a very large number of low-frequency signals must be telemetered, other methods such as pulse-duration modulation (PDM/FM) may be indicated. More sophisticated telemetry systems can be used over ranges of millions of miles, as in the space probes to Venus.

Radio telemetry is also useful over very short distances when the relative motion of the measuring device and the readout equipment prevent a suitable direct connection. Good examples of such situations are found in measurements on rotating machinery, where slip-ring techniques are not feasible because of high speeds or inaccessibility, and in physiological measurements on test animals or human beings, in which restriction of motion due to connecting wires is not tolerable. Miniaturization of the telemetry components and improvement of shock resistance by use of semiconductor devices now make possible many such applications formerly not feasible.

11.4

Pneumatic Transmission Transmission of pressure signals in industrial pneumatic control systems is regularly accomplished over distances of several hundred feet. Pneumatic-transmission-line dynamics is analogous to that of electrical cables but, of course, at a much lower frequency. Adequate simplified models employing a dead time equal to the acoustic transmission time and either a first-order or second-order system are available and were discussed in Chap. 6 [see Eqs. (6.89) and (6.90)].

11.5

Synchro Position Repeater Systems Figure 11.6 illustrates synchro position repeater systems used for transmitting low-power mechanical motion over considerable distances with only a three-wire interconnecting cable. Whenever the two angles θ_i and θ_o are not identical, an electromagnetic torque is exerted on the rotor of *each* machine, tending to bring the shafts into alignment. Thus, if the transmitter shaft θ_i is turned, the receiver shaft θ_o will follow accurately so long as there is no appreciable torque load on the θ_o shaft. Accuracy of such systems

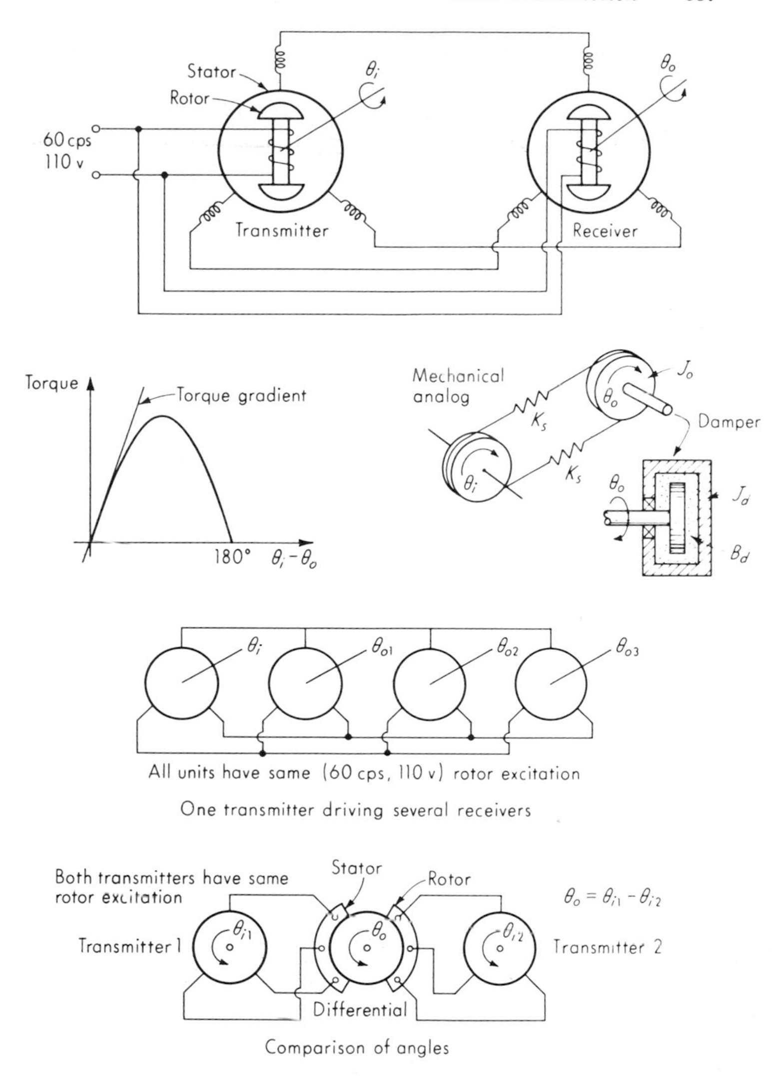

Fig. 11.6. *Torque-synchro angle transmission.*

depends on the torque gradient (torque per unit error angle) of the transmitter/receiver system. A typical value might be 0.35 in.-oz/degree. The electrical system serves only to transmit power from the θ_i to the θ_o shaft; all the mechanical work taken out at θ_o must be provided as mechanical work at θ_i. The torque gradient is reduced as the resistance of the connecting cable is increased; thus long cables result in reduced accuracy. In a typical unit, 10 ohms resistance in each of the three wires results in a 50 percent loss of torque. The dynamic response of these systems is essentially second-order, a mechanical analog being as shown in Fig. 11.6. A damper is sometimes put on the θ_o shaft to reduce oscillations since little inherent damping is present. When one transmitter drives several receivers (all units of identical size) the torque available at each receiver is $2/(N+1)$ times the torque for a single pair, where N is the number of receivers. The synchro differential shown in Fig. 11.6 is useful for comparing two rotations at a location remote from either. Its static and dynamic behavior is essentially the same as for a transmitter/receiver pair.

11.6

Slip Rings When transducers must be mounted on the rotating members of machines, some means must be provided to bring excitation power into the transducer and to take away the output signal. Some transducers (such as synchros) are themselves rotating "machines" in which such data and/or power transmission between a rotating and a stationary member is necessary. When only a small relative motion is involved, continuous flexible conductors (often in the form of light coil springs) can be used. In some cases of limited rotation through a few revolutions, the connecting wires can simply be allowed to wind or unwind on the rotating shaft. However, continuous high-speed rotation requires slip rings, radio telemetry, or some form of magnetic coupling between rotating and stationary parts.

Figure 11.7 shows the common forms of slip rings.[1] Rings are made of coin gold, silver, or other noble metals and alloys. Block-type brushes are often sintered silver graphite while wire-type brushes are alloys of platinum, gold, etc. An important consideration in slip rings used to transmit low-level instrumentation signals is the electrical noise produced at the sliding contact. One component of this noise is due to thermocouple action if the brush and ring are of different materials. The other main effect is a random variation of contact resistance due to surface roughness, vibration, etc. If the contact carries current, a variation in contact resistance causes a noise voltage to appear at the contact. A high-quality miniature sliding slip ring may exhibit a contact-resistance variation of the order of 0.05 ohm peak to peak and 0.005 ohm rms.[2]

While slip rings have been successfully operated at about 100,000 rpm, applications above 10,000 rpm generally require extreme care because of heating and vibration problems. A particular slip-ring assembly[3] usable

[1] A. J. Ferretti, Slip Rings, *Electromech. Design*, p. 145, July, 1964.

[2] E. J. Devine, Rolling Element Slip Rings for Vacuum Application, *NASA*, *Tech. Note* D-2261, p. 11, 1964.

[3] Ferretti, *op. cit.*, p. 159.

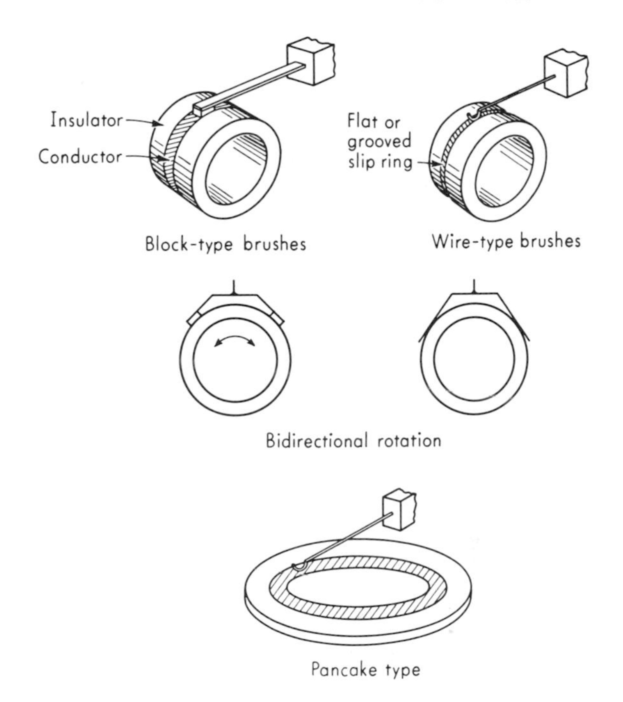

Fig. 11.7. *Slip-ring configurations.*

to 100,000 rpm and intended for strain-gage work had peak-to-peak noise voltage of 0.02 mv at 52,000 rpm and 0.40 mv at 100,000 rpm. This assembly used liquid cooling and lubrication of slip rings and bearings and gave a brush life of 30 hr at 35,000 rpm. At 52,000 rpm the noise level in a typical strain-gage circuit gave a signal/noise ratio of about 150:1. Hard gold rings of $\frac{1}{4}$-in.-diameter were used with two cantilevered wire-tuft brushes per ring.

When slip rings are used with strain-gage circuits, particular care must be taken since the resistance variation of the sliding contact may be comparable with the small strain-gage resistance change to be measured. If possible, a full bridge on the rotating member should be employed so that the sliding contacts can be taken out of the bridge circuit. This arrangement (Fig. 11.8) greatly reduces the effects of slip-ring resistance variations. For the most demanding applications, more complex schemes are available[1,2] to reduce noise to even lower levels.

[1] C. C. Perry and H. R. Lissner, "The Strain Gage Primer," p. 186, McGraw-Hill Book Company, New York, 1955.

[2] P. K. Stein, "Measurement Engineering," vol. II, chap. 29, Stein Engineering Services, Inc., Phoenix, Ariz.

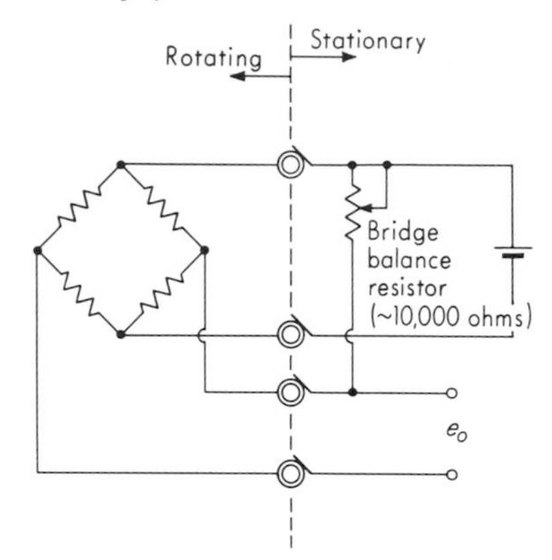

Fig. 11.8. Bridge-circuit slip-ring configuration.

A rotating disk dipping into a mercury pool (see Fig. 11.9) can perform the same function as a conventional slip ring. A commercially available device[1] is usable from 0 to 10,000 rpm, has a contact resistance of 0.005 ohm, contact-resistance variation of ± 0.00025 ohm for 0 to 600 rpm, and no measurable resistance variation from 600 to 10,000 rpm, is compensated for self-generated thermoelectric voltages, and can be made with 2 to 160 terminals.

Another alternative to slip rings is the rotary transformer.[2] Here signal or power voltages (they must be a-c) are transferred through an annular air gap between a concentrically rotatable primary and secondary coil.

Ordinary sliding slip rings may not operate properly in the high-vacuum environment of space. Preliminary research[3] using a thrust-type ball bearing as the signal-transfer mechanism indicates that rolling contact slip rings may provide a solution for such problems. A particular

[1] Meridian Laboratory, Lake Geneva, Wis.
[2] Data Tech, Cambridge, Mass.
[3] Devine, *op. cit.*

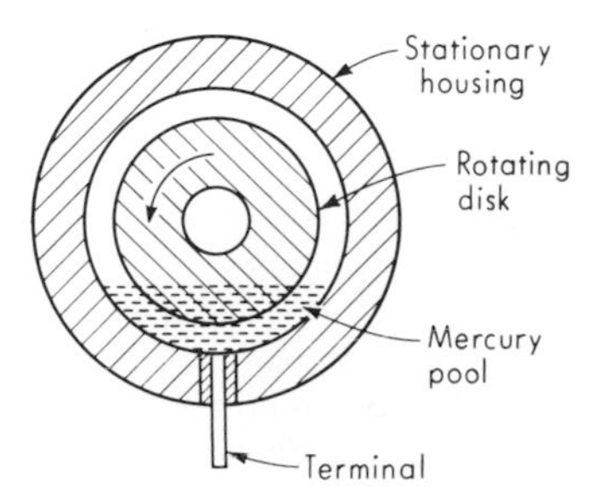

Fig. 11.9. Mercury-pool slip ring.

test at 2,000 rpm and vacuum of 2×10^{-9} torr gave operation for over 100 million revolutions at a resistance variation of 0.002 ohm rms.

Problems

11.1 Find a general expression for $(e_o/e_i)(D)$ for the system of Fig. 11.1*b*. If R = 20 ohms, C = 0.3 μf, and L = 0.2 millihenry, plot the logarithmic frequency-response curves.

11.2 A synchro repeater system has one transmitter and five receivers. The torque gradient of a single pair of devices with very short cable connections is 0.5 in.-oz/degree, and 10 percent of this is lost for each ohm of cable resistance. Each receiver drives a dial with 0.05 in.-oz of friction. If the allowable error is 0.5° and cable resistance is 0.05 ohm/ft, find the maximum allowable cable length.

Bibliography

1. R. H. Cerni and L. E. Foster: "Instrumentation for Engineering Measurement," chap. 5, John Wiley & Sons, Inc., New York, 1961.
2. J. D. Tate: Synchro Systems, *Machine Design*, p. 150, June 8, 1961.
3. R. J. Barber: 21 Ways to Pick Data Off Moving Objects, *Control Eng.*, p. 82, October, 1963; p. 61, January, 1964.
4. F. W. Hannula: Transmitting Test Information, *Control Eng.*, p. 173, September, 1959.
5. E. D. Lucas: Techniques for Radio Telemetry, *Control Eng.*, p. 71, December, 1962.
6. E. A. Ragland and D. E. Wassall: The Digital Answer to Data Telemetering, *Control Eng.*, p. 95, August, 1957.
7. E. H. Krause: Telemetering for Interplanetary Flight, *ISA J.*, p. 478, October, 1957.
8. C. I. Cummings and A. W. Newberry: Radio Telemetry, *ARS J.*, p. 141, May–June, 1953.
9. E. H. de Grey and J. G. Bayly: Measuring through Vessel Walls, *ISA J.*, p. 82, May, 1963.
10. L. W. Gardenhire: Evolution of PCM Telemetry, *Instr. Control Systems*, p. 87, April, 1965.
11. M. K. Stark: Short Range Telemetry System Provides Test Data on Rotating Parts, *Gen. Motors Eng. J.*, p. 23, January-February-March, 1965.
12. J. Valentich: Simple Slip Rings for Strain Gage Measurement, *Machine Design*, p. 154, Jan. 7, 1960.

12 Voltage indicating and recording devices

The majority of signals in measurement systems ultimately appear as voltages. Since voltage cannot be seen, it must be transduced into a form intelligible to a human observer. The form in which the data are presented is generally that of a pointer moving over a scale, a pen writing on a chart (including light beams writing on photosensitive paper and electron beams writing on cathode-ray tubes), visual presentation of a set of ordered digits, or printout of digital data by a typewriter or similar device. We shall consider the most common types of such indicating and/or recording devices.

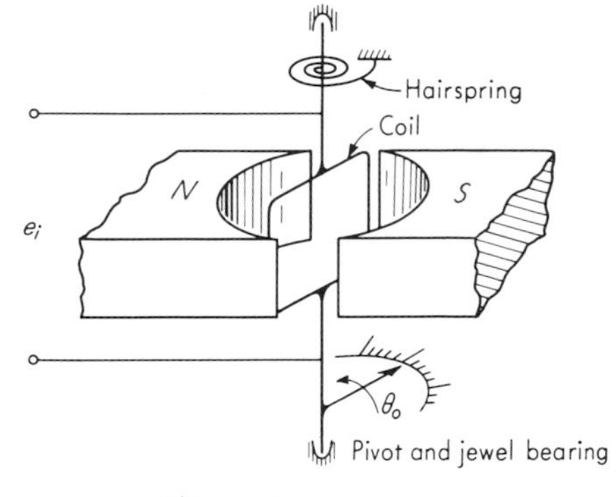

D'Arsonval meter movement

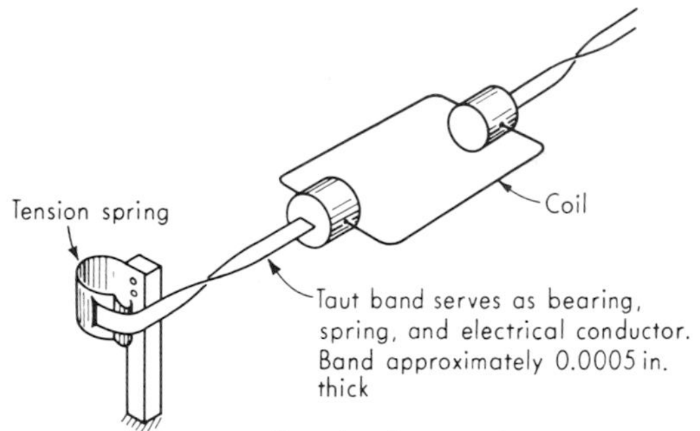

Taut-band suspension

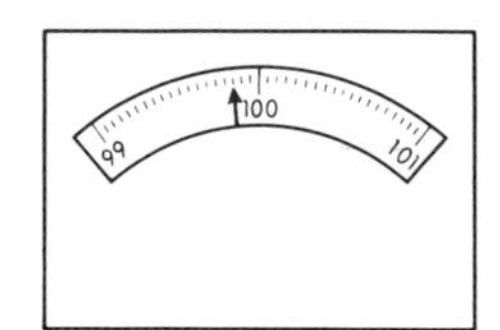

Expanded scale meter

Fig. 12.1. *D-C analog meters.*

12.1

Analog Voltmeters and Potentiometers The most widely used meter movement for d-c and (with rectifiers) a-c measurement in electronics and instrumentation work is the classical D'Arsonval movement (see Fig. 12.1). This is basically a current-sensitive device but is used to measure voltage by maintaining circuit resistance constant by means of compensating techniques (see Fig. 2.17*a*). Relatively recent improvements on this basic configuration include taut-band suspension (rather than pivot-and-jewel bearings), individually calibrated scale divisions, and expanded-scale instruments. Taut-band suspension completely eliminates bearing friction, reduces inertia and temperature effects, increases ruggedness, and results in less loading on the measured circuit since the reduced friction requires less power drain. The increased accuracy made possible by taut-band construction can be provided at reasonable cost through the use of automatic calibration systems which print an individual scale for each and every instrument. Expanded-scale instruments use a precision voltage-suppression circuit to measure a small variation around a larger voltage. Thus if one needs to measure a 100-volt signal accurately it is possible to do this with a meter whose scale goes from 99 to 101 volts. Static inaccuracies of 0.1 percent are attainable in a rugged and portable instrument with these methods.

Vacuum-tube voltmeters (see Fig. 3.23) still use the D'Arsonval meter movement but precede it by amplifier circuits. These increase the input impedance and overall sensitivity. Such instruments generally accept a wide range of d-c and a-c input voltages and have static error of the order of 1 to 3 percent of full scale.

When a-c (not necessarily sinusoidal) voltages are to be measured with a D'Arsonval movement, it is necessary to perform rectification. Depending on the circuitry used, a meter may be sensitive to the average, peak, or rms value of the input wave form. It is common practice to calibrate the scale of the meter to read rms value no matter what quantity is fundamentally sensed. This procedure is accurate only if pure sinusoidal wave forms are being measured since, in this case only, the peak, average, and rms values are all related by fixed constants and can thus be included in the scale calibration. For nonsinusoidal wave forms, peak- or average-sensing meters will not read the correct rms value. In some cases, peak or average value is actually what is wanted; however, rms is most often desired. A true rms voltmeter is complex and expensive; thus peak- and average-sensing meters calibrated to read rms are in wide use and are generally satisfactory except in the most critical applications.

Figure 12.2 shows circuits for peak, average, and rms meters. In the peak circuit the capacitor is charged to the peak value of a periodic input voltage. This charge cannot leak off rapidly because of the one-way conduction of the diodes and the high input impedance of the voltmeter (vacuum-tube voltmeters often use peak sensing). When the input reverses sign, the capacitor is additionally charged by an amount equal to the negative peak. The voltage across the meter thus stays near the peak-to-peak value of the input with only slight fluctuations due to diode reverse leakage and meter noninfinite impedance. In the average-reading circuit the input is full-wave-rectified, and the low-pass filtering characteristic of the meter movement is used to extract the

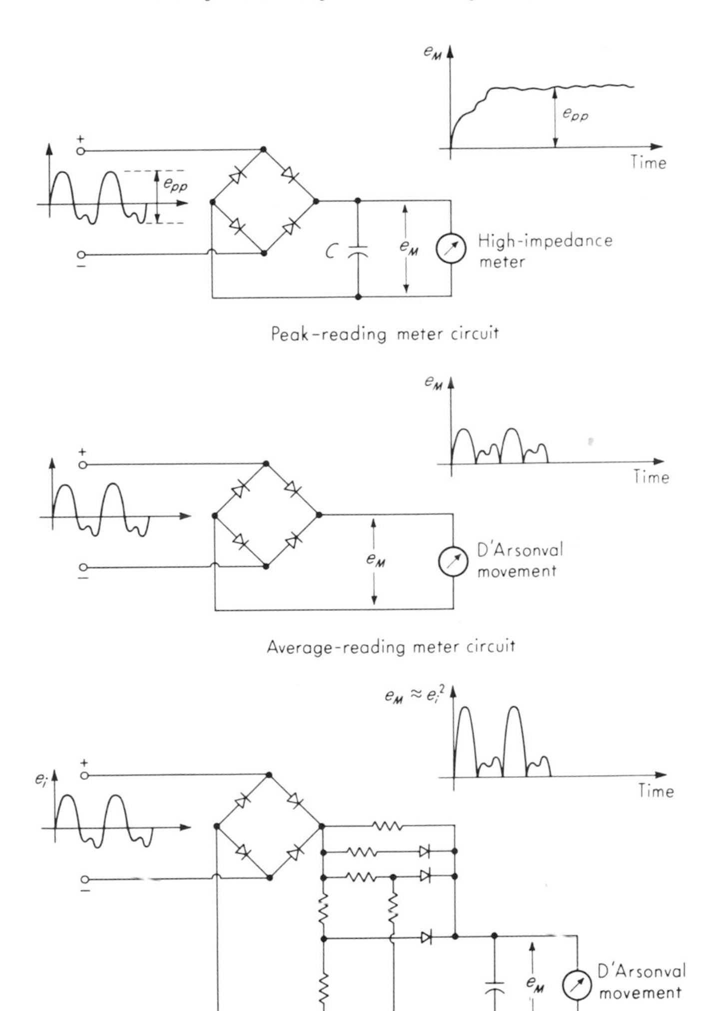

***Fig. 12.2.** Peak, average, and rms circuits.*

average value. The rms-reading circuit[1] approximates the required square-law parabola with a few straight-line segments in the fashion of a diode function generator. The average voltage on the capacitor is used to provide a variable bias on the diodes in the function generator,

[1] C. G. Wahrman, A True RMS Instrument, *B & K Tech. Rev.*, B & K Instruments, Cleveland, Ohio.

thereby obtaining higher accuracy than possible in a fixed-bias unit using the same number of diodes. The averaging required in obtaining an rms value is performed by the meter's low-pass filtering characteristic while the square-root operation is simply obtained by meter-scale distortion.

When highly accurate rms measurements of nonsinusoidal signals are required (random signals are a good example), methods based on the heating power of the wave form are employed since heating power is directly proportional to the mean-squared voltage. One such true rms voltmeter[1] amplifies the incoming signal and applies it to a highly stable resistance heating element bonded to silica and surrounded by an inert atmosphere. A thermopile of 45 copper/constantan junctions is attached to the heating element to measure its temperature. Conditions are such that highly linear heat-transfer processes occur and thus heater equilibrium temperature is closely proportional to the square of current, and therefore to the square of voltage, since heater resistance is constant. The thermopile output voltage is thus an accurate measure of rms input voltage and can easily be read on a D'Arsonval millivoltmeter. The described instrument accepts input signals in the frequency range 2 to 250,000 cps with response flat ± 0.2 db and has full-scale voltage ranges from 0.5 mv to 250 volts. The averaging time (related to the time over which the integration is carried out in the exact mathematical definition of mean-square value) of this meter is 16 sec. Since the *exact* determination of the mean-square value of a random signal requires an infinite integrating (averaging) time, long averaging times are needed for high accuracy. However, they also make the meter sluggish in reaching its final value; thus a switch-selectable 0.5-sec averaging time is provided for use in situations where the 16-sec value is not needed. Voltmeters for random signals must be able to handle peaks that are large compared with the rms value. This is specified by the peak factor of the meter. Large peak factors (ratio of peak to rms value) are desirable; the meter described above has a value of 10.

When the most accurate measurements of d-c voltage are required, potentiometers rather than deflection meters are employed. The potentiometer is a null-balance instrument in which the unknown voltage is compared with an accurate reference voltage which can be adjusted until the two are equal. Since, at the null point, no current flows, errors due to *IR* drops in lead wires are eliminated. Such *IR* drops are always present when a D'Arsonval-type meter is used to measure voltage directly. Figure 12.3*a* shows the basic potentiometer circuit. We see that a galvanometer (just a very sensitive D'Arsonval movement) is used as a

[1] Flow Corp., Cambridge, Mass., *Bull.* 59, 1960.

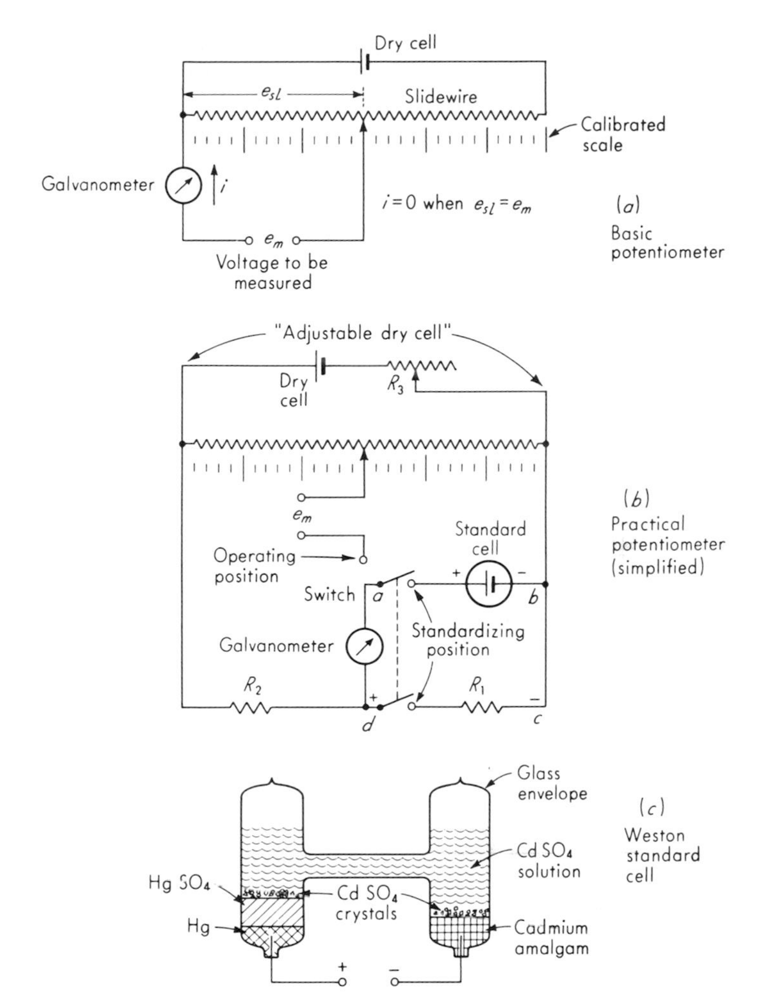

Fig. 12.3. *Manually-balanced potentiometer.*

null detector. It detects the presence or absence of current by deflecting whenever the unknown and reference voltages are unequal. However, it need not be calibrated since it must indicate only the presence of current, not its numerical value. The basic circuit of Fig. 12.3*a* is not practical since the accuracy of the reference voltage picked off the slide-wire is directly influenced by changes in the dry-cell voltage. Since the dry cell supplies power to the slide-wire, its voltage is bound gradually to drop off. This problem is solved in the practical circuit of Fig. 12.3*b* by inclusion of an additional component, the standard cell.

Figure 12.3*c* shows the Weston cadmium saturated standard cell which is the basic working standard of voltage. Its terminal voltage is 1.018636 volts and is reproducible to the order of 0.1 to 0.6 ppm. Its accuracy in terms of the fundamental mass, length, and time standards can be established to only about 10 ppm, however. Its temperature coefficient is -40 μv/C°; thus close temperature control must obviously be employed in the most exacting situations. A standard cell cannot be substituted for the dry cell of Fig. 12.3*a* since its accuracy is destroyed if any appreciable current is drawn from it over a time interval. It must thus be used as an intermittent reference against which the slide-wire excitation voltage can be checked whenever desired. The *unsaturated* Weston cell is used in practical instruments since it is more portable. Its terminal voltage varies from one unit to another. However its drift at constant temperature is only about -0.003 percent per year; thus it is perfectly adequate for most purposes. Its temperature coefficient is about -10 μv/C°.

The operation of the circuit of Fig. 12.3*b* is as follows: When the slide-wire scale on the potentiometer was originally calibrated at the factory, slide-wire excitation-adjusting resistor R_3 was set at a fixed value and resistor R_1 was adjusted until, when loop *abcd* was completed by the switch, no current flowed in the galvanometer. This means that the voltage drop across R_1 was just equal to the standard-cell voltage. Now the slide-wire, R_1, and R_2 are all fixed and stable resistors; thus if the voltage across R_1 is at its calibration value the slide-wire excitation voltage must also be at its calibration value. Thus, whenever we wish to check the calibration (this is called standardization) we merely complete the loop *abcd* momentarily (so as not to draw much current from the standard cell) and note whether the galvanometer deflects. If it does, we adjust the slide-wire excitation with R_3 until deflection ceases. We are then assured that the slide-wire excitation is at its original calibration value. The resistor R_2 is merely a current-limiting resistor to prevent drawing large current from the standard cell through the slide-wire path, which is fairly low-resistance.

Fairly common and inexpensive potentiometers which can be read to the nearest microvolt are in wide use. More sophisticated instruments intended for the most accurate calibration work provide greater accuracy and sensitivity. One such commercially available instrument[1] measures in three ranges: 0 to 1.611110 volts in steps of 1.0 μv, 0 to 0.1611110 in steps of 0.1 μv, and 0 to 0.01611110 in steps of 0.01 μv. The total parasitic thermoelectric voltage is less than 0.1 μv. The limit of error on the high range is ± 0.003 percent of reading ± 0.1 μv, while on the medium and low ranges it is ± 0.005 percent of reading ± 0.1 μv. These values

[1] Honeywell Inc., Philadelphia, Pa.

approach the level of the National Standards achieved by the National Bureau of Standards, which are about 0.001 percent from 0.01 to 1,000 volts.

12.2

Digital Voltmeters and Printers A digital voltmeter[1,2] accepts analog voltage inputs and produces a direct visual display of the voltage reading in decimal digits. All such instruments are essentially made up of some sort of analog-to-digital converter (see Sec. 10.12) plus some means of visual display. They are available in a wide range of capabilities to measure d-c and a-c voltages from less than a millivolt to more than 1,000 volts. Integrating digital voltmeters (see Sec. 10.11) are also included in this class. The time required to convert an analog voltage to digital form and display it varies widely with the type of conversion, ranging from about 1 μsec to several seconds. Inaccuracy of digital voltmeters is generally determined by the stability of the reference voltage (standard-cell or Zener-diode reference) plus the inherent error of ± 1 unit in the least significant digit. Five-digit units are available, thus giving a resolution of 0.001 percent of full scale. Accuracy of the order of 0.01 percent of full scale on a given range is attainable. Since coded digital signals must be produced in a digital voltmeter, many instruments provide these as outputs that can be sent to a digital printer or recorder.

Digital recorders or printers are electromechanical or electro-optical devices that print digital characters (usually the decimal digits 0 through 9 and a plus or minus sign) on a moving strip of paper. A line of printing may contain any number of digits, depending on the resolution of the equipment feeding the printer. Electromechanical parallel input printers[3] have a top speed at present of about 40 lines per second. Electro-optical printers[4] with serial input have speeds of 135 lines per second. When digitally recorded data must later be reproduced and/or processed, punched cards, punched paper tape, or digital magnetic tape may be employed.

12.3

Self-balancing Potentiometers and X-Y Recorders In considering analog indicating and recording instruments capable of producing

[1] Digital Voltmeters, *Electromech. Design*, p. 58, December, 1960.

[2] DVM, *Instr. Control Systems*, p. 101, June, 1964.

[3] Franklin Electronics, Bridgeport, Pa.

[4] Century Electronics, Tulsa, Okla.

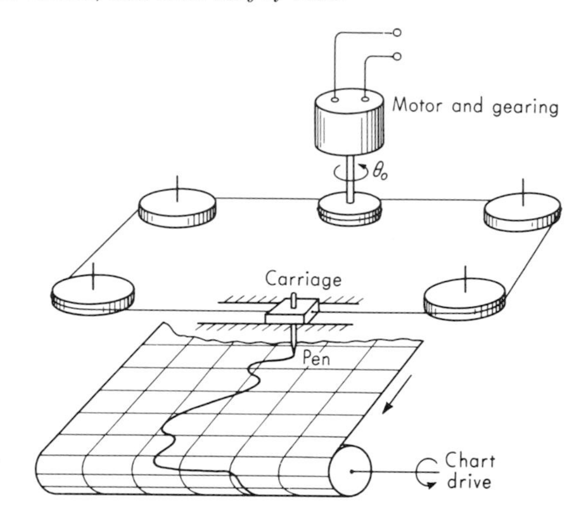

Fig. 12.4. *Self-balancing potentiometer.*

a permanent visual record of the time variation of a voltage, a classification with regard to static accuracy and speed of response is useful. The majority of recording tasks of this type are performed by three classes of instruments. These are:

Recorder	*Static error, percent of full scale*	*Frequency response, cps*
Self-balancing potentiometer	0.2	Less than 5
Galvanometer oscillograph	2	Up to 10,000
Cathode-ray oscilloscope	3	Up to 10^9

In this section we discuss self-balancing potentiometers and the closely related *X-Y* plotters.

Almost all self-balancing potentiometers use the instrument-servomechanism principle of Fig. 10.33. To make this into a recording instrument it is only necessary to convert the output angle θ_o to translation of a carriage to which a pen is attached (usually by means of a piano-wire and pulley drive) and then pass calibrated chart paper, from a roll under the pen, at a fixed and known speed to establish the time base (see Fig. 12.4). The pen will then trace the variation of the voltage e_i with time. Most such recorders now use a Zener-diode reference supply voltage rather than a standard cell. This semiconductor circuit provides an accurate and stable reference voltage and, unlike a standard cell, also can supply current to excite the slide-wire. With such units the inconvenience of periodic standardization is eliminated.

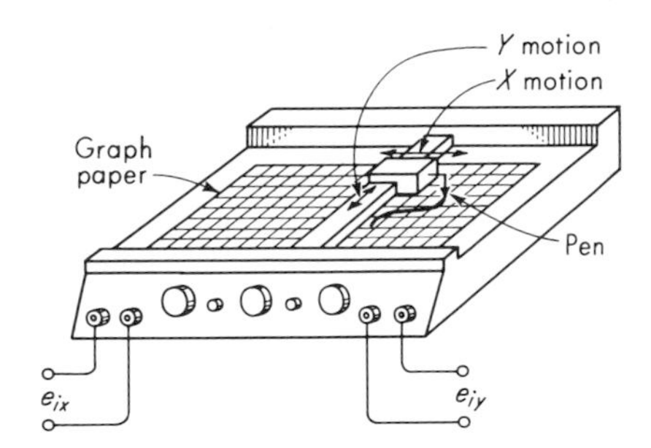

Fig. 12.5. *X-Y plotter.*

Self-balancing potentiometers come in a wide range of capabilities. A general-purpose laboratory type[1] exemplifies the limits of performance presently attainable with this class of instrument. Full-scale input voltage may be switch-selected from 0.1 mv to 100 volts in 19 ranges. Chart width is 6 in. with a static accuracy ± 0.25 percent of full scale or 1 μv, whichever is greater. Frequency response is flat within 1 percent to 5 cps for inputs that are 10 percent of full scale; full-scale inputs result in reduced frequency response. Time for full-scale pen travel is less than 0.5 sec. Chart speeds are thumbwheel-selected from 10 in the range 1 in./sec to 10 min/in.

When multichannel recording is required, two general approaches are used. For slowly varying data the pen is replaced with a print wheel having numbers, say, 1 to 24. A sampling switch connects each of 24 input signals to the potentiometer input in turn. When, for example, channel 9 is connected, the potentiometer drives to the correct chart position, prints a 9, and then goes on to channel 10, etc. For continuous recording, a separate servosystem and pen are provided for each channel but they all write on the same chart. To prevent mechanical interference and still allow each channel to use the full chart width, each pen is displaced by about $\frac{1}{16}$ in. from its neighbors. This causes a slight time displacement from channel to channel but this can be corrected for, if necessary. Up to four channels of such overlapping recording are available with 10-in. chart width.[2]

Often it is desired to cross-plot one variable against another rather than against time. This can be done by employing two independent servosystems to drive a pen over a stationary chart paper. Such arrangements are called *X-Y* plotters. Figure 12.5 shows their typical configuration. They are available for paper sizes up to several feet on a side; however the most common size accepts standard $8\frac{1}{2}$- by 11- or 11- by 17-in.

[1] Honeywell Inc., Philadelphia, Pa.
[2] Texas Instruments, Houston, Texas.

graph paper. Since each axis operates on essentially the same principle as the self-balancing potentiometer, their static and dynamic characteristics are quite similar. Many *X-Y* plotters also provide for making one of the axes a time base if it is desired to plot a single variable against time. This is done by generating a ramp-function voltage by charging a capacitor with a constant current, thereby making the capacitor voltage increase linearly with time. This voltage is then applied to the input of one of the servosystems and causes that axis to translate at a constant speed. One commercially available 11- by 17-in. instrument[1] has 17-d-c voltage ranges on each axis ranging from 0.1 mv/in. to 20 volts/in. and 12 a-c ranges from 5 mv/in. to 20 volts/in. Static error of d-c ranges is 0.2 percent of full scale and for a-c ranges (20 cps to 100 kc) it is 0.5 percent. Nonlinearity is 0.1 percent for d-c ranges and 0.2 for a-c. A time sweep giving eight speeds from 0.5 to 100 sec/in. may be applied to either axis.

12.4

Galvanometer Oscillographs While the D'Arsonval movement of Fig. 12.1 as applied to meters has a very limited frequency response (less than 1 cps), it is possible to miniaturize it so as to obtain rotational natural frequencies of the order of 10,000 cps. D'Arsonval movements of this type are called galvanometers and are the basic sensing elements of galvanometer oscillographs. We shall analyze the galvanometer to determine its performance characteristics.

Figure 12.6 shows schematically the construction of a typical galvanometer. Viscous damping B may or may not be present in a given design; we carry it along for generality. The galvanometer is an electromechanical transducer, and we shall write two equations, one electrical and one mechanical, to analyze its behavior. Basically, input voltage e_s from the signal source causes a current to flow in the coil. There is then a current-carrying conductor in a magnetic field; thus the coil experiences an electromagnetic force which, since it has a lever arm, causes a torque. This torque tends to rotate the coil until it is just balanced by the restoring torque of the torsion springs. For a constant e_s the output pointer comes to rest at a definite value of θ_o. By proper design the static relation between θ_o and e_s can be made quite linear.

Writing a Kirchhoff voltage-loop law for the electrical circuit, we have

$$i_g(R_s + R_g) + L_g \frac{di_g}{dt} + HNlb \frac{d\theta_o}{dt} - e_s = 0 \qquad (12.1)$$

[1] Hewlett-Packard Co., Moseley Div., Pasadena, Calif.

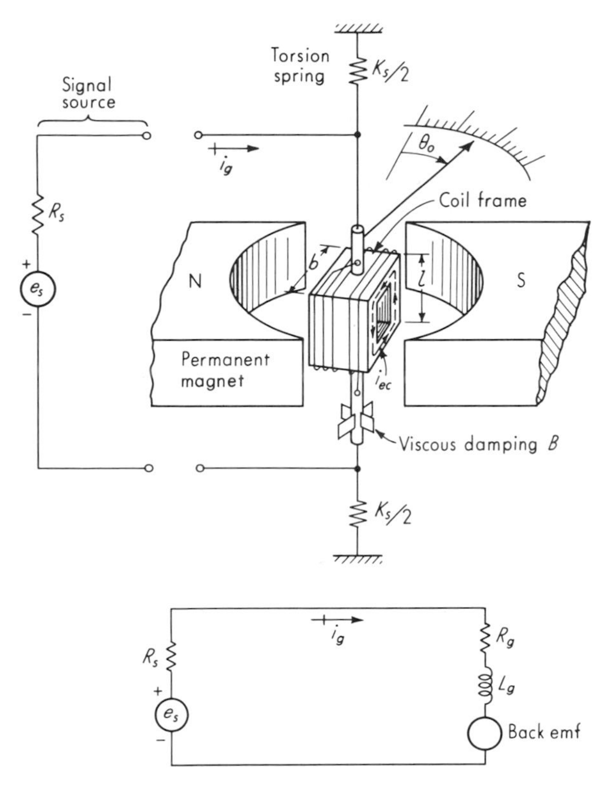

Fig. 12.6. *Galvanometer.*

where

$$HNlb(d\theta_o/dt) = \text{back emf of coil acting as generator}$$
$$H \triangleq \text{flux density}$$
$$N \triangleq \text{number of turns on coil}$$
$$l \triangleq \text{length of coil}$$
$$b \triangleq \text{breadth of coil}$$
$$L_g \triangleq \text{inductance of coil}$$
$$R_g \triangleq \text{resistance of coil}$$
$$R_s \triangleq \text{resistance of signal source}$$

We may note that this analysis is comparable to that of a d-c motor since the devices are basically the same. Now we can apply Newton's law to the rotational motion of the coil:

$$\Sigma \text{ torques} = J \frac{d^2\theta_o}{dt^2}$$

The main electromagnetic torque is given by $HNlbi_g$; however, a more subtle effect also produces an additional torque. If the *frame* on which the coil is wound is itself a conductor, a voltage $Hlb(d\theta_o/dt)$ will be induced in it (since it is a "one-turn coil") and an eddy current $i_{ec} = (Hlb/R_f)(d\theta_o/dt)$ will flow in the frame, where R_f is the resistance of the frame. This eddy current is also in the magnetic field and thus the frame feels an electromagnetic torque $-Hlbi_{ec}$. The minus sign is needed since such induced voltages always set up effects that oppose the original motion. Newton's law then reads

$$HNlbi_g - \frac{(Hlb)^2}{R_f}\frac{d\theta_o}{dt} - K_s\theta_o - B\frac{d\theta_o}{dt} = J\frac{d^2\theta_o}{dt^2} \tag{12.2}$$

where $B \triangleq$ viscous damping coefficient
$K_s \triangleq$ torsional spring constant
$J \triangleq$ moment of inertia of moving parts about axis of rotation

Equations (12.1) and (12.2) each contain two unknowns, i_g and θ_o; thus they must be solved simultaneously. Since our main interest is the transfer function relating input e_s to output θ_o, i_g is of little interest, and we do not bother to solve for it. A simplification is possible in Eq. (12.1) since experience shows that, in terms of frequency response, within the useful operating frequency range of the galvanometer the effect of inductance is invariably negligible. We thus drop this term and solve Eq. (12.1) for i_g.

$$i_g = \frac{e_s - HNlb(d\theta_o/dt)}{R_s + R_g} \tag{12.3}$$

This may now be substituted in Eq. (12.2), which gives, after some manipulation,

$$\frac{\theta_o}{e_s}(D) = \frac{K}{D^2/\omega_n^2 + 2\zeta D/\omega_n + 1} \tag{12.4}$$

where

$$K \triangleq \frac{HNlb}{K_s(R_s + R_g)} \qquad \text{rad/volt} \tag{12.5}$$

$$\omega_n \triangleq \sqrt{\frac{K_s}{J}} \qquad \text{rad/sec} \tag{12.6}$$

$$\zeta \triangleq \frac{B + (Hlb)^2/R_f + (HNlb)^2/(R_s + R_g)}{2\sqrt{K_sJ}} \tag{12.7}$$

We see that the galvanometer is a second-order instrument. One can increase sensitivity by increasing H, N, l, and b or by decreasing K_s, R_g, and R_s. The flux density H is usually at the maximum value possible with ordinary permanent magnets. Increases in N, l, and b result in increases in R_g since a longer total length of wire in the coil results; thus the net effect on sensitivity is not obvious. Decreases in K_s increase K directly; however, speed of response is lost since ω_n decreases. Note also that sensitivity varies with R_s; thus changing from one trans-

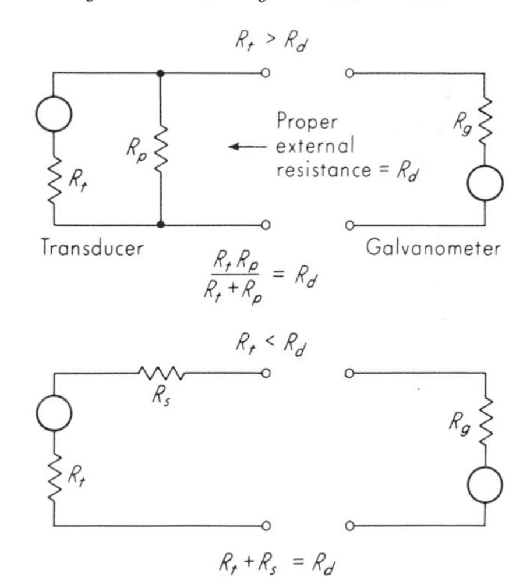

Fig. 12.7. *Damping networks.*

ducer to another with a different resistance results in a loss of calibration. This is due to the fact that the galvanometer is a current-sensitive device. Equation (12.7) shows that the mechanical viscous damping B can be completely eliminated and the system is still damped. The damping that remains is of electromagnetic origin and is caused by the back emf proportional to $d\theta_o/dt$ (mechanical-viscous-damping torque is also proportional to $d\theta_o/dt$). In low-frequency, high-sensitivity galvanometers the electrical damping is adequate to obtain the optimum ζ value of 0.65, and no intentional mechanical damping B is provided. Also such galvanometers may have the coil frame slotted so that $R_f = \infty$. Then

$$\zeta = \frac{(HNlb)^2}{2\sqrt{K_s J}\,(R_s + R_g)} \tag{12.8}$$

The most important feature of this result is that ζ depends on R_s, the source resistance. *Thus such a galvanometer can be properly damped for only one value of external resistance.* This does not mean it can be used only with transducers that have this value of R_s, but it does mean that a suitable resistance network may have to be interposed between the transducer and galvanometer. If the transducer resistance is too high, a shunt resistor is needed; if too low, a series resistor (see Fig. 12.7) is required. For high-frequency, low-sensitivity galvanometers the electromagnetic damping is inadequate, and intentional viscous damping is provided. The electromagnetic damping is a small percentage of the total, and so such galvanometers can generally be used with any external resistance in a wide range.

To construct a recording oscillograph it is necessary to provide chart paper moving at a known speed to give a time base, and a means of writing the galvanometer motion on this paper. A number of writing methods have been developed and are in use. The most obvious is simply to mount an ink pen at the end of an arm and let it move over the chart paper. To get straight-line motion from the rotation θ_o, special linkages have been developed. Another popular method also uses a mechanical arm but replaces the pen with a heated stylus. This requires use of a special heat-sensitive paper but eliminates the possible clogging and skipping of ink pens. Any oscillograph using a mechanical arm and a pen or stylus has so much inertia (due to the long arm) and friction (due to contact with the paper) that its frequency response is limited to 100 or 200 cps or less. This is adequate for many applications, and such instruments are in wide use. They generally come provided with amplifiers since their galvanometers are designed quite stiff to be accurate in the face of considerable friction.

In general, such instruments do not have interchangeable galvanometers of various sensitivities and response speeds. Since the input signal goes to an amplifier, the proper damping is provided there, and the signal source resistance usually has no effect on damping. Oscillographs of this type generally have a full-scale motion of 2 or 3 in. and a nonlinearity of 1 or 2 percent. Multichannel oscillographs of this class are generally limited to six or eight channels side by side since the mechanical arms cannot overlap. Recent developments utilizing pen-position feedback systems and pressurized inking systems have improved accuracy to about $\frac{1}{2}$ percent.

To realize the high frequency response (up to 10,000 cps) mentioned earlier, inertia and friction must be drastically reduced. This is accomplished by replacing the mechanical writing arm with a light beam. A tiny mirror is rigidly fastened to the moving coil and a light beam reflected from it. When the coil turns, the light beam, which is focused as a spot on the moving chart paper, deflects over the paper, leaving a trace (see Fig. 2.5). Until recently, ordinary photographic-type recording paper was used and records were not available until time-consuming darkroom work had been carried out to process the records. Today, there are available automatic processors attached directly to the oscillograph to make records quickly accessible. Even more convenient are the new recording processes using special papers and/or light sources that require no processing at all. For slowly varying signals at low chart speeds these processes give an immediately visible trace, just as in ink recording. For high-speed recording the trace becomes visible in 15 or 20 sec if the paper is exposed to an ordinary fluorescent lamp. These records will last for years if kept away from sunlight and can be subjected to a simple liquid fixing process if absolute permanence is required.

Oscillographs using light-beam galvanometers generally provide an

Undamped natural frequency, cps	Flat (±5%) frequency response, cps	External resistance for optimum damping, ohms	Coil resistance, ohms	Current sensitivity, in./ma	Maximum deflection for ±2% nonlinearity, in.
		Electromagnetic damped types			
40	0–24	120	20	136	8
40	0–24	350	35	225	8
100	0–60	120	32	91	8
100	0–60	350	67	160	8
200	0–120	120	53	44	8
400	0–240	120	116	12	8
		Fluid damped types			
1,000	0–600	20–1,000	37	0.356	8
1,650	0–1,000	20–1,000	25	0.107	8
3,300	0–2,000	20–1,000	31	0.039	6
5,000	0–3,000	20–1,000	37	0.023	3.5
8,000	0–4,800	20–1,000	33	0.027	2.0

Fig. 12.8. *Galvanometer family.*

entire family of interchangeable units covering the range from low frequency, high sensitivity to high frequency, low sensitivity. Figure 12.8 shows a typical selection.[1] The galvanometers can be easily and quickly removed from the magnet block and replaced by others suited to the particular job requirements. Since the light beams do not interfere with each other, as do mechanical arms, each channel of a multichannel instrument can use the entire chart width. Optical galvanometers also allow greater deflections, full scale being 4 to 8 in. in commercial instruments. Up to 60 channels may be recorded on one 12-in.-wide chart paper. Most instruments provide pushbutton selection of chart drive speeds; speeds up to about 160 in./sec are available. The time base provided by the paper drive is usually accurate only to about 3 to 5 percent. Thus if accurate time measurement is required, a timing trace (say the accurate 60-cps power-line frequency) can be put on one of the galvanometers. Accurate time-line generators internal to the oscillograph which print (optically) an accurate time grid at, say, 0.01-sec intervals on the paper are also available.

The optical galvanometer recording principle has been adapted[2]

[1] Honeywell Heiland Div., Denver, Colo.
[2] Sanborn Co., Waltham, Mass.

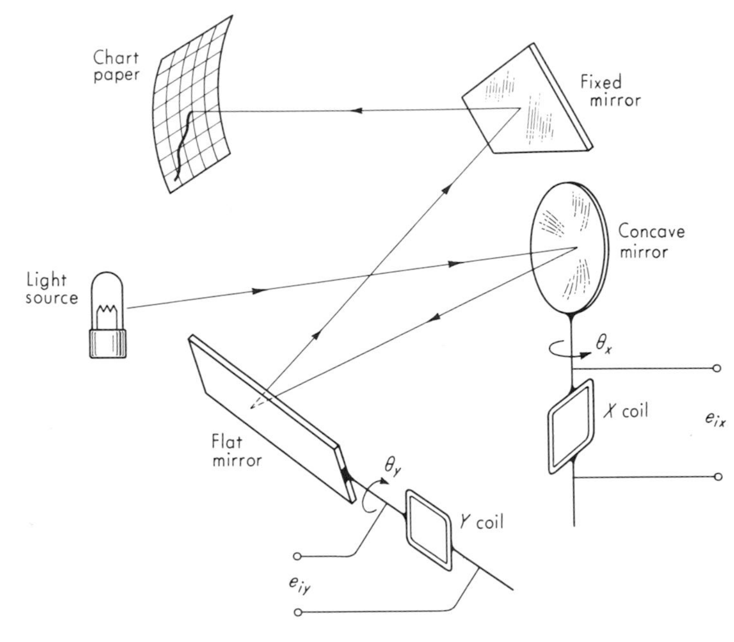

Fig. 12.9. Galvanometer X-Y plotter.

to X-Y recording as shown in Fig. 12.9. This instrument provides the X-Y recording function at speeds intermediate between self-balancing potentiometer types and cathode-ray oscilloscopes. The instrument shown employs ultraviolet-sensitive paper which requires no processing. Chart size is 8 by 8 in. with nonlinearity of 1 percent of full scale. Frequency response is flat from d-c to 100 cps at full-scale deflections.

12.5

Cathode-ray Oscilloscopes Figure 12.10 shows in simplified fashion the functional operation of a typical cathode-ray oscilloscope. A focused narrow beam of electrons is projected from an electron gun through a set of horizontal and vertical deflection plates. Voltages applied to these plates create an electric field which deflects the electron beam and causes horizontal and vertical displacement of its point of impingement on the phosphorescent screen. By proper design this displacement can be made closely linear with deflection-plate voltage. The phosphorescent screen emits light which is visible to the eye and may also be photographed for a permanent record.

The most common mode of operation is that in which one desires a plot of the input signal against time. This may be accomplished by

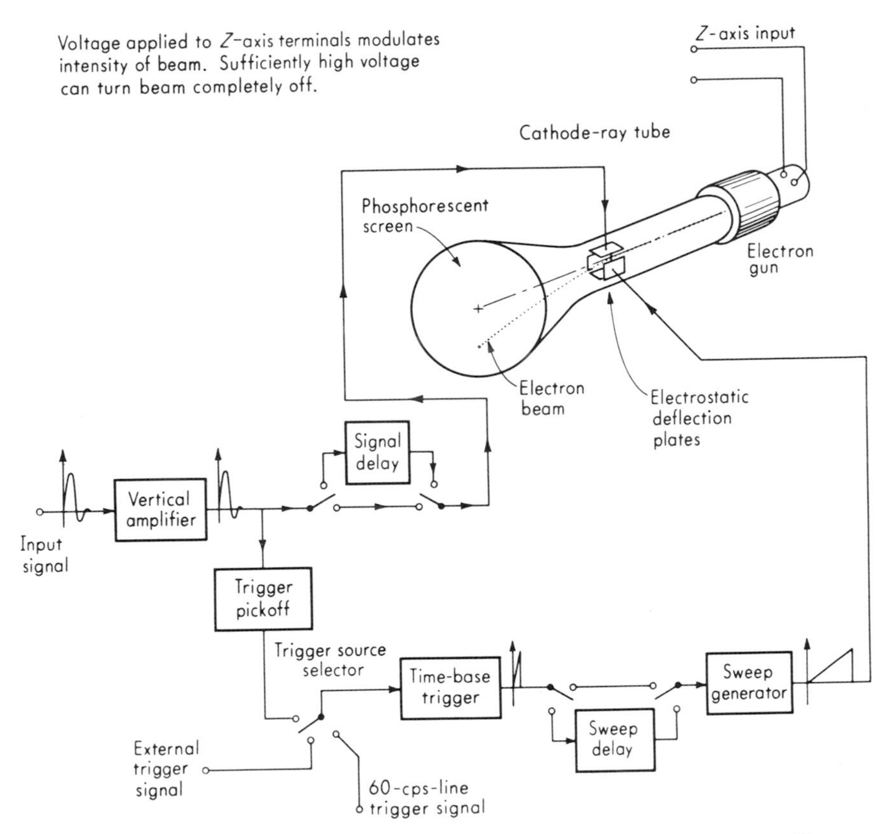

Fig. 12.10. *Cathode-ray oscilloscope.*

driving the horizontal deflection plates with a ramp voltage, thus causing the spot to sweep from left to right at a constant speed. To ensure that the sweep and the input signal applied to the vertical deflection plates are properly synchronized, the triggering of the sweep can be initiated by energizing the trigger circuit from the leading edge of the input signal itself (see Fig. 12.11). This results in a loss of the first instants of the input signal on the screen, but this is generally not serious since only about 1 mm of deflection is needed to cause triggering. In those cases where this loss is objectionable, oscilloscopes with signal delay are available. These delay the application of the input signal to the vertical deflection plates so that the sweep starts *before* the rise of the input signal on the screen. Thus the complete input signal is recorded. Most oscilloscopes also provide for triggering from either positive-going or negative-going voltages, and the instant of triggering can be adjusted from the minimum 1 mm level upward to any point on the input wave

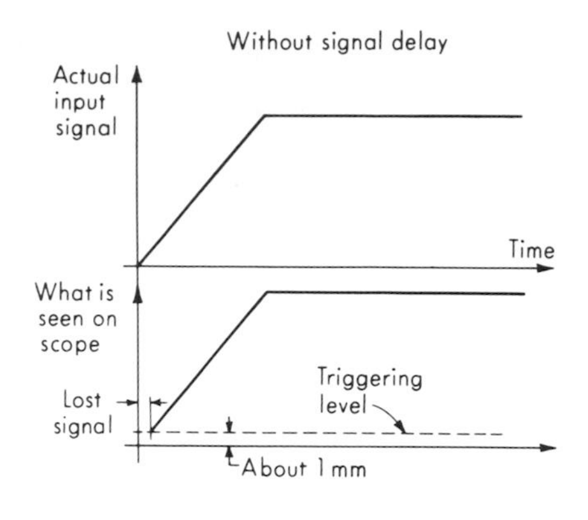

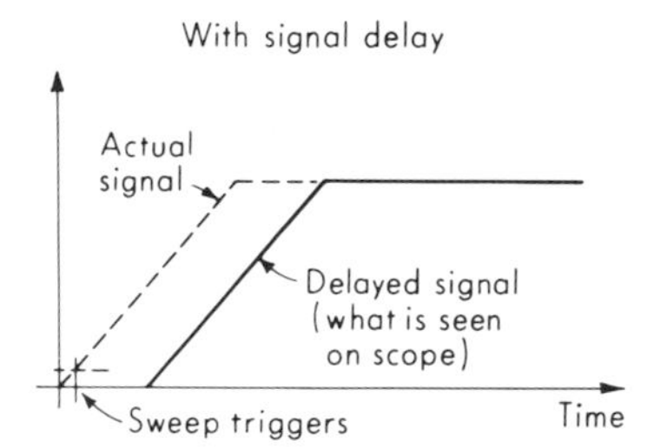

Fig. 12.11. *Signal delay.*

form. Triggering can also be controlled from external signals or the 60-cps power-line signal. When external trigger signals which are conveniently available occur somewhat before the input signal of interest, a sweep-delay feature may be useful; some instruments provide this capability. Since the deflection sensitivity of the cathode-ray tube itself is only of the order of 0.1 cm/volt, oscilloscopes generally include amplifiers so that the instrument can directly handle millivolt-level input signals. Oscilloscopes are also useful for X-Y plotting. For such operation the horizontal deflection plates are merely disconnected from the sweep generator and connected to an amplifier identical to the vertical amplifier.

Cathode-ray tubes are available with a number of different phosphors on the screen. The choice of phosphor controls the intensity of light available for visual observation or photographic recording and also the persistence of the trace after the electron beam has moved on. Both long- and short-persistence phosphors are available. Long-persistence phosphors are useful in visual observation of transients since the entire trace is visible long enough for an observer to note its characteristics. Persistence for several seconds is possible. When the moving-film method of trace photography is used, a very short-persistence phosphor is necessary to prevent blurring. (In this method the electron beam is deflected vertically only, while the film is moved horizontally in front of the screen at a fixed and known velocity.) Persistence of less than 1 μsec is available. The widely used P2 phosphor has a persistence of between 10^{-4} and 10^{-5} sec. For photography of the highest-speed traces the P11 phosphor produces more usable light and is often recommended. Dual-persistence phosphors (P7 is common) provide either a long or short persistence, depending on the color of the filter used over the scope screen. The most common method of photographing oscilloscope traces uses a still camera and the 10-sec Polaroid[1] film process. A common method of photographing transients uses a double exposure to record both the trace and the grid lines. With the grid-line illumination turned off and the camera shutter held open, the transient is triggered, thus recording its image on the film. The shutter is then closed. Now the grid lines are turned on and the shutter snapped in the normal manner (say $\frac{1}{25}$ sec at F:16) to superimpose the grid lines on the picture. This procedure is necessary since the illumination from the grid lines is so great that it would completely fog the film if left on during the long time that the shutter is left open to catch the transient. Because of the rapidity and ease of making trial runs with Polaroid film, it is generally best to determine camera settings by trial and error rather than attempting to calculate them.

To obtain multichannel capability in oscilloscopes, two approaches are used. The dual-beam oscilloscope has two separate electron beams in one cathode-ray tube, with separate deflection plates and amplifiers for each beam. In some units both beams use the same sweep system; thus the two traces are plotted against the same time base. Completely independent beams allowing different time bases on each trace are also available. The other approach uses a single-beam cathode-ray tube and a high-speed electronic switch to time-share the beam among several input signals. Such multitrace systems are available to give up to four traces on a single screen.

Versatility of operation is achieved in plug-in-type oscilloscopes by providing a wide variety of functional plug-in units for a single main frame. Typical plug-ins available include dual-trace and four-trace units, operational amplifiers, carrier amplifiers, spectrum analyzers, high-gain amplifiers, and time bases with special features.

While most laboratory oscilloscopes have a 5-in. diameter screen, larger screens up to about 21 in. are available. These are useful for viewing by large groups, presentation of many channels of data by bar-graph-type displays, etc. Large-screen scopes usually cannot attain the

[1] Polaroid Corp., Cambridge, Mass.

high-frequency capability of the 5-in. types. A special 17-in. display system achieving high accuracy by electronic generation of amplitude and time grid lines ("electronic graph paper") has been developed.[1] As many as eight variables can be plotted simultaneously against time by this system with an accuracy of 0.5 percent or better. The X-Y mode of operation is also possible with this equipment. By using suitable function generators in the grid-line system, useful distorted plotting scales (such as semilog, log-log, etc.) may be obtained.

While the Polaroid photography process makes recording very convenient, a *permanent* retention of the trace on the scope screen has certain advantages. This feature has recently become available in the so-called storage oscilloscope. Special cathode-ray tubes are now available that will retain a trace for long periods until it is erased electrically by pushing a button. Oscilloscopes using such storage tubes have essentially the same performance characteristics as normal types. Such oscilloscopes allow one to examine traces visually with ease; when a desired trace is noted, it can then be photographed.

The performance specifications of oscilloscopes cover such a wide range that one must really consult manufacturers' catalogs to appreciate the versatility of this instrument. We can, however, quote some limits of performance as presently available. Voltage sensitivities as high as 0.1 to 0.001 cm/μv with frequency response to 5,000 and 50,000 cps, respectively, can be obtained. Lower sensitivity results in higher frequency response, the upper limit around 1,000 Mc having a sensitivity of the order of 0.1 cm/volt. The accuracy of oscilloscope voltage and time scales is of the order of 2 or 3 percent at best, with operation at the limits of sensitivity and/or speed ranges resulting in reductions in these values. Increased accuracy of time measurements can be obtained in dual-beam or dual-trace scopes by applying an accurately known timing voltage, such as that from a crystal oscillator, to one channel. In single-trace instruments one can use the Z axis (intensity modulation) in a similar way to turn the beam on and off at known time intervals. This produces a dashed-line trace, with the dashes occurring at known time intervals. Increase in voltage (vertical deflection) accuracy is attainable by switching the beam rapidly between an accurately known reference voltage and the unknown voltage. An elaboration of this approach is actually what is done in the "electronic graph paper" system mentioned earlier.

To indicate the type of performance to be expected from a general-purpose oscilloscope suitable for a wide range of mechanical engineering studies, we quote the following specifications of a typical dual-beam

[1] G. A. Philbrick Researches, Boston, Mass.

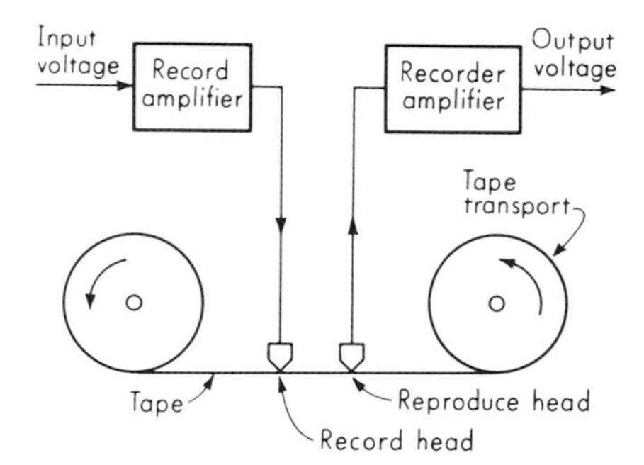

Fig. 12.12. *Tape recorder/reproducer.*

instrument.[1] The vertical-deflection factors (sensitivity) are selectable from 17 in the range 0.1 mv/cm to 20 volts/cm. Frequency response is from d-c to 50,000 cps at 0.1 mv/cm to d-c to 1 Mc at 0.2 volt/cm and higher. Both beams share the same time base, which is selectable from 21 in the range 5 sec/cm to 1 μsec/cm. The input amplifiers offer both single-ended and differential input and may be either a-c- or d-c-coupled. On a-c coupling the frequency response is flat beyond about 2 cps. The input impedance is 1 megohm paralleled by 47 pf. For high-sensitivity X-Y plotting only one beam is needed, and the vertical amplifier for the unused beam is "borrowed" for the horizontal-deflection system.

12.6

Magnetic Tape Recorder/Reproducers The magnetic tape recorder/reproducer has a number of unique features not shared by other recording devices discussed previously. These are derived mainly from its ability to record a voltage, store it for any length of time, and then reproduce it in electrical form essentially identical to its original occurrence. Recording methods used with tape recorders include the direct, FM, PDM (pulse-duration modulation), and digital techniques.[2] We shall consider briefly the direct and FM modes of operation.

Figure 12.12 shows a functional diagram of a tape recorder/reproducer, and Fig. 12.13 shows a closeup of the record and reproduce heads. A current i proportional to the input voltage is passed through the winding on the record head, producing a magnetic flux $\phi = K_\phi i$ at the recording gap. The tape (thin plastic coated with iron oxide particles) passes under the gap, and the oxide particles retain a state of permanent mag-

[1] Tektronix Inc., Beaverton, Ore.

[2] P. J. Weber, The Tape Recorder as an Instrumentation Device, Ampex Corp., Redwood City, Calif., 1963.

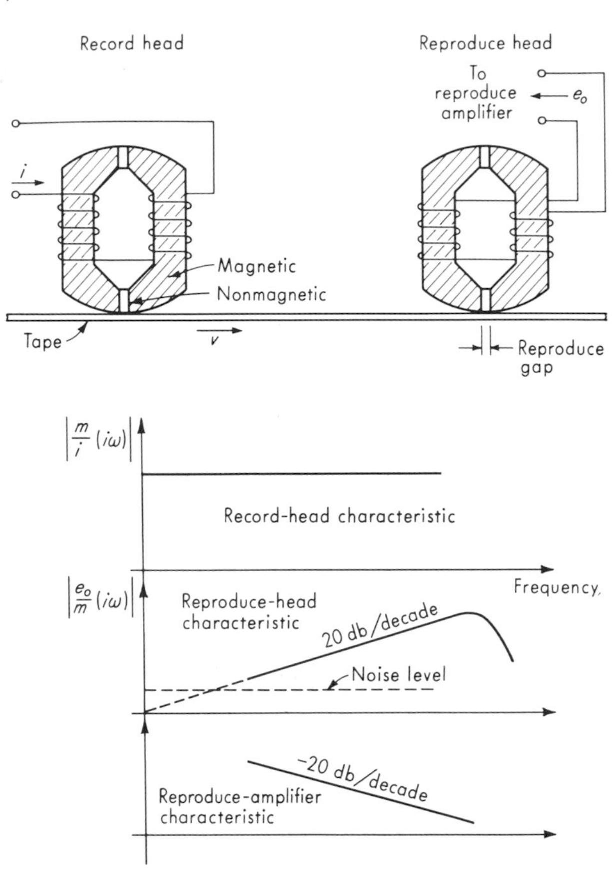

Fig. 12.13. *Record and reproduce heads.*

netization proportional to the flux existing at the instant the particle leaves the gap. (Actually the applied flux and induced magnetization are not proportional because of the nonlinearity of the magnetic-hysteresis curve. Effectively, however, a close linearity is obtained by a high-frequency bias technique.[1]) Thus, with a sinusoidal input signal $i = i_0 \sin 2\pi ft$ and a tape speed of v in./sec, the intensity of magnetization along the tape will vary sinusoidally with distance x according to

$$\text{Magnetization} \triangleq m = K_m K_\phi i_0 \sin\left(\frac{2\pi f}{v}x\right) \tag{12.9}$$

where $m = K_m\phi$

The wavelength of the magnetization variation is then v/f in. For example, a 60-cps signal at a 60-in./sec tape speed gives a wavelength

[1] *Ibid.*

of 1 in. If the tape with this signal on it is then passed under the reproduce head, a voltage proportional to the rate of change of flux bridging its gap will be generated in its coil. Note that since the output voltage depends on the rate of change of flux, if a d-c current at the input had produced a constant tape magnetization, the reproduce head would have given *zero* output. Thus the technique described above, the so-called direct recording process, can be used with varying input signals only, about 50 cps being the usual lower limit of frequency. Furthermore, since the reproducing head has a differentiating characteristic, the reproduce amplifier must have an integrating characteristic in order that the system output be proportional to the input. An upper frequency limit also exists because at sufficiently high frequencies, for a given reproduce gap and tape speed, one wavelength of magnetization will become equal to or less than the gap width. Then the average magnetization in the gap will be zero and no output voltage will be generated. For example, at the fastest currently available tape speed, 120 in./sec, and a gap width of 0.00008 in., this occurs at 1.5 Mc. Actually, the system is usable to only about half this frequency with reasonable accuracy. The frequency range of the direct recording process is approximately within the band 50 to 600,000 cps, with some machines going to several megacycles. The direct recording process does not give particularly high accuracy. This is essentially limited by the signal/noise ratio which is of the order of 25 db (about 18:1). The rather high noise level is due to minute defects in the tape surface coating to which the direct recording process is sensitive.

When more accurate recording and response to d-c voltages are required, the FM system is generally employed. Here the input signal is used to frequency-modulate a carrier which is then recorded on the tape in the usual way. Now, however, only the *frequency* of the recorded trace is significant, and tape defects causing momentary amplitude errors are of little consequence. The frequency modulators used here are similar in principle to those discussed under voltage-to-frequency converters and subcarrier oscillators in Chaps. 10 and 11. However the frequency deviation for tape recorders is ± 40 percent about the carrier frequency. The reproduce head reads the tape in the usual way and sends a signal to the FM demodulator and low-pass filter where the original input signal is reconstructed. The signal/noise ratio of an FM recorder may be of the order of 40 to 50 db (100:1 to 330:1), indicating the possibility of inaccuracies smaller than 1 percent. By using sufficiently high carrier frequencies (432 kc), the flat (± 1 db) frequency response of FM recorders may go as high as 80,000 cps at 120-in./sec tape speed. To conserve tape when high-frequency response is not needed, a range of tape speeds is generally provided. When the tape speed is changed, the carrier frequency is changed in direct proportion. This makes the recorded wavelength of a given d-c input signal the same, no matter what tape speed is being used, since ± 40 percent full-scale frequency deviation is used in all cases. Signals may be recorded at one tape speed and played back at any of the others without change in magnitude but with a compression or expansion of the time scale. A common set of specifications might be as follows:

Tape speed, in./sec	*Carrier frequency, kc*	*Flat frequency response* ± 0.5 *db, cps*	*RMS signal/noise ratio*
120	108	0–20,000	50
60	54	0–10,000	50
30	27	0–5,000	49
15	13.5	0–2,500	48
$7\frac{1}{2}$	6.75	0–1,250	47
$3\frac{3}{4}$	3.38	0–625	46
$1\frac{7}{8}$	1.68	0–312	45

Multichannel tape recorders are available with up to 14 channels on one 1-in.-wide tape. Input to tape recorders is generally at about a 1-volt level, and so most transducers require amplification before recording. The maximum time-base change of about 60:1 ($120/1\frac{7}{8}$) shown in the above table can be even further increased by rerecording the signal. For example, the original signal can be recorded at 120 in./sec. Then it is played back at $1\frac{7}{8}$ with the output of the reproduce amplifier feeding the input of the record amplifier of another machine running at 120 in./sec. If this tape is now played at $1\frac{7}{8}$, an overall slowdown of 4,096:1 is achieved. An example application of tape slowdown is the recording of a 20,000-cps signal on tape at 120 in./sec and playback at $7\frac{1}{2}$ in./sec into an optical oscillograph (frequency response to 2,000 cps) for a permanent record. Tape slowdown is also used when digital computations are to be performed on high-speed analog data. The digital equipment is very accurate but cannot handle rapidly varying inputs; thus one can analog-tape-record the data and then play it into the digital processing equipment at reduced speed. Tape speedup is useful in the processing (spectrum analysis, autocorrelation, etc.) of low-frequency signals in electronic equipment designed for high frequencies. The storage and playback feature of tape recording is widely used in simulation. A particular environmental condition, such as vibration in an aircraft, is measured and tape-recorded in the actual environment. The tape is then brought into the simulation laboratory where the environmental parameter (say vibration) is recreated by playing the tape into a vibration shaker.

Problems

12.1 Calculate the ratio peak value/rms value and average value/rms value for the following:
- *a.* Direct current
- *b.* A sine wave
- *c.* A square wave
- *d.* A half-wave-rectified sine wave
- *e.* A full-wave-rectified sine wave
- *f.* A train of rectangular pulses that are on 10 percent of the time and off 90 percent of the time.

12.2 Derive Eqs. (12.4) to (12.7).

12.3 Solve for $(i_g/e_s)(D)$ in the system of Fig. 12.6, neglecting inductance.

12.4 Taking inductance into account in the system of Fig. 12.6, find the following:
- *a.* $(\theta_o/e_s)(D)$
- *b.* $(i_g/e_s)(D)$

Show the possible shapes of frequency-response curves for these transfer functions.

Bibliography

1. Notes on the Julie Ratiometric Method of Measurement, Julie Research Laboratories, New York, 1964.
2. M. H. Aronson: "Handbook of Electrical Measurements," The Instruments Publishing Co., Pittsburgh, Pa., 1961.
3. L. W. Dean: Potentiometer Specifications, *Instr. Control Systems*, p. 73, January, 1965.
4. S. A. Davis: Analog Voltmeters, *Electromech. Design*, p. 48, November; p. 44, December, 1963.
5. R. Bergeson: Feedback Stiffens D'Arsonval Movement, *Control Eng.*, p. 121, September, 1964.
6. J. W. Martin: Error Analysis in Measuring RMS Voltages, *Electro-Technol. (New York)*, p. 38, April, 1965.
7. R. J. Erdman: DC Microvolt Measurements, *Instr. Control Systems*, p. 91, January, 1964.
8. R. T. Hood: Measuring Current in High-energy Arc Jets, *Instr. Control Systems*, p. 99, January, 1964.
9. J. F. Keithley: Electrometer Measurements, *Instr. Control Systems*, p. 74, January, 1962.
10. F. C. Martin: RMS Measurement of AC Voltages, *Instr. Control Systems*, p. 65, January, 1962.
11. W. H. Schaeffer: The Six-dial Thermofree Potentiometer, *Instr. Control Systems*, p. 283, February, 1961.
12. A. Miller: Design Considerations of D'Arsonval Galvanometer-Power Amplifier Systems, *The Right Angle*, The Sanborn Co., Waltham, Mass., August, November, 1958.
13. "Typical Oscilloscope Circuitry," Tektronix Inc., Beaverton, Ore., 1961.

14. A. L. Ispas: Interpretation of Magnetic Tape Recorder Specifications, *Instr. Control Systems,* p. 97, July, 1964.
15. P. J. Weber: The Tape Recorder as an Instrumentation Device, Ampex Corp., Redwood City, Calif., 1963.
16. P. A. Mohr: Magnetic Tape Systems for Data Recording, *Automation,* p. 72, February, 1958.
17. History of Magnetism, *Readout,* Ampex Corp., Redwood City, Calif., August–September, 1961.
18. R. E. Morley: Time Compression Disk, *Instr. Control Systems,* p. 108, July, 1964.
19. J. McElwain: Long-term Magnetic Tape Recording, *Instr. Control Systems,* p. 111, July, 1964.
20. E. D. Lucas: Miniature Tape Recorders, *Control Eng.,* p. 53, December, 1964.
21. G. H. Schulze: Tape Recording Errors, *ISA J.,* p. 61, May, 1964.

13 Large-scale Data systems

The measurement demands generated by experimental test programs of complex systems such as rockets have led to the development of sophisticated systems for the gathering and processing of measured data. In order to make test results quickly available, analog and/or digital computers may process measurements as they are made and present significant results to the test operator, who may then modify test conditions more nearly to meet requirements. Even if such on-line processing is not employed, the vast quantities of information generated by the transducers must be made available, in intelligible form, as quickly as possible. Such large-scale information-processing systems are also becoming more common in industrial-production operations. To illustrate the nature of these systems, we shall briefly examine a typical example. The example chosen is the central facility for recording and processing data from vibration, heat-transfer, and rocket-testing studies at the NASA Lewis Research Center in Cleveland, Ohio. The information presented below is based on the publication, A Central Facility for Recording and Processing Transient-type Data, *NASA*, *Tech. Note* D-1320, 1963. Since we shall discuss only the main features, those interested in more detail should consult this reference which gives a complete account of the problems involved in setting up such an operation.

Figure 13.1 shows a simplified overall block diagram of the data system. Transducer input signals at the 0 to 10-mv level may be accepted from six different test facilities at various scattered locations at the Lewis Research Center. When a test facility wishes to utilize the central data-processing system, a phone call is made; if the system is not in use, the necessary connections are established. Transducer signals are transmitted over a 200-pair shielded telephone cable for distances up to 7,000 ft at the 0 to 10-mv level. (Figure 11.1*d* and *e* show frequency-response tests of this cable.) Of the 200 available signal lines, 128 are used for a digital recording system and 24 for an analog system. The digital system features high accuracy and compatibility with digital-computer data processing but is of limited frequency response. It also provides for recording a large number of channels by use of time sharing. High frequency response is the main feature of the analog system. Various interconnections are possible between the analog and digital systems to provide a maximum of versatility and reliability.

Tracing through the digital system, we first encounter the premultiplexer. This is a rotating electromechanical sampling switch which samples the 128 input signals, 8 at a time, at a rate of 500 samples per second. If desired, it is also possible to sample 1, 2, 3, 4, 5, 6, 7, 8, 32, 64, or 128 inputs. The sample rate per signal is in each case 4,000/(number of signals) per second. Thus the maximum rate, if only one signal is sampled, is 4,000 per second. Sampling of signals can lead to difficulties if the number of samples per cycle of signal is not sufficiently great. For example, if a 60-cps sine-wave signal is sampled at exactly 60 samples per second, when the signal is reconstructed from the sampled points it will appear to be a d-c voltage. The Shannon sampling theorem[1] shows that the absolute minimum number of samples must be two per cycle of the highest frequency present in the signal. Actually, somewhat more than two samples per cycle must be used to achieve good accuracy when the signal is reconstructed. Rational methods of establishing the required

[1] L. W. Gardenhire, Selecting Sample Rates, *ISA J.*, p. 59, April, 1964.

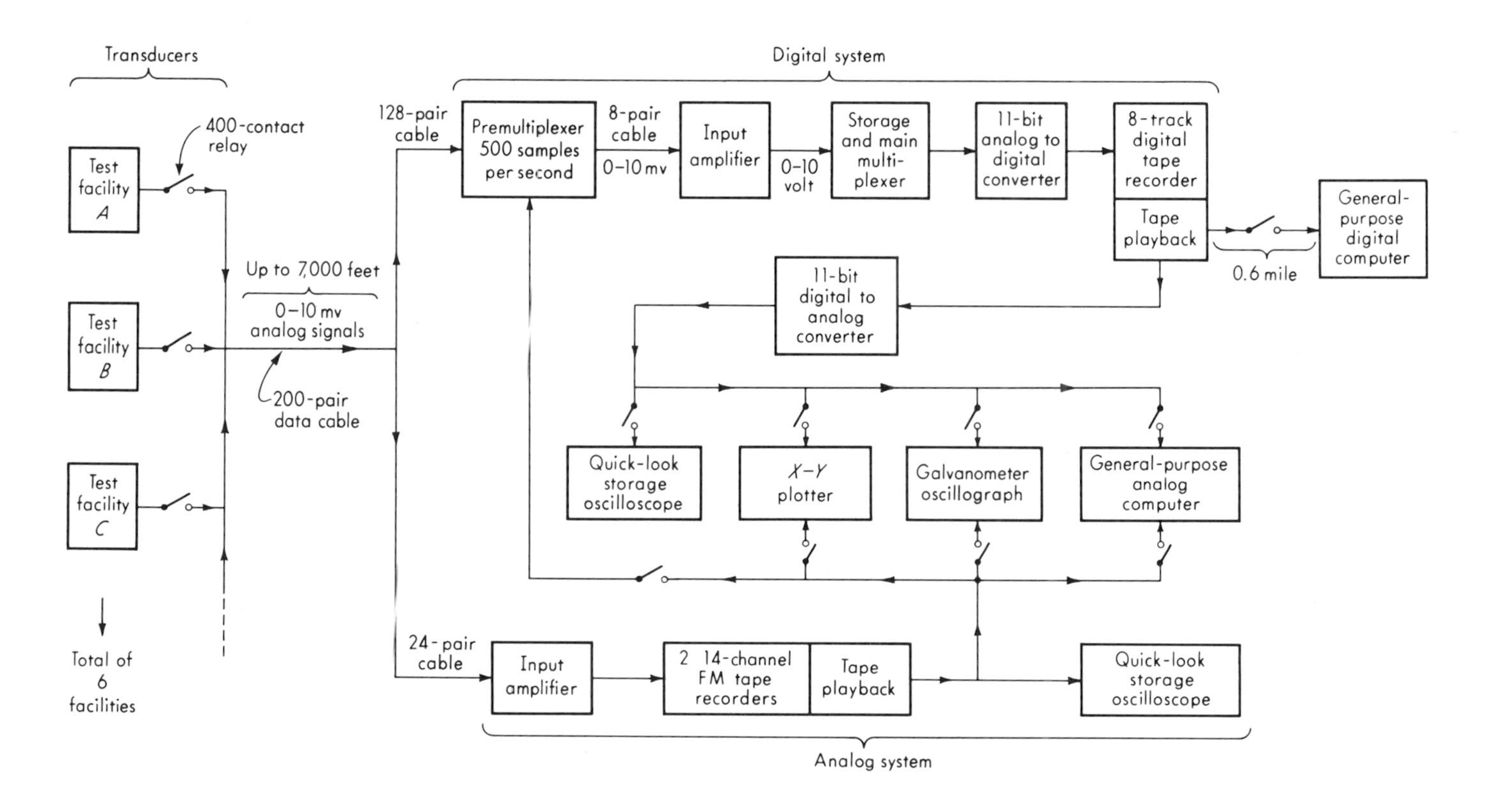

Fig. 13.1. *Large-scale data system.*

sampling rate have been developed[1] but their discussion is beyond the scope of this text. Depending on the accuracy required and the method of reconstructing the signal from the samples, the required sample rate varies over a wide range.[2] If one arbitrarily selects 10 samples per cycle as a reasonable rate, the above system can handle one signal of 400-cps frequency content, two signals of 200-cps content, etc.

The sampled 0 to 10-mv signals are boosted to 0 to 10-volt range by d-c amplifiers of the differential-input type. Their frequency response is flat (-3 db) to 3,000 cps, and the step response time to be within 0.1 percent of final value is 0.5 msec. Since signals are switched in a 2-msec (500 cps) cycle, the amplifiers are sufficiently fast to read correctly before the next sample is taken. Common-mode rejection at 60 cps is 10^6:1 for a 1,000-ohm unbalanced line, which is quite important in this system because of the extremely long cable.

The storage and main multiplexer unit stores simultaneously the eight sampled analog outputs of the amplifiers and feeds them one at a time to the analog-to-digital converter. This is an 11-bit successive-approximation type which can convert from analog to digital in 22 μsec. Conversion is accurate to 0.05 percent $\pm\frac{1}{2}$ the least-significant bit. Since 2^{11} is 2,048, the least-significant bit represents about 0.05 percent of full scale also. Digitally coded pulses are recorded on eight tracks of a tape recorder at either 15 or 60 in./sec. Playback from this recorder can be sent to a number of plotting or processing equipments. A standard telephone-type cable sends the digital pulses about 0.6 mile to a high-speed, stored-program, general-purpose digital computer. Pulse width at the sending end is about 25 μsec and degrades to 35 μsec at the receiving end. The digital computer can, of course, be programmed to perform any processing operations desired.

To provide quick-look checks on the performance of the digital system and readily available analog records of selected variables, the digital signal is converted back to analog in a digital-to-analog converter. This signal can then be sent to a storage-type oscilloscope, X-Y plotter, five-channel galvanometer oscillograph, or a general-purpose electronic analog computer. This computer has 12 integrator-summers, 24 summers, 8 electronic multipliers, 60 servoset coefficient potentiometers, and 6 servoset function generators. The storage oscilloscope may be used in four modes: One to four selected data channels may be plotted against time in a single sweep, any data channel may be displayed against any other channel in an X-Y plot, all data may be displayed at a fast-recurring

[1] L. W. Gardenhire, Selecting Sample Rates, *ISA J.*, p. 59, April, 1964.
[2] *Ibid.*

sweep to produce a vertical bar-graph display, and all data may be displayed during a single slow sweep.

In the analog system, the 0 to 10-mv transducer signals are boosted in d-c amplifiers with a frequency response of 0 to 10,000 cps. These signals are then applied to two 14-channel FM tape recorders. Twelve channels of each are used for data and two for identification and calibration information. The accuracy with which a d-c voltage can be recorded and played back is 1 percent of full scale. Frequency response of the overall system is good to 10,000 cps. Analog information recorded at maximum tape speed can later be played back at $\frac{1}{2}$, $\frac{1}{4}$, $\frac{1}{8}$, $\frac{1}{16}$, or $\frac{1}{32}$ of this speed into the digital system for desired processing. This is necessary since the digital system cannot handle high-speed data. Information from the analog system also may be sent to storage oscilloscopes, X-Y plotters, oscillographs, and the analog computer.

Problems

13.1 A 60-cps sinusoidal signal is sampled once every $\frac{1}{60}$ sec. If the sampled function is reconstructed by linear interpolation, describe the reconstructed function.

13.2 A 10-sec/cycle sine wave is sampled once every 11 sec. If the sampled function is reconstructed by linear interpolation, describe the reconstructed function. What fictitious frequency has been introduced by the sampling process?

Bibliography

1. A. T. Snyder: Airborne Recorder and Computer Speed Flight-test Data Processing, *ISA J.*, p. 44, July, 1958.
2. E. J. Kompass: Information Systems in Control Engineering, *Control Eng.*, p. 103, January, 1961.
3. J. P. Knight et al.: Low-level Data Multiplexing, *Instr. Control Systems*, p. 86, August, 1963.
4. L. W. Gardenhire: Selecting Sample Rates, *ISA J.*, p. 59, April, 1964.
5. W. T. Botner: Digital Data Gathering System for Blowdown Wind Tunnel, Sandia Corp., Albuquerque, N.Mex., *Rept.* SCR-23, 1958.
6. J. K. Slap: Recording and Processing Test Data, *Control Eng.*, p. 177, September, 1959.
7. E. Pacini: How Raytheon Cut Test Analysis Time from Days to Hours, *Instrumentation,* Honeywell Inc., Philadelphia, Pa., vol. 17, no. 1, 1964.
8. P. Westercamp: Computing Power Station Performance, *Control Eng.*, p. 72, December, 1963.
9. Digital Data System Takes 15,625 Engine Samples a Second in Saturn Rocket Static Tests, *Control Eng.*, p. 19, October, 1963.
10. W. C. Hixson: Instrumentation for the Pensacola Centrifuge Slow Rotation Room I Facility, *NASA*, *CR*-53341, 1964.
11. K. C. Sanderson: The X-15 Flight Test Instrumentation, *NASA*, *TM X*-56000, 1964.

12. J. D. Jones: High-speed Low-level Data Acquisition, *Instr. Control Systems*, p. 96, April, 1965.
13. H. Gruen and B. Olevsky: Increasing Information Transfer, *Space/Aeron.*, p. 40, February, 1965.
14. S. H. Boyd: Digital-to-Visible Character Generators, *Electro-Technol. (New York)*, p. 77, January, 1965.
15. W. Clifford: Digital Voltmeter Data Systems, *Instr. Control Systems*, p. 105, December, 1964.
16. W. E. Schilke: The Analysis of Transmission and Vehicle Field Test Data Using a Digital Computer, *Gen. Motors Eng. J.*, p. 19, October-November-December, 1964.

Index

Index